THE MECHANICS OF VIBRATION

THE MECHANICS OF VIBRATIONS

THE
MECHANICS OF
VIBRATION

BY

R. E. D. BISHOP

Kennedy Research Professor in the University of London and Fellow of University College.
Sometime Fellow of Pembroke College, Cambridge

AND

D. C. JOHNSON

Late Professor of Mechanical Engineering in the University of Bradford.
Sometime Fellow of Trinity Hall, Cambridge

CAMBRIDGE UNIVERSITY PRESS

CAMBRIDGE

LONDON . NEW YORK . MELBOURNE

CAMBRIDGE UNIVERSITY PRESS
Cambridge, New York, Melbourne, Madrid, Cape Town,
Singapore, São Paulo, Delhi, Tokyo, Mexico City

Cambridge University Press
The Edinburgh Building, Cambridge CB2 8RU, UK

Published in the United States of America by Cambridge University Press, New York

www.cambridge.org
Information on this title: www.cambridge.org/9781107402454

First published 1960
Reissued with minor revisions 1979
First paperback edition 2011

A catalogue record for this publication is available from the British Library

ISBN 978-0-521-04258-1 Hardback
ISBN 978-1-107-40245-4 Paperback

CONTENTS

CONTENTS

LIST OF TABLES

PREFACE

The understanding of the physical nature of a vibration problem is often both more difficult and more important than the calculation of frequencies and modal shapes. The latter can be tackled readily once a few techniques have been mastered; and modern computing machinery has removed the one difficulty which previously existed—that of excessive complexity. While, however, the study of methods of frequency calculation is often an important aid to understanding the physical nature of a problem, it is not always sufficient to enable the engineer to obtain the grasp that he requires. Further, the ease with which numerical work can now be handled makes a thorough understanding of *physical* problems even more necessary; without it the most important calculations may not be attempted and time may be wasted on others which are of little value.

It was when one of us was working in industry that he came to appreciate the value of the concept of receptance as an aid to understanding as well as to calculation. The concept was, at that time, being used in studying the vibration of aeroplane propellers in conjunction with the torsional vibration of the engine shafting on which they were mounted; it allowed the vibrational characteristics of the propellers and of the shafting to be discussed independently and this was helpful because—amongst other reasons—the propellers and engines were often made by different manufacturers! The receptance concept is used extensively throughout this book and, in particular, the classical theory of the small oscillation of a linear system is developed by means of it. We believe that the approach will be satisfying to the reader and that the familiarity with receptances which he will gain will enable him to apply them when examining new problems.

The notion of receptances is particularly useful when the vibration of one portion of a system, such as a single turbine blade, is being studied independently of the rest of the system. The justification for isolating a part of a system in this way, for the purposes of analysis, is commonly that of intuition and nothing more; the receptance idea should often enable the engineer to reconcile the intuitive results with sound reasoning or alternatively to reject them when they cannot be so reconciled. This revaluation of the results of intuition is one of the responsibilities of the academically trained engineer.

Since the aim of this book is to present ideas, rather than to describe the application of those ideas to particular engineering problems, much of the discussion is in terms of 'academic' systems only, that is to say of spring-connected rigid masses, taut strings and uniform elastic bodies. In Chapter 5, however, there is some discussion of the extent to which real systems may be analysed in terms of these ideal systems.

Some guidance may be needed by those who wish to study the receptance concept without working through the whole book, or by others who have other special needs. Chapter 1 introduces receptance ideas but the discussion is limited to very simple mechanical systems. Chapter 2 deals with dynamical methods for analysing more complex systems and Chapter 3 uses these methods in a general treatment of the vibration of multi-freedom systems. This latter chapter contains most of the essential ideas in receptance theory. After this, with the exception of Chapter 5, which has been mentioned above, the discussion is extended to include elastic bodies and the various chapter headings explain which types of system are treated in each. Systems with damping are then introduced, the general theory for these being developed in a similar way to that of Chapter 3. Finally the excitation of a system by transient forces is treated.

It will be seen that many items which are of interest to the engineer have been omitted. For instance, questions of non-linearity, instability, self-excitation, the vibration of rotating elastic bodies, as well as matters of analytical and experimental technique have not been dealt with. This is admittedly a somewhat arbitrary division; but it was our original intention to write more than one volume. Perhaps a second will be written, although we no longer enjoy the advantage of working in the same laboratory.

Much of the theory that is presented is capable of concise proof by matrix methods. These methods have acquired an important place in vibration analysis, particularly as a result of modern developments in high-speed computing. Nevertheless it is our belief that matrices should not be used by engineers when mastering the fundamentals of mechanical vibration theory, if only because matrices *are* so convenient. It is all too easy to become facile in the mathematical sense without acquiring a true understanding of the physical side of things.

Examples are provided at the ends of most of the sections in this book and we have used the 'absolute' system of units both in these and in the text. That is to say, the units of force, mass, length and time are the poundal, pound, foot and second respectively. In doing this we have made free use of a convenient approximation, namely that $g = 32$ ft./sec.^{2} (and that 32 poundals $= 1$ lb.wt.).

Finally we should like to mention a number of helpers, other than those whose names are mentioned later in connection with specific items. First we acknowledge our debt to Professor W. J. Duncan for his encouragement and for the benefit which we obtained from his pioneer work in the use of admittance methods for mechanical systems; without this our book would never have been started. Secondly we thank Miss P. L. A. Baker for typing the complete manuscript, and Messrs. B. Wood, S. Hother-Lushington and A. G. Parkinson for reading the proofs.

R. E. D. B.
D. C. J.

December 1957

PREFACE TO THE 1979 REISSUE

In preparing this book for reprinting, I have not had the help of my co-author Professor D. C. Johnson. His untimely death in 1969 not only deprived me of a friendship that I valued very greatly but robbed this country of one of its most accomplished engineers.

Apart from correcting a few misprints, I have slightly revised certain of the tabulated data. The previous presentation did not give rigid body modes sufficient prominence and, by way of a remedy, I have adopted detailed suggestions that were made to us by Professor G. M. L. Gladwell.

R. E. D. Bishop

London, June 1978

GENERAL NOTATION

See also the Supplementary Lists of Symbols at the ends of chapters

A	Compound oscillatory system.
$_kA_{rs}$	Numerator of kth partial fraction in series representation of α_{rs} [see equation (3.4.3)].
a_{rs}	Inertia coefficient [see equation (3.1.2)].
a_r	Inertia coefficient [see equation (3.6.9)].
$B, C, D, \ldots$	Sub-systems which, together, make up A.
b	Viscous damping coefficient of isolated damping element.
b_{rs}	Viscous damping coefficient [see equation (8.4.8)].
c_{rs}	Stability (or 'stiffness') coefficient [see equation (3.1.7)].
c_r	Stability (or 'stiffness') coefficient [see equation (3.6.9)].
D	Dissipation function [see equation (8.4.19)].
d_{rs}	Hysteretic damping coefficient [see equation (9.4.6)].
E	Young's modulus.
e	The exponential constant.
F	Amplitude of harmonic applied force or torque.
F_r	Amplitude of harmonic force or torque applied at x_r. If a letter (as well as a numerical) subscript is carried, this refers to a sub-system.
G	Shear modulus.
g	Gravitational constant (taken as 32 ft./sec. throughout this book).
h	Particular value of x defining a section of a taut string, shaft, bar, beam, distant h from origin $(0 \leqslant h \leqslant l)$; hysteretic damping coefficient of isolated damping element.
I	Moment of inertia of rigid disk; second moment of area of cross-section of beam about its neutral axis.
i	Imaginary operator [see equation (1.3.1)].
J	Second polar moment of area of circular shaft.
k	Stiffness.
l	Length of taut string, shaft, bar, beam.
M	Mass.
m	Mass of particle.
N	Magnification factor (see §8.2).
n	Magnification factor [see equation 9.2.10)].
P_r	Generalized force corresponding to p_r.
p_r	rth principal co-ordinate.
Q_r	Generalized force corresponding to q_r.
$_kQ_1, _kQ_2, \ldots, _kQ_n$	Set of generalized forces $Q_1, Q_2, \ldots, Q_n$ which produce distortion in kth principal mode only.

q_r	rth generalized co-ordinate.
$_kq_r$	Displacement at q_r in the kth principal mode.
R	Amplitude of x (a real constant).
S	Dissipation function for hysteretic damping [see equation (9.4.8)].
T	Kinetic energy.
t	Time.
U	$= V + S$.
V	Potential energy.
v	Lateral deflexion of taut string or beam.
v_r	Lateral deflexion of taut string or beam in rth principal mode.
W_r	Work done by forces of a system due to displacement at q_r.
w	Applied lateral load/unit length on beam or taut string.
w_r	Applied lateral load/unit length causing deflexion of a beam or taut string in its rth principal mode only.
X_r	Amplitude of displacement at x_r (may be complex). If a letter (as well as a numerical) subscript is carried, this refers to a sub-system.
x	Distance along taut string, shaft, bar, beam ($0 \leqslant x \leqslant l$).
x_r	rth co-ordinate.
α_{rs}	Receptance of system A.
α_r	Receptance at rth principal co-ordinate.
$\beta_{rs}, \gamma_{rs}, \delta_{rs}, \ldots$	Receptances of sub-systems $B, C, D, \ldots$.
Δ	Denominator of expressions for the receptances of a system having finite freedom [see equation (3.2.6), (8.5.9) or (9.5.5)], being the determinant of the coefficients of equations of motion.
Δ_{rs}	Determinant formed from Δ by omission of the rth column and sth row.
Ξ_r	Amplitude of P_r.
Π_r	Amplitude of p_r.
ρ	Mass density.
Φ_r	Amplitude of Q_r.
$\phi_r(x)$	rth characteristic function of taut string, shaft, bar, beam.
Ψ_r	Amplitude of q_r.
Ω_r	rth anti-resonance frequency (rad./sec.).
ω	Circular frequency of excitation (rad./sec.).
ω_r	rth natural frequency (rad./sec.).

CHAPTER 1

INTRODUCTION

The position of the moving parts of a steam-engine indicator clamped to an engine may be specified by giving the displacement of the pencil from some fixed point, or the displacement of the indicator piston, or the inclination of the pencil lever to the horizontal. Here again, the specification will be only a good approximation in actual fact, because, for example, the pencil lever will bend to some extent when rapidly moved.

B. HOPKINSON, *Vibrations of Systems having One Degree of Freedom* (1910)

1.1 Preliminary remarks

The subject of this book is the theory of the small oscillations of a dynamical system. This subject is important to the engineer because all materials that are used in the construction of machinery possess mass and the ability to store potential energy through the property of elasticity; the combination of these properties renders vibration possible. Gravity and other effects can also affect the potential energy of a system and so modify the vibration.

For the purpose of the study it will be convenient to classify systems by their 'degrees of freedom'. The number of degrees of freedom that a system possesses is the number of co-ordinates which must be specified in order to define its configuration. Thus a simple mathematical pendulum has one degree of freedom if its motion is confined to a single plane but it has two degrees of freedom if it can swing in more than one plane. Now all real systems have an infinite number of degrees of freedom and this renders impossible a complete vibration analysis. Even if such an analysis were possible the effort of making it would be mostly wasted because it would yield far more information than could be used. It is therefore necessary to make some simplifying assumptions about the motion of a system if an analysis is to be feasible.

Now it usually appears from inspection that some types of motion are likely to be unimportant; these can then be imagined to be suppressed entirely for the purpose of analysis. It will be shown later that the extent to which such restrictions can be justified depends on the frequencies of vibration to which the system is subjected. One of the most common simplifications of this type is to regard each vibrating link of a mechanism as a number of spring-connected rigid masses, the springs being treated as massless and therefore incapable of surging; it is sometimes possible to simplify further by restricting the rigid bodies to zero dimensions so that the links become systems of spring-connected particles. We may consider as an example the system that is formed by two flywheels which are joined by a shaft; the torsional oscillations of the system may be examined under the assumption that the wheels are rigid and the shaft massless. Simplifications of this type reduce the number of degrees of freedom of the system and thereby simplify the mathematical treatment.

We shall be concerned, in this book, with the analysis of systems after they have been simplified. The reader should be warned, however, that this is only part of the engineer's problem; the process of simplification of the system is often difficult and may need considerable knowledge and experience.

The vibrations which we shall discuss will be linear. That is to say, they will be governed by linear differential equations which have constant coefficients. In some problems this will be a result of the form taken by the system under discussion; thus, a linear equation governs the free oscillations of a mass attached to the free end of a massless spring. The form of the equation is due, in this case, to the nature of the spring—that is, it obeys Hooke's law—and to the nature of the oscillation—namely one of motion along the axis of the spring. But in order to arrive at a linear equation it is sometimes necessary, not only to have a suitable system, but also to restrict its motion to *small* oscillations. This is required when the geometry of the system would otherwise vary appreciably during the motion. For instance, a simple pendulum of length l oscillates according to the equation

$$\ddot{x} + \frac{g}{l}\sin x = 0, \tag{1.1.1}$$

where x is the angle made with the vertical. Only by limiting the angle of swing can we arrive at the linear equation

$$\ddot{x} + \frac{g}{l}x = 0. \tag{1.1.2}$$

All real mechanisms are affected by frictional forces, of which there are several causes. It is expedient to neglect these forces entirely during the development of the analysis. Their effects will be discussed in Chapters 8 and 9.

We shall begin our analysis by considering the motion of a system due to a sinusoidally varying force. This motion is of direct importance where the effects of lack of balance of rotating machinery are to be discussed. Such analysis also covers, through the medium of Fourier analysis, the effects of more complex periodic excitation. The motion of a system under transient loading, for instance pulse excitation, will be treated in Chapter 11 where it will be shown that the results for sinusoidal forcing provide useful data for the solution of the transient problem.

EXAMPLE 1.1

1. How many degrees of freedom have the following systems?
   (*a*) A particle in space.
   (*b*) A particle which is constrained to move along a fixed tortuous curve.
   (*c*) A lamina which can move in its own plane only.
   (*d*) A four-bar kinematic chain with one link fixed.
   (*e*) A four-bar kinematic chain with no links fixed.
   (*f*) A rigid gyroscope rotor in gimbals.
   (*g*) A tramcar.
   (*h*) A motor car.
   (*i*) An aeroplane.
For (*g*), (*h*) and (*i*) it is intended that the bodies should be treated as rigid.

1.2 Systems with one degree of freedom

The simplest type of vibrating system has a single degree of freedom. Consider, as an example, that shown in fig. 1.2.1, in which the rigid body of mass M is attached to a fixed abutment through the massless spring of stiffness k. Displacements of the mass from its equilibrium position will be denoted by x. The equation of motion of the system, when it is vibrating freely, is

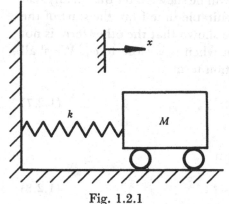

Fig. 1.2.1

$$\ddot{x} + \omega_1^2 x = 0, \tag{1.2.1}$$

where $\omega_1^2 = k/M$. The solution of (1.2.1) may be written in the forms

$$\left.\begin{aligned} x &= A\cos\omega_1 t + B\sin\omega_1 t \\ &= R\cos(\omega_1 t - \phi) \\ &= R\sin(\omega_1 t + \psi), \end{aligned}\right\} \tag{1.2.2}$$

where A, B, R, ϕ and ψ are constants which are determined by the initial conditions. The quantity ω_1 is sometimes called the 'circular frequency'; it will be convenient, when there is no danger of confusion, to refer to it simply as 'the frequency'.

Equation (1.2.1) is easily formed by the application of Newton's laws. With some single-degree-of-freedom systems, however, this method is somewhat laborious and an energy method may be used instead. Then, the kinetic energy T and the potential energy V of the system concerned are written as functions of the single co-ordinate. The equation of free motion can then be found from the relation

$$\frac{d}{dt}(T + V) = 0 \tag{1.2.3}$$

because the total energy of the system remains constant in the absence of external and frictional forces. Alternatively we may assume the motion to be harmonic and can then find the frequency from the relation

$$T_{\text{max.}} = V_{\text{max.}}, \tag{1.2.4}$$

that is, by equating the maximum values of T and V. This latter method requires the value of the potential energy to be taken as zero in the equilibrium position. These two energy methods can be tested on the system of fig. 1.2.1. A third, and much more powerful energy method—that of Lagrange—will be introduced in Chapter 2.

Now let a harmonic force $F\sin\omega t$ be applied to the mass M along the direction of the motion. The equation now becomes

$$\ddot{x} + \omega_1^2 x = \omega_1^2\left(\frac{F}{k}\right)\sin\omega t \tag{1.2.5}$$

and the general solution to this is the sum of the complementary function (1.2.2) and a particular integral. The latter corresponds, as may be expected, to a motion of the mass with the frequency ω of the disturbing force; the general solution is found to be

$$x = [A\cos\omega_1 t + B\sin\omega_1 t] + \frac{F/k}{1 - \dfrac{\omega^2}{\omega_1^2}}\sin\omega t. \qquad (1.2.6)$$

The terms within the bracket in this expression allow, by a suitable choice of A and B, for the initial conditions to be satisfied; it will be shown later that in any real system these terms are reduced to zero, after a suitable time from the start of the motion, by the action of damping. It will also be shown that the other term is not greatly affected by small damping forces except when ω is close to ω_1. We shall therefore confine our attention to the forced motion term

$$x = \frac{F/k}{1 - \dfrac{\omega^2}{\omega_1^2}}\sin\omega t. \qquad (1.2.7)$$

The solution (1.2.7) may be written in the form

$$x = \left(\frac{F}{k}\right) N\sin(\omega t - \zeta), \qquad (1.2.8)$$

where N is the 'magnification factor' which is defined by the relation

$$N = \left|\frac{1}{1 - \dfrac{\omega^2}{\omega_1^2}}\right|, \qquad (1.2.9)$$

and ζ is a phase angle which is zero for $\omega < \omega_1$ and π for $\omega > \omega_1$. These two quantities are functions of ω/ω_1 only and are shown sketched in fig. 1.2.2. It must be emphasized that any point on these curves represents a particular solution to equation

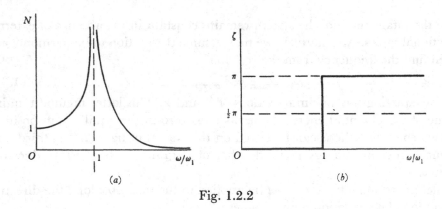

Fig. 1.2.2

(1.2.5) for the appropriate value of ω/ω_1; the curves do not imply that if, for a specified system, ω were varied, then the motion at any moment would be given by them. The forcing of the system at a *varying* frequency would imply transient excitation for which it is not permissible to neglect the complementary function.

Later on, we shall often meet the curves of fig. 1.2.2 in a slightly different form. This is obtained by writing (1.2.7) in the form

$$x = \left[\frac{1}{M(\omega_1^2 - \omega^2)} \right] F \sin \omega t \qquad (1.2.10)$$

and then plotting the quantity in the square bracket against ω. The curve is shown in fig. 1.2.3.

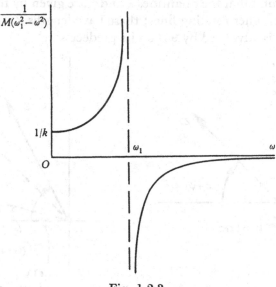

Fig. 1.2.3

As ω is taken closer and closer to ω_1, the amplitude of the motion increases without limit if the force amplitude is kept the same. Alternatively the force amplitude must be diminished indefinitely if the displacement amplitude is not to increase. The latter statement is usually preferable because an infinitely large amplitude is inconsistent with the assumption of small motion. The concept of reducing the force indefinitely as ω tends to ω_1 may be used to provide an alternative definition of natural frequency, namely that it is the frequency at which a finite response is produced by an infinitesimal force. We shall find it more convenient later to adopt this definition rather than to work explicitly with the equations of free vibration.

In using the expression (1.2.7) rather than the general solution to the differential equation the reader may prefer, at this stage, to suppose that the initial conditions were such as to make the constants A and B equal to zero. This will not impair the validity of the mathematics though it will restrict the physical application of it. Such a viewpoint avoids the apparent contradiction in the assumptions that the damping will eliminate the free motion but may be neglected in discussing the forced motion. This contradiction, and the important effect of damping on the forced motion in the immediate neighbourhood of the natural frequency, will be discussed in Chapters 8 and 9.

Both the free and forced motions of (1.2.2) and (1.2.7) are harmonic; we now introduce a method for representing such quantities graphically. A harmonically varying quantity

$$\xi = Z \sin (\Omega t + \psi) \qquad (1.2.11)$$

may be represented by a rotating line. Thus if a line of length Z rotates with angular velocity Ω about one end, as shown in fig. 1.2.4, then the value of ξ is given by the projection of the moving line on a fixed line; this is the vertical line in the figure. Further, it will be found that the quantities $\dot{\xi}$ and $\ddot{\xi}$ are given by the projections on the same fixed line of other rotating lines; these have lengths $Z\Omega$ and $Z\Omega^2$ respectively, and each one is advanced by $90°$ on its predecessor.

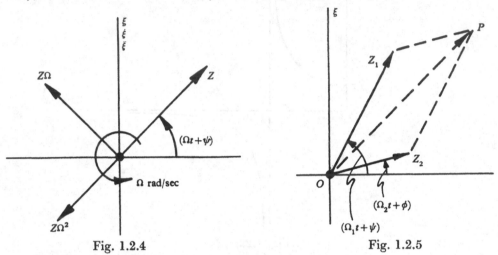

Fig. 1.2.4   Fig. 1.2.5

An important property of these rotating lines is that they may be added vectorially. Let some quantity ξ be given as the sum of two harmonic components so that we might write

$$\xi = Z_1 \sin (\Omega_1 t + \psi) + Z_2 \sin (\Omega_2 t + \phi). \qquad (1.2.12)$$

It may be seen from fig. 1.2.5 that ξ is given by the projection of the line OP, which is the vector sum of the lines whose projections give the two components of ξ. If the frequencies of the two components are not the same then the instantaneous value of ξ is still given by the projection of the vector-sum line; but under these conditions this line will not remain of constant length as it moves. In many problems only equal-frequency components are involved; when this is so the fixed line may be omitted from the figure because it is only the relative positions of the rotating lines, that is to say the shape of the diagram which depends on the relative phases, that concerns us.

When the rotating line representation is applied to the free vibration expression for the oscillator, then R of equation (1.2.2) will be identified with Z, ω_1 with Ω and x with ξ.

The forced vibration equation may also be represented in this way. The equation of motion may be written in the form

$$M\ddot{x} + kx = F \sin \omega t, \qquad (1.2.13)$$

so that the right-hand side will correspond to a rotating line of length F. We confine attention to the variation of x with the frequency ω so that ω is identified with Ω. The value of Z can now be deduced from the diagram of fig. 1.2.6. In order that the three lines in the figure should fit together as shown, which they must if the right-hand side of (1.2.13) is to be the vector sum of the left-hand side, it is necessary that

$$Z(k - M\omega^2) = F,$$

or

$$Z = \frac{F}{(k - M\omega^2)} = \frac{F/k}{1 - \dfrac{\omega^2}{\omega_1^2}}, \qquad (1.2.14)$$

which agrees with the analytical solution.

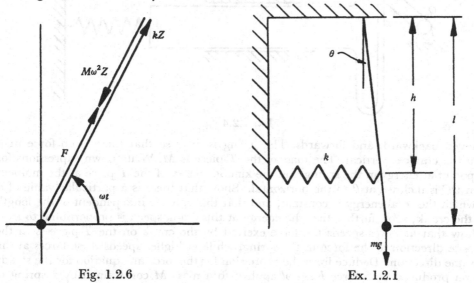

Fig. 1.2.6   Ex. 1.2.1

EXAMPLES 1.2

1. The figure (Ex.1.2.1) shows a simple pendulum, to the arm of which is attached a light spring of stiffness k. The system is in equilibrium when the arm is vertical. Find the frequency of free oscillations by

   (a) applying Newton's laws to the free system,

   (b) using an energy method,

   (c) finding the resonance condition when a harmonic, horizontal disturbing force is applied to the arm at a distance $\frac{1}{2}h$ below the point of suspension.

   [NOTE. While it is convenient to specify the point of application of the force, the choice cannot influence the value of the natural frequency which is found by this method.]

2. Two equal rollers, each of radius r, are placed with their axes at the same level and distant $2\cdot4r$ apart, in contact with a concave cylindrical surface of radius $3r$. They are maintained in this position by a third equal roller which is placed in contact with each of them, its axis thus coinciding with the axis of the concave surface. All the rollers are solid and homogeneous.

   Show that the periodic time of a small oscillation of the three rollers about the above position of equilibrium, assuming no slipping, will be $5\pi\sqrt{(r/g)}$.

   (C.U.M.S.T. Pt I, 1953)

3. A connecting rod weighing 5·10 lb. is suspended from a knife edge which is passed through the small-end bearing. It is found to perform 50 small oscillations in 75 sec. Find the moment of inertia of the rod about its centre of gravity, if the position of the latter is 10·1 in. from the point of suspension at the bearing.

4. In the mechanism shown the *T*-piece slides in the fixed guides *AB*, the movement being resisted by a spring which can be compressed or stretched unit length by an applied force *k*. The crank *OC* rotates with angular velocity ω about *O* so that the *T*-piece

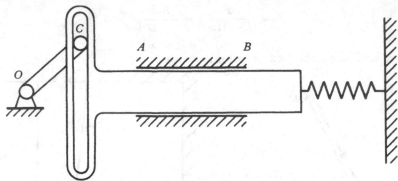

Ex. 1.2.4

is moved backwards and forwards. The spring is fixed so that there is no force in it when the crank is vertical. The mass of the *T*-piece is *M*. Write down expressions for the potential energy in the spring and the kinetic energy of the *T*-piece at the moment when *OC* is inclined at θ to the horizontal. Show that there is a particular value of ω for which the total energy is constant, and that this value is independent of the length *r* of the crank; show further that the energy at this crank speed is proportional to r^2.

Show that at lower speeds the force exerted by the crank on the *T*-piece is in the opposite direction to the force in the spring, while at higher speeds these forces are in the same direction. Deduce from the expression for the force an equation for the steady motion produced by a force $F \sin \omega t$ applied to a mass *M* constrained by a spring of stiffness *k*.

5. A mass *M* is attached to one end of a vertical light spring of stiffness *k*, the other end of which is fixed. The mass rests upon a table which has a vertical motion $a \sin \omega t$. If, in its mean position, the spring is compressed by an amount δ, over what frequency range will the mass rest permanently on the table?

6. A vibrating spring-mass system is governed by the equation

$$M\ddot{x} + kx = 0.$$

Show that if, at any instant, x and $\dot{x}$ are given, then the dynamical state of the system is completely known so that the motion may be fully represented by a point which moves in a plane so that its Cartesian co-ordinates are $(x, \dot{x})$ (the 'phase plane'). Show further, that the vibration is represented by a family of ellipses in the plane and express the lengths of the semi-axes in terms of the energy *E* of the system.

How will the ellipses be distorted if axes of x and $\dot{x}/\omega_1$ are used where $\omega_1 = \sqrt{(k/M)}$?

[For a further discussion of this type of representation see:

Andronow, A. A. and Chaikin, C. E., *Theory of Oscillations* (Moscow, 1937; English translation, Princeton University Press, 1949).

Lamoën, J., *Revue Universelle des Mines*, Series 8 (1935), vol. 11, no. 7, p. 213.

Bishop, R. E. D., *Proc. Inst. Mech. Engrs*, vol. 168 (1954), p. 299.

Jacobsen, L. S. and Ayre, R. S., *Engineering Vibrations* (McGraw-Hill, London, 1958).]

7. A uniform bar of length $2l$ and mass M is freely hinged at one end to a fixed support on level ground. The bar is held in equilibrium at an angle 2θ to the horizontal by means of an elastic cord which runs from the other end of the bar down to a point on the ground which is distant $2l$ from the hinge; the bar and cord are in a vertical plane. The elasticity of the cord is such that a force λ is required to produce unit extension.

Show that the equilibrium will be stable provided that

$$\frac{8l\lambda}{Mg} > \frac{\cos\theta}{\sin^3\theta}(1+2\sin^2\theta).$$

If this inequality holds find the natural frequency of small oscillations of the rod in the verical plane.

[NOTE. This system is non-linear since its geometry varies during the motion. As in the case of the pendulum, the solution given (p. 574) is a first approximation; it is accurate for very small oscillations.]

1.3 Representation of harmonic motion by the complex exponential

In this book we shall use the complex exponential $e^{i\psi}$ instead of the circular functions when expressing harmonic quantities. This is for mathematical convenience and it may be justified as follows.

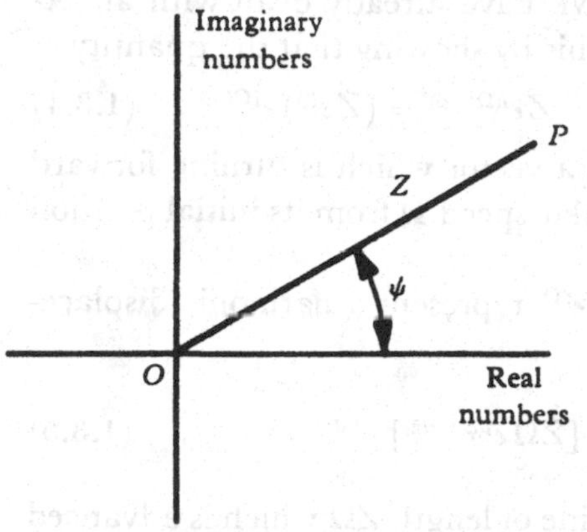

Fig. 1.3.1

Let z be a complex number whose real and imaginary parts are a and b respectively. z will be represented by the point P whose co-ordinates are (a,b) on the Argand diagram, fig. 1.3.1. The quantity may alternatively be written

$$z = a+ib = Z(\cos\psi + i\sin\psi) = Ze^{i\psi}, \tag{1.3.1}$$

where $Z = \sqrt{(a^2+b^2)}$; for it can be shown by expansion in series that

$$e^{i\psi} \equiv \cos\psi + i\sin\psi. \tag{1.3.2}$$

Now if ψ is replaced by the argument $\Omega t + \phi$, then the line OP will rotate about O with angular velocity Ω. The real and imaginary parts of z are obtained, at any instant, from the projections of OP on the horizontal and vertical axes; they are equal to $Z\cos(\Omega t + \phi)$ and $Z\sin(\Omega t + \phi)$ respectively.

It has been shown that a harmonically varying quantity, such as a displacement or velocity for instance, can be written in either of these forms. Such a quantity can therefore be represented by z if it is specified that the line joining the point z to the origin is to be projected onto some chosen fixed line. Thus the quantity $2\sin(3t+5)$ is given by the projection on the vertical axis, that is to say by the imaginary part, of $2e^{i(3t+5)}$. Similarly the real part is equal to $2\cos(3t+5)$. In this way the rotating line convention, which was explained in the last section, can be expressed in an analytical form.

It can be deduced from the above theory that the identity (1.3.2) ensures that addition and differentiation can be performed with the complex exponential, yielding correct results provided that the harmonic quantity is identified with the same part, be it either the real or the imaginary, throughout. It is desirable, however, to continue a little further with the geometrical interpretation of these processes because it correlates this method with the vector representation.

We first note that when two complex numbers are added, the real and imaginary parts of each are summed separately in order to give the real and imaginary parts of the whole. Inspection of fig. 1.2.5 shows that, if two lines are expressed in analytical form, the vector addition of the lines gives the same result as the analytical addition of the quantities.

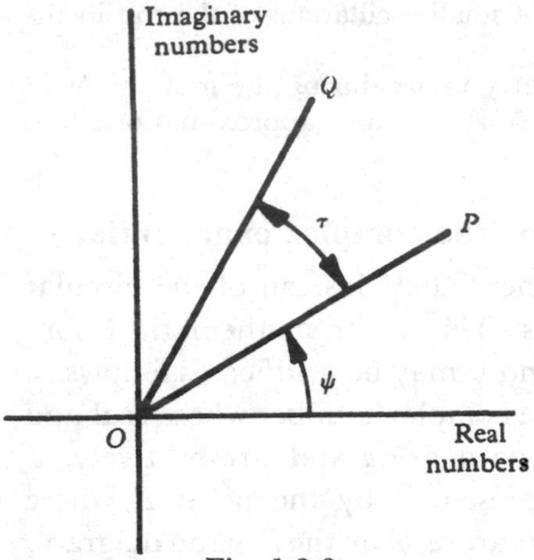

Let a new complex number z' be formed by rotating the line which corresponds to z through some angle τ, without changing its length, as shown in fig. 1.3.2. We have

$$z' = Z e^{i(\psi+\tau)} = z e^{i\tau} \qquad (1.3.3)$$

which shows that the formation of z' in this way is expressible analytically by multiplying z by $e^{i\tau}$. We have already dealt with an example of this by showing that the quantity

$$Z e^{i(\Omega t+\phi)} = (Z e^{i\phi}) e^{i\Omega t} \qquad (1.3.4)$$

represents a vector which is turning forward with angular speed Ω from its initial position $Z e^{i\phi}$.

Fig. 1.3.2

Let either part of a complex number $(Z e^{i\phi}) e^{i\Omega t}$ represent a harmonic displacement. By differentiation it is found that

$$\frac{d}{dt}[Z e^{i(\Omega t+\phi)}] = iZ\Omega\, e^{i(\Omega t+\phi)} = [Z\Omega\, e^{i(\phi+\pi/2)}]\, e^{i\Omega t}, \qquad (1.3.5)$$

so that the time derivative is represented by a line of length $Z\Omega$ which is advanced through $\frac{1}{2}\pi$ radians from the original vector and rotates with it. A second differentiation will bring about another multiplication by Ω and advancement by $\frac{1}{2}\pi$ radians, and so on. It will be seen from inspection of fig. 1.2.4 that this rule will give the vector representation of derivatives correctly provided that we adhere to either the real or the imaginary part of all the complex numbers.

Thus the vector manipulations of §1.2—addition and differentiation—can be performed symbolically by the rules of analysis without the need for drawing the vectors; and again the projection of the vectors on the axes is obtained by taking the real and imaginary parts separately. It must be noted that, in this analytical method,

$$i^2 = e^{i\pi} = -1 \qquad (1.3.6)$$

because, when it operates on a complex number, i^2 turns the vector through π radians without changing its length.

The complex exponential notation can now be applied to the problem of the simple oscillator that is shown in fig. 1.2.1. The solution (1.2.2) to this problem can be represented by means of the complex exponential; to do this, we identify R with Z and ω_1 with Ω. But it is not necessary to solve the equation of motion by means of trigonometric functions, to deduce from the solution a vector and finally to associate the vector with a complex number. Instead of this the differential equation can be solved immediately in terms of complex exponentials. By trying the solution $x = e^{\lambda t}$ in equation (1.2.1) we find

$$x = A\,e^{i\omega_1 t} + B\,e^{-i\omega_1 t}, \tag{1.3.7}$$

which can be regarded as two revolving lines in the Argand diagram, the directions of rotation being equal and opposite and A and B being constants that are fixed by the initial conditions.† If B is put equal to zero and A is identified with a vector z_0 (say) which represents the state of the system at the instant $t = 0$, then a single rotating vector is obtained of which either the real or imaginary part may be chosen for the harmonic quantity that is required.

Again the forced motion of the oscillator is given by the equation

$$M\ddot{x} + kx = F\sin \omega t \tag{1.3.8}$$

and this can be rewritten in the form

$$M\ddot{x} + kx = F e^{i\omega t} \tag{1.3.9}$$

so that the harmonic force corresponds to the imaginary part of the exponential. The trial solution $x = X e^{i\omega t}$ then gives

$$x = \frac{F e^{i\omega t}}{k - M\omega^2} = \frac{F/k}{1 - \dfrac{\omega^2}{\omega_1^2}} e^{i\omega t} \tag{1.3.10}$$

and the solution of equation (1.3.8) is the imaginary part of (1.3.10). If the right-hand side of (1.3.8) had been $F\cos \omega t$, then the real part of (1.3.10) would have been needed.

EXAMPLES 1.3

1. The values when $t = 0$ of a harmonic displacement x and velocity $\dot{x}$ are x_0 and $\dot{x}_0$ respectively. If the circular frequency of the motion is ω find:

    (a) the complex representation, the real part of which gives x, $\dot{x}$, etc.,

    (b) the complex representation from which x, $\dot{x}$, etc., are found by taking the imaginary part.

Explain the results by using the Argand diagram.

2. A point in the Argand diagram moves in such a way that its position at time t is given by

$$z = r_1 e^{i\omega t} + r_2 e^{-i\omega t}.$$

Show that the locus of the point is an ellipse, of which the major and minor axes are $r_1 + r_2$ and $r_1 - r_2$.

† This representation of harmonic motion is sometimes used in electrical work.

3. A point moves so that its position at time t is given by

$$z = r_1 e^{i\Omega t} + \frac{r_1}{3} e^{-i3\Omega t}.$$

Show that its locus is an epicycloid having four cusps. Show also that if the displacement of the point is measured relative to a set of axes which is rotating clockwise with angular velocity Ω, then the locus becomes an ellipse.

[NOTE. We propose to discuss the significance of these results when we come to problems of whirling shafts in a later volume.]

1.4 Systems with two degrees of freedom

The theory of systems with more than one degree of freedom involves several new concepts. In this article we shall discuss a two-degree-of-freedom system in order that the reader may become familiar with these before more general theory is introduced. The system shown in fig. 1.4.1 is similar to that of fig. 1.2.1 except that a pendulum element has been added to it. The pendulum is of length l and the bob is a particle of mass m; the pendulum is suspended from a frictionless pivot on the mass M.

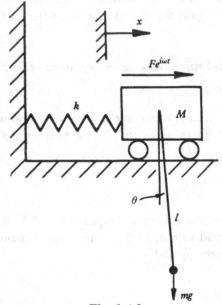

For this problem, and in the more general discussion subsequently, we shall adopt the definition of free vibration which was mentioned earlier; namely that it is the limiting case of forced vibration as resonance is approached. It will be necessary therefore to deal with equations of forced vibration and these will be handled with the complex exponential.

Small motions of the system are governed by the equations

Fig. 1.4.1

$$\left. \begin{array}{l} M\ddot{x} + kx - mg\theta = Fe^{i\omega t}, \\ \ddot{x} + l\ddot{\theta} + g\theta = 0, \end{array} \right\} \tag{1.4.1}$$

as may be found by direct application of Newton's laws. That is, it is necessary now to deal with simultaneous ordinary differential equations. We begin by trying a solution in which the co-ordinates x and θ vary harmonically with the impressed frequency; thus,

$$x = Xe^{i\omega t}, \quad \theta = \Theta e^{i\omega t}. \tag{1.4.2}$$

When these expressions are substituted in the equations they yield

$$\left. \begin{array}{l} (k - M\omega^2) X - mg\,\Theta = F, \\ -\omega^2 X + (g - l\omega^2)\,\Theta = 0, \end{array} \right\} \tag{1.4.3}$$

these being simultaneous algebraic equations which relate the amplitudes of the harmonic quantities. The solutions of these equations are

$$X = \frac{\begin{vmatrix} F & -mg \\ 0 & g - l\omega^2 \end{vmatrix}}{\begin{vmatrix} k - M\omega^2 & -mg \\ -\omega^2 & g - l\omega^2 \end{vmatrix}} = \frac{F(g - l\omega^2)}{(k - M\omega^2)(g - l\omega^2) - mg\omega^2},$$

$$\Theta = \frac{\begin{vmatrix} k - M\omega^2 & F \\ -\omega^2 & 0 \end{vmatrix}}{\begin{vmatrix} k - M\omega^2 & -mg \\ -\omega^2 & g - l\omega^2 \end{vmatrix}} = \frac{F\omega^2}{(k - M\omega^2)(g - l\omega^2) - mg\omega^2}.$$

$$(1.4.4.)$$

The excitation frequency at which finite displacements can be produced by a vanishingly small force is obtained by putting the denominator of the expressions for X and Θ equal to zero. We thus have

$$(k - M\omega^2)(g - l\omega^2) - mg\omega^2 = 0 \qquad (1.4.5)$$

which is the frequency equation. It is an equation in ω^2 which has the two roots

$$\omega_1^2 = \frac{1}{2}\left(\frac{k}{M} + \frac{g}{l} + \frac{mg}{Ml}\right) - \sqrt{\left[\frac{1}{4}\left(\frac{k}{M} + \frac{g}{l} + \frac{mg}{Ml}\right)^2 - \frac{kg}{Ml}\right]},$$

$$\omega_2^2 = \frac{1}{2}\left(\frac{k}{M} + \frac{g}{l} + \frac{mg}{Ml}\right) + \sqrt{\left[\frac{1}{4}\left(\frac{k}{M} + \frac{g}{l} + \frac{mg}{Ml}\right)^2 - \frac{kg}{Ml}\right]}.$$

$$(1.4.6)$$

It will be found that both roots are real and positive for all real positive values of the physical constants of the system. The corresponding frequencies are also real therefore. Subscripts have been affixed to the roots to distinguish them; they are in ascending order of magnitude, a system which will always be used hereafter. There are thus two natural frequencies at which the system will oscillate freely. The state of the system during this free vibration can now be examined.

For any value of the forcing frequency ω, equations (1.4.4) show that the amplitudes X and Θ are in the ratio

$$\frac{X}{\Theta} = \frac{g - l\omega^2}{\omega^2} = \chi \quad \text{(say)}. \qquad (1.4.7)$$

If ω is taken closer and closer to one of the natural frequencies, the magnitude of F being reduced in order to keep the motion small, then the ratio χ approaches one of the limits

$$\chi_1 = \left(\frac{X}{\Theta}\right)_{\omega=\omega_1} = \frac{g - l\omega_1^2}{\omega_1^2},$$

$$\chi_2 = \left(\frac{X}{\Theta}\right)_{\omega=\omega_2} = \frac{g - l\omega_2^2}{\omega_2^2}.$$

$$(1.4.8)$$

These limiting ratios depend only upon the constants of the system. It follows that two free harmonic oscillations are possible, namely

$$x = A\chi_1 e^{i\omega_1 t}, \quad \theta = A e^{i\omega_1 t},$$

and
$$x = B\chi_2 e^{i\omega_2 t}, \quad \theta = B e^{i\omega_2 t}. \tag{1.4.9}$$

The constants A and B, which may be complex, are introduced in order to satisfy any specified initial conditions.

Each pair of equations (1.4.9) defines motion in a 'principal mode', the form taken by the oscillation being dictated by the relevant relation (1.4.8); during such motion, the variations of x and θ have the same frequency and are in phase.

The equations of motion (1.4.1) are linear; it follows that if two solutions of them are found, then these may be added to give another solution. Thus a more general solution for free vibration is

$$x = A\chi_1 e^{i\omega_1 t} + B\chi_2 e^{i\omega_2 t},$$
$$\theta = A e^{i\omega_1 t} + B e^{i\omega_2 t}. \tag{1.4.10}$$

It will be found that these expressions satisfy equation (1.4.1) when F is zero. In general, therefore, free vibration is not simple harmonic, but is formed from the sum of motions in the principal modes. The superposition of the two motions allows four initial conditions to be satisfied; thus x, $\dot{x}$, θ and $\dot{\theta}$ can be specified and suitable values of the real and imaginary parts of A and B can then be chosen.

During free oscillation in a principal mode the ratio between the amplitudes of x and θ is fixed by the constants of the system. The magnitude, or intensity of the motion is free to take any value, however. Later on, we shall show that it is often necessary to define a unit magnitude for the distortion in a principal mode; such a choice is arbitrary and we could for instance take unit distortion in the first principal mode to correspond to unit displacement of θ so that the displacement of x would be χ_1. If we define unit distortion for both the principal modes in this way then the quantities A and B in (1.4.10) become direct measures of the amplitude and phase of the two modes.

In the following chapters several properties of the principal modes will be discussed; a particularly important one will be demonstrated now. This is that, once the form of a mode is known, the appropriate frequency can be calculated directly. This is important because, as we shall find later, the calculation of natural frequencies is an essential part of vibration analysis. The frequency which corresponds to a given mode can conveniently be found by an energy method; the reader can check that, if the numerical values of χ_1 and χ_2 are known, then ω_1 and ω_2 can be calculated (see example 1.4.5).

Now there are systems for which it is possible to obtain the principal modes by inspection. The system which we have been treating is not one of these (though it will be found subsequently that it is possible to deduce the modes if numerical values of M, m, k and l are known). The modes of the system of fig. 1.4.2 can be guessed, on the other hand, because of the symmetry. The system consists of two similar pendulums which are coupled together by a light spring. Small harmonic oscillations of the system are possible in which $\theta_1 = \theta_2 = \theta$ so that the spring remains unstrained; in this mode the energies are

$$T = ml^2\dot\theta^2, \quad V = mgl\theta^2, \tag{1.4.11}$$

and hence by using one of the energy methods of § 1.2 it may be shown that $\omega_1^2 = g/l$. Harmonic oscillations are also to be expected in a second mode for which $\theta_1 = -\theta_2$. In this mode the centre of the spring remains stationary, by symmetry, and the

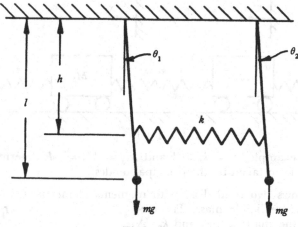

Fig. 1.4.2

oscillation of each half will be the same as it would be if the centre point of the spring were clamped. The energy expressions for the whole system are

$$T = ml^2\dot\theta^2, \quad V = mgl\theta^2 + 2kh^2\theta^2. \tag{1.4.12}$$

In these, the subscripts of θ have been omitted because the negative sign of θ_2 is lost when squaring. The expressions give the frequency as

$$\omega_2^2 = \frac{g}{l} + \frac{2kh^2}{ml^2}. \tag{1.4.13}$$

This may equally well be found by applying Newton's laws to one half of the system, assuming that the mid-point of the spring is anchored.

It will be shown in Chapter 3 that there are always as many principal modes as there are degrees of freedom. In general, each mode will have a different frequency but there are systems for which equal-frequency modes exist. It is possible for some of the modes to have degenerate forms.

EXAMPLES 1.4

1. The figure shows a system of two masses M_1 and M_2 which are coupled to each other and to fixed points by three springs of stiffness k_1, k_2 and k_3. Find the frequency equation of free vibration, using

(a) the equations of free motion which give the conditions for finite movement without an applied force,

(b) the equation of forced motion, assuming that a harmonic force is applied to one of the masses.

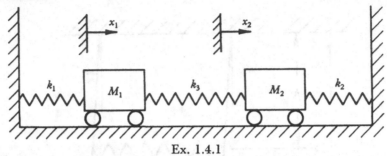

Ex. 1.4.1

2. If, in the last example, $k_1 = k_2 = k$ and $M_1 = M_2 = M$, determine the natural frequencies and the form taken by the principal modes.

3. The figure shows two rigid disks with moments of inertia I_1 and I_2 which are attached to shafts of negligible mass. The torsional stiffnesses of the shafts are k_1 and k_2. A sinusoidal torque $M\,e^{i\omega t}$ is applied to the inertia I_1.

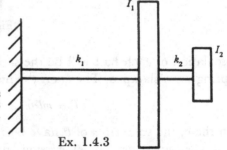

Considering only motion with the impressed frequency determine the value of ω for which the portion of the system formed by I_2 and k_2 acts as a tuned absorber—or 'reactor'—that is to say for which the inertia I_1 remains stationary and only I_2 moves.

Ex. 1.4.3

4. Find a rotating vector representation of the general motion, with the impressed frequency, of the system of the last example. What forms do the vectors take in the following conditions—?

(a) that of the tuned absorber described in the previous example,

(b) free oscillation, with infinitely small exciting force, at the two natural frequencies.

By inspection of the diagrams write down the solution to Ex. 3, and find the frequency equation.

5. For the system of fig. 1.4.1, let

$$k = 256 \text{ pdl./ft.,} \quad l = 1 \text{ ft.,}$$
$$M = 2 \text{ lb.,} \qquad m = 1 \text{ lb.}$$

(a) Find the natural frequencies of the system.

(b) Hence evaluate the quantities χ_1 and χ_2 of equation (1.4.8).

(c) Using these values of χ_1 and χ_2 find the natural frequencies by an energy method.

1.5 The concept of receptance

The procedure which has been used for examining the motion of a two-degree-of-freedom system can theoretically be extended to deal with systems having any number of degrees of freedom. If there are n degrees of freedom, then there will be n simultaneous equations of motion. These can usually be set up most conveniently by the method of Lagrange. The equations may be solved by trial solutions in which all the displacements vary harmonically at the disturbing frequency. The theory has been in existence for many years and has been thoroughly discussed;† unfortunately, however, direct application of the theory to numerical problems frequently becomes very unwieldy. Some simplification can often be gained in this respect by using the concept of 'receptance'.‡

Let a harmonic force $Fe^{i\omega t}$ act at some point of a dynamical system so that the system takes up a steady motion with the same frequency ω, such that the point of application of the force has the displacement

$$x = Xe^{i\omega t}. \tag{1.5.1}$$

Then, if the equations of motion are linear, this may be written

$$x = \alpha Fe^{i\omega t}, \tag{1.5.2}$$

where α depends upon the nature of the system and the frequency ω but not upon the amplitude F of the force. The quantity α is termed 'the direct receptance at x'.

If on the other hand x is the displacement at some point of the system other than that at which the force is applied, then equation (1.5.2) defines a 'cross-receptance' α. We shall discuss this type of receptance later.

We shall now derive some simple direct receptances in order to clarify the concept for the reader. The displacement x of the rigid mass M of fig. 1.5.1 is given by

$$M\ddot{x} = Fe^{i\omega t} \tag{1.5.3}$$

so that, if the displacement varies sinusoidally, we can write $x = Xe^{i\omega t}$ and it then follows that

$$-M\omega^2 X = F. \tag{1.5.4}$$

That is to say,

$$\alpha = -\frac{1}{M\omega^2} \tag{1.5.5}$$

which is the direct receptance at x.

---

† An account of this classical approach will be found in Lord Rayleigh's magnificent work, *The Theory of Sound* (Macmillan, London, 1894).

‡ In 1954, Professors W. J. Duncan and M. A. Biot and the authors proposed that the word 'receptance' should replace the term 'mechanical admittance' which had been used previously; see *J. Roy. Aero. Soc.* vol. 58 (April 1954), p. 305; *Applied Mechanics Reviews*, vol. 7 (June 1954), p. 238. This suggestion was made in order that confusion might be avoided between 'admittances' as applied to mechanical and electrical systems. The two uses of the word have related, but not identical, origins.

Much work had been done before 1954 on the uses and properties of mechanical admittances. The theory had been greatly extended by Professor Duncan and an account of his work had been published; see W. J. Duncan, *Mechanical Admittances and their Applications to Oscillation Problems, R. and M. 2000* (1947).

Again the displacement x at the free end of the massless spring of fig. 1.5.2 is given by

$$kx = Fe^{i\omega t}, \tag{1.5.6}$$

from which it follows that the direct receptance at x is

$$\alpha = \frac{1}{k}. \tag{1.5.7}$$

The receptance concept is not restricted to displacements of translation, but can also be applied to rotations. Indeed it will be shown in Chapter 2 that throughout all our mathematical theory there is no need to distinguish algebraically between forces and translational displacements on the one hand and torques and rotations

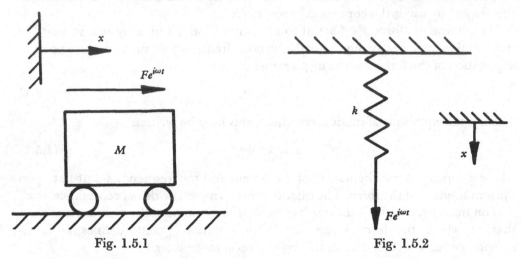

Fig. 1.5.1   Fig. 1.5.2

on the other; and for this reason, both harmonic disturbing forces and torques will usually be denoted in this chapter by $Fe^{i\omega t}$ and both types of harmonic displacements will usually be denoted by x $(= Xe^{i\omega t})$. Thus we may speak of the direct receptance at x of the freely-pinned rigid disk of fig. 1.5.3 (a) as

$$\alpha = -\frac{1}{I\omega^2} \tag{1.5.8}$$

when a torque $Fe^{i\omega t}$ acts as shown. Similarly the direct receptance at x of the massless shaft of torsional stiffness k in fig. 1.5.3 (b) is

$$\alpha = \frac{1}{k}. \tag{1.5.9}$$

Consider now the spring-mass system of fig. 1.5.4. The equation of motion is

$$M\ddot{x} + kx = Fe^{i\omega t}. \tag{1.5.10}$$

The trial solution for motion with the impressed frequency, is $x = Xe^{i\omega t}$ and this yields

$$X = \frac{F}{k - M\omega^2}, \tag{1.5.11}$$

so that the direct receptance at x is

$$\alpha = \frac{1}{k - M\omega^2}. \tag{1.5.12}$$

Fig. 1.2.3 showed a curve of α plotted vertically against ω horizontally. We shall find later that a second form of graphical representation, in which the reciprocal of the

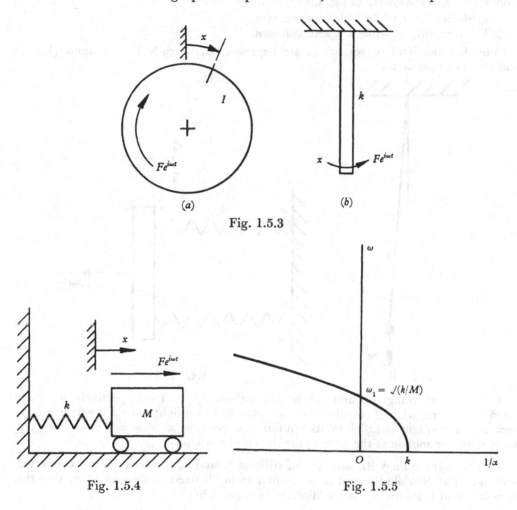

(a)　　　　　　　　　(b)

Fig. 1.5.3

Fig. 1.5.4　　　　　　　　　Fig. 1.5.5

receptance is plotted against ω, is also of use; such a curve is shown in fig. 1.5.5. The curve would clearly be symmetrical about the horizontal axis and there is therefore no need to draw it for negative values of ω. The condition of resonance occurs when $\omega^2 = k/M$ so that α becomes infinite. This is shown by the intersection of the curve of fig. 1.5.5 with the axis of ω so that this intersection represents a natural frequency of free vibration.

EXAMPLES 1.5

1. Find the direct receptance at x of the simple pendulum shown in the figure; thence deduce the natural frequency.

2. Trace the effects on the reciprocal receptance diagram of fig. 1.5.5 of the following modifications to the system of fig. 1.5.4:

   (a) decreasing k while M is kept constant,
   (b) decreasing M while k is kept constant.

Note that the limiting conditions are expressed analytically by equations (1.5.5) and (1.5.7) respectively.

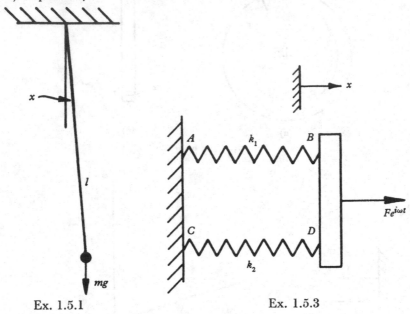

Ex. 1.5.1   Ex. 1.5.3

3. Two light springs AB and CD, having stiffnesses k_1 and k_2 respectively, are each fixed at one end while their other ends are attached to a light yoke BD which is constrained to remain parallel to its equilibrium position as shown. Show that the receptance for motion of the yoke in the direction x is $1/(k_1+k_2)$.

4. Two light springs AB and BC, of stiffness k_1 and k_2 respectively, are connected end-to-end at B while the end A of the first spring is fixed as shown. Show that the receptance at C for motion in the direction x is given by

$$\alpha = \frac{1}{k_1} + \frac{1}{k_2}.$$

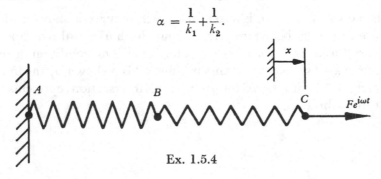

Ex. 1.5.4

1.6 Some properties of receptances

Some simple examples have been discussed which show how a receptance provides information about the response of a system to a sinusoidal force; also we have shown that the frequency at which a receptance becomes infinite is a natural frequency of the system. The expressions for α were found by direct substitution into the relevant equations of motion, and this technique can be extended to more complex problems; but if this were an essential step in every vibration analysis then little would be gained by introducing the receptance concept.

We shall now show that, by virtue of certain simple properties of the receptances, it is often possible to analyse a complex system into simpler parts whose receptances are known, are easy to find—possibly by experiment—or are tabulated; the receptances, the principal modes and the frequency equation of the complex system can then be calculated with this information. This method often saves much of the time and effort that is required for the determination of receptances by direct substitution into equations of motion.

In order to distinguish between direct and cross-receptances, and to specify the points at which receptances are to be calculated we shall add subscripts to the symbol α. The first subscript will indicate the co-ordinate at which the response is measured and the second will indicate that at which the disturbing force—or torque—is applied. Thus α_{mn} means that cross-receptance which gives the displacement at the mth co-ordinate due to an applied force at the nth. A direct receptance will therefore have identical subscripts; thus that at the mth co-ordinate will be written α_{mm}.

1.6.1 The direct receptance at a single co-ordinate that links two systems

Let a single co-ordinate x_1 be common to two systems which are otherwise separate. The systems are represented by the block diagrams B and C in fig. 1.6.1; the receptances of system B will be denoted by β's and those of C by γ's; this notation

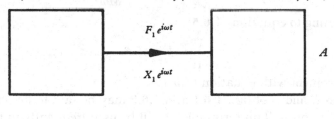

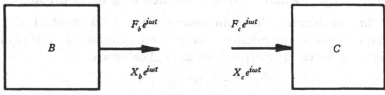

Fig. 1.6.1

will be used extensively hereafter. Omitting the factor $e^{i\omega t}$ from the forces and displacements, we have by definition

$$\beta_{11} = \frac{X_b}{F_b}, \quad \gamma_{11} = \frac{X_c}{F_c}, \tag{1.6.1}$$

where the applied forces and responses are as shown in the figure; the subscripts of F and X refer to the systems B and C. Again, using α to represent the receptance of the complete system which comprises B and C, we have,

$$\alpha_{11} = \frac{X_1}{F_1}. \tag{1.6.2}$$

Now because the systems B and C are connected it is necessary that

$$X_b = X_c = X_1 \tag{1.6.3}$$

and
$$F_1 = F_b + F_c, \tag{1.6.4}$$

from which it follows that
$$\frac{1}{\alpha_{11}} = \frac{1}{\beta_{11}} + \frac{1}{\gamma_{11}}. \tag{1.6.5}$$

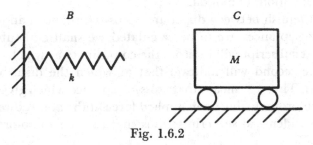

Fig. 1.6.2

As a simple example consider the system of fig. 1.5.4 which may be broken up as shown in fig. 1.6.2. The linking co-ordinate is x which we shall write x_1 in order to conform with our new system of notation. It was shown in § 1.5 that

$$\beta_{11} = \frac{1}{k}, \quad \gamma_{11} = -\frac{1}{M\omega^2} \tag{1.6.6}$$

so that, according to equation (1.6.5)

$$\frac{1}{\alpha_{11}} = k - M\omega^2. \tag{1.6.7}$$

This is in agreement with equation (1.5.12).

The systems B and C of figs. 1.6.1 and 1.6.2 may be said to be 'sub-systems' of their parent systems A. This terminology will be used frequently in this book.

1.6.2 The receptances when a force is applied to a mass through a spring

Consider the two-degree-of-freedom system of fig. 1.6.3 in which the applied force acts upon the end of the massless spring whose stiffness is k. The equations of motion in terms of the co-ordinates which are indicated are

$$\left. \begin{array}{l} F_1 e^{i\omega t} = k(x_1 - x_2), \\ M\ddot{x}_2 = k(x_1 - x_2). \end{array} \right\} \tag{1.6.8}$$

If the steady state amplitudes of x_1 and x_2 are X_1 and X_2 then direct substitution yields

$$X_1 = -\frac{(k - M\omega^2)}{kM\omega^2} F_1 = \alpha_{11} F_1,$$

$$X_2 = -\frac{F_1}{M\omega^2} \qquad = \alpha_{21} F_1. \tag{1.6.9}$$

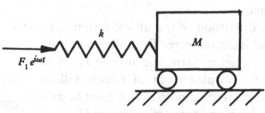

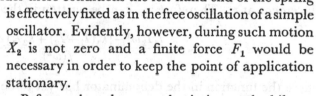

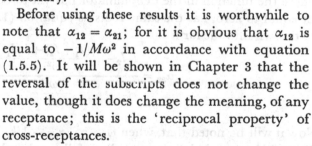

Fig. 1.6.3

When $\omega^2 = k/M$, $\alpha_{11} = 0$; under these conditions the left-hand end of the spring is effectively fixed as in the free oscillation of a simple oscillator. Evidently, however, during such motion X_2 is not zero and a finite force F_1 would be necessary in order to keep the point of application stationary.

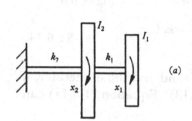

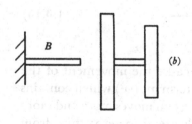

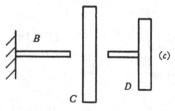

Fig. 1.6.4

Before using these results it is worthwhile to note that $\alpha_{12} = \alpha_{21}$; for it is obvious that α_{12} is equal to $-1/M\omega^2$ in accordance with equation (1.5.5). It will be shown in Chapter 3 that the reversal of the subscripts does not change the value, though it does change the meaning, of any receptance; this is the 'reciprocal property' of cross-receptances.

The use of these results may be illustrated by considering the torsional vibration of the system of shafts and disks shown in fig. 1.6.4; the disks are supposed to be rigid and the shafts massless. Let x_2 denote the rotation of that disk whose polar moment of inertia is I_2; we shall find α_{22} and thence the frequency equation of the system.

By equation (1.6.5) the direct receptance at x_2 is α_{22}, where

$$\frac{1}{\alpha_{22}} = \frac{1}{\beta_{22}} + \left(\frac{1}{\gamma_{22}} + \frac{1}{\delta_{22}}\right) \tag{1.6.10}$$

in which γ_{22} and δ_{22} are receptances of the systems C and D as indicated in fig. 1.6.4 (c); that is,

$$\frac{1}{\alpha_{22}} = \frac{1}{1/k_2} + \frac{1}{-1/I_2\omega^2} + \frac{1}{-(k_1 - I_1\omega^2)/k_1 I_1 \omega^2} \qquad (1.6.11)$$

or

$$\frac{1}{\alpha_{22}} = k_2 - I_2\omega^2 - \frac{k_1 I_1 \omega^2}{k_1 - I_1 \omega^2}. \qquad (1.6.12)$$

The frequency equation is obtained by equating $1/\alpha_{22}$ to zero so that the roots ω_1^2 and ω_2^2 of

$$k_2 - I_2\omega^2 - \frac{k_1 I_1 \omega^2}{k_1 - I_1 \omega^2} = 0 \qquad (1.6.13)$$

give the natural frequencies.

We can extend the discussion of the above system by considering the problem when the left-hand end is disconnected from the anchorage as shown in fig. 1.6.5. Let a harmonic torque $F_3 e^{i\omega t}$ be applied at this end, the displacement of which will be denoted by x_3. To find α_{33} we first note that it must be zero for those frequencies which are roots of equation (1.6.13). This follows because, as we saw previously, free vibration of the anchored system involves finite force but zero displacement at x_3. Therefore α_{33} can be written in the form

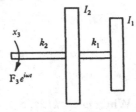

Fig. 1.6.5

$$\alpha_{33} = \frac{I_1 I_2 \omega^4 - [k_1(I_1 + I_2) + k_2 I_1]\omega^2 + k_1 k_2}{f(\omega^2)}, \qquad (1.6.14)$$

where the function in the denominator has yet to be found and the numerator, if set equal to zero, has the same roots as equation (1.6.13). Equation (1.6.14) can be thrown into the form

$$\alpha_{33} = \frac{I_1 I_2 - \dfrac{k_1(I_1 + I_2) + k_2 I_1}{\omega^2} + \dfrac{k_1 k_2}{\omega^4}}{\dfrac{f(\omega^2)}{\omega^4}}. \qquad (1.6.15)$$

Now it will be noted that, when $\omega \to \infty$, $\alpha_{33} \to 1/k_2$ because the movement of the disks will become indefinitely small. It follows that the term in $f(\omega^2)$ which contains the greatest power of ω is $I_1 I_2 k_2 \omega^4$. Again, if $\omega \to 0$ the system moves more and more as if it were a rigid body because the inertia forces also tend to zero; thus, from consideration of equation (1.6.14),

$$\alpha_{33} \to \frac{1}{-(I_1 + I_2)\omega^2} = \frac{k_1 k_2}{-(I_1 + I_2) k_1 k_2 \omega^2}. \qquad (1.6.16)$$

The complete receptance expression can now be written as

$$\alpha_{33} = \frac{I_1 I_2 \omega^4 - [k_1(I_1 + I_2) + k_2 I_1]\omega^2 + k_1 k_2}{I_1 I_2 k_2 \omega^4 - (I_1 + I_2) k_1 k_2 \omega^2}. \qquad (1.6.17)$$

1.6.3 The effect of applying a harmonic force to a system through a spring

Let the direct receptance α_{11} of a system A (fig. 1.6.6) be known; we now find the receptance α_{22} if a massless spring of stiffness k is attached to the system at x_1 as shown. For the modified system

$$F_2 e^{i\omega t} = k(x_2 - x_1) \tag{1.6.18}$$

while for the original one,

$$x_1 = \alpha_{11} F_1 e^{i\omega t} = \alpha_{11} F_2 e^{i\omega t}. \tag{1.6.19}$$

On eliminating x_1 it is found that

$$x_2 = \left(\frac{1 + k\alpha_{11}}{k} \right) F_2 e^{i\omega t} \tag{1.6.20}$$

and hence

$$\alpha_{22} = \frac{1}{k} + \alpha_{11}. \tag{1.6.21}$$

This relation is a special case of a more complex equation which holds when any type of system replaces the spring. We shall show how this more general relation may be found in § 1.6.4 below.

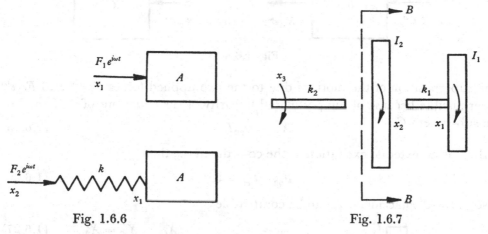

Fig. 1.6.6   Fig. 1.6.7

The problem of fig. 1.6.5 can now be solved more easily by dividing the system up as shown in fig. 1.6.7. The direct receptance β_{22} of the system B is seen from equation (1.6.5) to be given by

$$\frac{1}{\beta_{22}} = -I_2 \omega^2 - \frac{k_1 I_1 \omega^2}{k_1 - I_1 \omega^2}. \tag{1.6.22}$$

On applying equation (1.6.21) it is now found that

$$\alpha_{33} = \frac{1}{k_2} + \frac{1}{-I_2 \omega^2 - \dfrac{k_1 I_1 \omega^2}{k_1 - I_1 \omega^2}} \tag{1.6.23}$$

and this may be reduced to equation (1.6.17).

1.6.4 The effect upon a direct receptance of adding a remote system

Fig. 1.6.8 shows systems B and C linked by a single co-ordinate x_2. We shall now find the direct receptance α_{11} at x_1 of the composite system in terms of the receptances of the separate sub-systems B and C. With the forces and displacements as indicated in the diagram, motion of the sub-system B is such that

$$\left.\begin{aligned} X_1 &= \beta_{11}F_1 + \beta_{12}F_{b2}, \\ X_{b2} &= \beta_{21}F_1 + \beta_{22}F_{b2}. \end{aligned}\right\} \tag{1.6.24}$$

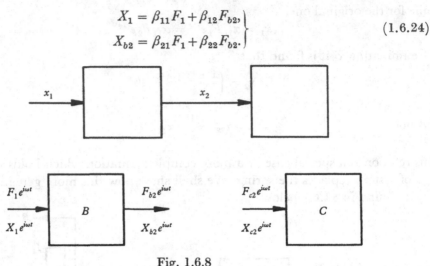

Fig. 1.6.8

The superposition of the motions due to the two applied forces $F_1 e^{i\omega t}$ and $F_{b2}e^{i\omega}$ is permissible because of the postulated linearity of the equations of motion. For the sub-system C,

$$X_{c2} = \gamma_{22}F_{c2}. \tag{1.6.25}$$

If there is no external excitation at the co-ordinate x_2, then

$$F_{b2} + F_{c2} = 0. \tag{1.6.26}$$

Also, if the displacements are to be compatible

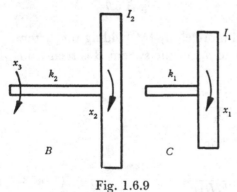

$$X_{b2} = X_{c2} = X_2. \tag{1.6.27}$$

These two conditions may be substituted into equations (1.6.24) and (1.6.25) and they then give, if the proof that $\beta_{12} = \beta_{21}$ is anticipated,

$$X_1 = \frac{\beta_{11}(\beta_{22}+\gamma_{22}) - \beta_{12}^2}{\beta_{22}+\gamma_{22}} F_1, \tag{1.6.28}$$

Fig. 1.6.9

so that $\quad \alpha_{11} = \beta_{11} - \dfrac{\beta_{12}^2}{\beta_{22}+\gamma_{22}}. \tag{1.6.29}$

This result permits a third solution of the problem of fig. 1.6.5. Let the system be divided as shown in fig. 1.6.9; if this is done, it is possible to use results which have already been obtained to write down α_{33} immediately. By analogy with the expression for α_{11} in equation (1.6.29), we have

$$\alpha_{33} = -\frac{k_2 - I_2\omega^2}{k_2 I_2 \omega^2} - \frac{\left(-\dfrac{1}{I_2\omega^2}\right)^2}{-\dfrac{1}{I_2\omega^2} - \dfrac{k_1 - I_1\omega^2}{k_1 I_1 \omega^2}}, \tag{1.6.30}$$

which yields equation (1.6.17) when it is simplified.

1.6.5 The effect upon a cross-receptance of adding a system at one of the co-ordinates concerned

From the relations (1.6.24), (1.6.25), (1.6.26) and (1.6.27) we can also obtain the result

$$X_2 = \left(\frac{\beta_{12}\gamma_{22}}{\beta_{22} + \gamma_{22}}\right) F_1, \tag{1.6.31}$$

so that

$$\alpha_{12} = \frac{\beta_{12}\gamma_{22}}{\beta_{22} + \gamma_{22}}, \tag{1.6.32}$$

where it has again been assumed that $\alpha_{12} = \alpha_{21}$ and $\beta_{12} = \beta_{21}$.

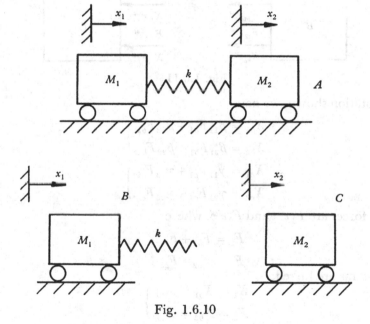

Fig. 1.6.10

This result may be illustrated with the system of fig. 1.6.10 for which the subsystem-receptances are

$$\beta_{12} = -\frac{1}{M_1 \omega^2}, \quad \beta_{22} = -\frac{k - M_1 \omega^2}{k M_1 \omega^2}, \\ \gamma_{22} = -\frac{1}{M_2 \omega^2}. \tag{1.6.33}$$

The cross-receptance α_{12} for the complete system is

$$\alpha_{12} = -\frac{\dfrac{1}{M_1\omega^2}\cdot\dfrac{1}{M_2\omega^2}}{\dfrac{k-M_1\omega^2}{kM_1\omega^2}+\dfrac{1}{M_2\omega^2}} = \frac{-k}{k(M_1+M_2)\,\omega^2 - M_1M_2\omega^4}. \qquad (1.6.34)$$

1.6.6 Systems linked by two co-ordinates

We have so far discussed systems which can be analysed in terms of sub-systems which are linked by a single co-ordinate. It is sometimes necessary to deal with systems which can only be so analysed if allowance is made for two linking co-ordinates. Let such a system be represented by the block diagrams of fig. 1.6.11.

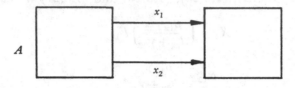

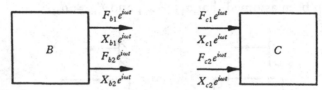

Fig. 1.6.11

Using the notation shown, we have

$$\left.\begin{aligned}X_{b1} &= \beta_{11}F_{b1}+\beta_{12}F_{b2},\\ X_{b2} &= \beta_{21}F_{b1}+\beta_{22}F_{b2},\end{aligned}\right\} \qquad (1.6.35)$$

and

$$\left.\begin{aligned}X_{c1} &= \gamma_{11}F_{c1}+\gamma_{12}F_{c2},\\ X_{c2} &= \gamma_{21}F_{c1}+\gamma_{22}F_{c2}.\end{aligned}\right\} \qquad (1.6.36)$$

The applied forces are $F_1 e^{i\omega t}$ and $F_2 e^{i\omega t}$, where

$$\left.\begin{aligned}F_1 &= F_{b1}+F_{c1},\\ F_2 &= F_{b2}+F_{c2}.\end{aligned}\right\} \qquad (1.6.37)$$

Since the systems are linked,

$$\left.\begin{aligned}X_1 &= X_{b1} = X_{c1},\\ X_2 &= X_{b2} = X_{c2}.\end{aligned}\right\} \qquad (1.6.38)$$

If $F_2 = 0$, so that excitation is applied at x_1 only, then it may be shown from the equations that

$$\alpha_{11} = \frac{X_1}{F_1} = \frac{\beta_{11}(\gamma_{11}\gamma_{22}-\gamma_{12}^2)+\gamma_{11}(\beta_{11}\beta_{22}-\beta_{12}^2)}{(\beta_{11}+\gamma_{11})\,(\beta_{22}+\gamma_{22})-(\beta_{12}+\gamma_{12})^2} \qquad (1.6.39)$$

and

$$\alpha_{21} = \frac{X_2}{F_1} = \frac{\beta_{12}(\gamma_{11}\gamma_{22}-\beta_{12}\gamma_{12})+\gamma_{12}(\beta_{11}\beta_{22}-\beta_{12}\gamma_{12})}{(\beta_{11}+\gamma_{11})\,(\beta_{22}+\gamma_{22})-(\beta_{12}+\gamma_{12})^2}, \qquad (1.6.40)$$

again assuming that the subscripts of the cross-receptances may be interchanged. Again if $F_1 = 0$, so that excitation is applied at x_2 only, then α_{12} is found to be given by the above expression for α_{21} and

$$\alpha_{22} = \frac{X_2}{F_2} = \frac{\beta_{22}(\gamma_{11}\gamma_{22} - \gamma_{12}^2) + \gamma_{22}(\beta_{11}\beta_{22} - \beta_{12}^2)}{(\beta_{11} + \gamma_{11})(\beta_{22} + \gamma_{22}) - (\beta_{12} + \gamma_{12})^2}. \qquad (1.6.41)$$

The frequency equation for the composite system is obtained, as usual, from the resonance condition at which all the receptances become infinite. This is when

$$(\beta_{11} + \gamma_{11})(\beta_{22} + \gamma_{22}) - (\beta_{12} + \gamma_{12})^2 = 0. \qquad (1.6.42)$$

The equation can be written in the determinantal form

$$\begin{vmatrix} \beta_{11} + \gamma_{11} & \beta_{12} + \gamma_{12} \\ \beta_{12} + \gamma_{12} & \beta_{22} + \gamma_{22} \end{vmatrix} = 0. \qquad (1.6.43)$$

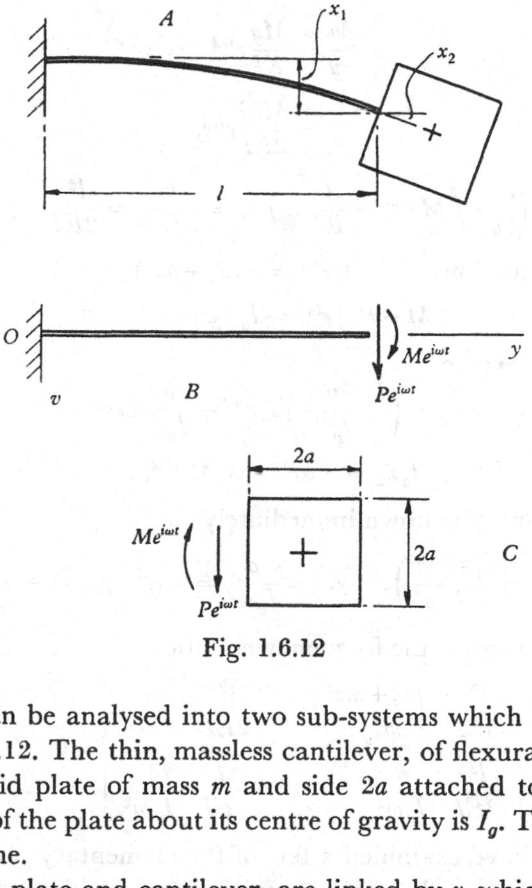

Fig. 1.6.12

A system that can be analysed into two sub-systems which have a double link is shown in fig. 1.6.12. The thin, massless cantilever, of flexural rigidity EI, has a square uniform rigid plate of mass m and side $2a$ attached to it as shown. The moment of inertia of the plate about its centre of gravity is I_g. The system oscillates in a horizontal plane.

The sub-systems, plate and cantilever, are linked by v_l which we shall identify with x_1 and $\left(\dfrac{\partial v}{\partial y}\right)_l$ which will be our x_2, v being the deflexion of the cantilever at any point. x_1 and x_2 will correspond to a disturbing force $P\,e^{i\omega t}$ and a couple $M\,e^{i\omega t}$ respectively. For the system B, with the couple absent,

$$\frac{\partial^2 v}{\partial y^2} = \frac{P(l-y)}{EI}\,e^{i\omega t}, \tag{1.6.44}$$

$$\frac{\partial v}{\partial y} = \frac{P}{EI}\left(ly - \frac{y^2}{2}\right)e^{i\omega t} \tag{1.6.45}$$

and

$$v = \frac{P}{EI}\left(\frac{ly^2}{2} - \frac{y^3}{6}\right)e^{i\omega t}, \tag{1.6.46}$$

the constants of integration being zero. We have therefore

$$\beta_{11} = \frac{v_l}{P\,e^{i\omega t}} = \frac{l^3}{3EI}, \quad \beta_{21} = \left(\frac{\partial v}{\partial y}\right)_l \bigg/ P\,e^{i\omega t} = \frac{l^2}{2EI}. \tag{1.6.47}$$

Similarly, in the absence of the force P,

$$\frac{\partial^2 v}{\partial y^2} = \frac{M}{EI}\,e^{i\omega t}, \tag{1.6.48}$$

$$\frac{\partial v}{\partial y} = \frac{My}{EI}\,e^{i\omega t} \tag{1.6.49}$$

and

$$v = \frac{My^2}{2EI}\,e^{i\omega t}, \tag{1.6.50}$$

whence

$$\beta_{22} = \left(\frac{\partial v}{\partial y}\right)_l \bigg/ M\,e^{i\omega t} = \frac{l}{EI}, \quad \beta_{12} = \frac{v_l}{M\,e^{i\omega t}} = \frac{l^2}{2EI} = \beta_{21}. \tag{1.6.51}$$

System C gives the equations

$$P\,e^{i\omega t} = m(\ddot{x}_1 + a\ddot{x}_2), \tag{1.6.52}$$

$$(M - Pa)\,e^{i\omega t} = I_g\,\ddot{x}_2. \tag{1.6.53}$$

These may be solved to give

$$m\ddot{x}_1 = \left(\frac{I_g + ma^2}{I_g}\right)P\,e^{i\omega t} - \frac{ma}{I_g}\,M\,e^{i\omega t}, \tag{1.6.54}$$

$$I_g\,\ddot{x}_2 = -aP\,e^{i\omega t} + M\,e^{i\omega t}, \tag{1.6.55}$$

and from these we can write down immediately

$$\gamma_{11} = -\left(\frac{I_g + ma^2}{mI_g\,\omega^2}\right), \quad \gamma_{12} = \frac{a}{I_g\,\omega^2} = \gamma_{21}, \quad \gamma_{22} = -\frac{1}{I_g\,\omega^2}. \tag{1.6.56}$$

Equation (1.6.43) now gives the frequency equation

$$\begin{vmatrix} \dfrac{l^3}{3EI} - \left(\dfrac{I_g + ma^2}{mI_g\,\omega^2}\right) & \dfrac{l^2}{2EI} + \dfrac{a}{I_g\,\omega^2} \\[4mm] \dfrac{l^2}{2EI} + \dfrac{a}{I_g\,\omega^2} & \dfrac{l}{EI} - \dfrac{1}{I_g\,\omega^2} \end{vmatrix} = 0. \tag{1.6.57}$$

In this article we have examined a few of the elementary properties of receptances; our illustrations have been taken mainly from problems of torsional oscillation of rigid disks and massless shafts and from the translational equivalent of these, namely the vibration of spring-connected masses. This type of problem, although particularly simple, is of great technical importance and the remainder of this chapter is devoted to it specifically.

It is not necessary to introduce Lagrange's equations for these problems so that they can be conveniently treated before we introduce more general methods into our analysis; the Lagrangian method will be used for this purpose later. In the problems of this chapter, however, we shall be concerned with many properties of receptances which are of general application.

EXAMPLES 1.6

1. Two masses M_1 and M_2 are joined by a light spring of stiffness k the masses being able to move freely parallel to the axis of the spring as shown in the diagram. Find the frequency equation of the system by splitting it up into two parts.

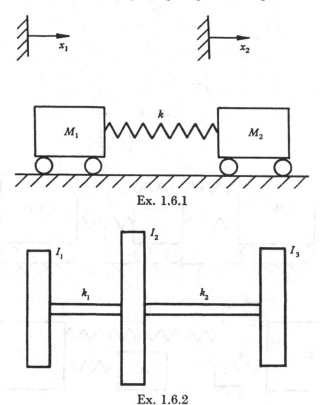

Ex. 1.6.1

Ex. 1.6.2

2. Three rigid disks, with polar moments of inertia I_1, I_2 and I_3 are connected by two light shafts with torsional stiffnesses k_1 and k_2 as shown in the figure. Show that the natural frequencies of torsional vibration are given by

$$\frac{k_1 I_1 \omega^2}{k_1 - I_1 \omega^2} + I_2 \omega^2 + \frac{k_2 I_3 \omega^2}{k_2 - I_3 \omega^2} = 0.$$

3. The crankshaft and flywheel of a particular 4-cylinder engine may be considered as equivalent, from the point of view of torsional vibration, to a set of four flywheels each with a moment of inertia I and a fifth flywheel with moment of inertia $4I$, connected

by a series of four shafts each with torsional stiffness k. Show that the natural frequencies of torsional vibration are given by

$$8 - 50x + 66x^2 - 29x^3 + 4x^4 = 0,$$

where

$$x = I\omega^2/k.$$

(C.U.M.S.T. Pt II, 1948, first part of question only)

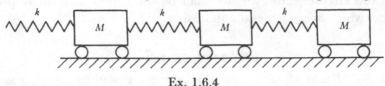

Ex. 1.6.4

4. Show that the direct receptance at the left-hand end of the system shown is given by

$$\frac{1}{k} + \cfrac{1}{-M\omega^2 + \cfrac{1}{\frac{1}{k} + \cfrac{1}{-M\omega^2 + \cfrac{1}{\frac{1}{k} - \frac{1}{M\omega^2}}}}}.$$

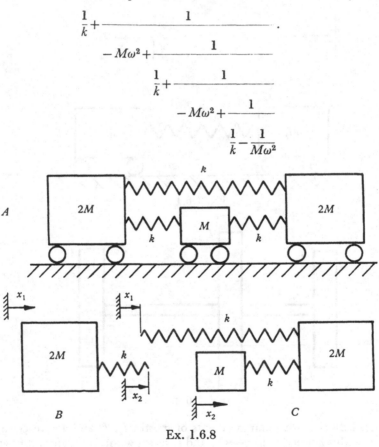

Ex. 1.6.8

5. Show that, in the system of fig. 1.6.4

$$\alpha_{12} = \frac{k_1}{k_1 k_2 - [k_1(I_1 + I_2) + k_2 I_1]\,\omega^2 + I_1 I_2 \omega^4}.$$

6. Determine the direct receptance at x of the system of fig. 1.4.1 and thence find the frequency equation by considering two systems—a simple spring-mass oscillator and a pendulum with a movable pivot. Notice that these systems are linked by *one* co-ordinate.

7. Derive equation (1.6.43) from the four pairs of equations which precede equation (1.6.39), using the fact that, for free vibration, F_1 and F_2 are zero and then examining the conditions for which the internal forces $F_{b1} e^{i\omega t}$, etc., may be non-zero.

8. Find the natural frequencies of the system shown by considering it as composed of the sub-systems B and C as indicated.

9. A mass of 10 lb. is suspended on a spring from a fixed point. A second mass, of 5 lb., is suspended by a second spring which is attached to the 10 lb. mass. With both masses supported in equilibrium the stretch of the upper spring is 0·15 in. and that of the lower 0·20 in.

Sketch, roughly to scale, the graph of the reciprocal of the direct receptance for vertical displacements of the lower mass. Obtain a rough estimate of the natural frequencies which the system would have if an additional mass of 5 lb. were fixed to the 10 lb. mass.

(Leeds B.Sc. Mech. Eng. 1958, part of question only)

1.7 The tabulation of receptances

We have shown how a complicated system can be regarded as if it were synthesized from simpler components linked at co-ordinates. The receptances and the equation which gives the natural frequencies of the composite system A are then obtainable from the receptances of the sub-systems $B, C, \dots$, etc. A few examples of the techniques involved have already been examined. The fact that, for this type of analysis, it is necessary to know only the receptances of the sub-systems suggests the possibility of constructing a set of tables.

Table 1 is a list of the receptances of some simple torsional systems. These results have been reached, in effect, by substitution into the relevant equations of motion in the manner of § 1.5; a convenient method of doing this is described in § 3.2. The systems may either be of immediate interest or they may be components of some more complicated systems.

By putting in numerical values of the stiffnesses k and inertias I, the response can be calculated of any system treated in the table to excitation at one of its co-ordinates.

It will be seen that all the receptances of any one system listed have the same denominator; this is denoted in the table by Δ. The condition

$$\Delta = 0 \qquad (1.7.1)$$

therefore implies that all the receptances are infinite† so that it is the condition of resonance. Equation (1.7.1) is a relation which governs ω^2 and is, as we have seen, the frequency equation. It has as many roots as there are degrees of freedom, all of which are real and positive, though one or more will be zero if the system is not anchored and is therefore capable of motion in the manner of a rigid body. These features will be established in Chapter 3 and it need only be noted at this stage that, by inserting numerical values into the various expressions Δ and equating them to zero, algebraic equations are obtained which give the squares of the natural frequencies.

† The possibility that both the numerators and denominators vanish for the same value of ω^2 is discussed in §3.11; this is a special case that can be disregarded at this stage.

TABLE 1

RECEPTANCES OF TORSIONAL SYSTEMS

Notation

I Polar moment of inertia of rigid disk about its axis.

k Torsional stiffness of light elastic shaft.

x_r Rotational displacement, from a position of rest, of the rth disk.

α_{rs} Receptance between x_r and x_s. If a harmonic torque $F_s e^{i\omega t}$ is applied to the sth disk, then the response at the rth disk is given by

$$x_r = \alpha_{rs} F_s e^{i\omega t}.$$

If $r \neq s$, α_{rs} is called a 'cross-receptance' and if $r = s$, it is called the 'direct receptance' at x_s. It is a general property of receptances of systems of the type under discussion that $\alpha_{rs} = \alpha_{sr}$.

Δ Function defined as required for any given system, as an abbreviation.

TABLE 1

(a) RECEPTANCES OF TORSIONAL SYSTEMS HAVING ONE DEGREE OF FREEDOM

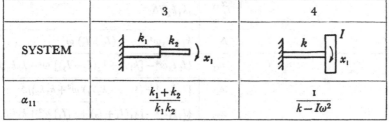

	1	2
SYSTEM		
α_{11}	$-\dfrac{1}{I\omega^2}$	$\dfrac{1}{k}$

	3	4
SYSTEM		
α_{11}	$\dfrac{k_1+k_2}{k_1 k_2}$	$\dfrac{1}{k-I\omega^2}$

(b) SYSTEMS HAVING TWO DEGREES OF FREEDOM

	SYSTEMS	α_{11}	α_{22}	$\alpha_{12}=\alpha_{21}$
5		$\dfrac{k-I\omega^2}{\Delta}$	$\dfrac{k}{\Delta}$	$\dfrac{k}{\Delta}$
		$\Delta=-kI\omega^2$		
6		$\dfrac{k_2}{\Delta}$	$\dfrac{k_1+k_2-I\omega^2}{\Delta}$	$\dfrac{k_2}{\Delta}$
		$\Delta=k_2(k_1-I\omega^2)$		
7		$\dfrac{k-I_2\omega^2}{\Delta}$	$\dfrac{k-I_1\omega^2}{\Delta}$	$\dfrac{k}{\Delta}$
		$\Delta=I_1 I_2\omega^4-k(I_1+I_2)\,\omega^2$		
8		$\dfrac{k_2-I_2\omega^2}{\Delta}$	$\dfrac{k_1+k_2-I_1\omega^2}{\Delta}$	$\dfrac{k_2}{\Delta}$
		$\Delta=I_1 I_2\omega^4-[k_1 I_2+k_2(I_1+I_2)]\,\omega^2+k_1 k_2$		
9		$\dfrac{k_2+k_3-I_2\omega^2}{\Delta}$	$\dfrac{k_1+k_2-I_1\omega^2}{\Delta}$	$\dfrac{k_2}{\Delta}$
		$\Delta=I_1 I_2\omega^4-[k_1 I_2+k_2(I_1+I_2)+k_3 I_1]\,\omega^2$ $+(k_1 k_2+k_2 k_3+k_3 k_1)$		

TABLE 1 (*cont.*)

(c) SYSTEMS HAVING THREE DEGREES OF FREEDOM

No.	Diagram		
10		α_{11}	$\{I_2 I_3 \omega^4 - [k_1 I_3 + k_2(I_2 + I_3)]\,\omega^2 + k_1 k_2\}/\Delta$
		α_{22}	$\{k_1(k_2 - I_3 \omega^2)\}/\Delta$
		α_{33}	$\{k_1(k_2 - I_2 \omega^2)\}/\Delta$
		α_{12} α_{21}	$\{k_1(k_2 - I_3 \omega^2)\}/\Delta$
		α_{13} α_{31}	$\{k_1 k_2\}/\Delta$
		α_{23} α_{32}	$\{k_1 k_2\}/\Delta$
		Δ	$k_1 I_2 I_3 \omega^4 - k_1 k_2(I_2 + I_3)\,\omega^2$
11		α_{11}	$\{I_2 I_3 \omega^4 - [k_1 I_3 + k_2(I_2 + I_3)]\,\omega^2 + k_1 k_2\}/\Delta$
		α_{22}	$\{I_1 I_3 \omega^4 - (k_1 I_3 + k_2 I_1)\,\omega^2 + k_1 k_2\}/\Delta$
		α_{33}	$\{I_1 I_2 \omega^4 - [k_1(I_1 + I_2) + k_2 I_1]\,\omega^2 + k_1 k_2\}/\Delta$
		α_{12} α_{21}	$\{k_1(k_2 - I_3 \omega^2)\}/\Delta$
		α_{13} α_{31}	$\{k_1 k_2\}/\Delta$
		α_{23} α_{32}	$\{k_2(k_1 - I_1 \omega^2)\}/\Delta$
		Δ	$-\{I_1 I_2 I_3 \omega^6 - [k_1(I_1 I_3 + I_2 I_3) + k_2(I_1 I_2 + I_1 I_3)]\,\omega^4 + k_1 k_2(I_1 + I_2 + I_3)\,\omega^2\}$
12		α_{11}	$\{I_2 I_3 \omega^4 - [k_2 I_3 + k_3(I_2 + I_3)]\,\omega^2 + k_2 k_3\}/\Delta$
		α_{22}	$\{I_1 I_3 \omega^4 - [k_1 I_3 + k_2 I_3 + k_3 I_1]\,\omega^2 + k_1 k_3 + k_2 k_3\}/\Delta$
		α_{33}	$\{I_1 I_2 \omega^4 - [k_1 I_2 + k_2(I_1 + I_2) + k_3 I_1]\,\omega^2 + k_1 k_2 + k_2 k_3 + k_3 k_1\}/\Delta$
		α_{12} α_{21}	$\{k_2(k_3 - I_3 \omega^2)\}/\Delta$
		α_{13} α_{31}	$\{k_2 k_3\}/\Delta$
		α_{23} α_{32}	$\{k_3(k_1 + k_2 - I_1 \omega^2)\}/\Delta$
		Δ	$-\{I_1 I_2 I_3 \omega^6 - [k_1 I_2 I_3 + k_2(I_1 I_3 + I_2 I_3) + k_3(I_1 I_2 + I_1 I_3)]\,\omega^4 + [I_1 k_2 k_3 + I_2(k_1 k_3 + k_2 k_3) + I_3(k_1 k_2 + k_2 k_3 + k_1 k_3)]\,\omega^2 - k_1 k_2 k_3\}$

TABLE 1 *(c)* *(cont.)*

13	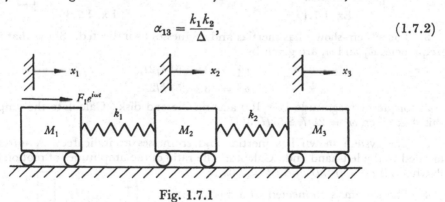	α_{11}	$\{I_2 I_3 \omega^4 - [k_2 I_3 + k_3 (I_2 + I_3) + k_4 I_2]\,\omega^2 + k_2 k_3 + k_3 k_4 + k_4 k_2\}/\Delta$
		α_{22}	$\{I_1 I_3 \omega^4 - [k_1 I_3 + k_2 I_3 + k_3 I_1 + k_4 I_1]\,\omega^2 + (k_1 + k_2)\,(k_3 + k_4)\}/\Delta$
		α_{33}	$\{I_1 I_2 \omega^4 - [k_1 I_2 + k_2 (I_1 + I_2) + k_3 I_1]\,\omega^2 + k_1 k_2 + k_2 k_3 + k_3 k_1\}/\Delta$
		α_{12} α_{21}	$\{k_2 (k_3 + k_4 - I_3 \omega^2)\}/\Delta$
		α_{13} α_{31}	$\{k_2 k_3\}/\Delta$
		α_{23} α_{32}	$\{k_3 (k_1 + k_2 - I_1 \omega^2)\}/\Delta$
		Δ	$-\{I_1 I_2 I_3 \omega^6 - [k_1 I_2 I_3 + k_2 (I_1 I_3 + I_2 I_3) + k_3 (I_1 I_2 + I_1 I_3) + k_4 I_1 I_2]\,\omega^4 + [I_1 (k_2 k_3 + k_3 k_4 + k_4 k_2) + I_2 (k_1 k_3 + k_3 k_2 + k_2 k_4 + k_4 k_1) + I_3 (k_1 k_2 + k_2 k_3 + k_3 k_1)]\,\omega^2 - [k_1 k_2 (k_3 + k_4) + k_3 k_4 (k_1 + k_2)]\}$

The results in Table 1 refer specifically to the torsional oscillations of systems of rigid disks and light shafts, these systems having been selected because of their practical importance. The receptances are, however, immediately applicable to the axial vibrations of spring-connected masses. For instance, the cross-receptance α_{13} of the system of fig. 1.7.1 is found from system 11 of the table; it is

$$\alpha_{13} = \frac{k_1 k_2}{\Delta}, \tag{1.7.2}$$

Fig. 1.7.1

where

$$\Delta = -\{M_1 M_2 M_3 \omega^6 - [k_1 (M_1 M_3 + M_2 M_3) + k_2 (M_1 M_2 + M_1 M_3)]\,\omega^4$$
$$+ k_1 k_2 (M_1 + M_2 + M_3)\,\omega^2\}. \tag{1.7.3}$$

This is found by replacing I_1, I_2 and I_3 by M_1, M_2 and M_3 and using k_1 and k_2 to represent extensional—rather than torsional—stiffness.

Receptances of systems other than those in the table may also be worked out and listed, either by direct calculation or by synthesis. Such receptances may then be used directly or as components of still more complicated systems which are to be analysed. The reader may thus extend the table to suit his own particular needs.

EXAMPLES 1.7

1. A stepped shaft is anchored at its thicker end and carries a rigid disk at its thinner end, the dimensions of the system being as shown in the figure. For the material of the shaft the shear modulus is $11 \cdot 5 \times 10^6$ psi and the density of the material of the disk is 480 lb./ft.^{3}. Estimate the natural frequency of torsional vibration under the assumption that the mass of the shaft can be neglected.

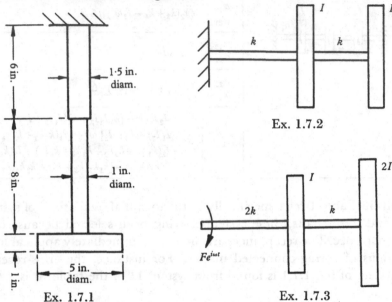

Ex. 1.7.1

Ex. 1.7.2

Ex. 1.7.3

2. The system shown has inertias and stiffnesses as indicated. Show that its natural frequencies ω_1 and ω_2 are given by

$$\omega_1^2 = (3 - \sqrt{5}) \, k/2I,$$
$$\omega_2^2 = (3 + \sqrt{5}) \, k/2I.$$

A torque of magnitude $k e^{i\omega t}/100$ acts on the end disk. Calculate the amplitudes at this disk when $\omega^2 = 2k/I$, $2 \cdot 5k/I$, $3k/I$.

3. The system shown has inertias and stiffnesses as indicated. A torque $F e^{i\omega t}$ is applied to the left-hand end. Calculate the ratio of the amplitudes of motion of the two flywheels if $\omega^2 = 0 \cdot 4k/I$ and $\omega^2 = 2 \cdot 0k/I$.

4. The system is connected to a mechanism at the left-hand end which gives this end a torsional motion with an amplitude of one degree and a frequency given by $\omega^2 = 3k/I$. What is the amplitude of the torque which the mechanism has to apply to the system in order to produce the motion and what will be the amplitudes of motion of the other two disks?

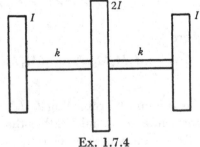

Ex. 1.7.4

5. Deduce α_{12} for system 9 of Table 1 by putting $I_3 = 0$ and $k_4 = \infty$ in α_{12} for system 13.

6. Show that, when a system of the type of fig. 1.6.1 oscillates freely at a natural frequency, it is necessary that
$$\beta_{11} = -\gamma_{11}$$
as a consequence of the third law of motion.

1.8 Receptances of composite systems

We have already shown examples of how a receptance of a composite system A can be written in terms of the receptances of its sub-systems $B, C, ...,$ etc.; equation (1.6.5) is a case in point. Table 2 is a selection of the more important relations for use in this process of synthesis. These relations are for general application and are not restricted to use with the systems of Table 1. We shall, however, continue to confine our attention to the latter until we reach the more general types of system in Chapter 3. The method of finding these relations has already been indicated, but here we shall review, in greater detail, the principles which are involved.

By hypothesis, only those systems are to be considered whose small forced oscillations are governed by linear equations with constant coefficients. It follows that the principle of superposition can be used throughout. Thus if harmonic excitations are applied at two or more points of a system, then the response at any co-ordinate is the sum of the responses which would be obtained if the forces were applied separately. This is a property of the differential equations of motion.

Let concentrated harmonic forces $F_1 e^{i\omega t}$, $F_2 e^{i\omega t}$, ..., $F_n e^{i\omega t}$ be applied at the co-ordinates $x_1, x_2, ..., x_n$ of a system, the latter representing the displacements in the appropriate directions of the points of application of the corresponding forces. If only those vibrations of the system are considered which have the excitation frequency ω, then the responses at the various co-ordinates are given by

$$\left.\begin{aligned}
x_1 &= \alpha_{11} F_1 e^{i\omega t} + \alpha_{12} F_2 e^{i\omega t} + ... + \alpha_{1n} F_n e^{i\omega t}, \\
x_2 &= \alpha_{21} F_1 e^{i\omega t} + \alpha_{22} F_2 e^{i\omega t} + ... + \alpha_{2n} F_n e^{i\omega t}, \\
&\cdots\cdots\cdots\cdots\cdots\cdots\cdots\cdots\cdots\cdots\cdots\cdots\cdots\cdots \\
x_n &= \alpha_{n1} F_1 e^{i\omega t} + \alpha_{n2} F_2 e^{i\omega t} + ... + \alpha_{nn} F_n e^{i\omega t}.
\end{aligned}\right\} \tag{1.8.1}$$

The quantities $x_1, x_2, ..., x_n$ are all proportional to $e^{i\omega t}$ so that this function may be omitted from the equations by writing the amplitudes $X_1, X_2, ..., X_n$ on the left-hand side. *When this is done, however, the restricted meaning of the equations—namely that they refer to steady harmonic motion with the frequency ω for which the receptances are to be calculated—must not be forgotten.*

Now equation (1.8.1) is not changed if some or all of the quantities $F_1 e^{i\omega t}$, $F_2 e^{i\omega t}$, ..., $F_n e^{i\omega t}$ represent harmonic torques rather than forces so that the appropriate quantities $x_1, x_2, ..., x_n$ are rotational displacements. We shall show in Chapter 3 that, algebraically, no distinction need be made between these two types of excitation.

Equations (1.8.1) hold good, not only for a complete system A but also for each sub-system $B, C, ...,$ etc. When these relations are written for a sub-system an

TABLE 2

RECEPTANCES OF COMPOSITE SYSTEMS

Notation

A — Composite oscillatory system which may be regarded, for the purpose of vibration analysis, as being composed of parts $B, C, D, \ldots.$

$B, C, D, \ldots$ — Sub-systems which, when fitted together (both as regards forces and displacements at the junctions) form the parent system A.

x_r — rth generalized co-ordinate being a co-ordinate at which two or more sub-systems are linked together. The symbol x_r usually denotes a linear displacement or a rotation.

α_{rs} — Receptance between x_r and x_s, these being co-ordinates of a system A. If a harmonic generalized force $F_s e^{i\omega t}$ corresponding to x_s (which is usually a simple force or a torque) acts at x_s, then the response at x_r is given by

$$x_r = \alpha_{rs} F_s e^{i\omega t}.$$

If $r \neq s$, α_{rs} is called a 'cross-receptance' and if $r = s$, it is a 'direct receptance'. It is a general property of the types discussed in this book that $\alpha_{rs} = \alpha_{sr}$.

β_{rs}, etc. — Receptance between x_r and x_s where x_r and x_s both refer to a sub-system B. This expression is defined in the same way as a receptance of A and possesses the same properties. Similarly γ_{rs} is a receptance of a sub-system C, δ_{rs} is a receptance of D and so on.

TABLE 2
RECEPTANCES OF COMPOSITE SYSTEMS

	SYSTEM	RECEPTANCE		

1 — (x_1 B x_2 C x_3)

α_{11}	α_{22}	α_{33}
$\beta_{11} - \dfrac{\beta_{12}^2}{\beta_{22}+\gamma_{22}}$	$\dfrac{\beta_{22}\gamma_{22}}{\beta_{22}+\gamma_{22}}$	$\gamma_{33} - \dfrac{\gamma_{23}^2}{\beta_{22}+\gamma_{22}}$
$\alpha_{12}=\alpha_{21}$	$\alpha_{13}=\alpha_{31}$	$\alpha_{23}=\alpha_{32}$
$\beta_{12} - \dfrac{\beta_{12}\beta_{22}}{\beta_{22}+\gamma_{22}}$	$\dfrac{\beta_{12}\gamma_{23}}{\beta_{22}+\gamma_{22}}$	$\gamma_{23} - \dfrac{\gamma_{22}\gamma_{23}}{\beta_{22}+\gamma_{22}}$

2 — (B x_1 C x_2 D)

α_{11}	$\dfrac{\beta_{11}[\gamma_{11}(\gamma_{22}+\delta_{22}) - \gamma_{12}^2]}{(\beta_{11}+\gamma_{11})\,(\gamma_{22}+\delta_{22}) - \gamma_{12}^2}$
α_{12} α_{21}	$\dfrac{\beta_{11}\gamma_{12}\delta_{22}}{(\beta_{11}+\gamma_{11})\,(\gamma_{22}+\delta_{22}) - \gamma_{12}^2}$

3 —

α_{11}	α_{44}	$\alpha_{14}=\alpha_{41}$
$\beta_{11} - \dfrac{n^2\beta_{12}^2}{n^2\beta_{22}+\gamma_{33}}$	$\gamma_{44} - \dfrac{\gamma_{34}^2}{n^2\beta_{22}+\gamma_{33}}$	$\dfrac{n\beta_{12}\gamma_{34}}{n^2\beta_{22}+\gamma_{33}}$

4 —

α_{11}	α_{44}
$\beta_{11} - \dfrac{\beta_{12}^2(n^2\gamma_{33}+m^2\delta_{55})}{\gamma_{33}\delta_{55}+\beta_{22}(n^2\gamma_{33}+m^2\delta_{55})}$	$\gamma_{44} - \dfrac{\gamma_{34}^2(n^2\beta_{22}+\delta_{55})}{\gamma_{33}\delta_{55}+\beta_{22}(n^2\gamma_{33}+m^2\delta_{55})}$
$\alpha_{14}=\alpha_{41}$	$\alpha_{46}=\alpha_{64}$
$\dfrac{m\beta_{12}\gamma_{34}\delta_{55}}{\gamma_{33}\delta_{55}+\beta_{22}(n^2\gamma_{33}+m^2\delta_{55})}$	$\dfrac{mn\beta_{22}\gamma_{34}\delta_{56}}{\gamma_{33}\delta_{55}+\beta_{22}(n^2\gamma_{33}+m^2\delta_{55})}$

5 — (B C, x_1, x_2)

α_{11}	$\{\beta_{11}(\gamma_{11}\gamma_{22}-\gamma_{12}^2)+\gamma_{11}(\beta_{11}\beta_{22}-\beta_{12}^2)\}/\Delta$
α_{22}	$\{\beta_{22}(\gamma_{11}\gamma_{22}-\gamma_{12}^2)+\gamma_{22}(\beta_{11}\beta_{22}-\beta_{12}^2)\}/\Delta$
α_{12} α_{21}	$\{\beta_{12}(\gamma_{11}\gamma_{22}-\beta_{12}\gamma_{12})+\gamma_{12}(\beta_{11}\beta_{22}-\beta_{12}\gamma_{12})\}/\Delta$
Δ	$(\beta_{11}+\gamma_{11})\,(\beta_{22}+\gamma_{22}) - (\beta_{12}+\gamma_{12})^2$

6 —

$$\alpha_{22}=n^2\beta_{11}$$

7 —

$$\alpha_{22}=\beta_{22}-\frac{\beta_{12}^2}{\beta_{11}}$$

appropriate letter is included with the subscripts of the displacements and forces concerned. Thus, while (1.8.1) applies to A, the comparable equations for B would be written

$$\left.\begin{array}{l} X_{b1} = \beta_{11}F_{b1} + \beta_{12}F_{b2} + \ldots + \beta_{1m}F_{bm}, \\ X_{b2} = \beta_{21}F_{b1} + \beta_{22}F_{b2} + \ldots + \beta_{2m}F_{bm}, \\ \cdots\cdots\cdots\cdots\cdots\cdots\cdots\cdots\cdots\cdots\cdots\cdots \\ X_{bm} = \beta_{m1}F_{b1} + \beta_{m2}F_{b2} + \ldots + \beta_{mm}F_{bm}, \end{array}\right\} \qquad (1.8.2)$$

where it is assumed that B has m degrees of freedom and where the factor $e^{i\omega t}$ has been omitted as mentioned above.

When a composite system A is built up from simpler systems it is necessary for the latter to fit together with regard both to the forces and the displacements at their junctions. This is in accordance with conditions which may be called those of 'equilibrium' and 'compatibility'.

1.8.1 Equilibrium

Consider the sub-systems B and C of fig. 1.8.1 which, when joined at x_1, form system 7 of Table 1. The torques which are exerted upon B and C at that co-ordinate must either be provided by an externally applied force $F_1 e^{i\omega t}$ or, in the absence of excitation at x_1, be equal and opposite. That is to say

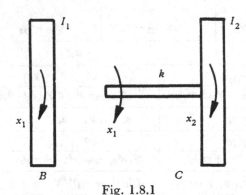

Fig. 1.8.1

$$F_{b1} + F_{c1} = F_1, \qquad (1.8.3)$$

where F_1 may be zero. This is an equilibrium condition.

System 3 of Table 2 has a gear ratio between B and C. The equilibrium condition in this case is, in the absence of excitation at the junction (x_2, x_3),

$$F_{b2} \times 1 + F_{c3} \times n = 0. \qquad (1.8.4)$$

For the branched geared system (number 4 in Table 2) equilibrium demands that

$$F_{b2} + mF_{c3} + nF_{d5} = 0 \qquad (1.8.5)$$

if there is no excitation at the branch point.

1.8.2 Compatibility

If B and C of fig. 1.8.1 are to remain coupled, then it is necessary that

$$X_{b1} = X_{c1}. \qquad (1.8.6)$$

This is a compatibility equation; it is concerned with the *continuity* of the compound system A.

The compatibility requirement of the geared system (number 3 of Table 2) is

$$nX_{b2} = X_{c3}, \qquad (1.8.7)$$

while that of the branched geared system (number 4) is

$$mX_{b2} = X_{c3},$$
$$nX_{b2} = X_{d5}.$$

$$(1.8.8)$$

By applying equilibrium and compatibility conditions to equations of the form (1.8.2) it is possible to find the receptances α of a system A in terms of the receptances $\beta, \gamma, \ldots$ of its component parts. A few instances of this procedure were discussed in § 1.6—equations (1.6.5), (1.6.29) and (1.6.32)—and a selection of these results is given in Table 2.

The object of expressing the receptances α in terms of the receptances $\beta, \gamma, \ldots$ is that the latter are more easily found since the systems $B, C, \ldots$ are simpler to analyse. Again it may be possible to measure the receptances of the sub-systems experimentally and then to combine the results to form the receptances of the composite system. Yet again some of the receptances of the sub-systems may be obtainable from a table such as Table 1.

It was stated in § 1.7 that the receptances of Table 1 were found by direct substitution in the relevant equations of motion. It was shown in § 1.6 that these receptances may also be built up by means of relations such as those of Table 2, starting with a very few elementary receptances. The first of these two methods is usually preferable when a purely analytical approach is required. If, however, numerical values are available from which the receptances of the subsystems can be readily evaluated, then much time and labour can often be saved by using results such as those of Table 2.

This table may be extended by the reader to fulfil his own particular needs.

EXAMPLES 1.8

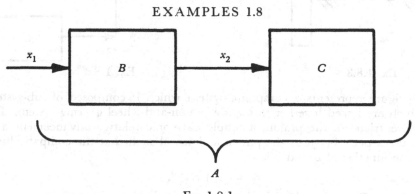

Ex. 1.8.1

1. Show that if a system A can be broken down into two sub-systems B and C as shown, then

$$\frac{\alpha_{12}}{\alpha_{22}} = \frac{\beta_{12}}{\beta_{22}}.$$

Explain carefully why this result is self-evident.

[NOTE. This relation is of great use in checking receptances.]

2. By writing down the equilibrium and compatibility equations for the *free* vibration of a system which is composed of two sub-systems B and C which are linked by two co-ordinates, deduce the frequency equation directly in its determinantal form. Show, by the same method, that if the systems are linked by j co-ordinates then the frequency equation becomes

$$\begin{vmatrix} \beta_{11}+\gamma_{11} & \beta_{12}+\gamma_{12} & \cdots & \beta_{1j}+\gamma_{1j} \\ \beta_{21}+\gamma_{21} & \beta_{22}+\gamma_{22} & \cdots & \beta_{2j}+\gamma_{2j} \\ \cdots & \cdots & \cdots & \cdots \\ \beta_{j1}+\gamma_{j1} & \beta_{j2}+\gamma_{j2} & \cdots & \beta_{jj}+\gamma_{jj} \end{vmatrix} = 0.$$

3. The arrangement shown in the figure consists of two shafts, each of which carries a gear wheel with n teeth on one end and one with m teeth on the other. The moments of inertia of the n-toothed wheels are each I_n and those of the m-toothed wheels are I_m. The torsional stiffnesses of each of the shafts is k. The shafts are mounted in bearings with the wheels at each end in mesh. Neglecting the effect of backlash, find an expression for the natural frequencies of torsional vibration of the system.

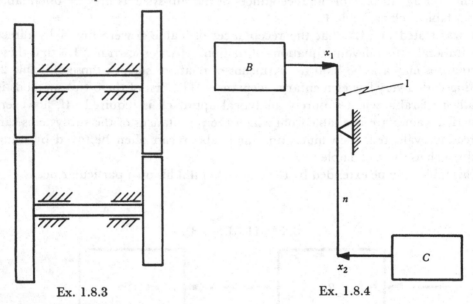

Ex. 1.8.3   Ex. 1.8.4

4. The figure represents a composite system which is composed of sub-systems B and C which are linked together through a pinion-and-wheel gearing system. Due to a slight eccentricity of the pinion, a simple harmonic relative advancement and retardation takes place between the pinion and wheel such that the compatibility condition at the junction of B and C is

$$x_2 = nx_1 + E\, e^{i\omega t},$$

where $E \ll nx_1$. Show that this 'displacement excitation' causes motions

$$x_1 = \frac{-n\beta_{11}E\, e^{i\omega t}}{n^2\beta_{11}+\gamma_{22}}, \quad x_2 = \frac{\gamma_{22}E\, e^{i\omega t}}{n^2\beta_{11}+\gamma_{22}},$$

and that the condition of resonance is the same as that of the given system when it is subjected to harmonic excitation by some externally applied force.

[NOTE. This type of excitation, and its technical importance, is discussed by H. G. Yates, *Proc. Inst. Mech. Engrs*, vol. 169 (1955), p. 611.]

5. The output shaft of a marine engine is connected to the propeller shaft through a pinion-and-wheel reduction gear whose ratio is $3:1$. The direct receptance β_{11} for torsional motion is measured at the pinion for the driving system only, with the wheel out of mesh. The direct receptance γ_{22} is measured at the wheel for the propeller system only, with the pinion out of mesh. The same range of driving frequencies is used in each test and the results are as follows:

Frequency (cyc./sec.)	16	18	20	22	24	26
$\beta_{11} \times 10^6$ (rad./in.lb.)	-40	-31	-23	-16	-9	-2
$\gamma_{22} \times 10^6$ (rad./in.lb.)	-21	-15	-8	0	$+8$	$+18$

When the wheel and pinion run in mesh, a simple harmonic relative advancement and retardation occurs between them due to a slight eccentricity of the pinion. Show that a condition of resonance may exist with a frequency within the test range, and find the engine speed at which this occurs.

(C.U.M.S.T. Pt II, 1957)

6. Draw up a table of the receptances of system 1 of Table 2, in the form

$$\begin{array}{ccc} \alpha_{11} & \alpha_{12} & \alpha_{13} \\ \alpha_{21} & \alpha_{22} & \alpha_{23} \\ \alpha_{31} & \alpha_{32} & \alpha_{33} \end{array}$$

for the case where the sub-systems B and C are identical with (i) dissimilar, and (ii) similar ends connected.

1.9 Frequency equations and free oscillations of composite systems

We have described a method for finding the frequency equation of a system, by using equation (1.7.1), namely $\Delta = 0$. That is to say the frequency equation is the condition that the denominator of the receptances of the system is zero so that the receptances are all infinite. A second method of finding the frequency equation is sometimes useful; it is based on the technique of analysing a system A in terms of its sub-systems $B, C, \ldots$.

The condition that a receptance α of A shall be infinite is that the denominator of an expression like those of Table 2 shall vanish. Thus the frequency equation of the geared system 3 is

$$n^2 \beta_{22} + \gamma_{33} = 0. \tag{1.9.1}$$

If therefore the variation of β_{22} and γ_{33} with ω are known, either by experiment or because they are given as explicit functions, then the natural frequency of the geared system can be found from equation (1.9.1). It may be convenient, for example, to plot $n^2 \beta_{22}$ and $-\gamma_{33}$ against ω so that the natural frequencies will be given by the intersections of the curves.

A number of frequency equations of this type are given in Table 3. The reader will find that in most cases the relations follow from the results in Table 2. For example, by putting $\gamma_{11} = \gamma_{22} = \gamma_{12} = 0$ in the quantity Δ of system 5 of Table 2 and then equating Δ to zero, the frequency equation of system 2 of Table 3 is found. It may be noted that these results are quite general and are not restricted to the type of system which is treated in Table 1.

TABLE 3

FREQUENCY EQUATIONS OF COMPOSITE SYSTEMS

Notation (as for Table 2)

The frequency equation of a system is the condition under which all its receptances are infinite. Thus the frequency equation of any system of Table 1 (*b*) or (*c*) is simply

$$\Delta = 0.$$

The roots of this equation are the squares of the natural frequencies of the system concerned.

A second way of finding the frequency equation of a system is to write down the condition that its receptances *expressed in terms of the receptances of its sub-systems* shall be infinite. Equations of this type are given in the following table. If, during free vibration of a composite system *A*, no forces are transmitted at the junctions of its sub-systems, the appropriate natural frequency will not be given by this second type of frequency equation; the important case is that of identical sub-systems symmetrically linked together. The missing frequency is then the common natural frequency of the sub-systems.

TABLE 3
FREQUENCY EQUATIONS OF COMPOSITE SYSTEMS

		SYSTEM	FREQUENCY EQUATION
I		x_1 B	$\beta_{11} = 0$
2		x_1 x_2 B	$\beta_{11}\beta_{22} - \beta_{12}^2 = 0$
3		B x_1 C	$\beta_{11} + \gamma_{11} = 0$
4		x_1 B x_2 C	$\beta_{11}(\beta_{22} + \gamma_{22}) - \beta_{12}^2 = 0$
5		x_1 B x_2 C x_3	$\beta_{11}\gamma_{33}(\beta_{22} + \gamma_{22}) - (\beta_{11}\gamma_{23}^2 + \beta_{12}^2\gamma_{33}) = 0$
6		B x_1 C x_2 D	$(\beta_{11} + \gamma_{11})(\gamma_{22} + \delta_{22}) - \gamma_{12}^2 = 0$

TABLE 3 (*cont.*)

	SYSTEM	FREQUENCY EQUATION
7		$\beta_{11}[(\beta_{22}+\gamma_{22})\,(\gamma_{33}+\delta_{33})-\gamma_{23}^2]-\beta_{12}^2(\gamma_{33}+\delta_{33})=0$
8		$\beta_{11}\{(\beta_{22}+\gamma_{22})\,[\delta_{44}(\gamma_{33}+\delta_{33})-\delta_{34}^2]-\gamma_{23}^2\delta_{44}\}$ $-\beta_{12}^2[\delta_{44}(\gamma_{33}+\delta_{33})-\delta_{34}^2]=0$
9		$(\beta_{11}+\gamma_{11})\,(\beta_{22}+\gamma_{22})-(\beta_{12}+\gamma_{12})^2=0$
10		$\begin{vmatrix} \beta_{11}+\gamma_{11} & \beta_{12}+\gamma_{12} & \cdots & \beta_{1j}+\gamma_{1j} \\ \beta_{21}+\gamma_{21} & \beta_{22}+\gamma_{22} & \cdots & \beta_{2j}+\gamma_{2j} \\ \cdots & \cdots & \cdots & \cdots \\ \cdots & \cdots & \cdots & \cdots \\ \beta_{j1}+\gamma_{j1} & \beta_{j2}+\gamma_{j2} & \cdots & \beta_{jj}+\gamma_{jj} \end{vmatrix}=0$
11		$n^2\beta_{11}+\gamma_{22}=0$
12		$(\gamma_{22}+m^2\beta_{11})\,(\delta_{44}+n^2\gamma_{33})-n^2\gamma_{23}^2=0$
13		$(m^2\beta_{11}+\gamma_{33})\,(n^2\beta_{22}+\gamma_{44})-(mn\beta_{12}+\gamma_{34})^2=0$
14		$\gamma_{22}\delta_{33}+\beta_{11}(n^2\gamma_{22}+m^2\delta_{33})=0$

The use of Table 3 may be illustrated by considering the branched system of fig. 1.9.1. Let the numerical values of the inertias and stiffnesses be

$$
\left.\begin{array}{ll}
k_1 = 8 \times 10^6 \,\text{pdl.ft./rad.}, & I_1 = 10 \,\text{lb.ft.}^2, \\
k_2 = 3 \times 10^6 \,\text{pdl.ft./rad.}, & I_2 = 40 \,\text{lb.ft.}^2, \\
k_3 = 6 \times 10^6 \,\text{pdl.ft./rad.}, & I_3 = 30 \,\text{lb.ft.}^2, \\
k_4 = 5 \times 10^6 \,\text{pdl.ft./rad.}, & I_4 = 30 \,\text{lb.ft.}^2,
\end{array}\right\} \tag{1.9.2}
$$

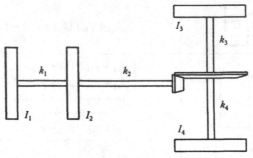

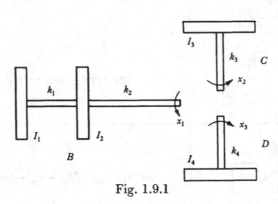

Fig. 1.9.1

and suppose that the reduction gear between the driving shaft and the axles is 4:1. Using the sub-systems B, C and D, as shown, the frequency equation is given by entry 14 of Table 3; it is

$$
\gamma_{22}\delta_{33} + \beta_{11}[(\tfrac{1}{4})^2\gamma_{22} + (\tfrac{1}{4})^2\delta_{33}] = 0. \tag{1.9.3}
$$

From entry 10 of Table 1, it is seen that

$$
\beta_{11} = \frac{10 \times 40\omega^4 - [3 \times 10^6 \times 10 + 8 \times 10^6(40+10)]\,\omega^2 + 8 \times 3 \times 10^{12}}{3 \times 10^6 \times 10 \times 40 \times \omega^4 - 8 \times 3 \times 10^{12}(40+10)\,\omega^2}, \tag{1.9.4}
$$

and from entry 5 of Table 1,

$$
\gamma_{22} = \frac{6 \times 10^6 - 30\omega^2}{-6 \times 10^6 \times 30\omega^2}, \tag{1.9.5}
$$

$$
\delta_{33} = \frac{5 \times 10^6 - 30\omega^2}{-5 \times 10^6 \times 30\omega^2}. \tag{1.9.6}
$$

On substituting these values into (1.9.3), the frequency equation in ω^2 is found. A convenient way of finding its roots will now be explained.

Equation (1.9.3) can be written in the alternative form

$$\frac{16}{\beta_{11}} = -\frac{1}{\gamma_{22}} - \frac{1}{\delta_{33}}. \tag{1.9.7}$$

The advantage of this is that an arithmetical trial-and-error process may be applied to it more readily than if it were left in form (1.9.3). The arithmetical equation now becomes

$$\frac{19\,200y(y-1)}{400y^2 - 430y + 24} = \frac{30y}{1 - 6y} + \frac{30y}{1 - 5y} \tag{1.9.8}$$

in which the notation $y = \omega^2/10^6$ has been adopted.

Now it will be seen that the two terms on the right-hand side are zero for $y = 0$ and that as y is increased each of them becomes positive and then infinite, afterwards becoming negative and asymptotic to a constant value. The values of y which correspond to the infinite values of the two terms are $0 \cdot 167$ and $0 \cdot 200$ respectively. The term on the left-hand side is zero for $y = 0$; it becomes first negative as y increases, passes through an infinite value for $y \doteqdot 0 \cdot 058$ and then, having become positive, it decreases to zero for $y = 1 \cdot 00$. There then follows a second infinite value at $y \doteqdot 1 \cdot 02$ after which it becomes positive and diminishes continuously, approaching the value 48. The reader should be able to deduce, by sketching a very rough graph of the two sides of equation (1.9.8), that one root lies a little below $0 \cdot 167$, another lies between $0 \cdot 167$ and $0 \cdot 200$ and the third lies slightly below $1 \cdot 02$. By trying suitable values he will then be able to find the correct roots quickly; they are $0 \cdot 15322$, $0 \cdot 18963$ and $1 \cdot 0035$. The corresponding values of ω are

$$\omega_1 = 391\,\text{rad./sec.}, \quad \omega_2 = 435\,\text{rad./sec.}, \quad \omega_3 = 1001\,\text{rad./sec.} \tag{1.9.9}$$

All the natural frequencies of a system may be found by the method of equation (1.7.1); that is, by finding a receptance and equating its denominator to zero. The second method, which involves equations of the form given in Table 3, leads to the same results but there is a special case when this method may fail to give all the natural frequencies. This arises when, during free oscillation, no forces are exerted between the sub-systems. The breakdown of the method is due to the fact that it is implicit in the results of Table 2 that the forces or torques at the junctions are not permanently zero during free motion. Suppose that such forces *were* zero; it would follow that, in the equations (1.8.2) for each sub-system, the various amplitudes X could be finite while the force amplitudes F were zero implying that the receptances of B, C, etc., would be infinite. If this were so, then it cannot be assumed that the equations of Table 3 give conditions for which the receptances of A become infinite.

The special conditions described above will occur if the systems B, C, etc., all have a common natural frequency and are linked together to form a chain or branched system in which adjacent sub-systems have one linking co-ordinate only. If any pair of sub-systems had two linking co-ordinates, or if part of the chain were closed to form a ring of sub-systems, then an additional equation of compatibility would be required and motion with zero forces in the links would not in general be possible. Special arrangements of this type, in which free motion with zero force in the links can occur, will arise if the modes of the doubly-linked sub-systems give the same ratio of displacements of the two linking co-ordinates. This can be brought about by the double coupling of identical systems.

With these special types of system the equations of Table 3 will not give all the natural frequencies. This fact does not detract seriously from the value of the table because the special cases can be recognized and calculated separately.

EXAMPLES 1.9

1. In the geared system shown,

$$I_1 = 2 \text{ lb.ft.}^2,$$
$$I_2 = 3 \text{ lb.ft.}^2,$$
$$I_3 = 5 \text{ lb.ft.}^2,$$
$$I_4 = 1 \text{ lb.ft.}^2,$$
$$I_5 = 0.5 \text{ lb.ft.}^2,$$
$$k_1 = k_2 = 2 \times 10^4 \text{ pdl.ft./rad.},$$
$$k_3 = 5 \times 10^3 \text{ pdl.ft./rad.}$$

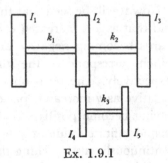

The ratio of the radii of the toothed wheels I_2 and I_4 is 2·5 : 1, I_2 having the larger radius, and the bearing friction may be neglected. Find the natural frequencies of the system.

Ex. 1.9.1

(C.U.M.S.T. Pt II, 1951, slightly altered; part of question only)

2. Show, from equations of the form (1.8.2) and the appropriate equilibrium and compatibility conditions, that entry 6 of Table 3 is the condition under which free oscillation of the given system can occur with finite forces appearing at the junctions.

3. Generalize the conclusions of the last problem by considering the adjacent sub-systems, B and C, connected at any number of co-ordinates through gears. That is, show that the frequency equation expressed in terms of receptances β and γ is the condition that free oscillation can occur with finite forces at the junctions.

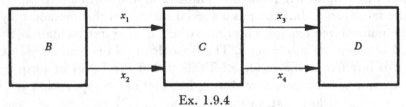

Ex. 1.9.4

4. (a) Deduce the frequency equation of entry 6 of Table 3 from entry 9.

(b) Show, by the use of entry 10 of Table 3, that the frequency equation of the system shown is

$$\begin{vmatrix} \beta_{11}+\gamma_{11} & \beta_{12}+\gamma_{12} & \gamma_{13} & \gamma_{14} \\ \beta_{21}+\gamma_{21} & \beta_{22}+\gamma_{22} & \gamma_{23} & \gamma_{24} \\ \gamma_{31} & \gamma_{32} & \delta_{33}+\gamma_{33} & \delta_{34}+\gamma_{34} \\ \gamma_{41} & \gamma_{42} & \delta_{43}+\gamma_{43} & \delta_{44}+\gamma_{44} \end{vmatrix} = 0.$$

5. The figure refers to a four-cylinder engine with a flywheel. The moment of inertia of the flywheel is 10^6 lb.in.^{2}, and the effective moment of inertia of the crank and

reciprocating parts for each cylinder is 10^5 lb.in.^{2}. The second polar moment of area of the cross-section of the shaft is 100 in.^{4}, and the shaft is of steel having modulus of rigidity 12.4×10^6 psi.

Determine the lowest natural frequency of torsional oscillation.

A high accuracy in the calculated frequency is not required.

(C.U.M.S.T. Section B, 1946)

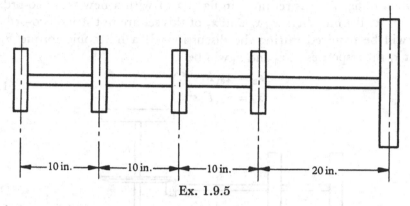

Ex. 1.9.5

6. It is known that a natural frequency of torsional vibration in a certain aero-engine lies between 450 and 650 cyc./sec. when the engine is driving a propeller. For the purposes of calculation the engine crankshaft assembly may be taken to consist of three disks, each of moment of inertia 10 lb.ft.^{2}, which are joined by two light shafts each of stiffness 4.0×10^8 pdl.ft./rad.; a third shaft of the same stiffness joins the end disk to the propeller.

The influence of the propeller on the vibration was found by experiment as follows. The propeller was removed from the engine and a torque $F \sin \omega t$ was applied to the hub, the amplitude X of the motion being measured for various values of ω. The following figures for the ratio F/X were found:

ω (sec.$^{-1}$)	3250	3500	3750	4000
F/X (pdl.ft./rad.)	$+0.10 \times 10^9$	-0.25×10^9	-0.50×10^9	-0.85×10^9

where the negative sign means that the motion was in antiphase with the torque. Estimate the natural frequency which is within the range mentioned.

(C.U.M.S.T. Pt II, 1950)

7. A closed gear-train in a gear-box is found to possess a troublesome natural frequency. It is proposed to alter its value by replacing a shaft connecting two wheels by one with a different stiffness. The original shaft is removed and the direct receptances α_{11} and α_{22} are measured at the co-ordinates x_1 and x_2, which are the angular displacements of the two wheels, measured in the same direction from a position of rest; the cross-receptance α_{12} between x_1 and x_2 is also measured with the shaft removed. The various test results are as follows:

Frequency (rad./sec.)	1200	1300	1400	1500
α_{11} (micro-rad./in.lb.)	-1.53	-0.59	-0.06	$+0.34$
α_{22} (micro-rad./in.lb.)	-8.31	-4.93	-3.19	-2.06
α_{12} (micro-rad./in.lb.)	$+5.33$	$+3.77$	$+3.06$	$+2.74$

For the purpose of vibration analysis, the new shaft may be regarded as if it is light. Find what range of values of stiffness must be avoided if the system is not to have a natural frequency between 1200 and 1500 rad./sec. (C.U.M.S.T. Pt II, 1955)

1.10 Free oscillations of a torsional system

In this section we shall examine the free oscillation of torsional systems more closely, developing a technique which was mentioned in §1.4. It is convenient to use an illustrative example in explaining the theory.

The system of fig. 1.9.1 is redrawn in fig. 1.10.1 with a new set of co-ordinates indicated on it; the members x_5, x_6 and x_7 of this set are not strictly co-ordinates, but they will be required during the discussion. If a harmonic torque $F_1 e^{i\omega t}$ is applied at x_1 the responses at x_3 and x_4 will be

$$x_3 = \alpha_{31} F_1 e^{i\omega t}, \\ x_4 = \alpha_{41} F_1 e^{i\omega t}. \bigg\} \tag{1.10.1}$$

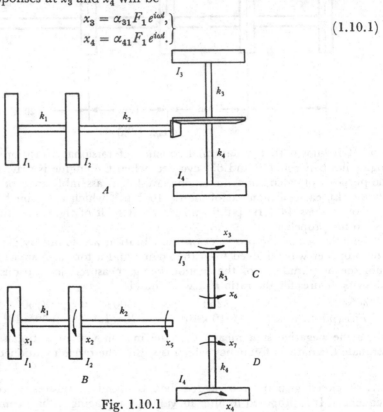

Fig. 1.10.1

The cross-receptances may be found from Table 2 (system 4) if the sub-systems B, C and D, which are shown in the figure, are used. The ratio of the two displacements of (1.10.1) is given by the ratio of the numerators of α_{31} and α_{41}; thus,

$$\frac{x_3}{x_4} = \frac{(\frac{1}{4}) \beta_{15} \gamma_{63} \delta_{77}}{(\frac{1}{4}) \beta_{15} \delta_{74} \gamma_{66}} = \frac{\gamma_{63} \delta_{77}}{\delta_{74} \gamma_{66}}. \tag{1.10.2}$$

The values of γ_{63}, γ_{66}, δ_{74} and δ_{77} are given by system 5 of Table 1. If the values given in §1.9 for the various parameters are used, this ratio becomes

$$\frac{x_3}{x_4} = \frac{\left(-\dfrac{1}{30\omega^2}\right)\left[-\dfrac{5 \times 10^6 - 30\omega^2}{5 \times 10^6 \times 30\omega^2}\right]}{\left(-\dfrac{1}{30\omega^2}\right)\left[-\dfrac{6 \times 10^6 - 30\omega^2}{6 \times 10^6 \times 30\omega^2}\right]} = \frac{10^6 - 6\omega^2}{10^6 - 5\omega^2}. \tag{1.10.3}$$

The ratio x_3/x_4 depends upon the excitation frequency ω and the effect may now be considered of taking ω closer and closer to one of the natural frequencies of the system. As we have already seen, this will mean that both α_{31} and α_{41} of equation (1.10.1) become indefinitely large so that the forcing amplitude F_1 must be made to approach zero if the motion is to remain finite. During this limiting process, however, the numerators of the receptances will remain finite and will continue to determine the ratio x_3/x_4 as in equation (1.10.3). It follows that during free vibration with the second natural frequency ω_2, for instance, the ratio in question will be given by

$$\frac{x_3}{x_4} = \frac{10^6 - 6 \times 0\cdot18963 \times 10^6}{10^6 - 5 \times 0\cdot18963 \times 10^6} = -2\cdot66, \tag{1.10.4}$$

this value for ω_2 being obtained from the second of the roots y which were given immediately above equation (1.9.9).

This result has been obtained by first considering excitation at x_1 and then letting this vanish as the natural frequency is approached. Now because the applied torque is to be made zero, the co-ordinate at which it is applied cannot affect the ratio (1.10.4). We leave it as an exercise for the reader to show that the same ratio is found by considering an applied torque at some other co-ordinate.

It is possible to obtain, by the above method, the ratio of displacements at any two co-ordinates during free vibration at any natural frequency. For instance, the ratio x_1/x_2 can be found from the ratio of the receptances α_{11} and α_{21}. This technique is applicable to any torsional system.† Thus Tables 1, 2 and 3 enable the ratio of the displacements to be determined for any two disks of a torsional system during free vibration.

The concept of a 'principal mode' was introduced in § 1.4; the reader will recall that it is the form (or *shape*) of the distortion which a system possesses when performing free vibration at one of its natural frequencies. The principal modes of a torsional system will not be discussed fully here; but it may be mentioned that the method used to find equation (1.10.4) provides a means of determining principal modes. Thus, using the second root ($\omega_2^2 = 435^2$) of the frequency equation for the system of fig. 1.10.1, and using the method already described, it is found that

$$\frac{x_1}{x_4} = \frac{0\cdot31}{1}, \quad \frac{x_2}{x_4} = \frac{0\cdot23}{1}, \quad \frac{x_3}{x_4} = -\frac{2\cdot66}{1}, \quad \frac{x_4}{x_4} = \frac{1}{1}. \tag{1.10.5}$$

During oscillation in this mode, therefore, at all instants‡

$$x_1 : x_2 : x_3 : x_4 :: 0\cdot31 : 0\cdot23 : -2\cdot66 : 1\cdot00. \tag{1.10.6}$$

We shall discuss in a later chapter other methods for finding principal modes; it is, however, common in practice for a single ratio, such as (1.10.4), to be required; when this is so the above method is sometimes useful and labour-saving.

It should be mentioned that in general, procedures of this sort become more difficult to follow with accuracy for a given system as higher natural frequencies

---

† A particular case in which it breaks down need not concern us here; it is dealt with fully in §3.11.
‡ When finding ratios of the type of x_2/x_4 the result quoted for Ex. 1.8.1 is often of great value. Here, however, the ratio is easily found from the ratio of α_{23} to α_{34}.

are used. This is a consequence of a property of vibrating systems which will be treated when we come to discuss Rayleigh's Principle in Chapters 3 and 5. It is to allow for this tendency that the roots y which precede equations (1.9.9) are quoted with such a high degree of accuracy.

The methods which have been explained in this chapter for dealing with torsional systems—that is, for finding natural frequencies, principal modes and responses to harmonic excitation—are tedious if the systems concerned are complicated. When this is so, it becomes necessary to adopt an arithmetical approach; we propose to discuss techniques of this sort in a later volume. The theory to be given in the present volume will be confined, generally speaking, to the fundamentals of vibration theory.

EXAMPLES 1·10

1. Find the ratio of the displacements x_3 and x_4 of the system of fig. 1.10.1 during free oscillation with the second natural frequency (see equation (1.9.9)). Use the method of this section, employing the receptances α_{44} and α_{34}. Compare the result with equation (1.10.4).

2. If the system of Ex. 1.9.1 oscillates freely in its first principal mode, find the ratio of the displacements x_3 and x_5 where the subscripts 3 and 5 are those of the inertias as shown. (The appropriate root of the frequency equation is $\omega^2 = 0.5025 \times 10^4$ rad.^{2}/sec.^{2}.)

NOTATION

Supplementary List of Symbols used in Chapter 1

A, B Arbitrary constants.

l Length of simple pendulum.

m, n Gear ratios at the connexions between sub-systems.

Z Amplitude of ξ (§ 1.2); modulus of z (§ 1.3).

z Complex number $= Z e^{i\psi} = Z e^{i(\Omega t + \phi)}$ (§ 1.3).

ξ Harmonically varying quantity (§ 1.2).

ϕ, ψ Phase angles.

Ω Circular frequency of ξ; angular speed of rotation of line in Argand diagram.

CHAPTER 2

GENERALIZED CO-ORDINATES AND LAGRANGE'S EQUATIONS

Mais si l'on cherche le mouvement de plusieurs corps qui agissent les uns sur les autres par impulsion ou par pression, soit immédiatement comme dans le choc ordinaire, ou par le moyen de fils ou de leviers inflexibles auxquels ils soient attachés, ou en général par quelque autre moyen que ce soit, alors la question est d'un ordre plus élevé, et les principes précédents sont insuffisants pour la résoudre. Car ici les forces qui agissent sur les corps sont inconnues, et il faut déduire ces forces de l'action que les corps doivent exercer entre eux, suivant leur disposition mutuelle. Il est donc nécessaire d'avoir recours à un nouveau principe qui serve à déterminer la force des corps en mouvement, eu égard à leur masse et à leur vitesse.   M. DE LA GRANGE, *Méchanique Analitique* (1788)

In order to develop a vibration theory which is not confined to any particular type of problem, we shall need a general method for forming the equations of motion of a mechanical system. The method of Lagrange, used with a system of 'generalized co-ordinates', fulfils this need.† This chapter is devoted to an explanation of these matters.

2.1 Generalized co-ordinates

The configuration of a given dynamical system—which will, in fact, be an idealization of a real system—can be expressed in terms of the values of a number of variables or 'co-ordinates'. These co-ordinates are the values of a number of independent quantities which are sufficient in number to specify the positions of all the parts of the system. The number of co-ordinates is thus equal to the number of degrees of freedom of the system.

Although only linear and angular displacements were used as co-ordinates in Chapter 1, there are many more possibilities. If, for instance, it is required to specify the position of a piston in its cylinder, as shown in fig. 2.1.1, we can choose as the co-ordinate the distance x between the piston crown and

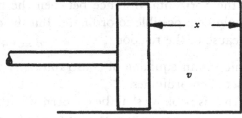

Fig. 2.1.1

the inside of the cylinder head. But equally the volume v that is enclosed can be used as a co-ordinate and this would be quite valid for our purposes.

For a given system several alternative co-ordinates are usually available, although they are not all independent. Thus, in the above example x and v cannot

† J. L. Lagrange, *Mécanique Analytique* (Paris, 1853). Section 5 of Part 2 of this famous book is concerned with the theory of vibration.

both be given arbitrary values because they are related in a known way. Therefore either one of them can be used as a co-ordinate, the other being calculated from the known 'equation of constraint' which relates them.

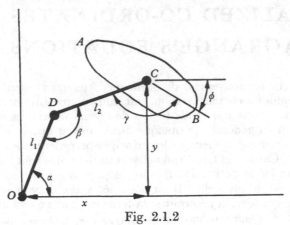

Fig. 2.1.2

If the link AB of fig. 2.1.2 is confined to plane motion then it has three degrees of freedom so that three co-ordinates will specify its position. One such set would be the Cartesian co-ordinates (x, y) of some point C within it and the angle ϕ between the horizontal and a line CB fixed in the link. Alternatively, for some purposes, it may be convenient to imagine that a point (C say) is joined to O, the fixed origin, by two hinged links of lengths l_1 and l_2 as shown by OD and DC respectively in the figure. The angles α, β and γ would then serve as a set of co-ordinates. The two sets, x, y, ϕ and α, β, γ are related by the three equations of constraint

$$\left. \begin{aligned} x &= l_1 \cos \alpha + l_2 \cos (\alpha + \beta - 180°), \\ y &= l_1 \sin \alpha + l_2 \sin (\alpha + \beta - 180°), \\ \phi &= 360° - (\alpha + \beta + \gamma). \end{aligned} \right\} \tag{2.1.1}$$

Another example can be taken from the system of fig. 1.4.1. The extension, d say, of the horizontal distance between the particle and the fixed end of the spring would be a possible co-ordinate. But the quantities x, θ and d are not independent because of the relation

$$d = x + l \sin \theta \tag{2.1.2}$$

which is an equation of constraint. Any two of these three variables will serve as a set of co-ordinates.

In this book we shall be concerned with the problem of subjecting a mechanical system to a force which is a function of time. It is, however, sometimes convenient to treat problems in which a system is subjected to a given *displacement* which varies with time. This introduces equations of constraint in which time appears explicitly or in which the time derivatives of the co-ordinates occur. These possibilities will not be dealt with here; but they are examined in books on general dynamics.†

All of the quantities which have been mentioned so far for use as co-ordinates

† See, for instance, S. P. Timoshenko and D. H. Young, *Advanced Dynamics* (McGraw-Hill, London, 1948).

have an evident physical meaning. It is not, however, essential that the co-ordinates should have immediate physical significance, and they may therefore be defined as algebraic combinations of the physical measurements. Thus two variables q_1 and q_2 can be used to describe the configuration of the system of fig. 1.4.1 if they are defined, for instance, by the relations

$$\left. \begin{array}{l} q_1 = 6x + 7\theta, \\ q_2 = 2x - 4\theta. \end{array} \right\} \tag{2.1.3}$$

This follows because, if q_1 and q_2 are known, x and θ can be found from the relations

$$\left. \begin{array}{l} x = (4q_1 + 7q_2)/38, \\ \theta = (q_1 - 3q_2)/19. \end{array} \right\} \tag{2.1.4}$$

This method of forming co-ordinates will not be used at present although it will be required later in the discussion of principal co-ordinates. It is sufficient at present to notice that any set of co-ordinates $q_1, q_2, q_3, \ldots$ which are defined as functions of some set of physically defined co-ordinates—such as x and θ in this problem—can be used. It is necessary that they shall be sufficient in number and be independent.

The configuration of a continuously deformable system, that is a system which contains one or more non-rigid bodies, cannot be specified by a finite number of co-ordinates. We shall leave this matter open until Chapter 4; but it may be remarked here that in such problems we adopt the idea described above.

The term 'generalized co-ordinates', which is used in the heading of this chapter, is applied to any set of co-ordinates which can be used to describe the configuration of a system. The adjective implies, not that the co-ordinates in question belong to any particular class, but rather that they are to be used in a particular manner in the analysis. This manner of use, which has been called the Lagrangian method, involves also the concept of generalized forces and this is discussed in the next section.

EXAMPLES 2.1

1. The configuration of a system of three particles can be specified by nine numbers which are the Cartesian co-ordinates of the particles referred to a set of rectangular axes; let these co-ordinates be written $x_1, y_1, z_1, x_2, y_2, z_2, x_3, y_3, z_3$.

If the motion of the particles is now limited in the ways described below, write down the relevant equations of constraint and state the number of degrees of freedom remaining in each case.

    (a) Particles restricted to motion in the plane $z = 1$.

    (b) Particles 1 and 2 are joined by a rigid link of length l.

2. Find the equations which govern the variation of q_1 and q_2 in equations (2.1.3) for the system of fig. 1.4.1.

3. The position of a lamina which can move in a plane is defined by three co-ordinates (x, y, θ), where x and y are the Cartesian co-ordinates of a point P in the body, referred to a pair of rectangular axes Ox and Oy, and θ is the angle between Ox and a

line PQ fixed in the body. The position of the body may alternatively be defined by the co-ordinates (X, Y, ϕ), where these co-ordinates are derived from the original ones by the equations

$$X = (x-a) \cos \alpha + (y-b) \sin \alpha,$$
$$Y = (y-b) \cos \alpha - (x-a) \sin \alpha,$$
$$\phi = \theta - \alpha.$$

Show that (X, Y) are the Cartesian co-ordinates of P with reference to axes $O'XY$ which are inclined at an angle α to Ox and Oy and which have their origin at the point (a, b) referred to Ox and Oy. Also show that ϕ is the angle between the original reference line in the lamina and the new axis $O'X$.

4. A number of straight bars are hinged to one another at the ends to form a chain which is suspended by one end from a fixed point. The lengths of the bars between the hinges are $a_1, a_2, a_3, \ldots$ starting from the top. Considering displacements of the bars in one plane only, the displacement from the vertical through the point of support of the lower end of the rth bar is denoted by q_r. $q_1, q_2, q_3, \ldots$ thus form a set of co-ordinates. It is to be assumed throughout this example that the values of the displacements q are small compared to the lengths a.

The configuration of the system can alternatively be defined by a set of co-ordinates $q'_1, q'_2, q'_3, \ldots$ which define the horizontal displacements of points on the bars other than the end-points. Show that these co-ordinates are related to the first set by the equations

$$q'_1 = K_1 q_1,$$
$$q'_2 = K_2 q_2 + (1 - K_2) q_1,$$
$$q'_3 = K_3 q_3 + (1 - K_3) q_2,$$
$$\text{etc.,}$$

where $K_r a_r$ is the distance of the reference point on the rth bar from its upper end.

Show further that it would be theoretically possible to construct a linkwork which could be attached to the hinges of the chain and which would have on it a set of reference points whose horizontal displacements would be

$$K'_1 q'_1,$$
$$K'_2 q'_2 + (1 - K'_2) q'_1,$$
$$K'_3 q'_3 + (1 - K'_3) q'_2,$$
$$\text{etc.,}$$

where $K'_1, K'_2, \ldots$ are a set of chosen constants. Deduce, by an extension of the argument, that a linkwork could be designed which carried reference points whose displacements would be given by any arbitrarily chosen independent linear functions of the original co-ordinates.

[NOTE. It is sometimes an aid to thought to imagine physical counterparts, such as this, to correspond to algebraic functions.]

2.2 Generalized forces

Lagrange's equations, towards the derivation of which we are working, are concerned with the variation of a set of generalized co-ordinates with time. For their use they require that the forces and couples which are applied to a system should be divided into components, each of which is associated with a particular co-ordinate. This leads to the concept of 'generalized forces' whose nature we shall now explain but whose use will be described later.

We shall derive sets of generalized forces which correspond to sets of applied forces and couples. Now, in dynamical systems, these latter are time-dependent so that the generalized forces also vary with time; but the generalized forces are defined in such a way that the relationships between the two sets (for a given system) do not depend upon time. It is therefore simplest to think of *static* systems of applied forces and couples $(F_1, F_2, \ldots,$ say), to find the appropriate generalized forces in terms of $F_1, F_2, \ldots$ and then to observe that, in fact, $F_1, F_2, \ldots$ may be time-dependent.

Let the configuration of a system with n degrees of freedom be defined by generalized co-ordinates $q_1, q_2, \ldots, q_n$. Since the co-ordinates q are independent, any one of them, q_m say, can be given a small increment δq_m while the others are kept constant. When this happens, the forces which act on and within the system will, in general, do a small amount of work δW_m. This can be expressed in the form

$$\delta W_m = Q_m \delta q_m, \tag{2.2.1}$$

which defines a certain function Q_m. This function is called the 'generalized force' which is associated with q_m.

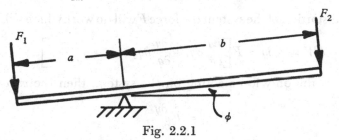

Fig. 2.2.1

As a first example, consider the light rigid rod shown in fig. 2.2.1. The small angle which the rod makes with the horizontal is denoted by ϕ, this being the single co-ordinate of the system. A small increase $\delta\phi$ of ϕ will be accompanied by an amount of work

$$\delta W_\phi = F_1 a . \delta\phi - F_2 b . \delta\phi \tag{2.2.2}$$

so that the generalized force associated with ϕ is $(F_1 a - F_2 b)$. It may be noted that this is the moment of the forces about the pivot.

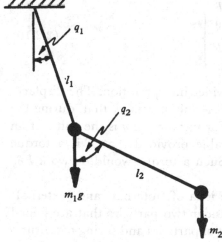

Fig. 2.2.2

For a second example, consider the double pendulum of fig. 2.2.2 in which particles of mass m_1 and m_2 are suspended by light strings of length l_1 and l_2. The angles marked q_1 and q_2 may be taken as generalized co-ordinates. If q_2 is increased by δq_2 while q_1 is kept constant, then work δW_2 will be done by gravity where

$$\delta W_2 = -m_2 g l_2 \delta q_2 \sin q_2. \tag{2.2.3}$$

Therefore $Q_2 = -m_2 g l_2 \sin q_2,$ (2.2.4)

which is the moment of the weight of the lower mass about the upper one. If, on the other hand, q_1 alone is increased (and this will involve relative movement at the pivot m_1) then the work will be

$$\delta W_1 = -(m_1 + m_2)\, gl_1\, \delta q_1 \sin q_1 \qquad (2.2.5)$$

so that
$$Q_1 = -(m_1 + m_2)\, gl_1 \sin q_1. \qquad (2.2.6)$$

This is no longer the moment of the forces $m_1 g$ and $m_2 g$ about the point of suspension because of the pivoting between the strings.

It is possible to express the generalized forces conveniently in terms of partial derivatives. Let a concentrated force F be applied to a system and let the displacement of the point of application, measured in the direction of the force, be y. If y is chosen as one of the generalized co-ordinates of the system then the associated generalized force will be simply F. But if y is not one of the chosen co-ordinates, then it will have to be expressed as a function of the generalized co-ordinates; that is to say,

$$y = y(q_1, q_2, \ldots, q_n). \qquad (2.2.7)$$

During a small distortion of the system the force F will do work which will be given by

$$\delta W = F\delta y = F\left[\frac{\partial y}{\partial q_1}\delta q_1 + \frac{\partial y}{\partial q_2}\delta q_2 + \ldots + \frac{\partial y}{\partial q_n}\delta q_n\right]. \qquad (2.2.8)$$

If, therefore, F is the only force acting on the system, then the mth generalized force will be

$$Q_m = F\frac{\partial y}{\partial q_m}. \qquad (2.2.9)$$

This is obtained from (2.2.8) by putting all the displacements $\delta q_1, \delta q_2, \ldots, \delta q_n$ zero except δq_m as is required by the definition of Q_m.

The system of fig. 1.4.1 provides an example of this process. It was pointed out that generalized co-ordinates could be chosen as in equation (2.1.3). The parts of the generalized forces which are due to a static force F that acts on the mass M in the direction x, can be found from equation (2.2.9) to be:

$$\left.\begin{aligned} Q_1 &= F\frac{\partial x}{\partial q_1} = \frac{2F}{19}, \\[2mm] Q_2 &= F\frac{\partial x}{\partial q_2} = \frac{7F}{38}. \end{aligned}\right\} \qquad (2.2.10)$$

Equation (2.2.9) can be given a somewhat wider interpretation. The displacement y may be a displacement of *any* type; for it is only necessary that, during the displacement, the work done should be $F\delta y$. For instance, if y is the value of an angle, then the above theory becomes applicable provided that F is a torque whose axis coincides with that of the angle. Such a torque would do work $F\delta y$ when the angle changed by δy.

It is expedient at this point to examine the idea of 'internal' and 'external' forces since we shall use this notion later. Consider two particles that are joined by a spring as in fig. 2.2.3 (a). The arrangement of particles and spring constitutes a dynamical system and the system may be imagined to be confined within a specified boundary B as shown. The spring is part of the system and the force in the spring is an internal force. But any other force which acts on the system, such as the force F shown, must act across the boundary B and is therefore an external force. Now when analysing a system it is necessary to define it by a boundary in this way thereby determining which are the internal—and which the external—forces; it should be realized that this division depends upon our initial choice of system and not upon any intrinsic property of the forces. In the above example,

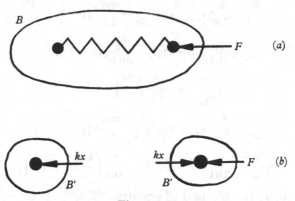

Fig. 2.2.3

we can alternatively choose to regard the particles as separate systems, in which case we provide boundaries B', B' as shown in fig. 2.2.3 (b). The force in the spring then acts across the boundaries of the systems and becomes an external force. Although the choice of the boundaries is arbitrary it must be adhered to throughout the subsequent analysis because it is the system inside the boundary for which the equations of motion will be formed; the initial choice will usually depend upon convenience in forming the equations of motion.

Now during a displacement of one co-ordinate q_m of an n-degree-of-freedom system, work will be done by both internal and external forces and both sets must be considered when finding Q_m. In the problems which we shall discuss first in this book, we shall be concerned with frictionless idealized systems in which the only internal forces which can do work will be those of elastic distortion. Later, in problems on damping, allowance will have to be made for internal dissipative forces. That part of Q_m which is produced by an internal elastic force is found by imagining the relevant spring to be severed and then proceeding as before.

For instance, if a static force F acts in the direction x upon the mass M of fig. 1.4.1, then the total work done will be

$$\delta W_x = F\delta x - kx\,\delta x \qquad (2.2.11)$$

during a small displacement δx. The generalized force corresponding to x will therefore be

$$Q_x = F - kx. \qquad (2.2.12)$$

Again the work done during a displacement $\delta\theta$ will be

$$\delta W_\theta = -mgl\theta\,.\,\delta\theta \qquad (2.2.13)$$

and the corresponding force will be

$$Q_\theta = -mgl\theta. \tag{2.2.14}$$

If, on the contrary, the co-ordinates q_1 and q_2 of equation (2.1.3) are to be used, then the expressions for the corresponding generalized forces will be

$$
\begin{aligned}
Q_1 &= \frac{2F}{19} - kx\frac{\partial x}{\partial q_1} - mg\frac{\partial}{\partial q_1}\left(\frac{l\theta^2}{2}\right) \\
&= \frac{2F}{19} - \frac{k}{38}(4q_1 + 7q_2)\frac{4}{38} - \frac{mgl}{19}(q_1 - 3q_2)\frac{1}{19} \\
&= \frac{2F}{19} - \frac{q_1}{361}(4k + mgl) - \frac{q_2}{361}(7k - 3mgl),
\end{aligned} \tag{2.2.15}
$$

$$
\begin{aligned}
Q_2 &= \frac{7F}{38} - kx\frac{\partial x}{\partial q_2} - mg\frac{\partial}{\partial q_2}\left(\frac{l\theta^2}{2}\right) \\
&= \frac{7F}{38} - \frac{k}{38}(4q_1 + 7q_2)\frac{7}{38} + \frac{mgl}{19}(q_1 - 3q_2)\frac{3}{19} \\
&= \frac{7F}{38} - \frac{q_1}{361}(7k - 3mgl) - \frac{q_2}{361}\left(\frac{49k}{4} + 9mgl\right),
\end{aligned} \tag{2.2.16}
$$

these being obtained by the method of equation (2.2.9). It should be noted how the vertical displacement of the pendulum bob, $\frac{1}{2}l\theta^2$, has to be brought into the equations but that subsequent differentiation leaves only first-degree terms.

It is sometimes convenient to form a 'potential function' from which the generalized forces may be obtained by differentiation. This involves an extension of the use of potential energy beyond those elementary problems where work can be done by gravity. The essential feature which a force must possess, if it is to be analysed in terms of a potential function, is that the work which it does during any displacement of the system upon which it acts shall depend only upon the initial and final configurations; it must not depend upon the 'route' which the system follows in moving from the one configuration to the other. This may be put another way by saying that the forces which act on the system, in any configuration, depend only upon the configuration and not upon the manner in which it was attained nor upon the instantaneous velocities. This applies to forces of gravitation because the work which they do on a body depends upon the drop of its centre of gravity and is not altered if the drop is accompanied by horizontal motion. Thus only the initial and final heights of the centre of gravity determine the work done and therefore the 'change of potential'. Frictional forces on the other hand, whether dry or viscous, do not fit the prescribed condition.

The existence of a potential function associated with a given set of forces is usually evident physically. For instance the forces of elasticity have potential because perfectly elastic bodies can have strain energy stored in them and this energy is a function only of the strain of the bodies. Thus if a spring of stiffness k is stretched from an extension x_1 to an extension x_2 then the work done by the spring will be

$$-\int_{x_1}^{x_2} kx\,dx = -\frac{k}{2}(x_2^2 - x_1^2), \tag{2.2.17}$$

which is seen to depend only upon the initial and final states. The quantities $\frac{1}{2}kx_1^2$ and $\frac{1}{2}kx_2^2$ give the energy which is stored in the spring in the initial and final states; the difference between these values is equal to the work done by the force during the displacement.

Both internal and external forces may have potential. If a force is known to have potential, then this may be expressed quantitatively. The *value* of the potential V is defined as the work that would be done by the force if the system to which it is applied were to move from its actual state to some previously chosen datum state. Thus the potential of the gravity force of a particle of mass M is

$$V = Mgh, \qquad (2.2.18)$$

where h is the vertical distance of the particle above a specified datum level. Again the potential of the force in a spring of stiffness k, which is extended by an amount x, is

$$V = \frac{kx^2}{2} + C \qquad (2.2.19)$$

if C is chosen as the value of the potential when the spring is not strained. It will be seen later that the value of C is not important because it vanishes on differentiation; it is usually convenient to choose C to be zero. The quantity V is sometimes called the 'potential energy' of the system.

The forces from which the generalized forces of a system are calculated may or may not have potential and in any particular problem the two kinds will be segregated. Consider now only those forces which have potential. The potential energy of a system depends only upon its configuration so that it will have the form

$$V = V(q_1, q_2, ..., q_n). \qquad (2.2.20)$$

The increase of V due to small increments of the generalized co-ordinates will therefore be

$$\delta V = \frac{\partial V}{\partial q_1} \delta q_1 + \frac{\partial V}{\partial q_2} \delta q_2 + ... + \frac{\partial V}{\partial q_n} \delta q_n. \qquad (2.2.21)$$

If only the co-ordinate q_m increases, then

$$\delta V = \delta V_m = \frac{\partial V}{\partial q_m} \delta q_m. \qquad (2.2.22)$$

Now since the forces have potential, this work is equal and opposite to the work which is done by them during the distortion so that

$$\delta W_m = -\frac{\partial V}{\partial q_m} \delta q_m. \qquad (2.2.23)$$

The generalized force is therefore given by

$$Q_m = -\frac{\partial V}{\partial q_m}. \qquad (2.2.24)$$

In future we shall often calculate the contributions of gravity and spring forces to generalized forces by differentiating their potential.

Returning to the system of fig. 1.4.1, suppose that a force F acts on the mass M in the direction x. The contributions of F to Q_x and Q_θ can be found by the direct method; they are F and zero respectively. The two other forces which act on the system are the weight and the spring force and both of these have potential. If the lowest position of m and the unstrained position of the spring are used as datum configurations, the potential energy of the distorted system becomes

$$V = \frac{mgl\theta^2}{2} + \frac{kx^2}{2},$$
(2.2.25)

provided that θ is small. The *complete* expressions for the generalized forces are therefore

$$\left. \begin{aligned} Q_x &= F - \frac{\partial V}{\partial x} = F - kx, \\ Q_\theta &= -\frac{\partial V}{\partial \theta} = -mgl\theta, \end{aligned} \right\}$$
(2.2.26)

which agree with previous results.

We leave it to the reader to rederive equations (2.2.15) and (2.2.16) by means of equation (2.2.25).

When generalized forces are calculated, account must be taken of forces within springs; these forces are sometimes internal to the system and sometimes external. Allowance must also be made for frictional forces within non-rigid members (these latter being forces which are not derivable from a potential function). But if a system contains *rigid* members then, although these members will have internal forces acting within them, no work can be done by these forces. Now if the equations for the equilibrium of a system are formed by applying the rules of statics to each member of the system separately, then the forces within the rigid members will appear in the equations because they produce the reactions between the different parts. These forces, which are not usually of immediate interest, have to be eliminated subsequently in the algebra.

The advantage of using generalized co-ordinates, in conjunction with the idea of generalized forces, is that the forces within rigid bodies are never introduced. The generalized forces are obtained from the work that is done in displacements which are specified in terms of the generalized co-ordinates; if, therefore, the co-ordinates are chosen so that an extension of some of the links in a system cannot be written in terms of them, then the generalized forces cannot contain any components due to forces within these links. In this way all the internal forces within rigid members are automatically removed from the analysis by choosing co-ordinates which imply the rigidity of the members.

Later on we shall refer to 'forces', 'displacements' and 'co-ordinates', and we shall mean by these terms, where there is no danger of confusion, generalized forces, changes in the values of generalized co-ordinates, and generalized co-ordinates respectively. It will thus be unnecessary, during general discussion, to distinguish between forces and torques on the one hand or between translations and rotations on the other. When it becomes essential to use the terms force and displacement in their more restricted senses, the context will prevent ambiguity.

EXAMPLES 2.2

1. Prove that the potential energy due to the weights of a number of particles, $m_1g, m_2g, ..., m_ng$ is equal to $hg \sum_{r=1}^{n} m_r$, where h is the height of the centre of mass of the particles above some datum level.

2. The figure shows a four-bar kinematic chain of which the link CD is fixed. The lengths of the links are as follows: $AB = 2a$, $BC = a$, $CD = 3a$, $DA = 2a$. AD and BC are light but the link AB has a uniformly distributed mass M. For the position of the chain in which the angle ADC is $60°$ find the generalized force Q_θ due to the weight of the top link, the co-ordinate θ being the angular displacement of AD.

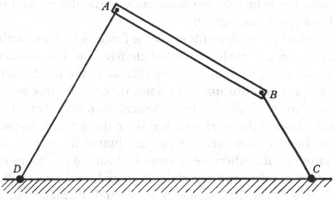

Ex. 2.2.2

3. Show how the generalized force Q_θ of the previous problem can be found graphically, by means of a displacement diagram, for any given value of θ.

4. A particle is supported by n concurrent coplanar springs the end of each spring which is remote from the particle being fixed. The axes of the springs make angles $\alpha_1, \alpha_2, ... \alpha_n$ with a certain fixed direction Ox and their stiffnesses are $k_1, k_2, ..., k_n$. A co-ordinate q_1 denotes the displacement of the particle in the direction Ox and q_2 denotes its displacement at right angles to this direction. Displacement out of the plane of the springs is prevented.

Write down an expression for the potential energy which will be stored in the springs due to the displacement (q_1, q_2) and derive from it an expression for the energy in terms of two new co-ordinates q_1' and q_2' which represent displacements in directions which are inclined at an angle α to those of the original co-ordinates. Deduce that there will be a certain value of α for which the potential energy expression contains no term involving the product $q_1'q_2'$ and hence show that if an external force is applied to the particle in the corresponding directions, then it will be deflected along these directions only.

5. Using direction cosines to specify the orientations of the springs, extend the discussion of the last example to three dimensions.

2.3 The principle of virtual displacements

The use of the method of virtual work, which involves the concept of virtual displacements, is primarily concerned with the analysis of static systems. We shall show later that the principle can also be applied to dynamic systems if these are approached by the method of d'Alembert. For the present, however, we shall restrict our attention to static systems in order to introduce the methods.

The method of virtual work permits direct use to be made of generalized co-ordinates when investigating the equilibrium conditions of a system. The principle may be stated thus: the work which is done by the forces (both internal and external) which are acting on a system that is in equilibrium, when the system undergoes a 'virtual' displacement, is zero.

It is first necessary to explain the significance of the adjective virtual. The word has been used by different writers in slightly different senses[†] and in recent years has become almost synonymous with 'small' or 'infinitesimal'. The ideas involved, however, when using the principle, are sufficiently particular to warrant the use of the word virtual in a specialized sense.

The equilibrium of a system depends upon the forces which are acting on it and not upon the way in which these forces would change in a real displacement; only the stability of the equilibrium is affected by this latter factor. Accordingly, if we are merely analysing the equilibrium conditions, then we are free to postulate that the forces do *not* change during a hypothetical displacement. This postulate facilitates the calculation of the work which is done during the displacement since a number of second-order terms are thereby eliminated from the expression; the neglect of these latter would otherwise have to be justified on the grounds of their small magnitude. When using the principle of virtual work, we shall employ this idea and we shall refer to the displacements as 'virtual' displacements.

This concept of a virtual displacement can be illustrated by the problem of the equilibrium of a mass m which is suspended on a light spring. The equilibrium condition, giving the static extension x of the spring, is that the force in the spring should be equal to the weight. If, in a real displacement, the mass is moved downwards by a distance δx, then the work done by the weight will be $mg\,\delta x$ and the work done by the spring will be

$$-\frac{k}{2}[(x+\delta x)^2 - x^2],   \tag{2.3.1}$$

where k is the spring stiffness and x is the initial extension. In a *virtual* displacement, on the other hand, the spring force remains constant during the displacement so that the work done by the spring is $-kx\,.\,\delta x$. If the net work done is now equated to zero, it is found that

$$mg\,\delta x - kx\,\delta x = 0.   \tag{2.3.2}$$

This is the equilibrium equation from which the condition of equilibrium is found by cancelling the non-zero quantity δx. This condition could alternatively have been reached by equating the total work in the real displacement to zero; thus

$$mg\,\delta x - \frac{k}{2}[(x+\delta x)^2 - x^2] = 0.   \tag{2.3.3}$$

The equilibrium equation can now be found by taking the limit as $\delta x \to 0$ so that the term containing $(\delta x)^2$ can be neglected in comparison with the δx terms and the

† See H. Lamb, *Statics* (Cambridge University Press, 3rd ed. 1928).

equation reduces to (2.3.2). The application of this limiting procedure to (2.3.3) is longer than the use of equation (2.3.2) directly and sometimes interferes with the proper understanding of the problem. We shall therefore always use virtual displacements when we are examining equilibrium conditions.

The above problem concerns a single-degree-of-freedom system. But the principle can be used for any number of degrees of freedom and an example of one possessing two degrees of freedom follows.

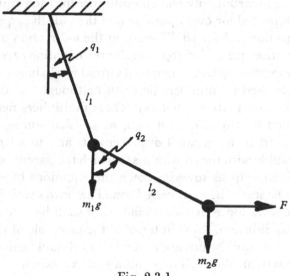

Fig. 2.3.1

Let it be required to find the equilibrium position of the double pendulum of fig. 2.3.1 when it is acted on by a horizontal force F at the lower mass. Consider first a virtual displacement involving the co-ordinate q_2 only. The virtual work during this displacement will be

$$\delta W_2 = F l_2 \delta q_2 \cos q_2 - m_2 g l_2 \delta q_2 \sin q_2, \tag{2.3.4}$$

and this will be zero if
$$\tan q_2 = \frac{F}{m_2 g}. \tag{2.3.5}$$

Again, for a virtual displacement which involves q_1 only

$$\delta W_1 = F l_1 \delta q_1 \cos q_1 - (m_1 + m_2) g l_1 \delta q_1 \sin q_1, \tag{2.3.6}$$

and if this is zero
$$\tan q_1 = \frac{F}{(m_1 + m_2) g}. \tag{2.3.7}$$

These two equations give the values of q_1 and q_2 in the equilibrium condition. In this problem, the two values are obtained independently, but in general the method will give simultaneous equations in the co-ordinates.

It is possible to derive the principle of virtual work from other statical theorems. We shall here, however, take it as an axiom because it is a safe and convenient starting point for the purposes of the engineer. It can alternatively be regarded in a different way, as a purely mathematical result. Imagine any system to be composed of a number of particles. At any instant, each separate particle is in equilibrium, which fact may be expressed by three equations; these equations state that the sums of all the forces in fixed directions Ox, Oy, Oz—both internal and external—which act upon the particle add up to zero. Imagine that the three equations are written down for a particular particle. Now the truth of these equations is not interfered with if each is multiplied throughout by some number; let this be done, using an arbitrarily chosen multiplier for each equation. Suppose that this process is repeated for every particle and then all the equations for equilibrium in the x-direction, all for equilibrium in the y-direction and all for equilibrium in the z-direction are added together. Three equations are found that can be regarded as those expressing the principle of virtual work since the terms in them represent forces multiplied by numbers that can be thought of as components of displacement (or, for that matter, velocity). These multipliers need not be small and are not restricted in any way although, as we shall show, it is sometimes advantageous to regard them as small displacements and to select them in such a way as to be compatible with the constraints to which the parent body is subjected.

If the principle is to help us to write down the equations of equilibrium of a system, then it must be stated in a converse form. That is to say, it is necessary that the vanishing of the work for a virtual displacement shall be a condition, and not merely a result, of equilibrium. Now in terms of the principle of virtual work, the complete conditions for equilibrium are that no work should be done in *any* virtual displacement of the system. We shall take this also as an axiom.

Consider a system with n degrees of freedom whose configuration is specified by the generalized co-ordinates $q_1, q_2, ..., q_n$. Virtual displacements of the system may be achieved by varying these co-ordinates by suitably chosen amounts $\delta q_1, \delta q_2, ..., \delta q_n$, the work done being

$$\delta W = Q_1 \delta q_1 + Q_2 \delta q_2 + ... + Q_n \delta q_n. \tag{2.3.8}$$

Now the only way in which δW can vanish *whatever the relative magnitudes* of δq_1, $\delta q_2, ..., \delta q_n$ is for each term on the right-hand side of (2.3.8) to vanish separately so that

$$\delta W_1 = \delta W_2 = ... = \delta W_n = 0. \tag{2.3.9}$$

That is, it is required that

$$Q_1 = Q_2 = ... = Q_n = 0. \tag{2.3.10}$$

It must be noticed that the generalized forces Q must necessarily include the internal—as well as the external forces (as in the above case of the spring-suspended mass).

These ideas can be applied to equations (2.2.15) and (2.2.16). If $Q_1 = Q_2 = 0$, then

$$\left.\begin{aligned}
(4k + mgl)\, q_1 + (7k - 3mgl)\, q_2 &= 38F, \\
(7k - 3mgl)\, q_1 + (\tfrac{49}{4}k + 9mgl)\, q_2 &= \frac{133F}{2}.
\end{aligned}\right\} \tag{2.3.11}$$

These equations give the equilibrium condition of the system of fig. 1.4.1 in terms of the co-ordinates of equation (2.1.3). If q_1 and q_2 are now replaced by the expressions (2.1.3) then this pair of equations reduces to

$$\left.\begin{aligned}2kx + mgl\theta &= 2F, \\ 7kx - 6mgl\theta &= 7F.\end{aligned}\right\} \tag{2.3.12}$$

These have the solutions $\qquad x = F/k, \quad \theta = 0,$ (2.3.13)

which we know of by other methods.

It is of interest, at this point, to discuss the equilibrium of systems that are acted on only by forces which have potential. In such a case the conditions (2.3.10) may be written

$$\frac{\partial V}{\partial q_1} = \frac{\partial V}{\partial q_2} \cdots = \frac{\partial V}{\partial q_n} = 0, \tag{2.3.14}$$

and this shows that the potential has a stationary value in the equilibrium condition. If the potential is a minimum, then small changes in the co-ordinates will increase the potential energy; this shows that, in any small displacement, the forces do negative work on the system. Under these conditions the equilibrium is said to be *stable*; thus stability is associated with positive second derivatives of V with respect to the co-ordinates of a system when it is in an equilibrium condition. Conversely the equilibrium is unstable if any of the second derivatives are negative.

As an example, consider the double pendulum system of fig. 2.2.2. It has been shown, in effect, that

$$\frac{\partial V}{\partial q_1} = (m_1 + m_2)\, gl_1 \sin q_1, \quad \frac{\partial V}{\partial q_2} = m_2 gl_2 \sin q_2, \tag{2.3.15}$$

from which the equilibrium positions may be derived by applying equations (2.3.14); they are
$$q_1 = 0 \text{ or } \pi, \quad q_2 = 0 \text{ or } \pi. \tag{2.3.16}$$

The second derivatives of the potential are

$$\frac{\partial^2 V}{\partial q_1^2} = (m_1 + m_2)\, gl_1 \cos q_1, \quad \frac{\partial^2 V}{\partial q_2^2} = m_2 gl_2 \cos q_2, \quad \frac{\partial^2 V}{\partial q_1 \partial q_2} = 0. \tag{2.3.17}$$

When $q_1 = q_2 = 0$, the equilibrium is stable; if, on the other hand, either co-ordinate has the value π, then the equilibrium is unstable.

EXAMPLES 2.3

1. A simple bridge is formed from two uniform rigid beams AB and BC which are of equal length and which weigh 400 lb. and 200 lb. respectively. They are hinged together at B and to abutments at the same level at A and C, the hinges being smooth. AC is 40 ft. and B is 10 ft. above AC.

Calculate the vertical and horizontal reactions at A and C when the bridge is loaded with 1000 lb. at B.

By how much will the hinge B descend if the length AC is increased by 1 in.?

2. A reciprocating engine is shown diagrammatically in the figure. For the position shown, find the thrust R on the piston which is necessary to overcome a resisting torque of 100 lb.in. applied to the crankshaft. Friction may be neglected.

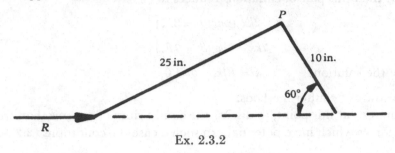

Ex. 2.3.2

3. If, in the system of the previous problem a friction couple if 5 lb.in. has to be overcome at P owing to tightness of the big-end bearing, show that the thrust R must be increased by about 0·58 lb.

4. A shaft $ABCDE$ is rigidly held at the ends A and E. Three equal disks of radius r are attached to the shaft at B, C and D and the torsional stiffness of the shaft between adjacent disks and between the disks and the end of the shaft is k. Each disk has a string wrapped round it with a mass of m lb. hanging on the end of it, the other end of each string being fixed to the rim of a disk. The three strings are wrapped round the disks in the same direction. Using co-ordinates which represent rotations of the disks measured from the position where there is no twist in the shaft, write down an expression for the potential energy which is stored in the shaft and for that which is lost due to the fall of the weights. Obtain the equilibrium equations by differentiating these expressions and confirm that these equations give the same conditions as are obtained by direct statics. Flexure of the shaft is to be neglected.

2.4 d'Alembert's principle

We have now to consider the means by which the methods of the last three sections can be applied to dynamics problems.

Let some particle which is within a mechanical system be acted on by forces X, Y, Z in the directions of three fixed co-ordinate axes $Oxyz$. If the mass of the particle is m then its three components of acceleration will be given by the equations

$$X = m\ddot{x}, \quad Y = m\ddot{y}, \quad Z = m\ddot{z}. \tag{2.4.1}$$

These equations can alternatively be written

$$X - m\ddot{x} = 0, \quad Y - m\ddot{y} = 0, \quad Z - m\ddot{z} = 0. \tag{2.4.2}$$

The trivial algebraic difference between the two sets of equations represents a significant difference of approach to the physical problem. In equations (2.4.1), the components of acceleration of the particle are regarded as resulting from the application of forces; but in (2.4.2), the particle is regarded as being *in equilibrium* under the combined action of the applied and the 'inertia' forces $(-m\ddot{x})$, $(-m\ddot{y})$, $(-m\ddot{z})$.

The advantage of the second method of approach is that it allows us to use any of the theorems of statics, including the method of virtual displacements, for the analysis of the forces. It is merely necessary to take the inertia forces into account, as well as the applied forces, when we are finding expressions for the work done during virtual displacements. If the problem concerns the motion of a single particle, or the irrotational motion of a single rigid body, then the methods implied in equations (2.4.1) and (2.4.2) are very similar and equally direct. When, however, a number of particles or rigid bodies are involved then the method of equation (2.4.2) has advantages.

As a first example, consider the system which is shown in fig. 2.4.1. The particles, which have masses m_1 and m_2, are connected by a light inextensible string which

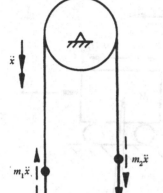

passes over a smooth pulley. Let P be the tension in the string and $\ddot{x}$ the acceleration of the mass m_1 downwards. The equations of motion, in the form (2.4.1), give

$$m_1 g - P = m_1 \ddot{x}, \Big\} \qquad (2.4.3)$$
$$P - m_2 g = m_2 \ddot{x}, \Big\}$$

and from these it is found that

$$\ddot{x} = \frac{(m_1 - m_2) g}{m_1 + m_2} \qquad (2.4.4)$$

by elimination of the tension P.

This solution may also be found by the use of d'Alembert's Principle, which is the name by which the method of equations (2.4.2) is known. We use the method of virtual displacements on the system, taking into account the work which is done by the forces $-m_1 \ddot{x}$ and $-m_2 \ddot{x}$ (which are shown dotted). This gives the equation

Fig. 2.4.1

$$(m_1 g - m_1 \ddot{x}) \,\delta x + (-m_2 \ddot{x} - m_2 g) \,\delta x = 0, \qquad (2.4.5)$$

from which the previous result may be obtained.

It will be seen that, even in this simple example, the method of (2.4.2) is slightly shorter because the tension P does not have to be introduced; it is avoided because the method allows us to use a virtual displacement in which this force does no work.

It is next necessary to show how the theory for a particle can be extended to cover the motion of a rigid body. A single rigid body may be thought of as being composed of a large number of massive particles, between which the total mass of the body is divided, together with a *massless* rigid skeleton to which the particles are attached. This skeleton serves to transmit any applied forces to the particles and, being massless, it is always *in equilibrium* under the combined action of these applied forces and the reactions from all the attached particles. The dynamical problem of the motion of the rigid body can be handled by the method of virtual work if it is treated as being the problem of the *static* equilibrium of the imaginary skeleton.

Consider a link or single body of which the skeleton is shown in fig. 2.4.2 (a) together with a single one of the attached particles, the mass of the particle being m.

Let the particle have an acceleration α in the direction shown, at some particular instant. If the particle is considered separately from the skeleton, as shown in fig. 2.4.2 (b), then it will be seen to be acted on by a force F as indicated. By equation (2.4.2) this force is equal to $m\alpha$ and is in the direction of α. The reaction on the link is equal and opposite to the force as shown in fig. 2.4.2 (c), and this reaction together with those of all the other particles, constitute the inertia loading on the skeleton link.

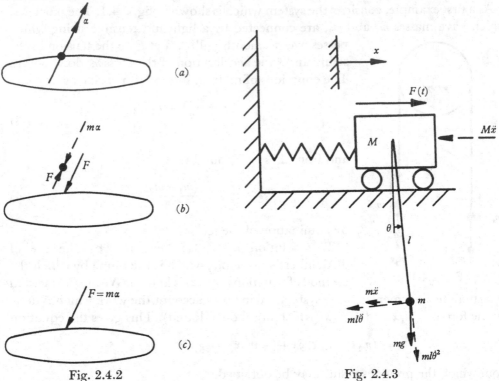

Fig. 2.4.2   Fig. 2.4.3

If a system contains a number of rigid bodies which are capable of relative movement then each one can be considered in terms of a skeleton with attached masses and the dynamic problem becomes that of finding the equilibrium conditions of the linked skeletons. This problem can always be treated by virtual work methods, once the resultant forces on each skeleton, due to the particles, have been found. Now for a single rigid body moving in two dimensions, the resultant of the inertia forces is known from elementary dynamics. Let the body have mass M and radius of gyration K about its centre of gravity G. Let the acceleration of G be $\ddot{x}$ and the angular acceleration of the body be $\ddot{\theta}$; then the resultant of the reactions from all the particles of the body, on the imaginary skeleton, will be the force $-M\ddot{x}$ acting at G and a couple $-MK^2\ddot{\theta}$.[†] These forces, for each body in a system, can be included with the applied forces when the generalized force components for the system are calculated.

† For example, see S. P. Timoshenko and D. H. Young, *Engineering Mechanics* (McGraw-Hill, 4th ed. London, 1956), §9.3.

Summarizing these ideas concerning the motion of rigid bodies, we note first that a particle may be thought of as being massless but in equilibrium under the action of its applied and its inertia forces; both these forces will contribute towards the generalized forces of any system of which the particle is a part. We have already shown how the generalized forces may be used in finding the conditions for equilibrium of the system. Secondly, a rigid body may be treated as if it were a massless rigid skeleton which is in equilibrium under the action of (a) the applied forces and (b) the inertia forces of the attached particles which represent the mass of the body. Both of the sets of forces contribute to the generalized forces of any system of which the body is a component part; and the equations of motion of the system are the equations of equilibrium of the system of skeletons. These latter equations state the conditions for the vanishing of all the generalized forces.

Fig. 2.4.3 shows the mechanism which was previously analysed in §1.4. It is shown here with the inertia forces added to it. Strictly, when the inertia forces are added in this way so that a system is shown in equilibrium, the figure must be taken to represent the skeleton system only. However, we shall not henceforth attempt to maintain the distinction between the two types of system because confusion is not likely to arise. The total generalized forces corresponding to the co-ordinates x and θ may be found, by the methods of §2.2, to be

$$\left.\begin{aligned} Q_x &= [F] + [-kx] + [-M\ddot{x} - m\ddot{x} - ml\ddot{\theta}], \\ Q_\theta &= [0] + [-mgl\theta] + [-ml^2\ddot{\theta} - ml\ddot{x}]. \end{aligned}\right\} \tag{2.4.6}$$

The first square bracket in each equation surrounds terms that correspond to forces which do not have potential, the second surrounds the terms which are derived from a potential function and the third surrounds those which are derived from the inertia forces. The conditions for the equilibrium of the static and inertia forces are, by equation (2.3.10)

$$\left.\begin{aligned} M\ddot{x} + m\ddot{x} + ml\ddot{\theta} + kx &= F, \\ \ddot{x} + l\ddot{\theta} + g\theta &= 0. \end{aligned}\right\} \tag{2.4.7}$$

These will be found to agree with equations (1.4.1).

As a second example consider the double pendulum shown in fig. 2.4.4 (a). G_1 and G_2 are the centres of gravity, K_1 and K_2 are the radii of gyration of the two bodies about these points and the masses are M_1 and M_2. The angles ϕ and ψ will be used as generalized co-ordinates. The applied forces, which include the components R and S of the reaction at the point of suspension, are shown in fig. 2.4.4 (b) as full arrows; inertia forces and couples are shown dotted. The generalized force Q_ψ corresponding to the co-ordinate ψ may be found from the work δW_ψ which is done by the forces when ψ is increased by a small amount, ϕ being held constant. This gives

$$\delta W_\psi = [-M_2 c^2 \ddot{\psi} - M_2 bc\ddot{\phi} \cos(\phi - \psi) - M_2 gc \sin \psi - M_2 K_2^2 \ddot{\psi}$$
$$+ M_2 bc\dot{\phi}^2 \sin(\phi - \psi)]\,\delta\psi, \quad (2.4.8)$$

whence
$$Q_\psi = -M_2[(c^2 + K_2^2)\,\ddot\psi + bc(\ddot\phi\cos\overline{\phi-\psi} - \dot\phi^2\sin\overline{\phi-\psi}) + cg\sin\psi]. \quad (2.4.9)$$

To find the other generalized force, we compute the work δW_ϕ which will be done when the angle ϕ is increased by $\delta\phi$ and ψ is held fixed. This gives

$$\delta W_\phi = [-M_1 a^2\ddot\phi - M_1 ga\sin\phi - M_1 K_1^2\ddot\phi - M_2 bc\ddot\psi\cos(\phi - \psi)$$
$$- M_2 b^2\ddot\phi - M_2 bc\dot\psi^2\sin(\phi - \psi) - M_2 gb\sin\phi]\,\delta\phi, \quad (2.4.10)$$

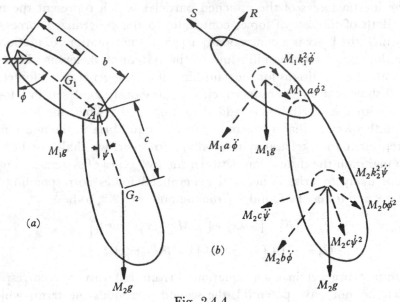

Fig. 2.4.4

so that the generalized force is

$$Q_\phi = -M_1[(a^2 + K_1^2)\,\ddot\phi + ag\sin\phi]$$
$$- M_2[b^2\ddot\phi + bc(\ddot\psi\cos\overline{\phi-\psi} + \dot\psi^2\sin\overline{\phi-\psi}) + bg\sin\phi]. \quad (2.4.11)$$

The equations of motion are now found by equating these generalized forces to zero. Doing this, and then modifying on the assumption that the angles ϕ and ψ remain small throughout the motion, we get

$$\left.\begin{aligned}(c^2 + K_2^2)\,\ddot\psi + bc\ddot\phi + cg\psi &= 0,\\ M_2 bc\ddot\psi + [M_1(a^2 + K_1^2) + M_2 b^2]\,\ddot\phi + (M_1 a + M_2 b)\,g\phi &= 0.\end{aligned}\right\} \quad (2.4.12)$$

When we treat dynamics problems in this way, we apply the principles of statics to a set of moving massless links; the fact that the links have velocity as well as acceleration does not invalidate the analysis.

When we apply d'Alembert's Principle, in conjunction with the method of virtual displacements, to a system which contains a large number of bodies then the summation of the inertia forces may become tedious or difficult. This is particularly so in three-dimensional problems. We shall show in the next section that Lagrange's equations provide a ready means of finding the contribution of the inertia forces to the generalized forces for each of the separate links.

EXAMPLES 2.4

1. The diagram shows a uniform rod of mass M and length $2a$ which is freely hinged at O so that it can swing in a vertical plane. The reaction at the hinge has components R and S as shown and the weight Mg is the only other applied force. The rod is initially at rest balanced upright when the hinge O is given a horizontal acceleration α to the right in the vertical plane containing the rod so that the angle θ increases from zero in the direction shown.

(a) Redraw the diagram and add the inertia forces for the position shown.

(b) Calculate the generalized force Q_θ corresponding to the angle θ taking into account both the applied and inertia force systems.

(c) By setting Q_θ equal to zero, find the equation of motion of the rod.

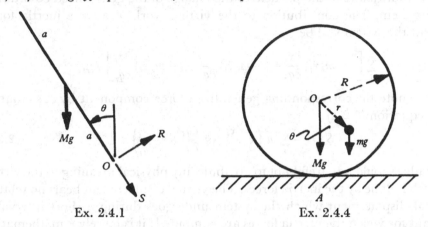

Ex. 2.4.1   Ex. 2.4.4

2. A connecting rod of uniform section has mass M/l lb. per ft., length l ft., and works with a crank of length r ft., rotating with angular velocity ω rad./sec. Assuming that the acceleration of the piston when the crank is perpendicular to the connecting rod is zero, show that the maximum bending moment in the connecting rod in this position is

$$Mlr\omega^2/9\sqrt{3} \text{ pdl.ft.}$$

3. A uniform rod of mass M and length l rests on a smooth horizontal plane. A steady force P is applied at one end in a horizontal direction at right angles to the rod. Calculate the greatest shearing force and bending moment set up.

4. A wheel of radius R, mass M and radius of gyration K has, attached to it at a radius r $(< R)$, a particle of mass m. The composite system rolls when placed upon a horizontal flat surface as shown. Using d'Alembert's principle and taking moments about A, obtain the equation of motion governing the angle θ.

2.5 Lagrange's equations

Consider a system which has n degrees of freedom and which consists of a number of rigid bodies, each one of which may be thought of as being equivalent to a large number of particles as described in the previous section. Let the system contain particles whose masses are $m_1, m_2, m_3, \ldots$ and let the Cartesian co-ordinates of the ith particle, with respect to a set of fixed axes $Oxyz$, be x_i, y_i, z_i. The inertia forces produced by the ith particle are $-m_i\ddot{x}_i$, $-m_i\ddot{y}_i$, $-m_i\ddot{z}_i$, in the directions of the respective axes. We now require to find the contribution of these inertia forces to a generalized force which is associated with some generalized co-ordinate q_r of the system.

Consider a virtual displacement of the system in which only one generalized co-ordinate—the rth—is increased by an amount δq_r; in this displacement the inertia forces of the ith particle will do an amount of work which is equal to

$$\left[(-m_i \ddot{x}_i) \frac{\partial x_i}{\partial q_r} + (-m_i \ddot{y}_i) \frac{\partial y_i}{\partial q_r} + (-m_i \ddot{z}_i) \frac{\partial z_i}{\partial q_r}\right] \delta q_r.$$

The terms $\partial x_i/\partial q_r$, $\partial y_i/\partial q_r$ and $\partial z_i/\partial q_r$ in this expression relate the changes of the Cartesian co-ordinates of the particle to the change of the generalized co-ordinate q_r of the system. The contribution to the virtual work of all the inertia forces throughout the system will be

$$\sum_i \left[(-m_i \ddot{x}_i) \frac{\partial x_i}{\partial q_r} + (-m_i \ddot{y}_i) \frac{\partial y_i}{\partial q_r} + (-m_i \ddot{z}_i) \frac{\partial z_i}{\partial q_r}\right] \delta q_r.$$

We shall denote the corresponding generalized force component by S_r so that we have, by equation (2.2.1),

$$S_r = -\sum_i m_i \left(\ddot{x}_i \frac{\partial x_i}{\partial q_r} + \ddot{y}_i \frac{\partial y_i}{\partial q_r} + \ddot{z}_i \frac{\partial z_i}{\partial q_r}\right). \tag{2.5.1}$$

The reader should be careful not to attribute any physical meaning to the virtual work which is done by the inertia forces. The virtual displacement bears no relation to the real displacement which the system undergoes during a short interval at that instant for which the inertia forces are computed; it is merely a mathematical device for analysing the force system.

Now the expression (2.5.1) contains the Cartesian co-ordinates x_i, etc., whereas we shall finally require forces in terms of the generalized co-ordinates $q_1, q_2, ..., q_n$. We may transform from one system to the other by the artifice of introducing a function which can be written down in terms of either. The necessary function is the kinetic energy T which, in terms of the Cartesian co-ordinates, can be written

$$T = \tfrac{1}{2} \sum_i m_i (\dot{x}_i^2 + \dot{y}_i^2 + \dot{z}_i^2). \tag{2.5.2}$$

We first note that the co-ordinates (x_i, y_i, z_i) must be functions of the co-ordinates $(q_1, q_2, ..., q_n)$. Thus we can write

$$\left. \begin{aligned} x_i &= x_i(q_1, q_2, ..., q_n), \\ y_i &= y_i(q_1, q_2, ..., q_n), \\ z_i &= z_i(q_1, q_2, ..., q_n). \end{aligned} \right\} \tag{2.5.3}$$

It follows that the rates of change of (x_i, y_i, z_i), that is to say the components of velocity of the various particles, can be written in terms of the rates of change of the co-ordinates q; thus, for the ith particle,

$$\dot{x}_i = \frac{dx_i}{dt} = \frac{\partial x_i}{\partial q_1} \dot{q}_1 + \frac{\partial x_i}{\partial q_2} \dot{q}_2 + ... + \frac{\partial x_i}{\partial q_n} \dot{q}_n. \tag{2.5.4}$$

The differentiation by which this relation is obtained is permissible because the q's are all independent. From the equation we may formally deduce, by differentiation with respect to $\dot{q}_r$, that

$$\frac{\partial \dot{x}_i}{\partial \dot{q}_r} = \frac{\partial x_i}{\partial q_r}. \tag{2.5.5}$$

This relation will be required shortly together with the corresponding equations for y_i and z_i which may be derived in the same way.

If the expression for T (in equation (2.5.2)) is differentiated with respect to $\dot{q}_r$, it is found that

$$\frac{\partial T}{\partial \dot{q}_r} = \sum_i m_i \left(\dot{x}_i \frac{\partial \dot{x}_i}{\partial \dot{q}_r} + \dot{y}_i \frac{\partial \dot{y}_i}{\partial \dot{q}_r} + \dot{z}_i \frac{\partial \dot{z}_i}{\partial \dot{q}_r} \right), \tag{2.5.6}$$

and from this it is found that

$$\frac{d}{dt}\left(\frac{\partial T}{\partial \dot{q}_r} \right) = \sum_i m_i \left(\ddot{x}_i \frac{\partial \dot{x}_i}{\partial \dot{q}_r} + \ddot{y}_i \frac{\partial \dot{y}_i}{\partial \dot{q}_r} + \ddot{z}_i \frac{\partial \dot{z}_i}{\partial \dot{q}_r} \right)$$

$$+ \sum_i m_i \left[\dot{x}_i \frac{d}{dt}\left(\frac{\partial \dot{x}_i}{\partial \dot{q}_r} \right) + \dot{y}_i \frac{d}{dt}\left(\frac{\partial \dot{y}_i}{\partial \dot{q}_r} \right) + \dot{z}_i \frac{d}{dt}\left(\frac{\partial \dot{z}_i}{\partial \dot{q}_r} \right) \right]. \tag{2.5.7}$$

Now if we substitute $\partial x_i/\partial q_r$, etc., for $\partial \dot{x}_i/\partial \dot{q}_r$, etc., in the first summation, in accordance with (2.5.5), then it yields the function $-S_r$. It follows, therefore, that

$$-S_r = \frac{d}{dt}\left(\frac{\partial T}{\partial \dot{q}_r} \right) - \sum_i m_i \left[\dot{x}_i \frac{d}{dt}\left(\frac{\partial \dot{x}_i}{\partial \dot{q}_r} \right) + \dot{y}_i \frac{d}{dt}\left(\frac{\partial \dot{y}_i}{\partial \dot{q}_r} \right) + \dot{z}_i \frac{d}{dt}\left(\frac{\partial \dot{z}_i}{\partial \dot{q}_r} \right) \right]. \tag{2.5.8}$$

The relation (2.5.5) can be substituted in the remaining summation also so that it becomes

$$\sum_i m_i \left[\dot{x}_i \frac{d}{dt}\left(\frac{\partial x_i}{\partial q_r} \right) + \dot{y}_i \frac{d}{dt}\left(\frac{\partial y_i}{\partial q_r} \right) + \dot{z}_i \frac{d}{dt}\left(\frac{\partial z_i}{\partial q_r} \right) \right]. \tag{2.5.9}$$

Further, the factors

$$\frac{d}{dt}\left(\frac{\partial x_i}{\partial q_r} \right), \quad \frac{d}{dt}\left(\frac{\partial y_i}{\partial q_r} \right), \quad \frac{d}{dt}\left(\frac{\partial z_i}{\partial q_r} \right)$$

are identical with

$$\frac{\partial \dot{x}_i}{\partial q_r}, \quad \frac{\partial \dot{y}_i}{\partial q_r}, \quad \frac{\partial \dot{z}_i}{\partial q_r},$$

respectively. This follows because, in each case, both of the expressions are equal to the same quantity; thus,

$$\frac{d}{dt}\left(\frac{\partial x_i}{\partial q_r} \right) = \frac{\partial \dot{x}_i}{\partial q_r} = \frac{\partial^2 x_i}{\partial q_1 \partial q_r} \cdot \dot{q}_1 + \frac{\partial^2 x_i}{\partial q_2 \partial q_r} \cdot \dot{q}_2 + \dots + \frac{\partial^2 x_i}{\partial q_n \partial q_r} \cdot \dot{q}_n, \tag{2.5.10}$$

as may be found by differentiating $\partial x_i/\partial q_r$ with respect to t and $\dot{x}_i$ with respect to q_r. We can now substitute

$$\frac{\partial \dot{x}_i}{\partial q_r}, \text{ etc.,} \quad \text{for} \quad \frac{d}{dt}\left(\frac{\partial x_i}{\partial q_r} \right), \text{ etc.,}$$

in equation (2.5.9) and so obtain a summation which is also found by differentiating T with respect to q_r as may be seen by comparison with (2.5.6); therefore,

$$-S_r = \frac{d}{dt}\left(\frac{\partial T}{\partial \dot{q}_r} \right) - \frac{\partial T}{\partial q_r}. \tag{2.5.11}$$

This expression for S_r is formed entirely in terms of the generalized co-ordinates, so that the equations of motion of the system can now be written without bringing in the Cartesian co-ordinates. They are

$$\frac{d}{dt}\left(\frac{\partial T}{\partial \dot{q}_r}\right) - \frac{\partial T}{\partial q_r} = Q_r \quad (r = 1, 2, \ldots, n). \tag{2.5.12}$$

If some of the forces have potential, then the equations may be written in the form

$$\frac{d}{dt}\left(\frac{\partial T}{\partial \dot{q}_r}\right) - \frac{\partial T}{\partial q_r} + \frac{\partial V}{\partial q_r} = Q_r \quad (r = 1, 2, \ldots, n), \tag{2.5.13}$$

where Q_r is retained with a new meaning; it now stands for that part of the generalized force at q_r which does not have potential. The quantities Q_r defined in this way are still loosely called the 'generalized forces' and we shall follow this convention here.

Let a force F act in the direction x on the mass M in the system of fig. 1.2.1. The displacement x may be used as a generalized co-ordinate and for this system

$$T = \frac{M\dot{x}^2}{2}, \quad V = \frac{kx^2}{2}, \quad Q_x = F. \tag{2.5.14}$$

Lagrange's equation then gives immediately

$$M\ddot{x} + kx = F. \tag{2.5.15}$$

There is of course no advantage in using the method for systems which are as simple as this, and indeed some understanding of the physical nature of such problems may be lost by so doing.

In the system of fig. 1.4.1 it is necessary, if the equations of motion are to be formed by Newton's laws directly, to imagine the system to be split into its component parts; these will, in fact, be the two masses. During the analysis the reactions between these parts will have to be allowed for and subsequently eliminated from the equations. If, however, Lagrange's equations are used to set up the equations of small motions about the equilibrium position, then the following functions are used as the starting point:

$$\left.\begin{array}{l} T = \frac{1}{2}[M\dot{x}^2 + m(\dot{x} + l\dot{\theta})^2], \\ V = \frac{1}{2}[kx^2 + mgl\theta^2], \\ Q_x = F, \quad Q_\theta = 0; \end{array}\right\} \tag{2.5.16}$$

and then, from the relations

$$\left.\begin{array}{l} \dfrac{d}{dt}\left(\dfrac{\partial T}{\partial \dot{x}}\right) - \dfrac{\partial T}{\partial x} + \dfrac{\partial V}{\partial x} = Q_x, \\[2mm] \dfrac{d}{dt}\left(\dfrac{\partial T}{\partial \dot{\theta}}\right) - \dfrac{\partial T}{\partial \theta} + \dfrac{\partial V}{\partial \theta} = Q_\theta, \end{array}\right\} \tag{2.5.17}$$

it is found that

$$\left.\begin{array}{l} M\ddot{x} + m\ddot{x} + ml\ddot{\theta} + kx = F, \\ \ddot{x} + l\ddot{\theta} + g\theta = 0. \end{array}\right\} \tag{2.5.18}$$

In this solution, it has not been necessary to introduce the reaction between the parts of the system.

EXAMPLES 2.5

1. A simple pendulum OG of length l and mass m_1 is modified as shown in the figure so as to cause a particle of mass m_2 to slide on a smooth horizontal surface which is a distance h below the point of suspension $O(h < l)$. Assuming that the light rod OG is rigid and that all friction may be neglected, find the equation of motion governing large oscillations.

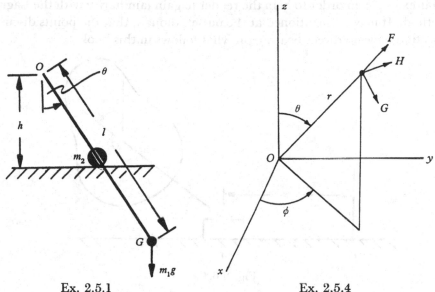

Ex. 2.5.1   Ex. 2.5.4

2. A uniform rod OA of length l is freely pivoted at O and hangs with A below O. The inclination of the rod to the vertical through O is ψ and the angle between a fixed vertical plane and the vertical plane containing OA is ϕ.

Using the angles ϕ and ψ as generalized co-ordinates, set up the equations of motion by Lagrange's method. Hence show from the ψ-equation that, if a state of uniform rotation exists such that $\dot{\phi}$ and ψ have constant values ω and θ respectively, then

$$2l\omega^2 \cos \theta = 3g.$$

What is the physical interpretation of the ϕ-equation?

3. A rigid bar of mass M is suspended in a horizontal position by two equal vertical strings of length l. Two particles of equal mass m are hung from points of the bar by two strings, each of length a. Find the frequency equation for small free oscillations in the vertical plane through the equilibrium position of the bar.

4. The position of a particle of mass m, relative to fixed co-ordinate axes $Oxyz$, is determined by polar co-ordinates (r, θ, ϕ) as shown. Forces F, G and H act on the particle, as indicated in the figure, in the directions δr, $r\,\delta\theta$ and $r \sin \theta\, \delta\phi$ respectively. Using Lagrange's method, show that the equations of motion governing the polar co-ordinates are

$$m(\ddot{r} - r\dot{\theta}^2 - r\dot{\phi}^2 \sin^2 \theta) = F,$$

$$m(r\ddot{\theta} + 2\dot{r}\dot{\theta} - r\dot{\phi}^2 \sin \theta \cos \theta) = G,$$

$$m(r\ddot{\phi} \sin \theta + 2\dot{r}\dot{\phi} \sin \theta + 2r\dot{\theta}\dot{\phi} \cos \theta) = H.$$

2.6 Remarks on Lagrange's equations

Lagrange's equations are an extremely valuable tool for the analysis of dynamic systems. In subsequent chapters, the equations will not be used in their full generality because we shall restrict ourselves to the study of *small* oscillations. In this article, however, we shall discuss the meanings of the various terms in a particular example, in order to help the reader to gain familiarity with the Lagrangian method. It may be mentioned at the outset, though, that the points discussed in this section have no direct bearing on what follows in this book.

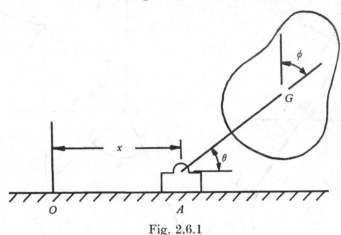

Fig. 2.6.1

We take as our example the system shown in fig. 2.6.1. A rigid body of mass M and radius of gyration K about its centre of gravity G, is freely pivoted at G to a light link of length l. The other end of the link is freely pivoted at A to a light block which slides without friction along OA. Generalized co-ordinates may be taken as the distance x and the angles θ and ϕ, which are marked in the figure.

The kinetic energy is given by

$$T = \tfrac{1}{2}M[(\dot{x} - l\dot{\theta}\sin\theta)^2 + (l\dot{\theta}\cos\theta)^2] + \tfrac{1}{2}MK^2(\dot{\theta} + \dot{\phi})^2. \tag{2.6.1}$$

In later problems on vibration, co-ordinates will be chosen which are all zero when the system concerned is in some known equilibrium configuration. To illustrate this possibility, we now choose three new generalized co-ordinates which are defined as the differences between the three original co-ordinates x, θ, ϕ and three specified values of these co-ordinates which correspond to some standard configuration of the system. Thus if the standard configuration is determined by the particular values x', θ', ϕ', then a new set of generalized co-ordinates q_1, q_2 and q_3 may be defined where

$$x = x' + q_1, \quad \theta = \theta' + q_2, \quad \phi = \phi' + q_3. \tag{2.6.2}$$

By substituting these expressions into the kinetic energy expression, T may be found in terms of the new co-ordinates. It is found that

$$2T = M\dot{q}_1^2 + M(l^2 + K^2)\,\dot{q}_2^2 + MK^2\dot{q}_3^2 - 2Ml\sin(\theta' + q_2)\,\dot{q}_1\dot{q}_2 + 2MK^2\dot{q}_2\dot{q}_3. \tag{2.6.3}$$

In a vibration problem of the type to be treated in this book, the motion would be restricted to small values of the co-ordinates q; here, however, we shall not restrict the treatment in this way at present.

If the kinetic energy is differentiated with respect to $\dot{q}_1$, $\dot{q}_2$ and $\dot{q}_3$, it is found that

$$\left. \begin{aligned} \frac{\partial T}{\partial \dot{q}_1} &= M\dot{q}_1 - Ml\sin(\theta' + q_2)\cdot\dot{q}_2, \\ \frac{\partial T}{\partial \dot{q}_2} &= -Ml\sin(\theta' + q_2)\cdot\dot{q}_1 + M(l^2 + K^2)\dot{q}_2 + MK^2\dot{q}_3, \\ \frac{\partial T}{\partial \dot{q}_3} &= MK^2\dot{q}_2 + MK^2\dot{q}_3. \end{aligned} \right\} \quad (2.6.4)$$

The first of these quantities will be found to be equal to the linear momentum of the system in the direction in which the co-ordinate q_1 is measured. The second is the moment of the momentum about the point A, round which the rotation q_2 takes place; this moment of momentum is positive in the direction of increasing q_2. The last expression is the moment of momentum, in the direction of increasing q_3, about the point G round which the displacement q_3 takes place. It is common, in fact, to refer to the quantity $\partial T/\partial \dot{q}_r$ as the 'generalized component of momentum' corresponding to q_r.

When the generalized momenta are differentiated with respect to time, they give the results

$$\left. \begin{aligned} \frac{d}{dt}\left(\frac{\partial T}{\partial \dot{q}_1}\right) &= M\ddot{q}_1 - Ml\sin(\theta' + q_2)\cdot\ddot{q}_2 - Ml\cos(\theta' + q_2)\cdot\dot{q}_2^2, \\ \frac{d}{dt}\left(\frac{\partial T}{\partial \dot{q}_2}\right) &= -Ml\sin(\theta' + q_2)\cdot\ddot{q}_1 + M(l^2 + K^2)\ddot{q}_2 + MK^2\ddot{q}_3 - Ml\cos(\theta' + q_2)\cdot\dot{q}_1\dot{q}_2, \\ \frac{d}{dt}\left(\frac{\partial T}{\partial \dot{q}_3}\right) &= MK^2\ddot{q}_2 + MK^2\ddot{q}_3. \end{aligned} \right\}$$

$$(2.6.5)$$

Now the linear momentum of a system of particles m_i, in the direction x is the sum

$$\sum_i m_i \dot{x}_i,$$

and the rate of change of this quantity with time is

$$\frac{d}{dt}\left(\sum_i m_i \dot{x}_i\right) = \sum_i m_i \ddot{x}_i. \quad (2.6.6)$$

This latter expression may be written in the form

$$\ddot{x}_g \sum_i m_i,$$

where x_g is the co-ordinate of the centre of gravity of the particles. This quantity is the sum of the *reversed* inertia forces (which is referred to as the 'effective force' in the direction x by some writers). In a similar way it will be found that the first expression (2.6.5), which is the rate of change of the generalized component of momentum in the direction q_1, is also the total reversed inertia force in the direction q_1 as is indicated in fig. 2.6.2; the Lagrangian equation is obtained by putting this quantity equal to the total applied force in this direction.

Again the moment of momentum of a system of particles m_i about some point O, which can be taken as the origin of the Cartesian co-ordinate system Oxy, is equal to

$$\sum_i m_i(\dot{y}_i x_i - \dot{x}_i y_i).$$

The rate of change of this quantity is

$$\frac{d}{dt}\Big[\sum_i m_i(\dot{y}_i x_i - \dot{x}_i y_i)\Big] = \sum_i m_i(\ddot{y}_i x + \dot{y}_i \dot{x}_i - \ddot{x}_i y_i - \dot{x}_i \dot{y}_i)$$

$$= \sum_i m_i(\ddot{y}_i x_i - \ddot{x}_i y_i). \qquad (2.6.7)$$

It will be seen that, because of the cancellation of the $m_i \dot{x}_i \dot{y}_i$ terms, this is equal to the moment of all the reversed inertia forces about O (sometimes called 'the moment of the effective force' about O). Now the third equation (2.6.5), which gives the rate of change of the generalized component of momentum corresponding to q_3, will be found to be equal to the moment of all the reversed inertia forces about G (see fig. 2.6.2).

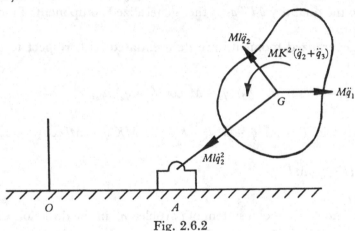

Fig. 2.6.2

The second equation (2.6.5), however, contains in addition to the moments of the reversed inertia forces about A (see fig. 2.6.2), the term

$$- Ml\cos(\theta' + q_2) \cdot \dot{q}_1 \dot{q}_2.$$

When the moment of the applied forces is equated to the moment of the inertia forces in the Lagrangian equation, this terms must evidently be subtracted from the $\dfrac{d}{dt}\Big(\dfrac{\partial T}{\partial \dot{q}_2}\Big)$ expression. This is done when the term $-\dfrac{\partial T}{\partial q_2}$ is included; for it will be found that this quantity is, in fact, identical with the term which has to be removed.

The reader may well ask why it is that the time rate of change of the generalized component of momentum does not (for the co-ordinate q_2) give the moment of the reversed inertia forces directly, especially in view of the fact that the theory by which we derived expression (2.6.7) appears to be perfectly general. The answer to this question is that in equation (2.6.7) it is assumed, implicitly, that the origin O is fixed. If the origin were moving then the moment of momentum about it would be changing by virtue of the change in the moment arms which the movement of the origin involves; the terms due to this effect are those which are removed by the term $(-\partial T/\partial q_r)$ in the Lagrangian equation for q_r. The need for this term here is thus due to the fact that the quantity $\dfrac{d}{dt}\left(\dfrac{\partial T}{\partial \dot{q}_2}\right)$ represents the instantaneous value of the moment of momentum about the point A which has a velocity $\dot{q}_1$. This may, perhaps, be made clearer by writing down the expression for the moment of momentum about the fixed point with which A coincides at time $t = 0$. The expression is

$$- Ml\sin (\theta' + q_2).\,\dot{q}_1 + Ml^2\dot{q}_2 + MK^2(\dot{q}_2 + \dot{q}_3) + Ml\cos (\theta' + q_2).\,\dot{q}_2 \int_0^t \dot{q}_1\,dt.$$

$$(2.6.8)$$

If this is now differentiated with respect to t and t is subsequently put equal to zero in order to make the instantaneous position of the point A correct, then it gives the expression

$$- Ml\sin (\theta' + q_2).\,\ddot{q}_1 + M(l^2 + K^2)\,\ddot{q}_2 + MK^2\ddot{q}_3 \qquad (2.6.9)$$

and this is equal to the moment of the reversed inertia forces about A, as is indicated in fig. 2.6.2.

EXAMPLES 2.6

1. In the above example the generalized component of momentum which corresponds to q_3 is expressed about the point G which is moving. Why is it that there is no term in the third of equations (2.6.5) which must be removed by a $\partial T/\partial q_3$ term in order that the expression shall equal the moment of the reversed inertia forces about G?

2. The system of fig. 2.6.1 is to be taken to lie in a horizontal plane and to be in equilibrium in the standard configuration previously defined. The equilibrium is maintained by the actions of three springs as follows: (a) a spring of stiffness k_1 which is connected between the block A and a fixed point on the slide, and is unstressed when x is equal to x', (b) a spring of torsional stiffness k_2 which is connected between the block and the link AG and which is unstressed when θ is equal to θ', and (c) a spring of torsional stiffness k_3 which is connected between the link AG and the mass M and is unstressed when ϕ is equal to ϕ'.

Write down the expression for the potential energy and hence deduce the equations of free motion. Show that, if it is assumed that a motion of the form

$$q_1 = \Psi_1\,e^{i\omega t}, \quad q_2 = \Psi_2\,e^{i\omega t}, \quad q_3 = \Psi_3\,e^{i\omega t}$$

is possible, during which the amplitudes of all the q's remain small, then the equations reduce to

$$(k_1 - M\omega^2)\,\Psi_1 \qquad\qquad + Ml\omega^2 \sin \theta'.\,\Psi_2 \qquad\qquad = 0,$$

$$Ml\omega^2 \sin \theta'.\,\Psi_1 + [k_2 - M(l^2 + K^2)\,\omega^2]\,\Psi_2 \qquad - MK^2\omega^2\Psi_3 = 0,$$

$$- MK^2\omega^2\Psi_2 + (k_3 - MK^2\omega^2)\,\Psi_3 = 0.$$

3. (*a*) Find the rate of change of moment of momentum of the lower body in the double compound pendulum of fig. 2.4.4 about the *fixed* point with which the pivot A coincides instantaneously using the method employed in deriving the expression (2.6.8). Verify, by comparison with the equation of free motion obtained from equation (2.4.9), that this is equal to the moment of the applied force $M_2 g$ about this point.

(*b*) Show that, for the system of fig. 2.4.4, the quantity $\dfrac{d}{dt}\left(\dfrac{\partial T}{\partial \dot{\psi}}\right)$ gives the rate of change of the moment of momentum about the moving pivot and show further that, by subtracting the expression for $\partial T/\partial \psi$, the result (*a*) is arrived at.

NOTATION

Supplementary List of Symbols used in Chapter 2

F	Generalized force applied to system.
m_i	Mass of ith particle.
S_r	Component of the generalized force corresponding to q_r which is due to inertia forces.
x_i, y_i, z_i	Co-ordinates, relative to fixed axes, of ith particle.
y	Generalized displacement of point of application of F in the direction of F.

CHAPTER 3

SYSTEMS HAVING ANY FINITE NUMBER OF DEGREES OF FREEDOM

The stroke given in the bell will cause response and some slight movement in another bell similar to itself; and the string of a lute as it sounds will cause response and movement in another similar string of like tone in another lute; and this you will see by placing a straw on the string similar to that which has sounded.   LEONARDO DA VINCI, *Selections from the Notebooks* ... (1500*c*)
(World's Classics Edition, O.U.P. 1952)

If one bows the bass string on a viola rather smartly and brings near it a goblet of fine, thin glass having the same tone as that of the string, this goblet will vibrate and audibly resound.

GALILEO GALILEI, *Dialogues Concerning Two New Sciences* (1638)
(Transl., Macmillan Co. 1914)

Lagrange's equations provide a method by which vibration theory may be developed in general terms so that reference need not constantly be made to specific systems. This approach is introduced in this chapter and the reader will see that, while particular systems are frequently referred to, they are introduced merely to illustrate features of the general theory.

3.1 The equations of motion

We shall form the equations of motion for the small oscillations of a system which has n degrees of freedom and which can vibrate without friction. The configuration of the system will be defined by the n generalized co-ordinates $q_1, q_2, ..., q_n$. In order that the Lagrange equations may be used, it is first necessary to find expressions for the kinetic and potential energies.

When the kinetic energy expression is formed for a system, a function is obtained which is the sum of a large number of terms, each of which has the form

$$\tfrac{1}{2}m_i(\dot{x}_i^2 + \dot{y}_i^2 + \dot{z}_i^2),$$

where m_i is the mass of the ith constituent particle. Now the components of velocity $\dot{x}_i, \dot{y}_i, \dot{z}_i$ of any particle of the system are expressible as functions of the generalized velocities as described in the last chapter so that, cf. equation (2.5.4),

$$\left.\begin{aligned}
\dot{x}_i &= \frac{\partial x_i}{\partial q_1}\dot{q}_1 + \frac{\partial x_i}{\partial q_2}\dot{q}_2 + ... + \frac{\partial x_i}{\partial q_n}\dot{q}_n, \\[4pt]
\dot{y}_i &= \frac{\partial y_i}{\partial q_1}\dot{q}_1 + \frac{\partial y_i}{\partial q_2}\dot{q}_2 + ... + \frac{\partial y_i}{\partial q_n}\dot{q}_n, \\[4pt]
\dot{z}_i &= \frac{\partial z_i}{\partial q_1}\dot{q}_1 + \frac{\partial z_i}{\partial q_2}\dot{q}_2 + ... + \frac{\partial z_i}{\partial q_n}\dot{q}_n.
\end{aligned}\right\} \qquad (3.1.1)$$

It follows therefore that the total kinetic energy will be a homogeneous second-degree function of the generalized velocities. The coefficients in this function will themselves be functions of the co-ordinates q since they will contain squares and products of the partial derivative expressions $\partial x_i/\partial q_r$, $\partial y_i/\partial q_r$, $\partial z_i/\partial q_r$.

We now restrict the motion so that the change of the configuration, away from some stable equilibrium state, is always small. It is convenient, further, to define the generalized co-ordinates in such a way that they are all zero in the equilibrium state; the value of all the co-ordinates during the motion will then be small. Let the coefficients in the kinetic energy expression be now expressed, by Taylor's theorem in its extended form, as a series of ascending powers of the co-ordinates q. If the q's are sufficiently small throughout the motion only the constant terms in these series need be taken when calculating the kinetic energy. In this event, the kinetic energy T is given by an expression of the form

$$2T = a_{11}\dot{q}_1^2 + a_{22}\dot{q}_2^2 + \dots + a_{nn}\dot{q}_n^2 + 2a_{12}\dot{q}_1\dot{q}_2 + \dots, \qquad (3.1.2)$$

where $a_{11}, a_{22}, \dots, a_{nn}, a_{12}, \dots$ are the aforesaid constants. They are known as the 'coefficients of inertia'. Since the kinetic energy of the system is essentially a positive quantity, no matter what values are given to the generalized velocities $\dot{q}_r$, the values of the coefficients of inertia are subject to certain restrictions;[†] for this reason the function T is said to be 'positive definite'.

If generalized co-ordinates q_1 and q_2 are identified with the quantities x and θ in the system of fig. 1.4.1, then the coefficients of inertia will be

$$a_{11} = M + m, \quad a_{22} = ml^2, \quad a_{12} = ml, \qquad (3.1.3)$$

as may be seen from equation (2.5.16).

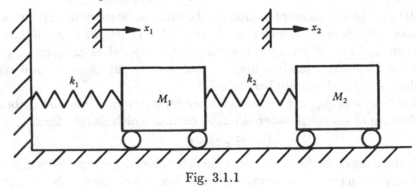

Fig. 3.1.1

Again, in the system of fig. 3.1.1, in which the masses M_1 and M_2 can move horizontally only, the kinetic energy will be given by

$$2T = M_1\dot{x}_1^2 + M_2\dot{x}_2^2, \qquad (3.1.4)$$

and if q_1 and q_2 are identified with x_1 and x_2 respectively, then

$$a_{11} = M_1, \quad a_{22} = M_2, \quad a_{12} = 0. \qquad (3.1.5)$$

† See, for instance, H. Lamb, *Higher Mechanics* (Cambridge University Press, 1920), §72.

When the equations of motion are formed, there are two types of forces which must be allowed for apart from those of inertia. First there are those which are inherent to the system, such as the spring forces and the weights, and secondly there are those which may be applied arbitrarily by some outside agency; the latter may include pulsating forces which are the ultimate source of the vibration which it is wished to study. Now the spring and weight forces will have potential and we shall choose the datum for the potential energy V in such a way that V is zero in the equilibrium configuration. It will be found that this was done when the system of fig. 1.4.1 was discussed (see equation (2.5.16)), and that also the co-ordinates were chosen to be zero in the equilibrium state in that example.

Returning to the general theory, let the system be slightly displaced from the equilibrium configuration, so that the generalized co-ordinates have small values. The potential energy may be written in the form

$$V = \left[\frac{\partial V}{\partial q_1}q_1 + \frac{\partial V}{\partial q_2}q_2 + \dots + \frac{\partial V}{\partial q_n}q_n\right]$$
$$+ \frac{1}{2}\left[\frac{\partial^2 V}{\partial q_1^2}q_1^2 + \frac{\partial^2 V}{\partial q_2^2}q_2^2 + \dots + \frac{\partial^2 V}{\partial q_n^2}q_n^2 + 2\frac{\partial^2 V}{\partial q_1 \partial q_2}q_1 q_2 + \dots\right]$$
$$+ \text{(higher-order terms)} \tag{3.1.6}$$

by virtue of the extended form of Taylor's series. In this expression the partial derivatives will have the values which hold in the equilibrium state. Now the first derivatives are zero in this state, as was shown in § 2.2. If, therefore, the co-ordinates q are sufficiently small for terms above the second degree to be neglected then the potential V can be expressed by the equation

$$2V = c_{11}q_1^2 + c_{22}q_2^2 + \dots + c_{nn}q_n^2 + 2c_{12}q_1 q_2 + \dots. \tag{3.1.7}$$

In this expression the coefficients c are constants and are usually known as the 'coefficients of stability', though they may be regarded, in the problems with which we are concerned, as 'coefficients of stiffness'. As the motion takes place about a *stable* equilibrium position only, the potential energy will be a minimum in that position; it follows that V is also a positive definite function of the q's, as T was of the $\dot{q}$'s. Restrictions on the values of the coefficients c are also present therefore;[†] removal of these restrictions introduces the possibility of instability which will not be discussed in the present volume.

Returning once more to the system of fig. 1.4.1, the coefficients of stability will be seen to be

$$c_{11} = k, \quad c_{22} = mgl, \quad c_{12} = 0 \tag{3.1.8}$$

if equation (2.5.16) is examined.

Again, in the example of fig. 3.1.1, if x_1 and x_2 are measured from the equilibrium positions of the masses, then the potential energy will be given by

$$2V = (k_1 + k_2)x_1^2 + k_2 x_2^2 - 2k_2 x_1 x_2, \tag{3.1.9}$$

from which we have $\quad c_{11} = k_1 + k_2, \quad c_{22} = k_2, \quad c_{12} = -k_2. \tag{3.1.10}$

† See Lamb, *op. cit.* §86.

It should be noted that no significance is attached to the order of the subscripts rs in the coefficients a_{rs} and c_{rs}. Thus a_{32} and a_{23} are the same quantity.

It is now necessary to consider the externally applied exciting forces which will provide generalized forces $Q_1, Q_2, ..., Q_n$. In this chapter, it will be assumed that these forces vary harmonically; indeed vibration analysis is largely concerned with the effects of harmonic forcing. The excitation will therefore be such that we can write

$$Q_1 = \Phi_1 e^{i\omega t}, \quad Q_2 = \Phi_2 e^{i\omega t}, \quad ..., \quad Q_n = \Phi_n e^{i\omega t}, \tag{3.1.11}$$

ω being the frequency of excitation. The amplitude Φ of the forces could, in the general case, be complex to allow for phase differences between forces which may be applied at different points; we shall usually, however, be concerned with excitation by a single force at a point and the generalized forces which are derived from such a source will then all be in phase.

Lagrange's equations may now be applied in the form

$$\frac{d}{dt}\left(\frac{\partial T}{\partial \dot{q}_r}\right) + \frac{\partial V}{\partial q_r} = Q_r \quad (r = 1, 2, ..., n), \tag{3.1.12}$$

where it will be seen that the term $-\partial T/\partial q_r$ of equation (2.5.13) has been omitted; this is done because all such terms would be zero by virtue of the fact that the a_{rs} coefficients are all constant. When the expressions (3.1.2) and (3.1.7) for T and V are substituted into equations (3.1.12), they yield

$$\left.\begin{aligned}
a_{11}\ddot{q}_1 + a_{12}\ddot{q}_2 + ... + a_{1n}\ddot{q}_n + c_{11}q_1 + c_{12}q_2 + ... + c_{1n}q_n &= \Phi_1 e^{i\omega t}, \\
a_{21}\ddot{q}_1 + a_{22}\ddot{q}_2 + ... + a_{2n}\ddot{q}_n + c_{21}q_1 + c_{22}q_2 + ... + c_{2n}q_n &= \Phi_2 e^{i\omega t}, \\
\cdots\cdots\cdots\cdots\cdots\cdots\cdots\cdots\cdots\cdots\cdots\cdots\cdots\cdots\cdots\cdots\cdots\cdots\cdots\cdots\cdots\cdots\cdots\cdots\cdots \\
a_{n1}\ddot{q}_1 + a_{n2}\ddot{q}_2 + ... + a_{nn}\ddot{q}_n + c_{n1}q_1 + c_{n2}q_2 + ... + c_{nn}q_n &= \Phi_n e^{i\omega t},
\end{aligned}\right\} \tag{3.1.13}$$

as the equations of motion. These are a set of n linear, second order, ordinary differential equations with constant coefficients. Such equations are obtained by restricting the motion to small amplitudes so that the approximations discussed above are valid; if these restrictions are not imposed, then the coefficients in the equations may become functions of the co-ordinates q and no general solution of them is possible; the study of such equations is part of the theory of non-linear vibration.

An example will illustrate the manner in which the restriction to small amplitudes affects the equations. Consider the system of fig. 3.1.2, which is the same as that of fig. 2.6.1 except that springs of stiffness k_1, k_2, k_3 have been added as shown. The system is to be considered as lying on its side in a horizontal position so that the springs are unstretched in equilibrium. The parts which are shown dotted are to be assumed to have negligible mass, as are the springs and likewise the arm AG. The lengths AG and GB are both l, the radius of the light wheel is r and the angle BGA is β in the equilibrium position.

The kinetic energy of the system is given by equation (2.6.3) in terms of the previously selected co-ordinates. If the motion is now restricted to small amplitudes then the expression reduces to

$$2T = M\dot{q}_1^2 + M(l^2 + K^2)\dot{q}_2^2 + MK^2\dot{q}_3^2 - 2Ml\sin\theta'.\dot{q}_1\dot{q}_2 + 2MK^2\dot{q}_2\dot{q}_3, \quad (3.1.14)$$

which has the form of the general expression (3.1.2).

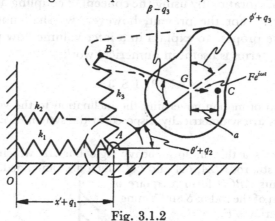

Fig. 3.1.2

The potential energy, which is composed entirely of elastic strain energy, will be given by

$$2V = k_1 q_1^2 + k_2(q_1 - rq_2)^2 + k_3 \left[2l\sin\left(\frac{\beta - q_3}{2}\right) - 2l\sin\left(\frac{\beta}{2}\right) \right]^2. \quad (3.1.15)$$

The restriction to small amplitudes reduces this to

$$2V = k_1 q_1^2 + k_2(q_1 - rq_2)^2 + k_3 l^2 \cos^2\left(\frac{\beta}{2}\right).q_3^2, \quad (3.1.16)$$

which has the form of equation (3.1.7).

Let an external force $Fe^{i\omega t}$ be applied at G as shown in the diagram. The corresponding generalized forces may be found by the method of equation (2.2.9). The displacement y of the point of application of the force $Fe^{i\omega t}$ is given by

$$y = q_1 - lq_2\sin(\theta' + q_2) \doteqdot q_1 - l\sin\theta'.q_2 \quad (3.1.17)$$

provided that the q's are small. Therefore

$$\left. \begin{aligned} Q_1 &= Fe^{i\omega t}\frac{\partial y}{\partial q_1} = Fe^{i\omega t} = \Phi_1 e^{i\omega t}, \\ Q_2 &= Fe^{i\omega t}\frac{\partial y}{\partial q_2} = -Fl\sin\theta'.e^{i\omega t} = \Phi_2 e^{i\omega t}, \\ Q_3 &= 0. \end{aligned} \right\} \quad (3.1.18)$$

Lagrange's equations (3.1.12) may now be used and they will be found to give

$$\left. \begin{aligned} M\ddot{q}_1 - Ml\sin\theta'.\ddot{q}_2 + (k_1 + k_2)q_1 - k_2 rq_2 &= Fe^{i\omega t}, \\ -Ml\sin\theta'.\ddot{q}_1 + M(l^2 + K^2)\ddot{q}_2 + MK^2\ddot{q}_3 - k_2 rq_1 + k_2 r^2 q_2 &= -Fl\sin\theta'.e^{i\omega t}, \\ MK^2\ddot{q}_2 + MK^2\ddot{q}_3 + k_3 l^2 \cos^2(\tfrac{1}{2}\beta).q_3 &= 0, \end{aligned} \right\} \quad (3.1.19)$$

which are of the general form (3.1.13). The method of solution of equations of this type will be treated in the next section.

It is possible for all the coefficients a_{rs} $(r \neq s)$ to be zero in a particular case;

equations (3.1.13) can then each contain one a term only, and the co-ordinates are then said to be 'dynamically uncoupled'. Similarly, if all the coefficients c_{rs} are zero, the co-ordinates are 'statically uncoupled'. The system of fig. 1.4.1 has dynamic coupling only while that of fig. 3.1.1 has only static coupling. It is possible to begin the analysis of vibration by using the concept of coupling and this has been done by some writers.† For the present, however, we shall not introduce this approach although we propose to explain in a later volume how the vanishing of either the a_{rs} or the c_{rs} terms is useful in numerical work.

EXAMPLES 3.1

1. Find the equation of motion governing the inclination to the vertical of the line about which the rollers are symmetrically disposed in the system of Ex. 1.2.2 by using Lagrange's equations.

2. The diagram shows a thin uniform heavy rigid bar AB which is supported by three light uniform elastic rods DA, CA and CB which are pin-ended. The mass of the bar is M and the points $ABCD$ form a square of side a. The cross-section of the rods is S and Young's modulus for their material is E.

Denoting the horizontal displacements to the right of A and B by q_1 and q_2 respectively, and the vertical downward displacement of B by q_3, find the equations of small free oscillations of the system. The effects of gravity are to be neglected.

3. Show that if $Q_1, Q_2, ..., Q_n$ are the static generalized forces which must be applied at the corresponding co-ordinates $q_1, q_2, ..., q_n$ of a system of the type discussed in this section to hold it at rest, then the potential energy V is given by

$$2V = q_1 Q_1 + q_2 Q_2 + ... + q_n Q_n.$$

4. Two parallel shafts, 15 in. apart, are mounted in bearings in a gear-casing and carry straight-toothed gears which are in mesh; one of the gears has 20 teeth and the other 50. The moment of

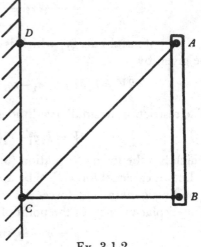

Ex. 3.1.2

inertia of the smaller gear is 3 lb.ft.² and that of the larger 16 lb.ft.². A flywheel of moment of inertia 7 lb.ft.² is keyed to the shaft carrying the smaller gear and the torsional stiffness of the shaft between the gear and the wheel is 12×10^4 pdl.ft./rad. A flywheel of inertia 18 lb.ft.² is keyed to the other shaft and the relevant stiffness of this is 60×10^4 in the same units.

The mass of the small gear and small flywheel, together with their shaft, is 15 lb. and that of the other shaft assembly 90 lb.

The gear-casing is mounted so that it can rotate about the axis of the low-speed shaft; its moment of inertia about this axis is 100 lb.ft.². Its mounting has a torsional stiffness of 2×10^6 pdl.ft./rad.

Form equations from which the natural frequencies for torsional motion of the shafting and gear-box can be found.

Show how these equations can be solved. The moments of inertia of the shafts may be assumed negligible.   (Leeds B.Sc. Mech. Eng. 1958)

† For instance, Th. von Kármán and M. A. Biot, *Mathematical Methods in Engineering* (McGraw-Hill, 1st ed. London, 1940), ch. 5.

3.2 Derivation of the receptances from the general equations

In this section we shall study the general equations (3.1.13) using a similar method to that used in Chapter 1. That is to say, we shall consider first steady forced motion which is at the frequency of the exciting force; by this means it will be possible to find the receptances of the system and these will be used to find the frequencies of free vibration. The steady forced motion which we shall study is, in fact, the important type of motion in practice, but justification of this statement must wait until the effects of damping have been examined.

Consider the trial solution

$$q_1 = \Psi_1 e^{i\omega t}, \quad q_2 = \Psi_2 e^{i\omega t}, \quad \ldots, \quad q_n = \Psi_n e^{i\omega t} \qquad (3.2.1)$$

to the set of equations (3.1.13), where the amplitudes Ψ_r (which may be complex) are at present unknown. If the trial solutions are substituted into the general equations, a set of *algebraic* equations emerge which relate the amplitudes of the forces and the displacements; these are

$$\left. \begin{aligned} (c_{11} - \omega^2 a_{11})\,\Psi_1 + (c_{12} - \omega^2 a_{12})\,\Psi_2 + \ldots + (c_{1n} - \omega^2 a_{1n})\,\Psi_n &= \Phi_1, \\ (c_{21} - \omega^2 a_{21})\,\Psi_1 + (c_{22} - \omega^2 a_{22})\,\Psi_2 + \ldots + (c_{2n} - \omega^2 a_{2n})\,\Psi_n &= \Phi_2, \\ \cdots\cdots\cdots\cdots\cdots\cdots\cdots\cdots\cdots\cdots\cdots\cdots\cdots\cdots\cdots\cdots\cdots\cdots\cdots\cdots\cdots\cdots\cdots \\ (c_{n1} - \omega^2 a_{n1})\,\Psi_1 + (c_{n2} - \omega^2 a_{n2})\,\Psi_2 + \ldots + (c_{nn} - \omega^2 a_{nn})\,\Psi_n &= \Phi_n. \end{aligned} \right\} \qquad (3.2.2)$$

If, in a particular case, the coefficients a and c, and also the force amplitudes Φ, are known then the displacement amplitudes Ψ can be calculated. But if the number of co-ordinates is large the work involved will be very arduous.

The receptance theory of Chapter 1 is concerned with the displacements (and rotations) of points of systems and with the amplitudes of concentrated forces (and couples) acting at these points. It is now necessary to extend these ideas to *generalized* co-ordinates and *generalized* forces. Thus, let a set of forces act on the system which do work during a virtual displacement q_r but which do no work during a virtual displacement of any of the other generalized co-ordinates. There will then be a generalized force Q_r (corresponding to the co-ordinate q_r) acting on the system, but there will be no generalized forces corresponding to any of the other co-ordinates. If the acting force Q_r is harmonic, then the displacements of the system will be harmonic and the displacements can be specified in terms of the generalized co-ordinates. There will then be a set of coefficients or receptances which relate the various generalized displacements with the one generalized force. In the previous notation, there will be constants

$$\alpha_{rr} = \frac{q_r}{Q_r}, \quad \alpha_{sr} = \frac{q_s}{Q_r}, \qquad (3.2.3)$$

where α_{rr} is the 'direct receptance at q_r' and α_{sr} is the 'cross-receptance between q_s and q_r'. We may also write

$$\alpha_{rr} = \frac{\Psi_r}{\Phi_r}, \quad \alpha_{sr} = \frac{\Psi_s}{\Phi_r} \qquad (3.2.4)$$

noting that whereas (3.2.4) is a relation between constants, (3.2.3) relates the instantaneous values of two harmonically varying quantities.

If only the generalized force Q_1 is acting on the system, so that all the terms other than Φ_1 on the right-hand side of equations (3.2.2) are zero, then by the rules for determinants the amplitude Ψ_1^c may be written in the form

$$\Psi_1^c = \frac{\begin{vmatrix} \Phi_1 & (c_{12}-\omega^2 a_{12}) & \cdots & (c_{1n}-\omega^2 a_{1n}) \\ 0 & (c_{22}-\omega^2 a_{22}) & \cdots & (c_{2n}-\omega^2 a_{2n}) \\ \cdots & \cdots & \cdots & \cdots \\ 0 & (c_{n2}-\omega^2 a_{n2}) & \cdots & (c_{nn}-\omega^2 a_{nn}) \end{vmatrix}}{\begin{vmatrix} (c_{11}-\omega^2 a_{11}) & (c_{12}-\omega^2 a_{12}) & \cdots & (c_{1n}-\omega^2 a_{1n}) \\ (c_{21}-\omega^2 a_{21}) & (c_{22}-\omega^2 a_{22}) & \cdots & (c_{2n}-\omega^2 a_{2n}) \\ \cdots & \cdots & \cdots & \cdots \\ (c_{n1}-\omega^2 a_{n1}) & (c_{n2}-\omega^2 a_{n2}) & \cdots & (c_{nn}-\omega^2 a_{nn}) \end{vmatrix}}. \tag{3.2.5}$$

This may be written in the form $\quad \Psi_1^c = \dfrac{\Delta_{11}}{\Delta}\Phi_1,$ (3.2.6)

where Δ represents the determinant of all the coefficients on the left-hand side of equations (3.2.2) and Δ_{11} is the determinant which is formed from Δ by omitting the first column and the first row. In the same way the amplitude Ψ_2^c is given by

$$\Psi_2^c = \frac{\begin{vmatrix} (c_{11}-\omega^2 a_{11}) & \Phi_1 & (c_{13}-\omega^2 a_{13}) & \cdots & (c_{1n}-\omega^2 a_{1n}) \\ (c_{21}-\omega^2 a_{21}) & 0 & (c_{23}-\omega^2 a_{23}) & \cdots & (c_{2n}-\omega^2 a_{2n}) \\ \cdots & \cdots & \cdots & \cdots & \cdots \\ (c_{n1}-\omega^2 a_{n1}) & 0 & (c_{n3}-\omega^2 a_{n3}) & \cdots & (c_{nn}-\omega^2 a_{nn}) \end{vmatrix}}{\Delta}, \tag{3.2.7}$$

and this is written $\qquad \Psi_2^c = -\dfrac{\Delta_{21}}{\Delta}\Phi_1,$ (3.2.8)

where Δ_{21} is formed from Δ by omitting the second column and the first row.

If, instead of Q_1, the only generalized force acting is Q_s then the amplitude of q_r is given by

$$\Psi_r^c = (-1)^{r+s}\frac{\Delta_{rs}}{\Delta}\Phi_s \tag{3.2.9}$$

so that the receptance α_{rs} is given by

$$\alpha_{rs} = (-1)^{r+s}\frac{\Delta_{rs}}{\Delta}. \tag{3.2.10}$$

For the direct receptances $r = s$ and for the cross-receptances $r \neq s$. In mathematical language, Δ_{rs} is the *minor* of that element of Δ which lies in the rth column and sth row. The quantity $(-1)^{r+s}\Delta_{rs}$ is called the *cofactor* of that element.

It has already been pointed out that the order of the subscripts of the coefficients a and c is immaterial. It follows that Δ_{rs} and Δ_{sr} are identical because the value of a determinant is not changed if its rows and columns are interchanged. It therefore follows that the cross-receptances α_{rs} and α_{sr} are equal.

The generalized co-ordinates of the system will not usually be so chosen that the applied force (or forces) will be equivalent to one generalized force only. This, however, does not lead to any more complex analysis because the effects of the different generalized forces can be superposed. Thus, in general,

$$\left.\begin{aligned}
\Psi_1' &= \alpha_{11}\Phi_1 + \alpha_{12}\Phi_2 + \ldots + \alpha_{1n}\Phi_n, \\
\Psi_2' &= \alpha_{21}\Phi_1 + \alpha_{22}\Phi_2 + \ldots + \alpha_{2n}\Phi_n, \\
&\cdots\cdots\cdots\cdots\cdots\cdots\cdots\cdots\cdots\cdots\cdots\cdots \\
\Psi_n' &= \alpha_{n1}\Phi_1 + \alpha_{n2}\Phi_2 + \ldots + \alpha_{nn}\Phi_n.
\end{aligned}\right\} \quad (3.2.11)$$

These equations are valid because of the linearity of the original set of differential equations of motion.

Consider as an example the system of fig. 3.1.2 for which the relations (3.1.19) are the equations of motion. The determinant Δ will be found to be

$$\Delta = \begin{vmatrix} k_1 + k_2 - M\omega^2 & -k_2 r + Ml\omega^2 \sin\theta' & 0 \\ -k_2 r + Ml\omega^2 \sin\theta' & k_2 r^2 - M\omega^2(l^2 + K^2) & -M\omega^2 K^2 \\ 0 & -M\omega^2 K^2 & k_3 l^2 \cos^2(\tfrac{1}{2}\beta) - M\omega^2 K^2 \end{vmatrix} \quad (3.2.12)$$

so that the receptances are:

$$\left.\begin{aligned}
\alpha_{11} &= \frac{\begin{vmatrix} k_2 r^2 - M\omega^2(l^2 + K^2) & -M\omega^2 K^2 \\ -M\omega^2 K^2 & k_3 l^2 \cos^2(\tfrac{1}{2}\beta) - M\omega^2 K^2 \end{vmatrix}}{\Delta}, \\[1em]
\alpha_{22} &= \frac{\begin{vmatrix} k_1 + k_2 - M\omega^2 & 0 \\ 0 & k_3 l^2 \cos^2(\tfrac{1}{2}\beta) - M\omega^2 K^2 \end{vmatrix}}{\Delta}, \\[1em]
\alpha_{33} &= \frac{\begin{vmatrix} k_1 + k_2 - M\omega^2 & -k_2 r + Ml\omega^2 \sin\theta' \\ -k_2 r + Ml\omega^2 \sin\theta' & k_2 r^2 - M\omega^2(l^2 + K^2) \end{vmatrix}}{\Delta}, \\[1em]
\alpha_{12} = \alpha_{21} &= -\frac{\begin{vmatrix} -k_2 r + Ml\omega^2 \sin\theta' & -M\omega^2 K^2 \\ 0 & k_3 l^2 \cos^2(\tfrac{1}{2}\beta) - M\omega^2 K^2 \end{vmatrix}}{\Delta}, \\[1em]
\alpha_{13} = \alpha_{31} &= \frac{\begin{vmatrix} -k_3 r + Ml\omega^2 \sin\theta' & k_2 r^2 - M\omega^2(l^2 + K^2) \\ 0 & -M\omega^2 K^2 \end{vmatrix}}{\Delta}, \\[1em]
\alpha_{23} = \alpha_{32} &= -\frac{\begin{vmatrix} k_1 + k_2 - M\omega^2 & -k_2 r + Ml\omega^2 \sin\theta' \\ 0 & -M\omega^2 K^2 \end{vmatrix}}{\Delta}.
\end{aligned}\right\} \quad (3.2.13)$$

With these, the steady motion at the forcing frequency may be obtained; it is given by the equations

$$\left.\begin{aligned}
\Psi_1 &= \alpha_{11}F - \alpha_{12}Fl\sin\theta', \\
\Psi_2 &= \alpha_{21}F - \alpha_{22}Fl\sin\theta', \\
\Psi_3 &= \alpha_{31}F - \alpha_{32}Fl\sin\theta'.
\end{aligned}\right\} \tag{3.2.14}$$

It is shown in § 2.2 that if y is the displacement in the direction of some applied force $Fe^{i\omega t}$ of the point Y at which the force is applied, then the generalized forces which result from the force $Fe^{i\omega t}$ are

$$Q_1 = F\frac{\partial y}{\partial q_1}e^{i\omega t}, \quad Q_2 = F\frac{\partial y}{\partial q_2}e^{i\omega t}, \quad \dots, \quad Q_n = F\frac{\partial y}{\partial q_n}e^{i\omega t}. \tag{3.2.15}$$

Suppose that it is wished to find the displacement z in some specified direction of a point Z in the system, when the force is acting at Y. This problem will involve a new receptance expression which can be found in terms of the original receptances as follows.

The displacement z will be a function of the co-ordinates q: since the latter are independent, small increments of the q's will give a total increment δz of z which can be found from the relation

$$\delta z = \left(\frac{\partial z}{\partial q_1}\right)\delta q_1 + \left(\frac{\partial z}{\partial q_2}\right)\delta q_2 + \dots + \left(\frac{\partial z}{\partial q_n}\right)\delta q_n. \tag{3.2.16}$$

As small motions only are contemplated, the derivatives may be taken as constants which are evaluated at the equilibrium position of the system. The equations may then be integrated to give

$$z = \left(\frac{\partial z}{\partial q_1}\right)q_1 + \left(\frac{\partial z}{\partial q_2}\right)q_2 + \dots + \left(\frac{\partial z}{\partial q_n}\right)q_n. \tag{3.2.17}$$

If equations (3.2.11) and (3.2.15) are now substituted into this expression, it is found that

$$\begin{aligned}
z = \Bigg\{ &\frac{\partial z}{\partial q_1}\left[\alpha_{11}F\frac{\partial y}{\partial q_1} + \alpha_{12}F\frac{\partial y}{\partial q_2} + \dots + \alpha_{1n}F\frac{\partial y}{\partial q_n}\right] \\
&+ \frac{\partial z}{\partial q_2}\left[\alpha_{21}F\frac{\partial y}{\partial q_1} + \alpha_{22}F\frac{\partial y}{\partial q_2} + \dots + \alpha_{2n}F\frac{\partial y}{\partial q_n}\right] \\
&\quad\dots\dots\dots\dots\dots\dots\dots\dots\dots\dots\dots\dots\dots\dots \\
&+ \frac{\partial z}{\partial q_n}\left[\alpha_{n1}F\frac{\partial y}{\partial q_1} + \alpha_{n2}F\frac{\partial y}{\partial q_2} + \dots + \alpha_{nn}F\frac{\partial y}{\partial q_n}\right]\Bigg\}e^{i\omega t}.
\end{aligned} \tag{3.2.18}$$

This relation may be written in the form

$$z = \alpha_{zy}Fe^{i\omega t}, \tag{3.2.19}$$

where the receptance α_{zy} is given by

$$\begin{aligned}
\alpha_{zy} = \alpha_{yz} = &\alpha_{11}\left(\frac{\partial y}{\partial q_1}\frac{\partial z}{\partial q_1}\right) + \alpha_{22}\left(\frac{\partial y}{\partial q_2}\frac{\partial z}{\partial q_2}\right) + \dots + \alpha_{nn}\left(\frac{\partial y}{\partial q_n}\frac{\partial z}{\partial q_n}\right) \\
&+ \alpha_{12}\left(\frac{\partial y}{\partial q_1}\frac{\partial z}{\partial q_2} + \frac{\partial z}{\partial q_1}\frac{\partial y}{\partial q_2}\right) + \dots.
\end{aligned} \tag{3.2.20}$$

The equality of α_{yz} and α_{zy} is to be expected because y and z might have been chosen as generalized co-ordinates of the system and then the cross-receptance would have been found by the method of equation (3.2.10).

If the points Y and Z are the same, and if the directions in which the displacements y and z are measured is also the same, then α_{yz} becomes a direct receptance and is given by

$$\alpha_{yy} = \alpha_{11}\left(\frac{\partial y}{\partial q_1}\right)^2 + \alpha_{22}\left(\frac{\partial y}{\partial q_2}\right)^2 + \ldots + \alpha_{nn}\left(\frac{\partial y}{\partial q_n}\right)^2 + 2\alpha_{12}\left(\frac{\partial y}{\partial q_1}\cdot\frac{\partial y}{\partial q_2}\right) + \ldots \quad (3.2.21)$$

The system of fig. 3.1.2 will illustrate these ideas. Let y be the displacement of G in the direction of the applied force $Fe^{i\omega t}$ and let z be the displacement of the point C in the direction perpendicular to and away from the slide OA. In terms of the generalized co-ordinates, y and z are given by

$$y = q_1 - lq_2\sin\theta', \quad z = lq_2\cos\theta' + a(q_2 + q_3), \quad (3.2.22)$$

provided that the motion is small. It follows that

$$\left.\begin{array}{l}\dfrac{\partial y}{\partial q_1} = 1, \quad \dfrac{\partial y}{\partial q_2} = -l\sin\theta', \quad \dfrac{\partial y}{\partial q_3} = 0, \\[2mm] \dfrac{\partial z}{\partial q_1} = 0, \quad \dfrac{\partial z}{\partial q_2} = l\cos\theta' + a, \quad \dfrac{\partial z}{\partial q_3} = a,\end{array}\right\} \quad (3.2.23)$$

so that, by equation (3.2.20),

$$\alpha_{yz} = -\alpha_{22}l\sin\theta'.(l\cos\theta' + a) + \alpha_{12}(l\cos\theta' + a) + \alpha_{13}.a - \alpha_{23}.la\sin\theta',$$
$$(3.2.24)$$

where α_{22}, α_{12}, α_{13}, α_{23} are given by equation (3.2.13). Similarly the direct receptance at y is, by equation (3.2.21),

$$\alpha_{yy} = \alpha_{11} + \alpha_{22}l^2\sin^2\theta' - 2\alpha_{12}l\sin\theta'. \quad (3.2.25)$$

All the direct and cross-receptances of a system will become infinite when

$$\Delta = 0. \quad (3.2.26)$$

This is the condition of resonance at which the system gives an infinite response to a finite excitation or a finite response with zero excitation—that is to say at which it can perform free oscillations at the corresponding frequencies.

It is worthwhile to point out a common alternative interpretation of the frequency equation (3.2.26). If a system performs free oscillation then its equations of motion are identical with equations (3.1.13) but with all the amplitudes Φ set equal to zero. The trial solutions (3.2.1) then lead to *homogeneous* algebraic equations (3.2.2). The frequency equation (3.2.26) is merely the condition that non-zero values for the amplitudes Ψ may exist in that event. In this book, however, we shall approach the subject of free oscillation through that of forced vibration.

If Δ is expanded it will be found to be a polynomial of the nth degree in ω^2, so that equation (3.2.26) is an algebraic equation of the nth degree. The roots of this equation in ω^2 give the natural frequencies of the system; these will be denoted, as in Chapter 1, by $\omega_1^2, \omega_2^2, \ldots, \omega_n^2$ in order of increasing magnitude.

The n roots of the frequency equation are all real and positive if motion occurs, as we have postulated, about a state of stable equilibrium. This may be shown analytically to be a consequence of the fact that the expressions for T and V are positive definite.† Rather than present here the mathematical argument to prove this, we merely point out that the fact may be deduced on physical grounds. For suppose that Δ had a negative or a complex root. The natural frequency deduced from such a root would involve an imaginary component and this would imply that the free motion would either increase or decrease indefinitely with time, as may be seen by inspection of equations (3.2.1). Such a motion would mean either that the system was unstable or that its energy was increasing or decreasing. The first of these implications has been ruled out by our initial postulate and the second is impossible because there is no source, and no means for the dissipation, of energy within the system.

Extraction of the roots of frequency equations of the type (3.2.26) is a matter of importance. Mathematical techniques exist for the purpose which, it is hoped, will be described in a later volume. This aspect will not be taken up here as we propose to restrict attention in this book to the *mechanics* of vibration. For information on this subject, the reader is commended to the mathematical literature.‡

It will be shown later that it is not uncommon for one or more of the roots of equation (3.2.26) to be zero; it will be sufficient for the present to regard such roots as extreme examples of real positive roots which have become indefinitely small. It is also possible for the system to have two or more equal roots; this matter is examined in some detail in § 3.11.

The arguments concerning the nature of the roots of equation (3.2.26) apply also to the roots of

$$\Delta_{rr} = 0 \tag{3.2.27}$$

since omission of the rth column and the rth row of Δ leaves a determinant of similar form to that of Δ. This determinant is in fact the Δ of a system which is obtained by restricting the motion of the original system in such a way that displacement of the rth co-ordinate q_r is prevented. The roots of equation (3.2.27) therefore give the natural frequencies of the system when it is constrained in this way. We shall suppose for the present that all the roots of (3.2.27) are unequal; this matter also will be examined later (in § 3.10). The frequencies of the system, when it is constrained so that one of the co-ordinates is fixed, are known as the anti-resonant frequencies in distinction to the *natural* or *resonance* frequencies of the unconstrained system. It will be seen that the anti-resonant frequencies corresponding to different co-ordinates will not be the same, in general, because Δ_{rr} and Δ_{ss} are different. There is no corresponding physical interpretation of the frequencies which can be found as roots of the equation

$$\Delta_{rs} = 0 \quad (r \neq s) \tag{3.2.28}$$

and such frequencies are not necessarily real.

---

† For example, see H. Lamb, *Higher Mechanics* (Cambridge University Press, 1st ed. 1920), §92 or A. S. Ramsey, *Dynamics*, Pt II (Cambridge University Press, 2nd ed. 1944), §10.4.

‡ For example, see Th. von Kármán and M. A. Biot, *Mathematical Methods in Engineering* (McGraw-Hill, 1939), ch. 5.

In this section we have introduced a more general concept of receptances. It is necessary to note, however, that such receptances can still be used in the formulae of Tables 2 and 3, because these formulae depend only upon the form of the equations of motion and not upon the physical meanings of the co-ordinates. It will be realized, though, that the physical meaning of the linking of two systems becomes less clear if the linking co-ordinate is not a simple translation or rotation. The following example shows a form which such linking can take.

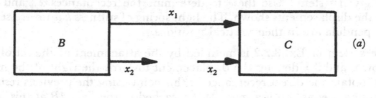

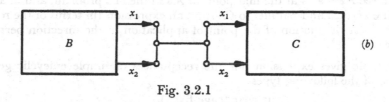

Fig. 3.2.1

Let two systems B and C each have two co-ordinates x_1 and x_2, these co-ordinates being the linear displacements of points of the systems. In the simple type of linking which we considered in Chapter 1 the two points whose displacements are x_1 might be joined so that B and C form a composite system as in fig. 3.2.1 (a); the displacement at x_1 of the two sub-systems will now be the same whereas the displacements x_2 of B and C may differ. Now suppose that generalized co-ordinates for each of the two systems, B and C, are chosen to be $x_1 + x_2$ and $x_1 - x_2$ respectively, these new co-ordinates being q_1 and q_2 respectively. Consider now the physical meaning of linking the co-ordinates q_1 (instead of x_1) for the two sub-systems so as to form a fresh composite system. In terms of the original co-ordinates this linking will mean that $x_1 + x_2$ will be the same for the two systems but that $x_1 - x_2$ may be different. Fig. 3.2.1 (b) indicates how such linking might be arranged if the original co-ordinates are represented by the arrows in the figure. The reader may note that the form of linking between two systems may influence the choice of co-ordinates for the systems, since a suitable choice may make the equations of motion easier to derive.

EXAMPLES 3.2

1. By means of the equation $\alpha_{rs} = (-1)^{r+s}\, \Delta_{rs}/\Delta$

find the cross-receptance α_{12} of the system shown in the figure.

Verify the result by considering the system as a combination of two simpler systems in the manner of §1.6.

2. Using the generalized co-ordinates q_1 and q_2 shown, find the receptances α_{11}, α_{22}, α_{12} for the given system. Use these to determine the receptances α_{yy} and α_{yz}, where y and z are the displacements shown. The light spring of stiffness k is unstressed when the two simple pendula are in their vertical positions.

3. If the system of Ex. 3.2.2 is modified by the attachment of the set of light rigid links as shown, and if x denotes the displacement towards the right of the mid-point of the link AB obtain the direct receptance at this point using the previous results. Hence find the frequency equation if a mass M is attached to the link AB at this point.

4. Find the receptances α_{11} and α_{12} of the system of Ex. 3.1.2.

5. If a force $F\,e^{i\omega t}$ acts at the mid-point of AB in the last problem, and in a direction at $45°$ to the vertical and parallel to CA, find an expression (in terms of the receptances α_{11}, α_{12}, etc.) for the motion of the point of application in the direction perpendicular to the force.

6. The table gives expressions for the receptances of simple epicyclic gears, these gears being of the following types:

> star gear (cage fixed);
> planetary gear (annulus fixed);
> solar gear (sun fixed).

Derive one of the direct and one of the cross-receptances.

TABLE OF RECEPTANCES OF EPICYCLIC GEARS

Star gear	$\alpha_{ss} = \dfrac{1}{\Delta}$	$\alpha_{aa} = \dfrac{(r_s/r_a)^2}{\Delta}$	$\alpha_{sa} = \alpha_{as} = \dfrac{-(r_s/r_a)}{\Delta}$
	$\Delta = -\omega^2\left[I_s + NI_p\left(\dfrac{r_s}{r_p}\right)^2 + I_a\left(\dfrac{r_s}{r_a}\right)^2 \right]$		
Planetary gear	$\alpha_{ss} = \dfrac{1}{\Delta}$	$\alpha_{cc} = \dfrac{(r_s/2r_c)^2}{\Delta}$	$\alpha_{sc} = \alpha_{cs} = \dfrac{(r_s/2r_c)}{\Delta}$
	$\Delta = -\omega^2\left[I_s + NI_p\left(\dfrac{r_s}{2r_p}\right)^2 + I_c\left(\dfrac{r_s}{2r_c}\right)^2 \right]$		
Solar gear	$\alpha_{cc} = \dfrac{1}{\Delta}$	$\alpha_{aa} = \dfrac{(2r_c/r_a)^2}{\Delta}$	$\alpha_{ca} = \alpha_{ac} = \dfrac{(2r_c/r_a)}{\Delta}$
	$\Delta = -\omega^2\left[I_c + NI_p\left(\dfrac{r_c}{r_p}\right)^2 + I_a\left(\dfrac{2r_c}{r_a}\right)^2 \right]$		

Symbols: I = polar moment of inertia; N = no. of planets; r = radius.

Subscripts: a = annulus; c = cage; p = planet; s = sun.

Note: (1) r_a, r_p, r_s = pitch circle radii; r_c = radius of circle containing planet axes. (2) I_p refers to single planet and its own axis. (3) I_c includes cage, planet spindles and allowance ($= NM_p r_c^2$) for translation of planets (where M_p = mass of planet).

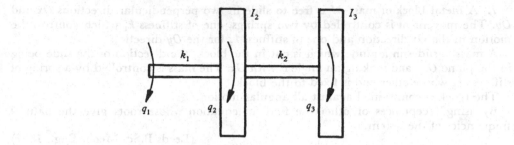

Ex. 3.2.1

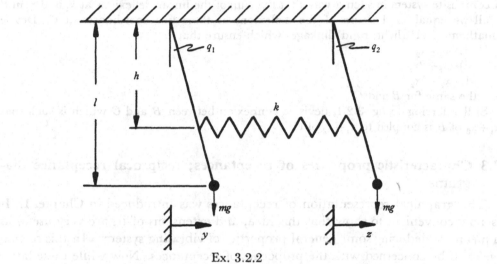

Ex. 3.2.2

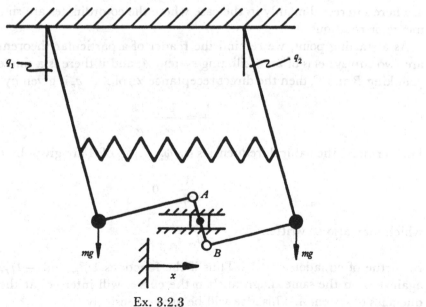

Ex. 3.2.3

7. A metal block of mass M is free to slide in two perpendicular directions Ox and Oy. The movement is controlled by two springs, one of stiffness k_1 which controls the motion in the Ox direction and one of stiffness k_2 for the Oy direction.

A mass m slides in a guide which is cut in the block, the direction of the slide being in the plane Oxy and making an angle θ with Ox. The mass is controlled by a spring of stiffness k_3 whose other end is fixed to the block.

The block is constrained against all angular motion.

By using receptances or otherwise find an equation whose roots give the natural frequencies of the system.

(Leeds B.Sc. Mech. Eng., 1957)

8. Fig. 3.2.1 (b) shows how two subsystems B and C may be connected so as to form a composite system in such a way that the sum of the linear deflexions at x_1 and x_2 in B shall be equal to the sum of the deflexions at these two coordinates at C. Devise 'mathematical' (light, rigid) linkages which ensure that

    (i) $x_1 - x_2$
    (ii) $x_1 + 2x_2$
    (iii) $x_1 - 2x_2$

are the same for B and C.

Still referring to fig. 3.2.1, devise a connexion between B and C which is such that $x_1 + x_2$ of B is coupled to $x_1 - x_2$ of C.

3.3 Characteristic properties of receptances; reciprocal receptance diagrams

The graphical representation of receptances was introduced in Chapter 1. It is now convenient to show that this idea, and extensions of it, are very useful as a means of deducing some general properties of vibrating systems. In this section we shall be concerned with the properties of receptances. Now while these latter can be regarded as being associated with *any* set of generalized co-ordinates q, we are here interested mainly in the case where the co-ordinates are simple displacements or rotations.

As a starting point, we remind the reader of a particular theorem; if B and C are two sub-systems of an oscillating system A, and if there is a single co-ordinate q_r linking B and C, then the direct receptance α_{rr} of A at q_r is given by the equation

$$\frac{1}{\alpha_{rr}} = \frac{1}{\beta_{rr}} + \frac{1}{\gamma_{rr}}. \tag{3.3.1}$$

Furthermore, the natural frequencies $\omega_1, \omega_2, ..., \omega_n$ of A are given by the equation

$$\frac{1}{\alpha_{rr}} = 0 \tag{3.3.2}$$

which may also be written
$$\frac{1}{\beta_{rr}} = -\frac{1}{\gamma_{rr}} \tag{3.3.3}$$

by virtue of equation (3.3.1). Thus if the functions $1/\beta_{rr}$ and $-1/\gamma_{rr}$ are plotted against ω on the same diagram then the curves will intersect at the natural frequencies of system A. This idea will be used extensively.

Two types of receptance diagram will be discussed subsequently. In one of these the reciprocals of receptances are plotted against ω, as already described in Chapter 1, and in the other the receptances are plotted directly against ω. The latter type will be referred to in the next section and we shall now examine some properties of the former.

Some idea of the form of the curve of the reciprocal of a direct receptance can be obtained from consideration of the function

$$\frac{1}{\alpha_{rr}} = \frac{\Delta}{\Delta_{rr}}. \tag{3.3.4}$$

If the system in question has n degrees of freedom, then there will be n frequencies for which Δ vanishes, thus making $1/\alpha_{rr}$ zero; these are the natural frequencies. Moreover, since Δ_{rr} is the determinant Δ for the system when one of its degrees of freedom is removed, there will be $(n-1)$ frequencies at which Δ_{rr} vanishes; these frequencies will make the function $1/\alpha_{rr}$ infinite and are called the 'anti-resonance' frequencies. It follows that the function $1/\alpha_{rr}$ has n zero values and $(n-1)$ infinite values.[†] Now when expanded, Δ_{rr} becomes a polynomial in ω^2 so that its sign changes when it passes through zero; the sign of $1/\alpha_{rr}$ also changes therefore as it passes through an infinite value.

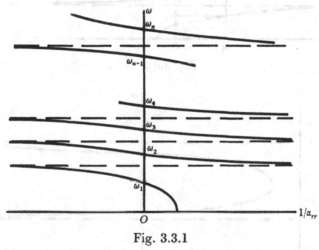

Fig. 3.3.1

The general form of curve shown in fig. 3.3.1 has the properties described above. It also has the property that the infinite and zero values of the abscissae $1/\alpha_{rr}$ occur alternately; it will be shown that this is necessary for the reciprocal of any direct receptance. This property and others will be deduced here by direct reasoning based upon our knowledge of physical systems; in the next section the same matters will be treated by analysis.

Let a system A, for which fig. 3.3.2 represents the reciprocal of the direct receptance at q_r, be modified by attaching to it (at q_r) a massless spring of stiffness k; the

† It is assumed here that none of the roots of equation (3.2.26) is equal to a root of (3.2.27); this special case will be treated in §3.10.

other end of this spring is to be assumed to be anchored. The original system and the spring now form two sub-systems which are linked by the co-ordinate q_r. The natural frequencies of the complete system are given by

$$\frac{1}{\alpha_{rr}} + \frac{1}{1/k} = 0 \qquad (3.3.5)$$

or

$$\frac{1}{\alpha_{rr}} = -k, \qquad (3.3.6)$$

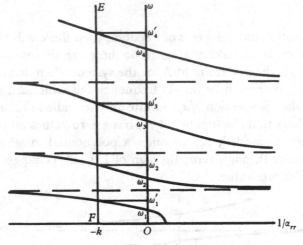

Fig. 3.3.2

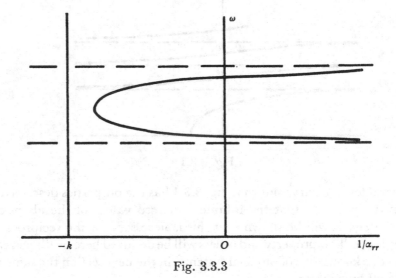

Fig. 3.3.3

which is a particular case of equation (3.3.3). The roots of equation (3.3.6) will correspond to the intersections of the original curve in fig. 3.3.2 with the line EF which is the graph of the reciprocal of the spring receptance, the sign being changed in accordance with equation (3.3.3).

From this diagram, it may be deduced that the roots and infinite values of $1/\alpha_{rr}$ occur alternately. For suppose that they did not, so that the curve could contain branches of the form shown in fig. 3.3.3. If the stiffness k of the spring were to be varied so that the vertical line of fig. 3.3.2 would be moved sideways, then it would be possible to find some values of k for which the line would not intersect the branch shown. For such values of the stiffness, two of the real roots of the frequency equation would disappear and would therefore have to be replaced by two complex or imaginary roots. But it has already been seen that all the roots of the frequency equation must be real and so the supposition that branches of the form shown in fig. 3.3.3 might exist must be wrong. Therefore no looped branch is possible which has the bottom of its loop on the left-hand side of the axis of ω.

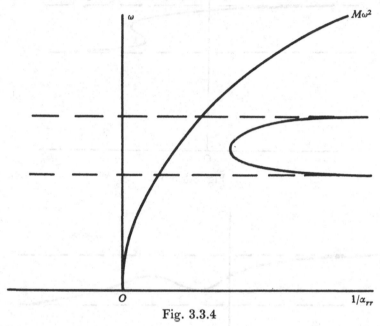

Fig. 3.3.4

By a similar argument, it can be shown that a looped branch with the bottom of its loop on the right-hand side of the axis of ω is also impossible. Suppose that the original system A is to be modified by attaching to it a concentrated mass M at q_r. The natural frequencies of the modified system will now be given by

$$\frac{1}{\alpha_{rr}} + \frac{1}{-1/M\omega^2} = 0 \qquad (3.3.7)$$

or

$$\frac{1}{\alpha_{rr}} = M\omega^2. \qquad (3.3.8)$$

The function $M\omega^2$ is represented by a parabola as shown in fig. 3.3.4. If the mass M is varied, then the parabola sweeps over the right-hand half of the diagram and can be arranged to cut, or alternatively to be clear of, the branch shown in the figure. It thus follows that a looped branch with the bottom of its loop on the right of the axis of ω cannot exist.

From the above facts, and from our knowledge that there are n roots and $(n-1)$ horizontal asymptotes to the curve of $1/\alpha_{rr}$, it follows that the graph of the function must have the general form shown by fig. 3.3.1. For it will be seen from arguments such as those above that bends of the curve as shown in fig. 3.3.5 are also impossible. Bends of the form shown in fig. 3.3.6 are ruled out by the fact that $1/\alpha_{rr}$ is single-valued for all values of ω, because Δ and Δ_{rr} are polynomials.

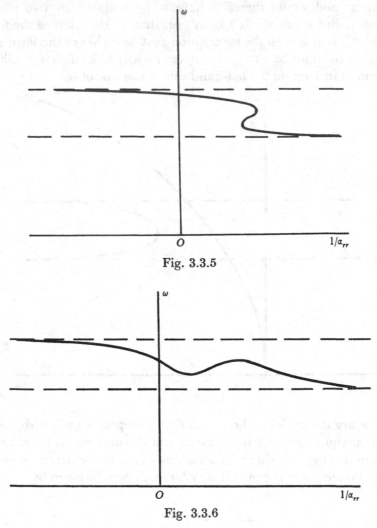

Fig. 3.3.5

Fig. 3.3.6

It will be noted that the toe of the reciprocal receptance diagram, at which $\omega = 0$, has been shown lying to the right of the origin. This must always be so because otherwise a *static* force Q_r at q_r would do a negative amount of work when applied to the system. This would be incompatible with the postulated stability of the system. If the toe of the diagram lies at the origin then there is no restoring force when a static deflexion is imposed; this means that the system is not anchored or, in other words, that it has a natural frequency of zero magnitude.

It can now be shown that the slopes of all the branches of the curves of fig. 3.3.2 are necessarily negative. For suppose that it were possible to find a system for which such slopes were positive. Fig. 3.3.7 represents the lowest branch of the diagram for such a system together with a parabola corresponding to an additional mass M in accordance with equation (3.3.8). The intersections of the parabola and the bottom branch may or may not exist depending upon the magnitude of M; and the disappearance of the intersections will not affect the number of intersections with the other branches. By the previous argument the disappearance of real roots cannot occur for a real system in this way. Hence it is not possible for such a system to have branches with positive slope.

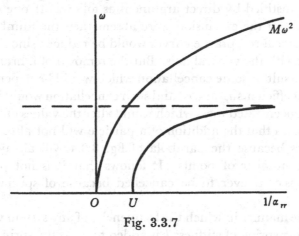

Fig. 3.3.7

The fact that the slopes are always negative can also be deduced as follows. Consider first a system for which the toe U lies to the right of O as in fig. 3.3.7; if the number of intersections of the curve with the axis of ω is to be one more than the number of horizontal asymptotes a negative slope for the branches is essential. Now if the toe coincides with the origin (so that the system is not anchored) then it may be turned into an anchored system by a spring of stiffness k which is attached to it at q_r and which has its other end fixed. The new direct receptance at q_r is α'_{rr}, where

$$\frac{1}{\alpha'_{rr}} = k + \frac{1}{\alpha_{rr}}. \tag{3.3.9}$$

This equation shows that the anchoring spring shifts the reciprocal receptance curves to the right without changing their form and, in particular, their slope. The previous argument can thus be used to show that the slopes are negative for an unanchored system just as they are for an anchored system.

Another method for examining the general character of the reciprocal receptance curves will be introduced in § 3.4, and we shall not therefore follow the matter any further here. The reader may note, however, that further arguments may be based on the possibility of attaching a simple spring and mass to a system, the point of attachment of the spring being remote from the mass; examples in which this idea has to be used are to be found at the end of this section.

The direct receptance α_{rr} which has been considered has been regarded as being measured at a particular generalized co-ordinate q_r. The arguments apply equally to any direct receptance measured at a point Y because it is always possible to choose the displacement y of such a point as one of the generalized co-ordinates of the system.

Before going on to use the properties of the reciprocal receptance curves which have just been deduced, we shall clear up a matter which was left open earlier. It was assumed that the frequency equation (3.2.26) would *always* be of the nth degree in ω^2, cancellation of the highest degree terms not being possible. The reader may justify this assumption by examination of the algebraic form of the expansion of Δ; but it may also be justified by direct argument as follows. If one or more of the highest degree terms of the expansion were absent then the number of branches of each of the reciprocal receptance curves would be reduced since there would be fewer intersections with the vertical axis. But the removal of a branch in this way could only be the result of some cancellation which would be dependent upon the *exact* values of the coefficients a_{rs} and c_{rs} and such cancellation would be impaired by the addition of a concentrated mass which would alter the values of the coefficients. But it has been shown that the addition of a particle will not alter the number of natural frequencies because the parabola of fig. 3.3.4 will always intersect the curve in the same number of points. It follows that it is not possible for the highest degree terms of Δ ever to be cancelled because of special values of the coefficients.

Consider now the manner in which the frequencies of any system will be changed if an extra anchoring spring of stiffness k is added to it. If the spring is attached at some point Y, then the new natural frequencies will be given by the intersections of the branches of $1/\alpha_{yy}$ with the line distant k on the left-hand side of the axis of ω. Since the slope of all the branches is negative, it can be deduced that all the natural frequencies of the system will be raised by the addition of the extra anchorage. However, it is possible for some of the branches to be horizontal straight lines so that the corresponding frequencies will not then be changed; we shall discuss this special case in § 3.10.

The argument may be extended to cover a modification of the system in which a spring is attached between any two points. This is because it is always possible, in theory, to attach a massless linkwork of rigid members between the two points concerned which is so arranged that the displacement of one point on the linkwork is equal to the stretch which would be produced in a spring between the points; being rigid and massless, this linkwork will not affect the energy coefficients a_{rs} and c_{rs} of the system and so will not alter Δ and hence the natural frequencies. Now if the spring were anchored at one end and the other end of it were fixed to this point on the imaginary linkwork, then the natural frequencies of the system would all be raised as before. If, on the other hand, the spring were attached between the points then the potential energy function would be modified in exactly the same way as it would be with the spring between the point of the linkwork and the anchorage. It follows that the addition of a spring between any two points, provided that they suffer relative motion, must raise the natural frequencies of the system.

A special case of the addition, at a co-ordinate q_r, of an extra anchoring spring may be noted. If such a spring has infinite stiffness then the intersections on the diagram of $1/\alpha_{rr}$ will be lifted up to lie on the asymptote which is immediately above. The natural frequencies will thus become the anti-resonance frequencies of the co-ordinate in question, a result which may be arrived at by simple argument.

The modification of a system by the addition of stiffness as described above alters the coefficients of stability; these coefficients may also be changed when the mass of a system is altered because of the effects of gravity. If gravity does not affect the values of the stability coefficients then it may be shown, by arguments similar to those above, that the addition of mass to the system will reduce all the natural frequencies. But there may be special cases in which some of the frequencies are unchanged.

Whereas the addition of stiffness will shift the relevant reciprocal receptance curves to the right without altering their form, the addition of mass will change the shape as well as the position of the curves. This is because the modified curves are obtained from the original ones by adding a parabola instead of a straight line.

In this section we have discussed several general properties of vibrating systems using a semi-graphical approach. The interested reader will find a further discussion of these matters, which is based on analytical reasoning, in Rayleigh's book.[†]

EXAMPLES 3.3

1. In the figure, A represents a mechanical system with any number of degrees of freedom, q_1 being one of its co-ordinates. The system is attached at q_1 to the mass M which is itself attached to the light spring of stiffness k, of which the remote end is anchored.

Show that, as a result of the coupling of the two systems, all the frequencies of A which are above $\sqrt{(k/M)}$ will be diminished and all those which are below this value will be increased.

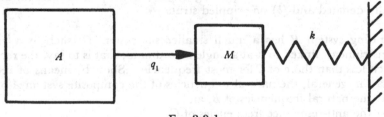

Ex. 3.3.1

2. The curves of the reciprocal receptance of two coupled systems can be formed by adding those for each system separately in accordance with equation (3.3.1). Consider graphically the case when the two sub-systems have receptances which vanish at very nearly the same frequency and deduce that, in the limit, when the receptances vanish at the same frequency, the curve for the coupled system has one branch which is a straight line.

[†] *Theory of Sound* (1894), §92a.

Show further that, if a third system is then coupled to the common point of the other two, there will be one frequency which will not be altered by the addition of the third system. What is the physical explanation of this?

(Problems such as this are treated in §3.10.)

3. A system B has attached to it, at a co-ordinate q_m, the spring-mass system C as shown in the diagram. B is anchored at some point and has natural frequencies $\omega_1, \omega_2, \ldots, \omega_n$ and anti-resonance frequencies $\Omega_1, \Omega_2, \ldots, \Omega_{n-1}$ at q_m. The anti-resonance frequency at q_m of the system C is Ω^*.

If $\Omega_r < \Omega^* < \Omega_s$ and $\Omega^* \neq \omega_s$ show that, due to the attachment of C,

    (i) the first s natural frequencies of B are diminished,

    (ii) the composite system has an extra natural frequency between ω_s and Ω_s,

    (iii) the remaining $(n-s)$ natural frequencies of B are increased.

If, further, any of the natural frequencies of B are much greater than Ω^*, show that the effect of the coupling upon these frequencies is approximately that of an anchored spring of stiffness k.

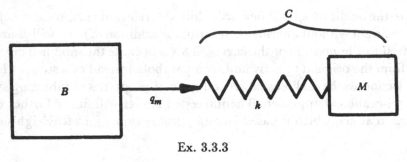

Ex. 3.3.3

4. Show that if, in the last problem, B is not anchored at any point, then the previous results are not altered except that the single natural frequency $\omega_1 = 0$ is left unchanged.

5. Assume that an aircraft behaves as a conservative system with a finite number of degrees of freedom for the purpose of vibration analysis.† When in flight, it possesses certain natural frequencies and it is required to measure these on the ground. Discuss the effects on these frequencies of supporting the aircraft during the test (a) on its own tyres, partially deflated and (b) on crippled struts.

6. A vibrating system B has a much smaller sub-system C (such as a vibration-measuring instrument) linked to it at a single co-ordinate; that is to say, the receptances of B are much less than those of C for most frequencies. Show by means of receptance diagrams that, in general, the natural frequencies of the composite system A occur

    (a) near the natural frequencies of B, and

    (b) near the anti-resonance frequencies of C.

What is the physical significance of these results?

7. In order to measure the direct receptance at a point of a structure at low frequencies, an exciter is used which involves the rigid attachment of a significant mass to the structure at that point. Explain how a correction of the measured receptances may be made for the mass of the exciter.

    † The validity of assumptions of this sort will be discussed in Chapters 5, 8 and 9.

3.4 Series form of receptances; principal modes of vibration

In §3.2, it was shown that the general expression for a receptance has the form

$$\alpha_{rs} = (-1)^{r+s}\frac{\Delta_{rs}}{\Delta}, \tag{3.4.1}$$

where Δ_{rs} is a polynomial in ω^2 of maximum degree $(n-1)$. This may be written in terms of the roots $\omega_1^2, \omega_2^2, \ldots, \omega_n^2$ of the frequency equation (3.2.26); thus

$$\alpha_{rs} = \frac{(-1)^{r+s}\Delta_{rs}}{K(\omega_1^2 - \omega^2)(\omega_2^2 - \omega^2)\ldots(\omega_n^2 - \omega^2)}, \tag{3.4.2}$$

where K is a constant which must be introduced because Δ will not, in general, be equal simply to the product of the n bracketed factors. This expression can be split up into partial fractions to give

$$\alpha_{rs} = \frac{{}_1A_{rs}}{\omega_1^2 - \omega^2} + \frac{{}_2A_{rs}}{\omega_2^2 - \omega^2} + \ldots + \frac{{}_nA_{rs}}{\omega_n^2 - \omega^2}, \tag{3.4.3}$$

where the numerators ${}_mA_{rs}$ are real constants. This form of the receptance enables us readily to deduce the nature of the free vibration at the various natural frequencies.

Let the system be acted on by a single generalized force $Q_s = \Phi_s e^{i\omega t}$ so that the motions at the various co-ordinates are given by

$$\left.\begin{aligned}
q_1 &= \alpha_{1s}Q_s = \left[\frac{{}_1A_{1s}}{\omega_1^2 - \omega^2} + \frac{{}_2A_{1s}}{\omega_2^2 - \omega^2} + \ldots + \frac{{}_nA_{1s}}{\omega_n^2 - \omega^2}\right]Q_s, \\
q_2 &= \alpha_{2s}Q_s = \left[\frac{{}_1A_{2s}}{\omega_1^2 - \omega^2} + \frac{{}_2A_{2s}}{\omega_2^2 - \omega^2} + \ldots + \frac{{}_nA_{2s}}{\omega_n^2 - \omega^2}\right]Q_s, \\
&\cdots\cdots\cdots\cdots\cdots\cdots\cdots\cdots\cdots\cdots\cdots\cdots\cdots \\
q_n &= \alpha_{ns}Q_s = \left[\frac{{}_1A_{ns}}{\omega_1^2 - \omega^2} + \frac{{}_2A_{ns}}{\omega_2^2 - \omega^2} + \ldots + \frac{{}_nA_{ns}}{\omega_n^2 - \omega^2}\right]Q_s.
\end{aligned}\right\} \tag{3.4.4}$$

Suppose now that the excitation frequency ω is made to approach the mth natural frequency ω_m while the magnitude of Φ_s is reduced in order that the amplitude of motion should remain small. That is to say, consider a limiting process by which

$$\lim_{\omega \to \omega_m}\left[\frac{\Phi_s}{\omega_m^2 - \omega^2}\right] = \text{some finite constant, } C_m. \tag{3.4.5}$$

When this limiting process is applied to the equations of motion (3.4.4), it leads to expressions for the *free* motion at the frequency ω_m; these are

$$\left.\begin{aligned}
q_1 &= C_m \cdot {}_mA_{1s}e^{i\omega_m t}, \\
q_2 &= C_m \cdot {}_mA_{2s}e^{i\omega_m t}, \\
&\cdots\cdots\cdots\cdots\cdots \\
q_n &= C_m \cdot {}_mA_{ns}e^{i\omega_m t}.
\end{aligned}\right\} \tag{3.4.6}$$

These equations show that a free harmonic motion with frequency ω_m is possible in which the amplitudes of $q_1, q_2, \ldots, q_n$ are governed by the set of ratios

$$\Psi_1 : \Psi_2 : \ldots : \Psi_n :: {}_mA_{1s} : {}_mA_{2s} : \ldots : {}_mA_{ns}. \tag{3.4.7}$$

This set of ratios defines the mth 'principal mode' of vibration; the mode is a characteristic of the system because the quantities $_mA_{rs}$ are dependent upon the coefficients of inertia and stability—the a's and c's from which Δ is formed. There is a principal mode corresponding to each natural frequency of the system, this mode being the *form* or shape of distortion in which the system vibrates freely at the frequency in question. It was shown in § 1.10 that the ratios which constitute the mode could be found from the ratios of the receptances at the corresponding frequency.

Equations (3.4.6) still contain the subscript s. It is a relic of the fact that the equations of free motion were reached by a limiting process from those of the forced motion due to Q_s. The subscript is no longer significant, however, once the forcing amplitude has been reduced to zero. The ratio of the quantities

$$_mA_{1t} : _mA_{2t} : \ldots : _mA_{nt}$$

must, for this reason, be equal to the ratios of

$$_mA_{1s} : _mA_{2s} : \ldots : _mA_{ns}$$

although the absolute values of the quantities may differ. The subscript s does not therefore have any effect on the modal shape.†

The system has n degrees of freedom and a like number of natural frequencies; we now see that it can vibrate freely in n ways. Owing to the linearity of the equations of motion, any two solutions of them may be added to give another solution. It follows that the system can vibrate in two or more of the principal modes simultaneously, provided that the motion in each mode occurs at its own natural frequency. The general expression for the free motion can therefore be written in the form

$$\left.\begin{aligned}
q_1 &= C_1 \cdot {_1A_1}\, e^{i\omega_1 t} + C_2 \cdot {_2A_1}\, e^{i\omega_2 t} + \ldots + C_n \cdot {_nA_1}\, e^{i\omega_n t}, \\
q_2 &= C_1 \cdot {_1A_2}\, e^{i\omega_1 t} + C_2 \cdot {_2A_2}\, e^{i\omega_2 t} + \ldots + C_n \cdot {_nA_2}\, e^{i\omega_n t}, \\
& \cdots\cdots\cdots\cdots\cdots\cdots\cdots\cdots\cdots\cdots\cdots\cdots\cdots\cdots\cdots \\
q_n &= C_1 \cdot {_1A_n}\, e^{i\omega_1 t} + C_2 \cdot {_2A_n}\, e^{i\omega_2 t} + \ldots + C_n \cdot {_nA_n}\, e^{i\omega_n t},
\end{aligned}\right\} \qquad (3.4.8)$$

where the subscript s has been omitted from the constants A for the reasons explained above. It will be shown in Chapter 10 how the arbitrary complex constants C in equations (3.4.8) allow for any initial conditions of displacement q and velocity $\dot{q}$ to be fulfilled. If, in particular, the system is released from rest when it is distorted in a principal mode then the free motion will be a vibration in that mode only and the motions of all the points of the system will have the appropriate natural frequency and will be in phase.

If the form of any one principal mode is known, then the corresponding energy expressions T and V can be written down. And since the oscillation in the mode is harmonic, and the displacements of all points are in phase, the natural frequency can be found from the relation‡

$$T_{\text{max.}} = V_{\text{max.}} \qquad (3.4.9)$$

† An exception to this, which need not concern us here, is discussed in §3.11.
‡ A simple example illustrating this technique will be found at the beginning of §3.9.

The concept of principal modes of vibration is important; we shall therefore consider here some simple examples in order that the reader may become more familiar with it.

Two masses, of magnitude $2M$ and M as shown in fig. 3.4.1, are suspended on springs of stiffness $2k$ and k. Let the vertical displacements of the masses, measured from the equilibrium position, be q_1 and q_2. The energy expressions are

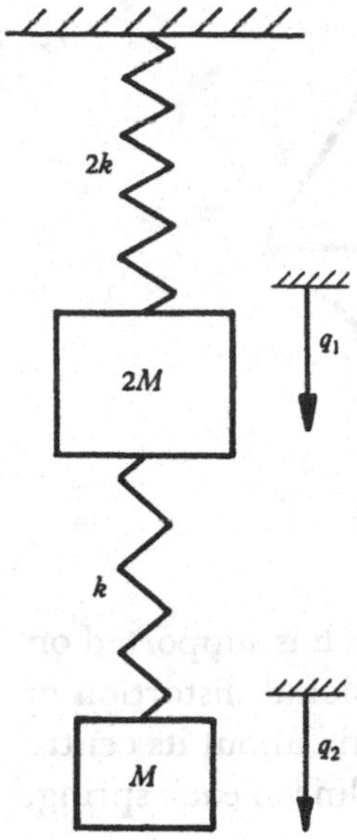

$$\left.\begin{aligned}2T &= 2M\dot{q}_1^2 + M\dot{q}_2^2, \\ 2V &= 2kq_1^2 + k(q_2 - q_1)^2 = 3kq_1^2 + kq_2^2 - 2kq_1q_2,\end{aligned}\right\} \quad (3.4.10)$$

so that

$$\left.\begin{aligned}a_{11} &= 2M, \quad a_{22} = M, \quad a_{12} = 0, \\ c_{11} &= 3k, \quad c_{22} = k, \quad c_{12} = -k.\end{aligned}\right\} \quad (3.4.11)$$

If the lower mass is acted on by a vertical force $Q_2 = \Phi_2 e^{i\omega t}$ then

$$\left.\begin{aligned}q_1 &= \alpha_{12}Q_2 = -\frac{\Delta_{12}}{\Delta}Q_2, \\ q_2 &= \alpha_{22}Q_2 = \frac{\Delta_{22}}{\Delta}Q_2,\end{aligned}\right\} \quad (3.4.12)$$

where

$$\Delta = \begin{vmatrix} 3k - 2M\omega^2 & -k \\ -k & k - M\omega^2 \end{vmatrix}. \quad (3.4.13)$$

Fig. 3.4.1

Equations (3.4.12) now become

$$\left.\begin{aligned}\alpha_{12} &= \frac{q_1}{Q_2} = \frac{k}{2k^2 - 5kM\omega^2 + 2M^2\omega^4} = \frac{\frac{2}{3}}{k - 2M\omega^2} + \frac{-\frac{1}{3}}{2k - M\omega^2}, \\ \alpha_{22} &= \frac{q_2}{Q_2} = \frac{3k - 2M\omega^2}{2k^2 - 5kM\omega^2 + 2M^2\omega^4} = \frac{\frac{4}{3}}{k - 2M\omega^2} + \frac{\frac{1}{3}}{2k - M\omega^2}.\end{aligned}\right\} \quad (3.4.14)$$

The natural frequencies of the system are

$$\omega_1 = \sqrt{(k/2M)}, \quad \omega_2 = \sqrt{(2k/M)}, \quad (3.4.15)$$

and the corresponding principal modes are defined by the ratios

$$\frac{q_1}{q_2} = \frac{1}{2}, \quad \frac{q_1}{q_2} = \frac{-1}{1}, \quad (3.4.16)$$

respectively. Thus, during free motion in the first principal mode, the displacements q_1 and q_2 occur in the same direction at any instant, the amplitude of q_2 being twice that of q_1. The second principal mode has twice the frequency of the first and involves equal displacements of the two masses in opposite directions at any instant. The form of these principal modes may be represented diagrammatically as in fig. 3.4.2. The amplitudes in the first mode are proportional to the lengths of the horizontal lines in (a) while those in the second mode are given in the same way by (*b*). Distances between the vertical axis and points on the inclined lines give the displacements of corresponding points on the springs although similar diagrams are sometimes drawn in which the inclined lines have no physical significance, being included merely as guides to the eye.

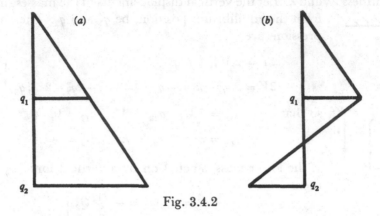

Fig. 3.4.2

As a second example, consider the motion of a rigid body which is supported on two springs as shown in fig. 3.4.3, the motion involving extensional distortion of the springs only. The body has mass M, and its moment of inertia about its centre of gravity G is $\frac{3}{4}Ma^2$, where a is the distance of G from the centre line of each spring.

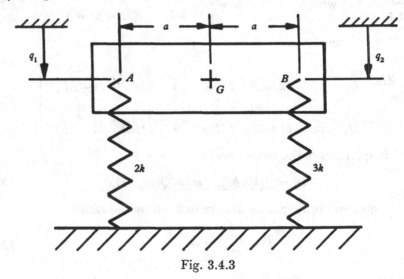

Fig. 3.4.3

The stiffnesses are $2k$ and $3k$. The vertical displacements of the tops of the springs A and B are a suitable pair of co-ordinates q_1 and q_2 as shown. The energy expressions are

$$2T = M\left(\frac{\dot{q}_1+\dot{q}_2}{2}\right)^2 + \frac{3Ma^2}{4}\left(\frac{\dot{q}_1-\dot{q}_2}{2a}\right)^2 = \frac{7M}{16}\dot{q}_1^2 + \frac{7M}{16}\dot{q}_2^2 + \frac{M}{8}\dot{q}_1\dot{q}_2, \left.\begin{matrix} \\ \\ \end{matrix}\right\} \quad (3.4.17)$$

$$2V = 2kq_1^2 + 3kq_2^2,$$

so that

$$a_{11} = \frac{7M}{16}, \quad a_{22} = \frac{7M}{16}, \quad a_{12} = \frac{M}{16},$$
$$c_{11} = 2k, \quad c_{22} = 3k, \quad c_{12} = 0$$

(3.4.18)

and therefore

$$\Delta = \begin{vmatrix} 2k - \dfrac{7M}{16}\omega^2 & -\dfrac{M\omega^2}{16} \\[2mm] -\dfrac{M\omega^2}{16} & 3k - \dfrac{7M}{16}\omega^2 \end{vmatrix}.$$

(3.4.19)

The receptances are

$$\alpha_{11} = \frac{\Delta_{11}}{\Delta} = \frac{3k - \dfrac{7M}{16}\omega^2}{6k^2 - \dfrac{35kM\omega^2}{16} + \dfrac{3M^2\omega^4}{16}} = \frac{2 \cdot 0054}{4 \cdot 409k - M\omega^2} + \frac{0 \cdot 3279}{7 \cdot 257k - M\omega^2},$$

$$\alpha_{12} = \frac{-\Delta_{12}}{\Delta} = \frac{\dfrac{M\omega^2}{16}}{6k^2 - \dfrac{35kM\omega^2}{16} + \dfrac{3M^2\omega^4}{16}} = \frac{0 \cdot 5161}{4 \cdot 409k - M\omega^2} + \frac{-0 \cdot 8494}{7 \cdot 257k - M\omega^2},$$

(3.4.20)

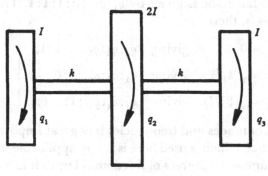

Fig. 3.4.4

and, from these, the natural frequencies and corresponding principal modes may be deduced; they are

$$\omega_1^2 = \frac{4 \cdot 409k}{M} \quad \text{giving} \quad \frac{q_1}{q_2} = \frac{2 \cdot 0054}{0 \cdot 5161} = 3 \cdot 885,$$
$$\omega_2^2 = \frac{7 \cdot 257k}{M} \quad \text{giving} \quad \frac{q_1}{q_2} = -\frac{0 \cdot 3279}{0 \cdot 8494} = -0 \cdot 3860.$$

(3.4.21)

The first mode consists of a combined turning and lifting motion in which the instantaneous centre of rotation lies on the line AB produced (see fig. 3.4.3) whereas for the second mode the instantaneous centre lies within AB. The oscillation of systems of this type is of interest in connexion with the study of motor vehicle suspensions.

The system of fig. 3.4.4, in which the vibration is torsional provides a third example. In this problem it is possible to guess the modes and thence to find the frequencies immediately by an elementary method; this, however, will be left to be done by the reader. The energy expressions are

$$2T = I(\dot{q}_1^2 + 2\dot{q}_2^2 + \dot{q}_3^2),$$
$$2V = k(q_2 - q_1)^2 + k(q_3 - q_2)^2,$$

$$(3.4.22)$$

and these give the receptances

$$\alpha_{11} = \frac{k^2 - 4kI\omega^2 + 2I^2\omega^4}{-2I\omega^2(k - I\omega^2)(2k - I\omega^2)} = \frac{\frac{1}{4}}{-I\omega^2} + \frac{\frac{1}{2}}{k - I\omega^2} + \frac{\frac{1}{4}}{2k - I\omega^2},$$

$$\alpha_{21} = \frac{k^2 - kI\omega^2}{-2I\omega^2(k - I\omega^2)(2k - I\omega^2)} = \frac{\frac{1}{4}}{-I\omega^2} + \frac{-\frac{1}{4}}{2k - I\omega^2},$$

$$\alpha_{31} = \frac{k^2}{-2I\omega^2(k - I\omega^2)(2k - I\omega^2)} = \frac{\frac{1}{4}}{-I\omega^2} + \frac{-\frac{1}{2}}{k - I\omega^2} + \frac{\frac{1}{4}}{2k - I\omega^2}.$$

$$(3.4.23)$$

These expressions differ from those in the previous examples in that they include the $1/\omega^2$ terms. These correspond to the free rotational motion of the shaft which has to be regarded, in this analysis, as a mode of vibration which has zero frequency and for which the modal shape is given by $q_1 : q_2 : q_3 :: 1 : 1 : 1$. The complete set of modes and frequencies is, therefore,

$$\omega_1 = 0 \qquad \text{giving} \quad q_1 : q_2 : q_3 :: 1 : 1 : 1,$$
$$\omega_2 = \sqrt{(k/I)} \qquad \text{giving} \quad q_1 : q_2 : q_3 :: 1 : 0 : -1,$$
$$\omega_3 = \sqrt{(2k/I)} \qquad \text{giving} \quad q_1 : q_2 : q_3 :: 1 : -1 : 1.$$

$$(3.4.24)$$

The determination of modes and frequencies is of great importance in vibration analysis. But the method which is used here is rarely applicable because it becomes unwieldy when the number of degrees of freedom is large; it is included here as an illustration of theory only. We shall show, in a later volume, that there are several different approaches to the calculation which reduce it to one of arithmetic.

We can now leave the study of free vibration and return to discuss the forced motion in terms of equation (3.4.4). Each term in every series of fractions has the form of the direct receptance of a simple spring-mass oscillator, as may be seen by comparison with equation (1.5.12). As the excitation frequency is raised from zero the system will pass successively through its various resonances near each of which one of the fractions of each series becomes predominant. Thus motion in each mode in turn rises to a maximum and then diminishes again. The total motion is made up from the sum of the contribution of the separate modes so that the amplitude-frequency curve for any point is obtained by adding a set of resonance curves.

The shape of the receptance diagrams, which matter was discussed in the previous section, can now be accounted for. In the first place, fig. 3.4.5 (a) shows the $1/\alpha$ curve for a simple system having one degree of freedom while fig. 3.4.5 (b) shows the curve of α for the same system; the first of these diagrams may be compared with fig. 1.5.5 while the second is, effectively, a new version of fig. 1.2.3. Consider now a system having two degrees of freedom. A direct receptance can evidently be expressed in the form

$$\alpha_{rr} = \frac{{}_{1}A_{rr}}{\omega_1^2 - \omega^2} + \frac{{}_{2}A_{rr}}{\omega_2^2 - \omega^2} \tag{3.4.25}$$

and each individual partial fraction can be represented by a curve of the form shown in fig. 3.4.5 (b). The dashed curves in fig. 3.4.6 (a) represent the two fractions so that the full-line curve (which gives their sum) represents the curve of α_{rr}. The

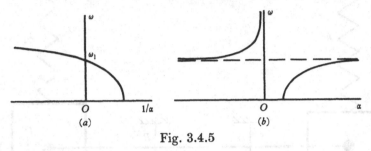

Fig. 3.4.5

reciprocal of this latter curve, giving $1/\alpha_{rr}$ is shown in fig. 3.4.6 (b) and its form will be found to be consistent with the ideas which were developed in the previous section. This idea can clearly be extended to indicate the form of the $1/\alpha_{rr}$ curve for a system with any number of degrees of freedom.

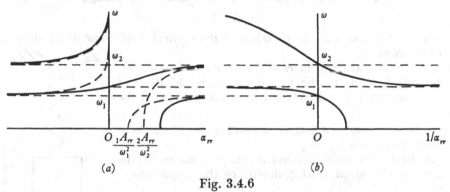

Fig. 3.4.6

It will be seen that none of the constants $_{m}A_{rr}$ can be negative because, if they were, loops would appear in the diagram of the reciprocal of a direct receptance and this has been shown to be impossible. If all the quantities $_{m}A_{rr}$ were negative then there would be no loops in the diagram but the slope of the $1/\alpha_{rr}$ curve would be positive; this has also been shown to be impossible.†

It is not impossible, however, for a cross-receptance curve—or a reciprocal cross-receptance curve—to possess a loop, since the reasoning of the previous section does not then apply. It will be found, for instance, that the signs of the numerators of the partial fractions are not all positive in the above examples of cross-receptances. We shall give later an algebraic treatment from which these rules of sign may be deduced.

† The possibility of some of the quantities $_{m}A_{rr}$ being zero will be discussed in §3.10.

EXAMPLES 3.4

1. The system shown in the figure consists of two masses M_1 and M_2 which are suspended on springs of stiffness k_1 and k_2 respectively. The masses are connected by a light rigid rod which is pinned to each mass at the centre of gravity of that mass. The

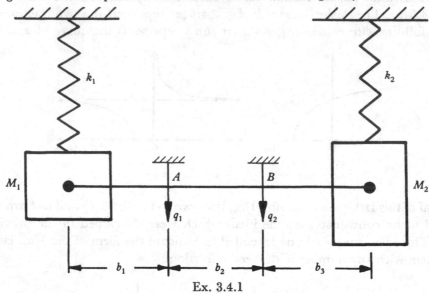

Ex. 3.4.1

co-ordinates of the system are to be taken as the vertical displacements of the points A and B as shown.

What are the natural frequencies of the system and the corresponding modes? Find, by the use of equations (3.2.20) and (3.2.21) or otherwise, the direct and cross-receptances of the system.

2. (a) Obtain the natural frequencies and principal modes of the system shown in the figure.

(b) If, during free oscillation in the first principal mode, the total energy of the system is $M\omega_1^2$, determine the amplitudes of q_1 and q_2.

(c) Find the ratio of the steady forces Q_1 and Q_2 which must be applied at q_1 and q_2 in order to produce *static* deflexions in the principal modes.

3. The figure shows a system which consists of two thin uniform beams of mass M which are hinged together and are supported in a horizontal position by vertical springs of which the stiffnesses are as shown. The co-ordinates are to be taken as the vertical downward displacements of the ends of the beams. Find, by inspection or otherwise, the natural frequencies and the principal modes.

Ex. 3.4.2

4. Show that, in the notation of this chapter, the ratios between the amplitudes $\Psi'_1, \Psi'_2, \ldots, \Psi'_n$ which determine the rth principal mode are given by

$$\frac{\Psi'_1}{\alpha_1} = \frac{\Psi'_2}{\alpha_2} = \ldots = \frac{\Psi'_n}{\alpha_n},$$

where $\alpha_1, \alpha_2, \ldots, \alpha_n$ are the cofactors of the elements in any row of Δ when ω^2 is given the value ω_r^2.

Verify this result using one of the systems discussed in this section.

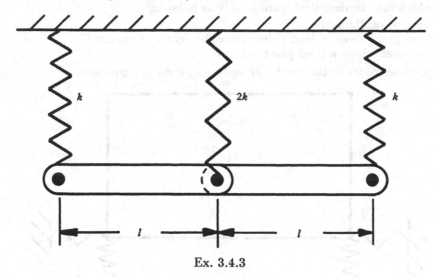

Ex. 3.4.3

5. (a) Show that K in equation (3.4.2) is given by

$$K = |a| = \frac{|c|}{\omega_1^2 \omega_2^2 \ldots \omega_n^2},$$

where $|a|$ and $|c|$ are the determinants of the inertia coefficients a_{rs} and stability coefficients c_{rs} respectively.

(b) Show, further, that in the notation of this section,

$$\Delta_{rr} = |a_{rr}| \, (\Omega_1^2 - \omega^2) \, (\Omega_2^2 - \omega^2) \ldots (\Omega_{n-1}^2 - \omega^2)$$

and hence that
$$|a_{rr}| = \frac{|c_{rr}|}{\Omega_1^2 \Omega_2^2 \ldots \Omega_{n-1}^2},$$

where $|a_{rr}|$ and $|c_{rr}|$ are the determinants of the coefficients a and c with the rth columns and rows removed and the Ω's are the anti-resonance frequencies at q_r.

(c) As the driving frequency ω is diminished to zero a receptance α_{rs} becomes a static *flexibility* f_{rs}. Noting that the direct flexibility at q_r is given by

$$f_{rr} = \frac{|c_{rr}|}{|c|}$$

show that†
$$\alpha_{rr} = f_{rr} \frac{\left(1 - \dfrac{\omega^2}{\Omega_1^2}\right)\left(1 - \dfrac{\omega^2}{\Omega_2^2}\right) \ldots \left(1 - \dfrac{\omega^2}{\Omega_{n-1}^2}\right)}{\left(1 - \dfrac{\omega^2}{\omega_1^2}\right)\left(1 - \dfrac{\omega^2}{\omega_2^2}\right) \ldots \left(1 - \dfrac{\omega^2}{\omega_n^2}\right)}.$$

6. Show that the results of the last example may be applied to one of the simple systems used in the text.

† See W. J. Duncan, *Mechanical Admittances and their Applications to Oscillation Problems, R. and M. 2000,* (1947), p. 31.

7. The figure shows a machine that is fixed to the ground through 'vibration-absorber' springs. Considering only motion in the plane of the diagram and taking reasonable values for the various parameters, show that the machine has three principal modes† which may be described qualitatively as follows:

(a) vertical oscillation;
(b) 'swinging' about a fixed point above the centre of gravity G;
(c) 'rocking' about a fixed point below G.

Arrange these modes in their order of increasing natural frequency.

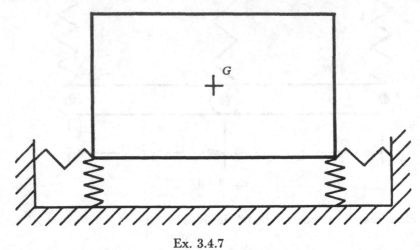

Ex. 3.4.7

3.5 Principal co-ordinates

It was pointed out in § 2.1 that it is possible to define generalized co-ordinates as algebraic combinations of variables which are physically recognizable (such as displacements or rotations) insteady of using the variables as generalized co-ordinates directly. Now a system with n degrees of freedom has n principal modes of vibration, and co-ordinates of the above type may be chosen such that each one corresponds to a distortion of the system in one of these modes. Thus the co-ordinates become parameters that determine the magnitude of distortion which the system has in each principal mode. Such co-ordinates are called *principal co-ordinates*. There will evidently be n of them and we shall denote them by $p_1, p_2, ..., p_n$.

Before discussing the properties of these new co-ordinates, it is necessary to define them precisely and thereby explain the meaning of unit value of any one of them. This involves a choice of scale factor as may be explained by means of an example.

Consider the system of fig. 3.4.4 for which the generalized co-ordinates q_1, q_2 and q_3 have simple physical meanings. The third principal mode is defined by the ratios

$$q_1 : q_2 : q_3 :: +1 : -1 : +1. \tag{3.5.1}$$

---

† It is assumed here that the machine is rigid and that the springs are light so that the whole system has but three degrees of freedom. As we shall show in later chapters, this idealization depends for its validity upon the range of frequency contemplated; we are here concerned with the lowest natural frequencies.

Let unit value of the third principal co-ordinate p_3 be chosen to make $q_2 = +2$, say. The three co-ordinates q are then given by

$$q_1 = -2, \quad q_2 = +2, \quad q_3 = -2 \tag{3.5.2}$$

when $p_3 = 1$. For any other value of p_3, the displacements at the original co-ordinates q will be

$$q_1 = -2p_3, \quad q_2 = +2p_3, \quad q_3 = -2p_3, \tag{3.5.3}$$

provided that the other two principal co-ordinates are zero.

Alternatively, we may choose scale factors such that unit values of all the principal co-ordinates make $q_1 = 1$ in the relevant modes. Then if distortion occurs in the first principal mode only,

$$q_1 = p_1, \quad q_2 = p_1, \quad q_3 = p_1. \tag{3.5.4}$$

If it occurs in the second mode only,

$$q_1 = p_2, \quad q_2 = 0, \quad q_3 = -p_2, \tag{3.5.5}$$

and for the third mode only,

$$q_1 = p_3, \quad q_2 = -p_3, \quad q_3 = p_3. \tag{3.5.6}$$

If a distortion of this system involves all the modes simultaneously then, by super-position,

$$\left. \begin{aligned} q_1 &= p_1 + p_2 + p_3, \\ q_2 &= p_1 + 0 - p_3, \\ q_3 &= p_1 - p_2 + p_3. \end{aligned} \right\} \tag{3.5.7}$$

These equations, which give the original co-ordinates q of the system as linear functions of the principal co-ordinates p, can be transformed by the rules of determinants to give

$$\left. \begin{aligned} p_1 &= \frac{q_1}{4} + \frac{q_2}{2} + \frac{q_3}{4}, \\ p_2 &= \frac{q_1}{2} + 0 - \frac{q_3}{2}, \\ p_3 &= \frac{q_1}{4} - \frac{q_2}{2} + \frac{q_3}{4}. \end{aligned} \right\} \tag{3.5.8}$$

The principal co-ordinates are thus expressed as functions of the q's. This shows that the p's, which were originally regarded as defining the magnitudes of the distortions in the principal modes, can be regarded mathematically merely as a set of special linear functions of the co-ordinates q. Further, any arbitrary displacement of the system, which corresponds to a given set of values of q_1, q_2 and q_3, can equally well be expressed as a set of values of p_1, p_2 and p_3; these values would be given uniquely by equations (3.5.8).

These results can evidently be expressed in more general terms. The expressions for the generalized co-ordinates q of any system can evidently be written in a form comparable with those of equations (3.5.7); thus

$$\left.\begin{aligned}
q_1 &= A_1 p_1 + B_1 p_2 + \dots + N_1 p_n, \\
q_2 &= A_2 p_1 + B_2 p_2 + \dots + N_2 p_n, \\
&\dots\dots\dots\dots\dots\dots\dots\dots\dots\dots \\
q_n &= A_n p_1 + B_n p_2 + \dots + N_n p_n.
\end{aligned}\right\} \qquad (3.5.9)$$

Here, the constants $A_1, A_2, \dots, A_n$ will be in the ratio of the values of $q_1, q_2, \dots, q_n$ in the first principal mode, while their absolute magnitudes will be fixed by the choice of unit value for p_1. Similar considerations govern the constants $B, C, \dots, N$. These equations may now be transformed to give the principal co-ordinates p as functions of the co-ordinates q; thus

$$\left.\begin{aligned}
p_1 &= A_1' q_1 + B_1' q_2 + \dots + N_1' q_n, \\
p_2 &= A_2' q_1 + B_2' q_2 + \dots + N_2' q_n, \\
&\dots\dots\dots\dots\dots\dots\dots\dots\dots\dots \\
p_n &= A_n' q_1 + B_n' q_2 + \dots + N_n' q_n.
\end{aligned}\right\} \qquad (3.5.10)$$

The constant coefficients $A_1', B_1', \dots,$ etc., in these equations are obtained from the coefficients $A_1, B_1, \dots,$ etc., by relations of the form

$$p_1 = \frac{\begin{vmatrix}
q_1 & B_1 & C_1 & \dots & N_1 \\
q_2 & B_2 & C_2 & \dots & N_2 \\
\multicolumn{5}{c}{\dots\dots\dots\dots\dots\dots} \\
q_n & B_n & C_n & \dots & N_n
\end{vmatrix}}{\begin{vmatrix}
A_1 & B_1 & . & \dots & N_1 \\
A_2 & B_2 & . & \dots & N_2 \\
\multicolumn{5}{c}{\dots\dots\dots\dots\dots\dots} \\
A_n & B_n & . & \dots & N_n
\end{vmatrix}}. \qquad (3.5.11)$$

These relations enable all the p's to be found if the values of the q's are known. We shall show later, in § 3.8, how the coefficients $A_1', B_1', \dots,$ etc., may be obtained from the coefficients $A_1, B_1, \dots,$ etc., without the necessity of solving the nth order determinants of equation (3.5.11).

It has been assumed that it will always be possible to transform the equations for the q's into a set for the p's. This is justifiable because the determinant of the coefficients of the p's in equations of the form (3.5.9) cannot vanish. If it were to do so then one line of the coefficients would be expressible as a linear function of the other lines and this would imply that the q's were not independent; this would be opposed to our original assumptions.

It may be helpful at this point if we pause to consider the physical meaning of the equations which relate the two sets of co-ordinates; in doing so we shall assume that the q's represent displacements at generalized co-ordinates which are directly measurable. The equations (3.5.9) mean that the displacement of any point of a distorted system can be regarded as being made up from a number of component displacements each one of which is due to distortion in a separate principal mode. Equations (3.5.10), on the other hand, mean that the distortion in the principal modes can be made up from components, each one of which is due to the (generalized) displacement of a separate point of the system. This idea, which is not so evident physically as the previous one, may be thought of by reference to a particular case of the previous one. For consider a distortion in which one generalized co-ordinate q of the system is given a finite value while all the other generalized co-ordinates remain zero; as may be seen from equations (3.5.9), this distortion can be expressed as the sum of distortions in the set of principal modes. It follows that in any distortion, the total displacement in any one principal mode will be made up of components, each one of which arises from a different particular distortion of the type mentioned, in which one of the co-ordinates q only is displaced.

It was shown in the previous section that the variation with time of the displacements in the principal modes are of a different character according to whether the motion is free or forced. If the vibration is free, then the displacement at each of the principal co-ordinates varies sinusoidally at the appropriate natural frequency; in forced motion, on the other hand, all the principal co-ordinates vary at the same frequency.

As an example of free motion, consider the system of fig. 3.4.4. If the principal co-ordinates which are defined in equations (3.5.4), (3.5.5) and (3.5.6) are used, and the amplitudes of the p's are the complex constants Π_1, Π_2 and Π_3 respectively, then

$$p_1 = \Pi_1 e^{i\omega_1 t}, \quad p_2 = \Pi_2 e^{i\omega_2 t}, \quad p_3 = \Pi_3 e^{i\omega_3 t}. \tag{3.5.12}$$

The displacements at the co-ordinates q can therefore be written as functions of time; thus, in accordance with equations (3.5.7),

$$\begin{aligned}
q_1 &= \Pi_1 e^{i\omega_1 t} + \Pi_2 e^{i\omega_2 t} + \Pi_3 e^{i\omega_3 t}, \\
q_2 &= \Pi_1 e^{i\omega_1 t} + \quad 0 \quad - \Pi_3 e^{i\omega_3 t}, \\
q_3 &= \Pi_1 e^{i\omega_1 t} - \Pi_2 e^{i\omega_2 t} + \Pi_3 e^{i\omega_3 t}.
\end{aligned} \tag{3.5.13}$$

The A, B, C coefficients in this example, as they are defined by equations (3.5.9), are

$$A_1 = A_2 = A_3 = 1, \quad B_1 = -B_3 = 1, \quad B_2 = 0, \quad C_1 = -C_2 = C_3 = 1. \tag{3.5.14}$$

Now consider the general problem of forced harmonic motion of a system when the force is applied at a single generalized co-ordinate q_s. The excitation is

$$Q_s = \Phi_s e^{i\omega t} \tag{3.5.15}$$

so that the displacements at the co-ordinates q are given by equations of the form

$$
\left.
\begin{aligned}
q_1 &= \frac{{}_1A_{1s}}{(\omega_1^2 - \omega^2)} \cdot \Phi_s e^{i\omega t} + \frac{{}_2A_{1s}}{(\omega_2^2 - \omega^2)} \cdot \Phi_s e^{i\omega t} + \dots + \frac{{}_nA_{1s}}{(\omega_n^2 - \omega^2)} \cdot \Phi_s e^{i\omega t}, \\
q_2 &= \frac{{}_1A_{2s}}{(\omega_1^2 - \omega^2)} \cdot \Phi_s e^{i\omega t} + \frac{{}_2A_{2s}}{(\omega_2^2 - \omega^2)} \cdot \Phi_s e^{i\omega t} + \dots + \frac{{}_nA_{2s}}{(\omega_n^2 - \omega^2)} \cdot \Phi_s e^{i\omega t}, \\
&\cdots\cdots\cdots\cdots\cdots\cdots\cdots\cdots\cdots\cdots\cdots\cdots\cdots\cdots \\
q_n &= \frac{{}_1A_{ns}}{(\omega_1^2 - \omega^2)} \cdot \Phi_s e^{i\omega t} + \frac{{}_2A_{ns}}{(\omega_2^2 - \omega^2)} \cdot \Phi_s e^{i\omega t} + \dots + \frac{{}_nA_{ns}}{(\omega_n^2 - \omega^2)} \cdot \Phi_s e^{i\omega t}.
\end{aligned}
\right\} \quad (3.5.16)
$$

The coefficients ${}_1A_{1s}, {}_1A_{2s}, \dots, {}_1A_{ns}$ determine the first mode and so on, independently of the value of s, thus showing that equations (3.5.16) are of the same form as (3.5.9). All the principal co-ordinates, however, now vary with the same frequency ω and they have amplitudes which are determined by the constants of the system and by the frequency and amplitude of the exciting force.

In general, systems will be acted on by more than one generalized force and it is left for the reader to show that the presence of two or more forces does not change the form of the equations (3.5.16).†

An illustrative example of this theory for forced vibration is furnished by the system of fig. 3.4.1 for which it has been shown that excitation by a force $Q_2 = \Phi_2 e^{i\omega t}$ will give

$$
\left.
\begin{aligned}
q_1 &= \left(\frac{\frac{2}{3}}{k - 2M\omega^2} \right) \Phi_2 e^{i\omega t} + \left(\frac{-\frac{1}{3}}{2k - M\omega^2} \right) \Phi_2 e^{i\omega t}, \\
q_2 &= \left(\frac{\frac{4}{3}}{k - 2M\omega^2} \right) \Phi_2 e^{i\omega t} + \left(\frac{\frac{1}{3}}{2k - M\omega^2} \right) \Phi_2 e^{i\omega t}.
\end{aligned}
\right\} \quad (3.5.17)
$$

If the principal co-ordinates are now selected so that, say,

    (i) $p_1 = 1$ when $q_1 = 2$ in the first mode, and
    (ii) $p_2 = 1$ when $q_1 = 4$ in the second mode,

then the two sets of co-ordinates are related by the equations

$$
\left.
\begin{aligned}
q_1 &= 2p_1 + 4p_2, \\
q_2 &= 4p_1 - 4p_2,
\end{aligned}
\right\} \quad (3.5.18)
$$

and by comparing these expressions with those of equations (3.5.17) it is seen that

$$
p_1 = \frac{1}{2} \left(\frac{\frac{2}{3}}{k - 2M\omega^2} \right) \Phi_2 e^{i\omega t}, \quad p_2 = \frac{1}{4} \left(\frac{-\frac{1}{3}}{2k - M\omega^2} \right) \Phi_2 e^{i\omega t}, \quad (3.5.19)
$$

while, in equations (3.5.9),

$$
A_1 = 2, \quad A_2 = 4, \quad B_1 = 4, \quad B_2 = -4. \quad (3.5.20)
$$

---

† It may be noted that, by a suitable choice of the generalized co-ordinates, only one generalized force need be produced by a specified applied force. This idea can be used in a simple proof of the equality of the sets of ratios ${}_1A_{1s} : {}_1A_{2s} : \dots : {}_1A_{ns}$ and ${}_1A_{1t} : {}_1A_{2t} : \dots : {}_1A_{nt}$.

EXAMPLES 3.5

1. Both principal co-ordinates of the system shown are to be taken to have unit value when $q_1 = 1$ in the appropriate principal mode. Determine the co-ordinates q_1 and q_2 as functions of time when the system performs free oscillations in which the principal co-ordinates p_1 and p_2 have amplitudes of 4 and 2 respectively.

2. Find the variations of p_1 and p_2 for the system of Ex. 3.5.1 if it performs steady oscillation under the action of a force $Q_2 = \Phi_2 e^{i\omega t}$.

3. Devise simple mathematical systems having (a) two-, (b) three-, (c) n-degrees of freedom for which the principal co-ordinates are directly measurable quantities.

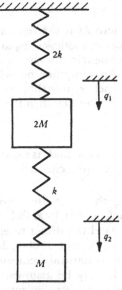

Ex. 3.5.1

4. Replace the co-ordinates q_1, q_2, q_3 used previously for the system of fig. 3.4.4. by q_1, q_1+q_2, q_1+q_3 respectively so that q_1 represents a rigid-body rotation of the complete system in its lowest principal mode, q_1 being the appropriate principal co-ordinate. Show that the direct receptance at the left-hand disk is given by

$$\alpha_{11} = -\frac{\left(1-\dfrac{\omega^2}{\Omega_1^2}\right)\left(1-\dfrac{\omega^2}{\Omega_2^2}\right)}{4I\omega^2\left(1-\dfrac{\omega^2}{\omega_2^2}\right)\left(1-\dfrac{\omega^2}{\omega_3^2}\right)},$$

where Ω_1 and Ω_2 are the anti-resonance frequencies measured at q_1 and ω_2 and ω_3 are the non-zero natural frequencies of equations (3.4.24).

(Note that the multiplier of ω^2 in the denominator of α_{11} is $4I$, the inertia coefficient a_{11}.)

5. Generalize the result of the previous example. That is to say consider a system such that q_1 can be chosen so that displacements at that co-ordinate do not affect the potential energy, thus giving

$$c_{11} = c_{12} = \ldots = c_{1n} = 0.$$

Following the steps indicated in Ex. 3.4.5, show that one natural frequency, ω_1 say, is zero and that, if $\Omega_1, \Omega_2, \ldots, \Omega_{n-1}$ are the anti-resonance frequencies measured at q_1, then

$$\alpha_{11} = -\frac{\left(1-\dfrac{\omega^2}{\Omega_1^2}\right)\left(1-\dfrac{\omega^2}{\Omega_2^2}\right)\ldots\left(1-\dfrac{\omega^2}{\Omega_{n-1}^2}\right)}{a_{11}\omega^2\left(1-\dfrac{\omega^2}{\omega_2^2}\right)\left(1-\dfrac{\omega^2}{\omega_3^2}\right)\ldots\left(1-\dfrac{\omega^2}{\omega_n^2}\right)}.$$

[This result, which has useful practical applications, is due to Biot† who considers aircraft propeller vibration. The co-ordinate q_1 is then the angle of rotation and a_{11} is the moment of inertia of the propeller (assumed rigid) about its axis; ω_2, ω_3, ... are the resonant frequencies for a harmonic exciting couple applied at the boss and Ω_1, Ω_2, ... are the anti-resonance frequencies at that point (or the resonant frequencies with the boss clamped). It is permissible to apply a result like this (for a system with n degrees of freedom) to a propeller which has infinite freedom if sufficient factors are retained in the numerator and denominator for the final factors to be very nearly equal to unity. The formula is useful because the parameters involved are all fairly easy to measure.]

† M. A. Biot, *Journ. Aero. Sci.* vol. 7 (July, 1940), p. 376. A proof of this formula can also be found in the work of Duncan, *R. and M. 2000* (1947), p. 32.

6. Show that, if a system has one zero frequency, the expression for a direct receptance can be written in the form

$$\alpha = \frac{\left(1 - \frac{\omega^2}{\Omega_1^2}\right)\left(1 - \frac{\omega^2}{\Omega_2^2}\right)\cdots}{-M\omega^2\left(1 - \frac{\omega^2}{\omega_1^2}\right)\left(1 - \frac{\omega^2}{\omega_2^2}\right)\cdots},$$

where M is the relevant effective mass of the system when it is moving in the zero-frequency mode, ω_1, ω_2 are the other natural frequencies, and Ω_1, Ω_2 are the anti-resonance frequencies for the co-ordinate in question.

An aeroplane propeller is mounted on a shaft which is clamped rigidly so close to the propeller hub that the latter may be considered fixed. The natural frequencies of the propeller are then found to be such that

$$\Omega_1^2 = 25, \quad \Omega_2^2 = 150, \quad \Omega_3^2 = 600.$$

When the shaft is mounted in bearings so that it is free to rotate but not to move endwise the frequencies are

$$\omega^2 = 60, \quad \omega_2^2 = 400,$$

together with the zero frequency. The moment of inertia of the propeller about the shaft axis is 1000 lb.ft.^{2}.

Find the direct receptance for angular motion at the hub. If the propeller shaft has a stiffness of 25000 pdl.ft./rad. and if the other end of it is clamped what will be the lowest natural frequency of the system?

It may be assumed that motion in any modes higher than those whose frequencies are given is sufficiently small to be neglected.   (Leeds B.Sc. Mech. Eng. 1957)

3.6 The principle of orthogonality; normalized principal co-ordinates

In order to appreciate the importance of principal co-ordinates it is necessary to be familiar with the property of *orthogonality*.†

Consider a system which is oscillating freely in two of its principal modes simultaneously so that two principal co-ordinates, p_l and p_m say, are varying harmonically at their corresponding natural frequencies while the other principal co-ordinates are all zero. The co-ordinates $q_1, q_2, \ldots, q_n$ will be related to p_l and p_m so that the energy expressions T and V, which have previously been quoted as functions of the q's, can also be expressed as functions of the p's by substituting equations of the form (3.5.9) into the energy expressions (3.1.2) and (3.1.7). These functions will also be of the second degree and for the particular system under consideration they may be written in the form

$$\left.\begin{aligned} 2T &= a_l\, p_l^2 + a_m\, p_m^2 + 2a_{(lm)}\, p_l\, p_m, \\ 2V &= c_l\, p_l^2 + c_m\, p_m^2 + 2c_{(lm)}\, p_l\, p_m. \end{aligned}\right\} \tag{3.6.1}$$

The coefficients a_l, etc., and c_l, etc., will depend upon the coefficients a and c of the original expressions for T and V and also upon the choice of the meaning of unit value for the two principal co-ordinates p_l and p_m.

---

† This name is derived from the idea of displacement vectors in a $2n$-dimensional space, a concept which can be used to discuss the analysis of n-degree-of-freedom systems; we shall not use this concept here.

Now during the free vibration the co-ordinates p_l and p_m will be given by expressions of the form

$$p_l = \Pi_l \sin (\omega_l t + \psi_l), \quad p_m = \Pi_m \sin (\omega_m t + \psi_m), \tag{3.6.2}$$

where the constants Π_l, Π_m, ψ_l, ψ_m are now real since we shall temporarily drop the complex exponential notation. If these are substituted into the energy expressions, then twice the total energy of the system, at any time t, is found to be

$$2(T + V) = \Pi_l^2[a_l \omega_l^2 \cos^2 (\omega_l t + \psi_l) + c_l \sin^2 (\omega_l t + \psi_l)]$$
$$+ \Pi_m^2[a_m \omega_m^2 \cos^2 (\omega_m t + \psi_m) + c_m \sin^2 (\omega_m t + \psi_m)]$$
$$+ 2\Pi_l \Pi_m[a_{(lm)} \omega_l \omega_m \cos (\omega_l t + \psi_l) \cos (\omega_m t + \psi_m)$$
$$+ c_{(lm)} \sin (\omega_l t + \psi_l) \sin (\omega_m t + \psi_m)]. \tag{3.6.3}$$

Now the right-hand side of this equation must be independent of t because there is no force acting on the system and no means by which energy may be dissipated or created within it. For this to be so, it is necessary that

$$a_l \omega_l^2 = c_l, \quad a_m \omega_m^2 = c_m \tag{3.6.4}$$

and

$$a_{(lm)} = c_{(lm)} = 0. \tag{3.6.5}$$

The first pair of equations demonstrates a property to which we have already alluded; namely that, if free motion occurs in one mode only, then the natural frequency can be found provided that the modal shape is known. We leave it to the reader to show that equation (3.6.4) is an alternative form of equation (3.4.9).

Equation (3.6.5) demonstrates the property of orthogonality. It may be stated in words in the following form. The energies T and V of a conservative system, when it is distorted in two or more principal modes simultaneously, are equal to the sums of the energies that the system would have if it were distorted in each of the modes separately; the modes are said to be *orthogonal* to one another. Although we have proved this result by considering free motion, the vanishing of the coefficients $a_{(lm)}$ and $c_{(lm)}$ shows that the principle still holds if the motion is forced at any frequency and in particular that it is true for a static distortion.

Consider as an example the system of fig. 3.4.1. If principal co-ordinates p_1 and p_2 are chosen so that $p_1 = 1$ when $q_1 = 1$ in the first mode and $p_2 = 1$ when $q_1 = 1$ in the second mode, then

$$\left. \begin{aligned} q_1 &= p_1 + p_2, \\ q_2 &= 2p_1 - p_2, \end{aligned} \right\} \tag{3.6.6}$$

because the modes are those of equation (3.4.16). If these values are now substituted into the energy expressions (3.4.10), it is found that

$$\left. \begin{aligned} 2T &= 2M(p_1 + p_2)^2 + M(2p_1 - p_2)^2 = 6Mp_1^2 + 3Mp_2^2, \\ 2V &= 3k(p_1 + p_2)^2 + k(2p_1 - p_2)^2 - 2k(p_1 + p_2)(2p_1 - p_2) = 3kp_1^2 + 6kp_2^2, \end{aligned} \right\} \tag{3.6.7}$$

from which it will be seen that

$$a_1 = 6M, \quad a_2 = 3M, \quad c_1 = 3k, \quad c_2 = 6k, \tag{3.6.8}$$

while $a_{(12)}$ and $c_{(12)}$ are zero as predicted.

In general, since the product terms are all zero when the energies are expressed in terms of the principal co-ordinates, the quantities T and V are expressible in the form

$$2T = a_1 p_1^2 + a_2 p_2^2 + \ldots + a_n p_n^2,$$
$$2V = c_1 p_1^2 + c_2 p_2^2 + \ldots + c_n p_n^2.$$

$$(3.6.9)$$

The single suffix will be retained instead of the double one (which was used for other types of co-ordinate) in order to distinguish this special set of coefficients; the vanishing of the product terms allows this to be done without danger of confusion.

The expressions (3.6.9) are used by some writers as a means of *defining* the principal co-ordinates; that is to say, they define them as a set of linear functions of the q's which transform the energy expressions into the sums of squares. It may be shown, by the following argument, that such a definition will lead to the same functions as we have discussed. Let a set of variables $p_1, p_2, \ldots, p_n$ be defined by the relations

$$q_1 = A_1^* p_1 + B_1^* p_2 + \ldots + N_1^* p_n,$$
$$q_2 = A_2^* p_1 + B_2^* p_2 + \ldots + N_2^* p_n,$$
$$\ldots\ldots\ldots\ldots\ldots\ldots\ldots\ldots\ldots\ldots\ldots$$
$$q_n = A_n^* p_1 + B_n^* p_2 + \ldots + N_n^* p_n.$$

$$(3.6.10)$$

Now the first column of quantities on the right-hand side determines a distortion of the system whose *shape* is defined by the set of ratios

$$q_1 : q_2 : \ldots : q_n :: A_1^* : A_2^* : \ldots : A_n^* \qquad (3.6.11)$$

and whose *intensity* is fixed by the parameter p_1. A meaning may be chosen for unit value of p_1 so that the value of one of the coefficients is fixed; for instance, it may be decided to make $q_1 = 1$ when $p_1 = 1$ and $p_2 = p_3 = \ldots = p_n = 0$ so that $A_1^* = 1$. Similarly the values of $B_1^*, C_1^*, \ldots, N_1^*$ may be selected. During the process of choosing unit magnitudes for the p's in this way, the values of n out of the original n^2 quantities (which are the starred coefficients) are fixed thus leaving $(n^2 - n)$ quantities to be fixed. The expressions (3.6.10) can now be substituted into the energy functions (3.1.2) and (3.1.7) and all the product terms containing $p_1 p_2$, $p_1 p_3$, ..., etc., and $p_1 p_2$, $p_1 p_3$, ..., etc., put equal to zero. This gives $n^2 - n$ equations because there are $\frac{1}{2}n(n-1)$ from the T expression and the same number from V. The coefficients in the equations (3.6.10) are thus determined uniquely. A set of co-ordinates is therefore obtained which must be identical with the set of principal co-ordinates which is found by the previous method provided that the choices of unit amplitude are the same for the two sets.

The coefficients $a_1, a_2, \ldots, a_n$ and $c_1, c_2, \ldots, c_n$ must all be positive if the system is stable; for otherwise it would be possible to select distortions for which the energy of the system would be negative.

The remainder of this section is devoted to a commonly-used process for choosing principal co-ordinates in a particular way. As this process will not be adopted in the remainder of this book (for a reason which is explained), the reader may wish to omit this matter during a first reading.

It has been pointed out that, when defining the principal co-ordinates, it is necessary to choose some arbitrary magnitude of distortion to be represented by unit value of each co-ordinate. In the previous examples, this magnitude has been chosen by reference to one of the original generalized co-ordinates; thus we have set $p_r = 1$, say, when some generalized displacement q_s is unity. There is another way of making the choice which does not involve reference to a particular generalized co-ordinate. The method is called 'normalization' and is used to bring the equations into a neater form.

Let a system be oscillating in its rth principal mode so that its kinetic and potential energies are given by

$$2T = a_r p_r^2, \quad 2V = c_r p_r^2, \tag{3.6.12}$$

where a_r and c_r depend not only upon the system but also upon the meaning which is to be given to unit value of p_r. Since these quantities are not yet fixed, it is permissible to make $a_r = 1$ arbitrarily. This will fix the meaning of unit value of p_r because the values of the q's which correspond to it must be such as to make the potential energy V equal to $\frac{1}{2}\omega_r^2$ when the system is distorted in its rth mode only, on account of equation (3.6.4). If this is done for all the principal co-ordinates, then the energy expressions assume the form which corresponds to *normalized* principal co-ordinates, namely

$$\begin{aligned} 2T &= p_1^2 + p_2^2 + \ldots + p_n^2, \\ 2V &= \omega_1^2 p_1^2 + \omega_2^2 p_2^2 + \ldots + \omega_n^2 p_n^2. \end{aligned} \tag{3.6.13}$$

The system of fig. 3.4.4. may be used to illustrate this. The principal modes of this system are known so that equations of the form (3.5.9) may be written down for it; they must be of the form

$$\begin{aligned} q_1 &= K_1 p_1 + K_2 p_2 + K_3 p_3, \\ q_2 &= K_1 p_1 + \quad 0 \quad - K_3 p_3, \\ q_3 &= K_1 p_1 - K_2 p_2 + K_3 p_3, \end{aligned} \tag{3.6.14}$$

where K_1, K_2 and K_3 are constants whose values will be found for the case where all the p's are normalized. The kinetic energies in the three modes are

$$\begin{aligned} \tfrac{1}{2}[IK_1^2 p_1^2 + 2IK_1^2 p_1^2 + IK_1^2 p_1^2] &= \tfrac{4}{2}IK_1^2 p_1^2, \\ \tfrac{1}{2}[IK_2^2 p_2^2 + \quad 0 \quad + IK_2^2 p_2^2] &= \tfrac{2}{2}IK_2^2 p_2^2, \\ \tfrac{1}{2}[IK_3^2 p_3^2 + 2IK_3^2 p_3^2 + IK_3^2 p_3^2] &= \tfrac{4}{2}IK_3^2 p_3^2. \end{aligned} \tag{3.6.15}$$

If the principal co-ordinates are to be normalized therefore, K_1, K_2 and K_3 must be given by

$$\tfrac{1}{2}p_1^2 = \tfrac{4}{2}IK_1^2 p_1^2, \quad \tfrac{1}{2}p_2^2 = \tfrac{2}{2}IK_2^2 p_2^2, \quad \tfrac{1}{2}p_3^2 = \tfrac{4}{2}IK_3^2 p_3^2, \tag{3.6.16}$$

so that

$$K_1 = \frac{1}{2\sqrt{I}}, \quad K_2 = \frac{1}{\sqrt{(2I)}}, \quad K_3 = \frac{1}{2\sqrt{I}}. \tag{3.6.17}$$

Thus none of the constants K of equations (3.6.14) has the value unity as has been the case before.

The process of normalization is convenient in the analytical theory of dynamics, but it will not be used here very much. Our reason for not employing it is that, whereas the unit values of the energy coefficients a simplify the algebraic form of the equations, they also conceal physical dimensions. Thus although all the a's become equal to unity it is necessary to remember that they still have the dimensions of inertia; also, while the c's take the numerical values of the quantities $\omega_1^2, \omega_2^2, \ldots, \omega_n^2$, they nevertheless retain the dimensions of stiffnesses and do not possess the dimensions of (frequency)^2. The reader may well be aware of other cases in physical theory where the choice of unity for the value of a physical constant has led to subsequent confusion in the physical interpretation of the equations.

EXAMPLES 3.6

1. Write down the co-ordinates q of the system of fig. 3.4.4 in terms of a set of principal co-ordinates. Substitute the expressions in the energy functions and hence show that the latter become sums of squares.

Show that the orthogonality of the first principal mode to the other two arises because the system has no net angular momentum during motion in the second and third modes.

2. Verify that, if the values (3.6.17) for the constants K_1, K_2, K_3 are used in equations (3.6.14) and the potential energy expression for the system of fig. 3.4.4 is then written in terms of the p's, then the coefficients of stability are given by

$$c_1 = \omega_1^2, \quad c_2 = \omega_2^2, \quad c_3 = \omega_3^2.$$

3. A system consists of three rigid disks which are fixed to a light elastic shaft, the whole being supported in frictionless bearings. The angular displacements of the disks are denoted by q_1, q_2, q_3 and the values of these quantities which correspond to unit distortion in the three principal modes are, respectively,

$$1, \; {}_1q_2, \; {}_1q_3 \quad \text{in mode 1,}$$
$$1, \; {}_2q_2, \; {}_2q_3 \quad \text{in mode 2,}$$
$$1, \; {}_3q_2, \; {}_3q_3 \quad \text{in mode 3.}$$

Show that

(a) ${}_1q_2 = {}_1q_3 = 1$;

(b) either ${}_2q_2$ or ${}_2q_3$ or both these quantities are negative and that the same applies to the displacements in the third mode;

(c) the arrangement of the signs of the displacements in the second mode cannot be the same as it is for the third mode.

4. A system consists of a single particle which is supported by light elastic members, motion in three dimensions being possible. Show that the orthogonality of the principal modes requires that the displacements in the modes shall be mutually perpendicular. Deduce that for any elastic system, at any point, there are three mutually perpendicular directions which are such that, if a force is applied at that point along any one of them then the deflexion of the point is also in that direction.

5. A rigid flat plate is supported on springs in such a way that it can move in the direction perpendicular to its plane and can also rotate about any axis in its plane. Show that each of its principal modes of vibration consists of an angular motion about some line in the plate and hence deduce that there will be three points on the plate such that, if a normal pulsating force acts at any one of them, it will produce motion in one principal mode only.

6. A rigid body is supported on a light elastic structure. Show that its six principal modes consist of screw motions in each of which the body rotates about an axis and simultaneously moves along the axis.

Show that it will be possible to select six wrenches, each one of which produces displacement in one principal mode only. (A wrench is a force system composed of a single force together with a couple which acts in a plane perpendicular to the force.)

3.7 The receptances at the principal co-ordinates

Lagrange's equations may be used with principal co-ordinates in just the same way that they can be used with any other generalized co-ordinates. The vibration equations (3.1.12) can therefore be written in the form

$$\frac{d}{dt}\left(\frac{\partial T}{\partial \dot{p}_r}\right) + \frac{\partial V}{\partial p_r} = P_r \quad (r = 1, 2, ..., n), \tag{3.7.1}$$

where the expressions for T and V will be functions of the co-ordinates p and where the quantities $P_1, P_2, ..., P_n$ are the generalized forces which correspond to the principal co-ordinates. Since the p's and the P's are merely special types of the q's and the Q's respectively, they obey the same rules. Thus, if a force (or couple) $F e^{i\omega t}$ acts at some point of the system, the deflexion (or rotation) of the point in the corresponding direction being y, then

$$P_1 = \left(F\frac{\partial y}{\partial p_1}\right)e^{i\omega t}, \quad P_2 = \left(F\frac{\partial y}{\partial p_2}\right)e^{i\omega t}, \quad ..., \quad P_n = \left(F\frac{\partial y}{\partial p_n}\right)e^{i\omega t}. \tag{3.7.2}$$

Now the energy functions T and V have the form given by equation (3.6.9) and Lagrange's equations therefore become

$$\left.\begin{array}{l} a_1\ddot{p}_1 + c_1 p_1 = P_1, \\ a_2\ddot{p}_2 + c_2 p_2 = P_2, \\ \cdots\cdots\cdots\cdots\cdots \\ a_n\ddot{p}_n + c_n p_n = P_n, \end{array}\right\} \tag{3.7.3}$$

in which each principal co-ordinate appears in a separate equation.

Forced motion will occur only in the rth mode if P_r is the only generalized force acting; in general a force system, or a single concentrated force or couple, will produce several different generalized forces P so that several modes will be excited together. If, on the other hand, the motion is free, then the right-hand side of each of the equations (3.7.3) will be zero and the motion will be given by equations of the form (3.5.12) because of the relations

$$\omega_1^2 = \frac{c_1}{a_1}, \quad \omega_2^2 = \frac{c_2}{a_2}, \quad ..., \quad \omega_n^2 = \frac{c_n}{a_n} \tag{3.7.4}$$

which follow from the results (3.6.4).

As an example of the equations of forced motion of a system, written in terms of the principal co-ordinates, consider the system of fig. 3.4.4. Let it be acted on by a torque $Q_3 = \Phi_3 e^{i\omega t}$ at the right-hand flywheel. The components of this generalized force which correspond to the various principal co-ordinates (as we selected them in equation (3.5.7)) are as follows:

$$\left. \begin{aligned} P_1 &= \Phi_3 \frac{\partial q_3}{\partial p_1} e^{i\omega t} = \Phi_3 e^{i\omega t}, \\ P_2 &= \Phi_3 \frac{\partial q_3}{\partial p_2} e^{i\omega t} = -\Phi_3 e^{i\omega t}, \\ P_3 &= \Phi_3 \frac{\partial q_3}{\partial p_3} e^{i\omega t} = \Phi_3 e^{i\omega t}. \end{aligned} \right\} \tag{3.7.5}$$

The energy expressions may be found by substituting equations (3.5.7) into equation (3.4.22) and this gives

$$\left. \begin{aligned} 2T &= 4Ip_1^2 + 2Ip_2^2 + 4Ip_3^2, \\ 2V &= 2kp_2^2 + 8kp_3^2. \end{aligned} \right\} \tag{3.7.6}$$

Equations (3.7.3) therefore become

$$\left. \begin{aligned} 4I\ddot{p}_1 &= \Phi_3 e^{i\omega t}, \\ 2I\ddot{p}_2 + 2kp_2 &= -\Phi_3 e^{i\omega t}, \\ 4I\ddot{p}_3 + 8kp_3 &= \Phi_3 e^{i\omega t}, \end{aligned} \right\} \tag{3.7.7}$$

so that the motion is given by

$$\left. \begin{aligned} p_1 &= \left(\frac{-1}{4I\omega^2} \right) \Phi_3 e^{i\omega t}, \\ p_2 &= \left(\frac{-1}{2k - 2I\omega^2} \right) \Phi_3 e^{i\omega t}, \\ p_3 &= \left(\frac{1}{8k - 4I\omega^2} \right) \Phi_3 e^{i\omega t}. \end{aligned} \right\} \tag{3.7.8}$$

If these results are now transformed back to the original co-ordinates, by means of equations (3.5.7), the result is found in the familiar receptance form; it is, in fact

$$\left. \begin{aligned} q_1 &= \left[\frac{-1}{4I\omega^2} + \frac{-1}{2k - 2I\omega^2} + \frac{1}{8k - 4I\omega^2} \right] \Phi_3 e^{i\omega t}, \\ q_2 &= \left[\frac{-1}{4I\omega^2} + \frac{-1}{8k - 4I\omega^2} \right] \Phi_3 e^{i\omega t}, \\ q_3 &= \left[\frac{-1}{4I\omega^2} + \frac{1}{2k - 2I\omega^2} + \frac{1}{8k - 4I\omega^2} \right] \Phi_3 e^{i\omega t}. \end{aligned} \right\} \tag{3.7.9}$$

When the generalized forces $P_1, P_2, \ldots, P_n$ are of the general types of equations (3.7.2), that is to say when they are harmonic with frequency ω, then the equations of motion (3.7.3) each have the form of that for a simple oscillator. This may be seen by comparing them with equation (1.5.10). The steady-motion relations can thus be written in the form

$$\left. \begin{aligned} p_1 &= \alpha_1 P_1, \\ p_2 &= \alpha_2 P_2, \\ &\cdots\cdots\cdots \\ p_n &= \alpha_n P_n, \end{aligned} \right\} \tag{3.7.10}$$

where the quantities $\alpha_1, \alpha_2, ..., \alpha_n$ are receptances which are given by

$$\alpha_1 = \frac{1}{c_1 - a_1 \omega^2}, \quad \alpha_2 = \frac{1}{c_2 - a_2 \omega^2}, \quad ..., \quad \alpha_n = \frac{1}{c_n - a_n \omega^2}. \qquad (3.7.11)$$

They are the receptances at the principal co-ordinates. In order to distinguish these functions from other receptances, we shall write them with a single subscript only; this does not produce any ambiguity because all the cross-receptances at the principal co-ordinates must be zero. The diagrams for these receptances will be discussed in § 3.10.

Suppose that it is required to find the displacement z of some point Z of a system, z being neither one of the selected generalized co-ordinates nor one of the principal co-ordinates. The quantity z will be a linear function of the q's and therefore, also, a linear function of the p's. A small increment of z can therefore be written in the form

$$\delta z = \frac{\partial z}{\partial p_1} \delta p_1 + \frac{\partial z}{\partial p_2} \delta p_2 + ... + \frac{\partial z}{\partial p_n} \delta p_n, \qquad (3.7.12)$$

where the partial derivatives are constants. Integration now gives

$$z = \frac{\partial z}{\partial p_1} p_1 + \frac{\partial z}{\partial p_2} p_2 + ... + \frac{\partial z}{\partial p_n} p_n. \qquad (3.7.13)$$

The value of the displacement z, due to the application of a force $F e^{i\omega t}$ at some other point Y, will therefore be

$$z = \frac{\partial z}{\partial p_1} \left(\alpha_1 F \frac{\partial y}{\partial p_1} \right) e^{i\omega t} + \frac{\partial z}{\partial p_2} \left(\alpha_2 F \frac{\partial y}{\partial p_2} \right) e^{i\omega t} + ... + \frac{\partial z}{\partial p_n} \left(\alpha_n F \frac{\partial y}{\partial p_n} \right) e^{i\omega t}. \qquad (3.7.14)$$

This may be written $\qquad\qquad z = \alpha_{zy} F e^{i\omega t}, \qquad\qquad (3.7.15)$

where the cross-receptance α_{zy} is given by

$$\alpha_{zy} = \alpha_{yz} = \alpha_1 \left(\frac{\partial y}{\partial p_1} \frac{\partial z}{\partial p_1} \right) + \alpha_2 \left(\frac{\partial y}{\partial p_2} \frac{\partial z}{\partial p_2} \right) + ... + \alpha_n \left(\frac{\partial y}{\partial p_n} \frac{\partial z}{\partial p_n} \right). \qquad (3.7.16)$$

If z and y are identical this expression becomes the direct receptance

$$\alpha_{yy} = \alpha_1 \left(\frac{\partial y}{\partial p_1} \right)^2 + \alpha_2 \left(\frac{\partial y}{\partial p_2} \right)^2 + ... + \alpha_n \left(\frac{\partial y}{\partial p_n} \right)^2. \qquad (3.7.17)$$

Equations (3.7.16) and (3.7.17) are special cases of equations (3.2.20) and (3.2.21) and can be derived directly from the latter.

It is often convenient to write the receptances $\alpha_1, \alpha_2, ..., \alpha_n$ in the form

$$\alpha_1 = \frac{1}{a_1 (\omega_1^2 - \omega^2)}, \quad \alpha_2 = \frac{1}{a_2 (\omega_2^2 - \omega^2)}, \quad ..., \quad \alpha_n = \frac{1}{a_n (\omega_n^2 - \omega^2)}. \qquad (3.7.18)$$

It may be noted that, since these quantities provide a means for finding the variations of $p_1, p_2, ..., p_n$ which are produced by some specified applied force, their values must be dependent on the choice of the meaning of unit value for each of the principal co-ordinates; the magnitudes of the coefficients $a_1, a_2, ..., a_n$ are in fact dependent on this choice. If this form of the expressions is used in equations (3.7.16) and (3.7.17), then it is found that

$$\alpha_{zy} = \alpha_{yz} = \frac{\dfrac{\partial y}{\partial p_1}\dfrac{\partial z}{\partial p_1}}{a_1(\omega_1^2 - \omega^2)} + \frac{\dfrac{\partial y}{\partial p_2}\dfrac{\partial z}{\partial p_2}}{a_2(\omega_2^2 - \omega^2)} + \dots + \frac{\dfrac{\partial y}{\partial p_n}\dfrac{\partial z}{\partial p_n}}{a_n(\omega_n^2 - \omega^2)} \quad (3.7.19)$$

and

$$\alpha_{yy} = \frac{\left(\dfrac{\partial y}{\partial p_1}\right)^2}{a_1(\omega_1^2 - \omega^2)} + \frac{\left(\dfrac{\partial y}{\partial p_2}\right)^2}{a_2(\omega_2^2 - \omega^2)} + \dots + \frac{\left(\dfrac{\partial y}{\partial p_n}\right)^2}{a_n(\omega_n^2 - \omega^2)}. \quad (3.7.20)$$

Now these expressions must be independent of the choice for the unit values of the co-ordinates p because they concern the displacements and forces at Y and Z only.

That they are so, may be seen by the following argument. Let the displacement y be given in terms of the co-ordinates q by the relation

$$y = Aq_1 + Bq_2 + \dots + Nq_n, \quad (3.7.21)$$

where $A, B, \dots, N$ are constants. Now if the q's are linear functions of the co-ordinates p, as given in equations (3.5.9), so that for distortion in the rth mode $q_1 = R_1 p_r,\ q_2 = R_2 p_r,\ \dots,\ q_n = R_n p_r$, then it follows that

$$\left(\frac{\partial y}{\partial p_r}\right)^2 = (AR_1 + BR_2 + \dots + NR_n)^2. \quad (3.7.22)$$

Also, by substituting equations (3.5.9) into (3.1.2), it is found that

$$a_r = a_{11}R_1^2 + a_{22}R_2^2 + \dots + a_{nn}R_n^2 + 2a_{12}R_1 R_2 + \dots. \quad (3.7.23)$$

The ratios $R_1 : R_2 : \dots : R_n$ are fixed by the shape of the rth principal mode but the absolute values of the R's depend upon the meaning of unit value of p_r. It follows that the expression

$$\frac{\left(\dfrac{\partial y}{\partial p_r}\right)^2}{a_r}$$

is independent of this choice since multiplication of all the R's by any factor leaves it unaltered. Therefore α_{yy} and (by a similar argument) α_{yz} are also independent of this choice. We shall show that equations (3.7.19) and (3.7.20) are of great practical value.

The use of the result (3.7.16) may be illustrated by the system of fig. 3.4.3. The ratios of the co-ordinates in the first and second modes were found to be

$$\frac{q_1}{q_2} = \frac{2 \cdot 0054}{0 \cdot 5161} = 3 \cdot 885, \quad \frac{q_1}{q_2} = -\frac{0 \cdot 3279}{0 \cdot 8494} = -0 \cdot 3860. \quad (3.7.24)$$

If the principal co-ordinates are defined such that $p_1 = 1$ when $q_2 = 1$ in the first mode and $p_2 = 1$ when $q_2 = 1$ in the second mode, then

$$\left.\begin{aligned} q_1 &= 3 \cdot 885 p_1 - 0 \cdot 386 p_2, \\ q_2 &= p_1 + p_2. \end{aligned}\right\} \quad (3.7.25)$$

On substituting these relations into the energy expressions (3.4.17), it is found that

$$2T = 7 \cdot 528 M p_1^2 + 0 \cdot 4544 M p_2^2,$$
$$2V = 33 \cdot 18 k p_1^2 + 3 \cdot 298 k p_2^2. \tag{3.7.26}$$

The receptances at the principal co-ordinates are therefore

$$\alpha_1 = \frac{1}{33 \cdot 18 k - 7 \cdot 528 M \omega^2}, \quad \alpha_2 = \frac{1}{3 \cdot 298 k - 0 \cdot 4544 M \omega^2}. \tag{3.7.27}$$

These expressions may be used to find, for instance, the angle θ between the line AB of fig. 3.4.3 and the horizontal when a vertical force $F e^{i\omega t}$ is applied at A. The angle θ is given in terms of the co-ordinates q_1 and q_2 by

$$\theta = \frac{q_1 - q_2}{2a} \tag{3.7.28}$$

and therefore in terms of the principal co-ordinates it is, by equations (3.7.25),

$$\theta = \frac{1 \cdot 442 p_1 - 0 \cdot 693 p_2}{a}. \tag{3.7.29}$$

The relevant cross-receptance is now given by equation (3.7.16); it is

$$\alpha_1 \left(\frac{3 \cdot 885 \times 1 \cdot 442}{a} \right) + \alpha_2 \left(\frac{0 \cdot 386 \times 0 \cdot 693}{a} \right) = \frac{5 \cdot 604}{a} \alpha_1 + \frac{0 \cdot 2675}{a} \alpha_2$$

so that $\quad \theta = \left[\dfrac{5 \cdot 604}{a(33 \cdot 18 k - 7 \cdot 528 M \omega^2)} + \dfrac{0 \cdot 2675}{a(3 \cdot 298 k - 0 \cdot 4544 M \omega^2)} \right] F e^{i\omega t}. \tag{3.7.30}$

This result may also, of course, be found by substituting the expressions for q_1 and q_2, which are derived from equations (3.1.20) into equation (3.7.28).

In some systems the principal modes may be found easily; this may arise as the result of previous analysis of similar systems or it may be due to symmetry in the system. In such systems, the co-ordinates q may be written down as functions of the co-ordinates p immediately. Now if the system is slightly modified, the principal modes will be altered; but it may be convenient to retain, as co-ordinates, the principal co-ordinates of the *unmodified* system (although these will no longer be principal co-ordinates after the modification is made). It is assumed here that the modification allows the co-ordinates to be chosen in this way; this would not be the case if, for instance, an extra degree of freedom were introduced by the modification.

This technique can be illustrated by means of a simple example. Consider the coupled-pendulum system of fig. 3.7.1 (*a*) in which the springs are unstrained in the equilibrium position. These springs have stiffnesses k and λ as indicated. The system is regarded as a modification of that of fig. 3.7.1 (*b*), for which the principal modes consist of equal in-phase and equal anti-phase motions of the pendula. The angles of swing of the pendula are denoted by q_1 and q_2 as shown and if principal co-ordinates p are chosen such that $p_1 = 1$ when $q_1 = 1$ in the first mode and $p_2 = 1$ when $q_1 = 1$ in the second mode, then

$$q_1 = p_1 + p_2,$$
$$q_2 = p_1 - p_2. \tag{3.7.31}$$

The natural frequencies of these modes, that is for the system of fig. 3.7.1 (b), are given by

$$\omega_1^2 = \frac{g}{l}, \qquad \omega_2^2 = \left(\frac{g}{l} + \frac{2kh^2}{ml^2}\right). \tag{3.7.32}$$

The kinetic energy function is

$$2T = ml^2\dot{q}_1^2 + ml^2\dot{q}_2^2 = 2ml^2\dot{p}_1^2 + 2ml^2\dot{p}_2^2 \tag{3.7.33}$$

so that the receptances α_1 and α_2 can be formed; they are

$$\left.\begin{aligned}
\alpha_1 &= \frac{1}{a_1(\omega_1^2 - \omega^2)} = \frac{1}{2ml^2[(g/l) - \omega^2]}, \\[2mm]
\alpha_2 &= \frac{1}{a_2(\omega_2^2 - \omega^2)} = \frac{1}{2ml^2\left[\left(\dfrac{g}{l} + \dfrac{2kh^2}{ml^2}\right) - \omega^2\right]}.
\end{aligned}\right\} \tag{3.7.34}$$

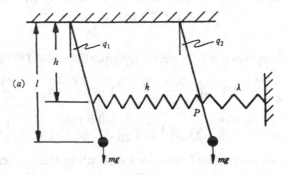

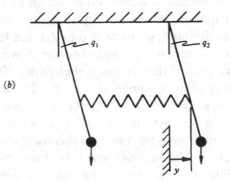

Fig. 3.7.1

The displacement of the point P, in terms of the principal co-ordinates, is

$$y = hq_2 = h(p_1 - p_2) \tag{3.7.35}$$

so that the direct receptance at this point is

$$\alpha_{yy} = \frac{h^2}{2ml^2\left[\left(\dfrac{g}{l}\right) - \omega^2\right]} + \frac{h^2}{2ml^2\left[\left(\dfrac{g}{l} + \dfrac{2kh^2}{ml^2}\right) - \omega^2\right]}. \tag{3.7.36}$$

If a horizontal force $Fe^{i\omega t}$ is applied at P then the displacement there will be

$$y = \frac{h^2 F e^{i\omega t}}{2ml^2\left[\left(\frac{g}{l}\right) - \omega^2\right]} + \frac{h^2 F e^{i\omega t}}{2ml^2\left[\left(\frac{g}{l} + \frac{2kh^2}{ml^2}\right) - \omega^2\right]}. \tag{3.7.37}$$

This displacement is made up from components due to the two principal modes of the unmodified system, these being

$$p_1 = \frac{hF e^{i\omega t}}{2ml^2\left[\left(\frac{g}{l}\right) - \omega^2\right]},$$

$$p_2 = \frac{-hF e^{i\omega t}}{2ml^2\left[\left(\frac{g}{l} + \frac{2kh^2}{ml^2}\right) - \omega^2\right]}, \tag{3.7.38}$$

as may be seen by comparison with equations (3.7.35). All these results have been obtained very simply for the unmodified system; the modified system can now be analysed by the same process.

Entry number 3 of Table 3 gives the frequency equation of the modified system; for the unmodified system of fig. 3.7.1 (b) may be regarded as one sub-system and the spring which is to be introduced as the other. The equation is

$$\frac{1}{\lambda} + \alpha_{yy} = 0, \tag{3.7.39}$$

where α_{yy} is given in equation (3.7.36). It is convenient, at this stage, to introduce numerical values for the various parameters; suppose $m = 16$ lb.; $l = 2$ ft.; $h = 1$ ft.; $k = 1024$ pdl./ft.; $\lambda = 513$ pdl./ft. When used in the frequency equation, these quantities give

$$\omega_1^2 = 19\cdot5 \text{ rad.}^2/\text{sec.}^2, \quad \omega_2^2 = 52\cdot5 \text{ rad.}^2/\text{sec.}^2. \tag{3.7.40}$$

The principal modes of the modified system can be found by observing that its free motion involves forced motion of the original system at the frequencies ω_1 and ω_2 of equations (3.7.40). The first principal mode is given by equations (3.7.38) in terms of p_1 and p_2 and is seen to be such that

$$\frac{p_1}{p_2} = \frac{-\left[\left(\frac{g}{l} + \frac{2kh^2}{ml^2}\right) - 19\cdot5\right]}{\left[\left(\frac{g}{l}\right) - 19\cdot5\right]} = -\frac{(48 - 19\cdot5)}{(16 - 19\cdot5)} = 8\cdot12. \tag{3.7.41}$$

Thus the ratio of the co-ordinates q_1 and q_2 for this mode will be

$$\frac{q_1}{q_2} = \frac{p_1 + p_2}{p_1 - p_2} = 1\cdot28. \tag{3.7.42}$$

Similarly, for the second mode,

$$\frac{p_1}{p_2} = \frac{-(48 - 52\cdot5)}{(16 - 52\cdot5)} = -0\cdot12, \tag{3.7.43}$$

which gives

$$\frac{q_1}{q_2} = -0\cdot78. \tag{3.7.44}$$

In a problem such as this, where there are only two degrees of freedom, the gain to be had from this method is small. However, the example illustrates the principles involved and, in more complex problems, the gain may be considerable. This is particularly true of systems which have an infinite number of degrees of freedom. We shall discuss these in later chapters.

EXAMPLES 3.7

1. A thin uniform rigid bar of mass M is supported symmetrically by three springs as shown. Obtain the principal modes by inspection and thence find the direct receptance for vertical movement at one end of the bar.

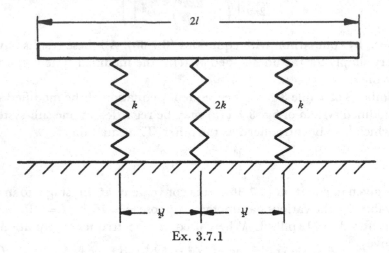

Ex. 3.7.1

2. If the system in the last example is modified by attaching a fourth vertical spring of stiffness $2k$ to the left-hand end of the bar, the lower end of this spring being anchored, obtain the equation for the natural frequencies and find these with the corresponding principal modes.

3. Express the combined rotational and linear displacements, which together constitute the principal modes, as simple rotations about particular points on the axis of the bar which is dealt with in Ex. 3.7.2. Find the combinations of torque and vertical force at the centre of the bar which will produce static displacements in each of the principal modes separately. Verify that the force system which is found in this way for either mode can be reduced to a single force which acts at that point on the bar which remains fixed during motion in the other principal mode. Explain this result by reference to the principle of orthogonality.

4. A uniform rigid bar of length 8 ft., and mass 100 lb., is supported in a horizontal position by springs at points 1 ft. from each end; the stiffness of each of the springs is 500 lb.wt./in.

Find the direct receptance for vertical displacement at a point P, 2 ft. from one end of the bar, and find also the cross-receptance between this point and the point which is 2 ft. from the other end of the bar.

If an additional mass of 25 lb. is suspended from P by a spring of stiffness 20 lb.wt./in., find the three natural frequencies of the system.   (C.U.M.S.T. Pt II, 1954)

5. Show that if the co-ordinate q_1 of a certain system is constrained to be zero then the direct receptance at another co-ordinate q_2 is given by

$$\frac{\alpha_{11}\alpha_{22} - \alpha_{12}^2}{\alpha_{11}},$$

where the α's are the receptances of the unconstrained system.

Hence, or otherwise, show that if the system has two degrees of freedom only the above expression may be written

$$\frac{\left[\dfrac{\partial q_1}{\partial p_1}\dfrac{\partial q_2}{\partial p_2} - \dfrac{\partial q_2}{\partial p_1}\dfrac{\partial q_1}{\partial p_2}\right]^2}{\left(\dfrac{\partial q_1}{\partial p_1}\right)^2 a_2(\omega_2^2 - \omega^2) + \left(\dfrac{\partial q_1}{\partial p_2}\right)^2 a_1(\omega_1^2 - \omega^2)}.$$

6. Two cantilevers, B and C, are mounted in the same horizontal plane and a short distance apart. A light rigid beam joins the free ends of the cantilevers, the connexions being by ball joints at each end.

The vibration characteristics, in a vertical plane, are found for each cantilever; the free motion in this plane is, in each case, independent of any horizontal motion.

For the cantilever B the natural frequencies are found to be given by $\omega_1 = 200$ rad./sec. and $\omega_2 = 300$ rad./sec., and the mass, which would give the same kinetic energy as the cantilever if it were concentrated at the tip of the cantilever, was estimated to be 5 lb. for the first mode and 3 lb. for the second. The corresponding figures for C are 150 and 250 for frequencies and 7 lb. and 4 lb. for the masses.

Assuming that motion in the higher modes may be neglected, estimate the lowest natural frequency for vertical motion which the system would have if a concentrated mass of 6 lb. were attached to the linking beam at a point one-third of the way along it from the tip of B. (Leeds B.Sc. Mech. Eng. 1958)

7. A uniform rectangular block has its lower face $ABCD$ horizontal and is supported by four equal vertical springs, each of stiffness 200 lb.wt./in., at the lower corners. The edges AB and CD are 10 in. long.

The block is constrained so that the only displacements which it can undergo are

    (a) vertical translation,

    (b) horizontal translation in a direction parallel to AB,

    (c) rotation about an axis parallel to BC.

The block weighs 50 lb.; the centre of gravity is 2 in. from the lower face and central in plan.

Find the direct receptance at the mid-point of BC for movement in a vertical plane parallel to AB, and in a direction 45° to the horizontal. Hence, or otherwise, find the natural frequencies if this point is connected, in this direction, by a spring of stiffness 100 lb.wt./in. to a fixed point. (Leeds B.Sc. Mech. Eng. 1957)

3.8 Alternative forms of the orthogonality condition

So far, the orthogonal property of the principal modes has been expressed symbolically merely as the vanishing of the product coefficients when the energy expressions are written in terms of the principal co-ordinates. It is possible to state this property in terms of *any* set of generalized co-ordinates. In order to do this, we shall require to give a symbol to the contribution of the sth principal mode to the mth generalized co-ordinate; the symbol $_sq_m$ will be used for this quantity, so that, in the notation of equation (3.5.9) we should have

$$_sq_m = S_m p_s. \tag{3.8.1}$$

During a distortion which is a combination of the rth and sth modes, the energies will be given by T and V where

$$
\left.\begin{aligned}
2T &= a_{11}(_r\dot{q}_1 + {_s}\dot{q}_1)^2 + a_{22}(_r\dot{q}_2 + {_s}\dot{q}_2)^2 + \cdots \\
&\quad + a_{nn}(_r\dot{q}_n + {_s}\dot{q}_n)^2 + 2a_{12}(_r\dot{q}_1 + {_s}\dot{q}_1)\,(_r\dot{q}_2 + {_s}\dot{q}_2) + \cdots, \\
2V &= c_{11}(_rq_1 + {_s}q_1)^2 + c_{22}(_rq_2 + {_s}q_2)^2 + \cdots \\
&\quad + c_{nn}(_rq_n + {_s}q_n)^2 + 2c_{12}(_rq_1 + {_s}q_1)\,(_rq_2 + {_s}q_2) + \cdots.
\end{aligned}\right\} \quad (3.8.2)
$$

Since the total energies are the sums of those due to each of the principal modes separately, it follows that

$$
\left.\begin{aligned}
a_{11}\cdot{_r}\dot{q}_1\cdot{_s}\dot{q}_1 + a_{22}\cdot{_r}\dot{q}_2\cdot{_s}\dot{q}_2 + \cdots + a_{nn}\cdot{_r}\dot{q}_n\cdot{_s}\dot{q}_n + a_{12}(_r\dot{q}_1\cdot{_s}\dot{q}_2 + {_s}\dot{q}_1\cdot{_r}\dot{q}_2) + \cdots = 0, \\
c_{11}\cdot{_r}q_1\cdot{_s}q_1 + c_{22}\cdot{_r}q_2\cdot{_s}q_2 + \cdots + c_{nn}\cdot{_r}q_n\cdot{_s}q_n + c_{12}(_rq_1\cdot{_s}q_2 + {_s}q_1\cdot{_r}q_2) + \cdots = 0.
\end{aligned}\right\} \quad (3.8.3)
$$

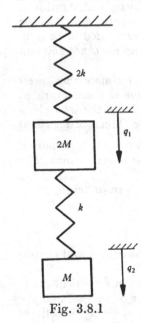

Fig. 3.8.1

The first of this pair of equations may be rewritten with the dots over the q's omitted; this follows because each of the terms in it contains the product $p_r p_s$ which may be cancelled out and replaced by $p_r p_s$ without interfering with the relation. That is to say,

$$
a_{11}\cdot{_r}q_1\cdot{_s}q_1 + a_{22}\cdot{_r}q_2\cdot{_s}q_2 + \cdots + a_{nn}\cdot{_r}q_n\cdot{_s}q_n
$$
$$
+ a_{12}(_rq_1\cdot{_s}q_2 + {_s}q_1\cdot{_r}q_2) + \cdots = 0, \quad (3.8.4)
$$

so that this equation and the second of equations (3.8.3) may be regarded as a statement of the orthogonality relations.

As an illustration of this form of the statement, consider the system of fig. 3.4.1 which is reproduced for convenience in fig. 3.8.1. Let the system be acted on by the force $Q_2 = \Phi_2 e^{i\omega t}$. It will be found from equations (3.4.11) and (3.4.14) that the expressions on the left-hand sides of (3.8.4) and the second of the relations (3.8.3) are, respectively,

$$
2M\left(\frac{\frac{2}{3}}{k - 2M\omega^2}\right)\left(\frac{-\frac{1}{3}}{2k - M\omega^2}\right)Q_2^2 + M\left(\frac{\frac{4}{3}}{k - 2M\omega^2}\right)\left(\frac{\frac{1}{3}}{2k - M\omega^2}\right)Q_2^2 - 0
$$

and

$$
3k\left(\frac{\frac{2}{3}}{k - 2M\omega^2}\right)\left(\frac{-\frac{1}{3}}{2k - M\omega^2}\right)Q_2^2 + k\left(\frac{\frac{4}{3}}{k - 2M\omega^2}\right)\left(\frac{\frac{1}{3}}{2k - M\omega^2}\right)Q_2^2
$$

$$
- k\left[\left(\frac{\frac{2}{3}}{k - 2M\omega^2}\right)\left(\frac{\frac{1}{3}}{2k - M\omega^2}\right) + \left(\frac{-\frac{1}{3}}{2k - M\omega^2}\right)\left(\frac{\frac{4}{3}}{k - 2M\omega^2}\right)\right]Q_2^2.
$$

The reader may confirm by expanding these expressions that they do in fact vanish.

Consider now the *free* oscillation of the system of fig. 3.8.1. If the principal co-ordinates are chosen so that $p_1 = 1$ when $_1q_2 = 3$ and $p_2 = 1$ when $_2q_1 = -2$ (say), then

$$\left.\begin{aligned}q_1 &= \tfrac{3}{2}p_1 - 2p_2, \\ q_2 &= 3p_1 + 2p_2.\end{aligned}\right\} \tag{3.8.5}$$

During free oscillation p_1 and p_2 will be of the form shown in equation (3.5.12) so that

$$\left.\begin{aligned}_1q_1 &= \tfrac{3}{2}\Pi_1 e^{i\omega_1 t}, & _2q_1 &= -2\Pi_2 e^{i\omega_2 t}, \\ _1q_2 &= 3\Pi_1 e^{i\omega_1 t}, & _2q_2 &= 2\Pi_2 e^{i\omega_2 t}.\end{aligned}\right\} \tag{3.8.6}$$

By using the appropriate constants a_{rs} and c_{rs} from equation (3.4.11), it will be found that the relations (3.8.3) and (3.8.4) are satisfied by the displacements (3.8.6).

Returning now to the general theory, consider a set of forces $_rQ_1, {}_rQ_2, ..., {}_rQ_n$ which, when applied to the system at $q_1, q_2, ..., q_n$ respectively, will produce a *static* distortion in the rth principal mode only. The values of the forces will be given by equations (3.1.13) if the inertia terms are omitted from them; that is,

$$\left.\begin{aligned}c_{11} \cdot {}_rq_1 + c_{12} \cdot {}_rq_2 + ... + c_{1n} \cdot {}_rq_n &= {}_rQ_1, \\ c_{21} \cdot {}_rq_1 + c_{22} \cdot {}_rq_2 + ... + c_{2n} \cdot {}_rq_n &= {}_rQ_2, \\ \cdots\cdots\cdots\cdots\cdots\cdots\cdots\cdots\cdots\cdots\cdots\cdots\cdots \\ c_{n1} \cdot {}_rq_1 + c_{n2} \cdot {}_rq_2 + ... + c_{nn} \cdot {}_rq_n &= {}_rQ_n.\end{aligned}\right\} \tag{3.8.7}$$

Now let us multiply the first of the expressions by $_sq_1$, the second by $_sq_2$ and so on, and then add. The left-hand side of the resulting equation will be identical with the left-hand side of the second of the relations (3.8.3) so that it vanishes. The right-hand side must therefore be equated to zero and this gives

$$_rQ_1 \cdot {}_sq_1 + {}_rQ_2 \cdot {}_sq_2 + ... + {}_rQ_n \cdot {}_sq_n = 0. \tag{3.8.8}$$

This equation states that if the set of forces which will produce a static distortion in the rth mode is applied to the system and that if the system is then given, by some means, a slow distortion in the sth mode, no work will be done by the set of forces. This is a further statement of the orthogonality relation.

In the static problem which has been considered, the formal algebraic proof which is given is not necessary. This is because the final statement which was deduced can be seen to follow directly from the fact that the total potential energy is the sum of the separate contributions which would be produced in each of the principal modes separately. We have given the proof because we shall shortly consider the corresponding relation for a set of harmonic forces for which the direct argument cannot be used so easily. Before doing this, however, we illustrate the steady force relation with an example.

Consider again the system of fig. 3.8.1. A pair of distortions which correspond to the first and second principal modes are respectively:

$$\left.\begin{aligned}_1q_1 &= 3, & _2q_1 &= 2, \\ _1q_2 &= 6, & _2q_2 &= -2.\end{aligned}\right\} \tag{3.8.9}$$

The steady forces which would produce these displacements may be found either from equation (3.8.7) or from simple statics; they are

$$_1Q_1 = 3k, \quad _2Q_1 = 8k, \atop _1Q_2 = 3k, \quad _2Q_2 = -4k. \bigg\}$$

$$(3.8.10)$$

The application of equation (3.8.8) to these results gives

$$_1Q_1 \cdot _2q_1 + _1Q_2 \cdot _2q_2 = _2Q_1 \cdot _1q_1 + _2Q_2 \cdot _1q_2 = 0 \qquad (3.8.11)$$

which relations will be found to be correct.

To return again to the general case, consider now forces which are varying harmonically. Such a set of forces, which would produce oscillation in the rth mode only, will be given by

$$\left. \begin{aligned} &- (a_{11} \cdot _rq_1 + a_{12} \cdot _rq_2 + \dots + a_{1n} \cdot _rq_n)\, \omega^2 \\ &\quad + (c_{11} \cdot _rq_1 + c_{12} \cdot _rq_2 + \dots + c_{1n} \cdot _rq_n) = {}_rQ_1, \\ &- (a_{21} \cdot _rq_1 + a_{22} \cdot _rq_2 + \dots + a_{2n} \cdot _rq_n)\, \omega^2 \\ &\quad + (c_{21} \cdot _rq_1 + c_{22} \cdot _rq_2 + \dots + c_{2n} \cdot _rq_n) = {}_rQ_2, \\ &\quad\dots\dots\dots\dots\dots\dots\dots\dots\dots\dots\dots\dots\dots\dots \\ &- (a_{n1} \cdot _rq_1 + a_{n2} \cdot _rq_2 + \dots + a_{nn} \cdot _rq_n)\, \omega^2 \\ &\quad + (c_{n1} \cdot _rq_1 + c_{n2} \cdot _rq_2 + \dots + c_{nn} \cdot _rq_n) = {}_rQ_n, \end{aligned} \right\}$$

$$(3.8.12)$$

where, it must be remembered, all the q's and all the Q's now vary harmonically with frequency ω. These equations can be manipulated in the same way as before and equations (3.8.8) are again arrived at; but that equation will now have been shown to hold when the forces and displacements are sinusoidally varying quantities instead of constants as they were previously. From the point of view of the physical implication of the equations, it may be noted that the inertial forces, as determined by Lagrange's equations, have merely been added to the forces which were present previously.

As an example let there be a forced motion of the system of fig. 3.8.1 in which

$$_1q_1 = 2e^{i\omega t}, \quad _2q_1 = -5e^{i\omega t}, \atop _1q_2 = 4e^{i\omega t}, \quad _2q_2 = 5e^{i\omega t}. \bigg\}$$

$$(3.8.13)$$

The forces which would be necessary to produce this motion can be found from equations of the type of (3.8.12); in fact, by using the energy coefficients a and c of equations (3.4.11) it is found that

$$_1Q_1 = 2(k - 2M\omega^2)\, e^{i\omega t}, \quad _2Q_1 = -10(2k - M\omega^2)\, e^{i\omega t}, \atop _1Q_2 = 2(k - 2M\omega^2)\, e^{i\omega t}, \quad _2Q_2 = 5(2k - M\omega^2)\, e^{i\omega t}. \bigg\}$$

$$(3.8.14)$$

The reader can now verify that, again,

$$_1Q_1 \cdot _2q_1 + _1Q_2 \cdot _2q_2 = _2Q_1 \cdot _1q_1 + _2Q_2 \cdot _1q_2 = 0. \qquad (3.8.15)$$

It will be shown later that the concept of a set of forces which will produce a distortion in one particular mode only is of use when calculating frequencies of complicated systems. For the present we shall not require to follow up this idea, but it is convenient at this stage to mention one property of such sets of forces.

Let a set of harmonically varying forces, which would produce distortion in the rth mode only, be defined by the equations (3.8.12), their frequency being ω. It is evidently possible to solve the equations so as to give the generalized co-ordinates as functions of the forces; the coefficients which we shall obtain will in fact be the receptances at the q co-ordinates as may be seen by comparison with the theory of §3.2. That is to say

$$_r q_u = \alpha_{u1} \cdot _r Q_1 + \alpha_{u2} \cdot _r Q_2 + \ldots + \alpha_{un} \cdot _r Q_n. \tag{3.8.16}$$

Now it has been shown that the displacement q_u at the uth generalized co-ordinate can be expressed as the sum of a number of components, each one of which is due to the distortion in one of the principal modes. It is permissible to write, therefore,

$$q_u = _1 q_u + _2 q_u + \ldots + _n q_u. \tag{3.8.17}$$

Each of the terms on the right-hand side may now be expressed by equations of the form (3.8.16); it thus follows that

$$\left. \begin{aligned} q_u = \alpha_{u1}[_1 Q_1 + _2 Q_1 + \ldots + _n Q_1] \\ + \alpha_{u2}[_1 Q_2 + _2 Q_2 + \ldots + _n Q_2] \\ + \ldots\ldots\ldots\ldots\ldots\ldots\ldots\ldots\ldots \\ + \alpha_{un}[_1 Q_n + _2 Q_n + \ldots + _n Q_n]. \end{aligned} \right\} \tag{3.8.18}$$

By comparing this relation with equation (3.2.11) it will be found that

$$Q_v = _1 Q_v + _2 Q_v + \ldots + _n Q_v. \tag{3.8.19}$$

This result is comparable with that of equation (3.8.17). It implies that any set of forces which acts on a system may be expressed as the sum of a number of other sets, each one of which would produce a distortion in one principal mode only if it acted alone. Thus the force at any particular co-ordinate q may be regarded as being made up from a number of components, each component being the contribution of one of these special sets.

The orthogonality property, as it is stated in equation (3.8.8) is useful for solving equations (3.5.9) to obtain equations (3.5.10). By its use, expressions may be found for principal co-ordinates p in terms of other generalized co-ordinates q without the necessity of solving n simultaneous equations. This is advantageous when n is large and essential if n is infinite. Suppose that a system is distorted in some way and it is required to express the displacement at the rth principal co-ordinate p_r in terms of the displacements $q_1, q_2, \ldots, q_n$. Evidently expressions are wanted for the coefficients $A_r', B_r', \ldots, N_r'$ in the relation

$$p_r = A_r' q_1 + B_r' q_2 + \ldots + N_r' q_n \tag{3.8.20}$$

which is the rth equation (3.5.10); these expressions are required in terms of the constants $A_1, B_1, \ldots$, etc., of equations (3.5.9).

In the notation of §3.5, the contributions to displacements at $q_1, q_2, ..., q_n$ which are due to a distortion in the rth mode are

$$_rq_1 = R_1 p_r, \quad _rq_2 = R_2 p_r, \quad ..., \quad _rq_n = R_n p_r, \tag{3.8.21}$$

in accordance with equation (3.8.1). The *static* generalized forces $_rQ_1, _rQ_2, ..., _rQ_n$ which will produce these contributions to a *static* distortion, and which have values given by (3.8.7), may thus be written in the form

$$\left.\begin{aligned}
_rQ_1 &= [c_{11}R_1 + c_{12}R_2 + ... + c_{1n}R_n]\, p_r, \\
_rQ_2 &= [c_{21}R_1 + c_{22}R_2 + ... + c_{2n}R_n]\, p_r, \\
&\cdots\cdots\cdots\cdots\cdots\cdots\cdots\cdots\cdots\cdots\cdots \\
_rQ_n &= [c_{n1}R_1 + c_{n2}R_2 + ... + c_{nn}R_n]\, p_r.
\end{aligned}\right\} \tag{3.8.22}$$

If these forces are imagined to be applied and then the deflexions $q_1, q_2, ..., q_n$ to be imposed by some means, then the work done by these forces is evidently

$$W_r = {}_rQ_1 \cdot q_1 + {}_rQ_2 \cdot q_2 + ... + {}_rQ_n \cdot q_n. \tag{3.8.23}$$

Now these displacements $q_1, q_2, ..., q_n$ can be written in the form of equation (3.8.17). Moreover, the principle of orthogonality in the form (3.8.8) states that the forces $_rQ_1, _rQ_2, ..., _rQ_n$ only do work as they move through the deflexions $_rq_1, _rq_2, ..., _rq_n$. An alternative expression for the work W_r is, therefore,

$$W_r = {}_rQ_1 \cdot {}_rq_1 + {}_rQ_2 \cdot {}_rq_2 + ... + {}_rQ_n \cdot {}_rq_n. \tag{3.8.24}$$

The two expressions, (3.8.23) and (3.8.24), for W_r can now be equated giving the relation

$$_rQ_1 \cdot q_1 + {}_rQ_2 \cdot q_2 + ... + {}_rQ_n \cdot q_n = {}_rQ_1 \cdot {}_rq_1 + {}_rQ_2 \cdot {}_rq_2 + ... + {}_rQ_n \cdot {}_rq_n. \tag{3.8.25}$$

If the expressions for the force contributions $_rQ_1$, etc., and for the displacement contributions $_rq_1$, etc., are now substituted from equations (3.8.21) and (3.8.22) it is found that

$$\left.\begin{aligned}
p_r = {}&\left[\frac{c_{11}R_1 + c_{12}R_2 + ... + c_{1n}R_n}{c_{11}R_1^2 + c_{22}R_2^2 + ... + c_{nn}R_n^2 + 2c_{12}R_1R_2 + ...}\right] q_1 \\
&+ \left[\frac{c_{21}R_1 + c_{22}R_2 + ... + c_{2n}R_n}{c_{11}R_1^2 + c_{22}R_2^2 + ... + c_{nn}R_n^2 + 2c_{12}R_1R_2 + ...}\right] q_2 \\
&+ \cdots\cdots\cdots\cdots\cdots\cdots\cdots\cdots\cdots\cdots\cdots \\
&+ \left[\frac{c_{n1}R_1 + c_{n2}R_2 + ... + c_{nn}R_n}{c_{11}R_1^2 + c_{22}R_2^2 + ... + c_{nn}R_n^2 + 2c_{12}R_1R_2 + ...}\right] q_n.
\end{aligned}\right\} \tag{3.8.26}$$

Comparison of this result with equation (3.8.20) reveals that the expressions that are enclosed by square brackets are the required constants $A_r', B_r', ..., N_r'$.

The use of equation (3.8.26) may be illustrated by reference to the system of fig. 3.8.1 for which

$$c_{11} = 3k, \quad c_{22} = k, \quad c_{12} = -k, \tag{3.8.27}$$

by equation (3.4.11). It has been shown (see equation (3.5.18)) that principal co-ordinates may be chosen for this system such that

$$\left.\begin{aligned}
q_1 &= 2p_1 + 4p_2, \\
q_2 &= 4p_1 - 4p_2.
\end{aligned}\right\} \tag{3.8.28}$$

By the use of (3.8.26), it follows that

$$p_1 = \left[\frac{3k \times 2 - k \times 4}{3k \times 2^2 + k \times 4^2 - 2k \times 2 \times 4} \right] q_1$$

$$+ \left[\frac{-k \times 2 + k \times 4}{3k \times 2^2 + k \times 4^2 - 2k \times 2 \times 4} \right] q_2$$

$$= \tfrac{1}{8} \times q_1 + \tfrac{1}{8} \times q_2 \tag{3.8.29}$$

and

$$p_2 = \left[\frac{3k \times 4 + k \times 4}{3k \times 4^2 + k \times 4^2 + 2k \times 4^2} \right] q_1$$

$$+ \left[\frac{-k \times 4 - k \times 4}{3k \times 4^2 + k \times 4^2 + 2k \times 4^2} \right] q_2$$

$$= \tfrac{1}{6} \times q_1 - \tfrac{1}{12} \times q_2. \tag{3.8.30}$$

These results are those that are found by solving equations (3.8.28) directly.

In this section, we have introduced the notion of a set of forces $_rQ_1, {}_rQ_2, \dots, {}_rQ_n$ which produce distortion in the rth principal mode only. Now it is evident that such a distortion will also be produced by the generalized force P_r at the rth principal co-ordinate, as may be seen by reference to equations (3.7.3). The set of forces is thus related to P_r. Now we shall not examine the nature of this relation in detail here as it is not of immediate importance; the interested reader may wish, however, to work out examples 3.8.4 and 3.8.5 which deal with this point.

EXAMPLES 3.8

1. Find the pairs of forces $_1Q_1, {}_1Q_2$ and $_2Q_1, {}_2Q_2$ which, when applied to the system of fig. 3.8.1, produce the same motions in the principal modes as would the pair of applied forces $Q_1 = \Phi_1 e^{i\omega t}$ and $Q_2 = \Phi_2 e^{i\omega t}$ and show that the solutions agree with equation (3.8.19). Show further that the solutions satisfy the orthogonality requirement in the form (3.8.8).

2. Rework Ex. 3.6.3 using the results of this section.

3. A suitably supported beam is to be used to carry some additional masses and it is decided to investigate its vibration characteristics in the vertical plane. Co-ordinates are chosen as the vertical displacements of three points on the beam and are denoted by q_1, q_2, q_3. The first three natural frequencies are found, experimentally, to be given by

$$\omega^2 = 2 \cdot 5 \times 10^4, \quad \omega_2^2 = 12 \cdot 0 \times 10^4, \quad \omega_3^2 = 70 \cdot 0 \times 10^4, \quad (\text{rad.}^2/\text{sec.}^2)$$

and the ratios of $q_1 : q_2 : q_3$ at these frequencies are

$$+1 \cdot 0 : +2 \cdot 0 : +1 \cdot 5,$$

$$+1 \cdot 0 : +0 \cdot 5 : -0 \cdot 5,$$

$$+1 \cdot 0 : -0 \cdot 8 : +1 \cdot 2,$$

respectively.

The beam is subsequently loaded statically at q_1, q_2 and q_3 in such a way that the deflexion has the form of the first principal mode; the loading is then changed to give the second and third modes. Denoting the steady forces to produce the first mode deflexion by $_1Q_1$, $_1Q_2$, etc., the three sets were found to be

$$_1Q_1 = -50 \text{ lb.wt.,} \quad _1Q_2 = +150 \text{ lb.wt.,} \quad _1Q_3 = +150 \text{ lb.wt.,}$$

$$_2Q_1 = +400 \text{ lb.wt.,} \quad _2Q_2 = +300 \text{ lb.wt.,} \quad _2Q_3 = -700 \text{ lb.wt.,}$$

$$_3Q_1 = +1820 \text{ lb.wt.,} \quad _3Q_2 = -2660 \text{ lb.wt.,} \quad _3Q_3 = +2340 \text{ lb.wt.}$$

In each case the deflexion q_1 was equal to $0 \cdot 010$ in.

Obtain expressions for the receptances α_{11}, α_{12}, and α_{22}.

Estimate the first natural frequency if a concentrated mass of 300 lb. is attached to the bar at q_2. Obtain an equation for the natural frequencies if two masses, each of 150 lb. are attached at q_1 and q_2 in place of the single 300 lb. mass.

(Leeds B.Sc. Mech. Eng. 1956)

4. A system is given a static distortion $_rq_1$, $_rq_2$, ..., $_rq_n$ in its rth principal mode only. The static force distribution $_rQ_1$, $_rQ_2$, ..., $_rQ_n$ which must be applied is given by equations (3.8.7) and (3.8.22) while the value of P_r the generalized force at the rth principal co-ordinate p_r is $c_r p_r$ by equation (3.7.3). Show that

(i) whereas the set of forces $_rQ_1$, etc., is known completely, the value of P_r depends on the choice that is made for the meaning of unit value of p_r;

(ii) the set of forces $_rQ_1$, etc., can be derived from P_r and that the derivation is such that the ambiguity referred to does not affect the result.

Illustrate these results by reference to the system of fig. 3.8.1, taking more than one possible set of principal co-ordinates.

5. Repeat Ex. 3.8.4, making reference to harmonic forces and displacements.

3.9 Rayleigh's Principle

It is stated in § 3.4 that, if the form of one of the principal modes of a system is known, then the corresponding natural frequency can be calculated immediately. Since this is of great practical importance and is the starting-point of discussion of Rayleigh's Principle, a simple illustrative example will help to fix the reader's ideas.

In the system of fig. 3.9.1, the following numerical values are used:

$$M = 2 \text{ lb.,} \quad m = 1 \text{ lb.,} \quad l = 1 \text{ ft.,} \quad k = 256 \text{ pdl./ft.} \tag{3.9.1}$$

By the methods of § 3.4, it may be shown that the motion caused by the exciting force $F e^{i\omega t}$ pdl. indicated in the figure is of the form

$$x = \left[\frac{0 \cdot 018}{27 \cdot 6 - \omega^2} + \frac{0 \cdot 482}{148 \cdot 4 - \omega^2} \right] F e^{i\omega t} \text{ (ft.),}$$

$$\theta = \left[\frac{0 \cdot 114}{27 \cdot 6 - \omega^2} - \frac{0 \cdot 614}{148 \cdot 4 - \omega^2} \right] F e^{i\omega t} \text{ (rad.).} \tag{3.9.2}$$

It follows that *free* vibration in the first principal mode is of the form

$$x = 0 \cdot 018 A\, e^{i\omega_1 t} \text{ (ft.),}$$

$$\theta = 0 \cdot 114 A\, e^{i\omega_1 t} \text{ (rad.),} \tag{3.9.3}$$

where A is a constant and $\omega_1^2 = 27 \cdot 6$ rad.^{2}/sec.^{2}.

Let us suppose now that by some means (possibly by guessing) we can find the modal shape of equations (3.9.3), namely

$$\frac{x}{\theta} = \frac{0\cdot018}{0\cdot114} = \frac{1}{6\cdot33} \tag{3.9.4}$$

(in units of ft. and radians). With this data it is possible to calculate the corresponding natural frequency ω_1. Free vibration in the first mode is known to be of the form

$$\left.\begin{aligned} x &= A\sin\omega_1 t \quad \text{(ft.)}, \\ \theta &= 6\cdot33A\sin\omega_1 t \ \text{(rad.)}. \end{aligned}\right\} \tag{3.9.5}$$

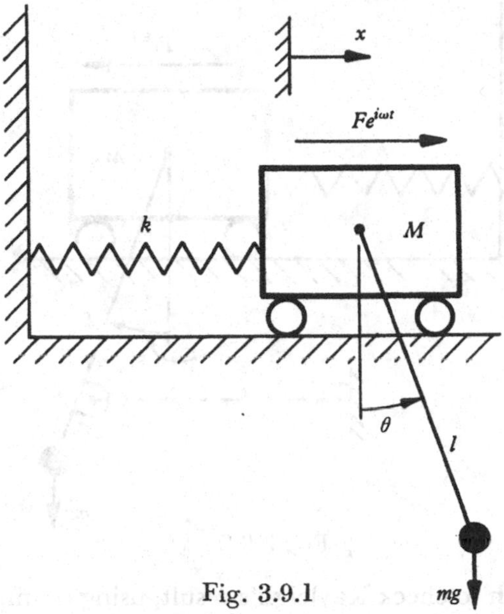

Fig. 3.9.1

The energies are

$$\left.\begin{aligned} T &= \tfrac{1}{2}[M\dot{x}^2 + m(\dot{x}+l\dot{\theta})^2], \\ V &= \tfrac{1}{2}[kx^2 + lmg\theta^2] \end{aligned}\right\} \tag{3.9.6}$$

and when the values (3.9.1) and (3.9.5) are substituted into these expressions, they become

$$\left.\begin{aligned} T &= 27\cdot86A^2\omega_1^2\cos^2\omega_1 t, \\ V &= 769\cdot1A^2\sin^2\omega_1 t. \end{aligned}\right\} \tag{3.9.7}$$

But the maximum values of T and V must be equal since the motion is harmonic so that

$$\omega_1^2 = \frac{769\cdot1}{27\cdot86} = 27\cdot6 \ \text{rad.}^2/\text{sec.}^2. \tag{3.9.8}$$

Now the derivation of the modal shape as a preliminary step in a frequency calculation is usually tedious and if the number of degrees of freedom of the system is large the labour may well be prohibitive unless a computing machine is available. But there are many problems in which a rapid approximate calculation of the lowest natural frequency of a system is required; these problems can be dealt with by using Rayleigh's Principle.† This is a theorem which states that a small error

† *Theory of Sound* (1894), §88; see also *Rayleigh's Principle* by G. Temple and W. G. Bickley (Oxford University Press, 1933).

in the assumed modal shape will produce a second-order error only in the calculated frequency. The principle also applies to frequencies other than the lowest but the direct use of it when calculating such frequencies is rather limited. It is, nevertheless, of great importance in general theory and extends into other fields, notably that of elastic instability. It also provides a justification for the idealization of a physical system into a simpler form in which it may be analysed as will be shown in Chapter 5.

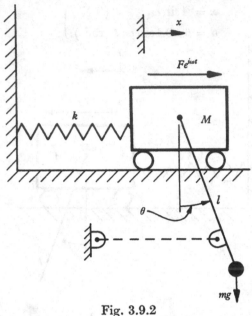

Fig. 3.9.2

The reader may wish to check Rayleigh's result, using a simple example, before proceeding. Suppose that the ratio (3.9.4) has been *guessed* and that instead of the correct figure 6.33, the erroneous figure 8 has been assumed. By following the procedure used before, it will be found that the frequency ω_1 would then be given by the approximate relation

$$\omega_1^2 \doteqdot 27 \cdot 8 \,\text{rad.}^2/\text{sec.}^2. \qquad (3.9.9)$$

The error in the modal shape was substantial whereas that in the frequency is quite small.

Before examining the principle, it will be useful to describe the concept of a 'constrained mode' on which it is based. This may be introduced by means of an example.

The system of fig. 3.9.1 has two degrees of freedom. By the addition of the dotted massless rigid link shown in fig. 3.9.2, the system is transformed into one with a single degree of freedom only. Now the presence of the link does not alter the values of the coefficients in the energy expressions because the added member has neither mass nor elasticity; its only effect is to impose a fixed ratio between the values of x and θ. When the link is incorporated, the system can oscillate in one particular modal shape only and this is said to be a 'constrained mode' of the original system of fig. 3.9.1.

The idea of a 'constraint' can be extended to systems with any number of degrees of freedom, suitable mechanisms of light rigid links being imagined to constrain all the freedoms but one. A possible arrangement for a three-degree-of-freedom system is shown in fig. 3.9.3. The use of the idea of a constrained mode is not limited by the practicability of constructing a real constraining mechanism nor yet by the engineer's ability to imagine the form of a theoretical one. The physical idea which we have presented, namely that of light rigid links, is merely intended as an aid to the reader in grasping the significance of a purely mathematical operation, to wit, the imposition of a set of ratios between the displacements at different co-ordinates of a system.

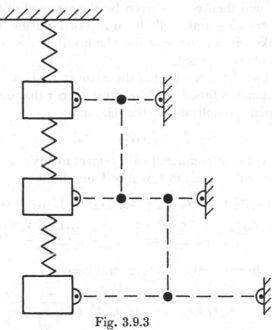

Fig. 3.9.3

Let the set of ratios be chosen so that the n co-ordinates $q_1, q_2, ..., q_n$ are related to a single variable q by the equations

$$q_1 = \tau_1 q, \quad q_2 = \tau_2 q, \quad ..., \quad q_n = \tau_n q, \tag{3.9.10}$$

the τ's being the quantities which define the constrained modal shape.† The energy expressions for the constrained system are

$$
\left.
\begin{aligned}
2T &= (a_{11}\tau_1^2 + a_{22}\tau_2^2 + ... + a_{nn}\tau_n^2 + 2a_{12}\tau_1\tau_2 + ...)\, \dot{q}^2, \\
2V &= (c_{11}\tau_1^2 + c_{22}\tau_2^2 + ... + c_{nn}\tau_n^2 + 2c_{12}\tau_1\tau_2 + ...)\, q^2.
\end{aligned}
\right\} \tag{3.9.11}
$$

---

† We have left undefined the meaning of unit value of q. This choice of scale is conveniently made by setting one of the τ's equal to unity. Thus if $\tau_1 = 1$, q has unit value when $q_1 = 1$ and the *shape* of the distortion may be altered by adjusting $\tau_1, \tau_3, ..., \tau_n$; the shape is then defined by the $n-1$ quantities

$$\tau_2 = \frac{q_2}{q} = \frac{q_2}{q_1}, \quad \tau_3 = \frac{q_3}{q_1}, \quad ..., \quad \tau_n = \frac{q_n}{q_1}.$$

This method of fixing the scale of the distortion will be used later; but it is convenient at this stage to retain the n constants τ and to bear in mind that the *shape* of the distortion is determined by $(n-1)$ quantities.

During free oscillation in the constrained mode, q must vary harmonically since the constrained system has only one degree of freedom. Thus the natural frequency ω_c can be found by equating the maximum values of the kinetic and potential energies. This gives

$$\omega_c^2 = \frac{c_{11}\tau_1^2 + c_{22}\tau_2^2 + \ldots + c_{nn}\tau_n^2 + 2c_{12}\tau_1\tau_2 + \ldots}{a_{11}\tau_1^2 + a_{22}\tau_2^2 + \ldots + a_{nn}\tau_n^2 + 2a_{12}\tau_1\tau_2 + \ldots}. \qquad (3.9.12)$$

If the quantities τ are chosen so that equation (3.9.10) corresponds to a distortion in one of the principal modes of the unconstrained system—the mth say—then the value of ω_c as found from equation (3.9.12) will be identical with ω_m. The free vibration will then be unaltered by the removal of the constraints and evidently the constraining links will be unstressed during the vibration. The removal of the links will, of course, allow the possibility of other free vibrations which were previously suppressed.

It is clear, from the above result, that the natural frequency of any constrained mode of a given system is a function of the parameters τ that define the mode. This fact may be expressed symbolically by the relation

$$\omega_c^2 = \omega_c^2(\tau_1, \tau_2, \ldots, \tau_n). \qquad (3.9.13)$$

The function ω_c^2 may be differentiated with respect to any one of the parameters τ. Thus the derivative with respect to τ_r will be found to be

$$\frac{\partial \omega_c^2}{\partial \tau_r} = \frac{\left\{ \begin{array}{l} 2(a_{11}\tau_1^2 + a_{22}\tau_2^2 + \ldots + a_{nn}\tau_n^2 + 2a_{12}\tau_1\tau_2 + \ldots)(c_{1r}\tau_1 + c_{2r}\tau_2 + \ldots + c_{nr}\tau_n) \\ -2(c_{11}\tau_1^2 + c_{22}\tau_2^2 + \ldots + c_{nn}\tau_n^2 + 2c_{12}\tau_1\tau_2 + \ldots)(a_{1r}\tau_1 + a_{2r}\tau_2 + \ldots + a_{nr}\tau_n) \end{array} \right\}}{(a_{11}\tau_1^2 + a_{22}\tau_2^2 + \ldots + a_{nn}\tau_n^2 + 2a_{12}\tau_1\tau_2 + \ldots)^2},$$
$$\qquad (3.9.14)$$

and by substituting from equation (3.9.12) this becomes

$$\frac{\partial \omega_c^2}{\partial \tau_r} = \frac{2[(c_{1r} - \omega_c^2 a_{1r})\tau_1 + (c_{2r} - \omega_c^2 a_{2r})\tau_2 + \ldots + (c_{nr} - \omega_c^2 a_{nr})\tau_n]}{a_{11}\tau_1^2 + a_{22}\tau_2^2 + \ldots + a_{nn}\tau_n^2 + 2a_{12}\tau_1\tau_2 + \ldots}. \qquad (3.9.15)$$

Now let $\tau_1, \tau_2, \ldots, \tau_n$ have the values which correspond to the sth principal mode so that

$$\omega_c = \omega_s. \qquad (3.9.16)$$

The motion is free and there are no forces in the constraining links; it follows that the numerator of equation (3.9.15) is zero at all instants because it is identical with the rth line of equation (3.2.2) with Φ_r zero. The denominator of equation (3.9.15) does not vanish at all instants because it is twice the kinetic energy of the motion. Therefore

$$\frac{\partial \omega_c^2}{\partial \tau_r} = 0 \quad (r = 1, 2, \ldots, n) \qquad (3.9.17)$$

when $\omega_c^2 = \omega_1^2, \omega_2^2, \ldots, \omega_n^2$.

This is the mathematical statement of Rayleigh's Principle. The immediate conclusion which may be drawn from it is that if, in calculating the frequency of a mode of oscillation, we make a small error in guessing the shape of the mode, then the resulting frequency will be only slightly affected; by this we mean that an error of the first order of small quantities in any of the parameters τ will involve an error of the second order in the value of ω_c^2.

Suppose, for instance, that a guess were made as to the shape of the sth principal mode so that values are assigned to the constants τ. Equation (3.9.12) would then give an estimate of ω_s^2. The value of ω_c^2 that is obtained in this way will not be identical with ω_s^2 owing to the errors $\delta\tau_1$ in τ_1, $\delta\tau_2$ in τ_2, ..., $\delta\tau_n$ in τ_n. The extended form of Taylor's series shows that the error is given by

$$\omega_c^2 - \omega_s^2 = \left[\frac{\partial \omega_c^2}{\partial \tau_1} \delta\tau_1 + \frac{\partial \omega_c^2}{\partial \tau_2} \delta\tau_2 + \ldots + \frac{\partial \omega_c^2}{\partial \tau_n} \delta\tau_n \right]$$

$$+ \frac{1}{2} \left[\frac{\partial^2 \omega_c^2}{\partial \tau_1^2} (\delta\tau_1)^2 + \frac{\partial^2 \omega_c^2}{\partial \tau_2^2} (\delta\tau_2)^2 + \ldots + \frac{\partial^2 \omega_c^2}{\partial \tau_n^2} (\delta\tau_n)^2 \right.$$

$$\left. + 2 \frac{\partial^2 \omega_c^2}{\partial \tau_1 \partial \tau_2} (\delta\tau_1)(\delta\tau_2) + \ldots \right] + \ldots. \qquad (3.9.18)$$

The derivatives of ω_c^2 are all evaluated at $\omega_c^2 = \omega_s^2$ so that Rayleigh's Principle in the form (3.9.17) shows that the contents of the first square brackets are zero; thus the error $\omega_c^2 - \omega_s^2$ depends upon the second and higher powers of the small quantities $\delta\tau$.

Rayleigh used the principle extensively in the development of vibration theory and, in particular, employed it in the proofs of some of the theorems which have been presented here with the aid of receptance diagrams. We shall now discuss various aspects and approaches to the idea in order to familiarize the reader with it.

First of all we repeat the above theory using the principal co-ordinates. By adding constraints, the co-ordinates $p_1, p_2, \ldots, p_n$ are made proportional to a single variable, p say. Thus the motion in the constrained mode is such that

$$p_1 = \rho_1 p, \quad p_2 = \rho_2 p, \quad \ldots, \quad p_n = \rho_n p, \qquad (3.9.19)$$

where the constants ρ determine the shape of the mode. The energy expressions become

$$2T = (a_1\rho_1^2 + a_2\rho_2^2 + \ldots + a_n\rho_n^2) \, \dot{p}^2, \\ 2V = (c_1\rho_1^2 + c_2\rho_2^2 + \ldots + c_n\rho_n^2) \, p^2, \Big\} \qquad (3.9.20)$$

where the coefficients a and c are those of the original unconstrained system. The frequency of free vibration of the constrained system is therefore given by

$$\omega_c^2 = \frac{c_1\rho_1^2 + c_2\rho_2^2 + \ldots + c_n\rho_n^2}{a_1\rho_1^2 + a_2\rho_2^2 + \ldots + a_n\rho_n^2}. \qquad (3.9.21)$$

Now let the constraints be chosen so that the mode differs only slightly from the sth principal mode of the unconstrained system; that is to say, let all the parameters ρ_r ($r \neq s$) be small compared with ρ_s. In these circumstances the algebraic form of (3.9.21) shows that ω_c^2 will differ from ω_s^2 by a small quantity of the second order if the ratios ρ_1/ρ_s, ρ_2/ρ_s, etc., are small quantities of the first order. This is because the parameters ρ in (3.9.21) are all squared; Rayleigh's original exposition was based on this argument. The reader may confirm by differentiation that, when $\omega_c^2 = \omega_1^2, \omega_2^2, \ldots, \omega_n^2$

$$\frac{\partial \omega_c^2}{\partial \rho_r} = 0 \quad (r = 1, 2, \ldots, n). \qquad (3.9.22)$$

This is another statement of the same fact.

The choice of the meaning of unit amplitude for each of the principal co-ordinates will not affect the proofs of Rayleigh's Principle. Suppose therefore that all have been normalized. Equation (3.9.21) then becomes

$$\omega_c^2 = \frac{\omega_1^2 \rho_1^2 + \omega_2^2 \rho_2^2 + \ldots + \omega_n^2 \rho_n^2}{\rho_1^2 + \rho_2^2 + \ldots + \rho_n^2} \tag{3.9.23}$$

and this may be written in the alternative forms

$$\omega_c^2 = \omega_1^2 + \frac{(\omega_2^2 - \omega_1^2)\, \rho_2^2 + (\omega_3^2 - \omega_1^2)\, \rho_3^2 + \ldots + (\omega_n^2 - \omega_1^2)\, \rho_n^2}{\rho_1^2 + \rho_2^2 + \ldots + \rho_n^2} \tag{3.9.24}$$

and

$$\omega_c^2 = \omega_n^2 - \frac{(\omega_n^2 - \omega_{n-1}^2)\, \rho_{n-1}^2 + (\omega_n^2 - \omega_{n-2}^2)\, \rho_{n-2}^2 + \ldots + (\omega_n^2 - \omega_1^2)\, \rho_1^2}{\rho_1^2 + \rho_2^2 + \ldots + \rho_n^2}. \tag{3.9.25}$$

Now the fractions in these expressions are essentially positive so that the frequency of the constrained mode must be greater than the lowest natural frequency and less than the highest natural frequency; that is to say,

$$\omega_1^2 \leqslant \omega_c^2 \leqslant \omega_n^2. \tag{3.9.26}$$

This fact, like Rayleigh's Principle, can also be deduced directly from the algebraic form of (3.9.21).

It should be useful at this stage to exemplify the principle by calculating the constrained mode frequency of the system of fig. 3.9.1. The constraining link which is shown in fig. 3.9.2, introduces the relation

$$x = \chi\theta \tag{3.9.27}$$

between the co-ordinates x and θ of the unconstrained system, where χ is a constant. The energy expressions are

$$\left.\begin{aligned} 2T &= [M\chi^2 + m(\chi + l)^2]\, \theta^2, \\ 2V &= (k\chi^2 + mgl)\, \theta^2, \end{aligned}\right\} \tag{3.9.28}$$

and for harmonic motion in the constrained mode the frequency will thus be given by

$$\omega_c^2 = \frac{k\chi^2 + mgl}{(M + m)\, \chi^2 + 2ml\chi + ml^2}. \tag{3.9.29}$$

As the single parameter χ defines the constraint, in this problem a graph may be plotted relating χ and the frequency. This is shown in fig. 3.9.4, where the curve is drawn for the values of the constants which are given in equation (3.9.1). The maximum and minimum values of the curve correspond to the conditions of constraint for which

$$\chi = \chi_1 = 0 \cdot 159 \,\text{ft.}, \quad \chi = \chi_2 = -0 \cdot 784 \,\text{ft.} \tag{3.9.30}$$

The corresponding ordinates are

$$\left.\begin{aligned} \omega_c^2 &= 27 \cdot 6 \,\text{rad.}^2/\text{sec.}^2 = \omega_1^2, \\ \omega_c^2 &= 148 \cdot 4 \,\text{rad.}^2/\text{sec.}^2 = \omega_2^2. \end{aligned}\right\} \tag{3.9.31}$$

These then are the natural frequencies and it will be seen that for all other values of χ the curve is such that the inequality

$$\omega_1^2 < \omega_c^2 < \omega_2^2 \tag{3.9.32}$$

is satisfied in accordance with the result (3.9.26).

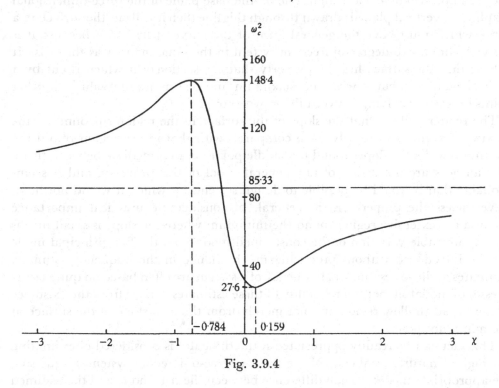

Fig. 3.9.4

This graphical representation of Rayleigh's Principle can be extended to a three-degree-of-freedom system if the graph is made three-dimensional. Two constraints will be required for such a system and they will prescribe the ratios q_2/q_1 and q_3/q_1 of the generalized co-ordinates. Let the two ratios be measured along perpendicular axes in a horizontal plane so that any point on the plane corresponds to particular values of the ratios and therefore to a particular constrained mode of the system. Now if vertical ordinates are erected from each point of the plane, the heights of the ordinates being measures of the appropriate values of ω_c^2, then the tops of the ordinates will define a surface. Rayleigh's Principle states that there will be three points on the surface at which the tangent plane is horizontal and these three points represent the natural frequencies and the corresponding principal modes.

Fig. 3.9.5 is a photograph of a surface which was constructed to show the frequencies of the constrained modes of the system of Example 3.1.2. The lowest natural frequency of the system is given by the point (1) in the figure, the second by the point (2) and the third by (3). It will be seen that (1) is a minimum,

(3) is a maximum while (2) is a saddle point; moreover, the heights of all the points on the surface are intermediate between the maximum and minimum at (3) and (1).

If one constraint only is placed upon this system then it will retain two degrees of freedom and there will be a linear relation between the quantities q_2/q_1 and q_3/q_1. This relation will define a straight line on the base plane of the three-dimensional graph. If a vertical plane is drawn through this line then it will cut the surface in a curve which must have the general form of the curve of fig. 3.9.4 because it is derived from a two-degree-of-freedom system in the same manner as this was. It follows that the surface has the property that *any* section of it which is cut by a vertical plane is a curve with one maximum and one minimum value, all other points on the curve lying between these two extremes.

The reader will see that the slope of the surface in the region surrounding the maximum value is extremely steep compared with that in the region round the minimum while the slopes round the saddle point are intermediate between these. The authors are not aware of any general proof of this property, and it seems probable that it would be possible to design systems for which it would not hold. Nevertheless the property is, in general, of considerable practical importance because the extensive region round the minimum where the slope is small means that appreciable variation of the constrained mode from the first principal mode can be introduced without producing much change in the frequency estimate. Estimates of the lowest natural frequency of a system are often based on quite crude guesses of modal shape; the reliability of these estimates derives from the existence of the broad shallow region and not merely from the zero slope of the surface at the minimum point.

This fact can be readily appreciated if the difficulty is considered of estimating the highest natural frequency of the three-degree-of-freedom system by guessing the appropriate mode; a small difference between the true mode and the assumed mode may then produce a very large frequency error and it would be most improbable that a crude guess would correspond to the small flat region on top of the sharp peak shown in the figure. This fact is not inconsistent with the original statement of Rayleigh's Principle, but it shows that the term 'small error' is of very restricted meaning when a higher mode of the system is being considered.

In a later volume, we propose to describe more fully the technique of frequency estimation by Rayleigh's method and to explain methods by the use of which it is possible to obtain second and higher order approximations successively. These techniques, when applied to n-degree-of-freedom systems are analogous to a process of finding, for the three-degree-of-freedom system, a sequence of points on the surface which progressively approach one of the flat regions. This idea is illustrated in the next paragraph by an example for a two-degree-of-freedom system. In this elementary example no special technique for the handling of the algebra is necessary. The convergence of the process, however, is ensured by the same fact as in the more complex problems, namely that the rate of change of frequency with modal shape tends to zero as a principal mode is approached.

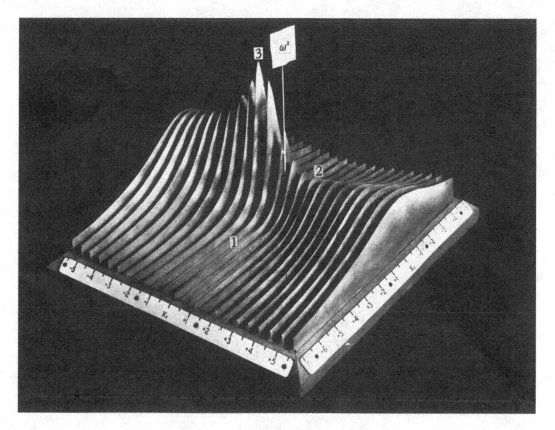

Fig. 3.9.5

Consider the system of fig. 3.9.1 for which the constants are as in equation (3.9.1). The principal modes are defined by the quantities χ_1 and χ_2 as seen from equations (3.9.30). If these modes were unknown and are to be found, we can begin by writing the equations of motion for the system which are

$$\left.\begin{aligned} M\ddot{x} + kx - mg\theta &= 0, \\ \ddot{x} + l\ddot{\theta} + g\theta &= 0. \end{aligned}\right\} \tag{3.9.33}$$

From these the equations for steady free vibration of the constrained system can be deduced, namely

$$\left.\begin{aligned} (k - M\omega^2)\,\chi - mg &= 0, \\ \chi\omega^2 + (l\omega^2 - g) &= 0. \end{aligned}\right\} \tag{3.9.34}$$

These are correct either when $\chi = \chi_1$ and $\omega^2 = \omega_1^2$ or when $\chi = \chi_2$ and $\omega^2 = \omega_2^2$. Suppose now that the expression

$$x = \frac{l\theta}{2} \quad \text{or} \quad \chi = \frac{l}{2} \tag{3.9.35}$$

is taken as a trial value of χ defining a principal mode. Equation (3.9.29) then gives

$$\omega_c^2 = 34 \cdot 91 \,\text{rad.}^2/\text{sec.}^2 \tag{3.9.36}$$

for the corresponding frequency. Now this frequency will not allow both of equations (3.9.34) to be satisfied simultaneously. We therefore take one of them only, in this case the first, and substitute in it for ω^2. This provides a second trial value for χ; thus

$$(256 - 2 \times 34 \cdot 91)\,\chi - 32 = 0 \tag{3.9.37}$$

or

$$\chi = 0 \cdot 172. \tag{3.9.38}$$

For this value, the constrained mode frequency (as given by equation (3.9.29)), is found to be

$$\omega_c^2 = 27 \cdot 62 \,\text{rad.}^2/\text{sec.}^2 \tag{3.9.39}$$

and this is a close approximation to ω_1^2. By substituting it into either of the equations (3.9.34), the first mode is found approximately. The process may then be repeated if great accuracy is required.

The reader may note than when one of the modes has been found, together with its natural frequency, there is no immediate way of finding out which one it is. In many problems experience and insight into the physical nature of the system will allow a sufficiently good guess to be made for the lowest mode to ensure that the process converges towards it.

By starting with a different trial mode, the other principal mode may be found. It is left to the reader to begin with the value

$$x = -\frac{l\theta}{2} \quad \text{or} \quad \chi = -\frac{l}{2} \tag{3.9.40}$$

in the above problem. If the second of the two equations (3.9.34) is used it will be found that the second principal mode will be reached.

The question of how to arrange the process so that a particular principal mode is reached will be discussed more fully later.

Before leaving the subject of Rayleigh's Principle, it will be useful to show how closely it is related to the equations of motion. If the expression (3.9.12) is written in the form

$$\omega_c^2 = \frac{V(\tau_1, \tau_2, \ldots, \tau_n)}{T(\tau_1, \tau_2, \ldots, \tau_n)} \qquad (3.9.41)$$

then its derivative with respect to one of the τ's becomes

$$\frac{\partial \omega_c^2}{\partial \tau_r} = \frac{1}{T}\left[\frac{\partial V}{\partial \tau_r} - \frac{V}{T}\frac{\partial T}{\partial \tau_r}\right] = \frac{1}{T}\left[\frac{\partial V}{\partial \tau_r} - \omega_c^2\frac{\partial T}{\partial \tau_r}\right]. \qquad (3.9.42)$$

Now by comparing equations (3.9.11) and (3.9.41) it will be seen that the quantity $\partial V/\partial \tau_r$ may be expanded in the form

$$\frac{\partial V}{\partial \tau_r} = (c_{r1}\tau_1 + c_{r2}\tau_2 + \ldots + c_{rn}\tau_n)\, q^2 \qquad (3.9.43)$$

so that, by equations (3.9.10),

$$\frac{\partial V}{\partial \tau_r} = (c_{r1}q_1 + c_{r2}q_2 + \ldots + c_{rn}q_n)\, q \qquad (3.9.44)$$

and similarly for $\partial T/\partial \tau_r$. If these results are substituted into the bracket of expression (3.9.42), then it becomes

$$\{(c_{r1} - \omega_c^2 a_{r1})\, q_1 + (c_{r2} - \omega_c^2 a_{r2})\, q_2 + \ldots + (c_{rn} - \omega_c^2 a_{rn})\, q_n\}\, q. \qquad (3.9.45)$$

But this is known to be zero when $\omega_c^2 = \omega_1^2, \omega_2^2, \ldots, \omega_n^2$ because the quantity within the curly brackets, when equated to zero, is one of the equations of free motion of the system as found by Lagrange's method (cf. equation (3.2.2)). It therefore follows that

$$\frac{\partial \omega_c^2}{\partial \tau_r} = 0 \quad (r = 1, 2, \ldots, n), \qquad (3.9.46)$$

when $\omega_c^2 = \omega_1^2, \omega_2^2, \ldots, \omega_n^2$.

In this section we have discussed the effects of imposing constraints on a system until it is left with but one degree of freedom. It is possible to relax the conditions of constraint to leave more degrees of freedom than one, but fewer than the number possessed by the original system. This problem is dealt with by Rayleigh[†] who deduces, in this way, many general properties of vibrating systems. We have examined some of the latter by the alternative methods of § 3.3.

EXAMPLES 3.9

1. Two equal disks of moment of inertia I are attached to a shaft. One end of the shaft is fixed and the torsional stiffness between that end and the first disk is k; the stiffness of the portion of shaft between the two disks is also k. The angular displacement of the first disk is q_1 and of the second is q_2.

Obtain an expression for the frequency of torsional vibration of the system in a mode in which the ratio of q_1 to q_2 is constrained to have the value x. Plot a graph of the frequency against x and check that the minimum and maximum values correspond with the natural frequencies of the unconstrained system.

† *Theory of Sound* (1894), §92*a*.

2. A flywheel of moment of inertia $10I$ is attached to the end of a uniform shaft of stiffness k, the other end of the shaft being fixed. Two other flywheels, each of moment of inertia I, are fixed to the shaft at points one-third and two-thirds of the way along it. Find an approximate value for the lowest natural frequency of torsional vibration by considering a constrained mode in which the amplitudes of motion of the three wheels are proportional to the distances of the wheels from the fixed end of the shaft. Compare your results with the true natural frequency.

3. In an attempt to estimate the rth natural frequency of a system, a modal shape is guessed such that
$$\rho_1, \ \rho_2, \ ..., \ \rho_{r-1}, \ \rho_{r+1}, \ ..., \ \rho_n \ll \rho_r,$$
where the ρ's relate the normalized principal co-ordinates to a single variable p as in equation (3.9.19). Show by the use of Taylor's series that the error in the square of the frequency is given approximately by
$$\omega_c^2 - \omega_r^2 \doteqdot (\omega_1^2 - \omega_r^2) \left(\frac{\rho_1}{\rho_r}\right)^2 + (\omega_2^2 - \omega_r^2) \left(\frac{\rho_2}{\rho_r}\right)^2 + ... + (\omega_n^2 - \omega_r^2) \left(\frac{\rho_n}{\rho_r}\right)^2.$$

4. Show that, in the notation of equations (3.9.19) and (3.9.21),
$$\omega_c^2 = \omega_r^2 - \left[\frac{(\omega_r^2 - \omega_1^2) \, \rho_1^2 + (\omega_r^2 - \omega_2^2) \, \rho_2^2 + ... + (\omega_r^2 - \omega_{r-1}^2) \, \rho_{r-1}^2}{\rho_1^2 + \rho_2^2 + ... + \rho_n^2}\right]$$
$$+ \left[\frac{(\omega_{r+1}^2 - \omega_r^2) \, \rho_{r+1}^2 + (\omega_{r+2}^2 - \omega_r^2) \, \rho_{r+2}^2 + ... + (\omega_n^2 - \omega_r^2) \, \rho_n^2}{\rho_1^2 + \rho_2^2 + ... + \rho_n^2}\right],$$
where $r = 1, 2, ..., n$. In view of the fact that the contents of both of the pairs of square brackets are essentially positive, what can be deduced concerning the effects of distortions in extraneous modes in an estimation of ω_r by Rayleigh's method?

3.10 Special forms of the receptances

In dealing with the properties of receptances, we have made certain assumptions as to their algebraic form. For instance, we have only discussed systems whose frequency equations have no repeated roots; again the discussion of direct receptances α_{rr} (as given by equation (3.2.10)) has been based upon the supposition that equations (3.2.26) and (3.2.27) have no common root. This section, and the one following, is concerned with the special cases which arise when these assumptions are not valid. Now the reader may well wish to omit these two sections on a first reading and to regard these special cases as being of mathematical interest only since they can never be realized *exactly* in practical systems. The arrangement of this book is such that the development of the theory will not be interrupted by this omission. These special cases are not without importance to the engineer however.

The special forms which receptances can possess are related to particular properties of the system concerned and are of interest when it is wished to examine the effects of modifications to the system. They produce branches in the receptance diagrams which are straight lines. These effects will be dealt with at some length because they will help the reader to appreciate the connexions which exist between the receptance curves, the coefficients in the expressions for receptances and the physical characteristics of the system.

Consider first the concept of a node. It often happens that, during motion in one of the principal modes, there are one or more points of a system which do not move.

If for instance the system consists of two flywheels which are joined to the ends of a light shaft, then there will be a mode in which the two wheels move in anti-phase; in this mode there must be a section of the shaft which has zero amplitude. It is usual to call this section a *node* or to say that a node exists at this section.

Now the presence of a node will not affect the form of the receptances unless it occurs at a point whose displacement is one of the co-ordinates of the system. Suppose then that there is a node, at a point of which the displacement is the co-ordinate q_r, when the system is vibrating in the kth principal mode. The existence of this mode will evidently be manifested by the relation

$$\frac{\partial q_r}{\partial p_k} = K_r = 0, \tag{3.10.1}$$

where K_r is the appropriate constant of equations (3.5.9). Since the presence of a node will only become evident in the algebra by virtue of a relation of the form (3.10.1), it is convenient to *define* a node by this relation. This is not usual in vibration theory but it is convenient here and should not cause confusion with the more generally accepted notion of a node; it is a mathematical generalization of that concept. It must be realized that, whereas if q_r denotes a translational displacement of a point or the angular displacement at a section, this definition accords with the usual physical idea, this definition is nevertheless of wider significance. For it has been shown that a generalized co-ordinate may be any linear function of displacements at a number of points; the vanishing of such a co-ordinate for motion in a principal mode may not therefore have any obvious physical significance although, by our definition, a node will be present.

In order to examine the algebraic consequences of the vanishing of one of the K_r terms it is useful to consider a limiting process. A state in which a node occurs at one co-ordinate of a given system can be approached by the artifice of changing the point on the system at which the co-ordinate is measured, or alternatively by changing the values of the parameters in the linear function which defines the co-ordinate. Thus in the system of two flywheels and a shaft, which is mentioned above, the angular displacement of some particular section of the shaft may be used as a co-ordinate; it is then possible to study the manner in which the receptances change as the section is taken closer and closer to the node.

Suppose that, in some way such as this, a state is approached in which a system has a node at the rth co-ordinate q_r when it is moving in the kth mode. The receptances at q_r are, by equations (3.4.4),

$$\left.\begin{aligned}
\alpha_{r1} &= \frac{_1A_{r1}}{\omega_1^2 - \omega^2} + \ldots + \frac{_kA_{r1}}{\omega_k^2 - \omega^2} + \ldots + \frac{_nA_{r1}}{\omega_n^2 - \omega^2}, \\[2mm]
\alpha_{r2} &= \frac{_1A_{r2}}{\omega_1^2 - \omega^2} + \ldots + \frac{_kA_{r2}}{\omega_k^2 - \omega^2} + \ldots + \frac{_nA_{r2}}{\omega_n^2 - \omega^2}, \\[2mm]
&\cdots\cdots\cdots\cdots\cdots\cdots\cdots\cdots\cdots\cdots\cdots\cdots\cdots\cdots\cdots\cdots \\[2mm]
\alpha_{rn} &= \frac{_1A_{rn}}{\omega_1^2 - \omega^2} + \ldots + \frac{_kA_{rn}}{\omega_k^2 - \omega^2} + \ldots + \frac{_nA_{rn}}{\omega_n^2 - \omega^2}.
\end{aligned}\right\} \tag{3.10.2}$$

Suppose now, that the system is excited harmonically at $q_1, q_2, ..., q_n$ in turn, the frequency being very close to ω_k. The amplitude at q_r becomes vanishingly small as q_r approaches the node, no matter where the exciting force acts. Evidently, then, all the coefficients $_kA_{r1}, \, _kA_{r2}, ..., \, _kA_{rn}$ tend to zero during the process.

These ideas may be illustrated by the system of fig. 3.10.1. If the moment of inertia about G is $\frac{3}{4}Ma^2$, this system has a second principal mode in which

$$\frac{q_1}{q_2} = -0.386 \tag{3.10.3}$$

as found in § 3.4, q_1 and q_2 being the downward displacements of the tops of the springs A and B. It will be found by simple proportion that a point C which is

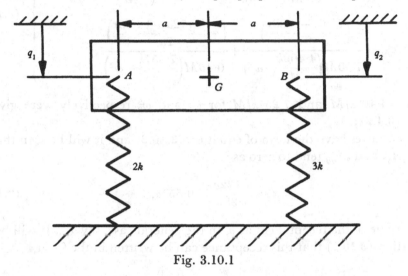

Fig. 3.10.1

distant $0.557a$ from A along AB remains at rest during motion in the second mode. If the system is analysed by the usual methods, using the data of § 3.4, then it is found that

$$2T = \frac{7M}{16}\dot{q}_1^2 + \frac{7M}{16}\dot{q}_2^2 + \frac{M}{8}\dot{q}_1\dot{q}_2. \tag{3.10.4}$$

Let principal co-ordinates be chosen such that each has unit value when $q_2 = 1$ in the relevant mode; this gives, according to our previous results,

$$\left.\begin{aligned} q_1 &= 3.88p_1 - 0.386p_2, \\ q_2 &= p_1 + p_2, \end{aligned}\right\} \tag{3.10.5}$$

so that equation (3.10.4) becomes

$$2T = 7.50Mp_1^2 + 0.45Mp_2^2. \tag{3.10.6}$$

That is, in the usual notation,

$$a_1 = 7.50M, \quad a_2 = 0.45M. \tag{3.10.7}$$

Let y be the downward displacement of a point distant x from A along AB in fig. 3.10.1; then

$$y = \frac{2a-x}{2a}q_1 + \frac{x}{2a}q_2 \tag{3.10.8}$$

or, in terms of the principal co-ordinates,

$$y = \left(\frac{7\cdot76a - 2\cdot88x}{2a}\right)p_1 + \left(\frac{-0\cdot772a + 1\cdot386x}{2a}\right)p_2. \tag{3.10.9}$$

The cross-receptance α_{y1} between y and q_1 and the direct receptance α_{yy} at y can now be formed; from equations (3.7.19) and (3.7.20) they are found to be

$$\left.\begin{array}{l} \alpha_{y1} = \dfrac{\left(\dfrac{7\cdot76a - 2\cdot88x}{2a}\right)3\cdot88}{7\cdot50M\left(\dfrac{4\cdot409k}{M} - \omega^2\right)} + \dfrac{\left(\dfrac{-0\cdot772a + 1\cdot386x}{2a}\right)(-0\cdot386)}{0\cdot45M\left(\dfrac{7\cdot257k}{M} - \omega^2\right)}, \\[6mm] \alpha_{yy} = \dfrac{\left(\dfrac{7\cdot76a - 2\cdot88x}{2a}\right)^2}{7\cdot50M\left(\dfrac{4\cdot409k}{M} - \omega^2\right)} + \dfrac{\left(\dfrac{-0\cdot772a + 1\cdot386x}{2a}\right)^2}{0\cdot45M\left(\dfrac{7\cdot257k}{M} - \omega^2\right)}. \end{array}\right\} \tag{3.10.10}$$

The values $4\cdot409k/M$ and $7\cdot257k/M$ for ω_1^2 and ω_2^2 respectively were given in equations (3.4.21).

The receptances have the form of equation (3.10.2) and it will be seen that, for instance, $_2A_{y1}$ and $_2A_{yy}$ tend to zero as

$$x \to \frac{0\cdot772a}{1\cdot386} = 0\cdot557a. \tag{3.10.11}$$

Consider one of the receptances at q_r in the general case, say α_{rs}. It will be seen from equation (3.10.2) that this receptance can be written in the form

$$\alpha_{rs} = \frac{_kA_{rs}}{\omega_k^2 - \omega^2} + \frac{f(\omega^2)}{(\omega_1^2 - \omega^2) \dots (\omega_{k-1}^2 - \omega^2)(\omega_{k+1}^2 - \omega^2) \dots (\omega_n^2 - \omega^2)}$$

$$= \frac{_kA_{rs}}{\omega_k^2 - \omega^2} + \frac{f(\omega^2)}{g(\omega^2)} \quad \text{(say)}. \tag{3.10.12}$$

Now when the limiting process is used so that the co-ordinate q_r approaches the nodal condition in the kth mode, the first term of equation (3.10.12) is very small for all values of ω other than ω_k. If therefore the applied frequency is not equal to ω_k the receptance is determined almost entirely by the second term only and this will give rise to a receptance diagram of the usual type. When the frequency becomes very close to ω_k however, the first term will become important eventually, no matter how small $_kA_{rs}$ has been made in the previous limiting process. The term will, in fact, quickly grow from a negligible value to an infinite one and will then change sign and diminish again almost to zero; all this will occur in the immediate vicinity of ω_k. In order to represent this term on the receptance diagram, therefore, a horizontal line must be drawn across it at the point where $\omega = \omega_k$. This straight line represents a branch of the diagram which becomes degenerate during the limiting process.

The receptance given in equation (3.10.12) can be written in a slightly different form which will be useful later. We first rewrite the expression as

$$\alpha_{rs} = \frac{{}_kA_{rs}.g(\omega^2) + (\omega_k^2 - \omega^2).f(\omega^2)}{(\omega_k^2 - \omega^2).g(\omega^2)}$$

(3.10.13)

and then note that this may be put in the form

$$\alpha_{rs} = \frac{(\omega_k^2 + \epsilon - \omega^2).f(\omega^2)}{(\omega_k^2 - \omega^2).g(\omega^2)},$$

(3.10.14)

where

$$\epsilon = {}_kA_{rs}\frac{g(\omega^2)}{f(\omega^2)}.$$

(3.10.15)

The expression

$$\frac{\omega_k^2 + \epsilon - \omega^2}{\omega_k^2 - \omega^2}$$

is very nearly equal to unity if ϵ is small and ω is not very close to ω_k; under these conditions its presence does not significantly affect the value of α_{rs} which therefore reduces to $f(\omega^2)/g(\omega^2)$. The above factor, however, passes through all values between $+\infty$ and $-\infty$ as ω passes through the value ω_k, *no matter how small ϵ may be.* The factor may be seen in this way to add a horizontal line to the diagram which is otherwise given by $f(\omega^2)/g(\omega^2)$.

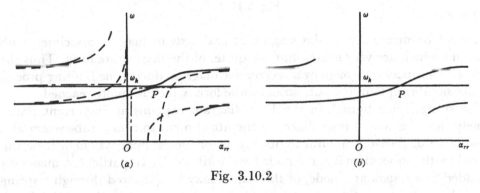

Fig. 3.10.2

As $\epsilon \to 0$ and the horizontal line appears in the receptance diagram, so the receptance approaches the form

$$\alpha_{rs} = \left[\frac{(\omega_k^2 - \omega^2)}{(\omega_k^2 - \omega^2)}\right]\frac{f(\omega^2)}{g(\omega^2)}$$

(3.10.16)

in which the factor $(\omega_k^2 - \omega^2)$ is common to the numerator and the denominator. In equation (3.10.16) we have placed the multiplier in additional square brackets to indicate that it has been arrived at in the way described and is therefore significant when $\omega = \omega_k$. We shall require this convention again later.

This result can be reached directly from consideration of the manner in which receptance diagrams change during the limiting process, without specific mention of the algebraic function. It may be deduced, for instance, from the diagram of α_{rr}. Such a diagram is shown in fig. 3.10.2 (a) where the separate curves from which the full diagram is constructed (in the manner of fig. 3.4.6 (a)) are shown dotted. The straight line branch, which is shown in fig. 3.10.2 (b), is produced by that member of the family of dotted curves which is chain-dotted and which is very close to the ω axis at all points except where $(\omega - \omega_k)$ is small. This attenuated curve arises from the fact that, when q_r is very close to a node in the kth mode, then this mode is scarcely excited by a force which is applied at q_r. Now it will be seen that a straight line in the α_{rr} diagram must produce a corresponding one in the $1/\alpha_{rr}$ diagram as in fig. 3.10.3. The argument is equally valid for any of the cross-receptance diagrams.

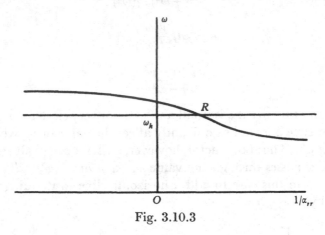

Fig. 3.10.3

It must be mentioned at this stage that real systems may be associated with diagrams which are very nearly, but not quite, of the degenerate form. Thus the stage which is passed through by a receptance diagram during the limiting process is more significant than the pure straight line form which is finally reached.

The straight line branch in the $1/\alpha_{rr}$ diagram explains an important result, namely that if a system is modified by the attachment to it of a sub-system at a co-ordinate q_r then the natural frequency ω_k of the original system will be unaltered by the sub-system if q_r is a node for the kth mode. In particular, mass may be added to a system at a node, or the system may be anchored through a spring at a node, without affecting the relevant frequency. This matter arises in the theory of vibration dampers and absorbers.

The reader will now be able to see that the limiting process which was introduced is an essential step in the discussion. This is because, without it, the straight line would be lost from the reciprocal receptance diagram and the intersection of the line with the ω axis or with the $-1/\alpha_{rr}$ curves of an attached sub-system would therefore be lost also. If this happened, one of the natural frequencies would not be shown on the diagram. This may be seen from the algebra. For if the discussion on nodes is started simply by putting $_kA_{rs}$ equal to zero, then the corresponding term in the receptance series for α_{rs} would be lost; this would mean that the degree of the denominator, if the receptance were re-expressed as a single algebraic fraction, would be reduced by one. There would therefore be one root lost from any frequency equation which might be formed from the receptance. This lost root would be that which arises from the intersection of the straight line with the axis of ω in the diagram.

We shall now examine some special cases of a receptance diagram which contains a straight line branch, beginning the discussion with an example.

The system shown in fig. 3.10.4 has co-ordinates q as indicated. It follows from the symmetry that there will be a mode which has a node at q_2. And since the system is not anchored there will also be a rigid body mode which has zero frequency.

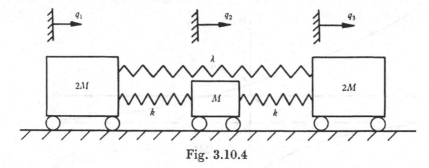

Fig. 3.10.4

The third mode will involve equal in-phase motions of the end masses while the centre mass moves in anti-phase with them. This set of modes and frequencies, together with a suitable set of principal co-ordinates, is shown by the table. It has been assumed, in deciding the order in which the frequencies are numbered, that $(2\lambda + k) < 5k$; the table then corresponds to the usual convention of numbering

	q_1	q_2	q_3	
$p_1 = 1$	$\overrightarrow{1}$	$\overrightarrow{1}$	$\overrightarrow{1}$	$\omega_1^2 = 0$
$p_2 = 1$	$\overleftarrow{1}$	0	$\overrightarrow{1}$	$\omega_2^2 = \dfrac{2\lambda + k}{2M}$
$p_3 = 1$	$\overrightarrow{1}$	$\overleftarrow{4}$	$\overrightarrow{1}$	$\omega_3^2 = \dfrac{5k}{2M}$

the natural frequencies in order of increasing magnitude. The coefficients of inertia for these principal co-ordinates are

$$a_1 = 5M, \quad a_2 = 4M, \quad a_3 = 20M, \tag{3.10.17}$$

and, by using equation (3.7.20), the direct receptance at q_2 can thus be written down. It will be found to be

$$\alpha_{22} = -\frac{1}{5M\omega^2} + \frac{0}{4M\left(\dfrac{2\lambda + k}{2M} - \omega^2\right)} + \frac{16}{20M\left(\dfrac{5k}{2M} - \omega^2\right)}, \tag{3.10.18}$$

$$= \frac{0}{4M\left(\dfrac{2\lambda + k}{2M} - \omega^2\right)} + \frac{2\omega^2 - \left(\dfrac{k}{M}\right)}{\omega^2(5k - 2M\omega^2)}, \tag{3.10.19}$$

in which the term with the zero numerator is retained since it is to be interpreted in the light of the previous discussion. The reader will be able to deduce that other receptances at q_2 also have the term with the zero numerator.

The reciprocal receptance diagram is shown in fig. 3.10.5. It will be seen to show that α_{22} vanishes when $\omega = \sqrt{(k/2M)}$. Now the height of the horizontal line in the diagram is given by $\omega_2 = \sqrt{[(2\lambda + k)/2M]}$ and it can therefore be altered by changing the value of λ. The lowest position of the line, when $\lambda = 0$, will be $\sqrt{(k/2M)}$

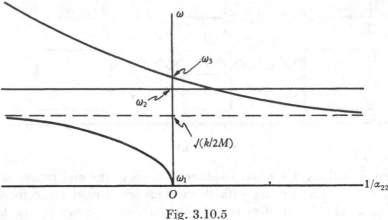

Fig. 3.10.5

and the line may be raised as much above this position as is desired. Evidently then it will be possible, by a suitable choice of λ, to arrange that the line passes through the point of intersection of the curve and the axis of ω. If this is done, the system develops special characteristics which we shall examine in the next section. First of all, however, we note that when $\lambda = 0$ the straight line coincides with an asymptote of the curve; it is necessary to examine the physical significance of this.

When a straight line in a diagram of $1/\alpha_{rr}$ coincides with an asymptote of the remainder of the curve, it means that the natural frequency ω_k which is given by the straight line is equal to an anti-resonance frequency of the system. This arises when the point P of fig. 3.10.2 (b) lies on the vertical axis and R of fig. 3.10.3 lies at infinity. It implies that if the system is excited at q_r by a force of frequency ω_k then the motion at q_r which is due to the combination of all modes other than the kth is zero because of the anti-resonance condition, while that due to the kth mode is also zero because a node exists at q_r for this mode. This is illustrated by equation (3.10.19).

This result may be expressed generally in analytical terms as follows. The direct receptance has the form

$$\alpha_{rr} = \frac{{}_kA_{rr}}{\omega_k^2 - \omega^2} + \frac{f(\omega^2)}{(\omega_1^2 - \omega^2) \dots (\omega_{k-1}^2 - \omega^2)(\omega_{k+1}^2 - \omega^2) \dots (\omega_n^2 - \omega^2)}, \quad (3.10.20)$$

and this may be rewritten as

$$\alpha_{rr} = \frac{{}_kA_{rr}}{\omega_k^2 - \omega^2} + \frac{(\omega_k^2 - \omega^2) \cdot h(\omega^2)}{g(\omega^2)} \quad (3.10.21)$$

if the contribution of all the modes to the motion at q_r is to be zero when $\omega = \omega_k$. Now, before proceeding to the limit by which $_kA_{rr} \to 0$, α_{rr} can be written in the form

$$\alpha_{rr} = \frac{_kA_{rr} \cdot g(\omega^2) + (\omega_k^2 - \omega^2)^2 \cdot h(\omega^2)}{\Delta}. \tag{3.10.22}$$

Thus when $_kA_{rr} \to 0$ there remains the expression

$$\alpha_{rr} = \frac{(\omega_k^2 - \omega^2)^2 \cdot h(\omega^2)}{\Delta} = \left[\frac{(\omega_k^2 - \omega^2)}{(\omega_k^2 - \omega^2)}\right] \frac{(\omega_k^2 - \omega^2) \cdot h(\omega^2)}{g(\omega^2)}, \tag{3.10.23}$$

where we again use the additional brackets as in equation (3.10.16). This may be compared with equation (3.10.16) for α_{rs}.

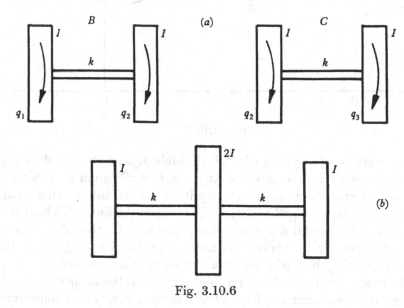

Fig. 3.10.6

The conclusion to be drawn from this result is that, whereas the presence of a straight line in the reciprocal receptance diagram corresponds to the existence of a factor which is common to Δ and Δ_{rr}, this factor is repeated in Δ_{rr} when the straight line coincides with an asymptote.

In order to illustrate this, the reader may care to find all the receptances at q_2 of the system of fig. 3.10.4. When λ is put equal to zero it will be found that α_{12} and α_{23} retain the form (3.10.16) whereas the direct receptance at q_2 takes the form (3.10.23).

This state of affairs arises, particularly, in a way which is of practical importance. This is when a system may be divided into two sub-systems, these being linked at the co-ordinate q_r only, and q_r is a node. An example of this situation is shown in fig. 3.10.6 (b) where the sub-systems into which the system may be divided are given in fig. 3.10.6 (a). They are linked by the co-ordinate q_2 and it is found (see equation (3.4.24)) that there is a node at q_2 in the second principal mode.

Now the reciprocal receptance curves for the complete system can be formed by adding those for the sub-systems in accordance with equation (3.3.1). Fig. 3.10.7 (*a*) depicts part of the curves for the sub-systems by broken lines and the combined curve by the full lines; it has been drawn for a system in which the anti-resonance frequencies for the two portions are slightly different. It can be seen that, as these two frequencies approach one another, the branch PQ of the combined curve becomes straighter and, in the limit, it produces a straight line branch which is also an asymptote.

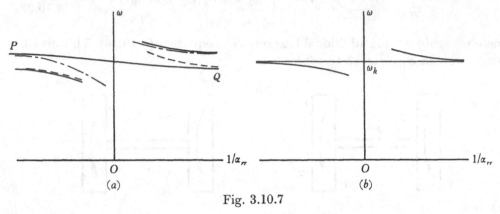

Fig. 3.10.7

In the general case of a system which is divisible at a single co-ordinate q_r with equal anti-resonance frequencies ω_k of its parts, free vibration may occur at the frequency ω_k. During the motion the amplitudes of the sub-systems must be in such a ratio as to produce equal and opposite forces at the link. This balance of the forces determines the modal shape and this shape may be altered, without interfering with the specialized character of the system, by changing one of the sub-systems but preserving the anti-resonance frequency of that sub-system at q_r.

It may be readily seen that this 'divisible' system has the properties which have already been deduced from the form of the receptances. Let a harmonic force Q_r act on the system at q_r, the frequency of the force being that of the anti-resonance of the parts. Since q_r is a node in the corresponding principal mode of the complete system, there will be no displacement at q_r due to motion in this mode. But the total displacement at the point is zero because of the properties of the sub-systems; it follows that the sum of the motions due to all the other modes must be zero also. For example, in the system of fig. 3.10.6 (*b*),

$$\alpha_{22} = \frac{\frac{1}{4}}{-I\omega^2} + \frac{0}{k - I\omega^2} + \frac{\frac{1}{4}}{2k - I\omega^2} \qquad (3.10.24)$$

so that there is a node in the second principal mode, the frequency being $\omega_2 = \sqrt{(k/I)}$. If the driving frequency ω has this value, then the first and third terms on the right-hand side of equation (3.10.24) cancel. A second example of this has already been discussed, namely the system of fig. 3.10.4 when $\lambda = 0$.

We mentioned earlier that a node, giving a straight line branch in a receptance diagram, might have no obvious physical significance if the co-ordinate at which the receptance is measured is not a simple displacement or rotation. Now, in equation (3.7.18) the direct receptance at a principal co-ordinate p_r was written as

$$\alpha_r = \frac{1}{a_r(\omega_r^2 - \omega^2)}. \tag{3.10.25}$$

The corresponding $1/\alpha_r$ curve is parabolic. It can now be seen that the parabola is not, itself, the full receptance curve but that it must be supplemented by a set of straight line branches, each one of which will correspond to one of the principal modes. In particular, if the system is not anchored, so that it has one or more zero frequencies, the $1/\alpha_r$ axis will be one of the straight branches. Thus the receptance at p_r should, strictly, be written in the form

$$\alpha_r = \frac{0}{a_1(\omega_1^2 - \omega^2)} + \frac{0}{a_2(\omega_2^2 - \omega^2)} + \cdots + \frac{1}{a_r(\omega_r^2 - \omega^2)} + \cdots + \frac{0}{a_n(\omega_n^2 - \omega^2)}. \tag{3.10.26}$$

Having examined the case where a straight line branch of the receptance diagram coincides with one of the anti-resonance frequencies of the system, we must now turn to the state of affairs where the branch coincides with one of the natural frequencies. This obtains when the point P of fig. 3.10.2 (b) lies at infinity and R of fig. 3.10.3 lies on the vertical axis. Now we shall show that this occurs when the equation

$$\Delta = 0 \tag{3.10.27}$$

has two equal roots. This condition is important and needs to be discussed at some length; it has therefore been included in a separate section.

EXAMPLES 3.10

1. Show that, when expressed in the form

$$\alpha_{rs} = (-1)^{r+s} \frac{\Delta_{rs}}{\Delta}$$

all the receptances at q_2 for the system of fig. 3.10.4 contain a factor common to Δ_{rs} and Δ in conformity with equation (3.10.16). Examine and explain the results that relate to the special case which arises when $\lambda = 0$.

2. Do the receptances α_{11} and α_{13} of the system of Ex. 3.10.1 each contain a factor common to the appropriate numerator (Δ_{11} or Δ_{13}) and Δ when they are expressed in the determinantal form? Do either of these receptances assume the form of equation (3.10.16) or equation (3.10.23) when $\lambda = 0$? Explain these results.

3.11 Systems with two or more equal frequencies

We shall here take up a matter which was referred to in the preceding section. It is that of a straight line branch in a direct receptance diagram whose height corresponds, not to an anti-resonance frequency, but to a natural frequency.

Let the diagram for $1/\alpha_{rr}$, referring to the direct receptance at a co-ordinate q_r of a particular system, contain a straight line branch at $\omega = \omega_k$ so that q_r is a node for the kth principal mode. Let another branch of the diagram cut the ω axis at ω_l, so that ω_k and ω_l are two natural frequencies. Consider now the effects of modifying the system so that these two frequencies become indefinitely close to one another. The receptance may be written in the form

$$\alpha_{rr} = \frac{{}_1A_{rr}}{\omega_1^2 - \omega^2} + \ldots + \frac{{}_kA_{rr}}{\omega_k^2 - \omega^2} + \frac{{}_lA_{rr}}{\omega_l^2 - \omega^2} + \ldots + \frac{{}_nA_{rr}}{\omega_n^2 - \omega^2} \qquad (3.11.1)$$

$$= \frac{(\omega_k^2 - \omega^2)\,(\omega_l^2 - \omega^2)f_1(\omega^2) + [{}_kA_{rr}(\omega_l^2 - \omega^2) + {}_lA_{rr}(\omega_k^2 - \omega^2)]f_2(\omega^2)}{\Delta}$$

$$(3.11.2)$$

and, for the present, we shall not introduce the limiting process by which ${}_kA_{rr} \to 0$. Instead we begin by modifying the system so that $\omega_l \to \omega_k$. The receptance (3.11.2) may then be put in the form

$$\alpha_{rr} = \left[\frac{(\omega_k^2 - \omega^2)}{(\omega_k^2 - \omega^2)}\right]\left[\frac{f_3(\omega^2)}{(\omega_1^2 - \omega^2) \ldots (\omega_k^2 - \omega^2) \ldots (\omega_n^2 - \omega^2)}\right]. \qquad (3.11.3)$$

Now the factor inside the second pair of square brackets has the usual form of a receptance for a system with $n - 1$ degrees of freedom. The factor inside the first pair of square brackets is unity for all values of ω other than those very close to ω_k, while for these particular values it gives a straight line branch on the receptance diagram in the manner which was discussed in § 3.10. This branch cuts the ω axis at the same point as one of the branches given by the bracketed expression, so that the state of affairs has been obtained which we wished to examine. Moreover, this position has been reached without our ever having used the earlier limiting process by which ${}_kA_{rr} \to 0$. Thus the characteristic diagram for a node has been obtained without our introducing the nodal state explicitly.

Since the above argument has been based solely on the equality of two natural frequencies of a system, it must apply equally well to the receptance at any co-ordinate, and that receptance may be direct or cross. There is thus a straight branch in every receptance diagram; all these branches are at the same height ω_k, and in each diagram another branch of the curve also cuts the ω axis at ω_k.

These ideas may be illustrated and extended by the use of an example. Let $\lambda = 2k$ in the system of fig. 3.10.4, so that Δ has two roots which are both equal to $5k/2M$, and consider all the receptances at q_1 and q_2. At q_1, they are

$$\alpha_{11} = \left[\frac{(5k - 2M\omega^2)}{(5k - 2M\omega^2)}\right]\left[\frac{-\frac{1}{5}}{M\omega^2} + \frac{\frac{3}{5}}{5k - 2M\omega^2}\right],$$

$$\alpha_{12} = \left[\frac{(5k - 2M\omega^2)}{(5k - 2M\omega^2)}\right]\left[\frac{-\frac{1}{5}}{M\omega^2} + \frac{-\frac{2}{5}}{5k - 2M\omega^2}\right], \qquad (3.11.4)$$

$$\alpha_{13} = \left[\frac{(5k - 2M\omega^2)}{(5k - 2M\omega^2)}\right]\left[\frac{-\frac{1}{5}}{M\omega^2} + \frac{-\frac{2}{5}}{5k - 2M\omega^2}\right]$$

while, at q_2,

$$\alpha_{21} = \left[\frac{(5k-2M\omega^2)}{(5k-2M\omega^2)}\right]\left[\frac{-\frac{1}{5}}{M\omega^2}+\frac{-\frac{2}{5}}{5k-2M\omega^2}\right],$$

$$\alpha_{22} = \left[\frac{(5k-2M\omega^2)}{(5k-2M\omega^2)}\right]\left[\frac{-\frac{1}{5}}{M\omega^2}+\frac{\frac{8}{5}}{5k-2M\omega^2}\right], \qquad (3.11.5)$$

$$\alpha_{23} = \left[\frac{(5k-2M\omega^2)}{(5k-2M\omega^2)}\right]\left[\frac{-\frac{1}{5}}{M\omega^2}+\frac{-\frac{2}{5}}{5k-2M\omega^2}\right].$$

All these receptances will be seen to have the straight line factor

$$\left[\frac{(5k-2M\omega^2)}{(5k-2M\omega^2)}\right].$$

If the second of the fractions within the brackets is used to determine the modal shape which is associated with the frequency $\sqrt{(5k/2M)}$ then, from the receptances at q_1 it will be seen that

$$q_1:q_2:q_3::3:-2:-2 \qquad (3.11.6)$$

while, from the receptances at q_2,

$$q_1:q_2:q_3::-1:4:-1. \qquad (3.11.7)$$

The modal shape thus depends upon the point at which the indefinitely small exciting force is imagined to be applied; this property does not exist in systems without multiple roots.

The special characteristics which are found in systems that have two or more equal roots of Δ are, then, that all the receptance diagrams have straight line branches at the value of the repeated root and that the modal shape associated with the repeated root is not unique. These two characteristics are related, the first being a consequence of the second. The reason for the lack of uniqueness in the modes is that, if two principal modes have the same frequency, the system can vibrate freely in any form which is a linear combination of these modes. Such a linear combination can, itself, be regarded as a new principal mode and there are thus an infinite number of possible principal modes. Further it will always be possible to choose a form for one of the modes which will give a node at some specified co-ordinate; the receptance diagrams for that co-ordinate, and therefore for any co-ordinate must have a straight line branch.

A simple example of these ideas is provided by a particle which is constrained to move in a plane and is controlled by two light springs which lie in the plane and are mutually perpendicular. This is shown in fig. 3.11.1. If the spring stiffnesses k and λ are different, then the system has two natural frequencies, $\sqrt{(k/m)}$ and $\sqrt{(\lambda/m)}$, the corresponding principal modes being motions along the axes of the two

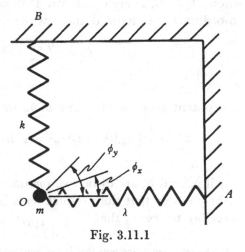

Fig. 3.11.1

springs. But if $k = \lambda$, then any straight line motion of the particle in the plane becomes possible and may be regarded as a principal mode. If the receptance diagram is drawn for the displacement in any specified direction, then free motion along a line perpendicular to this direction will be possible and this motion will give a node, in accordance with our definition, at the co-ordinate which is measured along the specified direction.†

If, in the above example, a single force in a specified direction produces forced motion, then the principal mode which is obtained by the usual limiting process (by which the amplitude of the force is diminished as its frequency is adjusted to approach the natural frequency) involves motion in the direction of the force. This agrees with the result which was found previously, whereby the modal form depended upon the co-ordinate at which the vanishingly small force was imagined to be applied.

If the forced motion of a system having multiple roots is to be analysed, and if principal co-ordinates are to be used, there will be an infinite choice of such co-ordinates since there is an infinite choice of principal modes. It is necessary to note, however, that if one principal co-ordinate, corresponding to a double root of Δ, has been chosen then the second co-ordinate corresponding to this root is thereby determined; this is apart from the usual matter of scale factor. This follows because the principal modes which are chosen must be made orthogonal if their co-ordinates are to yield equations which are independent.

Let p_k and p_l be two *orthogonal* principal co-ordinates corresponding to the same frequency ω_k; further, let the associated stability coefficients be c_k and c_l so that the potential energy will be given by

$$2V = c_k p_k^2 + c_l p_l^2. \tag{3.11.8}$$

Now suppose that two more principal co-ordinates are defined by the relations

$$p_x = A p_k + B p_l, \quad p_y = C p_k + D p_l, \tag{3.11.9}$$

where A, B, C, D are constants. If these equations are solved to give the original co-ordinates p in terms of the new ones, it is found that

$$p_k = \frac{D p_x - B p_y}{z}, \quad p_l = \frac{A p_y - C p_x}{z}, \tag{3.11.10}$$

where

$$z = AD - BC. \tag{3.11.11}$$

The potential energy, in terms of p_x and p_y, will therefore be given by

$$2V = \frac{1}{z^2} \left[(c_k D^2 + c_l C^2) p_x^2 + (c_k B^2 + c_l A^2) p_y^2 - 2(c_k BD + c_l AC) p_x p_y \right]. \tag{3.11.12}$$

But if p_x and p_y are to be two *consistent* principal co-ordinates then they must be orthogonal so that the product term in the above expression must vanish. It is therefore necessary that

$$c_k BD + c_l AC = 0. \tag{3.11.13}$$

---

† The reader may also note that linear combinations of two straight line displacements can give circular or elliptical motions if complex coefficients are used. We intend to develop this idea in a later volume.

If, then, A and B have been chosen to define p_x, the ratio of C to D which fixes the shape of the other principal mode, is determined.

The advantage of choosing *orthogonal* principal co-ordinates, even in a very simple problem, will be seen if these ideas are applied to the spring-controlled particle of fig. 3.11.1. It has been shown that, when the two spring stiffnesses are both equal to k, principal modes can be taken as straight line displacements in any direction. We shall first choose *any* two directions to define the principal modes and hence the principal co-ordinates and show subsequently how the equations are simplified by choosing perpendicular directions. Let p_x and p_y be the principal co-ordinates which are displacements measured in directions that make angles ϕ_x and ϕ_y respectively with the direction OA of one of the springs; this is shown in fig. 3.11.1. Using these co-ordinates, we now wish to determine the motion produced by the force $F e^{i\omega t}$ which makes an angle θ with OA.

Displacements of the particle along the directions OA and OB will be taken as corresponding to p_k and p_l of the above argument. They are given in terms of p_x and p_y by the equations

$$\left.\begin{aligned} p_k &= p_x \cos \phi_x + p_y \cos \phi_y, \\ p_l &= p_x \sin \phi_x + p_y \sin \phi_y, \end{aligned}\right\} \tag{3.11.14}$$

these being equivalent to the relations (3.11.10). The energy expressions, in terms of p_x and p_y will therefore be

$$\begin{aligned} 2V &= k[(p_x \cos \phi_x + p_y \cos \phi_y)^2 + (p_x \sin \phi_x + p_y \sin \phi_y)^2] \\ &= k[p_x^2 + p_y^2 + 2 p_x p_y \cos (\phi_x - \phi_y)], \end{aligned} \tag{3.11.15}$$

and similarly

$$2T = m[\dot{p}_x^2 + \dot{p}_y^2 + 2 \dot{p}_x \dot{p}_y \cos (\phi_x - \phi_y)]. \tag{3.11.16}$$

The components of the applied force in the p_x and p_y directions are

$$\cos (\phi_x - \theta) F e^{i\omega t}, \quad \cos (\phi_y - \theta) F e^{i\omega t}$$

and these are the generalized forces Q_x and Q_y as may be verified by the use of equation (2.2.9). The equations of motion are, by Lagrange's method,

$$\left.\begin{aligned} m\ddot{p}_x + m\ddot{p}_y \cos (\phi_x - \phi_y) + kp_x + kp_y \cos (\phi_x - \phi_y) &= F \cos (\phi_x - \theta) e^{i\omega t}, \\ m\ddot{p}_y + m\ddot{p}_x \cos (\phi_x - \phi_y) + kp_y + kp_x \cos (\phi_x - \phi_y) &= F \cos (\phi_y - \theta) e^{i\omega t}. \end{aligned}\right\} \tag{3.11.17}$$

If solutions are now sought of the form

$$p_x = X e^{i\omega t}, \quad p_y = Y e^{i\omega t}, \tag{3.11.18}$$

where X and Y are constants, it is found that

$$\left.\begin{aligned} (k - m\omega^2) X + (k - m\omega^2) \cos (\phi_x - \phi_y) Y &= F \cos (\phi_x - \theta), \\ (k - m\omega^2) \cos (\phi_x - \phi_y) X + (k - m\omega^2) Y &= F \cos (\phi_y - \theta). \end{aligned}\right\} \tag{3.11.19}$$

These equations can be solved and produce the results

$$\left.\begin{aligned} X &= \frac{-F}{(k - m\omega^2)} \frac{\sin (\phi_y - \theta)}{\sin (\phi_x - \phi_y)}, \\ Y &= \frac{F}{(k - m\omega^2)} \frac{\sin (\phi_x - \theta)}{\sin (\phi_x - \phi_y)}. \end{aligned}\right\} \tag{3.11.20}$$

From these, the displacements in any specified direction can be found. But if the directions in which p_x and p_y are measured are chosen to be perpendicular, so that

$$\phi_x - \phi_y = \tfrac{1}{2}\pi \tag{3.11.21}$$

the equations of motion (3.11.19) reduce to

$$\left.\begin{aligned} (k - m\omega^2)\, X &= F\cos(\phi_x - \theta), \\ (k - m\omega^2)\, Y &= F\cos(\phi_y - \theta), \end{aligned}\right\} \tag{3.11.22}$$

so that the solution can be written down immediately.

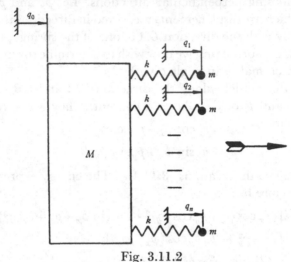

Fig. 3.11.2

It has already been implied that the characteristics found in a system which has two equal roots are also present if the number of equal roots is greater than two. If this is so, the choice of a set of mutually orthogonal modes may be difficult. A system of this sort is shown in fig. 3.11.2; here, the motion of each of the n small masses m, and of the large mass M, can only occur in the direction shown by the arrow. This type of system is met with in the analysis of a turbine in which a rotor carries a number of similar blades. A method of selecting a set of orthogonal principal modes for this system has, in fact, been explained.†

EXAMPLES 3.11

1. A rigid body is free to move in three dimensions so that it has six natural frequencies all of which are zero. How may six principal co-ordinates be chosen if they are to be mutually orthogonal?

2. A uniform rigid bar of mass M and length $2l$ is supported horizontally on two springs each of stiffness k, each of the springs being distant a from the mid-point. Considering vibration in the vertical plane what must be the value of a/l if the system is to have two equal frequencies?

If one principal mode is taken to be an angular displacement of the bar about one end what must the second principal mode be if it is orthogonal to the first?

† D. C. Johnson and R. E. D. Bishop, 'The Modes of Vibration of a Certain System having a Number of Equal Frequencies', *Journ. Appl. Mech.* vol. 23 (1956), p. 379.

NOTATION

Supplementary List of Symbols used in Chapter 3

A_r, B_r, etc.	Constants relating co-ordinates p to co-ordinates q; see equations (3.5.9).
A'_r, B'_r, etc.	Constants relating co-ordinates q to co-ordinates p; see equations (3.5.10).
C_m	Constant introduced in limiting process of equation (3.4.5).
x_i, y_i, z_i	Co-ordinates, relative to fixed axes, of ith particle.
ρ_r	Quantities defining the shape of a constrained mode; see equation (3.9.19).
τ_r	Quantities defining the shape of a constrained mode; see equations (3.9.10).
ω_c	Natural frequency of constrained system.

CHAPTER 4

THE TAUT STRING

The philosophical theory of harmonics, or of the combinations of sounds, was considered by the ancients as affording one of the most refined employments of mathematical speculation; nor has it been neglected in modern times, but it has been in general either treated in a very abstruse and confused manner, or connected entirely with the practice of music, and habitually associated with ideas of mere amusement.

THOMAS YOUNG, *A Course of Lectures on Natural Philosophy and the Mechanical Arts* (1807)

4.1 Introduction

In Chapter 3 we were concerned with systems which have a finite number of degrees of freedom. In this chapter, and in some subsequent ones, we shall treat systems in which the freedom is infinite so that the configuration cannot be expressed in terms of a finite number of co-ordinates. Real systems may be thought of as being composed of elastic bodies and therefore of possessing infinite freedom; they may be idealized, for the purposes of analysis, either into finite- or infinite-freedom systems depending upon their nature and upon the convenience with which the equations of motion may be handled. The reader will be able to judge the relative merits of the two methods when he has studied the analysis for systems having infinite freedom.

The small transverse vibrations of a uniform stretched string form a suitable introduction to the general theory of elastic bodies which is to follow. The string theory is not of much immediate engineering importance; it is presented here because it involves particularly simple analytical functions and because the physical meaning of the equations is readily seen. The analysis of more complicated systems, for instance that of the torsional and flexural vibrations of bars, will be found to have much in common with the string theory.

It may appear at first that the finite-freedom theory of the last chapter can be made to cover the analysis of the string merely by increasing the number of co-ordinates sufficiently. Thus a uniform string can be idealized into a light string to which a number of equal massive particles are fixed at equal intervals. The displacements of the particles can be used as co-ordinates and their number n can be increased without limit so as to represent the motion of a uniform string with sufficient exactness. Such an analysis is possible and can be found, for instance, in Rayleigh's *Sound*.†

This procedure has two drawbacks however. First of all, it introduces a mathematical difficulty if we attempt to extend the process to a limit in which n becomes infinite, because it then becomes impossible to solve the simultaneous equations or to postulate that they can be solved. It also introduces a practical difficulty

† *Theory of Sound* (1894), §120.

because the analogous process for other elastic bodies is unmanageably complicated. The other means for analysing the motion of a string, and later the motion of shafts and beams, is to form the partial differential equation for the motion of an element and to seek solutions of the equation which give the same characteristics to the oscillatory motion as those which we have described for finite-freedom systems. The advantage of this process is that it avoids cumbersome sets of simultaneous equations and also sheds new light on vibration theory generally. We shall show that the ideas of principal modes and natural frequencies, and of the orthogonality of the modes, are again valid and useful.

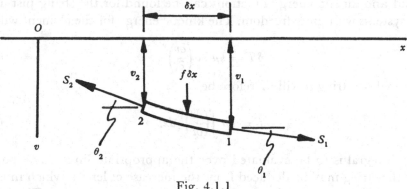

Fig. 4.1.1

Consider the short element of string 2,1 shown in fig. 4.1.1. The x-axis is drawn along the undeflected position of the string and the mass per unit length of the string is μ; S_1 and S_2 are the tensions on the two sides of the element and v is the mean displacement of the element at right angles to the x axis. The equation of motion for the element is

$$S_1 \sin \theta_1 - S_2 \sin \theta_2 + f(x, t)\, \delta x = \mu\, \delta x \frac{\partial^2 v}{\partial t^2}, \tag{4.1.1}$$

where $f(x, t)$ is the magnitude per unit length of any externally applied force.

If the motion is restricted to small amplitudes only the slopes θ_1 and θ_2 are small and are given approximately by

$$\left(\frac{\partial v}{\partial x}\right)_1$$

and

$$\left(\frac{\partial v}{\partial x}\right)_2$$

respectively. We may also put S_1 and S_2 equal to the initially applied tension S. If equation (4.1.1) is divided by the quantity δx, these changes are made, and δx is taken as indefinitely small, then the equation reduces to

$$S \frac{\partial^2 v}{\partial x^2} + f(x, t) = \mu \frac{\partial^2 v}{\partial t^2}. \tag{4.1.2}$$

This is the partial differential equation governing the deflexions v.

This problem has been treated as a two-dimensional one although a string is usually free to vibrate in any direction perpendicular to its axis. The treatment is sufficient for most purposes because any vibration of the string can be resolved into components in two planes, preferably at right angles, and each of the components will be governed by an equation like (4.1.2). This matter is mentioned here because it shows that the string is a system which has pairs of equal frequencies and is therefore comparable with the systems which have one pair of equal frequencies and which are described in § 3.11. Moreover, although the ability to vibrate in more than one plane is not of any great importance here, it is of great significance during the flexural vibration of shafts which are rotating. We shall treat this subject in a later volume.

Potential and kinetic energy functions can be found for the string just as they could for systems with finite freedom. The kinetic energy for the element will be

$$\delta T = \tfrac{1}{2}\mu\,\delta x \left(\frac{\partial v}{\partial t}\right)^2, \tag{4.1.3}$$

and for the whole string it will therefore be

$$T = \tfrac{1}{2}\mu \int \left(\frac{\partial v}{\partial t}\right)^2 dx, \tag{4.1.4}$$

where the integral is to be evaluated over the appropriate limits. The potential energy of the string may be deduced from the increase of length which must take place when it is deflected without moving the points of anchorage. The increase of length will be given approximately by

$$\int \left[\sqrt{\left\{ 1 + \left(\frac{\partial v}{\partial x}\right)^2 \right\}} - 1 \right] dx = \frac{1}{2} \int \left(\frac{\partial v}{\partial x}\right)^2 dx,$$

and if the string is sufficiently low in elastic modulus for the tension to be sensibly unchanged, as we have assumed earlier, then the potential energy will be

$$V = \tfrac{1}{2}S \int \left(\frac{\partial v}{\partial x}\right)^2 dx. \tag{4.1.5}$$

Again appropriate limits are to be used for the integral.

EXAMPLES 4.1

1. By considering the deflected form of a taut string as having been brought about by a static lateral applied force distribution, and by equating the work which the force would do during the deflexion to the potential energy of the string, show that the latter is given by

$$V = -\tfrac{1}{2}S \int_0^l v\,\frac{\partial^2 v}{\partial x^2}\,dx.$$

Show that this agrees with equation (4.1.5).

2. A taut string is deflected by a single concentrated force which acts at a distance h from one end. If the deflexion under the force is v_1, find the energy stored in the string and the magnitude of the force, by using equation (4.1.5).

3. Having noted that Ex. 4.1.2 cannot be done by using the formula of Ex. 4.1.1, consider a string that is deflected by a force which is uniformly distributed over a very short length ϵ of the string, the ends of the loaded segment being distant h and $h+\epsilon$ from one end. Using the formula of Ex. 4.1.1 and assuming that the whole of the segment ϵ is equally deflected, form an expression for the potential energy using equation (4.1.2) to find $\partial^2 v/\partial x^2$; show that the expression is equal to half the product of the deflexion of the segment and the total force.

4.2 The receptances of a taut string

It is now necessary to investigate the response of strings to harmonically varying forces and, in doing this, we shall show that the concept of receptance is again useful.

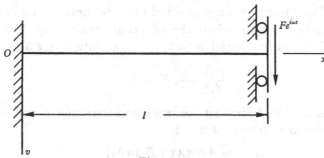

Fig. 4.2.1

So far nothing has been said about the nature of the supports to which the string is attached at its ends. These supports may be rigidly fixed or they may be subject to some displacement in a direction perpendicular to the axis of the string; it is of course essential to the simple theory that they should not move in the direction of the string axis since the tension S is constant. For the purpose of analysis, it is convenient to think of two idealized forms of support. One of these is the rigid support which we shall indicate diagrammatically as shown on the left of fig. 4.2.1, while the other is a support which has zero lateral stiffness and which will be depicted as shown on the right of fig. 4.2.1. The vibration of a string which is supported at a point which has a finite lateral stiffness can be analysed by using the receptance for the freely supported end together with the receptance for the spring of the actual support. We shall next find the receptances for strings having the various possible sets of supports.

4.2.1 The fixed-free string

Let a string of length l be supported as shown in fig. 4.2.1. and let a force $F e^{i\omega t}$ be applied to the right-hand end. It is assumed that there are no other external forces perpendicular to the string.

The tensioning device at the right-hand end is assumed to be light. It follows that, at the right-hand end,

$$F e^{i\omega t} = S \frac{\partial v}{\partial x},$$

(4.2.1)

while at the left-hand end, $v = 0$. It is now assumed that the string will have a steady forced motion similar to that of a system with finite freedom; thus a solution of the differential equation must be found which fits the end conditions and which consists of the product of a function of x and the factor $e^{i\omega t}$. Such a solution can be written in the form

$$v = X(x)\, e^{i\omega t}. \tag{4.2.2}$$

It implies, if X is real, that the vibrations of all the points of the string are in phase with the driving force $F e^{i\omega t}$.

The equation (4.1.2) may be written

$$a^2 \frac{\partial^2 v}{\partial x^2} = \frac{\partial^2 v}{\partial t^2}, \tag{4.2.3}$$

where $a^2 = S/\mu$. The f term has been omitted because there are no applied lateral forces except those at the ends, where the effects are treated by choosing suitable end conditions. Substituting equation (4.2.2) into equation (4.2.3) gives

$$\frac{d^2 X}{dx^2} + \frac{\omega^2}{a^2} X = 0, \tag{4.2.4}$$

the factor $e^{i\omega t}$ being omitted because it does not affect the vanishing of the expression. The general solution of equation (4.2.4) is

$$X = A \cos \lambda x + B \sin \lambda x, \tag{4.2.5}$$

where A and B are arbitrary constants and

$$\lambda = \frac{\omega}{a} = \omega \sqrt{\frac{\mu}{S}}. \tag{4.2.6}$$

If axes of x and v are used as shown in the figure, it will be found that this solution satisfies the end conditions when

$$A = 0, \quad B = \frac{F}{S\lambda \cos \lambda l}.$$

That is to say, the motion of the string which is produced by the force is

$$v = \frac{F \sin \lambda x}{S\lambda \cos \lambda l} e^{i\omega t}. \tag{4.2.7}$$

This expressions gives the displacement at any point on the string and thereby gives the receptance between the end and any point. We shall write for the cross-receptance

$$\alpha_{xl} = \frac{\sin \lambda x}{S\lambda \cos \lambda l} \tag{4.2.8}$$

and, for the direct receptance at the end $x = l$,

$$\alpha_{ll} = \frac{\tan \lambda l}{S\lambda}. \tag{4.2.9}$$

The suffices of α follow a slightly different convention from that which was used previously because we cannot now number the co-ordinates at which the receptances are measured, there being an infinite number of them. The new suffices denote, in the order of writing them, the distances along the string at which the displacement is to be measured and at which the harmonic force is applied.

The receptances can again be used to find a frequency equation because the amplitude will be infinite at natural frequencies. Here the frequency equation is

$$\lambda l \cos \lambda l = 0. \tag{4.2.10}$$

This has the roots $\lambda l = \tfrac{1}{2}\pi, \tfrac{3}{2}\pi, \tfrac{5}{2}\pi, \ldots$, and so on indefinitely together with the zero root for which the string remains straight as it swings about the fixed end.[†] The natural frequencies which correspond to these solutions are

$$\omega_1 = \frac{\pi a}{2l}, \quad \omega_2 = \frac{3\pi a}{2l}, \quad \ldots, \quad \omega_r = \frac{(2r-1)\,\pi a}{2l}, \quad \ldots. \tag{4.2.11}$$

The zero root has been omitted from these in the numbering of the ω's. It will be noted that there is an infinite number of natural frequencies.

If the reciprocal of the receptance α_{ll} is plotted against ω, in the manner discussed in § 3.3, then it gives a curve with the characteristics that were found previously except that now the number of branches is infinite. Again each branch ends at an asymptote which corresponds to an anti-resonance frequency. These frequencies, which occur alternatively with the natural frequencies, are given by

$$\sin \lambda l = 0. \tag{4.2.12}$$

This has solutions $\lambda l = 0, \pi, 2\pi, \ldots$, where the zero root must be disregarded as it does not correspond to a zero value of the receptance; the relevant frequencies are

$$\omega = \frac{\pi a}{l}, \quad \frac{2\pi a}{l}, \quad \ldots. \tag{4.2.13}$$

These frequencies are, by the same arguments as were used for finite-freedom systems, the natural frequencies of the string when both its ends are fixed. We shall examine this condition again later.

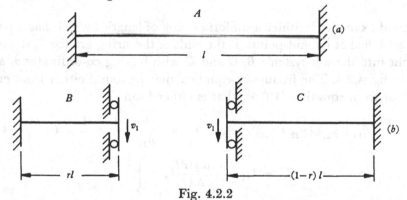

Fig. 4.2.2

The receptances (4.2.8) and (4.2.9) can be used with the formulae of Tables 2 and 3; this may be illustrated by the following example. Let a string with two fixed ends, shown in fig. 4.2.2 (a), be regarded as being built up from two fixed-free

† Motion of this type would only be possible in reality if the surface on which the end of the string is supported were circular with centre at O.

strings as indicated in fig. 4.2.2 (*b*), so that some fraction *r* of the total length is the length of one sub-system. In the usual notation, the sub-systems being denoted by *B* and *C* as shown, the frequency equation for the whole string is

$$\beta_{11} + \gamma_{11} = 0, \qquad (4.2.14)$$

where, by equation (4.2.9),

$$\beta_{11} = \frac{\tan \lambda rl}{S\lambda}, \qquad \gamma_{11} = \frac{\tan \lambda (1-r) l}{S\lambda}. \qquad (4.2.15)$$

The frequency equation is thus

$$\tan \lambda rl + \tan \lambda (1-r) l = 0,$$

and since

$$\tan A + \tan B = \frac{\sin (A+B)}{\cos A \cos B}$$

it may also be written as

$$\sin \lambda l = 0. \qquad (4.2.16)$$

This agrees with equation (4.2.12).

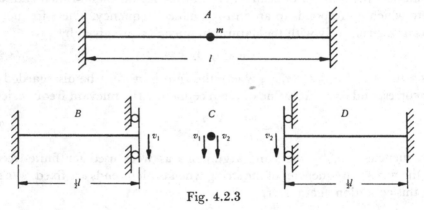

Fig. 4.2.3

As a second example, consider a uniform string of length *l* which has a particle of mass *m* attached at its mid-point. If the ends of the string are fixed, the system may be split into the sub-systems *B*, *C* and *D* with linking co-ordinates v_1 and v_2 as shown in fig. 4.2.3. The frequency equation may be found either from entry 6 of Table 3 or from equation (1.6.5), that is either from

$$(\beta_{11} + \gamma_{11}) (\gamma_{22} + \delta_{22}) - \gamma_{12}^2 = 0 \quad \text{or} \quad \frac{1}{\beta_{11}} + \frac{1}{\gamma_{11}} + \frac{1}{\delta_{22}} = 0, \qquad (4.2.17)$$

where

$$\left.\begin{array}{l} \beta_{11} = \delta_{22} = \dfrac{\tan \frac{1}{2}(\lambda l)}{S\lambda}, \\[2mm] \gamma_{11} = \gamma_{22} = \gamma_{12} = -\dfrac{1}{m\omega^2}. \end{array}\right\} \qquad (4.2.18)$$

Substitution into either of equations (4.2.17) gives the frequency equation

$$\frac{m\omega a}{2S} \tan \left(\frac{\lambda l}{2}\right) = 1, \qquad (4.2.19)$$

which gives those natural frequencies for which the corresponding mode involves movement of the particle. The reader may examine the motion for which the particle remains at rest in the light of our discussion of such special cases in Chapter 3.

The problem of the vibration of a particle on a taut string is discussed by Rayleigh.†

4.2.2 The free-free string

Let both ends of a uniform string be supported at points which are free to move perpendicular to the string, as in fig. 4.2.4. Let an external force $Fe^{i\omega t}$ be applied at the right-hand end so that the end conditions are

$$S\frac{\partial v}{\partial x} \quad \text{or} \quad \frac{\partial v}{\partial x} = 0 \quad \text{when} \quad x = 0, \\ S\frac{\partial v}{\partial x} = Fe^{i\omega t} \quad \text{when} \quad x = l.$$

(4.2.20)

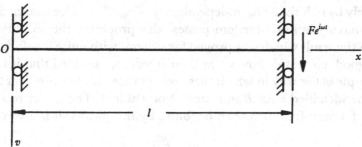

Fig. 4.2.4

The constants A and B which are required for the function (4.2.5) are now

$$A = -\frac{F}{S\lambda \sin \lambda l}, \quad B = 0$$

so that the displacements at all points along the string are given by

$$v = -\frac{F \cos \lambda x}{S\lambda \sin \lambda l} e^{i\omega t}.$$

(4.2.21)

From this the receptances are found to be

$$\alpha_{xl} = -\frac{\cos \lambda x}{S\lambda \sin \lambda l},$$

(4.2.22)

$$\alpha_{0l} = -\frac{1}{S\lambda \sin \lambda l},$$

(4.2.23)

$$\alpha_{ll} = -\frac{\cot \lambda l}{S\lambda}.$$

(4.2.24)

The suffix 0 is now found in the receptances whereas previously there were only l and x. This is because there is now a displacement at the end where $x = 0$, whereas previously this end was fixed.

† *Theory of Sound* (1894), §136.

The natural frequencies of the free-free string are given by

$$\lambda l \sin \lambda l = 0 \qquad (4.2.25)$$

this being the condition under which all the receptances become infinite. The frequencies are

$$\omega_0 = 0, \quad \omega_1 = \frac{\pi a}{l}, \quad \omega_2 = \frac{2\pi a}{l}, \quad \dots, \quad \omega_r = \frac{r\pi a}{l}, \quad \dots . \qquad (4.2.26)$$

The zero frequency reflects the string's ability to move bodily. Again it will be found that the receptance diagrams which can be drawn for the functions $1/\alpha_{ll}$, etc., are of the usual form and have an infinite number of branches.

The receptances may be used with Table 2 and 3 as before, but this statement requires a note to justify it. The tables were derived on the assumption that the cross-receptances α_{rs} and α_{sr} are equal and, while this may be readily proved for a system with finite freedom, we have so far not proved it for infinite freedom. It may, however, be deduced from the symmetry of the expressions in this instance or alternatively by rederiving α_{l0} independently of α_{0l}. It will be assumed hereafter that systems having infinite freedom possess this property; the validity of this is inferred from the result which was proved for systems with finite freedom since that result holds good, no matter how great their freedom, provided that it is finite.

As an example of the way in which these receptances may be used, let the system of fig. 4.2.4 be identified with B in system 7 of Table 2. The direct receptance at the free end of a fixed-free string can be found by this means; it is given by

$$\alpha_{ll} = \beta_{22} - \frac{\beta_{12}^2}{\beta_{11}}, \qquad (4.2.27)$$

where

$$\beta_{11} = \beta_{22} = -\frac{\cot \lambda l}{S\lambda} \qquad (4.2.28)$$

by equation (4.2.24) and

$$\beta_{12} = -\frac{\csc \lambda l}{S\lambda} \qquad (4.2.29)$$

by equation (4.2.23). Substitution of these values gives equation (4.2.9).

Consider now the system of a free-free string with a massive particle attached to it at one end as shown in fig. 4.2.5. The above results can be used in conjunction with the tables to find the direct receptance of the string at the opposite end from the particle. The system may be split up as shown and the direct receptance at the left-hand end of the system A is then given by

$$\alpha_{11} = \beta_{11} - \frac{\beta_{12}^2}{\beta_{22} + \gamma_{22}} \qquad (4.2.30)$$

in accordance with entry 1 of Table 2. Here,

$$\beta_{11} = \beta_{22} = -\frac{\cot \lambda l}{S\lambda}, \quad \beta_{12} = -\frac{\csc \lambda l}{S\lambda} \qquad (4.2.31)$$

and

$$\gamma_{22} = -\frac{1}{m\omega^2}. \qquad (4.2.32)$$

On substituting these values back into equation (4.2.30), it is found that

$$\alpha_{11} = \frac{m\omega^2 - S\lambda \cot \lambda l}{S\lambda(m\omega^2 \cot \lambda l + S\lambda)}. \qquad (4.2.33)$$

It will be seen that if m is made zero in this expression it reduces to that of equation (4.2.24) as required.

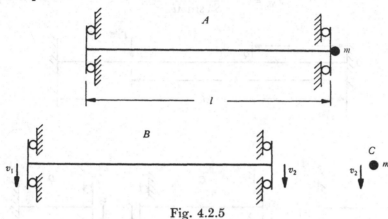

Fig. 4.2.5

4.2.3 The fixed-fixed string

If the ends of a string are fixed then there will be no deflexions at the ends $x = 0$ and $x = l$ and we shall be concerned therefore only with receptances at points other

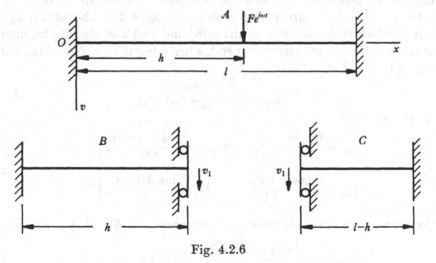

Fig. 4.2.6

than the ends. Let a force $Fe^{i\omega t}$ act at a point distant h from the left-hand end as shown in fig. 4.2.6 and let the system be split into sub-systems as shown. The direct receptance α_{hh} can now be written in terms of the receptances of the sub-systems; it is

$$\alpha_{hh} = \frac{\beta_{11}\gamma_{11}}{\beta_{11} + \gamma_{11}}, \qquad (4.2.34)$$

where, from equation (4.2.9),

$$\beta_{11} = \frac{\tan \lambda h}{S\lambda}, \quad \gamma_{11} = \frac{\tan \lambda(l-h)}{S\lambda}. \tag{4.2.35}$$

From these relations it is found that

$$\alpha_{hh} = \frac{\sin \lambda h \sin \lambda(l-h)}{S\lambda \sin \lambda l}. \tag{4.2.36}$$

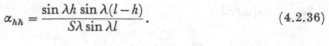

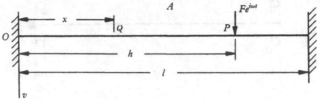

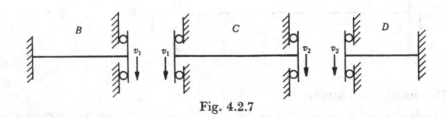

Fig. 4.2.7

Finding the cross-receptances between two points on the string requires the string to be split into three sub-systems as shown in fig. 4.2.7. The force is applied at a point P which is distant h from the left-hand end and the displacement is required at a point Q distant x from that end, x being less than h. Now from entry 2 of Table 2, it is seen that

$$\alpha_{xh} = \frac{\beta_{11}\gamma_{12}\delta_{22}}{(\beta_{11}+\gamma_{11})(\gamma_{22}+\delta_{22}) - \gamma_{12}^2}, \tag{4.2.37}$$

where, in this problem

$$\left.\begin{array}{l} \beta_{11} = \dfrac{\tan \lambda x}{S\lambda}, \quad \gamma_{11} = \gamma_{22} = -\dfrac{\cot \lambda(h-x)}{S\lambda}, \\[2mm] \gamma_{12} = -\dfrac{\operatorname{cosec} \lambda(h-x)}{S\lambda}, \quad \delta_{22} = \dfrac{\tan \lambda(l-h)}{S\lambda}. \end{array}\right\} \tag{4.2.38}$$

When these expressions are substituted into equation (4.2.37), they give

$$\alpha_{xh} = \frac{\sin \lambda x \sin \lambda(l-h)}{S\lambda \sin \lambda l} \quad (0 \leqslant x \leqslant h). \tag{4.2.39}$$

By a similar process it may be shown that the displacements at points to the right of P, where $x > h$, are given by the receptance

$$\alpha_{xh} = \frac{\sin \lambda h \sin \lambda(l-x)}{S\lambda \sin \lambda l} \quad (h \leqslant x \leqslant l). \tag{4.2.40}$$

It may be seen to follow from equations (4.2.39) and (4.2.40) that the two cross-receptances α_{xh} and α_{hx} between points P and Q are equal because each may be written in the form

$$\frac{\sin \lambda a \sin \lambda b}{S \lambda \sin \lambda l},$$

where a and b are the distances of the points from the ends of the string as shown in fig. 4.2.8.

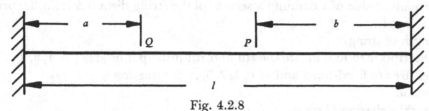

Fig. 4.2.8

The receptances (4.2.36), (4.2.39) and (4.2.40) give the frequency equation (4.2.12) which has already been found for this type of string and which has been found also to govern the frequencies of the free-free string. Some interesting comments on the receptances can be found in Duncan's monograph.†

4.2.4 Table of receptances

The receptances for all possible end conditions and points of deflexion can be found by similar methods to those used above. Such receptances may then be used to synthesize those of more complicated systems. Table 4 contains a number of these receptances.

For all the string systems it may be shown, as has been done for the fixed-fixed string, that the two cross-receptances relating to any two points are equal. This may be done without using equations like those of Table 2 which presuppose this property.

When receptances for systems having infinite freedom are expressed (as are those in this section) in terms of trigonometrical or other functions, they will be referred to as being 'in closed form'. This distinguishes them from functions which are expressed as infinite series and which are referred to as being 'in open form' or 'in series form'. This latter form will be introduced shortly.

EXAMPLES 4.2

1. Find from first principles the two expressions for α_{xh}, of a fixed-free string, which are given in Table 4.

2. Show that the sth principal mode of the system of fig. 4.2.3, *in which the particle moves* has a natural frequency which approaches the value

$$\omega_s = \frac{2(s-1)\,\pi a}{l}$$

when s becomes large.

† W. J. Duncan, *R. and M. 2000* (1947), p. 40.

TABLE 4

LATERAL VIBRATION OF TAUT UNIFORM STRINGS

Notation

a $= \sqrt{(S/\mu)}$.

h Particular value of x defining a section of the string distant h from the origin $(0 \leqslant h \leqslant l)$.

l Length of string.

r, s Subscripts used to indicate the rth and sth principal modes $(= 1, 2, 3, \ldots$ for fixed-free or fixed-fixed and $= 0, 1, 2, 3, \ldots$ for free-free).

S Tension.

v Lateral deflexion of string.

x Distance along the string $(0 \leqslant x \leqslant l)$.

α_{xh} Receptance between the sections x and h. If a harmonic point force $F e^{i\omega t}$ is applied laterally to a taut string at the section $x = h$, then the deflexion at any point x of the string is given by

$$v = \alpha_{xh} F e^{i\omega t}.$$

λ $= \omega \sqrt{(\mu/S)} = \omega/a$.

μ Mass per unit length of string.

Equation of motion: $S \dfrac{\partial^2 v}{\partial x^2} = \mu \dfrac{\partial^2 v}{\partial t^2}$.

Boundary conditions: $v = 0$ at a fixed end,

$$\frac{\partial v}{\partial x} = 0 \text{ at a free end.}$$

By a 'free end' of a taut string is meant an end at which no *lateral* restraint is applied; there is, however, some means of maintaining the tension in the string.

Energy expressions:

$$T = \frac{\mu}{2} \int_0^l \left(\frac{\partial v}{\partial t}\right)^2 dx, \quad V = \frac{S}{2} \int_0^l \left(\frac{\partial v}{\partial x}\right)^2 dx.$$

In the series representation of deflexion the scale of measurement of distortions in the principal modes is such that

$$v = \sum_{r=0}^{\infty} p_r \cdot \phi_r(x) \text{ (free-free)},$$

or

$$v = \sum_{r=1}^{\infty} p_r \cdot \phi_r(x) \quad \text{(fixed-free or fixed-fixed)}.$$

TABLE 4

LATERAL VIBRATION OF TAUT UNIFORM STRINGS

Quantity	Fixed-free $v=0$ when $x=0$ $\partial v/\partial x=0$ when $x=l$	Free-free $\partial v/\partial x=0$ when $x=0$ $\partial v/\partial x=0$ when $x=l$	Fixed-fixed $v=0$ when $x=0$ $v=0$ when $x=l$
α_{xh} $0\leqslant x\leqslant h$ $0\leqslant h\leqslant l$	$\dfrac{\sin \lambda x \cos \lambda(l-h)}{S\lambda \cos \lambda l}$	$-\dfrac{\cos \lambda x \cos \lambda(l-h)}{S\lambda \sin \lambda l}$	$\dfrac{\sin \lambda x \sin \lambda(l-h)}{S\lambda \sin \lambda l}$
α_{xh} $h\leqslant x\leqslant l$ $0\leqslant h\leqslant l$	$\dfrac{\sin \lambda h \cos \lambda(l-x)}{S\lambda \cos \lambda l}$	$-\dfrac{\cos \lambda h \cos \lambda(l-x)}{S\lambda \sin \lambda l}$	$\dfrac{\sin \lambda h \sin \lambda(l-x)}{S\lambda \sin \lambda l}$
α_{xh} $0\leqslant x\leqslant l$ $0\leqslant h\leqslant l$	$\dfrac{2}{\mu l}\sum\limits_{r=1}^{\infty}\dfrac{\phi_r(x)\,\phi_r(h)}{\omega_r^2-\omega^2}$	$\dfrac{2}{\mu l}\sum\limits_{r=0}^{\infty}\dfrac{\phi_r(x)\,\phi_r(h)}{\omega_r^2-\omega^2}$	$\dfrac{2}{\mu l}\sum\limits_{r=1}^{\infty}\dfrac{\phi_r(x)\,\phi_r(h)}{\omega_r^2-\omega^2}$
$\omega_r=\sqrt{\left(\dfrac{c_r}{a_r}\right)}$	$\dfrac{(2r-1)\,\pi a}{2l}$	$\dfrac{r\pi a}{l}$	$\dfrac{r\pi a}{l}$
$\phi_r(x)$	$\sin\dfrac{(2r-1)\,\pi x}{2l}$	$\dfrac{1}{\sqrt{2}}\ (r=0),$ $\cos\dfrac{r\pi x}{l}\ (r=1,2,3,\ldots)$	$\sin\dfrac{r\pi x}{l}$
$\displaystyle\int_0^l \phi_r(x)\,\phi_s(x)\,dx$	$0\ \ (r\neq s)$ $l/2\ \ (r=s)$	$0\ \ (r\neq s)$ $l/2\ \ (r=s)$	$0\ \ (r\neq s)$ $l/2\ \ (r=s)$
a_r	$\mu l/2$	$\mu l/2$	$\mu l/2$
c_r	$\dfrac{(2r-1)^2\,\pi^2 S}{8l}$	$\dfrac{r^2\pi^2 S}{2l}\ (r=0,1,2,3,\ldots)$	$\dfrac{r^2\pi^2 S}{2l}$

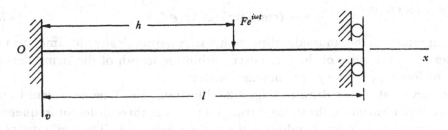

Fixed end   Free end

3. Show that the first principal mode of the system of fig. 4.2.3, in which the particle moves, has a frequency that tends towards the value $\omega = 2\sqrt{(S/ml)}$ as the mass of the string becomes indefinitely small. Show also that this value may be obtained for the light string by direct analysis.

4. A string of length l is stretched between two fixed supports; the central portion of the string, which has length rl, is of mass per unit length μ' while the two end portions have mass per unit length μ. Find the frequency equation and show that, if $r \to 0$ while the product $\mu' rl$ remains constant, the equation gives the natural frequencies for a string with a particle as shown in fig. 4.2.3.

5. A uniform disk of mass M is held by two strings which are attached to it at the ends of a diameter, the two strings each having a length l and being fixed at the ends remote from the disk. The diameter of the disk is a. Obtain the equation which gives the natural frequencies of the system for vibrations which are wholly in the plane of the disk and in which the disk has:

(a) angular motion about its axis;
(b) transverse motion without rotation.

The usual notation for the mass and tension in the string should be used.

4.3 The principal modes

It is possible to find principal modes of vibration for systems with infinite freedom, and, in particular, for strings. To show this, we shall use a method which is comparable with that used for finite-freedom systems. But, instead of using receptances expressed in the series form, as in § 3.4, we shall use them in the closed form which we have already examined; the procedure is like that required to solve Ex. 3.4.4.

If a fixed-free string is acted on by a harmonic force at its free end, the steady forced motion is such that a displacement

$$v = \alpha_{xl} F e^{i\omega t} = \frac{F \sin \lambda x}{S\lambda \cos \lambda l} e^{i\omega t} \tag{4.3.1}$$

is produced at any point x. The expression (4.3.1) shows that the form (in space) of the forced mode of vibration is sinusoidal, since the expression for the displacement may be written as

$$v = (\text{constant}) \sin \lambda x \,.\, e^{i\omega t}. \tag{4.3.2}$$

The magnitude of the sinusoidal distortion varies harmonically (in time) and the number of wavelengths of the sine curve within the length of the string depends upon the forcing frequency and increases with it.

The curves which are drawn in fig. 4.3.1 illustrate this. They represent instantaneous displacements of the string during vibration at three different frequencies, the lowermost curve corresponding to the highest frequency. The ordinates of the curves have been drawn much larger than would be permissible if they were to represent directly the displacements of a real string if the simple theory is to be applicable; this has been done for clarity. The tensioning device at the right-hand end of the string, which was shown on previous diagrams, has also been omitted for simplicity; it will not be shown in string diagrams henceforward.

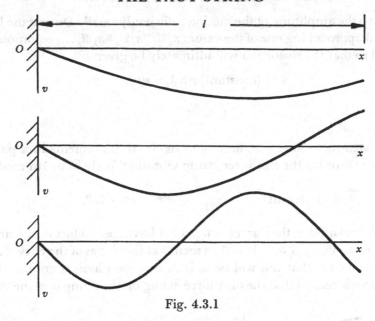

Fig. 4.3.1

Now the natural frequencies of the string are given by $\lambda l = \frac{1}{2}\pi, \frac{3}{2}\pi, \frac{5}{2}\pi, \ldots$. We shall therefore examine the distorted form of the string as the frequency approaches one of these values just as we did for the finite-freedom systems. As before, the amplitude of the force, F, must be diminished as the denominator of α_{xl} decreases

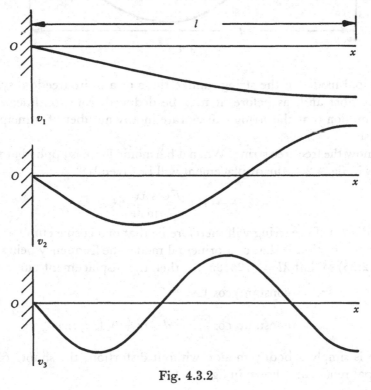

Fig. 4.3.2

in order to keep the amplitude of the motion sufficiently small. During the limiting process, with λ approaching one of the values $\pi/2l, 3\pi/2l, 5\pi/2l, \ldots$, equation (4.3.2) remains valid so that the distortion will ultimately be given by

$$v = (\text{constant}) \sin \lambda_r x \,.\, e^{i\omega_r t}, \tag{4.3.3}$$

where

$$\lambda_r = \frac{\omega_r}{a}. \tag{4.3.4}$$

It will be convenient to affix a suffix to v to signify displacements in a particular principal mode; thus for the fixed-free string vibrating freely in its rth mode,

$$v_r = (\text{constant}) \sin \frac{(2r-1)\,\pi x}{2l} \,.\, e^{i\omega_r t} \quad (r = 1, 2, \ldots). \tag{4.3.5}$$

For these principal modes, the curves of fig. 4.3.1 have special forms, the first three of which are shown in fig. 4.3.2. It will be seen that the slopes of the curves are zero at the right-hand end; that this will be so is evident on physical grounds because the applied force is zero so that the only force acting on the string is in the direction of the x axis.

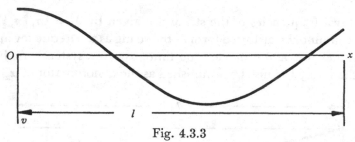

Fig. 4.3.3

The principal modes of the string, unlike those of a finite-freedom system, are infinite in number and, as before, it may be deduced from the linearity of the equation of motion that the string can vibrate in any number of principal modes simultaneously.

Consider now the free-free string. When a harmonic force is applied at the right-hand end, as in fig. 4.2.4, the displacement will be given by

$$v = \alpha_{xl} F e^{i\omega t} = -\frac{F \cos \lambda x}{S\lambda \sin \lambda l} e^{i\omega t}. \tag{4.3.6}$$

The distorted form of the string will therefore be that of a cosine curve as shown in fig. 4.3.3. If the motion is that of a principal mode, the frequency being given by equation (4.2.25) so that $\lambda l = \pi, 2\pi, 3\pi, \ldots$, then the displacement curve takes the special form

$$v_r = (\text{constant}) \cos \lambda_r x \,.\, e^{i\omega_r t}$$

$$= (\text{constant}) \cos \frac{r\pi x}{l} \,.\, e^{i\omega_r t} \quad (r = 0, 1, 2, \ldots). \tag{4.3.7}$$

The 0-mode is simply a bodily motion without distortion; the shapes of the first three principal modes are shown in fig. 4.3.4.

The principal modes of the fixed-fixed string can be found by using the receptances which were derived in the previous section. These are

$$v = \frac{F \sin \lambda x \sin \lambda(l-h)}{S\lambda \sin \lambda l} e^{i\omega t} \quad (0 \leqslant x \leqslant h),$$
$$v = \frac{F \sin \lambda h \sin \lambda(l-x)}{S\lambda \sin \lambda l} e^{i\omega t} \quad (h \leqslant x \leqslant l).$$

(4.3.8)

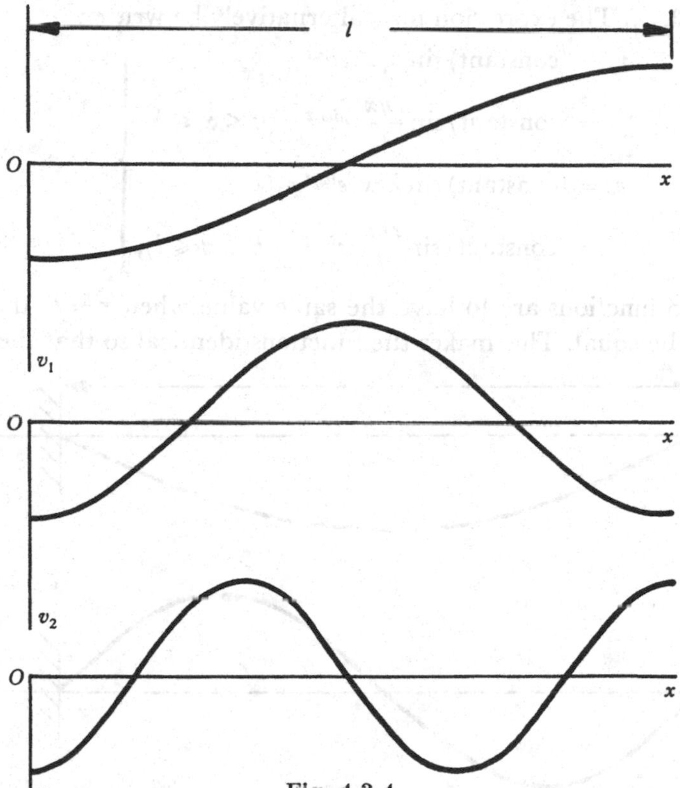

Fig. 4.3.4

They show that the string has a sinusoidal distortion on either side of the point of excitation. In general the two sine curves do not have the same amplitude although they have the same wavelength. Each has the same displacement at $x = h$, where

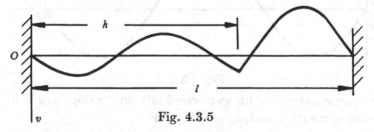

Fig. 4.3.5

they meet, but the slope at this point will not be continuous unless the applied force is zero. At frequencies other than natural frequencies, the instantaneous form of the string will be similar to that shown in fig. 4.3.5.

It was deduced from equation (4.2.12) that the natural frequencies are given by $\lambda l = \pi, 2\pi, 3\pi, \ldots$. These give the deflexions

$$\left.\begin{aligned}
v_r &= \text{(constant)} \sin\frac{r\pi x}{l}\sin r\pi\left(1-\frac{h}{l}\right).e^{i\omega_r t} \quad (0 \leqslant x \leqslant h),\\
v_r &= \text{(constant)} \sin\frac{r\pi h}{l}\sin r\pi\left(1-\frac{x}{l}\right).e^{i\omega_r t} \quad (h \leqslant x \leqslant l),
\end{aligned}\right\} \tag{4.3.9}$$

where $r = 1, 2, 3, \ldots$. The expression may alternatively be written

$$\left.\begin{aligned}
v_r &= \text{(constant)} \sin\lambda_r x . e^{i\omega_r t}\\
&= \text{(constant)} \sin\frac{r\pi x}{l} . e^{i\omega_r t} \quad (0 \leqslant x \leqslant h),\\
v_r &= \text{(constant)} \sin\lambda_r x . e^{i\omega_r t}\\
&= \text{(constant)} \sin\frac{r\pi x}{l} . e^{i\omega_r t} \quad (h \leqslant x \leqslant l),
\end{aligned}\right\} \tag{4.3.10}$$

and if these two functions are to have the same value when $x = h$, then the two constants must be equal. This makes the functions identical so that the distortion

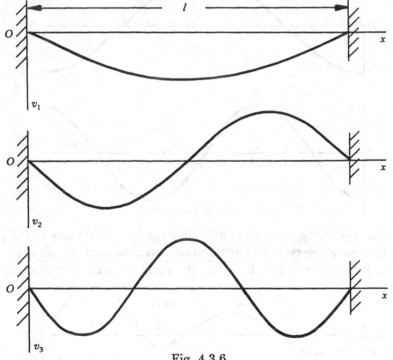

Fig. 4.3.6

becomes one continuous wave with no discontinuity in the slope at $x = h$. The form of free vibration in the rth principal mode is thus

$$\begin{aligned}
v_r &= \text{(constant)} \sin\lambda_r x . e^{i\omega_r t}\\
&= \text{(constant)} \sin\frac{r\pi x}{l} . e^{i\omega_r t} \quad (r = 1, 2, \ldots) \tag{4.3.11}
\end{aligned}$$

over the whole length of the string $(0 \leqslant x \leqslant l)$. The shapes of the first three modes are shown in fig. 4.3.6.

The principal modes and their frequencies, for the fixed-free, free-free and fixed-fixed string will be found in Table 4; the modes are specified in terms of 'characteristic functions' whose significance is explained in the next section. The strings are able to vibrate in any number of the modes simultaneously, each with its appropriate frequency, as do systems with finite freedom.

EXAMPLE 4.3

1. Deduce the principal modes of a free-free string from the two receptances

$$\alpha_{xh} = -\frac{\cos \lambda x \cos \lambda (l-h)}{S\lambda \sin \lambda l} \quad (0 \leqslant x \leqslant h),$$

$$\alpha_{xh} = -\frac{\cos \lambda h \cos \lambda (l-x)}{S\lambda \sin \lambda l} \quad (h \leqslant x \leqslant l),$$

where $h \neq 0$ and $h \neq l$.

4.4 Principal co-ordinates and characteristic functions

The treatment in this chapter of the vibration theory of elastic bodies does not begin, as does that in Chapter 3 for finite-freedom systems, with the choice of a set of generalized co-ordinates. Indeed it is not clear at this stage how a set of co-ordinates could be selected. It will be shown later that, in fact, it is still possible to use the Lagrangian method; but for the present we may note that the principal modes which have been found can be used to define a set of principal co-ordinates. The type of co-ordinate so defined is of great value in the analysis of systems having infinite freedom.

If a fixed-free string is given a static deflexion in its rth principal mode then the distortion is given by

$$v_r(x) = C_r \sin \lambda_r x = C_r \sin \frac{(2r-1)\pi x}{2l}, \tag{4.4.1}$$

where C_r is a constant. Similarly during vibration with frequency ω in this principal mode the distortion is, at any instant,

$$v_r(x, t) = C_r \sin \lambda_r x . e^{i\omega t}. \tag{4.4.2}$$

A principal co-ordinate p_r can now be selected, whose value will specify the magnitude of the distortion in this particular mode.

The selection of a principal co-ordinate will involve, as explained in §3.5, a choice of meaning for unit magnitude of the co-ordinate. For uniform strings a natural choice is to make $p_r = 1$ correspond to unit amplitude of the sine wave deflexion; thus for the fixed-free string equations (4.4.1) and (4.4.2) become

$$v_r = p_r \sin \lambda_r x, \tag{4.4.3}$$

where p_r is a constant for a steady deflexion and is a harmonically varying quantity during harmonic vibration. We shall select co-ordinates in this way; but the scaling of the $r = 0$ mode for the free-free string we leave open for the moment. Thus

$$
\left.
\begin{aligned}
v_r(x, t) &= p_r \sin \lambda_r x = p_r \sin \frac{(2r-1)\,\pi x}{2l} && \text{(fixed-free)} \\[4pt]
v_0(x, t) &= p_0 C_0; \quad v_r(x, t) = p_r \cos \lambda_r x = p_r \cos \frac{r\pi x}{l} && \text{(free-free)} \\[4pt]
v_r(x, t) &= p_r \sin \lambda_r x = p_r \sin \frac{r\pi x}{l} && \text{(fixed-fixed)}
\end{aligned}
\right\} (r = 1, 2, 3, \ldots).
$$

$$(4.4.4)$$

If the relations (4.4.4) are compared with equations (4.3.5), (4.3.7) and (4.3.11), then it will be seen that during *free* vibration

$$p_r = \Pi_r e^{i\omega_r t} \quad (r = 1, 2, \ldots), \tag{4.4.5}$$

where Π_r is an arbitrary complex constant. This result is valid for all types of string and is the same as that found in § 3.5 for systems having finite freedom.

It is now convenient to introduce the function $\phi_r(x)$, which is defined as

$$
\left.
\begin{aligned}
\phi_r(x) &= \sin \lambda_r x = \sin \frac{(2r-1)\,\pi x}{2l} && \text{(fixed-free)} \\[4pt]
\phi_0(x) &= C_0; \quad \phi_r(x) = \cos \lambda_r x = \cos \frac{r\pi x}{l} && \text{(free-free)} \\[4pt]
\phi_r(x) &= \sin \lambda_r x = \sin \frac{r\pi x}{l} && \text{(fixed-fixed)}
\end{aligned}
\right\} (r = 1, 2, 3, \ldots).
$$

$$(4.4.6)$$

It is called a 'characteristic function' and its form depends upon the end conditions of the string to which it relates. By using this notation it will be possible to express, in the form of a single equation, an important mathematical relation which holds in different forms for different strings. The characteristic function for a particular type of string defines the modal shape for the string and may be compared with the sets of constants $_rA_{uv}$ which were introduced in § 3.4 and which defined the modal shapes of finite-freedom systems. For many purposes the characteristic function is more convenient to use than the sets of constants.

Equations (4.4.4), which give the motion of a string in its rth principal mode, may now be written in the form

$$v_r(x, t) = p_r(t) \cdot \phi_r(x). \tag{4.4.7}$$

Further, since motion may occur in any number of principal modes simultaneously, a more general expression for the deflexion is

$$v(x, t) = \sum_{r=0 \text{ or } 1}^{\infty} v_r(x, t) = \sum_{r=0 \text{ or } 1}^{\infty} p_r(t) \cdot \phi_r(x). \tag{4.4.8}$$

If the motion is harmonically forced, then it will occur in all the modes with the driving frequency. But if the motion is free, each modal component will have its own frequency in accordance with equation (4.4.5).

It was shown in Chapter 3 that any distortion of a finite-freedom system could be regarded as a combination of deflexions in the n principal modes, the magnitudes of the deflexions being given by the values of principal co-ordinates $p_1, p_2, ..., p_n$ (see equation (3.5.9)). We may expect, on physical grounds, that this will hold also for infinite-freedom systems. Thus, for taut strings in particular it should be possible to represent *any* distorted form by a series of the form (4.4.8), assuming that the distortion is consistent with the physical conditions under which the analysis is valid. We shall show shortly how the magnitudes of the various terms in the series may be chosen to correspond to a specified distortion; for this purpose we shall require the orthogonality relations and the nature of these relations is discussed in the next section.

EXAMPLES 4.4

1. Express the displacement of a fixed-free string in the first principal mode in terms of the principal co-ordinate p_1 if $p_1 = 1$ when the deflexion half-way along the string is unity.

2. The fifth principal co-ordinate p_5 of a fixed-fixed string is selected to have unit value when the deflexion is the negative of a sine wave of amplitude 2. Find an expression for the deflexion of the string in this mode.

4.5 Orthogonality

It may be remembered that the orthogonal property of the principal modes for a finite-freedom system was deduced from the form of the energy expressions when these were written in terms of the principal co-ordinates. We shall treat orthogonality for infinite freedom by again starting with the energy expressions, and begin with a particular case.

Let a fixed-fixed string be oscillating in two of its principal modes simultaneously these being the rth and the sth. The deflexion at any point will be

$$v = p_r \sin \frac{r\pi x}{l} + p_s \sin \frac{s\pi x}{l}. \tag{4.5.1}$$

The kinetic energy of the whole string can be calculated from equation (4.1.4); thus

$$T = \frac{\mu}{2} \int_0^l \left(\frac{\partial v}{\partial t}\right)^2 dx$$

$$= \frac{\mu}{2} \left\{ \dot{p}_r^2 \int_0^l \sin^2 \frac{r\pi x}{l} dx + \dot{p}_s^2 \int_0^l \sin^2 \frac{s\pi x}{l} dx \right.$$

$$\left. + 2\dot{p}_r \dot{p}_s \int_0^l \sin \frac{r\pi x}{l} \sin \frac{s\pi x}{l} dx \right\}. \tag{4.5.2}$$

This reduces to

$$T = \frac{\mu l}{4} (\dot{p}_r^2 + \dot{p}_s^2) \tag{4.5.3}$$

provided that $r \neq s$ because, since r and s are integers,

$$\int_0^l \sin \frac{r\pi x}{l} \sin \frac{s\pi x}{l} dx = 0. \tag{4.5.4}$$

It may be noted here that our choice for unit value of the p's results in the same inertia coefficient (a_r or a_s in our previous notation) for all the co-ordinates.

Again, the potential energy expression may be formed from equation (4.1.5); it is

$$V = \frac{S}{2} \int_0^l \left(\frac{\partial v}{\partial x}\right)^2 dx$$

$$= \frac{S}{2}\left\{ p_r^2 \frac{r^2\pi^2}{l^2} \int_0^l \sin^2 \frac{r\pi x}{l} dx + p_s^2 \frac{s^2\pi^2}{l^2} \int_0^l \sin^2 \frac{s\pi x}{l} dx \right.$$

$$\left. + 2p_r p_s \frac{rs\pi^2}{l^2} \int_0^l \sin \frac{r\pi x}{l} \sin \frac{s\pi x}{l} dx \right\} \tag{4.5.5}$$

and this reduces to

$$V = \frac{S\pi^2}{4l} (r^2 p_r^2 + s^2 p_s^2). \tag{4.5.6}$$

The form of this and of equation (4.5.3) shows that the modes are orthogonal, as they are in finite-freedom systems. That is to say, the total kinetic energy during any motion is the sum of the separate energies due to each of the modes which are present, and similarly for the potential energy. The proof of these facts has been entirely algebraic and is therefore not dependent upon whether the motion is free or forced. The full energy expressions for the fixed-fixed string may now be written

$$2T = a_1 \dot{p}_1^2 + a_2 \dot{p}_2^2 + a_3 \dot{p}_3^2 + \dots, \tag{4.5.7}$$

$$2V = c_1 p_1^2 + c_2 p_2^2 + c_3 p_3^2 + \dots, \tag{4.5.8}$$

where

$$a_r = \frac{\mu l}{2}, \quad c_r = \frac{S\pi^2 r^2}{2l}.$$

The above proof for the orthogonality of the modes of a fixed-fixed string depends upon the vanishing of the integral

$$\int_0^l \sin \frac{r\pi x}{l} \sin \frac{s\pi x}{l} dx \tag{4.5.9}$$

when $r \neq s$. Now in other elastic-body problems, the expressions within the corresponding integral may not be so simple and the mathematical proof may then become arduous. It should be noted therefore that it is not necessary to adopt this type of proof because orthogonality may be deduced from the argument which was used in §3.6; in that method we assume simultaneous free vibration in two modes and then show that the total energy cannot remain constant throughout unless there are no product terms in the T and V expressions. Although the argument is based on free motion the algebraic relation which it gives is applicable to any motion.

The reader may verify that the orthogonal relations hold for strings with other end-fixings, using the mathematical process; for the fixed-free string

$$a_r = \frac{\mu l}{2}, \quad c_r = \frac{S(2r-1)^2\pi^2}{8l} \quad (r = 1, 2, 3, \dots). \tag{4.5.10}$$

and for the free-free string

$$a_r = \frac{\mu l}{2}, \quad c_r = \frac{S\pi^2 r^2}{2l} \quad (r = 0, 1, 2, 3, \dots), \tag{4.5.11}$$

provided the constant C_0 in equation (4.4.4) is given the value $1/\sqrt{2}$.

If the principal modes of a string are known they may be used to find the natural frequencies, again following the technique for systems having finite freedom. Thus the relation $T_{\text{max.}} = V_{\text{max.}}$ for harmonic motion in a single mode gives, for the fixed-free string,

$$\omega_r^2 = \frac{c_r}{a_r} = \frac{(2r-1)^2\pi^2 S}{4\mu l^2} = \left[\frac{(2r-1)\pi a}{2l}\right]^2 \qquad (4.5.12)$$

which agrees with the result found previously.

In §3.8, certain results were shown to be corollaries of the orthogonality relation. These results apply also when the system concerned has infinite freedom as we shall now show, confining our treatment to a fixed-fixed string and leaving the reader to examine the other types.

Consider a new function $f_r(x, t)$ which represents an applied force per unit length which would produce deflexion in the rth principal mode only. Now the applied *static* force which would produce a *static* deflexion

$$v_r(x) = C_r \sin\frac{r\pi x}{l} \qquad (4.5.13)$$

in the rth mode may be found from equation (4.1.2). This gives

$$f_r(x) = -S\frac{d^2 v_r}{dx^2} = C_r\left(\frac{Sr^2\pi^2}{l^2}\right)\sin\frac{r\pi x}{l} \qquad (4.5.14)$$

which shows that a harmonic force distribution *in space* is required to give a harmonic displacement in space. The work which would be done by the force distribution $f_r(x)$ if, while it was acting, the string were to be displaced in the sth mode, is

$$W = \int_0^l f_r(x) \cdot v_s(x)\, dx = C_r C_s\left(\frac{Sr^2\pi^2}{l^2}\right)\int_0^l \sin\frac{r\pi x}{l}\sin\frac{s\pi x}{l}\, dx = 0. \qquad (4.5.15)$$

Thus the force distribution which will produce a static deflexion in the rth mode will do no work during a distortion in the sth mode. This is true for all the strings, whatever may be their end-fixings. The general result is

$$\int_0^l f_r(x) \cdot v_s(x)\, dx = 0 \quad (r \neq s). \qquad (4.5.16)$$

This may be compared with the relation given by equation (3.8.8).

As with systems having finite freedom, comparable results hold for the work which would be done against the inertia forces; but the physical significance of this is less evident.

EXAMPLE 4.5

1. Remove the restriction in the derivation of equation (4.5.16) to static forces by showing that

$$\int_0^l f_r(x, t) \cdot v_s(x, t)\, dx = 0,$$

where both f_r and v_s are proportional to $e^{i\omega t}$.

[NOTE. This result may be used at several points in the ensuing theory. We shall prefer, however, to restrict ourselves to the simpler result (4.5.16) which will be sufficient for our purposes.]

4.6 Receptances at the principal co-ordinates; receptances in series form

It is often convenient to use the receptances at the principal co-ordinates in calculations on the vibration of taut strings. These receptances relate the variation of the principal co-ordinates to the generalized forces which are associated with them. We begin by examining the nature of these generalized forces, following the reasoning of § 2.2.

Let a static force distribution $f(x)$ be applied to a taut string, thereby producing a deflexion. Consider the work which will be done by the force system during a

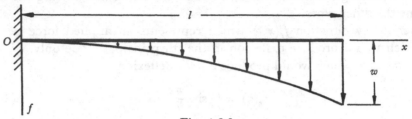

Fig. 4.6.1

small increment δp_r of the deflexion at the rth principal co-ordinate p_r; this will imply a small increase in the deflexion in the rth principal mode. The work will be

$$\delta W_r = \int_0^l f(x) \,.\, \delta p_r \, \phi_r(x) \, dx = \delta p_r \int_0^l f(x) \, \phi_r(x) \, dx \qquad (4.6.1)$$

and the generalized force P_r corresponding to p_r will therefore be

$$P_r = \int_0^l f(x) \,.\, \phi_r(x) \, dx. \qquad (4.6.2)$$

As an illustration of this result, consider the force distribution shown in fig. 4.6.1. The loading is parabolic, of the form

$$f(x) = \frac{wx^2}{l^2} \qquad (4.6.3)$$

and the string is of the fixed-free type. Equation (4.6.2) shows that the generalized force at the rth principal co-ordinate is

$$P_r = \frac{w}{l^2} \int_0^l x^2 \sin \frac{(2r-1)\,\pi x}{2l} \, dx$$

$$= \frac{8wl}{(2r-1)^2 \pi^2} \left[\sin \frac{(2r-1)\,\pi}{2} - \frac{2}{(2r-1)\,\pi} \right]. \qquad (4.6.4)$$

As a second example, consider the fixed-free string shown in fig. 4.6.2 in which a concentrated static force F acts at the right-hand end. It is not possible now to use the integral of equation (4.6.2) in its simple form to calculate P_r since $f(x)$ is now zero in the range $0 \leqslant x < l$ and infinite at the section $x = l$. It is therefore necessary to reconsider the definition of P_r and to observe that an increment δp_r in the deflexion at p_r causes a displacement

$$\delta p_r \sin \frac{(2r-1)\,\pi}{2}$$

at the end $x = l$. Thus the work δW_r of equation (4.6.1) is now

$$\delta W_r = F.\,\delta p_r \sin \frac{(2r-1)\,\pi}{2} \tag{4.6.5}$$

and

$$P_r = F \sin \frac{(2r-1)\,\pi}{2}. \tag{4.6.6}$$

If now the applied force distribution varies with time, its form in space remaining constant, then equation (4.6.2) gives the generalized force P_r at any instant. Thus if the loading is harmonic of the form $f(x)\, e^{i\omega t}$ then

$$P_r(t) = e^{i\omega t} \int_0^l f(x).\,\phi_r(x)\,dx. \tag{4.6.7}$$

Let the loading shown in fig. 4.6.1 be time-dependent such that the quantity w is an *amplitude* and the loading varies harmonically with frequency ω. In this event, it is necessary to include the factor $e^{i\omega t}$ in the expression (4.6.4) for P_r. Again, if the static force F of fig. 4.6.2 is replaced by a harmonic force $F e^{i\omega t}$, then the factor $e^{i\omega t}$ must be included in the expression (4.6.6) for P_r.

The generalized forces P_r may be used in Lagrange's equations, as in § 3.7. Now certain mathematical objections are sometimes raised to the use of Lagrange's method with infinite-freedom systems. We shall disregard these objections here believing that they are not significant in the treatment of real problems; an alternative approach will be shown later, however, in which this point does not arise. We now therefore treat the string problem by the Lagrangian method.

It has been assumed that any displacement of a taut string can be expressed in the form of a series of characteristic functions as in equation (4.4.8). If the deflexion is static, the coefficients p_r in this series are constants. But if v is time-dependent, then the series holds at every instant, the coefficients p_r being functions of time. The coefficients p_r will serve as a set of generalized co-ordinates in the manner of § 2.1 since, by suitably adjusting their values, the deflexion v can be given any form. If the functions $\phi_r(x)$ in the series (4.4.8) are the true characteristic functions which correspond to the string's end-fixings, then these co-ordinates are *principal* co-ordinates; it is this case which is dealt with in this chapter.

The kinetic and potential energies T and V are of the form of equations (4.5.7) and (4.5.8) where the coefficients a and c are given in Table 4. Thus the equations of motion may be set up in accordance with Lagrange's equations

$$\frac{d}{dt}\left(\frac{\partial T}{\partial \dot{p}_r}\right) + \frac{\partial V}{\partial p_r} = P_r \quad (r = 1, 2, \ldots \text{ or } = 0, 1, 2, \ldots). \tag{4.6.8}$$

There is now an infinite number of these relations and they are of the form

$$a_r \ddot{p}_r + c_r p_r = P_r \quad (r = 1, 2, \ldots \text{ or } = 0, 1, 2, \ldots), \tag{4.6.9}$$

or

$$\ddot{p}_r + \omega_r^2 p_r = \frac{P_r}{a_r} \quad (r = 1, 2, \ldots \text{ or } = 0, 1, 2, \ldots). \tag{4.6.10}$$

If P_r varies harmonically with frequency ω and if a solution of equation (4.6.10) is sought in which p_r varies with this frequency then it is found that

$$p_r = \frac{P_r}{a_r(\omega_r^2 - \omega^2)} = \alpha_r P_r, \tag{4.6.11}$$

where α_r is the receptance at the rth co-ordinate. This receptance may be expressed in the alternative forms

$$\alpha_r = \frac{1}{a_r(\omega_r^2 - \omega^2)} = \frac{1}{c_r - a_r \omega^2} \tag{4.6.12}$$

and, since a_r has the value $\frac{1}{2}\mu l$ for all the three types of string which we have considered, α_r has the value

$$\alpha_r = \frac{2}{\mu l(\omega_r^2 - \omega^2)}. \tag{4.6.13}$$

Since the p's are principal co-ordinates, there are no cross-receptances.

It will now be seen that other receptances can be found in series form. Consider, for instance, the motion of the fixed-free string of fig. 4.6.2 when the force at the right-hand end of the string is harmonic of magnitude $F e^{i\omega t}$. It has been shown that, in that case,

$$P_r = F e^{i\omega t} \sin \frac{(2r-1)\pi}{2} \quad (r = 1, 2, \ldots) \tag{4.6.14}$$

so that the variation of p_r is

$$p_r = \frac{2F e^{i\omega t}}{\mu l(\omega_r^2 - \omega^2)} \sin \frac{(2r-1)\pi}{2} \quad (r = 1, 2, \ldots) \tag{4.6.15}$$

by equation (4.6.11). If this result is now substituted into equation (4.4.8) it is found that

$$v(x, t) = \left\{ \frac{2}{\mu l} \left[\frac{\phi_1(x)}{\omega_1^2 - \omega^2} - \frac{\phi_2(x)}{\omega_2^2 - \omega^2} + \frac{\phi_3(x)}{\omega_3^2 - \omega^2} - \cdots \right] \right\} F e^{i\omega t}, \tag{4.6.16}$$

where

$$\omega_r = \frac{(2r-1)\pi}{2l} \sqrt{\left(\frac{S}{\mu}\right)}, \quad \phi_r(x) = \sin \frac{(2r-1)\pi x}{2l}. \tag{4.6.17}$$

Thus the receptance α_{xl} of a fixed-free string is represented by the quantity within the curly brackets, which is an infinite series.

If the response (4.6.16) is compared with the general expression for a deflexion in terms of the principal co-ordinates (see equation (4.4.8)), then it will be seen that in general all the modes are excited by the force. All the principal co-ordinates vary with the impressed frequency in accordance with equation (4.6.15). Furthermore, those modes whose frequencies are close to the driving frequency ω will tend to be the most strongly stimulated. The series form of receptances gives more insight into the nature of forced motion by bringing out these points. We shall show later that the series form also has other advantages.

In practical problems we are not usually concerned directly with the variations of the co-ordinates p and the forces P, and indeed it is rarely necessary to find the latter specifically. However, the receptances α_r provide a convenient means for treating harmonic distortions. This is because other receptances can readily be found in terms of them. Receptances α_{xh} have already been introduced which relate the displacement at a single point distant x along the string with an applied force which acts at a point distant h along it. Moreover it has been shown in §3.7 how a receptance can be found in the form of a series which gives the displacement of a co-ordinate z due to a force $F e^{i\omega t}$ at a co-ordinate y, such that

$$z = \alpha_{zy} F e^{i\omega t}. \tag{4.6.18}$$

In that instance, the series had n terms only because finite-freedom systems only were considered. The same technique can again be used, this time giving an infinite series, to find receptances of the α_{zy} type; the appropriate relation is merely an extended form of equation (3.7.16), being

$$\alpha_{zy} = \alpha_{yz} = \sum_{r=0\,\text{or}\,1}^{\infty} \alpha_r \frac{\partial y}{\partial p_r} \frac{\partial z}{\partial p_r} \tag{4.6.19}$$

and, when y and z are identical,

$$\alpha_{yy} = \sum_{r=0\,\text{or}\,1}^{\infty} \alpha_r \left(\frac{\partial y}{\partial p_r}\right)^2. \tag{4.6.20}$$

Consider as an example the cross-receptance α_{xh} of a fixed-fixed string which has a force $F e^{i\omega t}$ applied to it at the section where $x = h$. In using equation (4.6.12), the expressions are needed for y and z, namely

$$\left.\begin{aligned} y = v(h, t) = \sum_{r=1}^{\infty} p_r \phi_r(h), \\ z = v(x, t) = \sum_{r=1}^{\infty} p_r \phi_r(x). \end{aligned}\right\} \tag{4.6.21}$$

From these, the receptance α_{xh} may be written down; it is

$$\alpha_{xh} = \alpha_{hx} = \sum_{r=1}^{\infty} \frac{2\phi_r(h) \cdot \phi_r(x)}{\mu l(\omega_r^2 - \omega^2)} = \frac{2}{\mu l} \sum_{r=1}^{\infty} \frac{\sin\dfrac{r\pi h}{l} \sin\dfrac{r\pi x}{l}}{\omega_r^2 - \omega^2}, \tag{4.6.22}$$

where, for this type of string $\qquad \omega_r^2 = \dfrac{r^2\pi^2}{l^2} \sqrt{\left(\dfrac{S}{\mu}\right)}. \tag{4.6.23}$

It should be noted that this series form of the receptance applies to the displacement at any point along the string whereas two different functions were necessary to give the 'closed-form' receptance according to whether x was greater or less than h (see Table 4).

EXAMPLES 4.6

1. Find the static deflexion $v_r(x)$ in the rth principal mode of the taut fixed-fixed string loaded as shown.

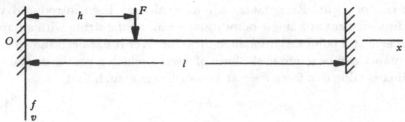

Ex. 4.6.1

2. Write down the receptance α_r for a fixed-fixed string and thence the cross-receptance α_{xh} in series form.

By considering a force of magnitude $w\,dx\,e^{i\omega t}$ acting at a distance h from the left-hand end of the string shown in the figure, find the deflexion due to the loading indicated.

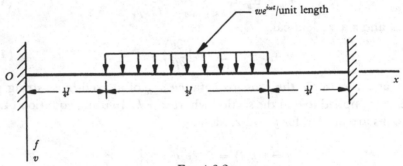

Ex. 4.6.2

3. A fixed-free taut string has a uniformly distributed harmonic loading applied to it of magnitude $w\,e^{i\omega t}$/unit length. Find the deflexion of the string at a point distant x from the fixed end, expressing the result in series form.

4. A taut fixed-fixed string has applied to it a harmonic distributed force whose magnitude is $w\,e^{i\omega t}$ per unit length. Show that the steady response of the string is given by

$$v(x, t) = \frac{4w\,e^{i\omega t}}{\pi\mu} \sum_{r=1,\,3,\,5,\,\ldots}^{\infty} \frac{1}{r(\omega_r^2 - \omega^2)} \sin\frac{r\pi x}{l}.$$

5. Show that the result of the previous example can be written in the form

$$v(x, t) = \frac{4w\,e^{i\omega t}}{\pi\mu} \left[\sum_{r=1,\,3,\,5,\,\ldots}^{\infty} \frac{1}{r\omega_r^2} \sin\frac{r\pi x}{l} \right] + \frac{4w\omega^2\,e^{i\omega t}}{\pi\mu} \left[\sum_{r=1,\,3,\,5,\,\ldots}^{\infty} \frac{1}{r\omega_r^2(\omega_r^2 - \omega^2)} \sin\frac{r\pi x}{l} \right].$$

Show further that the sum of the first of these series may be found by consideration of *static* loading and is of magnitude

$$\frac{wx(l-x)}{2S} e^{i\omega t}.$$

[Note that the second series converges very rapidly and is thus in a useful form for providing approximate solutions.]

4.7 Series representation of specified deflexions

The harmonically varying deflexion $v(x, t)$ of a string which is caused by a harmonic applied force distribution $f(x, t)$ has been shown to be expressible in the form of a series. The method of finding that series was, briefly, to form the generalized force P_r at the rth co-ordinate p_r and then to extract the deflexion series term by term by using the equations of motion in the form (4.6.9). This method is, of course, still valid if the conditions are static; equation (4.6.9) then shows that

$$p_r = \frac{P_r}{c_r}. \tag{4.7.1}$$

From this relation the constant coefficients p in the deflexion series

$$v(x) = p_1\phi_1(x) + p_2\phi_2(x) + \dots \tag{4.7.2}$$

can be found. The free-free string requires that $P_0 = 0$ and so $p_0\phi_0 = p_0/\sqrt{2}$ is indeterminate. We shall now consider a related problem, namely that of finding the coefficients p in the deflexion series if the *deflexion* is given as a function of x. The problem now before us is to calculate the coefficients p in equation (4.7.2) for any specified function $v(x)$.

Now a static force distribution $f_r(x)$ will produce a deflexion of the string concerned in its rth mode only, such that (say)

$$v_r(x) = C_r\phi_r(x), \tag{4.7.3}$$

and this force distribution can be found from equation (4.1.2). In fact,

$$f_r(x) = -S\frac{d^2v_r(x)}{dx^2} = -SC_r\frac{d^2\phi_r}{dx^2} = (\text{constant}) \times \phi_r(x) \ (r \neq 0), \tag{4.7.4}$$

where the last relation holds by virtue of the particular analytical form of all the characteristic functions which have been introduced. Let this force distribution be applied to the string and let the string be then deflected to the form of the specified function $v(x)$. During the deflexion the only work which will be done by the distributed force will be that due to the displacement in the rth mode, because of the relation (4.5.16). It therefore follows that

$$\int_0^l f_r(x).v(x)\,dx = 0 + 0 + \dots + p_r\int_0^l f_r(x).\phi_r(x)\,dx + 0 + \dots. \tag{4.7.5}$$

If the expression (4.7.4) for $f_r(x)$ is now substituted into this equation, an explicit equation is obtained for p_r, namely

$$p_r = \frac{\displaystyle\int_0^l v(x).\phi_r(x)\,dx}{\displaystyle\int_0^l [\phi_r(x)]^2\,dx} \ (r \neq 0). \tag{4.7.6}$$

Consider a fixed-free string whose deflected shape is the straight line AB of fig. 4.7.1 (*a*), the displacement at the free end being Y. The deflexion is

$$v(x) = \frac{Yx}{l}. \tag{4.7.7}$$

For these end-fixings
$$\phi_r(x) = \sin\frac{(2r-1)\,\pi x}{2l} \tag{4.7.8}$$

so that equation (4.7.6) yields the result

$$p_r = \frac{\dfrac{Y}{l}\displaystyle\int_0^l x\sin\dfrac{(2r-1)\,\pi x}{2l}\,dx}{\displaystyle\int_0^l \sin^2\dfrac{(2r-1)\,\pi x}{2l}\,dx} = \frac{8Y}{\pi^2(2r-1)^2}\sin\frac{(2r-1)\,\pi}{2} \quad (r = 1, 2, \ldots).$$
$$\tag{4.7.9}$$

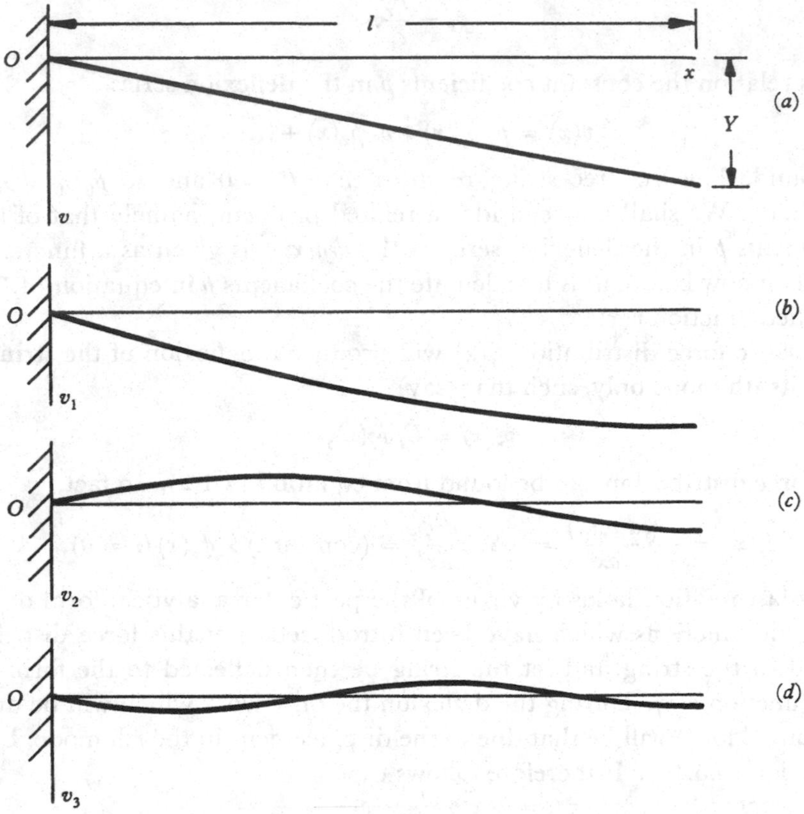

Fig. 4.7.1

The string deflexion that is shown in fig. 4.7.1 (a) may therefore be written

$$v(x) = \frac{8Y}{\pi^2}\left[\sin\frac{\pi x}{2l} - \frac{1}{9}\sin\frac{3\pi x}{2l} + \frac{1}{25}\sin\frac{5\pi x}{2l} - \ldots\right]. \tag{4.7.10}$$

The first three terms of this infinite series represent displacement components which are shown sketched in fig. 4.7.1 (b), (c) and (d). The reader may find it instructive to add these curves graphically, noting how the addition of each new curve brings the sum of them closer to the initial straight-line deflexion.

As a second example, consider the symmetrical deflexion of a fixed-fixed string, shown in fig. 4.7.2 (a). The integral which determines the displacement components must be evaluated over the two halves of the string separately so that

$$p_r = \frac{\dfrac{2Y}{l}\displaystyle\int_0^{l/2} x\sin\dfrac{r\pi x}{l}\,dx + 2Y\displaystyle\int_{l/2}^{l}\left(1-\dfrac{x}{l}\right)\sin\dfrac{r\pi x}{l}\,dx}{\displaystyle\int_0^{l}\sin^2\dfrac{r\pi x}{l}\,dx} = \frac{8Y}{r^2\pi^2}\sin\frac{r\pi}{2}. \qquad (4.7.11)$$

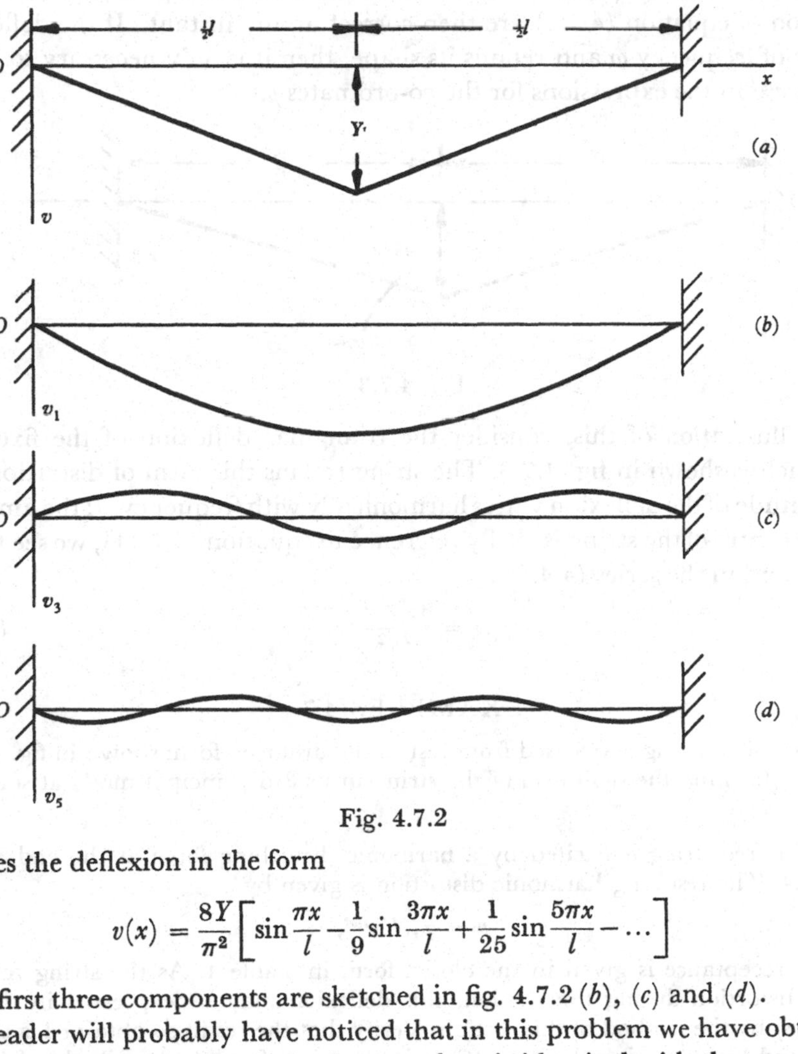

Fig. 4.7.2

This gives the deflexion in the form

$$v(x) = \frac{8Y}{\pi^2}\left[\sin\frac{\pi x}{l} - \frac{1}{9}\sin\frac{3\pi x}{l} + \frac{1}{25}\sin\frac{5\pi x}{l} - \ldots\right] \qquad (4.7.12)$$

and the first three components are sketched in fig. 4.7.2 (b), (c) and (d).

The reader will probably have noticed that in this problem we have obtained a Fourier series, and that the algebraic procedure is identical with that used in the ordinary Fourier expansion.

At this stage it is desirable to mention the matter of the convergence of the series (4.7.2), although a full treatment of this is beyond the scope of this book and involves some mathematical difficulty. We have seen that there are *physical* grounds for assuming that the arbitrary displacement function $v(x)$ can be expressed as a series of characteristic functions as in equation (4.7.2); indeed this important point was made by Rayleigh.† It may also be inferred that when the series is found it will be convergent; for if this were not so then the sign of equality in equation (4.7.2) would be out of place since the right-hand side would represent an infinite quantity. This would not be consistent with the results which we found previously for systems having finite freedom.

Suppose now that it is wished to represent a specified time-dependent deflexion $v(x, t)$ in the series form (4.4.8). The values of the coefficients p which are found by the method of equation (4.7.6) are then correct at any instant. If the deflexion is harmonic of frequency ω and retains its shape, then it is only necessary to include the factor $e^{i\omega t}$ in the expressions for the co-ordinates p.

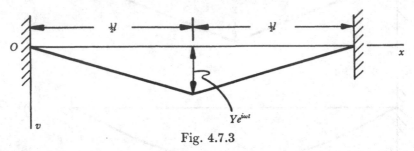

Fig. 4.7.3

As an illustration of this, consider the triangular deflexion of the fixed-fixed string which is shown in fig. 4.7.3. The string retains this form of distortion while the magnitude of the deflexion varies harmonically with frequency ω; the amplitude of v at the centre of the string is Y. By reference to equation (4.7.11), we see that the rth coefficient in the series (4.4.8) is

$$p_r(t) = \frac{8Y e^{i\omega t}}{r^2\pi^2} \sin\frac{r\pi}{2}. \tag{4.7.13}$$

EXAMPLES 4.7

1. A fixed-free string is released from rest in the distorted form shown in fig. 4.7.1 (a) at time $t = 0$. Find the deflexion of the string in its 3rd principal mode at some later time t.

2. A free-free string is excited by a harmonic disturbing force at the end $x = l$ as in fig. 4.2.4. The resulting harmonic distortion is given by

$$v = \alpha_{xl} F e^{i\omega t},$$

where the receptance is given in the closed form in Table 4. As the string retains its shape of distortion during the motion, this deflexion may be expressed in the series form (4.4.8) by the method of this section. Verify that the result so obtained is the same as that found by expressing the receptance in the series form as described in §4.6.

4.8 Series representation of applied forces and other functions

We have shown in §§4.6 and 4.7 how the deflexions of strings may be found in the form of series. In this section, we shall show that other functions, and in particular applied force distributions, can be expanded in this way.

† *Theory of Sound* (1894), §92.

Fig. 4.8.1 shows a fixed-fixed taut string which is deflected to the parabolic form

$$v(x) = \frac{4Yx(l-x)}{l^2}.$$ (4.8.1)

The component of this deflexion in the rth mode may be found from equation (4.7.6) and is, in fact,

$$v_r(x) = p_r \cdot \phi_r(x) = \left\{ \frac{8Y}{l^3} \int_0^l x(l-x) \sin \frac{r\pi x}{l} \, dx \right\} \sin \frac{r\pi x}{l}$$

$$= \frac{16Y}{r^3\pi^3}(1 - \cos r\pi) \sin \frac{r\pi x}{l} \quad (r = 1, 2, \ldots).$$ (4.8.2)

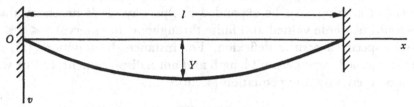

Fig. 4.8.1

The force distribution $f_r(x)$ which would be necessary to produce this component of deflexion can be found from equation (4.1.2), and is

$$f_r(x) = -S \frac{d^2 v_r(x)}{dx^2} = \frac{16SY(1 - \cos r\pi)}{rl^2\pi} \sin \frac{r\pi x}{l} \quad (r = 1, 2, \ldots).$$ (4.8.3)

The sum of contributions such as this produces the deflexion (4.8.1) so that the total applied load is obtained as the sum of an infinite series; that is

$$f(x) = \sum_{r=1}^{\infty} f_r(x) = \frac{16SY}{l^2\pi} \sum_{r=1}^{\infty} \frac{1 - \cos r\pi}{r} \sin \frac{r\pi x}{l}.$$ (4.8.4)

The reader should compare this result with that of equation (3.8.19).

Without pursuing the matter further at this stage, we notice that the series (4.8.4) converges.† Indeed, it is not clear that the sign of equality could be used in equation (4.8.4) were this not so.

The static deflexion of the fixed-free taut string of fig. 4.7.1 (a) in its rth mode can be written down on inspection of equation (4.7.9); it is

$$v_r(x) = \frac{8Y}{\pi^2(2r-1)^2} \sin \frac{(2r-1)\pi}{2} \sin \frac{(2r-1)\pi x}{2l}.$$ (4.8.5)

The corresponding function $f_r(x)$, found from equation (4.1.2) is

$$f_r(x) = \frac{2SY}{l^2} \sin \frac{(2r-1)\pi}{2} \sin \frac{(2r-1)\pi x}{2l}.$$ (4.8.6)

Now the applied force distribution which will produce the static deflexion shown in fig. 4.7.1 (a) evidently consists of a concentrated force F applied at the end $x = l$ of the string. The magnitude of this force is

$$F = \frac{SY}{l}.$$ (4.8.7)

† This may be proved by applying Dirichlet's test; e.g. see G. H. Hardy, *A Course of Pure Mathematics* (Cambridge University Press, ed. 10, 1952), § 196.

It follows that the required applied loading $f(x)$—which quantity is a force per unit length—is everywhere nil except at the end $x = l$ of the string, where it is infinite. As we shall show, it is this circumstance which accounts for the fact that the sum of all the components (4.8.6) forms a *divergent* series.

We have seen that there are *physical* grounds for assuming that the arbitrary displacement function $v(x)$ can be expressed as a series of characteristic functions as in equation (4.7.2) and that it may be inferred that when the series is found it will converge. Now an important aspect of the series representation of the arbitrary displacement $v(x)$ is that *any* function of x, which may not relate to a real physical displacement of a string, can be expanded in this way. It is necessary that the function should be single valued and finite throughout the interval $0 \leqslant x \leqslant l$ and that it could specify a *possible* deflexion. For instance the function $f(x)$, which represents an applied force per unit length and not a displacement, can be written in the form of a series by using equation (4.7.6).†

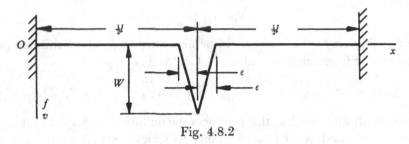

Fig. 4.8.2

This idea may be illustrated by the function whose graph is sketched in fig. 4.8.2. The curve is symmetrical about the point $x = \frac{1}{2}l$ and is finite only over a length 2ϵ, the maximum value of the function being W. This curve is a possible form of deflexion of a taut fixed-fixed string and a convergent series can be found to represent it. The coefficients in the series are, by equation (4.7.6),

$$p_r = \frac{\int_{\frac{1}{2}l-\epsilon}^{\frac{1}{2}l} \sin\frac{r\pi x}{l}\left[\frac{x-(\frac{1}{2}l-\epsilon)}{\epsilon}\right]W\,dx + \int_{\frac{1}{2}l}^{\frac{1}{2}l+\epsilon} \sin\frac{r\pi x}{l}\left[\frac{\frac{1}{2}l+\epsilon-x}{\epsilon}\right]W\,dx}{\int_0^l \sin^2\frac{r\pi x}{l}\,dx}$$

$$= \frac{4Wl\sin\frac{1}{2}r\pi}{\epsilon r^2\pi^2}\left(1-\cos\frac{r\pi\epsilon}{l}\right). \tag{4.8.8}$$

Now suppose that the curve represents a static force distribution $f(x)$ applied to the string, rather than a deflected form, with W the maximum value of the force per unit length. Evidently the series representation of the force distribution will be

$$f(x) = \frac{4Wl}{\epsilon\pi^2}\sum_{r=1}^{\infty}\left\{\frac{1}{r^2}\sin\frac{r\pi}{2}\left(1-\cos\frac{r\pi\epsilon}{l}\right)\sin\frac{r\pi x}{l}\right\}. \tag{4.8.9}$$

† It is sometimes useful to express the deflexion of one string in terms of the characteristic functions of another. If this kind of thing is done, however, there is no physical reason why the resulting series should converge unless the function concerned is a *possible* deflected form of the string whose characteristic functions are used.

As this force distribution has the shape of a possible deflected form of a fixed-fixed string, the series converges.

The total applied force is

$$F = \int_0^l f(x)\,dx = \tfrac{1}{2}.W.2\epsilon = W\epsilon. \tag{4.8.10}$$

Consider now a limiting process in which $\epsilon \to 0$ so that the region in which the force is applied becomes indefinitely small, while at the same time the product $W\epsilon$, which is the total force, is kept constant. When this is done the rth coefficient (4.8.8) becomes

$$\frac{2W\epsilon}{l}\sin\frac{r\pi}{2} \tag{4.8.11}$$

and it can now be seen that the series representation of $f(x)$ will not converge, for it becomes

$$\frac{2F}{l}\left[\sin\frac{\pi x}{l} - \sin\frac{3\pi x}{l} + \sin\frac{5\pi x}{l} - \ldots\right]. \tag{4.8.12}$$

This loss of convergence, which occurs when the length ϵ is made indefinitely small, is associated with the fact that the function ceases to represent a possible deflected form of the string; such a deflected form is ruled out because our theory assumes no longitudinal displacements of the string. The reader should not infer from these remarks that, if a function does not represent a possible string deflexion, then its series representation is *necessarily* divergent; it is merely that we have no physical assurance to the contrary.

It is of interest to find the deflected form which would be produced by the above force distribution after it has been concentrated. The rth term of the series representation of force is

$$\frac{2F}{l}\sin\frac{r\pi}{2}\sin\frac{r\pi x}{l} \quad (r = 1, 2, \ldots), \tag{4.8.13}$$

while that of the deflexion series is

$$p_r\sin\frac{r\pi x}{l} \quad (r = 1, 2, \ldots). \tag{4.8.14}$$

If the force and deflexion series are substituted into equation (4.1.2) it is now found that

$$\sum_{r=1}^{\infty}\left\{\left[\frac{2F}{l}\sin\frac{r\pi}{2} - \frac{r^2\pi^2 S p_r}{l^2}\right]\sin\frac{r\pi x}{l}\right\} = 0 \tag{4.8.15}$$

so that a fresh series is arrived at. The sum of this latter series is zero for all values of x so that each term must vanish separately; that is to say, the contents of the square brackets can be equated to zero so that

$$p_r = \frac{2Fl}{r^2\pi^2 S}\sin\frac{r\pi}{2} \quad (r = 1, 2, \ldots). \tag{4.8.16}$$

The deflected form is therefore given by

$$v(x) = \frac{2Fl}{\pi^2 S}\left[\sin\frac{\pi x}{l} - \frac{1}{9}\sin\frac{3\pi x}{l} + \frac{1}{25}\sin\frac{5\pi x}{l} - \ldots\right]. \tag{4.8.17}$$

This series converges and gives the correct form for the string when it is loaded by a concentrated force F at its mid-point. This may be verified by considering the statics of the string at its mid-point as indicated in fig. 4.8.3; at that point,

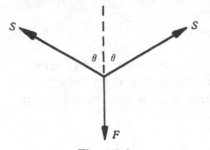

$$F = 2S\cos\theta = 2S\left(\frac{Y}{\frac{1}{2}l}\right) = \frac{4SY}{l}, \quad (4.8.18)$$

where Y is the central deflexion. If this relation is used to eliminate F in the series (4.8.17), then the latter becomes identical with that of equation (4.7.12).

Fig. 4.8.3

We have seen that it may not be possible to find converging series to represent certain force distributions. If the functional form of $f(x)$ is not such that it could represent a deflexion of the string whose characteristic functions are used, then we have no assurance that the loading series will converge. But in the above example, a divergent series for $f(x)$ produces the correct converging series for $v(x)$. This apparent paradox may be resolved by using the converging series (4.8.9) instead of (4.8.12) to find a deflexion series and only then to let $\epsilon \to 0$. Thus problems of this sort do not raise the question of diverging series if the standpoint is adopted that the concept of a concentrated force is too artificial and that in any real system the force must necessarily be spread over a finite length of string, no matter how short.

It is not wrong, however, to use the terms in a diverging loading series provided that these terms are not summed. We have seen that, under certain conditions, we have no assurance on physical grounds that the sum of a loading series will be finite; but we do not really require the series to have a finite sum. This may be explained by reference to the above problem. The generalized force P_r due to a concentrated load F at the centre of a fixed-fixed string is

$$P_r = F\sin\frac{r\pi}{2} \quad (r = 1, 2, \ldots) \quad (4.8.19)$$

(see § 4.6). If this quantity is used in equations (4.6.9), it is found that

$$p_r = \frac{P_r}{c_r} = \frac{2Fl}{r^2\pi^2 S}\sin\frac{r\pi}{2} \quad (r = 1, 2, \ldots) \quad (4.8.20)$$

which agrees with the result (4.8.16). There is no question of adding the forces P_r together and the reader may readily verify that the process of deriving these quantities in any string problem is effectively the same as that of finding the coefficients of the various terms of the loading series.

The ability to find a convergent deflexion series, corresponding to any specified force function, is of great practical importance. The first few terms of the series, and sometimes a single term only, may be used as an approximation to the distorted shape.

It was shown in § 4.6 that receptances have a series form. One way of finding such a series is suggested by the above theory, namely to treat the receptance in closed form as if it were a deflexion. Consider for instance the receptance α_{xh} of a fixed-free string. The series form of it may be found by applying equation (4.7.6) to the following function which could clearly represent a possible deflected form of a fixed-free string:

$$\alpha_{xh} = \begin{cases} \dfrac{\sin \lambda x \cos \lambda (l-h)}{S\lambda \cos \lambda l} & (0 \leqslant x \leqslant h) \\[3mm] \dfrac{\sin \lambda h \cos \lambda (l-x)}{S\lambda \cos \lambda l} & (h \leqslant x \leqslant l) \end{cases}. \tag{4.8.21}$$

This gives

$$\alpha_{xh} = \frac{1}{S\lambda \cos \lambda l} \sum_{r=1}^{\infty} \left\{ \frac{\left[\cos \lambda (l-h) \displaystyle\int_0^h \sin \lambda x \cdot \phi_r(x)\, dx + \sin \lambda h \displaystyle\int_h^l \cos \lambda (l-x) \cdot \phi_r(x)\, dx \right]}{\displaystyle\int_0^l [\phi_r(x)]^2\, dx} \phi_r(x) \right\}$$

$$\tag{4.8.22}$$

which reduces to
$$\alpha_{xh} = \frac{2}{\mu l} \sum_{r=1}^{\infty} \frac{\phi_r(h) \cdot \phi_r(x)}{\omega_r^2 - \omega^2}. \tag{4.8.23}$$

This result can be found readily by the methods of § 4.6; the technique was suggested in Ex. 4.7.2.

In this section we have examined *static* problems in some detail. The reader can readily demonstrate that the various manipulations which have been illustrated have a counterpart in dynamical theory when forces and deflexions are harmonic. Series can then be found for deflexions and applied loadings in a similar way to that used here. It is, however, necessary to include the inertia term in equation (4.1.2) when using that equation to obtain one type of series from the other.

EXAMPLES 4.8

1. A uniformly distributed load of w per unit length is applied to the middle half of a taut fixed-fixed string as shown in the figure.

(a) Find, by the method of this section, the load $f_r(x)$ which causes the deflexion of the string in its rth principal mode.

(b) Using this value find the deflexion $v_r(x)$ in the rth principal mode.

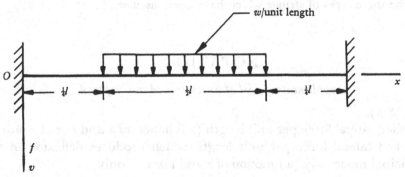

Ex. 4.8.1

2. The figure shows a taut free-free string to which is applied a balanced force distribution.

(a) Find the deflexion, relative to the end $x = 0$ of the string by direct integration.

(b) Express the deflexion found in (a) in series form.

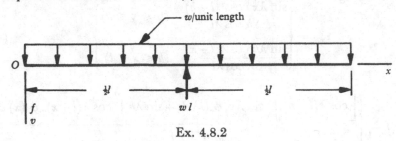

w/unit length

O

x

$\tfrac{1}{2}l$ $w\,l$ $\tfrac{1}{2}l$

f
v

Ex. 4.8.2

3. In this section, two methods have been described for finding a series representation for a given loading $f(x)$ applied to a taut string. They are, briefly,

(i) to start from the series for the deflexion that it produces, and

(ii) to treat the force distribution as if it were a deflexion and thus to find its series directly.

Show that the series found in these ways are identical.

4. From the series representation of the harmonic deflexion of the fixed-fixed string shown in fig. 4.7.3 (see equation (4.7.13)), show that the force distribution $f_r(x)$ which would cause the motion in the rth mode is

$$f_r(x, t) = \frac{8Y\mu(\omega_r^2 - \omega^2)\,e^{i\omega t}}{r^2\pi^2} \sin\frac{r\pi}{2} \cdot \sin\frac{r\pi x}{l}.$$

5. It has been shown that an applied loading $f(x)$ can be expressed in the form of a series by treating the function as if it were a deflexion. Let the series found in this way be

$$f(x) = \sum_{r=1}^{\infty} F_r \phi_r(x),$$

where the constants F_r are given by equation (4.7.6).

Show that, whereas the value of P_r, the generalized force at the rth principal coordinate, depends upon the particular magnitude of distortion in the rth mode which is associated with unit value of p_r, the value of F_r does not. What amplitude of deflexion $v_r(x)$ *in space* must be associated with unit value of p_r if these two quantities are to be the same for the three types of strings which have been discussed?

NOTATION

Supplementary List of Symbols used in Chapter 4

a $\quad = \sqrt{(S/\mu)}$.

f $\quad$ Applied lateral force per unit length (a function of x and t or of x only).

f_r $\quad$ Applied lateral force per unit length which produces deflexion in the rth principal mode only (a function of x and t or of x only).

S $\quad$ Tension in string.

v_r Lateral deflexion in rth mode (a function of x and t or of x only).†

W_r See equation (4.6.1).

α_{xh} Receptance between the sections x and h. If a harmonic point force $F e^{i\omega t}$ is applied laterally to a taut string at the section $x = h$, then the deflexion at any point x of the string is given by

$$v = \alpha_{xh} F e^{i\omega t}.$$

λ $= \omega/a.$

λ_r $= \omega_r/a.$

μ Mass per unit length of string.

† A numerical subscript is also used in some places to signify a co-ordinate v at which sub-systems are linked together (for example, see fig. 4.2.6.)

CHAPTER 5

THE ANALYSIS OF REAL SYSTEMS; APPROXIMATE METHODS

> We replace the arbitrary functions by a convergent series of harmonic vibrations. Taking a finite number of terms as an approximation, we have a perfectly continuous solution whose initial conditions differ but slightly from those of the proposed problem. This difference is less and less, the more terms of the series are included in the solution.
>
> E. J. ROUTH, *Dynamics of a System of Rigid Bodies*, Pt II (1892)

So far, in this book, we have examined the properties of certain highly-idealized oscillating systems. The first three chapters refer to systems having only finite freedom—systems, that is, which do not exist in the physical world. The vibrations of taut strings, which were dealt with in Chapter 4, were introduced in order to exemplify the theory of a continuous elastic system; but again the postulated systems could never be manufactured with complete precision.

In this chapter, we shall discuss the nature of this necessary process of idealization when it is used in the analysis of real systems. This can, of course, be done only in general terms. But several of the results can be illustrated conveniently by reference to the theory of taut strings. It must be emphasized, however, that the theory of strings is only used for convenience of presentation. The results, in their generality, apply equally well to beams, shafts, structures and so on.

The extent to which the vibration theory here developed applies to real systems is a matter, ultimately, for experiment. It is possible, however, to predict to some extent the discrepancies which will be found and the range of conditions within which the discrepancies will be small enough to be unimportant. This matter will be developed in this chapter.

5.1 The differences between real and theoretical systems

The most important difference between real systems and those which have been treated in our analysis so far is that all real systems are subject to damping; that is to say, energy is dissipated within them, whereas our theoretical systems have been conservative. Although this difference is vital it will not be discussed in this chapter because the effects of damping will be introduced into the theory later. We shall at present be concerned with discrepancies which would exist even if there were no damping in real systems. It is necessary to say that the effects of those differences which are to be discussed here cannot be completely explained without mention of damping. Nevertheless it is convenient to deal with some of the matters now and to clear up those which concern damping later.

There are three types of assumption which have been made in our development of the theory so far. These are as follows:

(a) It has been assumed that systems can be isolated from their surroundings, being supported on perfectly rigid anchorages or existing without any support and having no connexion with the external world (even through the air) or else being free to slide along smooth guides. The first essential in any vibration analysis is a definition of the system to which reference is made. Thus, when we discussed the behaviour of a fixed-fixed taut string in Chapter 4 we referred to an ideal elastic body stretched between two *rigidly*-fixed points, the whole being *in vacuo* since no allowance was made for the effects of surrounding air. The end-fixings and vacuum play an essential part in the analysis by forming the boundary which defines the system. Ideal boundaries such as these do not exist in nature and this, at the outset, is a main difficulty in the analysis of many engineering systems.

(b) It is further assumed that systems can be made of perfectly homogeneous materials and that their geometry is exact. Thus, the wires which are discussed in Chapter 4 are supposed to be absolutely uniform. Now it is well known that metallic machine parts are not perfectly homogeneous, isotropic and elastic; they are composed of large numbers of crystals. It is, however, experimentally observed that, provided these crystals possess a random orientation and are sufficiently numerous, they exhibit average properties which approximate to those of a homogeneous, isotropic and elastic body such as is envisaged in the theory of Strength of Materials.

(c) Finally, some systems have been assumed to possess finite freedom only so that they must be constructed of rigid bodies and massless springs.

Assumptions of these three types are commonly made in the analysis of static problems, but it is not usually considered necessary to discuss very fully the limitations which they impose on the application of theory. The reason for this is that the effects of approximations which are involved in the static problems can usually be seen to be small by quite simple arguments. For instance the effect of slight yielding of the supports of a structure, when theory assumes the supports rigid, can often be seen by superposing the displacements due to the yielding on the calculated displacements. Again the extension of a slightly non-uniform rod under tension can be seen to differ from that of a uniform rod only because of the small extra forces which would be applied to the rod if the thinner portions were filled out and the fatter portions were reduced. Arguments of this kind, often buttressed by energy considerations, are sufficient to dispose of any doubts about the applicability of analysis. We have now to determine whether similar arguments can be used which will dispose equally easily of the difficulties in vibration theory.

The difference between static problems and vibration problems arises because the latter introduce the frequency parameter. If the analysis is required to be valid for all values of this parameter between zero and infinity, then the idealized theory will be insufficient.

Consider, for instance, the assumption of a rigid support in the vibration theory of taut strings. This is an assumption of zero (or, rather, negligible) receptance.

But while we know that the receptance at a point may be very small over a limited frequency range, we also know that there will be some frequencies at which it will be large and, in undamped systems, infinite! Evidently our approximations cannot hold under these latter conditions.

Again, the assumption that a crystalline machine-part may be treated as if it were homogeneous, isotropic and perfectly elastic, rests on the assumption that irregularities are *small*. This, in turn, implies that adjacent crystals are subjected to almost the same mechanical loading at any given instant. A limit is thus placed on the frequencies with which the system concerned may be vibrated although this limit is usually far greater than any practically interesting driving frequency. For the higher frequencies are found, in general, to produce distortion shapes whose associated displacements vary rapidly *in space*; the taut strings provide a simple instance of this.

It has been shown by Lord Rayleigh† that this question of the small divergencies of a real—from a theoretical—system can be elucidated to a great extent on *theoretical* grounds. We shall refer later in this chapter to this argument which is based upon Rayleigh's Principle.

In general all the motions of which even a perfectly determined system is capable are not known. This raises considerations which may be illustrated in terms of taut strings.

Let a transverse harmonic force be applied to the 'free' end of a length of wire which is in tension and suppose that it may be regarded as being perfectly elastic, exactly uniform and ideally mounted. The vibration of the wire which is produced by the force can usually be calculated by using the theory of Chapter 4 although that theory refers to the motion of an infinitely thin and perfectly flexible string and not directly to a wire. Now the ideal string and the wire differ significantly in their behaviour only when motion in the higher modes is considered. Thus, in a mode whose associated natural frequency is very high, the wavelength in space of the appropriate characteristic function becomes comparable with the wire diameter; under these conditions the shear force and bending moment in the wire become important. Further, at still higher frequencies there will be modes of vibration in which the surface of the wire becomes rippled. Thus in the higher modes the replacement of the wire by the imaginary string is no longer even approximately valid. It follows that the substitution of the ideal string for the wire at low frequencies involves the assumption that the effects of all the high-frequency modes (which are generally unknown) may be neglected.

The whole question of the response of a given system in higher modes (that is, in modes whose frequencies are much higher than the driving frequency or would raise questions of non-homogeneity, etc.) is a crucial one in deciding how that system may be analysed. In general, these modes are very difficult to find and to specify. But experiments show that the process of simplifying systems for the purposes of analysis rests on the assumption that motions in these modes are small. It

† *Theory of Sound* (1894), §§88, 90 and 91.

transpires that these small motions can, in fact, be neglected. This is axiomatic in practical vibration analysis.

In this chapter we shall discuss the limitations of frequency within which a given analysis is valid. We shall also show how some receptance expressions may be further simplified if a lower range of frequencies for their applicability is acceptable.

5.2 The curtailment of receptance series

It has been pointed out that the fact that vibration theory can be applied to real systems implies that, as a rule, very small motions in high-frequency modes may be neglected in vibration analysis. The analysis is performed by using some idealized form of the real system which does not possess even approximately these high-frequency modes (whose forms are not known anyway). Now each term of a receptance series of the ideal version of the system gives the response in one of its principal modes. This suggests a further simplification, namely that we should be able to curtail their series, obtaining useful approximations from a limited, rather than an infinite, number of terms.

Suppose that a transverse harmonic force acts at some point on a taut fixed-fixed wire. We have seen that, provided the driving frequency is not too high, the wire may be treated as a taut string since it is intuitively obvious that the low-frequency modes of the wire will closely resemble those of a string. The modal shapes will be in error and the magnitude of the discrepancy becomes progressively more marked as the order of the mode is increased; moreover, the possibility of motions in other (unknown) modes will be disregarded if this idealization is made. We now enquire to what extent the motions in the higher modes of the *string* can be neglected as well. This is a problem in the theory of taut strings and reference need not be made to the wire when examining it. In dealing with it, we shall adopt the standpoint that a receptance in closed form is 'exact' while the appropriate curtailed series is an 'approximation' to it. It must be emphasized however that, from the *physical* standpoint we have no means of saying which is likely to be the more accurate although we should certainly expect the closed form to provide a good approximation to the truth.

It was shown in § 4.6 that if P_r, the generalized force corresponding to the rth principal co-ordinate p_r varies sinusoidally with frequency ω, then

$$p_r = \frac{P_r}{a_r(\omega_r^2 - \omega^2)} \tag{5.2.1}$$

(see equation (4.6.11)). This holds for both finite- and infinite-freedom systems. It is seen from this equation that a mode whose frequency ω_r is very close to ω will be excited strongly; but all the modes with higher frequencies ω_{r+1}, ω_{r+2}, etc., will contain larger frequency differences in their denominators so that their amplitudes will, in general, be successively reduced. The quantity P_r in the numerators depends upon the position of the applied force and varies from term to term but it does not increase regularly as the denominators do. This may be confirmed by examining the functions from which these generalized forces are derived.

Now the falling off of the amplitude of the higher modes is of great practical significance because it allows some of the higher modes to be neglected in calculations. That is to say, approximate receptances may be used which contain a small number of terms in their series form instead of either a large or an infinite number.

When dealing with infinite-freedom systems it is not usually possible, except in certain special cases, to prove mathematically that the receptance series is convergent, still less that the sum of all the terms beyond some specified number is small compared with the sum of the terms which precede them. The assumption that the sum of the higher terms is small or negligible can only be based on reasoning about the physical meaning of the problem.

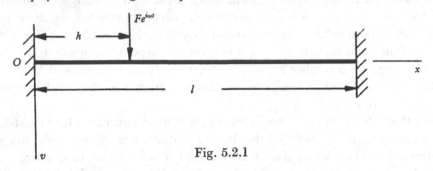

Fig. 5.2.1

Consider the motion of the fixed-fixed string of fig. 5.2.1. The harmonic deflexion of a point distant x from the left-hand support is given by the relation

$$v(x, t) = \alpha_{xh} F e^{i\omega t}, \tag{5.2.2}$$

the external force being applied at a point distant h from the left-hand end. Now the receptance α_{xh} is known in closed form (see Table 4). Its numerical value can therefore be calculated in any particular case. This receptance is also given in the series form in the table so that an approximate solution may be obtained from the series by excluding all the terms beyond some chosen value. In this problem, therefore, we can obtain a figure for the error which is introduced by the curtailment of the series.

As an example, let x/l be equal to 0.3 and h/l be equal to 0.4. The closed-form receptance is then

$$\alpha_{xh} = \frac{\sin\left[0.3l\omega\sqrt{(\mu/S)}\right]\sin\left[0.6l\omega\sqrt{(\mu/S)}\right]}{S\omega\sqrt{(\mu/S)}\sin\left[l\omega\sqrt{(\mu/S)}\right]}. \tag{5.2.3}$$

If we now take the value $\omega = 2.5(\pi/l)\sqrt{(S/\mu)}$ for the driving frequency, this gives an exact value $\alpha_{xh} = -0.090l/S$. If the series form of the receptance is used and all the terms for which ω_r is more than three times as great as ω are excluded, then the following approximate expression is arrived at:

$$\alpha_{xh} \doteq \frac{2l}{S\pi^2}\left[\frac{\sin 0.3\pi \sin 0.4\pi}{1 - 6.25} + \frac{\sin 0.6\pi \sin 0.8\pi}{4 - 6.25} + \frac{\sin 0.9\pi \sin 1.2\pi}{9 - 6.25} + \frac{\sin 1.2\pi \sin 1.6\pi}{16 - 6.25}\right.$$
$$\left. + \frac{\sin 1.5\pi \sin 2.0\pi}{25 - 6.25} + \frac{\sin 1.8\pi \sin 2.4\pi}{36 - 6.25} + \frac{\sin 2.1\pi \sin 2.8\pi}{49 - 6.25}\right]. \tag{5.2.4}$$

The sum of this is $-0.085l/S$, representing a discrepancy of 5.6%.

It will be seen that, in the string problem, there is little to be gained by curtailing the series in order to obtain an approximation to its sum because the closed form of the receptance is a convenient mathematical function. Subsequently we shall meet problems of elastic bodies in which the closed form of the receptance is far less easily handled and the approximate method then becomes valuable. Further, there are many problems in which it is not practicable to deduce a receptance in the closed form.

Having found that it is possible to obtain useful results by employing curtailed receptance series, we shall next examine a little more closely the nature of the approximation that this involves. It is desirable to be able to judge the magnitude of the error which is involved by a specified curtailment of a receptance series since the more modes that are neglected in a vibration analysis, the shorter do the necessary calculations become.

In assessing the error that curtailment of the series introduces the engineer must take into account two factors; these are the frequencies of the neglected modes and the static deflexion which would be produced by distortion of the system in those modes. If the frequencies of all the neglected modes are much higher than the exciting frequency, then the contribution that the corresponding partial fractions would make to the receptance differs only slightly from that which they would make at zero frequency (i.e. with ω set equal to zero). Thus if the lowest neglected mode has a frequency which is more than five times greater than the forcing frequency, then it will be found that the difference between the actual contribution and the contribution at zero frequency to a receptance will be less than 4 %. Whether or not the static deflexion in a set of modes is negligible can usually be estimated without great difficulty since the mass of the system does not enter into the calculation. In some problems the static deflexion is not negligible although all the frequencies are high. In these cases the neglected terms may be replaced by an estimate of the static deflexion, instead of being merely omitted.

We may use as an example of this process the deflexion of the string which was treated above. The full static cross-flexibility of the string may be found from the quantity

$$\lim_{\lambda \to 0} \left(\frac{\sin 0 \cdot 3\lambda l \, \sin 0 \cdot 6\lambda l}{S\lambda \sin \lambda l} \right) = \frac{0 \cdot 18l}{S}. \tag{5.2.5}$$

Now the first seven terms of the receptance series give, at zero frequency,

$$\frac{2l}{S\pi^2} \left[\frac{\sin 0 \cdot 3\pi \, \sin 0 \cdot 4\pi}{1} + \frac{\sin 0 \cdot 6\pi \, \sin 0 \cdot 8\pi}{4} + \frac{\sin 0 \cdot 9\pi \, \sin 1 \cdot 2\pi}{9} \right.$$

$$+ \frac{\sin 1 \cdot 2\pi \, \sin 1 \cdot 6\pi}{16} + \frac{\sin 1 \cdot 5\pi \, \sin 2 \cdot 0\pi}{25} + \frac{\sin 1 \cdot 8\pi \, \sin 2 \cdot 4\pi}{36}$$

$$\left. + \frac{\sin 2 \cdot 1\pi \, \sin 2 \cdot 8\pi}{49} \right]. \tag{5.2.6}$$

This has the sum $0 \cdot 185l/S$. Thus that part of the static flexibility which arises from the terms beyond the seventh is equal to $-0 \cdot 005l/S$. At a frequency ω for which

$$\omega \ll \omega_7, \tag{5.2.7}$$

the term beyond the seventh may be replaced therefore by $-0\cdot005l/S$; it will be seen that this correction nullifies the error (to this accuracy) of the approximation represented by equation (5.2.4).

This use of the static deflexion shows that the neglect of high-frequency terms in calculations can be justified even when the exact magnitudes of these terms are not known. Provided that the static deflexion can be found either by theory or experiment, a limit may be set to the error which is introduced by neglect of them.

The nature of the error that is introduced by the curtailment of a receptance series may be examined in another way. Consider first a problem of static deflexion. It was shown in §4.7 that if a taut string is given a displacement $v(x)$ from its equilibrium position then $v(x)$ may be represented by a series, thus:

$$v(x) = p_1\phi_1(x) + p_2\phi_2(x) + \ldots, \tag{5.2.8}$$

where the coefficients p are not functions of x and have the values

$$p_r = \frac{\displaystyle\int_0^l v(x)\,.\,\phi_r(x)\,dx}{\displaystyle\int_0^l [\phi_r(x)]^2\,dx}. \tag{5.2.9}$$

Such series are often useful in numerical work, for which it is necessary to use a limited number of the terms only. It is now useful to examine the nature of the approximation which is introduced when a series is curtailed. This discussion is presented in terms of the deflexions of a taut wire as usual, but is applicable to many different problems. In particular it may be applied to receptances, which are, after all, displacement functions with certain multiplying factors; the connexion between the general problem of an arbitrary displacement (5.2.8) and that of receptances is illustrated by the expansion given in §4.8 of the receptance of equation (4.8.21).

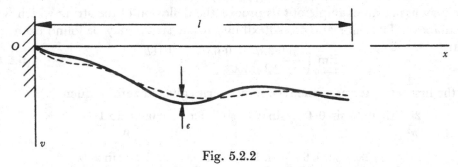

Fig. 5.2.2

In fig. 5.2.2 the full-line curve represents a fixed-free taut string which has been given some arbitrary static deflexion $v(x)$. This deflexion can be expressed in the form of equation (5.2.8) and if the number of terms taken is large enough, the representation may be made to approach any specified degree of accuracy. Since the series converges, the higher terms may always be excluded when approximate results only are required. At this stage we may examine the process of approximation in a different manner,† by seeking the best approximation which may be obtained from a series of m characteristic functions; thus

$$\bar{v}(x) = A_1 \phi_1(x) + A_2 \phi_2(x) + \ldots + A_m \phi_m(x). \tag{5.2.10}$$

The coefficients A in this expression have yet to be determined and it cannot be assumed *a priori* that they are identical with the coefficients p given by equation (5.2.9). The approximate function $\bar{v}(x)$ will have some form which is represented by the dotted curve of fig. 5.2.2.

Before setting about the task of finding the constants, it is necessary to define the meaning of the term *best approximation*; we shall define it by means of the least mean square error. This means that if ϵ is the error between the dotted and the full lines then the constants of the series will be selected in such a way that the expression

$$I = \frac{1}{l} \int_0^l \epsilon^2 \, dx \tag{5.2.11}$$

is a minimum. The notion of the 'least mean square error' has several uses in statistical theory since it is often essential to choose an even power of the error. This is because, if an odd power were chosen, positive and negative errors would tend to cancel each other out.

It is now required that the A's be chosen in such a way that the integral

$$I = \frac{1}{l} \int_0^l [v(x) - \bar{v}(x)]^2 \, dx = \frac{1}{l} \int_0^l \left[v(x) - \sum_{s=1}^m A_s \phi_s(x) \right]^2 dx \tag{5.2.12}$$

is a minimum. The quantity I is a function of the multipliers A alone and the necessary condition is therefore given by

$$\frac{\partial I}{\partial A_1} = 0, \quad \frac{\partial I}{\partial A_2} = 0, \quad \ldots, \quad \frac{\partial I}{\partial A_m} = 0. \tag{5.2.13}$$

It is therefore necessary that

$$\frac{2}{l} \int_0^l \left[v(x) - \sum_{s=1}^m A_s \phi_s(x) \right] \phi_r(x) \, dx = 0 \quad (r = 1, 2, \ldots, m) \tag{5.2.14}$$

which result is obtained by differentiating with respect to A_r within the integral of equation (5.2.12). The equation gives

$$\int_0^l v(x) \cdot \phi_r(x) \, dx = \int_0^l \left[\sum_{s=1}^m A_s \cdot \phi_s(x) \cdot \phi_r(x) \right] dx. \tag{5.2.15}$$

Now it has already been shown (see §4.7) that, owing to the property of orthogonality, the right-hand side of this equation is equal to

$$A_r \int_0^l [\phi_r(x)]^2 \, dx. \tag{5.2.16}$$

† A treatment is given of Fourier series by I. S. Sokolnikoff using the method to be discussed; see his book *Advanced Calculus* (McGraw-Hill, London, 1st ed. 1939), ch. 11.

Therefore,

$$A_r = \frac{\int_0^l v(x) \cdot \phi_r(x) \, dx}{\int_0^l [\phi_r(x)]^2 \, dx}. \tag{5.2.17}$$

It is seen that this expression for A_r is precisely the same as that of equation (5.2.9). If therefore a deflexion is expanded into an infinite series of characteristic functions and if the series is then curtailed, then the resulting function is the best approximation obtainable with the number of terms used. In this statement, 'the best' is to be understood according to the restricted sense which has been defined above. While this proof refers to a fixed-free string it could be equally well interpreted for any of the three types of string which we have discussed and also for other types of system, such as beams.

EXAMPLES 5.2

1. A fixed-free taut string has the following details:

$$\text{length} \quad l = 1 \text{ ft.}$$
$$\text{tension} \quad S = 100 \text{ lb.wt.}$$
$$\text{mass} \quad \mu = 0 \cdot 0137 \text{ lb./ft.}$$

A disturbing force of amplitude $F = 2$ lb.wt., and frequency 100 cyc./sec. is applied to the free end.

(*a*) Calculate the amplitude of the lateral motion at a point 3 in. from the fixed end of the string by using 1-, 2- and 3-term approximations of the cross-receptance series.

(*b*) Find the 'correct' result by using the receptance in closed form.

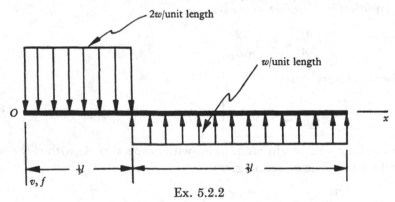

Ex. 5.2.2

2. The figure shows a taut free-free string to which is applied the force distribution shown. The tension is S.

(*a*) Find, by integration of the equation of equilibrium, the deflexions at the points $x = \frac{1}{3}l, \frac{2}{3}l, l$ measured relative to the left-hand end.

(*b*) Find the approximate solutions to (*a*) by using a curtailed series of the characteristic functions of a free-free string.

5.3 Rayleigh's Principle

The discussion of Rayleigh's Principle in § 3.9 was based on systems with finite freedom. It was shown that, by stating the principle in the form

$$\frac{\partial \omega_c^2}{\partial \tau_r} = 0 \quad (r = 1, 2, ..., n), \quad (5.3.1)$$

when $\omega_c^2 = \omega_1^2, \omega_2^2, ..., \omega_n^2$, then a set of equations may be deduced which turn out to be identical with the equations of motion previously found by Lagrange's method. From this it follows that Rayleigh's Principle may be taken as an alternative starting point in deducing equations of motion. Lagrange's equations and Rayleigh's Principle are evidently connected. We showed further, in § 4.6, that Lagrange's method could be applied to the string problem and that it gives identical results with those obtained from the partial differential equations of motion. In view of these facts, Rayleigh's Principle will now be applied to the string problem. In doing this, we first note that our original justification of the principle involved the use of the Rayleigh quotient (see equation (3.9.12)) whereas with a system having infinite freedom, such as the string, no useful meaning can be applied to the quotient.

In this section we shall show how Rayleigh's Principle may be applied to ideal systems having infinite freedom in approximate calculations of natural frequencies. That is, we shall apply it to systems all of whose principal modes are known. We shall return later to the question of the wider significance of the principle and the way it provides some justification of our whole approach to the vibration analysis of real systems.

The application of Rayleigh's Principle to infinite-freedom systems involves a matter which does not arise with systems having only finite freedom; this is the possibility of building up any chosen constrained mode from principal-mode components. When the freedom is finite we have already shown that any set of ratios between the n co-ordinates can be obtained by a suitable combination of the principal modes. This was explained in § 3.5, by considering the set of linear equations which has to be satisfied. Now with infinite freedom, no such mathematical argument can be used and we have regarded it as axiomatic that any distortion of an elastic system can be regarded as the sum of distortions in its various principal modes (see § 4.7).

Let some constrained mode of an elastic body be chosen and suppose that the body is distorted in the chosen mode by a set of static forces (rather than by a suitable constraining mechanism). This way of visualising a constrained mode may be used when deciding whether or not a particular choice of mode is likely to give a good approximation to a natural frequency. For it is clearly necessary that the set of static forces should have finite values everywhere. Thus, if we are considering a string with a fixed end, we must assume a modal shape which gives zero deflexion at the fixed end. Again, if we are considering the motion of a cantilever then we again choose a modal shape which gives both zero deflexion and zero slope at the built-in end. These limitations in the choice of a constrained mode may be referred to as the 'geometric boundary conditions'.

There are also other limitations, which we shall refer to as the 'force boundary conditions', which it is commonly an advantage to comply with but which, if neglected, do not completely invalidate the analysis. An example of a force boundary condition is provided by the free end of a taut string. At such an end there is no applied transverse force and therefore the string must have zero slope. If a constrained mode is chosen which has finite slope at the end of such a string then it cannot be built up from a finite set of principal modes all of which have zero slope at the end. The apparent difficulty is resolved by the fact that although the slope at the end is zero for all principal modes it is nevertheless possible to find a principal mode which gives finite slope at a point on the string as close to the end as we wish, without the amplitude of that mode being excessively large, provided that the frequency of the mode is sufficiently high. This suggests that if a force boundary condition is not complied with, it means that the constrained mode contains high-frequency components. An estimate of the lowest frequency based on such a mode is therefore likely to produce a poorer approximation than would otherwise be the case.

The method of applying Rayleigh's Principle to an elastic body may be illustrated by estimating the first natural frequency of a fixed-free taut string. The true first principal mode of this system has already been found, being given by

$$v_1(x, t) = p_1 \sin \frac{\pi x}{2l}, \tag{5.3.2}$$

p_1 being the principal co-ordinate. If the true mode is used as the constrained mode, then the energy expressions are found to be

$$\left. \begin{aligned} 2T &= \mu p_1^2 \int_0^l \sin^2 \frac{\pi x}{2l}\, dx = \frac{\mu l}{2} p_1^2 = a_1 p_1^2, \\ 2V &= S p_1^2 \left(\frac{\pi}{2l}\right)^2 \int_0^l \cos^2 \frac{\pi x}{2l}\, dx = \frac{S\pi^2}{8l} p_1^2 = c_1 p_1^2. \end{aligned} \right\} \tag{5.3.3}$$

The natural frequency of the one-degree-of-freedom system defined by this mode is

$$\omega_1^2 = \frac{\pi^2 S}{4\mu l^2} \quad \text{or} \quad \omega_1 = \frac{1 \cdot 571 a}{l} \tag{5.3.4}$$

as previously calculated.

Now let it be supposed that the true first principal mode is not known. An estimated value for the lowest frequency may be found by taking as a constrained mode the parabolic form given by

$$v = Y \frac{x}{l} \left(2 - \frac{x}{l}\right). \tag{5.3.5}$$

During vibration in this constrained mode, the deflexion would be given by

$$v(x, t) = p_1 \frac{x}{l} \left(2 - \frac{x}{l}\right). \tag{5.3.6}$$

In this expression p_1 is a principal co-ordinate whose value gives the deflexion of the string at its free end. Fig. 5.3.1 shows this together with the modal shape. The energy expressions corresponding to this shape are

$$\left.\begin{aligned} 2T &= \mu p_1^2 \int_0^l \left[\frac{x}{l}\left(2-\frac{x}{l}\right)\right]^2 dx = \frac{8\mu l}{15} p_1^2 = a_1 p_1^2, \\ 2V &= S p_1^2 \left(\frac{4}{l^2}\right) \int_0^l \left(1-\frac{x}{l}\right)^2 dx = \frac{4S}{3l} p_1^2 = c_1 p_1^2. \end{aligned}\right\} \quad (5.3.7)$$

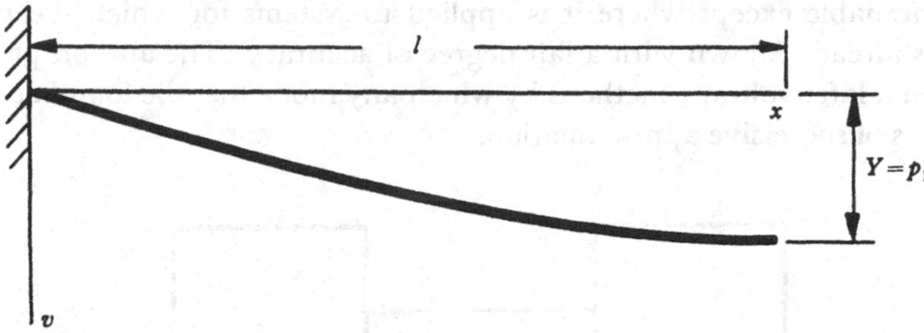

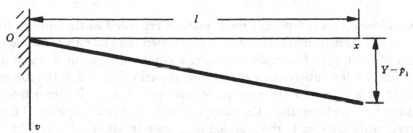

Fig. 5.3.1

These expressions give a new natural frequency

$$\omega_1^2 \doteqdot \omega_c^2 = \frac{5S}{2\mu l^2} \quad \text{or} \quad \omega_1 \doteqdot \frac{1{\cdot}581a}{l}. \quad (5.3.8)$$

This is a close approximation to the previous, exact, result. It is slightly too high as would be expected from finite-freedom theory. The error is only 0·64 %.

The parabolic shape assumed above fulfils both the geometric boundary condition (namely $v = 0$ when $x = 0$) and the force boundary condition ($\partial v/\partial x = 0$ when $x = l$). We now estimate the frequency by assuming a constrained modal

Fig. 5.3.2

shape which satisfies the geometric condition only. The shape shown in fig. 5.3.2 is suitable for this purpose and is given by the relation

$$v(x, t) = p_1 \frac{x}{l}. \quad (5.3.9)$$

In this expression p_1 again represents the value of the end deflexion. The energy expressions are now

$$\left.\begin{aligned} 2T &= \mu p_1^2 \int_0^l \left(\frac{x}{l}\right)^2 dx = \frac{\mu l}{3} p_1^2 = a_1 p_1^2, \\ 2V &= S p_1^2 \left(\frac{1}{l}\right)^2 \int_0^l dx = \frac{S}{l} p_1^2 = c_1 p_1^2 \end{aligned}\right\} \quad (5.3.10)$$

and the corresponding natural frequency is

$$\omega_1^2 \doteqdot \omega_c^2 = \frac{3S}{\mu l^2} \quad \text{or} \quad \omega_1 \doteqdot \frac{1 \cdot 732a}{l}. \tag{5.3.11}$$

The estimated value is about 10 % too high.

It may be mentioned here that it is occasionally possible to use Rayleigh's Principle for estimating the natural frequency of one of the higher modes of an elastic system. But the necessity of guessing the modal shape closely makes the process unreliable except where it is applied to systems for which the mode in question is already known with a fair degree of accuracy. The authors propose to describe, in a later volume, methods by which any mode may be found accurately by a process of successive approximation.

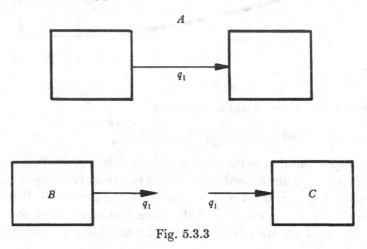

Fig. 5.3.3

The curtailment of a receptance series may be regarded as the application of a particular type of constraint on the motion of a system and its effects can be viewed in the light of Rayleigh's Principle. Consider a composite system A which may be broken down into the sub-systems B and C as shown in fig. 5.3.3, the sub-systems being linked at a single co-ordinate q_1. When system A is vibrating freely in a principal mode (which we shall here suppose to be the first), system B is undergoing forced vibration due to the reaction of C upon it; similarly C is being forced by the reaction of B. Now if either or both of the systems B and C have infinite freedom, then an approximate value for their direct receptances at q_1 may be found by curtailing the appropriate receptance series. If the series is curtailed for system B then it is implied that the higher modes of B are scarcely excited by forcing at the first natural frequency of A; equally, it may be convenient to curtail the series for C. The curtailed receptance series can be used in the equation

$$\frac{1}{\beta_{11}} + \frac{1}{\gamma_{11}} = 0 \quad \text{or} \quad \beta_{11} + \gamma_{11} = 0 \tag{5.3.12}$$

which will then give an approximate value for the lowest natural frequency of A. This approximation will be seen to be obtained from a constrained mode which differs from the principal mode in that the components due to the higher modes of B (and perhaps C) have been removed. Since motions in these higher modes are of small amplitude, Rayleigh's Principle ensures that the calculated frequency will be close to the true frequency of A.

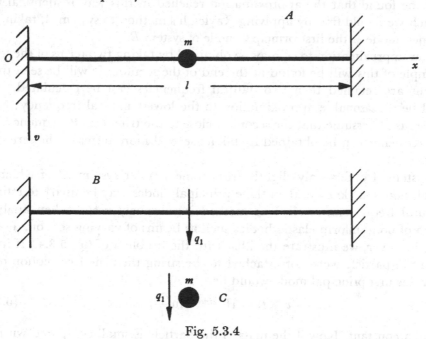

Fig. 5.3.4

The ideas above may be illustrated by considering the system shown in fig. 5.3.4 in which a particle of mass m is attached to the centre of a taut fixed-fixed string. This system may be analysed into the sub-systems B and C as shown in the figure; B represents the string only and C the particle only. The two sub-systems are linked by the co-ordinate q_1 which defines the central deflexion of the string. The frequency equation may be taken in the second form of equations (5.3.12) and will be found to be

$$\frac{2}{\mu l}\left[\frac{1}{\omega_{b1}^2 - \omega^2} + \frac{1}{\omega_{b3}^2 - \omega^2} + \frac{1}{\omega_{b5}^2 - \omega^2} + \dots\right] - \frac{1}{m\omega^2} = 0, \qquad (5.3.13)$$

where ω_{br} represents the rth natural frequency of system B. It will be seen that in the receptance series for system B the terms which correspond to the even numbered frequencies are omitted. This is because the corresponding modes give zero displacements at the centre of the string. The series is taken from Table 4 which also gives the values of the natural frequencies ω_{br}.

When seeking an approximate solution to equation (5.3.13), it should be remembered that the addition of the particle to the string will decrease all the odd numbered natural frequencies of B (that is, ω_{b1}, ω_{b3}, etc.); this property was pointed out in § 3.3 and was based on the known shape of receptance diagrams. Thus ω_1 (the first natural frequency of A) will be less than ω_{b1}. As a first approximation, therefore, all the terms of the series beyond the first may be neglected. This gives the expression

$$\omega_1^2 \doteqdot \frac{\mu l \omega_{b1}^2}{2m + \mu l} = \frac{\pi^2 S}{l(2m + \mu l)}. \tag{5.3.14}$$

It will be found that the approximation reached in this way is identical with that which we should find by applying Rayleigh's method to system A, taking the constrained mode as the first principal mode of system B.

A better approximation to ω_1 may be obtained by taking two terms of the series. An example of this will be found at the end of the section. It will be seen that if two terms are retained then two natural frequencies can be calculated. The first will be a reasonable approximation to the lowest natural frequency; but it would be rash to assume that the second is close to the true second frequency. The second frequency can be obtained by taking several more terms of the series into account.

If a system A differs only slightly from some simpler system A' of which the principal modes are known, then these principal modes may be used† to estimate the natural frequencies of A. This method is extremely useful when analysing vibration of non-uniform elastic bodies such as beams of varying section. For the present, however, we illustrate the idea with the problem of fig. 5.3.4. If in this problem the particle were not attached to the string then the free motion of the latter in this first principal mode would be

$$v(x, t) = \Pi_1 e^{i\pi at/l} \sin \frac{\pi x}{l}, \tag{5.3.15}$$

Π_1 being a constant. Now if the mass of the particle is small compared with that of the string, then the first principal mode of the system A will not greatly differ from that of the string alone. It is therefore possible to estimate this natural frequency in what is effectively the same manner as was described in the derivation of equation (5.3.14). Thus the same result as before may be arrived at though a different viewpoint is adopted in achieving it.

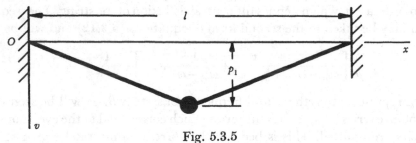

Fig. 5.3.5

The new viewpoint may also be used, however, if the mass of the particle is large compared with that of the string. For the estimate may be regarded as differing only slightly from an estimate for a light string. The principal mode of a particle on a light string is shown in fig. 5.3.5. In this case (where the mass of the particle is relatively great) the approximation is not easily reached by consideration of

† This powerful method is due to Rayleigh and is developed in *Theory of Sound* (1894), §§ 90, 91.

linked systems as the single co-ordinate linking the string's mass to the light string is not immediately available.

Rayleigh's Principle for finding the lowest natural frequency of a system is sometimes used by choosing a static deflexion as the constrained mode. This method is frequently applied in beam problems. It may be discussed here by referring again to the system of fig. 5.3.4. Consider the static deflexion of the system which would be produced by a concentrated static force F. It should be mentioned that the gravitational force is being neglected and the displacement may be imagined if desired as taking place in the horizontal plane. Now the advantage of using a static deflexion is that the potential energy expression may be obtained from the product of the static force and the displacement of the point at which it is applied. In this problem,

$$2V = p_1 F = p_1\left(\frac{4Sp_1}{l}\right) = \frac{4S}{l}p_1^2 = c_1 p_1^2. \tag{5.3.16}$$

Here, p_1 is the central deflexion and the relation between F and p_1 was found in equation (4.8.18). The deflexion of any point of the string is given by

$$v(x, t) = \begin{cases} p_1 \dfrac{2x}{l} & (0 \leqslant x \leqslant \tfrac{1}{2}l), \\[2mm] p_1 2\left(1 - \dfrac{x}{l}\right) & (\tfrac{1}{2}l \leqslant x \leqslant l) \end{cases} \tag{5.3.17}$$

and, from this, the following kinetic energy expression may be written down:

$$2T = 2\int_0^{l/2} \left(p_1\frac{2x}{l}\right)^2 \mu\, dx + m\dot p_1^2 = \left(\frac{\mu l}{3} + m\right)\dot p_1^2 = a_1 \dot p_1^2. \tag{5.3.18}$$

This gives the square of the natural frequency as

$$\omega_1^2 \doteqdot \frac{12S}{l(3m + \mu l)} \tag{5.3.19}$$

which may be compared with the previous estimate (see equation (5.3.14)). It will be found that the approximation is equivalent to treating the system as if one-third of the mass of the string were concentrated at the mid-point. This follows from the form of T.

Instead of using the static deflexion which is produced by a single concentrated force, we might take that which would be produced by a uniform distributed force or again that which would be produced by a combination of the two in some chosen ratio. This would give better results for problems in which the mass of the string is not much smaller than that of the particle.

A convenient check on the approximate values given in equations (5.3.14) and (5.3.19) is obtained when it is noted that the former becomes exact when m is zero and the latter becomes exact when μ is zero. It will be found that this follows immediately from the assumptions.

EXAMPLES 5.3

1. A taut uniform fixed-fixed wire of length 1 ft. has a tension $S = 100$ lb. and total mass 0·0137 lb. A small body weighing 4 oz. is attached to its mid-point (as in fig. 5.3.4). Considering only those principal modes which involve motion of the body, make the following calculations of the natural frequencies.

(a) Estimate ω_1 using a single term of the series of equation (5.3.13) (i.e. by means of equation (5.3.14)).

(b) Estimate ω_1 and ω_2 from two terms of the series in equation (5.3.13).

(c) Estimate ω_1 by the use of equation (5.3.19).

(d) Calculate the 'exact' values of ω_1 and ω_2 by the method of §4.2.

2. Show by reference to graphs of receptance that, if the mass of the attached particle in Ex. 5.3.1 (a) is diminished, the approximation to ω_1, afforded by one term of the series (5.3.13), will be improved.

3. The book entitled *Rigid Frame Formulas* by A. Kleinlogel is concerned with simple frames. These indeterminate structures are composed of two or more uniform beams, rigidly connected together. The treatment of each frame includes the presentation of vertical and horizontal reactions and bending moments at all points, corresponding to various conditions of static loading. These data are given in terms of the constants relating to the various members. Deflexions are not quoted.

Discuss the relevance of Kleinlogel's book to the *dynamics* of structures.

5.4 The use of finite-freedom analysis

The treatments given in §§ 5.1, 5.2 and 5.3 place us in a position to take up certain questions concerning the vibration analysis of real engineering systems. In particular, we can now discuss the validity and usefulness of finite-freedom analysis.

It was pointed out in § 5.3 that Rayleigh's Principle permits us to determine approximately the natural frequencies of a system A which differs only slightly from some simpler system A' whose principal modes are known. This was illustrated by means of the system of fig. 5.3.4 in which both the loaded string A and the unloaded string A' were 'ideal'. But it is not essential that the system A should be an ideal one; that is to say, A may be a real system and A' a simplified 'model' of it, which is used for the purposes of analysis.

The determination of modes and natural frequencies of an engineering system would usually be a matter of great difficulty if it had to be performed rigorously. This is because it is only for such simple ideal elastic bodies as taut strings and prismatic bars that characteristic functions are tabulated. It is therefore usual to rely upon the type of idealization that we have referred to; thus a simple (ideal) system is treated in detail and the properties of a real system are thence deduced under the assumption that it differs but little from the simpler one. The reasoning is due to Rayleigh and it is put on a quantitive basis in his book.†

When a mathematical treatment of a physical problem is undertaken, it is always necessary to idealize the real system in order to make the mathematics possible. The manner in which a real system is idealized is important because it

† *Theory of Sound* (1894), §90.

places a limit on the usefulness of accuracy in subsequent calculations. It is thus necessary, in particular, that we should justify the use of the finite-freedom analysis which was developed in Chapters 1 to 3. Clearly such analysis cannot be applied immediately to real systems without some previous investigation of the approximations involved, since rigid bodies and light elastic links do not exist in the physical world.

One method of reducing an infinite-freedom system to one having finite freedom is frequently used in the analysis of problems involving beam flexure and the like. The method is to divide the beam into a number of segments and to imagine the total mass of each segment to be concentrated at its centre of gravity. This simplification is, however, merely an aid to calculation and is used without regard to the physical nature of the actual system concerned; it will not therefore be discussed further here, though we hope to return to it in a later volume. The problems with which we are at present concerned are those in which the finite-freedom system closely resembles the real physical system.

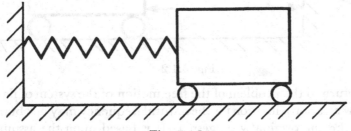

Fig. 5.4.1

Let an elastic steel block be connected, through an elastic steel spring to an elastic steel abutment as shown in fig. 5.4.1. The engineer will intuitively treat this problem in the manner of §1.2; that is, he will assume that the block and abutment are rigid and that the spring is light. He may subsequently make an extra allowance for the mass of the spring if he finds that it is not very small compared with that of the block; this correction will be discussed later.

Now this intuitive idealization does not conflict with the properties which have been shown to be possessed by continuous elastic systems. Thus, although the system has an infinite number of degrees of freedom its lowest natural frequency is known to correspond to a principal mode in which the distortion is confined almost entirely to the spring; for, with such a mode, a relatively large value of a_1 and a relatively small value of c_1 are obtained. Therefore by Rayleigh's Principle it is to be expected that a close approximation to ω_1 may be had by supposing no distortion to occur outside the spring.

The intuitive idealization of the system of fig. 5.4.1 is also valid if a harmonic force $Fe^{i\omega t}$, acting parallel to the axis of the spring, is applied to some point of the steel block. But this is true only if ω is sufficiently small. This may be shown from the fact that the receptance at the point of application of the force may be expanded in series form and by noting that if ω is much less than $\omega_2, \omega_3, \ldots$, etc., then only the first term in the series is of importance. This means that the forced motion will take place almost entirely in the first mode, and this mode involves little distortion elsewhere than in the spring. It is for this reason that no regard need be paid to the question of how the force is applied to the block. The engineer must be aware, however, that his intuitive idealization will become invalid if the forcing frequency becomes comparable with one or more of the higher natural frequencies of the system. In these higher modes, the mass of the spring becomes important with the result that 'surging' takes place in the spring. Such a motion may produce high stresses and failure of the spring in practice.

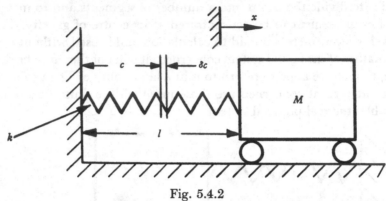

Fig. 5.4.2

We now return to the problem of the free motion of the system of fig. 5.4.1 and investigate how an approximation for the natural frequency may be obtained which is better than the one originally suggested. It is based upon the assumption that the steel block and the abutment are rigid but an allowance is made for the mass of the spring.

The system is redrawn in fig. 5.4.2 which shows an element of the spring distant c from the abutment. If the spring does not surge, then its stretch will be uniform and the displacement of the element will be cx/l, where x is the displacement of the mass and l is the length of the spring. If m is the mass of the spring and M that of the block then (assuming uniform stretch) the kinetic energy expression is

$$2T = M\dot{x}^2 + \int_0^l \left(\frac{c\dot{x}}{l}\right)^2 \frac{m\,dc}{l} = \left(M + \frac{m}{3}\right)\dot{x}^2 = a_1\dot{x}^2. \tag{5.4.1}$$

The potential energy expression is

$$2V = kx^2 = c_1 x^2 \tag{5.4.2}$$

so that the natural frequency is

$$\omega_1^2 = \frac{c_1}{a_1} = \frac{k}{M + \frac{1}{3}m}. \tag{5.4.3}$$

This is a Rayleigh estimate of the first natural frequency which corresponds to a modal shape that would hold for a massless spring. It is a good approximation to the true natural frequency provided that the spring mass is fairly small compared with the mass of the block. It may be noted here that the 'one-third rule' for allowing for the mass of the spring occurs also in other problems in which the displacement of points on a uniform bar are proportional to the distances of the points from one end. The rule is related to the moment of inertia expression for a uniform bar about one end.

Not only is it possible to simplify the equation of a system by replacing elastic bodies by rigid ones but also by replacing whole sections of mechanical systems by single masses and sometimes by infinite masses. An example will show the type of simplification which is possible.

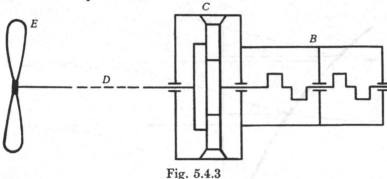

Fig. 5.4.3

Fig. 5.4.3 shows a system which consists of an engine B which drives a propeller E through a reduction gear C and a line-shaft D. By arguing as before, it may be concluded that certain types of motion of the system will be important because they are associated with low natural frequencies. Thus, when torsional vibration of the system is considered, motions are of interest which involve (a) twisting of the crankshaft of B, (b) twisting of the shaft D, (c) flexure of the blades of E, (d) angular motion of the whole engine and gear-case assembly in its mounting.

The presence of the last of these effects is due to the torque reaction which the reduction gear experiences from the shafting. This reaction will cause vibration of the engine in the same manner as would a harmonic torque that was applied to the engine casing when the shafting was absent. There will also be other types of distortion in, for instance, the gear wheels themselves and in the engine-casing. These distortions, however, are likely to remain small at the excitation frequencies which occur in practice. So far, we have idealized the system by disregarding certain very small motions in the same way as was suggested in previous discussions.

The simplification may be extended further in such a way that certain of the motions mentioned above can be examined separately. This action can be justified by referring to a diagram of reciprocal receptances. In fig. 5.4.4, the full line represents the reciprocal of the direct receptance for torsional motion of the engine and gear-casing on its mounting, the shafting not being included in the engine inertia; the system has one degree of freedom since all distortions are being neglected. The dotted lines in the diagram represent the reciprocal of the direct receptance for torsional motion of the rotating parts of the system; we shall imagine this receptance to be measured at the pitch circle on the reduction-gear annulus. The dotted curves are sketched with the sign reversed as usual. It is not necessary for the purposes of this argument that the curves should be known accurately; but some immediate features of the principal modes may be deduced from them.

The intersections of the full line and dotted lines represent natural frequencies of torsional vibration of the system. Now it will be seen that if the natural frequency of the engine and gear-casing on its mounting, which is given by ω_b in the figure, is very much lower than the lowest natural frequency which the rotating parts would have if the engine-casing were fixed, this being ω_c, and if also the inertia of the casing is much larger than that of the shafting, then the lowest frequency of

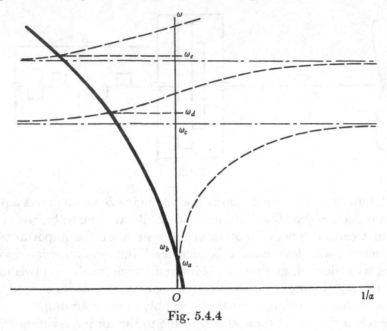

Fig. 5.4.4

the composite system will be very close to that of the empty casing. The two frequencies are ω_a and ω_b respectively and they will be very close together if the intersection between the curves is close to the axis of ω. In these circumstances the rotating parts may be neglected when the frequency of the engine on its mounting is calculated.

Again, it will be seen that the higher natural frequencies of the system, namely $\omega_d, \omega_e, \ldots$, etc., are very close to the anti-resonance frequencies of the rotating parts. Thus, when calculating these frequencies, the engine and casing may be treated as if fixed.

The argument may be extended further by considering vibration of the rotating parts only. Let the system be divided into two parts at the hub of the propeller; a reciprocal receptance curve for the shafting is represented by the full lines of fig. 5.4.5. The corresponding curves for the propeller, drawn with the sign reversed, are shown dotted. If the lowest natural frequency of the propeller with its hub clamped (ω_i in fig. 5.4.5) is greater than the lowest natural frequencies (ω_g and ω_h say) of the shafting alone, then the propeller will behave as if it were rigid over a considerable range of frequency. The appropriate receptance, measured at its hub, may then be taken as

$$- (1/I\omega^2),$$

where I is the moment of inertia; this may be regarded as the first term of the series representation of the receptance. The corresponding line on the diagram is shown chain-dotted. The error that is introduced into the calculated values of the lowest natural frequencies are given by the small differences between the ordinates of a and a' and of b and b'. The approximation is rarely permissible for aircraft propellers but is widely used for marine propellers.

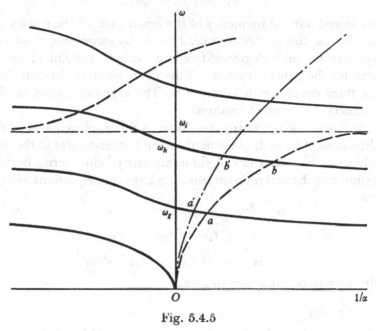

Fig. 5.4.5

5.5 Dunkerley's method

A part of this chapter has been devoted to Rayleigh's Principle and we now return to the use of it for estimating the lowest natural frequency of an ideal system (in the manner of § 5.3). The remainder of this chapter is concerned with two other approximate methods which may be used for this particular purpose. They are related to Rayleigh's Principle and it is thus convenient to introduce them at this point. The first of these methods was originally used by Dunkerley[†] on a purely empirical basis.

Dunkerley was concerned with the lowest natural frequency of flexural vibration of a beam carrying a number of concentrated masses. We shall first limit our discussion to this specific problem but we shall examine Dunkerley's method from a rather more general point of view later on. The discussion which follows refers to *beams* merely for historical reasons; the reader may prefer, at this stage, to think of the problem of a loaded taut string, for which the argument is identical. Dunkerley's method consisted of finding a number of frequencies, each one of which corresponded to the attachment of one of the concentrated masses only to the beam;

† S. Dunkerley, 'On the Whirling and Vibration of Shafts', *Phil. Trans. Roy. Soc.* A, vol. 185, Pt 1 (1894), pp. 279–360.

thus if there were n masses he obtained the n natural frequencies of the n one-degree-of-freedom systems obtained by attaching each mass in turn to the beam (which is supposed to be of negligible mass). Calling these frequencies $_1\omega_1, _2\omega_1, \ldots, _n\omega_1$, Dunkerley then used the formula

$$\frac{1}{\omega_1^2} \doteqdot \frac{1}{_1\omega_1^2} + \frac{1}{_2\omega_1^2} + \ldots + \frac{1}{_n\omega_1^2}, \qquad (5.5.1)$$

where ω_1 is the lowest natural frequency of the beam with all the masses.

The justification for this procedure may be seen by examining the form of the frequency equation for an n-degree-of-freedom system. Instead of forming the necessary terms for the general case it will be convenient to develop them for a system having three degrees of freedom only. The arguments used are identical for greater numbers of degrees of freedom.

Let the co-ordinates for the system be taken as the displacements of the three masses, the directions of these displacements being perpendicular to the axis of the beam. This choice of co-ordinates will eliminate any product terms in the kinetic energy expression and the determinant from the Lagrange equations will therefore be of the form

$$\Delta = \begin{vmatrix} c_{11} - \omega^2 a_{11} & c_{12} & c_{13} \\ c_{12} & c_{22} - \omega^2 a_{22} & c_{23} \\ c_{13} & c_{23} & c_{33} - \omega^2 a_{33} \end{vmatrix}. \qquad (5.5.2)$$

This determinant may be expanded to give

$$\Delta = \begin{vmatrix} c_{11} & c_{12} & c_{13} \\ c_{12} & c_{22} & c_{23} \\ c_{13} & c_{23} & c_{33} \end{vmatrix} - \omega^2 [a_{11}(c_{22}c_{33} - c_{23}^2) + a_{22}(c_{11}c_{33} - c_{13}^2) + a_{33}(c_{11}c_{22} - c_{12}^2)]$$

$$+ \omega^4 (a_{11}a_{22}c_{33} + a_{11}c_{22}a_{33} + c_{11}a_{22}a_{33}) - \omega^6 a_{11}a_{22}a_{33}. \qquad (5.5.3)$$

Now if the natural frequencies of the system are ω_1, ω_2, ω_3, then the frequency equation which is obtained from this expression, namely

$$\Delta = 0 \qquad (5.5.4)$$

must be equivalent to

$$(\omega_1^2 - \omega^2)(\omega_2^2 - \omega^2)(\omega_3^2 - \omega^2) = 0. \qquad (5.5.5)$$

This latter form can be expanded to

$$-\omega^6 + \omega^4(\omega_1^2 + \omega_2^2 + \omega_3^2) - \omega^2(\omega_1^2\omega_2^2 + \omega_2^2\omega_3^2 + \omega_3^2\omega_1^2) + \omega_1^2\omega_2^2\omega_3^2 = 0. \qquad (5.5.6)$$

By comparing the coefficients of ω^2 in equations (5.5.3) and (5.5.6) it is found that

$$\frac{1}{\omega_1^2} + \frac{1}{\omega_2^2} + \frac{1}{\omega_3^2} = \frac{a_{11}(c_{22}c_{33} - c_{23}^2) + a_{22}(c_{11}c_{33} - c_{13}^2) + a_{33}(c_{11}c_{22} - c_{12}^2)}{\begin{vmatrix} c_{11} & c_{12} & c_{13} \\ c_{12} & c_{22} & c_{23} \\ c_{13} & c_{23} & c_{33} \end{vmatrix}}. \qquad (5.5.7)$$

Now consider the natural frequency $_1\omega_1$ which the system would have if only the mass at q_1 were present. The determinant (5.5.2) would then reduce to

$$\Delta = \begin{vmatrix} c_{11} & c_{12} & c_{13} \\ c_{12} & c_{22} & c_{23} \\ c_{13} & c_{23} & c_{33} \end{vmatrix} - \omega^2 a_{11}(c_{22}c_{33} - c_{23}^2) \tag{5.5.8}$$

and this gives,

$$\frac{1}{_1\omega_1^2} = \frac{a_{11}(c_{22}c_{33} - c_{23}^2)}{\begin{vmatrix} c_{11} & c_{12} & c_{13} \\ c_{12} & c_{22} & c_{23} \\ c_{13} & c_{23} & c_{33} \end{vmatrix}}. \tag{5.5.9}$$

Expressions for the frequencies $_2\omega_1$ (with a mass at q_2 only) and $_3\omega_1$ (with a mass at q_3 only) will be similar. By adding the three expressions for these frequencies, it will be seen that

$$\frac{1}{_1\omega_1^2} + \frac{1}{_2\omega_1^2} + \frac{1}{_3\omega_1^2} = \frac{1}{\omega_1^2} + \frac{1}{\omega_2^2} + \frac{1}{\omega_3^2}. \tag{5.5.10}$$

Now for the problems with which Dunkerley was concerned the second and higher frequencies of the complete system were considerably greater than the lowest frequency. Consequently it was possible to neglect all the terms except the first on the right-hand side of equation (5.5.10) thus leaving Dunkerley's relation

$$\frac{1}{_1\omega_1^2} + \frac{1}{_2\omega_1^2} + \frac{1}{_3\omega_1^2} \doteq \frac{1}{\omega_1^2}. \tag{5.5.11}$$

The Dunkerley method of frequency calculation may be approached in another manner which shows that the idea is more general than might appear from the above discussion. Consider any system A which may have finite or infinite freedom and let systems $1, 2, ..., r$ be formed from it by dividing its total mass distribution into r different distributions and supposing that each were present separately. Let this division of the mass of A be made in such a way that the first principal modes of systems $1, 2, ..., r$ are all approximately similar to the first principal mode of the parent system A. The latter modal shape may be used to obtain Rayleigh approximations to the lowest natural frequencies of systems $1, 2, ..., r$. Thus, using the lowest mode of the complete system A as a constrained mode for system 1, let the kinetic energy be

$$\tfrac{1}{2} \cdot {_1}a_1 \cdot \dot{p}_1^2,$$

where p_1 is the first principal co-ordinate of A. Again with the second distribution of mass let the kinetic energy be

$$\tfrac{1}{2} \cdot {_2}a_1 \cdot \dot{p}_1^2$$

and so on. The potential energy will, in each case, be

$$\tfrac{1}{2} c_1 p_1^2,$$

where c_1 is the first stability coefficient of the complete system A. Now since the modal shape of the complete system A is approximately the same as those of the systems $1, 2, \ldots, r$, it follows from Rayleigh's Principle that the true lowest frequencies $_1\omega_1, _2\omega_1, \ldots, _r\omega_1$ of the latter are given by

$$_1\omega_1^2 \doteqdot \frac{c_1}{_1a_1}, \quad _2\omega_1^2 \doteqdot \frac{c_1}{_2a_1}, \quad \ldots, \quad _r\omega_1^2 \doteqdot \frac{c_1}{_ra_1}. \tag{5.5.12}$$

Moreover, these estimates will be generally somewhat high. With all the partial mass distributions present together the natural frequency of the complete system will be given exactly by

$$\omega_1^2 = \frac{c_1}{a_1} = \frac{c_1}{_1a_1 + _2a_1 + \cdots + _ra_1} \tag{5.5.13}$$

or

$$\frac{1}{\omega_1^2} = \frac{_1a_1}{c_1} + \frac{_2a_1}{c_1} + \cdots + \frac{_ra_1}{c_1}, \tag{5.5.14}$$

and from this it follows that

$$\frac{1}{\omega_1^2} \doteqdot \frac{1}{_1\omega_1^2} + \frac{1}{_2\omega_1^2} + \cdots + \frac{1}{_r\omega_1^2}. \tag{5.5.15}$$

This estimate of ω_1 in terms of $_1\omega_1, _2\omega_1, \ldots, _r\omega_1$ will be somewhat low, or, more accurately, the estimated value of ω_1 represents a lower limit for the true value.

This approach shows that Dunkerley's formula for the lowest natural frequency of a system is a good approximation if the fundamental modal shapes associated with the different mass distributions are similar to one another and to the fundamental modal shape of the complete system. It follows that, for the type of system to which Dunkerley originally applied the formula, the similarity of the modal shapes must correspond to the fact that the higher natural frequencies of the complete system are very high compared with the fundamental frequency.

The connexion between these two properties can be readily seen for a system which carries two concentrated masses only, being otherwise without mass. Consider such a system in which the two masses m_1 and m_2 are to be attached at co-ordinates q_1 and q_2. The system will possess two natural frequencies and two modal shapes. The complete arrangement may be regarded as a system A which may be divided into a sub-system B comprising the massless portion only and sub-system C comprising only the masses m_1 and m_2. Now if the mass m_1 only is attached to B then the modal shape for the natural frequency will be defined by the ratio of the displacements at q_1 and q_2. This ratio will be equal to the ratio

$$\beta_{11}/\beta_{12}$$

between the receptances of system B (or, more accurately since B is massless, its flexibilities). Similarly the modal shape with the mass m_2 present will be given by the ratio

$$\beta_{12}/\beta_{22}.$$

If the two modal shapes are identical then these two ratios must be equal so that

$$\beta_{11}\beta_{22} - \beta_{12}^2 = 0. \tag{5.5.16}$$

Now the frequency equation of the complete system is given by entry number 9 of Table 3, namely

$$(\beta_{11} + \gamma_{11})(\beta_{22} + \gamma_{22}) - (\beta_{12} + \gamma_{12})^2 = 0. \tag{5.5.17}$$

In this case, the receptances of C are

$$\gamma_{11} = \frac{-1}{m_1 \omega^2}, \quad \gamma_{22} = \frac{-1}{m_2 \omega^2}, \quad \gamma_{12} = 0, \tag{5.5.18}$$

so that the natural frequencies of A are given by

$$(\beta_{11}\beta_{22} - \beta_{12}^2) - \frac{1}{\omega^2}\left(\frac{\beta_{11}}{m_2} + \frac{\beta_{22}}{m_1}\right) + \frac{1}{\omega^4 m_1 m_2} = 0. \tag{5.5.19}$$

If equation (5.5.16) holds, so that the first term of equation (5.5.19) vanishes, then the natural frequencies of A are given by

$$\frac{1}{\omega^2}\left(\frac{1}{\omega^2 m_1 m_2} - \frac{\beta_{11}}{m_2} - \frac{\beta_{22}}{m_1}\right) = 0. \tag{5.5.20}$$

The roots of this relation are

$$\frac{1}{\omega^2} = \beta_{11} m_1 + \beta_{22} m_2, \quad \frac{1}{\omega^2} = 0. \tag{5.5.21}$$

Now let $_1\omega_1$ and $_2\omega_1$ be the natural frequencies of B when m_1 and m_2 are attached to it singly. Then, according to the frequency equations of these subsidiary systems,

$$\frac{1}{\beta_{11}} - m_1 \cdot _1\omega_1^2 = 0, \quad \frac{1}{\beta_{22}} - m_2 \cdot _2\omega_1^2 = 0. \tag{5.5.22}$$

Thus the first of the roots (5.5.21) is the Dunkerley relation, which is exact here,

$$\frac{1}{\omega^2} = \frac{1}{\omega_1^2} = \frac{1}{_1\omega_1^2} + \frac{1}{_2\omega_1^2}. \tag{5.5.23}$$

The second root shows that the other natural frequency ω_2 of A is infinite.

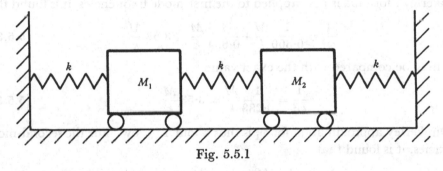

Fig. 5.5.1

All the above discussion has been concerned with the lowest natural frequencies only. The argument which is based on the similarity of the modal shapes of the different mass distributions can be applied, however, to the second and higher modes. In some systems good numerical results can be obtained in this way. However, it will be seen that, while an estimate of a fundamental frequency by this method may be shown by Rayleigh's Principle to be either exactly correct or (more generally) low, this will no longer be necessarily true for higher frequencies; if the system has finite freedom, then Rayleigh's Principle shows that an estimate of the highest frequency will either be correct or (more generally) somewhat high. The following example illustrates the application of Dunkerley's method for both the first and second principal modes of a system.

The complete system A is shown in fig. 5.5.1, the values of the masses M_1 and M_2 being shown in the following table. Let the total mass of the system be divided into two distributions which are indicated by the lines marked 1 and 2 of the table.

System	Value of mass M_1	Value of mass M_2
A	$4M$	$3M$
1	$2M$	$2M$
2	$2M$	M

First, the natural frequencies of the complete system A may be calculated using the frequency equation for the system which may be obtained from entry 9 of Table 1. It will be found to be

$$12M^2\omega^4 - k(3M + 7M + 4M)\,\omega^2 + 3k^2 = 0 \qquad (5.5.24)$$

which has the roots $\quad \omega_1^2 = 0\cdot283k/M, \quad \omega_2^2 = 0\cdot884k/M.$ $\qquad (5.5.25)$

Now, for the system 1, the frequency equation is

$$4M^2\omega^4 - k(2M + 4M + 2M)\,\omega^2 + 3k^2 = 0 \qquad (5.5.26)$$

which has the roots $\quad {}_1\omega_1^2 = 0\cdot500k/M, \quad {}_1\omega_2^2 = 1\cdot500k/M.$ $\qquad (5.5.27)$

Again, for the system 2, the frequency equation is

$$2M^2\omega^4 - k(M + 3M + 2M)\,\omega^2 + 3k^2 = 0 \qquad (5.5.28)$$

which has the roots $\quad {}_2\omega_1^2 = 0\cdot634k/M, \quad {}_2\omega_2^2 = 2\cdot366k/M.$ $\qquad (5.5.29)$

If Dunkerley's formula is now applied to the first-mode frequencies, it is found that

$$\frac{1}{\omega_1^2} \doteqdot \frac{1}{0\cdot500}\frac{M}{k} + \frac{1}{0\cdot634}\frac{M}{k} = 3\cdot58\frac{M}{k} \qquad (5.5.30)$$

which may be compared with the exact value

$$\frac{1}{\omega_1^2} = \frac{1}{0\cdot283}\frac{M}{k} = 3\cdot53\frac{M}{k}. \qquad (5.5.31)$$

The Dunkerley approximation for ω_1^2 is thus about 1 % low. For the second-mode frequencies, it is found that

$$\frac{1}{\omega_2^2} \doteqdot \frac{1}{1\cdot500}\frac{M}{k} + \frac{1}{2\cdot366}\frac{M}{k} = 1\cdot09\frac{M}{k}, \qquad (5.5.32)$$

whereas the exact value is $\quad \dfrac{1}{\omega_2^2} = \dfrac{1}{0\cdot884}\dfrac{M}{k} = 1\cdot13\dfrac{M}{k}$ $\qquad (5.5.33)$

and the Dunkerley approximation for ω_2^2 is therefore somewhat less than 4 % high.

In the above discussion, the use of the formula

$$\frac{1}{\omega_s^2} \doteqdot \frac{1}{{}_1\omega_s^2} + \frac{1}{{}_2\omega_s^2} + \ldots + \frac{1}{{}_r\omega_s^2} \qquad (5.5.34)$$

to estimate the sth natural frequency of a system has been referred to as Dunkerley's method whether or not the problem to which it was being applied was of the type with which Dunkerley was originally concerned. The practical utility of the formula is probably limited to the Dunkerley type of problem and to related ones, but it is convenient to refer to Dunkerley's method in the more general sense. The approach to the problem here has been made general in order to emphasize the principles involved.

EXAMPLE 5.5

1. Show, by an extension of the argument of this section, e.g. see equation (5.5.19), that Dunkerley's method will give a good approximation to the lowest principal mode of a system which has three degrees of freedom provided that the three principal modes which are obtained from three separate mass distributions are similar.

5.6 Southwell's theorem

A method for estimating the lowest natural frequencies of systems, which is comparable with that of Dunkerley, was introduced by Southwell.[†] It, too, has its basis on Rayleigh's Principle and it has acquired the name of 'Southwell's Theorem'.

Southwell's process is complementary to Dunkerley's and the two are dependent upon different aspects of the same mathematical relationship. Whereas Dunkerley divided the total mass distribution into a number of partial distributions, in Southwell's method it is the stiffness distribution which is so divided. The mathematical justification for the method is identical with that for Dunkerley's; this may be seen as follows.

The natural frequency equation of a system having n degrees of freedom may be written

$$\begin{vmatrix} c_{11} - \omega^2 a_{11} & c_{12} - \omega^2 a_{12} & \cdots & c_{1n} - \omega^2 a_{1n} \\ c_{21} - \omega^2 a_{21} & c_{22} - \omega^2 a_{22} & \cdots & c_{2n} - \omega^2 a_{2n} \\ \cdots\cdots\cdots\cdots\cdots\cdots\cdots\cdots\cdots\cdots\cdots \\ c_{n1} - \omega^2 a_{n1} & c_{n2} - \omega^2 a_{n2} & \cdots & c_{nn} - \omega^2 a_{nn} \end{vmatrix} = 0 \qquad (5.6.1)$$

and, if we now write $x^2 = 1/\omega^2$, this equation may be rearranged to give

$$\begin{vmatrix} a_{11} - x^2 c_{11} & a_{12} - x^2 c_{12} & \cdots & a_{1n} - x^2 c_{1n} \\ a_{21} - x^2 c_{21} & a_{22} - x^2 c_{22} & \cdots & a_{2n} - x^2 c_{2n} \\ \cdots\cdots\cdots\cdots\cdots\cdots\cdots\cdots\cdots\cdots\cdots \\ a_{n1} - x^2 c_{n1} & a_{n2} - x^2 c_{n2} & \cdots & a_{nn} - x^2 c_{nn} \end{vmatrix} = 0. \qquad (5.6.2)$$

This second equation has the same mathematical form as the first but the coefficients c and a are interchanged, and ω^2 is replaced by x^2 or $1/\omega^2$. It follows that if the stiffness of a system is divided into a number of component distributions $1, 2, \ldots, r$ and if the corresponding values of x are found for each component separately, then there will be a relation between the various x's which is similar to that between the ω's in the Dunkerley method. Thus, if the x's for the first modes of the systems $1, 2, \ldots, r$ with the partial distributions of stiffness are $_1x_1, _2x_1, \ldots, _rx_1$ and if the

[†] H. Lamb and R. V. Southwell, 'The Vibration of a Spinning Disk', *Proc. Roy. Soc.* A, vol. 99 (1921), pp. 272–80.

stiffness components are so chosen that these first modes are all of similar shape to the fundamental mode of the complete system, then the relation will be

$$\frac{1}{_1x_1^2} + \frac{1}{_2x_1^2} + \dots + \frac{1}{_rx_1^2} \doteqdot \frac{1}{x_1^2}, \tag{5.6.3}$$

where x_1 is the lowest root of the equation (5.6.2) for the complete system. We may therefore write

$$_1\omega_1^2 + _2\omega_1^2 + \dots + _r\omega_1^2 \doteqdot \omega_1^2, \tag{5.6.4}$$

where $_1\omega_1^2 = 1/_1x_1^2$, etc. It is this relation which is referred to as Southwell's Theorem.

The theorem may be derived in a way that is comparable with that used above for Dunkerley's method and the sign of the error introduced in the approximation may then be seen. Consider a system A whose total distribution of stiffness is divided into r different distributions. We consider the lowest natural frequencies of systems $1, 2, \dots, r$, which systems have each separately one of the partial distributions of stiffness. Let the first principal mode of each of these systems $1, 2, \dots, r$ be of similar form, this form being similar to that of the first principal mode of A. The first mode of A may then be regarded as a constrained mode for the systems $1, 2, \dots, r$. If the magnitude of the distortion in this constrained mode is given by p_1, let the appropriate potential energy expression for system 1 be

$$\tfrac{1}{2} \cdot {_1}c_1 \cdot p_1^2.$$

With the second stiffness distribution, let the potential energy expression for system 2 be

$$\tfrac{1}{2} \cdot {_2}c_1 \cdot p_1^2,$$

and so on. The mass distribution is the same for all the systems so that the kinetic energy is

$$\tfrac{1}{2} \cdot a_1 \cdot p_1^2$$

for all of them. By Rayleigh's Principle, the true lowest frequencies $_1\omega_1, {_2}\omega_1, \dots, {_r}\omega_1$ of the systems $1, 2, \dots, r$ are given by

$$_1\omega_1^2 \doteqdot \frac{_1c_1}{a_1}, \quad _2\omega_1^2 \doteqdot \frac{_2c_1}{a_1}, \quad \dots, \quad _r\omega_1^2 \doteqdot \frac{_rc_1}{a_1}; \tag{5.6.5}$$

these estimates being, in general, somewhat high.

If all the partial stiffness distributions are present together the lowest natural frequency ω_1 of the complete system A is given exactly by

$$\omega_1^2 = \frac{c_1}{a_1} = \frac{_1c_1 + _2c_1 + \dots + _rc_1}{a_1}. \tag{5.6.6}$$

It follows therefore that $\qquad \omega_1^2 \doteqdot {_1}\omega_1^2 + {_2}\omega_1^2 + \dots + {_r}\omega_1^2, \tag{5.6.7}$

where the estimate of ω_1^2 is, in general, somewhat low. That is to say, if $_1\omega_1^2, {_2}\omega_1^2, \dots, {_r}\omega_1^2$ are determined by some means (either analytically or experimentally), their sum gives a lower limit to the possible value of ω_1^2.

As with Dunkerley's method, this argument may be applied to *any* natural frequency of A provided that there is a sufficient degree of similarity between the corresponding modal shapes of A on the one hand and of systems 1, 2, ..., r on the other. However, it is not then possible to predict easily what will be the sign of the error in the approximation unless the frequency be the highest of a system with finite freedom, when it will be high.

The following example, which is similar to that used for Dunkerley's method, illustrates the use of Southwell's theorem.

A uniform rigid bar has mass M and length l and is supported at its ends by two light springs, one of stiffness k and one of stiffness $3k$. To the ends of the bar there are also attached two *massless* strings each of length l; the other ends of the strings are fixed and the tension in each is S. The mid-point of the beam is constrained by a pin which slides in smooth guides, the arrangement being as shown in fig. 5.6.1.

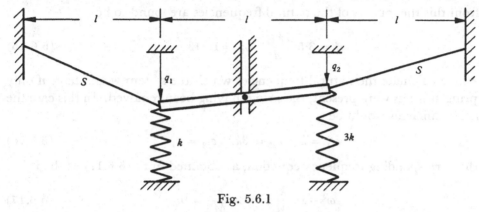

Fig. 5.6.1

The system will have two principal modes of vibration in the plane of the diagram and the frequencies will depend both upon the spring stiffness and the string tension. Consider first the two natural frequencies allowing for all the stiffness, taking as co-ordinates the displacements q_1 and q_2 of the ends of the bar as shown in the figure.

The total potential energy V of the system will be given by

$$2V = kq_1^2 + 3kq_2^2 + \frac{2S}{l}(q_1^2 + q_2^2 - q_1 q_2), \tag{5.6.8}$$

where the contribution which depends upon S is found from the products of the tension and the small extensions of the two strings. The total kinetic energy is T where

$$2T = M\left(\frac{\dot{q}_1 + \dot{q}_2}{2}\right)^2 + \frac{Ml^2}{12}\left(\frac{\dot{q}_1 - \dot{q}_2}{l}\right)^2 = \frac{M}{3}(\dot{q}_1^2 + \dot{q}_2^2 + \dot{q}_1 \dot{q}_2). \tag{5.6.9}$$

We shall examine the problem numerically when the tension of the string is such that

$$\frac{S}{l} = \frac{k}{4}. \tag{5.6.10}$$

The natural frequencies will be given by the equation

$$\begin{vmatrix} c_{11} - \omega^2 a_{11} & c_{12} - \omega^2 a_{12} \\ c_{12} - \omega^2 a_{12} & c_{22} - \omega^2 a_{22} \end{vmatrix} = 0 \tag{5.6.11}$$

and, here, the stiffness coefficients are

$$c_{11} = \frac{3k}{2}, \quad c_{22} = \frac{7k}{2}, \quad c_{12} = -\frac{k}{4}, \tag{5.6.12}$$

while the inertia coefficients are

$$a_{11} = \frac{M}{3}, \quad a_{22} = \frac{M}{3}, \quad a_{12} = \frac{M}{6}. \tag{5.6.13}$$

These figures give a frequency equation

$$\omega^4 - \frac{21k}{M}\omega^2 + 62 \cdot 25\frac{k^2}{M^2} = 0 \tag{5.6.14}$$

and from this the squares of the natural frequencies are found to be

$$\omega_1^2 = 3 \cdot 57\frac{k}{M}, \quad \omega_2^2 = 17 \cdot 43\frac{k}{M}. \tag{5.6.15}$$

We now calculate the natural frequencies which the system would have if only the spring stiffness were present, the strings having been removed. In this case the stiffness coefficients would be

$$c_{11} = k, \quad c_{22} = 3k, \quad c_{12} = 0, \tag{5.6.16}$$

and the corresponding frequency equation, as obtained from (5.6.11), is then

$$\omega^4 - 16\frac{k}{M}\omega^2 + 36\frac{k^2}{M^2} = 0. \tag{5.6.17}$$

This has the roots $\quad _1\omega_1^2 = 2 \cdot 71\frac{k}{M}, \quad _1\omega_2^2 = 13 \cdot 29\frac{k}{M}.$ \hfill (5.6.18)

If on the other hand the springs were removed and only the string tension were available to provide the restoring forces, then the stiffness coefficients would be

$$c_{11} = \frac{k}{2}, \quad c_{22} = \frac{k}{2}, \quad c_{12} = -\frac{k}{4}, \tag{5.6.19}$$

and the corresponding frequency equation then becomes

$$\omega^4 - 5\frac{k}{M}\omega^2 + 2 \cdot 25\frac{k^2}{M^2} = 0. \tag{5.6.20}$$

This has the roots $\quad _2\omega_1^2 = 0 \cdot 50\frac{k}{M}, \quad _2\omega_2^2 = 4 \cdot 50\frac{k}{M}.$ \hfill (5.6.21)

If Southwell's theorem is now used to obtain an approximation for the natural frequencies with both springs and string tension present, then it is found that

$$\left.\begin{aligned} \omega_1^2 &\doteqdot (2 \cdot 71 + 0 \cdot 50)\frac{k}{M} = 3 \cdot 21\frac{k}{M}, \\[2mm] \omega_2^2 &\doteqdot (13 \cdot 29 + 4 \cdot 50)\frac{k}{M} = 17 \cdot 79\frac{k}{M}. \end{aligned}\right\} \tag{5.6.22}$$

If these results are compared with the true figures calculated above, the first will be found to be 10 % low and the second 2 % high.

It may be mentioned here that Southwell's theorem has been used for calculating natural frequencies of flexure of the blades of rotating propellers. In this problem some of the stiffness arises from elastic effects while some arises from the effects of centrifugal forces when the propeller is rotating. The frequencies due to each of these effects separately may be calculated and an approximation then obtained for the frequencies when both effects are present. The theorem is very useful in this problem because the natural frequencies of the blades when the propeller is not rotating can be easily checked by experiment.[†]

EXAMPLE 5.6

1. The figure shows an elastic system comprising two equal masses M and three light springs whose stiffnesses are k_1, k_2 and k_3. Three systems A, B and C are formed, as shown in the following table, by giving particular values to the stiffnesses such that the

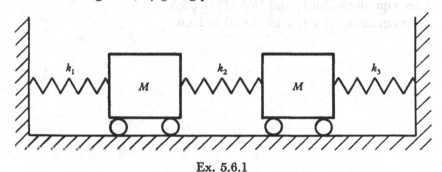

Ex. 5.6.1

values of A are the sums of those for B and C. Show that the square of each natural frequency of A is approximately equal to the sum of the squares of the corresponding natural frequencies of B and C. Explain how this result follows from Rayleigh's Principle and comment on the sign of the error obtained in this approximation.

System	Value of k_1	Value of k_2	Value of k_3
A	k	2k	2k
B	k	k	k
C	o	k	k

Explain how the lowest natural frequency of a rotating propeller can be estimated by dividing the potential energy into two parts in a comparable way.

(C.U.M.S.T. Pt II, 1955)

† R. V. Southwell and Barbara S. Gough, 'On the Transverse Vibrations of Airscrew Blades', *R. and M.* 766 (1921). This matter is discussed by J. P. den Hartog, *Mechanical Vibrations* (McGraw-Hill, London, 4th ed. 1956), p. 270.

NOTATION

Where reference is made to taut strings, see the Notation for Chapter 4.

Supplementary List of Symbols used in Chapter 5

A_r — Coefficient of $\phi_r(x)$ in approximating series (see equation (5.2.10)).

$_1a_1$, etc. — Approximate inertia coefficient (see the text between equations (5.5.11) and (5.5.12)).

$_1c_1$, etc. — Approximate stability coefficient (see the text between equations (5.6.4) and (5.6.5)).

$\bar{v}(x)$ · — Approximation to a given curve $v(x)$ (see equation (5.2.10)).

x — $= 1/\omega$; a substitution made in equation (5.6.1).

$_1x_1$, etc. — $= 1/_1\omega_1$, etc. (see equation (5.6.3)).

ϵ — Error introduced by $\bar{v}(x)$ in approximation for $v(x)$.

ω_c — Natural frequency of constrained system.

$_1\omega_1$, etc. — $\begin{cases} \text{See equations (5.5.1) and (5.5.12) in § 5.5.} \\ \text{See equations (5.6.4) and (5.6.5) in § 5.6.} \end{cases}$

CHAPTER 6

TORSIONAL AND LONGITUDINAL VIBRATION OF UNIFORM SHAFTS AND BARS

The statement that every small motion of the system can be represented as the result of superposed motions in normal modes is equivalent to a theorem, viz: that any arbitrary displacement (or velocity) can be represented as the sum of a finite or infinite series of normal functions. Such theorems concerning the expansions of functions are generalizations of Fourier's theorem, and, from the point of view of a rigorous analysis, they require independent proof. Every problem of free vibrations suggests such a theory of expansion.

A. E. H. LOVE, *A Treatise on the Mathematical Theory of Elasticity* (1892)

This chapter is devoted to a treatment of vibrations which can be dealt with by the same mathematical methods as were used for taut strings, but which are of more direct importance in engineering. As a result of this mathematical similarity, the string theory will be used extensively after the basic differential equations for torsional and longitudinal motion have been derived. We shall restrict the discussion of torsional motion to that of circular shafts; the torsion of non-circular shafts requires a more complete theory but fortunately is not often required in engineering practice. The longitudinal vibration theory of this chapter applies to uniform bars of any cross-section.

6.1 The basic equations of torsional vibration of a uniform circular shaft

Consider a uniform shaft whose circular cross-section is of radius R, and let distance along the shaft, measured from some suitable origin, be denoted by x.

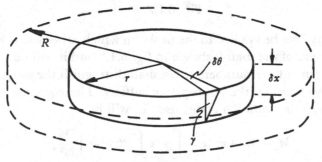

Fig. 6.1.1

A thin element of the shaft is shown in fig. 6.1.1. If this element is subjected to a torque which acts about the axis of the shaft, then the shear strain γ at a radius r is given by

$$r\,\delta\theta = \gamma\,\delta x$$

as shown in the figure, θ being the rotation at any cross-section. It is here assumed that lines which were radial in the shaft before torque was applied remain straight after the shaft has been twisted; this assumption may be justified by arguments based on symmetry. If the shear stress at radius r is τ then $\gamma = \tau/G$, where G is the shear modulus of the material of the shaft. It follows that, in the limit when $\delta x \to 0$,

$$\tau = rG\frac{d\theta}{dx}. \tag{6.1.1}$$

This shear stress is provided by the total torque M which has to be applied to the element. The magnitude of this twisting moment is therefore

$$M = \int_0^R \tau \,.\, 2\pi r^2\, dr = G\frac{d\theta}{dx}\int_0^R 2\pi r^3\, dr = GJ\frac{d\theta}{dx}, \tag{6.1.2}$$

where J represents the value of the definite integral, being the second polar moment of area of the cross-section.

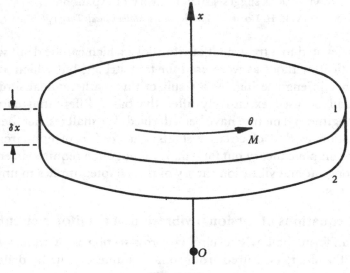

Fig. 6.1.2

These ideas may now be applied to the problem which arises when θ is a function of time. The element of the shaft is shown in fig. 6.1.2 and it will be noted that the two faces of the element are numbered; we shall distinguish the torques which are applied to the two faces by the appropriate suffices. The equation of motion for the element, considering rotation about its axis, will be

$$M_1 - M_2 + m\,\delta x = \left[\rho\,\delta x\int_0^R 2\pi r^3\, dr\right]\frac{\partial^2\theta}{\partial t^2}, \tag{6.1.3}$$

where ρ is the mass density of the material. In this equation the quantity m, which we may write $m(x,t)$ because it can vary with time and with position along the bar, represents an externally applied distributed torque per unit length. Such an applied torque will not be present in real problems because there will usually be no means by which it could act. It is included here because we shall require to use it in the later development of the theory; it is comparable with the force distributions which were required in the string theory.

The values of M_1 and M_2 may be substituted from equation (6.1.2). This gives

$$GJ\frac{\left[\left(\frac{\partial\theta}{\partial x}\right)_1 - \left(\frac{\partial\theta}{\partial x}\right)_2\right]}{\delta x} + m = \rho J \frac{\partial^2\theta}{\partial t^2}$$

and, in the limit when $\delta x \to 0$, the equation of motion becomes

$$G\frac{\partial^2\theta}{\partial x^2} + \frac{m}{J} = \rho\frac{\partial^2\theta}{\partial t^2}. \tag{6.1.4}$$

This equation may be regarded as valid for practical engineering purposes. Its derivation, however, does involve certain assumptions as, for instance, that the shear stress τ is the only significant stress present. The validity of the equation in general can be examined in detail by comparing it with the results which are obtainable directly from the mathematical theory of elasticity;† but even the latter theory does not include the distributed applied torque per unit length m.

In order to develop the vibration theory of shafts along the same lines that were used previously, we shall require the energy expressions. The kinetic energy of the element of fig. 6.1.2 is given by

$$\delta T = \frac{1}{2}\left[\rho\,\delta x \int_0^R 2\pi r^3\,dr\right]\left(\frac{\partial\theta}{\partial t}\right)^2 = \frac{1}{2}J\rho\left(\frac{\partial\theta}{\partial t}\right)^2 \delta x. \tag{6.1.5}$$

The kinetic energy of a finite length of the shaft will therefore be

$$T = \frac{1}{2}J\rho \int\left(\frac{\partial\theta}{\partial t}\right)^2 dx, \tag{6.1.6}$$

where the integral is to be evaluated between appropriate limits.

The potential, or elastic strain energy of the disk element is

$$\delta V = \frac{1}{2}M\delta\theta = \frac{1}{2}\left(GJ\frac{\partial\theta}{\partial x}\right)\delta\theta = \frac{1}{2}GJ\left(\frac{\partial\theta}{\partial x}\right)^2 \delta x \tag{6.1.7}$$

and hence the potential energy of a finite length of the shaft is

$$V = \frac{1}{2}GJ \int\left(\frac{\partial\theta}{\partial x}\right)^2 dx, \tag{6.1.8}$$

where again appropriate limits must be used.

6.2 Receptances in closed form for shafts in torsion; natural frequencies

Receptances in 'closed form' may now be derived by the method which was used for taut strings. Consider a clamped-free shaft as shown in fig. 6.2.1 to which a torque

$$M = Fe^{i\omega t} \tag{6.2.1}$$

† The theory is due to L. Pochhammer, *J.f. Math. (Crelle)*, vol. 81 (1876), p. 324; an account of it is given by A. E. H. Love, *Mathematical Theory of Elasticity* (Cambridge University Press, 4th ed. 1927), p. 288.

is applied at the free end where $x = l$. It must be remembered that the method of application of this torque in a real problem may be complicated and probably incompletely known. Nevertheless this will not invalidate the subsequent theory by virtue of the arguments which were advanced in Chapter 5.

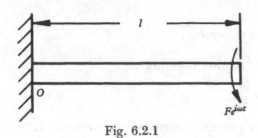

Fig. 6.2.1

The differential equation of motion for the shaft is

$$a^2 \frac{\partial^2 \theta}{\partial x^2} = \frac{\partial^2 \theta}{\partial t^2}, \tag{6.2.2}$$

where $a^2 = G/\rho$ since m is now zero in equation (6.1.4). It is necessary to seek solutions of the differential equations which are of the form

$$\theta = X(x)\, e^{i\omega t}. \tag{6.2.3}$$

These solutions will be required to satisfy the boundary conditions

$$\left. \begin{aligned} \theta &= 0 && \text{when} \quad x = 0, \\ \frac{\partial \theta}{\partial x} &= \frac{F}{GJ} e^{i\omega t} && \text{when} \quad x = l. \end{aligned} \right\} \tag{6.2.4}$$

The second of these conditions follows from equation (6.1.2). The solution is found exactly as in § 4.2. It is

$$\theta = \frac{F \sin \lambda x}{GJ\lambda \cos \lambda l} e^{i\omega t}, \tag{6.2.5}$$

where

$$\lambda = \frac{\omega}{a}. \tag{6.2.6}$$

This result can be rewritten in the usual receptance forms

$$\alpha_{xl} = \frac{\sin \lambda x}{GJ\lambda \cos \lambda l} \tag{6.2.7}$$

and

$$\alpha_{ll} = \frac{\tan \lambda l}{GJ\lambda}. \tag{6.2.8}$$

It will be seen that the derivation of these receptances begins with an equation of motion which is mathematically identical with that for the string, equation (4.2.3). Further, the boundary conditions for the clamped-free shaft are mathematically identical with those for the fixed-free string. The receptance functions are for this reason also identical. There is thus an exact analogy between the problem of the small transverse vibration of a uniform taut string and that of the torsional vibration of a uniform circular shaft.

A table of receptances may now be constructed either directly or by using the analogous results for the taut string. Such a table for clamped-free, free-free and clamped-clamped shafts, which are subject to an applied torque at any particular section is given (Table 5, pp. 256–257).

It will be found that the reciprocal relation still holds for all the cross-receptances in the table.

Those values of ω which make the receptances infinite will be the natural frequencies ω_r of the three types of shaft referred to. These frequencies are

$$\omega_r = \frac{(2r-1)\,\pi a}{2l} \quad \text{(clamped-free)},$$

$$\omega_0 = 0; \quad \omega_r = \frac{r\pi a}{l} \quad \text{(free-free)}, \quad \left.\right\} \quad (r = 1, 2, 3, \ldots). \quad (6.2.9)$$

$$\omega_r = \frac{r\pi a}{l} \quad \text{(clamped-clamped)},$$

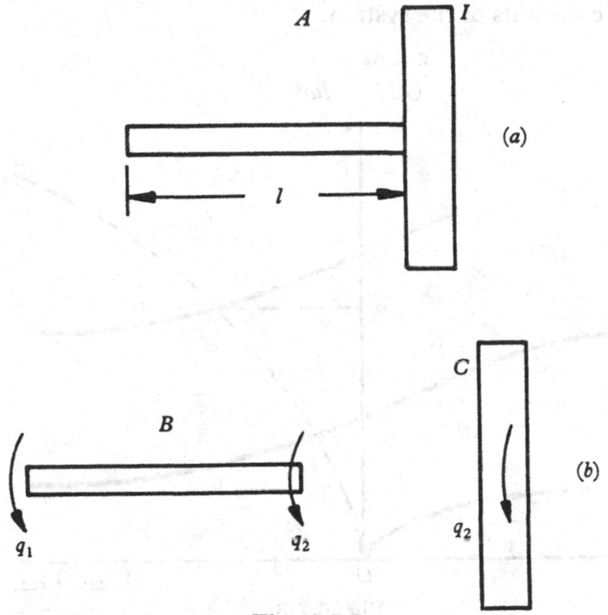

Fig. 6.2.2

The use of Table 5 will now be illustrated by finding the direct receptance at the free end of the shaft in the system of fig. 6.2.2 (a). The system may be divided into sub-systems as in fig. 6.2.2 (b) for which entry 1 in Table 2 gives

$$\alpha_{11} = \beta_{11} - \frac{\beta_{12}^2}{\beta_{22} + \gamma_{22}}. \quad (6.2.10)$$

The subscripts 1 and 2 in this equation refer to the co-ordinates shown in the figure. Table 5 now shows that

$$\beta_{11} = \beta_{22} = -\frac{\cot \lambda l}{GJ\lambda}, \quad \beta_{12} = -\frac{1}{GJ\lambda \sin \lambda l}, \quad (6.2.11)$$

while from Table 1 it is found that

$$\gamma_{22} = -\frac{1}{I\omega^2}. \quad (6.2.12)$$

These values may be substituted into equation (6.2.10) to give

$$\alpha_{11} = \frac{I\omega^2}{GJ\lambda \sin^2 \lambda l[I\omega^2 \cot \lambda l + GJ\lambda]} - \frac{\cot \lambda l}{GJ\lambda}. \quad (6.2.13)$$

The system of fig. 6.2.2 and system 5 of Table 1 differ in that the shaft in the former has mass while that of the latter is light. We leave it as an exercise for the reader to derive the receptance α_{11} for the system having the light shaft by treating it as the limiting case, as the density is decreased indefinitely, of the receptance which is given by equation (6.2.13).

The frequency equation for the system of fig. 6.2.2 can be written down once one of the receptances has been found. The equation is

$$\beta_{22} + \gamma_{22} = 0 \qquad (6.2.14)$$

or, in terms of the constants of the system,

$$\frac{\cot \lambda l}{GJ\lambda} + \frac{1}{I\omega^2} = 0. \qquad (6.2.15)$$

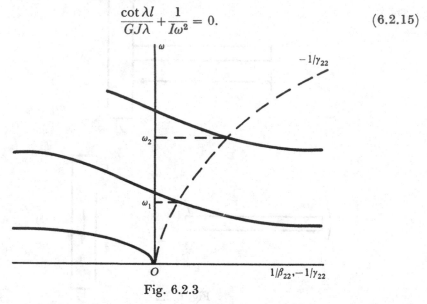

Fig. 6.2.3

This is an equation in ω because $\lambda = \omega/a$. The solution to this equation may be illustrated by using the curves of *reciprocal* receptance and writing the frequency equation in the form

$$\frac{1}{\beta_{22}} = -\frac{1}{\gamma_{22}}. \qquad (6.2.16)$$

The curves of $1/\beta_{22}$ and $-1/\gamma_{22}$ are sketched in fig. 6.2.3, the latter being shown dotted. The first two intersections, which give the first two natural frequencies, and also the frequency of zero value, are shown. It will be seen that at very high frequencies the intersections will lie very close to the asymptotes of the curve of $1/\beta_{22}$. These asymptotes occur at regular intervals as we know from the properties of the tangent function. Thus the natural frequencies will also occur at regular intervals.

EXAMPLES 6.2

1. A rigid flywheel of moment of inertia I is attached to one end of a uniform circular shaft, the other end of the shaft being rigidly clamped. Show that the natural frequencies are given by

$$\frac{1}{J\sqrt{(G\rho)}} \tan [\omega l \sqrt{(\rho/G)}] = \frac{1}{I\omega}$$

in the notation of this section.

Solve the equation by plotting the appropriate curves of $1/\alpha$ for the shaft and for the wheel separately for the case where

diameter of shaft = 1 in.,
length of shaft l = 2 ft.,
moment of inertia of wheel = 20 lb.in.²

The material of the shaft is steel for which

$$\rho = 0\cdot28 \text{ lb./in.}^3, \quad G = 12 \times 10^6 \text{ lb./in.}^2.$$

Obtain the two lowest natural frequencies.

2. Calculate the natural frequency of the system of the last example under the assumption that the mass of the shaft may be neglected.

3. Show that the natural frequencies of the system shown are given by

$$\frac{\dfrac{a}{\omega J_L G}\cot\dfrac{\omega l_L}{a}+\dfrac{1}{I_L\omega^2}}{\dfrac{a^2}{\omega^2 J_L^2 G^2}-\dfrac{a}{I_L\omega^3 J_L G}\cot\dfrac{\omega l_L}{a}}+\frac{\dfrac{a}{\omega J_N G}\cot\dfrac{\omega l_N}{a}+\dfrac{1}{I_N\omega^2}}{\dfrac{a^2}{\omega^2 J_N^2 G^2}-\dfrac{a}{I_N\omega^3 J_N G}\cot\dfrac{\omega l_N}{a}}-I_M\omega^2 = 0,$$

where $a^2 = G/\rho$.

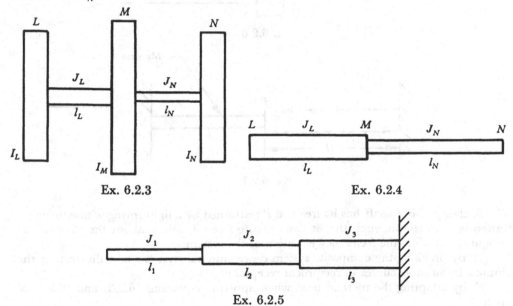

Ex. 6.2.3   Ex. 6.2.4

Ex. 6.2.5

4. Show that the direct receptance at the end L of the stepped shaft shown is

$$\frac{\dfrac{a}{\omega G J_L}\left[\dfrac{1}{J_L}-\dfrac{1}{J_N}\cot\dfrac{\omega l_L}{a}\cot\dfrac{\omega l_N}{a}\right]}{\dfrac{1}{J_N}\cot\dfrac{\omega l_N}{a}+\dfrac{1}{J_L}\cot\dfrac{\omega l_L}{a}},$$

where $a^2 = G/\rho$.

5. The shaft shown has portions whose lengths are l_1, l_2 and l_3 and whose polar second moments of area are J_1, J_2 and J_3. The material is uniform throughout. Neglecting

any local actions at the shoulders, show that the natural frequencies for torsional vibration of the stepped shaft when one end is clamped as shown, are given by

$$\frac{J_1 \tan \dfrac{\omega l_1}{a} + J_2 \tan \dfrac{\omega l_2}{a}}{J_2 - J_1 \tan \dfrac{\omega l_1}{a} \tan \dfrac{\omega l_2}{a}} = \frac{J_3}{J_2} \cot \frac{\omega l_3}{a},$$

where $a^2 = G/\rho$.

6. Two shafts LM and MN are geared together at M as shown. The gear-wheels may be regarded as light and the ratio is $1:r$. If details of the shafts are as shown and they are of the same material, find the frequency equation for torsional vibration, assuming that bending vibrations are prevented by suitable bearings.

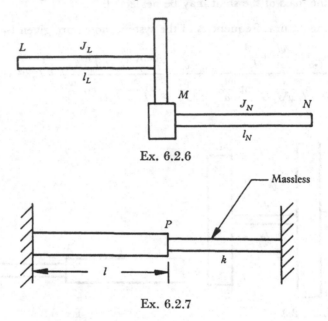

Ex. 6.2.6

Ex. 6.2.7

7. A clamped-free shaft has its free end P restrained by a light spring whose torsional stiffness is k, as shown, such that it can execute free vibrations about the 'untwisted' configuration. Find the frequency equation of the system

(a) by breaking the composite system down into sub-systems and illustrating the solutions by suitable curves of reciprocal receptance;

(b) by adapting the method used when applying equations (6.2.3) and (6.2.4) of this section.

8. A horizontal cylindrical shaft is mounted in bearings and carries a flywheel at one end. The shaft is of length l and is made of metal which has density ρ and modulus of rigidity G; the flywheel may be treated as a rigid body. The flywheel and the shaft have moments of inertia about the axis of rotation I and I' respectively.

(a) If the free end of the shaft is acted upon by a torque $F e^{i\omega t}$, derive an expression for the motion of the flywheel.

(b) Show that, if $\omega = a\pi/2l$, where $a = \sqrt{(G/\rho)}$, then the amplitude of the flywheel oscillation is independent of I, and determine its magnitude.

(C.U.M.S.T. Pt II, 1954, slightly altered)

6.3 Torsional vibration. Principal modes; principal co-ordinates; characteristic functions; orthogonality

The method of §4.3 may readily be adapted to find the principal modes in torsion of clamped-free, free-free and clamped-clamped shafts. Let a torque $M = Fe^{i\omega t}$ be applied at some section $x = h$ of a uniform circular shaft; the rotational displacement at any section x of the shaft is then given by

$$\theta = \alpha_{xh} Fe^{i\omega t}. \tag{6.3.1}$$

If the exciting frequency ω is now allowed to approach the rth natural frequency ω_r, the receptance α_{xh} becomes very large so that it is necessary for F to tend to zero if the amplitude of motion is to be kept finite. In this way a condition of free vibration is approached. For the three types of shaft with which §6.2 is concerned, expressions can be written down for the motion at any section x during free vibration at the rth natural frequency. The expressions are

$$\left.\begin{aligned}
\theta_r &= A_r \sin\frac{(2r-1)\pi x}{2l} e^{i\omega_r t} \quad \text{(clamped-free),} \\
\theta_0 = A_0; \quad \theta_r &= A_r \cos\frac{r\pi x}{l} e^{i\omega_r t} \qquad \text{(free-free),} \\
\theta_r &= A_r \sin\frac{r\pi x}{l} e^{i\omega_r t} \qquad \text{(clamped-clamped),}
\end{aligned}\right\} (r = 1, 2, 3, \ldots) \tag{6.3.2}$$

where, in each case, the multiplying factor A_r is independent of x.

It may be noted here that the free-free shaft has a mode for which r is zero, the motion in this mode being rotation of the shaft as a rigid body. It is important, in some problems, that this mode should not be forgotten.

The principal modes which are defined by the functions θ_r are all sinusoidal. It is therefore again possible to choose the principal co-ordinates so that each has unit value when the maximum amplitude of the shaft is unity. Thus a distortion in the rth mode may be expressed in the form

$$\theta_r(x, t) = p_r(t) \cdot \phi_r(x), \tag{6.3.3}$$

where the characteristic function $\phi_r(x)$ is given by

$$\left.\begin{aligned}
\phi_r(x) &= \sin\frac{(2r-1)\pi x}{2l} \quad \text{(clamped-free),} \\
\phi_0(x) = \frac{1}{\sqrt{2}}; \quad \phi_r(x) &= \cos\frac{r\pi x}{l} \qquad \text{(free-free),} \\
\phi_r(x) &= \sin\frac{r\pi x}{l} \qquad \text{(clamped-clamped),}
\end{aligned}\right\} (r = 1, 2, 3, \ldots). \tag{6.3.4}$$

The similarity between present theory and that of the taut string is again apparent.

Now any torsional deflexion θ of a shaft may be regarded as being composed of deflexions in the various principal modes. Therefore the deflexion may be expressed in the form

$$\theta = \sum_{r=0 \text{ or } 1}^{\infty} \theta_r = \sum_{r=0 \text{ or } 1}^{\infty} p_r(t) \cdot \phi_r(x). \tag{6.3.5}$$

If one of these series is substituted into the energy expressions (6.1.6) and (6.1.8), and if the integrals are then evaluated over the shaft length, then it will be found that

$$2T = a_1 p_1^2 + a_2 p_2^2 + \ldots, \tag{6.3.6}$$

$$2V = c_1 p_1^2 + c_2 p_2^2 + \ldots, \tag{6.3.7}$$

where

$$a_r = \frac{J\rho l}{2}, \quad c_r = \frac{(2r-1)^2 \pi^2 GJ}{8l} \quad \text{(clamped-free)},$$

$$a_r = \frac{J\rho l}{2}, \quad c_r = \frac{r^2 \pi^2 GJ}{2l} \quad \text{(free-free)},$$

$$a_r = \frac{J\rho l}{2}, \quad c_r = \frac{r^2 \pi^2 GJ}{2l} \quad \text{(clamped-clamped)}. \tag{6.3.8}$$

Notice that if the scaling of p_0 for the free-free shaft is such that $\phi_0(x) = 1/\sqrt{2}$, then

$$c_0 = 0, \quad a_0 = \frac{J\rho l}{2} \quad \text{(free-free)}. \tag{6.3.9}$$

These results show that the total kinetic energy during any motion is equal to the sum of the kinetic energies in each of the principal modes of which the motion is composed. This is true also of the potential energy. The principle of orthogonality holds good, therefore, this being an alternative way of stating that there are no 'cross-terms' in the energy expressions. In deriving the energy series, it is found that the vanishing of the 'cross-terms' is due to the fact that

$$\int_0^l \phi_r(x) \cdot \phi_s(x)\, dx = 0 \quad (r \neq s), \tag{6.3.10}$$

whereas the comparable quantity for the squared terms is

$$\int_0^l [\phi_r(x)]^2\, dx = \frac{l}{2}. \tag{6.3.11}$$

The values of the integrals (6.3.10) and (6.3.11) may be checked readily for each type of shaft; this however is not necessary because the characteristic functions are identical with those of the taut string and their properties have been dealt with already. But it has been shown previously that the orthogonality conditions may be expected also on physical grounds and the vanishing of the integrals (6.3.10) may therefore be assumed.

It was shown in § 4.5 that the orthogonality property could also be expressed in a different way. This was to find the force distribution which would produce deflexion of the system in one particular mode only and then to write down the condition that this force distribution would do no work during a displacement of the system in some other mode. In order to apply this idea here, consider the applied torque m per unit length (see equation (6.1.4)) which will produce a static distortion $\theta_r(x)$ in the rth principal mode of a shaft. The distribution of torque $m_r(x)$ is found by substituting the displacement function $\theta_r(x)$ into equation (6.1.4). This gives

$$m_r(x) = -GJ\frac{d^2\theta_r}{dx^2} = (\text{constant}) \times \phi_r(x), \tag{6.3.12}$$

where the characteristic function will be chosen to suit the end conditions of the shaft in question. Now let the distributed torque $m_r(x)$ be applied to the shaft, so as to cause a static deflexion $\theta_r(x)$, and then let a static deflexion $\theta_s(x)$ in some other principal mode be given to the shaft by some suitable means. The work done by the torque distribution $m_r(x)$ during the second displacement will then be equal to

$$\int_0^l m_r(x) \cdot \theta_s(x)\, dx = (\text{constant}) \times \int_0^l \phi_r(x) \cdot \phi_s(x)\, dx = 0. \qquad (6.3.13)$$

This is zero by virtue of the orthogonality property.

As indicated in equation (6.3.5), any torsional deflexion of a shaft can be expressed in the form of a series of the characteristic functions which are appropriate to the end conditions of the shaft. If the distortion is static then the principal co-ordinates, which appear as coefficients in the series, will have constant values. These coefficients can be obtained from the function $\theta(x)$ by using the orthogonality relations. Thus, if a given static deflexion is expressed in the form

$$\theta(x) = p_1 \phi_1(x) + p_2 \phi_2(x) + \dots \qquad (6.3.14)$$

the argument of § 4.7 can be used to show that

$$p_r = \frac{\displaystyle\int_0^l \theta(x)\, \phi_r(x)\, dx}{\displaystyle\int_0^l [\phi_r(x)]^2\, dx} \qquad (6.3.15)$$

by the result of equation (6.3.13).

Some of the results in the present section have been collected together in Table 5 for ease of reference.

EXAMPLES 6.3

1. Show that unit value of the rth *normalized* principal co-ordinate p_r for a clamped-free, a free-free or a clamped-clamped shaft corresponds to a sinusoidal variation of θ with x of amplitude $\sqrt{(2/J\rho l)}$.

2. Using the methods of this section, find the variation of the rotation θ with distance x from the left-hand end of the shaft in Ex. 6.2.7 when it is distorted in its rth principal mode.

3. Show, analytically, that the principal modes of the system referred to in Ex. 6.3.2, are orthogonal. If unit value of the rth principal co-ordinate corresponds to a sinusoidal deflexion in space of unit amplitude, find the values of the energy coefficients a_r and c_r. Show that the relation $\omega_r^2 = c_r/a_r$ gives the frequency equation of the system.

6.4 Receptances in series form for shafts in torsion; variation of principal co-ordinates

The receptance at the rth principal co-ordinate p_r, for any of the shafts which have been considered may be found by the method explained in § 4.6. It is given in each case by the expression

$$\alpha_r = \frac{1}{c_r - a_r \omega^2} = \frac{1}{a_r(\omega_r^2 - \omega^2)}. \qquad (6.4.1)$$

TABLE 5

TORSIONAL VIBRATION OF UNIFORM CIRCULAR SHAFTS

Notation

a $= \sqrt{(G/\rho)}$.

G Shear modulus.

h Particular value of x defining a section of the shaft distant h from the origin $(0 \leqslant h \leqslant l)$.

J Second polar moment of area of shaft cross-section.

l Length of shaft.

r, s Subscripts used to indicate the rth and sth principal modes ($= 1, 2, 3, \ldots$ for clamped-free or clamped-clamped and $= 0, 1, 2, 3, \ldots$ for free-free).

x Distance along the shaft $(0 \leqslant x \leqslant l)$.

α_{xh} Receptance between the sections x and h. If a harmonic torque $F e^{i\omega t}$ is applied to a shaft at the section $x = h$, then the twist at any point x of the shaft is given by
$$\theta = \alpha_{xh} F e^{i\omega t}.$$

θ Angle of twist of shaft.

λ $= \omega \sqrt{(\rho/G)} = \omega/a$.

ρ Mass density.

Equation of motion: $\quad G \dfrac{\partial^2 \theta}{\partial x^2} = \rho \dfrac{\partial^2 \theta}{\partial t^2}.$

Boundary conditions: $\quad \theta = 0$ at a clamped end,

$$\frac{\partial \theta}{\partial x} = 0 \text{ at a free end.}$$

Energy expressions:
$$T = \frac{J\rho}{2} \int_0^l \left(\frac{\partial \theta}{\partial t}\right)^2 dx, \quad V = \frac{GJ}{2} \int_0^l \left(\frac{\partial \theta}{\partial x}\right)^2 dx.$$

In the series representation of twist, the principal co-ordinates p_r have been chosen so that
$$\theta = \sum_{r=1}^{\infty} p_r \cdot \phi_r(x) \text{ (clamped-free or clamped-clamped),}$$
or
$$\theta = \sum_{r=0}^{\infty} p_r \cdot \phi_r(x) \quad \text{(free-free).}$$

TABLE 5

TORSIONAL VIBRATION OF UNIFORM CIRCULAR SHAFTS

Quantity	Clamped-free $\theta=0$ when $x=0$ $\partial\theta/\partial x=0$ when $x=l$	Free-free $\partial\theta/\partial x=0$ when $x=0$ $\partial\theta/\partial x=0$ when $x=l$	Clamped-clamped $\theta=0$ when $x=0$ $\theta=0$ when $x=l$
α_{xh} $0\leqslant x\leqslant h$ $0\leqslant h\leqslant l$	$\dfrac{\sin\lambda x\cos\lambda(l-h)}{GJ\lambda\cos\lambda l}$	$-\dfrac{\cos\lambda x\cos\lambda(l-h)}{GJ\lambda\sin\lambda l}$	$\dfrac{\sin\lambda x\sin\lambda(l-h)}{GJ\lambda\sin\lambda l}$
α_{xh} $h\leqslant x\leqslant l$ $0\leqslant h\leqslant l$	$\dfrac{\sin\lambda h\cos\lambda(l-x)}{GJ\lambda\cos\lambda l}$	$-\dfrac{\cos\lambda h\cos\lambda(l-x)}{GJ\lambda\sin\lambda l}$	$\dfrac{\sin\lambda h\sin\lambda(l-x)}{GJ\lambda\sin\lambda l}$
α_{xh} $0\leqslant x\leqslant l$ $0\leqslant h\leqslant l$	$\dfrac{2}{J\rho l}\displaystyle\sum_{r=1}^{\infty}\dfrac{\phi_r(x)\,\phi_r(h)}{\omega_r^2-\omega^2}$	$\dfrac{2}{J\rho l}\displaystyle\sum_{r=0}^{\infty}\dfrac{\phi_r(x)\,\phi_r(h)}{\omega_r^2-\omega^2}$	$\dfrac{2}{J\rho l}\displaystyle\sum_{r=1}^{\infty}\dfrac{\phi_r(x)\,\phi_r(h)}{\omega_r^2-\omega^2}$
$\omega_r=\sqrt{\left(\dfrac{c_r}{a_r}\right)}$	$\dfrac{(2r-1)\,\pi a}{2l}$	$\dfrac{r\pi a}{l}$	$\dfrac{r\pi a}{l}$
$\phi_r(x)$	$\sin\dfrac{(2r-1)\,\pi x}{2l}$	$\dfrac{1}{\sqrt{2}}\ \ (r=0)$ $\cos\dfrac{r\pi x}{l}\ \ (r=1,2,3,\ldots)$	$\sin\dfrac{r\pi x}{l}$
$\displaystyle\int_0^l\phi_r(x)\,\phi_s(x)\,dx$	$0\quad(r\neq s)$ $l/2\quad(r=s)$	$0\quad(r\neq s)$ $l/2\quad(r=s)$	$0\quad(r\neq s)$ $l/2\quad(r=s)$
a_r	$J\rho l/2$	$J\rho l/2$	$J\rho l/2$
c_r	$\dfrac{(2r-1)^2\pi^2 GJ}{8l}$	$\dfrac{r^2\pi^2 GJ}{2l}\ (r=0,1,2,3,\ldots)$	$\dfrac{r^2\pi^2 GJ}{2l}$

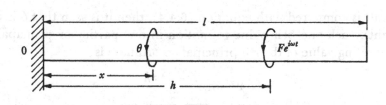

Clamped end   Free end

Since the values of a_r, c_r and ω_r^2 are known for each type of shaft, these receptances may be written down immediately. In most real problems, however, we are concerned with the receptances at, and between, different sections of a shaft rather than with receptances at principal co-ordinates. In order to find the former, the series discussed in § 4.6 may be used; thus, in the absence of a rigid mode,

$$\alpha_{zy} = \alpha_{yz} = \alpha_1\left(\frac{\partial y}{\partial p_1}\frac{\partial z}{\partial p_1}\right) + \alpha_2\left(\frac{\partial y}{\partial p_2}\frac{\partial z}{\partial p_2}\right) + \cdots, \tag{6.4.2}$$

$$\alpha_{yy} = \alpha_1\left(\frac{\partial y}{\partial p_1}\right)^2 + \alpha_2\left(\frac{\partial y}{\partial p_2}\right)^2 + \cdots. \tag{6.4.3}$$

Let it be required to find the variation of θ at any section x of a clamped-free shaft to which a torque $Fe^{i\omega t}$ is applied at the section $x = h$, as shown in fig. 6.4.1.

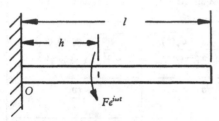

Fig. 6.4.1

The co-ordinate y of equation (6.4.2) may be identified with the rotation θ at the section x, and similarly z may be associated with the rotation at the section where $x = h$. Thus if

$$\theta = p_1(t).\phi_1(x) + p_2(t).\phi_2(x) + \cdots \tag{6.4.4}$$

as in equation (6.3.5), then

$$\frac{\partial y}{\partial p_r} = \phi_r(x), \quad \frac{\partial z}{\partial p_r} = \phi_r(h). \tag{6.4.5}$$

The series (6.4.2) now yields

$$\alpha_{xh} = \frac{\phi_1(x).\phi_1(h)}{a_1(\omega_1^2 - \omega^2)} + \frac{\phi_2(x).\phi_2(h)}{a_2(\omega_2^2 - \omega^2)} + \cdots. \tag{6.4.6}$$

On putting in the values for the characteristic functions and for the natural frequencies, it is now found that

$$\theta = \alpha_{xh}Fe^{i\omega t} = \left\{\frac{2}{J\rho l}\sum_{r=1}^{\infty}\frac{\sin\dfrac{(2r-1)\pi x}{2l}\sin\dfrac{(2r-1)\pi h}{2l}}{\left[\dfrac{(2r-1)^2\pi^2 a^2}{4l^2} - \omega^2\right]}\right\}Fe^{i\omega t}. \tag{6.4.7}$$

If this result is compared with equation (6.4.4), then it is seen that θ is the sum of components, each one being due to motion in one particular principal mode. The corresponding value of the rth principal co-ordinate is

$$p_r(t) = \frac{2F\sin\dfrac{(2r-1)\pi h}{2l}}{J\rho l\left[\dfrac{(2r-1)^2\pi^2 a^2}{4l^2} - \omega^2\right]}e^{i\omega t}. \tag{6.4.8}$$

The principal co-ordinates vary with the frequency of the exciting force, as is usual for steady forced harmonic motion.

Table 5 gives the receptances in series form for clamped-free, free-free and clamped-clamped shafts.

EXAMPLES 6.4

1. Repeat the calculation of the lowest natural frequency in Ex. 6.2.1. assuming that the shaft distorts only in the first principal mode that it would possess in the absence of the attached flywheel. That is to say, use the curtailed direct receptance series at the free end of the shaft in setting up the frequency equation.

2. The system of Ex. 6.2.7 vibrates freely in a principal mode with the appropriate frequency Ω. The amplitude of θ is unity at the end $x = l$ (that is, at the point of attachment of the torsion spring). Show that the variation of p_r, the rth principal co-ordinate of the shaft in the absence of the torsion spring, is given by

$$p_r = \frac{-k\phi_r(l)\, e^{i\Omega t}}{a_r(\omega_r^2 - \Omega^2)},$$

where $p_r(t)$, $\phi_r(x)$, a_r and ω_r refer to the clamped-free shaft as in Table 5.

6.5 Principal modes of composite systems of shafts

The receptances in either closed or series form may be used to find the natural frequencies of a variety of systems containing shafts as is discussed in § 6.2. They may also be used to find the corresponding principal modes of such systems. The process may be explained by means of an example.

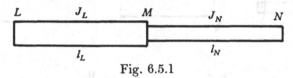

Fig. 6.5.1

Let two shafts LM and MN be joined together at M while ends L and N are free as in fig. 6.5.1. The natural frequency of the complete system is obtained by equating to zero the total reciprocal receptance at the point M. Thus, the frequency equation is

$$\frac{GJ_L\lambda}{\cot \lambda l_L} + \frac{GJ_N\lambda}{\cot \lambda l_N} = 0, \qquad (6.5.1)$$

where the suffixes L and N distinguish quantities which are associated with LM and MN respectively. Alternatively, the frequency equation can be written as the sum of the direct receptances at M of the two portions LM and MN (see entry 3 of Table 3); the two forms of the equation lead to the same results.

Now when the complete system is vibrating freely with some natural frequency ω_r, which is a root of equation (6.5.1), then the shaft LM may be regarded as undergoing forced vibration from an applied torque at the end M. This applied torque is due to the reaction from the shaft MN. Equally, the shaft MN is subjected to a forced vibration from the equal and opposite reaction. Now let the amplitude of motion at M be unity; then the amplitude of the torque which is being applied to LM will be given by the reciprocal of the direct receptance, namely

$$-\frac{GJ_L\lambda_r}{\cot \lambda_r l_L},$$

where $\lambda_r = \omega_r \sqrt{(\rho/G)}$. Using the cross-receptance α_{xl} of LM in the closed form, the amplitude of motion at some other section of LM distant x from L will be

$$\left(-\frac{GJ_L\lambda_r}{\cot\lambda_r l_L}\right)\left(-\frac{\cos\lambda_r x}{GJ_L\lambda_r\sin\lambda_r l_L}\right) = \frac{\cos\lambda_r x}{\cos\lambda_r l_L}. \qquad (6.5.2)$$

Similarly, the amplitude at any point in MN distant y from M is given by the receptance α_{x0} for that portion of the shaft. Thus the amplitude is

$$\left(-\frac{GJ_N\lambda_r}{\cot\lambda_r l_N}\right)\left(-\frac{\cos\lambda_r(l_N-y)}{GJ_N\lambda_r\sin\lambda_r l_N}\right) = \frac{\cos\lambda_r(l_N-y)}{\cos\lambda_r l_N}. \qquad (6.5.3)$$

The functions (6.5.2) and (6.5.3) define the principal mode of the composite system of fig. 6.5.1 for the natural frequency ω_r.

The principal modes of composite systems can also be written down by using the receptances in series form. This is sometimes preferable in numerical work.

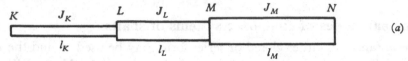

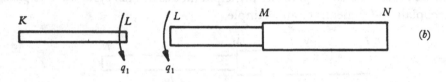

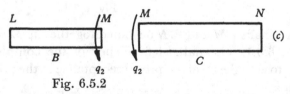

Fig. 6.5.2

Consider a system which is composed of three shafts which are connected end-to-end. Let the system be $KLMN$ for which the components KL, LM and MN are uniform shafts with different cross-sections as shown in fig. 6.5.2 (a). The ideas which are utilized above may again be used though the treatment is somewhat longer.

It is first necessary to find the natural frequencies of the composite system. If the system is split at L, as in fig. 6.5.2 (b), then the direct receptance of KL at L is

$$-\frac{\cot\lambda l_K}{GJ_K\lambda} \qquad (6.5.4)$$

as before. The direct receptance of LMN at L can be found by using the formula for α_{11} from entry 1 of Table 2 together with the relevant receptances for LM (sub-system B) and MN (sub-system C). This gives the receptance

$$\left(-\frac{\cot\lambda l_L}{GJ_L\lambda}\right) - \frac{\left(\dfrac{-1}{GJ_L\lambda\sin\lambda l_L}\right)^2}{\left(-\dfrac{\cot\lambda l_L}{GJ_L\lambda}\right)+\left(-\dfrac{\cot\lambda l_M}{GJ_M\lambda}\right)} = \frac{1}{GJ_L\lambda}\left[\frac{J_M\tan\lambda l_L\tan\lambda l_M-J_L}{J_L\tan\lambda l_L+J_M\tan\lambda l_M}\right]. \qquad (6.5.5)$$

The frequency equation for the complete system $KLMN$ is found by equating the sum of the reciprocals of the receptances (6.5.4) and (6.5.5) to zero; it is

$$\frac{GJ_K\lambda}{\cot \lambda l_K} + GJ_L\lambda \left[\frac{J_L \tan \lambda l_L + J_M \tan \lambda l_M}{J_L - J_M \tan \lambda l_L \tan \lambda l_M} \right] = 0. \tag{6.5.6}$$

Let ω_r be some root of this equation, and suppose that it is required to find the corresponding modal shape. Consider free vibration at this frequency in which the amplitude of angular motion at L is unity. The corresponding torque, acting on KL at L is of amplitude

$$F_1 = -\frac{GJ_K\lambda_r}{\cot \lambda_r l_K} \tag{6.5.7}$$

by the same argument as before. The amplitude of motion at any section of the shaft KL can thus be written down as in the previous problem. The amplitudes of motion at any section of the shaft LM however cannot be written down in this way because this portion is acted on by reaction torques at both L and M and the magnitude of the reaction torque at M is not yet known. This also makes it impossible to write down an expression for the amplitudes of motion at any section of the shaft MN.

The next step is to find the amplitude of motion at M. This may be found by using the formulae for α_{11} and α_{12} of system 1 in Table 2, together with the relevant receptances of the shafts LM and MN. If the shaft LM is identified with the subsystem B and if MN is identified with C (see fig. 6.5.2 (c)), then the amplitude at M when the amplitude at L is unity will be given by

$$\frac{\alpha_{21}}{\alpha_{11}} = \frac{\beta_{12}\gamma_{22}}{\beta_{11}\beta_{22} + \beta_{11}\gamma_{22} - \beta_{12}^2}$$

$$= \frac{\left(\dfrac{-\operatorname{cosec} \lambda_r l_L}{GJ_L\lambda_r} \right) \left(\dfrac{-\cot \lambda_r l_M}{GJ_M\lambda_r} \right)}{\left(-\dfrac{\cot \lambda_r l_L}{GJ_L\lambda_r} \right)^2 + \left(-\dfrac{\cot \lambda_r l_L}{GJ_L\lambda_r} \right) \left(-\dfrac{\cot \lambda_r l_M}{GJ_M\lambda_r} \right) - \left(\dfrac{-\operatorname{cosec} \lambda_r l_L}{GJ_L\lambda_r} \right)^2}$$

$$= \Theta_M \text{ (say)}. \tag{6.5.8}$$

Therefore the amplitude of the reactive torque applied at M to the shaft MN in this free vibration is $F_2 = \Theta_M/\gamma_{22}$ or

$$F_2 = -\frac{GJ_M\lambda_r}{\cot \lambda_r l_M} \Theta_M. \tag{6.5.9}$$

The amplitude at a section of the shaft MN distant z from M is now seen to be

$$\left(-\frac{GJ_M\lambda_r}{\cot \lambda_r l_M} \Theta_M \right) \left(-\frac{\cos \lambda_r(l_M - z)}{GJ_M\lambda_r \sin \lambda_r l_M} \right) = \Theta_M \frac{\cos \lambda_r(l_M - z)}{\cos \lambda_r l_M}. \tag{6.5.10}$$

In order to find the amplitude at any section in LM, we first note that this portion is acted upon by torques at L and M. Now if a torque of amplitude F_1 is acting on LM at L and if a torque of amplitude F_2 is acting on it at M, then the amplitude of motion at a section distant y from L will be given by

$$-\left(\frac{\cos \lambda_r(l_L - y)}{GJ_L\lambda_r \sin \lambda_r l_L} \right) F_1 - \left(\frac{\cos \lambda_r y}{GJ_L\lambda_r \sin \lambda_r l_L} \right) F_2. \tag{6.5.11}$$

If the values for F_1 and F_2 are now substituted from equations (6.5.7) and (6.5.9), then the amplitude at any point of LM can be found. In this way, the rth principal mode is found for the composite system.

The method by which the principal modes of the shaft of fig. 6.5.2 have been found can be extended further, but it will be seen that the algebra becomes very heavy; alternative procedures may be devised, however, of which we shall describe two. The first of these is essentially an extension of the processes already explained and we shall merely outline the method here. The second represents a quite different, and more powerful, approach; it will be explained in § 6.6.

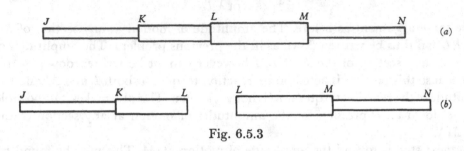

Fig. 6.5.3

Consider a system of four shafts $JKLMN$ (see fig. 6.5.3 (a)) whose principal modes and natural frequencies it is required to find. Let the system be split, as in fig. 6.5.3 (b), at the point L and let the natural frequencies of the complete system be found by equating the sum of the receptances at L (or their reciprocals) to zero. Instead of determining immediately the amplitudes at K and M which correspond to unit amplitude at L, the natural frequencies of the sub-systems JKL and LMN are found. A knowledge of these frequencies enables the principal modes of these sub-systems to be derived also; this process will be identical with that for the system of fig. 6.5.1 as described above. All the necessary data is now available for substituting in the formula $\alpha_r = 1/[a_r(\omega_r^2 - \omega^2)]$ except for the magnitude of the coefficients a_r, and this may evidently be found by integration. These receptances α_r for the portion JKL can now be used to determine receptances for this sub-system. Thus the amplitudes at sections of the shafts JK and KL can now be found for unit amplitude at L when the driving frequency ω is a natural frequency of the entire system $JKLMN$. The same sort of analysis can be performed for the sub-system LMN and in this way the principal modes of $JKLMN$ may be calculated. The convergence of the receptance series is usually sufficiently rapid to make the arithmetical work reasonably short.

EXAMPLES 6.5

1. A uniform shaft which is free to rotate in bearings has attached to it at one end a rigid flywheel. The moment of inertia about its axis is twice as great as the total moment of inertia of the shaft. Determine the lowest natural frequency for torsional vibration other than the zero frequency, in terms of the constants G, ρ and l of the shaft.

Find also the ratio of the amplitude at the flywheel to that at the other end of the shaft during motion in the corresponding mode.

2. The two portions of the stepped shaft shown are of the same material, for which

$$G = 11 \cdot 5 \times 10^6 \text{ psi}, \quad \rho = 480 \text{ lb./ft.}^3$$

If LM is of length 6 in. and diameter 0·5 in., and MN is of length 9 in. and diameter 0·375 in., calculate the first two natural frequencies of torsional vibration. Deduce and plot the corresponding principal modes.

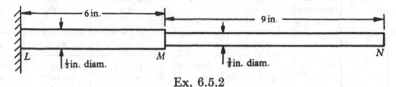

Ex. 6.5.2

3. A rigid disk whose polar moment of inertia is 20 lb.in.^{2} is attached to one end of a circular steel shaft of diameter 2 in. and length 4 ft. Derive the frequency equation for torsional oscillation of this system and determine the lowest natural frequency.

When this system oscillates freely in its first principal mode, and the amplitude at the disk is 0·001 rad., what is the amplitude at the mid-point of the shaft?

(For steel: modulus of rigidity = $11 \cdot 5 \times 10^6$ psi, density = 480 lb./ft.^{3})

(C.U.M.S.T. Pt II, 1955)

6.6 Tabular calculation of the receptances at the end of a chain of systems

The methods of finding natural frequencies and principal modes which are described in § 6.5 are very cumbersome if the shafts concerned consist of many portions. This section is concerned with an alternative method which is essentially a means of calculating receptances; from these, the modes and frequencies may be derived.

The nature of the calculations will first be explained in fairly general terms. The reader will thus see that this approach is one of wide application. A numerical example on torsional vibration will then follow.

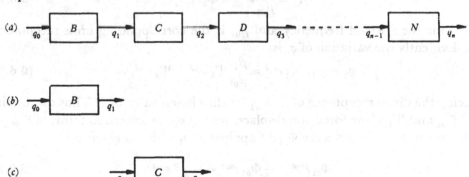

Fig. 6.6.1

Consider a system A which is composed of sub-systems $B, C, ..., N$ that are linked together, as shown in fig. 6.6.1 (a), by the co-ordinates $q_0, q_1, q_2, ..., q_n$. The co-ordinate q_0 is introduced here simply for convenience of notation. The object of our calculations will now be to determine the receptances $\alpha_{0n}, \alpha_{1n}, \alpha_{2n}, ..., \alpha_{nn}$ of

TABLE A

1	2	3	4	5	6	7	8
Sub-system	Subscript of left-hand co-ordinate r	Φ_r	Ψ_r	Direct receptance at q_r	Part of Ψ_r due to Φ_r	Part of Ψ_r due to Φ_{r+1}	Cross-receptance q_r to q_{r+}
		−9	13		3×5	4−6	
		pdl.ft.	(entry) ×10⁻⁴ rad.	(entry) ×10⁻⁴ rad./pdl.ft.	(entry) ×10⁻⁴ rad.	(entry) ×10⁻⁴ rad.	(entry) ×10⁻⁶ rad./pdl
$\lambda = 1{\cdot}300$ rad./ft.	B 0	0·000	1·000	3·121	0·000	1·000	− 8·4
	C 1	0·118	−0·368	− 0·431	−0·051	−0·317	− 1·6
	D 2	−0·197	−0·275	− 2·183	0·430	−0·705	− 8·1
$\lambda = 1{\cdot}068$ rad./ft.	B 0	0·000	1·000	0·296	0·000	1·000	− 9·5
	C 1	0·105	−0·031	− 1·038	−0·109	0·078	− 2·1
	D 2	0·036	−0·189	− 5·253	−0·189	0·000	− 10·8
$\lambda = 0{\cdot}362$ rad./ft.	B 0	0·000	1·000	−46·685	0·000	1·000	−54·5
	C 1	0·018	0·840	−14·709	−0·265	1·105	−15·7
	D 2	0·070	0·747	−74·419	−5·209	5·956	−79·5

A in terms of the receptances $\beta, \gamma, \dots, \nu$ of $B, C, \dots, N$ respectively, for a selected value of ω, the driving frequency.

Let the amplitude of q_0 be Ψ_0 and consider the motion of B (see fig. 6.6.1 (b)). Since no applied force acts at q_0, by hypothesis, the force acting at q_1 is

$$Q_{b1} = \frac{\Psi_0}{\beta_{01}} e^{i\omega t} = \Phi_{b1} e^{i\omega t}, \qquad (6.6.1)$$

where ω is the selected frequency and β_{01} is the corresponding cross-receptance of B. Evidently the variation of q_1 is

$$q_1 = \beta_{11} \Phi_{b1} e^{i\omega t} = \frac{\beta_{11}}{\beta_{01}} \Psi_0 e^{i\omega t} = \Psi_1 e^{i\omega t}, \qquad (6.6.2)$$

β_{11} being the direct receptance of B at q_1 for the chosen value of ω. Thus the amplitudes Φ_{b1} and Ψ_1 of the force and displacement at q_1 are known in terms of Ψ_0.

The sub-system C has a force $\Phi_{c1} e^{i\omega t}$ applied at q_1 which is given by

$$\Phi_{c1} e^{i\omega t} = -\Phi_{b1} e^{i\omega t} = -\frac{\Psi_0}{\beta_{01}} e^{i\omega t} \qquad (6.6.3)$$

in view of the equilibrium that is maintained at the junction of B and C. This force contributes to the motion at q_1 an amount $\gamma_{11} \Phi_{c1} e^{i\omega t}$ so that the displacement $(\Psi_1 - \gamma_{11} \Phi_{c1}) e^{i\omega t}$ must be accounted for by the force $\Phi_{c2} e^{i\omega t}$. That is,

$$Q_{c2} = \Phi_{c2} e^{i\omega t} = \frac{(\Psi_1 - \gamma_{11} \Phi_{c1}) e^{i\omega t}}{\gamma_{12}}. \qquad (6.6.4)$$

TABLE A

9	10	11	12	13	
r+1	Direct receptance at q_{r+1}	Part of Ψ_{r+1} due to Φ_r	Part of Ψ_{r+1} due to Φ_{r+1}	Ψ_{r+1}	
/8		3×8	9×10	11+12	
l.ft.	(entry) ×10⁻⁴ rad./pdl.ft.	(entry) ×10⁻⁴ rad.	(entry) ×10⁻⁴ rad.	(entry) ×10⁻⁴ rad.	
·118	3·121	0·000	−0·368	−0·368	$\alpha_{33}=\dfrac{1·414}{0·087}\times 10^{-4}=16·3\times 10^{-4}$ rad./pdl.ft.
·197	−0·431	−0·190	−0·085	−0·275	
·087	−2·183	1·604	−0·190	1·414	$\dfrac{1}{\alpha_{33}}=620$ pdl.ft./rad.
·105	0·296	0·000	−0·031	−0·031	$\alpha_{33}=\infty$
·036	−1·038	−0·226	0·037	−0·189	
·000	−5·253	−0·392	0·000	−0·392	$\dfrac{1}{\alpha_{33}}=0$
·018	−46·685	0·000	0·840	0·840	$\alpha_{33}=\dfrac{0·011}{-0·075}\times 10^{-4}=-0·15\times 10^{-4}$ rad./pdl.ft.
·070	−14·709	−0·283	1·030	0·747	
·075	−74·419	−5·570	5·581	0·011	$\dfrac{1}{\alpha_{33}}=-66\,700$ pdl.ft./rad.

The variation of q_2 can now be found since

$$q_2 = (\gamma_{21}\Phi_{c1} + \gamma_{22}\Phi_{c2})\, e^{i\omega t} = \Psi_2 e^{i\omega t}. \tag{6.6.5}$$

Thus systems B and C have been 'crossed' and displacements and forces found in terms of Ψ_0, ω and the receptances β and γ.

By similar means it is possible to work towards q_n, crossing one sub-system at a time. The result of the calculations is a knowledge of q_n ($= \Psi_n e^{i\omega t}$) and Q_n ($= \Phi_n e^{i\omega t}$), both of the amplitudes being found in the form:

$$\text{(a numerical factor)} \times \Psi_0. \tag{6.6.6}$$

Unless the selected value of ω happens to be a natural frequency, the value of Φ_n will be non-zero so that

$$\alpha_{nn} = \frac{\Psi_n}{\Phi_n} \tag{6.6.7}$$

since the force Q_n must be applied by some outside agency if the postulated amplitude at q_0 is to be maintained. Equally the cross-receptances are given by

$$\alpha_{0n} = \frac{\Psi_0}{\Phi_n}, \quad \alpha_{1n} = \frac{\Psi_1}{\Phi_n}, \quad \dots. \tag{6.6.8}$$

Since $\Psi_1, \Psi_2, \dots, \Psi_n$ (as well as Φ_n) are all of the form (6.6.6), there is clearly no point in carrying the symbol Ψ_0 through the calculation and it can conveniently be taken as unity or as some convenient power of 10.

If the value of ω happens to be an anti-resonance frequency at q_n, Ψ_n will be zero and the receptance (6.6.7) will be nil.

If various values of ω are used, a curve may be plotted of $1/\alpha_{nn}$ against ω and it will give the natural frequencies. Let one of these be ω_r and let one of the above calculations be made for this value of ω so that $\Phi_n = 0$. Evidently the corresponding mode is given by the values of $\Psi_0, \Psi_1, \Psi_2, ..., \Psi_n$. The amplitudes within any one sub-system may readily be found from its receptances and from the amplitudes of force at its junctions with neighbouring sub-systems.

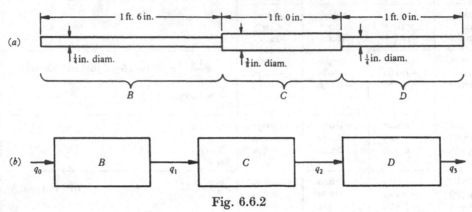

Fig. 6.6.2

Stepped shafts may be regarded as chains of the type shown in fig. 6.6.1 (a), each of the uniform portions being one sub-system. Thus a numerical example may now be given for such a system. Consider the shaft shown in fig. 6.6.2 (a) which consists of three parallel portions, for the material of which

$$G = 11 \cdot 5 \times 10^6 \, \text{p.s.i.} \quad \text{and} \quad \rho = 480 \, \text{lb./ft.}^3 \qquad (6.6.9)$$

This system is too simple to justify use of the tabular method since it does not contain a sufficient number of parallel portions, but it will serve for the purposes of explanation. The arrangement of sub-systems is shown in fig. 6.6.2 (b) and the receptances at q_3 will be found for the whole system. The calculation will be illustrated for the particular value $\lambda = 1 \cdot 3$ rad./ft. This corresponds to the driving frequency

$$\omega = \lambda a = 1 \cdot 3 \sqrt{\left(\frac{11 \cdot 5 \times 10^6 \times 32 \times 144}{480}\right)} = 13\,660 \, \text{rad./sec.} \qquad (6.6.10)$$

but it is simpler to work with λ than with ω.

It is first necessary to evaluate the receptances of the portions B, C and D. This can best be done in tabular form if a range of values of λ are to be used, by means of the expressions given in Table 5. The values corresponding to $\lambda = 1 \cdot 3$ rad./ft., are found to be

$$\left.\begin{array}{ll}
\beta_{00} = \beta_{11} = 3 \cdot 121 \times 10^{-4} \, \text{rad./pdl.ft.}, & \beta_{01} = -8 \cdot 446 \times 10^{-4} \, \text{rad./pdl.ft.}, \\
\gamma_{11} = \gamma_{22} = -0 \cdot 431 \times 10^{-4} \, \text{rad./pdl.ft.}, & \gamma_{12} = -1 \cdot 609 \times 10^{-4} \, \text{rad./pdl.ft.}, \\
\delta_{22} = \delta_{33} = -2 \cdot 183 \times 10^{-4} \, \text{rad./pdl.ft.}, & \delta_{23} = -8 \cdot 140 \times 10^{-4} \, \text{rad./pdl.ft.}
\end{array}\right\}$$

$$(6.6.11)$$

These values may now be entered in a table and the calculations mentioned above may then be performed in a systematic fashion. A convenient layout, which is thought to be self-explanatory, is given in Table A where calculations are shown for the system of fig. 6.6.2; the usual f.p.s. system of units is used and the first calculation corresponds to the value $\lambda = 1 \cdot 3$ rad./ft. It may be mentioned that all the calculations referred to in this section have been made with reasonable, but not high, accuracy; three places of decimals have been retained in all tabular working, after removal of the factor 10^{-4}.

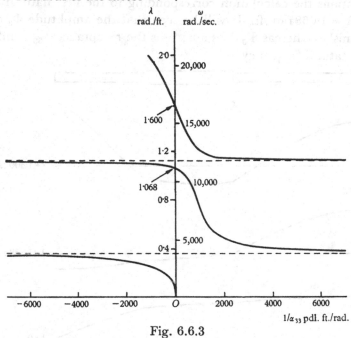

Fig. 6.6.3

From the results given in Table A, the receptances at q_3 may be found. They are

$$
\begin{aligned}
\alpha_{03} &= +\frac{10^{-4}}{0 \cdot 087} = +11 \cdot 5 \times 10^{-4}\,\text{rad./pdl.ft.,} \\
\alpha_{13} &= -\frac{0 \cdot 368}{0 \cdot 087} \times 10^{-4} = -4 \cdot 23 \times 10^{-4}\,\text{rad./pdl.ft.,} \\
\alpha_{23} &= -\frac{0 \cdot 275}{0 \cdot 087} \times 10^{-4} = -3 \cdot 16 \times 10^{-4}\,\text{rad./pdl.ft.,} \\
\alpha_{33} &= +\frac{1 \cdot 414}{0 \cdot 087} \times 10^{-4} = +16 \cdot 3 \times 10^{-4}\,\text{rad./pdl.ft.}
\end{aligned}
\tag{6.6.12}
$$

If tabular calculations of this sort are made with a series of values of λ, receptance curves can be plotted. Fig. 6.6.3 shows the curve of $1/\alpha_{33}$ for the shaft of fig. 6.6.2 and its intersections with the axis of ω (or λ) show that the first two natural frequencies (apart from the zero frequency) are

$$
\begin{aligned}
\lambda_1 &= 1 \cdot 068\,\text{rad./ft.} \quad \text{or} \quad \omega_1 = 11\,200\,\text{rad./sec.,} \\
\lambda_2 &= 1 \cdot 600\,\text{rad./ft.} \quad \text{or} \quad \omega_2 = 16\,800\,\text{rad./sec.}
\end{aligned}
\tag{6.6.13}
$$

The positions of the asymptotes of this curve reveal that the anti-resonance frequencies at q_3 are given by

$$\left.\begin{array}{l} \lambda = 0\cdot362 \text{ rad./ft.} \quad \text{or} \quad \Omega_1 = 3\,800 \text{ rad./sec.,} \\ \lambda = 1\cdot126 \text{ rad./ft.} \quad \text{or} \quad \Omega_2 = 11\,800 \text{ rad./sec.} \end{array}\right\} \qquad (6.6.14)$$

These latter are the two lowest natural frequencies that the shaft would have were it clamped at q_3.

Table A contains the calculation corresponding to the first natural frequency ω_1, that is to $\lambda = 1\cdot068$ rad./ft. It will be seen that the amplitude Φ_3 of applied torque at q_3 vanishes whereas Ψ_3 does not; thus the receptance α_{33} is infinite as it should be at a natural frequency.

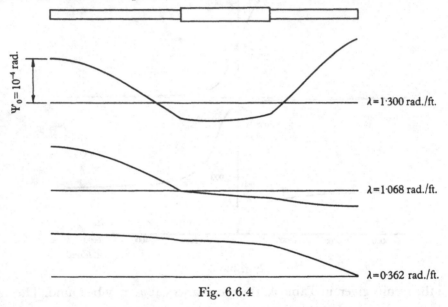

Fig. 6.6.4

Also given in the table is the calculation for the first anti-resonance frequency at q_3, that is for $\lambda = 0\cdot362$ rad./ft. Here, the amplitude Ψ_3 of motion at q_3 should be zero while Φ_3 is non-zero so that α_{33} vanishes. In fact, Ψ_3 is not zero and the receptance is finite though small; this is an outcome merely of the degree of accuracy to which the computations have been made.

It is now necessary to show how the deflexion shapes can be deduced from the tabular calculations. The table for $\lambda = 1\cdot3$ rad./ft. shows that the sub-system B of the shaft of fig. 6.6.2 (a) is subjected to a torque of amplitude $-0\cdot118$ pdl.ft. at its right-hand end. The portion C is excited by torques at each end, their amplitudes being $0\cdot118$ and $0\cdot197$ pdl.ft. Finally D experiences torques at each end whose magnitudes are $-0\cdot197$ and $0\cdot087$ pdl.ft. The deflected forms of B, C and D can thus be found separately and, when the appropriate curves are placed end-to-end, they give the distortion-shape of the entire shaft.

Fig. 6.6.4 shows the distortion curves for the shaft of fig. 6.6.2 (a) corresponding to the three values of λ that are used in Table A. The first curve shows the variation of amplitude Ψ with distance along the shaft when the system is excited by a harmonic torque of frequency 13 660 rad./sec. (i.e. $\lambda = 1\cdot3$ rad./ft.). The second curve shows the first principal mode of the free-free shaft. The third curve gives the first principal mode that the shaft would have were it clamped at its right-hand end.

It has been assumed in this section that the sub-system B (fig. 6.6.1 (a)) is free at the co-ordinate q_0. This is not a necessary condition where the use of the tabular method is concerned. Suppose that B is clamped at q_0 so that, while Ψ_0 is always zero, Φ_0 is not. In this case, it is merely necessary to assume a unity value of the force amplitude Φ_0 rather than of the displacement amplitude Ψ_0 as in Table A. The calculations then proceed in the usual way.

EXAMPLES 6.6

1. The figure shows a system composed of two uniform mild steel shafts (for which $G = 11\cdot5 \times 10^6$ psi and $\rho = 480$ lb./ft.^{3}) and a rigid disk. Find the reciprocal of the direct receptance at the right-hand end for some selected value of the driving frequency, using the tabular method of this section and treating the disk as one of the sub-systems.

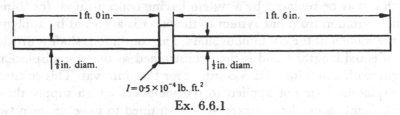

Ex. 6.6.1

2. Repeat Ex. 6.6.1 for the case where the left-hand end of the composite system is clamped. Deduce the second principal mode.

[Note that the method described in this section requires the left-hand portion of the system to be treated as if it is free-free; the left-hand end of it is then constrained to have zero deflexion. If this portion is regarded as being clamped-free and the direct receptance at the free end is calculated, sufficient information is obtained for that line in the tabular calculations corresponding to this sub-system to be dispensed with.]

3. Calculate the receptances of equation (6.6.12) using those of equation (6.6.11) by means of system 1 of Table 2.

6.7 Comparison of continuous shafts and torsional systems having concentrated (discrete) inertias

Consider a light elastic shaft to which are attached a number of heavy rigid disks, the disks being spaced at intervals along the shaft. Such a system will be capable of torsional vibration and methods by which the modes and frequencies of vibration can be found have been described in Chapters 1 and 3. If the number of disks is very large then the methods of Chapters 1 and 3 may become arduous and a modification of the tabular method of § 6.6 is found to be useful.† If there are many disks

† The tabular method, when applied to this problem, is known as Holzer's Method; see, for example, R. E. D. Bishop, *J. Roy. Aero. Soc.* vol. 58 (1954), p. 703.

and the system is such that all the disks are identical and are equally spaced along the shaft, then the difficulty can be avoided. This type of system may be treated for many purposes as if it were a continuous shaft with distributed mass. That is to say the inertia of each disk may be regarded as being spread over the two half-lengths of shaft which are adjacent to it. Thus if each of the disks has an inertia I and if they are spaced along the shaft at intervals of length l then the system can be treated as a continuous shaft which has an inertia I/l per unit length; this quantity replaces the quantity $J\rho$ of the previous analysis.

This replacement of an n-degree-of-freedom system by an infinite-freedom system may be shown to be justified provided that the distance between adjacent nodes, which are present during the vibration, is large compared with the intervals between the disks. This restriction means that there is a limit to the vibration frequencies for which the fictitious system may be used, and indeed it is evident that the continuous system cannot be applicable at very high frequencies because an infinite set of high-frequency modes has been introduced by the substitution.

The justification for the process may be seen by regarding it in reverse. Let us, therefore, begin with the continuous-shaft system and then examine the conditions under which it may be replaced by a system having concentrated- (or 'lumped-') inertia. The transition from one system to the other is achieved by applying constraints to the motion of the continuous shaft. The continuous shaft is divided into a number of equal lengths l and is then constrained so that the displacement of all the inertia within any interval is constant for that interval. This constraint on the mass displacement is not applied to the members which supply the elastic forces; these members on the contrary are constrained to have uniform twist per unit length within any one interval.

This arrangement of constraints is essentially different from those which were contemplated in Chapter 5. The latter never brought about distortions in which displacements of the mass were made different from the displacements of the associated elastic members. The new type of constraint evidently produces distortions which cannot be built up from components of the principal modes of the unconstrained system; for such modes will never involve different displacements of masses and elastic members when these are one and the same in the real physical system. It follows that Rayleigh's Principle is no longer applicable; nor, in particular, can it be assumed that the lowest constrained frequency will be greater than the lowest natural frequency of the unconstrained system.

The adoption of this new type of constraint may be justified by showing that the energy coefficients a_r and c_r for the modes concerned will not be altered greatly by the constraint. Now if the intervals between adjacent disks of the original system are small compared with the distance between successive nodes in some particular principal mode, then the twist per unit length of the imaginary uniform shaft will be almost constant over each interval. From this it follows that the corresponding coefficient c for the constrained system will be very close to that of the unconstrained system; this is evident because the strain energy in the shaft will be very little altered by the constraints. Further the kinetic energy of the shaft will be concentrated mainly in those portions of the shaft where the amplitude is large. Now if the amplitudes of two adjacent disks of the original shaft are Θ_1 and Θ_2 and if these disks are not close to a node then the difference between Θ_1 and Θ_2 will be small compared with the separate amplitudes. This means that the replacement of the disk-inertia by an inertia which is distributed along the interval will not greatly alter the kinetic energy, and hence the coefficient a, of the system.

These arguments show that the natural frequencies of a lumped-inertia system will be close to those of the corresponding continuous shaft system within the range for which the approximations discussed are valid. This being so, the reverse procedure of substitution can clearly be used; that is to say, the system having lumped-inertia can be replaced by the continuous system in calculations of the natural frequencies and principal modes.

Let seven disks, each having a polar moment of inertia I, be mounted on a light shaft with equal intervals between them and let the stiffness of the portion of the shaft between adjacent disks be k. If such a system is replaced by a uniform massive shaft, then the inertia of each disk will be replaced by an inertia which is distributed uniformly along the shaft. The inertia of any particular disk will evidently be represented in the new system by that inertia which lies between the mid-points of the two portions of shaft to which the particular disk is attached. If this arrangement is to preserve the original polar moment of inertia, then a uniform shaft is required for which the product $J\rho l$ is equal to I where l is the distance between adjacent disks. In order to allow for all the inertia of the end disks, it is necessary for the uniform shaft to extend a distance $\frac{1}{2}l$ beyond each end of the real shaft. This idea of extending the imaginary shaft is only permissible if the ends of the real shaft are free.

It is now necessary to examine the question of stiffness. If the torsional stiffness k between adjacent disks is to be retained between the corresponding points of the continuous shaft, then it is necessary that GJ/l for the latter shall be equal to k. This is because the twisting moment M can be expressed as

$$M = GJ\left(\frac{\partial\theta}{\partial x}\right) = k\theta, \quad \text{or} \quad kl\left(\frac{\theta}{l}\right). \tag{6.7.1}$$

The natural frequencies of the imaginary free-free shaft are given by

$$\lambda \tan 7\lambda l = 0. \tag{6.7.2}$$

The lowest root of this equation (apart from zero) is $7\lambda l = \pi$, where

$$\lambda = \omega\sqrt{\left(\frac{\rho}{G}\right)} = \omega\sqrt{\left(\frac{I}{kl^2}\right)}, \tag{6.7.3}$$

whence,
$$\omega_1 = \frac{\pi}{7}\sqrt{\left(\frac{k}{I}\right)} = 0.449\sqrt{\left(\frac{k}{I}\right)}. \tag{6.7.4}$$

The natural frequency for the corresponding mode of the original light shaft carrying the disks is given by

$$I^3\omega^6 - 5kI^2\omega^4 + 6k^2I\omega^2 - k^3 = 0. \tag{6.7.5}$$

This equation is obtained by considering half the shaft only and setting $\Delta = 0$ for system 12 of Table 1, because the mode must involve a central node. The first root of equation (6.7.5) is

$$\omega_1 = 0{\cdot}445 \sqrt{\left(\frac{k}{I}\right)}. \tag{6.7.6}$$

Suppose now that the shaft, which we have discussed above, has attached to it at one end a large disk having a moment of inertia I_2 in place of one of the original equal disks, as in fig. 6.7.1 (a). If the shaft and disks are replaced by a uniform shaft with distributed inertia, then it is not now permissible to extend the imaginary shaft the extra distance $\frac{1}{2}l$ at the end at which the inertia I_2 is attached and to retain a *uniform* equivalent shaft. Instead, the system may be replaced by that shown in fig. 6.7.1 (b). The large inertia on the end of the shaft is replaced here by the inertia $I_2 - \frac{1}{2}I$ so that there is an inertia $\frac{1}{2}I$ which is represented by the length $\frac{1}{2}l$ of shaft which adjoins the large disk. The free end of the shaft is extended as before.

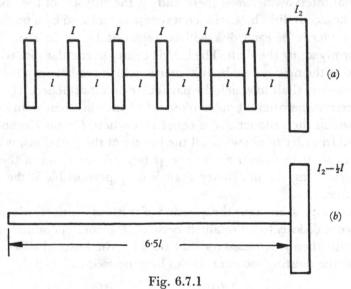

Fig. 6.7.1

The natural frequencies of the new system may be found by equating the sum of the reciprocal receptances at the disk to zero; they are given by the equation

$$\frac{-kl\omega\sqrt{(I/kl^2)}}{\cot[6{\cdot}5l\omega\sqrt{(I/kl^2)}]} - \left(I_2 - \frac{I}{2}\right)\omega^2 = 0. \tag{6.7.7}$$

For the special case where $I_2 = 4I$, this equation reduces to

$$k\omega\sqrt{(I/k)}\tan[6{\cdot}5\omega\sqrt{(I/k)}] + 3{\cdot}5I\omega^2 = 0 \tag{6.7.8}$$

and the lowest root is $\omega_1 = 0{\cdot}348\sqrt{(k/I)}$. The corresponding root of the 'lumped-inertia' system of fig. 6.7.1 (a) is $0{\cdot}345\sqrt{(k/I)}$ as may be found by splitting the system at the central disk and using receptances from Table 1.

The above discussion has been concerned with the replacement of one ideal type of system by another. In real systems the concentrated inertias will not be absolutely rigid, nor will the flexible members be massless. In many problems the distribution of mass and stiffness may be much more complex than in the disk-and-shaft system described above; a particular instance is provided by the crankshaft assembly of an in-line engine. It is common to replace the crankshaft by an imaginary system of disks in order to analyse it, the disks being coincident with the central planes of the cylinders. Whereas the flexibility between adjacent disks is represented in such an idealization by that of a uniform shaft, in the crankshaft itself the flexibility will be distributed along the journal and in the crankpins and webs in a manner which is difficult to determine accurately. Many empirical formulae have been devised for calculating the total flexibility of a crankshaft however, and static tests on a particular shaft can always be used to determine the total flexibility between one end and the other.† Provided, then, that all the throws are identical this flexibility may be regarded as being uniformly distributed along the shaft, and the total inertia may be similarly distributed. We thus arrive again at the replacement of a non-uniform system by a uniform one.

EXAMPLES 6.7

1. Five disks are attached to a straight uniform light shaft at equal intervals so that the torsional stiffness of the shaft between adjacent disks is k. The moment of inertia of the two end disks is $\frac{1}{2}I$ and that of the other disks is I. Calculate the lowest natural frequency of the system for which a node exists at the centre and also the natural frequency of the principal mode having two nodes; the latter will have an anti-node at the centre.

Find the frequencies (a) by replacing the system by a uniform shaft and (b) by direct calculation.

(Note that the modes with a central node may be found by considering the half system only, so that only 2 degrees of freedom are required. Note also that the mode having two-nodes may be found by splitting the system at the centre so that the two halves consist of three disks and are thus symmetrical.)

2. Modify the tabular method given in §6.6 so as to obtain a means of calculating the receptances at the end of a light shaft to which a number of rigid disks are attached.

6.8 Longitudinal vibration of uniform bars; basic equations and general results

In this section, we shall discuss the extensional vibration of elastic bars, confining the treatment to bars of uniform (but not necessarily of circular) cross-section. It will be seen therefore that the restrictions are not quite so great in this respect as they were for the discussion of torsional vibration.

If a bar (whose axis lies along the direction Ox and whose cross-sectional area is A) is subjected to a steady axial force P, compressive or tensile stress is produced at every cross-section. Any cross-section which was initially plane will remain so under stress. The axial strain in the bar will be the same all over a cross-section and will be equal to du/dx where u is the axial displacement of the section in the direction x. This strain will be related to the stress P/A by Hooke's law, so that

$$P = AE\frac{\partial u}{\partial x}, \qquad (6.8.1)$$

† See, for instance, the British Internal Combustion Engine Research Association's *Handbook on Torsional Vibration*, compiled by E. J. Nestorides (Cambridge University Press, 1958).

where E is Young's modulus. The partial derivative notation has here been introduced because we assume that this relation also holds when the stress varies with time.

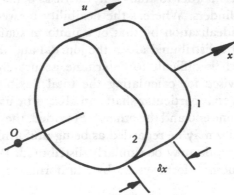

Fig. 6.8.1

Consider an element of the bar between plane faces 1 and 2, the faces being perpendicular to the centroidal axis Ox of the bar as shown in fig. 6.8.1. The equation of motion for this element of bar can be written in the form

$$AE\left[\left(\frac{\partial u}{\partial x}\right)_1 - \left(\frac{\partial u}{\partial x}\right)_2\right] + f\delta x = A\rho\,\delta x\frac{\partial^2 u}{\partial t^2}, \tag{6.8.2}$$

where ρ is the mass density of the elastic material. The first term which is enclosed within square brackets represents the difference between the forces which are due to the stress at the two ends of the element. The second term represents an axial force $f(x, t)$ per unit length which is imagined to be applied in some way to the material of the bar, and which we introduce for purposes of subsequent analysis as we did in the torsional problem. When $\delta x \to 0$, the equation of motion becomes

$$E\frac{\partial^2 u}{\partial x^2} + \frac{f}{A} = \rho\frac{\partial^2 u}{\partial t^2}. \tag{6.8.3}$$

The derivation of this equation involves several assumptions and it is only true approximately. The physical conditions which the equation represents do not obtain exactly in practice because of the existence of effects which are associated with Poisson's ratio; these produce displacements which are perpendicular to the axis of the bar. A more complete theory which takes these effects into account is known for bars of circular cross-section; the equation is derived by using the mathematical theory of elasticity.[†] The related two-dimensional problem of the extensional vibration of a flat plate has also been solved.[‡]

There is experimental evidence[§] to show that the simple theory is insufficient in some cases but it may be applied provided that the length of the bar is much

† L. Pochhammer, *J. f. Math.* (*Crelle*), vol. 81 (1876), p. 324; see A. E. H. Love, *Mathematical Theory of Elasticity*, Cambridge University Press, 4th ed. (1927), p. 289.

‡ H. Lamb, *Proc. Roy. Soc.* A, vol. 93 (1917), p. 114.

§ R. M. Davies, *Phil. Trans. Roy. Soc.* A, vol. 240 (1948), p. 375.

greater than any of its cross-sectional dimensions; the ratio of these quantities should be at least ten to one. Even so the theory is applicable only for the lower modes of vibration. These restrictions are not usually important for engineering purposes because any natural frequencies of longitudinal vibration are usually very high and consequently the higher modes are rarely excited appreciably.

It may be mentioned that it is possible to increase the range of principal modes for which the elementary equation may be used. This is done by applying corrections to the simple equation without introducing the full treatment which is necessary for an exact solution.[†]

The kinetic energy of the element of bar shown in fig. 6.8.1 is given by

$$\delta T = \tfrac{1}{2} A \rho \, \delta x \left(\frac{\partial u}{\partial t}\right)^2. \tag{6.8.4}$$

The total kinetic energy of a finite length of the bar is therefore

$$T = \tfrac{1}{2} A \rho \int \left(\frac{\partial u}{\partial t}\right)^2 dx. \tag{6.8.5}$$

The integral in this expression must be evaluated between appropriate limits.

The potential energy which is stored in the element is

$$\delta V = \tfrac{1}{2} P \delta u = \tfrac{1}{2} A E \frac{\partial u}{\partial x} \delta u = \tfrac{1}{2} A E \left(\frac{\partial u}{\partial x}\right)^2 \delta x \tag{6.8.6}$$

and the potential energy for a finite length of the bar is therefore

$$V = \tfrac{1}{2} A E \int \left(\frac{\partial u}{\partial x}\right)^2 dx. \tag{6.8.7}$$

Now it will be seen that the differential equation for longitudinal vibration is mathematically identical with that for the torsional vibration of a circular shaft. The boundary conditions which must be applied to the equation are:

$$\left. \begin{aligned} u &= 0 \quad \text{for a clamped end,} \\ \frac{\partial u}{\partial x} &= 0 \quad \text{for a free end (see equation (6.8.1)).} \end{aligned} \right\} \tag{6.8.8}$$

Again, if a harmonic force $P = F e^{i\omega t}$ is applied at some cross-section, then

$$\frac{\partial u}{\partial x} = \frac{F}{AE} e^{i\omega t} \tag{6.8.9}$$

at that section. It will be seen that not only the differential equation, but also the boundary conditions are mathematically identical with those for torsional vibration. It follows that all the previous results may be applied directly to longitudinal vibration problems.

Table 6 gives some results for longitudinal vibration. These, as the reader may readily verify, are analogous to the previous results for torsional vibration.

[†] R. E. D. Bishop, *Aero. Quart.* vol. 3, (1952), p. 280.

TABLE 6

LONGITUDINAL VIBRATION OF UNIFORM BARS

Notation

A Area of cross-section.

a $= \sqrt{(E/\rho)}$.

E Young's modulus.

h Particular value of x defining a section of the bar distant h from the origin ($0 \leqslant h \leqslant l$).

l Length of bar.

r, s Subscripts used to indicate the rth and sth principal modes ($= 1, 2, 3, \ldots$ for clamped-free or clamped-clamped and $= 0, 1, 2, 3, \ldots$ for free-free).

u Longitudinal displacement in the direction of x.

x Distance along the bar ($0 \leqslant x \leqslant l$).

α_{xh} Receptance between the sections x and h. If a harmonic longitudinal force $F e^{i\omega t}$ is applied at the section $x = h$ (and in practice this can only be at an otherwise free end), then the response at any point x is given by

$$u = \alpha_{xh} F e^{i\omega t}.$$

λ $= \omega \sqrt{(\rho/E)} = \omega/a$.

ρ Mass density.

Equation of motion: $E\dfrac{\partial^2 u}{\partial x^2} = \rho \dfrac{\partial^2 u}{\partial t^2}$.

Boundary conditions: $u = 0$ at a clamped end,

$\dfrac{\partial u}{\partial x} = 0$ at a free end.

Energy expressions:

$$T = \frac{A\rho}{2} \int_0^l \left(\frac{\partial u}{\partial t}\right)^2 dx, \quad V = \frac{AE}{2} \int_0^l \left(\frac{\partial u}{\partial x}\right)^2 dx.$$

In the series representation of displacement, the principal co-ordinates p_r have been chosen so that

$$u = \sum_{r=1}^{\infty} p_r . \phi_r(x) \quad \text{(clamped-free or clamped-clamped)},$$

or

$$u = \sum_{r=0}^{\infty} p_r . \phi_r(x) \quad \text{(free-free)}.$$

TABLE 6

LONGITUDINAL VIBRATION OF UNIFORM BARS

Quantity	Clamped-free	Free-free	Clamped-clamped
	$u = 0$ when $x = 0$	$\partial u / \partial x = 0$ when $x = 0$	$u = 0$ when $x = 0$
	$\partial u / \partial x = 0$ when $x = l$	$\partial u / \partial x = 0$ when $x = l$	$u = 0$ when $x = l$
α_{xh} $0 \leqslant x \leqslant h$ $0 \leqslant h \leqslant l$	$\dfrac{\sin \lambda x \cos \lambda(l-h)}{AE\lambda \cos \lambda l}$	$-\dfrac{\cos \lambda x \cos \lambda(l-h)}{AE\lambda \sin \lambda l}$	$\dfrac{\sin \lambda x \sin \lambda(l-h)}{AE\lambda \sin \lambda l}$
α_{xh} $h \leqslant x \leqslant l$ $0 \leqslant h \leqslant l$	$\dfrac{\sin \lambda h \cos \lambda(l-x)}{AE\lambda \cos \lambda l}$	$-\dfrac{\cos \lambda h \cos \lambda(l-x)}{AE\lambda \sin \lambda l}$	$\dfrac{\sin \lambda h \sin \lambda(l-x)}{AE\lambda \sin \lambda l}$
α_{xh} $0 \leqslant x \leqslant l$ $0 \leqslant h \leqslant l$	$\dfrac{2}{A\rho l} \displaystyle\sum_{r=1}^{\infty} \dfrac{\phi_r(x)\,\phi_r(h)}{\omega_r^2 - \omega^2}$	$\dfrac{2}{A\rho l} \displaystyle\sum_{r=0}^{\infty} \dfrac{\phi_r(x)\,\phi_r(h)}{\omega_r^2 - \omega^2}$	$\dfrac{2}{A\rho l} \displaystyle\sum_{r=1}^{\infty} \dfrac{\phi_r(x)\,\phi_r(h)}{\omega_r^2 - \omega^2}$
$\omega_r = \sqrt{\left(\dfrac{c_r}{a_r}\right)}$	$\dfrac{(2r-1)\pi a}{2l}$	$\dfrac{r\pi a}{l}$	$\dfrac{r\pi a}{l}$
$\phi_r(x)$	$\sin \dfrac{(2r-1)\pi x}{2l}$	$\dfrac{1}{\sqrt{2}} \quad (r = 0)$ $\cos \dfrac{r\pi x}{l} \quad (r = 1, 2, 3, \ldots)$	$\sin \dfrac{r\pi x}{l}$
$\displaystyle\int_0^l \phi_r(x)\,\phi_s(x)\,dx$	$0 \quad (r \neq s)$ $l/2 \quad (r = s)$	$0 \quad (r \neq s)$ $l/2 \quad (r = s)$	$0 \quad (r \neq s)$ $l/2 \quad (r = s)$
a_r	$A\rho l/2$	$A\rho l/2$	$A\rho l/2$
c_r	$\dfrac{(2r-1)^2 \pi^2 AE}{8l}$	$\dfrac{r^2 \pi^2 AE}{2l} (r = 0, 1, 2, 3, \ldots)$	$\dfrac{r^2 \pi^2 AE}{2l}$

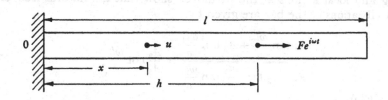

Clamped end   Free end

EXAMPLES 6.8

1. The longitudinal deflexion of a free-free bar may be written in terms of the principal co-ordinates $p_r(t)$ and characteristic functions $\phi_r(x)$; thus

$$u = \frac{1}{\sqrt{2}}\, p_0(t) + \sum_{r=1}^{\infty} p_r(t) \cdot \phi_r(x),$$

where the 0th principal co-ordinate $p_0(t)$ represents motion as a rigid body. By substituting this series into equations (6.8.5) and (6.8.7) show that, for this type of bar

$$2T = a_0\,\dot{p}_0^2 + a_1\,\dot{p}_1^2 + a_2\,\dot{p}_2^2 + \cdots,$$

$$2V = c_0\,p_0^2 + c_1\,p_1^2 + c_2\,p_2^2 + \cdots,$$

where
$$\left.\begin{aligned} a_r &= A\rho l/2 \\ c_r &= r^2\pi^2 AE/2l \end{aligned}\right\} (r = 0, 1, 2, 3, \ldots).$$

Using these values of the coefficients a, find the receptance α_{00} in series form for a free-free bar.

[NOTE. The first term of this series represents motion in the zero-frequency mode and the approximation to the true value of α_{00} that is obtained by rejecting all but the first term is that which is found by treating the bar as if it were rigid.]

2. Show that the error which is obtained in α_{00} for a free-free bar by treating the bar as if it were rigid is

$$\left(1 - \frac{\sin \lambda l}{\lambda l}\right) \times 100\,\%,$$

where $\lambda = \omega\sqrt{(\rho/E)}$.

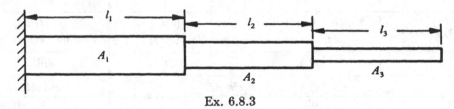

Ex. 6.8.3

3. The composite bar shown has lengths l_1, l_2, l_3 and cross-sectional areas A_1, A_2, A_3, the material being uniform throughout.

Neglecting any local actions at the shoulders, show that the natural frequencies for longitudinal vibrations of the bar are given by

$$\frac{A_1 \tan \dfrac{\omega l_2}{a} + A_2 \tan \dfrac{\omega l_1}{a}}{A_1 - A_2 \tan \dfrac{\omega l_1}{a} \tan \dfrac{\omega l_2}{a}} = \frac{A_2}{A_3} \cot \frac{\omega l_3}{a},$$

where $a = \sqrt{(E/\rho)}$.

(C.U.M.S.T. Pt II, 1947)

6.9 Steady wave motion in uniform shafts and bars

In all the problems on longitudinal or torsional motion which we have discussed so far, the motions at all points along the bar have been in phase. This has been the result of considering excitation by a single torque or by two torques which are in phase (referring specifically to *torsional* systems). In this section, we shall examine the conditions when two exciting torques, having the same frequency but differing phase, act on a shaft at different sections. In doing this, we shall consider the particular problem of torsional motion of a free-free shaft of length l; but, while the description will be in terms of torsion, the theory is applicable to longitudinal motion also. It is convenient to begin by describing the motion and subsequently to find the torques which would be required to produce it.

Let the angular displacement of a section of a free-free shaft, distant x from the end, be given by

$$\theta = A\left(\cos\frac{\omega x}{a} + i\sin\frac{\omega x}{a}\right)e^{i\omega t}.$$ (6.9.1)

This function satisfies the differential equation

$$\frac{\partial^2\theta}{\partial x^2} = \frac{1}{a^2}\frac{\partial^2\theta}{\partial t^2}$$ (6.9.2)

and consists of two components. One of these is a real multiple of $e^{i\omega t}$, namely,

$$A\cos\frac{\omega x}{a}e^{i\omega t}$$ (6.9.3)

and one an imaginary multiple. The latter may be written in the alternative form

$$A\,e^{i\frac{1}{2}\pi}\sin\frac{\omega x}{a}.e^{i\omega t} = A\sin\frac{\omega x}{a}.e^{i(\omega t+\frac{1}{2}\pi)}.$$ (6.9.4)

This shows that the motion is $\frac{1}{2}\pi$ rad. in advance of the motion given by the component (6.9.3).

The instantaneous deflexion of the shaft at any point is made up from these two sinusoidal components of equal frequency; it is therefore itself sinusoidal. The nature of the motion can be seen more easily by writing equation (6.9.1) in the form

$$\theta = A[e^{i\omega x/a}]\,e^{i\omega t} = A\,e^{i\omega(t+x/a)}.$$ (6.9.5)

This shows that the amplitude of the deflexion at any point remains constant and equal to A but that the points of maximum deflexion move continuously along the shaft.

The deflexion at any section x_1 differs from that at the section $x = 0$ by being in advance by the phase angle $\omega x_1/a$. This means that a sinusoidal wave moves from the section $x = x_1$ to the section $x = 0$ with a speed a, the time taken being x_1/a and the phase change in that time being $\omega x_1/a$. This type of travelling wave motion can occur at any frequency, the speed of the wave being independent of ω.

It will be seen that another possible solution of equation (6.9.2) is of the form

$$\theta = A\left(\cos\frac{\omega x}{a} - i\sin\frac{\omega x}{a}\right)e^{i\omega t}.$$ (6.9.6)

This may be written $$\theta = A\,e^{i\omega(t-x/a)}$$ (6.9.7)

and represents a similar motion to that of equation (6.9.5), but the wave now travels in the direction in which x increases instead of that in which x decreases.

If both motions ((6.9.1) and (6.9.6)) are present together, A being the same for each, then the net motion will be

$$\theta = 2A \cos \frac{\omega x}{a} . e^{i\omega t}. \tag{6.9.8}$$

This represents a steady forced motion of the familiar form. Thus such a motion may be regarded as the sum of two wave motions which have the same amplitude and wavelength and which travel in opposite directions.

Suppose that the motion of the shaft is that given in equations (6.9.1) and (6.9.5). The torque M which must be applied at the end $x = 0$ is

$$-GJ\frac{\partial \theta}{\partial x} \tag{6.9.9}$$

in accordance with equation (6.1.2); the negative sign is included here since M is to be counted positive in the positive direction of θ. It follows then that, when $x = 0$,

$$\left. \begin{array}{l} \theta = A\,e^{i\omega t}, \\[2mm] M = -i\dfrac{GJA\omega}{a} e^{i\omega t} = \dfrac{GJA\omega}{a} e^{i(\omega t - \frac{1}{2}\pi)}, \end{array} \right\} \quad (x = 0). \tag{6.9.10}$$

The deflexion and torque at the end $x = l$ are

$$\left. \begin{array}{l} \theta = A\,e^{i\omega(t+l/a)}, \\[2mm] M = i\dfrac{GJA\omega}{a} e^{i\omega(t+l/a)} = \dfrac{GJA\omega}{a} e^{i[\omega(t+l/a)+\frac{1}{2}\pi]}, \end{array} \right\} \quad (x = l). \tag{6.9.11}$$

In the latter expression the torque has been found by direct application of equation (6.1.2) as indicated in fig. 6.1.2.

It will be seen that the amplitude of the torque is the same at each end of the shaft and that the phase of the torque is also $\frac{1}{2}\pi$ different from that of the displacement. It follows that this results in energy being continuously fed into the bar at the end $x = l$ and removed from the bar at $x = 0$. There is a transfer of energy in the same direction as the wave motion. In order to find this energy, the instantaneous values of torque and angular velocity are required in real terms. At the end $x = l$, these are given by the real parts of

$$+\frac{GJA\omega}{a} e^{i[\omega(t+l/a)+\frac{1}{2}\pi]} \quad \text{and} \quad A\omega\, e^{i[\omega(t+l/a)+\frac{1}{2}\pi]},$$

respectively, and are

$$-\frac{GJA\omega}{a} \sin\left(\omega t + \frac{\omega l}{a}\right) \quad \text{and} \quad -A\omega \sin\left(\omega t + \frac{\omega l}{a}\right).$$

These quantities are in phase. The instantaneous rate of working is

$$\frac{GJA^2\omega^2}{a} \sin^2\left(\omega t + \frac{\omega l}{a}\right) \tag{6.9.12}$$

which is always positive and varies sinusoidally with frequency 2ω. While the magnitude of the rate of working is independent of l, the phase of the variation depends on l.

These results show that energy is supplied to the shaft at the end where $x = l$. It may be shown by a similar process that it is withdrawn from the shaft at $x = 0$.

The wave motion thus implies a continuous flow of energy along the shaft in the direction of motion of the waves of deflexion. Although the mean rate of supply of energy is equal to the mean rate of withdrawal, the instantaneous rates are not necessarily equal. This is explained by the fact that at any instant some energy is stored in the shaft, and this quantity is not constant.

It is of interest to note that if the end of the bar at $x = 0$ is attached to a damping device which exerts a torque proportional and opposite to its velocity, the constant of proportionality being GJ/a, then the damper will produce the required torque and only the torque at $x = l$ will have to be applied from outside to maintain the motion.

We shall discuss the idea of wave motions again in connexion with the effect of distributed damping forces within elastic bodies. The interested reader may wish to refer to an introductory book on the subject of wave motions and will find that of Coulson† helpful.

EXAMPLES 6.9

1. Find the velocity of propagation of harmonic waves of axial displacement along a uniform bar for the material of which

$$E = 30 \times 10^6 \text{ psi}, \quad \rho = 480 \text{ lb./ft.}^3$$

2. Show that free torsional vibration of a uniform free-free shaft may be regarded as the sum of the motions due to suitably chosen harmonic waves travelling in the two directions along the shaft.

3. A uniform fixed-free taut string (see fig. 4.2.1) oscillates freely in its rth principal mode, the maximum amplitude being $2A$. Show that the motion may be regarded as being due to the two waves

$$A \, e^{i\omega_r(t + x/a)} \quad \text{and} \quad A \, e^{i[\omega_r(t - x/a) + \pi]}.$$

NOTATION

Supplementary List of Symbols used in Chapter 6

A Area of cross-section of bar.

a Propagation speed of waves, equal to $\sqrt{(G/\rho)}$ for torsional motion or $\sqrt{(E/\rho)}$ for longitudinal motion.

f Distributed longitudinal force per unit length applied to bar.

M Twisting moment.

m Distributed torque per unit length applied to shaft.

R Radius of shaft.

r Radius.

P Longitudinal force.

u Axial displacement in direction x.

θ Angle of twist.

θ_r Angular deflexion in rth principal mode.

λ $= \omega/a$.

λ_r $= \omega_r/a$.

† C. A. Coulson, *Waves* (Oliver and Boyd, London, 5th ed. 1949).

CHAPTER 7

FLEXURAL VIBRATION OF
UNIFORM BEAMS

> It has been shewn in this chapter that the theory of bars, even when simplified
> to the utmost by the omission of unimportant quantities, is decidedly more
> complicated than that of perfectly flexible strings.
>
> Lord Rayleigh, *The Theory of Sound* (1894)

The vibration of continuous systems, that is of systems having infinite freedom, was introduced in Chapter 4 where taut strings were discussed. These systems were introduced for simplicity of explanation and, on account of mathematical similarity, they led us to the treatment of shafts and bars in Chapter 6. Now, although these latter systems have greater technical importance than strings they have seldom raised practical vibration problems. The reason for this is that even the lowest natural frequencies of these systems are usually much higher than the exciting frequencies which are commonly met. The reader will recall that, under these circumstances, receptances are nearly equal to the corresponding flexibilities; this may be seen from the series form of receptances.

The natural frequencies of beam flexure are generally lower, for members of comparable size, than those of shafts in torsion and bars in longitudinal vibration; this type of motion, then, is more liable to raise vibration problems. The analysis is, however, somewhat more complicated.

7.1 The equation of motion

The treatment of beam flexure which is developed here is based on simple bending theory such as is normally used for engineering purposes. The method of analysis is known as the Bernoulli–Euler theory and is based upon the assumption that plane cross-sections of a beam remain plane during flexure and that the radius of curvature of a bent beam is large compared with the beam's depth.†

Fig. 7.1.1 shows a short element of a beam. It is of length δx and is bounded by plane faces which are perpendicular to the axis; the faces are identified by the numbers 1 and 2. The forces and couples which act upon the element are also shown in the figure; they are the shear forces S_1, S_2, the bending moments M_1, M_2 and the applied lateral load $w \, \delta x$. If the deflexion of the beam is small, as the theory presupposes, then the inclination of the beam element from the unstrained position is also small. Under these conditions, the equation of motion perpendicular to the axis Ox of the undeflected beam is

$$S_1 - S_2 + w \, \delta x = (\rho A \, \delta x) \frac{\partial^2 v}{\partial t^2}, \qquad (7.1.1)$$

† See, for instance, S. P. Timoshenko, *Strength of Materials*, vol. 1 (Van Nostrand, 2nd ed. 1941), ch. 4.

where A is the area of cross-section, ρ is the mass density and v is the deflexion. If the equation is divided by δx and if the length of the element is then made indefinitely short, the equation of motion is obtained in the form

$$\frac{\partial S}{\partial x} + w = \rho A \frac{\partial^2 v}{\partial t^2}. \tag{7.1.2}$$

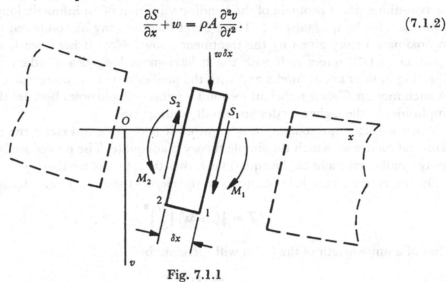

Fig. 7.1.1

If it is further assumed that the rate of change of the moment of momentum of the element about its centre of gravity can be neglected, then the equation for the rotational motion of the element may be written

$$S\,\delta x + M_1 - M_2 = 0. \tag{7.1.3}$$

When the length of the element is made small, this equation gives

$$S = -\frac{\partial M}{\partial x}. \tag{7.1.4}$$

Now from the simple bending theory we have the relation

$$M = EI\frac{\partial^2 v}{\partial x^2}, \tag{7.1.5}$$

where E is Young's modulus of the material and I is the second moment of area of the cross-section about the neutral axis through its centroid. If this is combined with equation (7.1.4) it gives

$$S = -EI\frac{\partial^3 v}{\partial x^3}. \tag{7.1.6}$$

This expression for S may be substituted in equation (7.1.2) which then gives the differential equation of motion in the form

$$\frac{\partial^2 v}{\partial t^2} + \frac{EI}{A\rho}\frac{\partial^4 v}{\partial x^4} = \frac{w}{A\rho}. \tag{7.1.7}$$

It will be seen that this is a partial differential equation of the fourth order.

Equation (7.1.7) is approximate only and it is important that we should be aware in what respects it is not reliable. If the term on the right-hand side is disregarded for the present, then it is possible to obtain, by using the mathematical theory of elasticity, an exact solution for the three-dimensional problem of the flexural vibration of an infinitely long uniform circular cylinder.† The solution of the two-dimensional problem of the bending vibration of an infinitely long elastic plate has also been obtained.‡ These exact theories may be compared with the approximate theory given by the treatment above. Now it has been found that equation (7.1.7) agrees well with the behaviour of long beams when they are vibrating in their lowest modes and with the predictions of the more exact theories for such motion. This is sufficient for most engineering purposes because the small amplitudes in the higher modes are usually unimportant.

When a better approximation is required it is possible to extend the range of principal modes for which the simple theory is adequate. The procedure is then to apply small corrections to the equations rather than to replace them.§

During vibration the kinetic energy of the element shown in fig. 7.1.1 is

$$\delta T = \tfrac{1}{2}(A\rho\,\delta x)\left(\frac{\partial v}{\partial t}\right)^2. \tag{7.1.8}$$

That of a finite length of the beam will therefore be

$$T = \frac{A\rho}{2}\int\left(\frac{\partial v}{\partial t}\right)^2 dx. \tag{7.1.9}$$

The integration for this expression must be performed between appropriate limits.

Again the potential energy due to the bending of the element will be

$$\delta V = \tfrac{1}{2}M\frac{\partial}{\partial x}\left(\frac{\partial v}{\partial x}\right)\delta x = \frac{EI}{2}\frac{\partial^2 v}{\partial x^2}\frac{\partial}{\partial x}\left(\frac{\partial v}{\partial x}\right)\delta x = \frac{EI}{2}\left(\frac{\partial^2 v}{\partial x^2}\right)^2 \delta x \tag{7.1.10}$$

and this may be integrated over a length of beam to give

$$V = \frac{EI}{2}\int\left(\frac{\partial^2 v}{\partial x^2}\right)^2 dx. \tag{7.1.11}$$

EXAMPLE 7.1

1. In deriving equation (7.1.3) it was assumed that the rate of change of the moment of momentum of the elemental slice of the beam (shown in fig. 7.1.1) could be neglected. Remove this restriction and show that, without it, the equation of motion becomes‖

$$\frac{\partial^2 v}{\partial t^2}+\frac{EI}{A\rho}\frac{\partial^4 v}{\partial x^4} = \frac{I}{A}\frac{\partial^4 v}{\partial x^2\partial t^2}+\frac{w}{A\rho}.$$

If this equation of motion is used, what must be the expression for S in subsequent theory (to replace that of equation (7.1.6))?

† L. Pochhammer, *J. f. Math.* (*Crelle*), vol. 81 (1876), p. 324; see A. E. H. Love, *Mathematical Theory of Elasticity* (Cambridge University Press, 4th ed. 1927), p. 287.

‡ H. Lamb, *Proc. Roy. Soc.* A, vol. 93 (1917), p. 114; see also R. E. D. Bishop, *Q.J.M.A.M.*, vol. 6 (1953), p. 250.

§ See S. P. Timoshenko, *Vibration Problems in Engineering*, 3rd ed. van Nostrand (1955), p. 329; also, see J. Prescott, *Phil. Mag.* vol. 33 (1942) p. 707.

‖ This correction was first made by Lord Rayleigh in his *The Theory of Sound* (2nd ed. 1894), §186.

7.2 Receptances in closed form

The receptances of uniform beams may be obtained in closed form by using equation (7.1.7) with the term $w/A\rho$ omitted. The equation is

$$\frac{\partial^2 v}{\partial t^2} + \frac{EI}{A\rho}\frac{\partial^4 v}{\partial x^4} = 0. \qquad (7.2.1)$$

Those solutions of this equation which are of the same general type as those found for other types of system when they are excited harmonically with frequency ω will have the form

$$v = X(x)\,e^{i\omega t}. \qquad (7.2.2)$$

If this expression is substituted in equation (7.2.1), it gives an ordinary differential equation from which the function $X(x)$ may be deduced. This equation is

$$\frac{d^4 X}{dx^4} - \frac{\omega^2 A\rho}{EI}X = 0, \qquad (7.2.3)$$

and the general form of the function $X(x)$ is therefore

$$X = A\cos\lambda x + B\sin\lambda x + C\cosh\lambda x + D\sinh\lambda x, \qquad (7.2.4)$$

where

$$\lambda^4 = \frac{\omega^2 A\rho}{EI} \qquad (7.2.5)$$

and where A, B, C and D are constants which are determined by the end conditions.

To find the receptances, it is necessary to specify the end-conditions suitably. The geometric boundary conditions relate to the deflexion v and slope $\partial v/\partial x$ while the force boundary conditions are found from equations (7.1.5) and (7.1.6); they are

$$
\left.
\begin{aligned}
v &= \frac{\partial v}{\partial x} = 0 \quad \text{at a clamped end,} \\[4pt]
v &= \frac{\partial^2 v}{\partial x^2} = 0 \quad \text{at a pinned end,} \\[4pt]
\frac{\partial v}{\partial x} &= \frac{\partial^3 v}{\partial x^3} = 0 \quad \text{at a sliding end,} \\[4pt]
\frac{\partial^2 v}{\partial x^2} &= \frac{\partial^3 v}{\partial x^3} = 0 \quad \text{at a free end.}
\end{aligned}
\right\}
\qquad (7.2.6)
$$

If a harmonic force $F\,e^{i\omega t}$, whose positive direction coincides with that of S in fig. 7.1.1, acts at an otherwise free end of a beam then, at that section

$$\frac{\partial^3 v}{\partial x^3} = -\frac{F}{EI}e^{i\omega t} \qquad (7.2.7)$$

by equation (7.1.6). Again, if a harmonic bending couple $H e^{i\omega t}$ acts at an otherwise free end, then

$$\frac{\partial^2 v}{\partial x^2} = \frac{H}{EI}e^{i\omega t} \qquad (7.2.8)$$

at that section, by equation (7.1.5), provided that the positive direction of the couple is the same as that of M in fig. 7.1.1. The following examples illustrate the procedure for finding receptances.

The clamped-free beam shown in fig. 7.2.1 is excited by a harmonic force of frequency ω which acts at the end as shown. The appropriate boundary conditions are

$$v = \frac{\partial v}{\partial x} = 0 \qquad \text{when} \quad x = 0,$$

$$\frac{\partial^2 v}{\partial x^2} = 0 \quad \text{and} \quad \frac{\partial^3 v}{\partial x^3} = -\frac{F}{EI}e^{i\omega t} \qquad \text{when} \quad x = l. \tag{7.2.9}$$

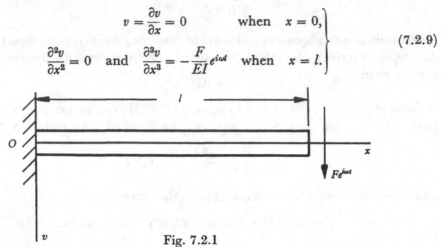

Fig. 7.2.1

The values of the constants A, B, C and D which correspond to these end conditions are

$$A = -C = \frac{-F(\sin \lambda l + \sinh \lambda l)}{2EI\lambda^3(1 + \cos \lambda l \cosh \lambda l)},$$

$$B = -D = \frac{F(\cos \lambda l + \cosh \lambda l)}{2EI\lambda^3(1 + \cos \lambda l \cosh \lambda l)}. \tag{7.2.10}$$

If these values for the constants are substituted back into equation (7.2.4) they give for the displacement v at any section, the expression

$$v = -\left\{ \frac{(\sin \lambda l + \sinh \lambda l)(\cos \lambda x - \cosh \lambda x) - (\cos \lambda l + \cosh \lambda l)(\sin \lambda x - \sinh \lambda x)}{2EI\lambda^3(1 + \cos \lambda l \cosh \lambda l)} \right\} F e^{i\omega t}. \tag{7.2.11}$$

The slope at any section may be found by differentiating this expression; it is

$$\frac{\partial v}{\partial x} = \left\{ \frac{(\cos \lambda l + \cosh \lambda l)(\cos \lambda x - \cosh \lambda x) + (\sin \lambda l + \sinh \lambda l)(\sin \lambda x + \sinh \lambda x)}{2EI\lambda^2(1 + \cos \lambda l \cosh \lambda l)} \right\} F e^{i\omega t}. \tag{7.2.12}$$

Let the harmonic force in the previous example be replaced by a couple $H e^{i\omega t}$ as shown in fig. 7.2.2. The boundary conditions with this arrangement are

$$v = \frac{\partial v}{\partial x} = 0 \quad \text{at} \quad x = 0,$$

$$\frac{\partial^2 v}{\partial x^2} = \frac{H}{EI}e^{i\omega t} \quad \text{and} \quad \frac{\partial^3 v}{\partial x^3} = 0 \quad \text{at} \quad x = l. \tag{7.2.13}$$

The same method as before now gives for the displacement

$$v = -\left\{\frac{(\cos \lambda l + \cosh \lambda l)(\cos \lambda x - \cosh \lambda x) + (\sin \lambda l - \sinh \lambda l)(\sin \lambda x - \sinh \lambda x)}{2EI\lambda^2(1 + \cos \lambda l \cosh \lambda l)}\right\} H e^{i\omega t},$$

(7.2.14)

and again for the slope

$$\frac{\partial v}{\partial x} = -\left\{\frac{(\sin \lambda l - \sinh \lambda l)(\cos \lambda x - \cosh \lambda x) - (\cos \lambda l + \cosh \lambda l)(\sin \lambda x + \sinh \lambda x)}{2EI\lambda(1 + \cos \lambda l \cosh \lambda l)}\right\} H e^{i\omega t}.$$

(7.2.15)

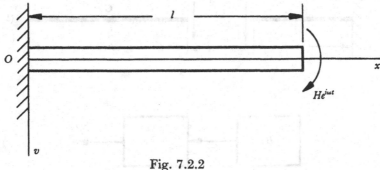

Fig. 7.2.2

Equations (7.2.11), (7.2.12), (7.2.14) and (7.2.15) can be written in the usual receptance form; thus,

$$\begin{pmatrix} \text{response} \\ v \text{ or } \dfrac{\partial v}{\partial x} \end{pmatrix} = (\text{a receptance}) \times \begin{pmatrix} F \\ \text{or} \\ H \end{pmatrix} e^{i\omega t}.$$

(7.2.16)

From this, it will be seen that there are four cross-receptances between any two sections of a beam instead of only the one as in longitudinal or torsional vibration. The reason for this is that we are now concerned with two co-ordinates at any section of the member, these being the deflexion and the slope (or possibly some functions of these quantities). In the theory of torsional vibration, on the other hand, the angular displacement of a section is the only co-ordinate required.

In order to emphasize this, the above problem may be contrasted with that of a fixed-free taut string which has a harmonic force applied at its end, as shown in fig. 7.2.3 (a). To find the deflexion v at some point along the string which is distant x from the fixed end, the system may be divided into two sub-systems B and C as shown in fig. 7.2.3 (b). The sub-systems are represented in block form in fig. 7.2.3 (c) where they will be seen to be linked by the co-ordinate v. The single cross-receptance α_{xl} can now be found by the method of §4.2.

The problem of the clamped-free beam shown in fig. 7.2.4 (a) may be contrasted with this. The system may be divided into sub-systems B and C as shown in fig. 7.2.4 (b). For convenience we introduce the notation

$$\left. \begin{aligned} v &= q_1, \\ \frac{\partial v}{\partial x} &= q_2, \end{aligned} \right\} \quad \text{at the section } x.$$

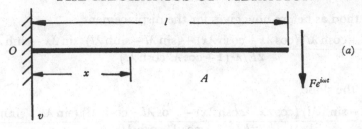

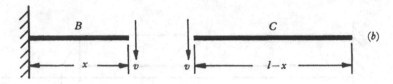

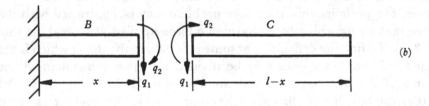

Fig. 7.2.3

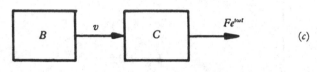

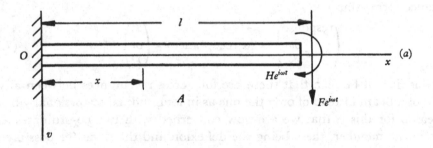

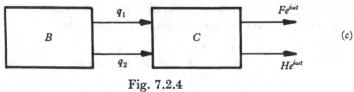

Fig. 7.2.4

The block diagram for the system is shown in fig. 7.2.4 (c), where q_1 and q_2 are the co-ordinates linking B and C. Now, in general, displacements at both q_1 and q_2 will be caused both by the force $F e^{i\omega t}$ and by the couple $H e^{i\omega t}$ so that there are four cross-receptances between the sections at x and l of the clamped-free beam.

The existence of *two* linking co-ordinates at any section where a beam is thought of as being severed introduces a need for suitable notation. The notation which we shall adopt may be explained under headings of the various quantities between which it is necessary to distinguish.

(i) *Systems.* The symbol α will still be used for the receptances of a complete system A and β, γ, etc., will continue to be used for the receptances of its component sub-systems B, C, ...

(ii) *Section of beam.* Subscripts to α, β, γ, ... will be used as before to indicate the section of the beam at which the response is measured and to which the excitation is applied. Thus α_{xh} gives the response at the section x of a beam which is excited harmonically at the section where $x = h$.

(iii) *Type of co-ordinate and type of excitation.* It is now necessary to distinguish between deflexions and slopes on the one hand and between forces and couples on the other. This will be done by the use of a prime (') on the subscript of a receptance whenever a slope or a couple is implied. Thus $\alpha_{x'h}$ will be the cross-receptance giving the slope at a section x when a harmonic force is applied at h; again $\alpha_{xh'}$ will give the deflexion v at x when a harmonic couple is applied at h.

The notation may be illustrated by writing the results (7.2.11), (7.2.12), (7.2.14), (7.2.15) in receptance form. They become

$$v = \alpha_{xl} F e^{i\omega t}, \qquad \frac{\partial v}{\partial x} = \alpha_{x'l} F e^{i\omega t} \tag{7.2.17}$$

for the system of fig. 7.2.1 and

$$v = \alpha_{xl'} H e^{i\omega t}, \qquad \frac{\partial v}{\partial x} = \alpha_{x'l'} H e^{i\omega t} \tag{7.2.18}$$

for that of fig. 7.2.2.

The direct and cross-receptances at the end of a clamped-free beam may be found by putting $x = l$ in these results. In this way it is found that

$$\alpha_{ll} = \frac{\sin \lambda l \cosh \lambda l - \cos \lambda l \sinh \lambda l}{EI\lambda^3(1 + \cos \lambda l \cosh \lambda l)}, \tag{7.2.19}$$

$$\alpha_{l'l} = \frac{\sin \lambda l \sinh \lambda l}{EI\lambda^2(1 + \cos \lambda l \cosh \lambda l)} \tag{7.2.20}$$

from equation (7.2.11) and equation (7.2.12) respectively. Again

$$\alpha_{ll'} = \frac{\sin \lambda l \sinh \lambda l}{EI\lambda^2(1 + \cos \lambda l \cosh \lambda l)}, \tag{7.2.21}$$

$$\alpha_{l'l'} = \frac{\sin \lambda l \cosh \lambda l + \cos \lambda l \sinh \lambda l}{EI\lambda(1 + \cos \lambda l \cosh \lambda l)} \tag{7.2.22}$$

from equation (7.2.14) and equation (7.2.15) respectively.

The reciprocal relation holds good for the cross-receptances of a beam as for the other systems that we have mentioned. Thus equations (7.2.20) and (7.2.21) show that

$$\alpha_{l'l} = \alpha_{ll'}. \tag{7.2.23}$$

It will be seen that all the receptances become infinite for the same values of frequency. These are the values for which

$$1 + \cos \lambda l \cosh \lambda l = 0. \tag{7.2.24}$$

Evidently this must be the frequency equation of a clamped-free beam.

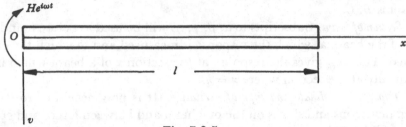

Fig. 7.2.5

Fig. 7.2.5 shows a free-free beam to which a couple of magnitude $He^{i\omega t}$ is applied at the end where $x = 0$. It is important to note that an applied couple is considered positive if it acts in the direction of increasing slope of the beam at the point of application; the couple indicated in the diagram is therefore positive. The implication of this sign convention is that if the slope is taken as a generalized co-ordinate of the system then the applied couple will be the corresponding generalized force and must therefore act 'in the direction of' the co-ordinate. In this problem, then, the force boundary condition at $x = 0$ is

$$\frac{\partial^2 v}{\partial x^2} = -\frac{H}{EI} e^{i\omega t}. \tag{7.2.25}$$

The remaining conditions are

$$\left. \begin{aligned} \frac{\partial^3 v}{\partial x^3} &= 0 \quad \text{when} \quad x = 0, \\[2mm] \frac{\partial^2 v}{\partial x^2} = \frac{\partial^3 v}{\partial x^3} &= 0 \quad \text{when} \quad x = l. \end{aligned} \right\} \tag{7.2.26}$$

When applied to equation (7.2.4), these relations give

$$v = \left\{ \frac{(F_1 - F_3) \cos \lambda x + (F_1 + F_3) \cosh \lambda x - F_6 (\sin \lambda x + \sinh \lambda x)}{-2EI\lambda^2 F_3} \right\} He^{i\omega t}, \tag{7.2.27}$$

where the quantities F_1, F_3, etc., are defined in Table 7.1 (a) (p. 359). It is almost essential to introduce this extra notation because otherwise the expressions become too unwieldy. The quantity within the curly brackets will be seen to be the receptance $\alpha_{x0'}$.

By differentiating equation (7.2.27), the slope at any section may be obtained in the form

$$\frac{\partial v}{\partial x} = \left\{ \frac{F_6(\cos \lambda x + \cosh \lambda x) + (F_1 - F_3)\sin \lambda x - (F_1 + F_3)\sinh \lambda x}{2EI\lambda F_3} \right\} H e^{i\omega t}.$$

$$(7.2.28)$$

In this expression, the function within the brackets is the receptance $\alpha_{x'0'}$ for a free-free beam.

The receptances α_{x0} and $\alpha_{x'0}$ may be found in a similar way for a free-free beam; the boundary conditions which then apply are

$$\left. \begin{array}{l} \dfrac{\partial^2 v}{\partial x^2} = 0, \quad \dfrac{\partial^3 v}{\partial x^3} = \dfrac{F}{EI} e^{i\omega t} \quad \text{when} \quad x = 0, \\[3mm] \dfrac{\partial^2 v}{\partial x^2} = \dfrac{\partial^3 v}{\partial x^3} = 0 \qquad \text{when} \quad x = l. \end{array} \right\}$$

$$(7.2.29)$$

The system illustrating this is shown in fig. 7.2.6.

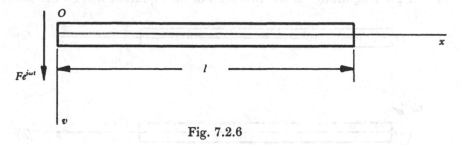

Fig. 7.2.6

Table 7.1 (b) (p. 360) is a list of receptances for uniform beams with various end-fixings, each of which may be found by methods similar to those used above. The notation is used which is given in Table 7.1 (a). The use of these receptances will be discussed in the next section.

It has been shown in previous chapters that it is useful to obtain receptances which refer to co-ordinates measured at the *ends* of strings, shafts and bars. Such receptances, which we shall refer to as 'tip' receptances, allow us to analyse systems which are built up by joining individual shafts, etc., by their ends. Examples of this type of problem will be found, for instance, at the end of § 6.2. Now it is also useful to obtain tip receptances for beams in flexure because they may be used in the analysis of systems, such as frames, which are formed by joining a number of beams together at their ends. A list of tip receptances for beams is given in Table 7.1 (c) (p. 363). The entries in this table are obtained by putting $x = 0$ or $x = l$ in the appropriate receptances of Table 7.1 (b) and they are included in a separate table for easy reference.

The frequency equation for any one of the beams to which Tables 7.1 (b) and 7.1 (c) refer may be found from the condition that the appropriate receptances shall become infinite. For instance, the natural frequencies of free-pinned and of clamped-pinned beams will be seen to be given by the condition $F_5 = 0$. That is, the natural frequencies of these types of beam are given by the equation

$$\tan \lambda l = \tanh \lambda l. \tag{7.2.30}$$

The nature of the roots of this type of frequency equation will be discussed later.

The double headings to the columns on Table 7.1 (c) arise by virtue of the reciprocal property of cross-receptances which makes, for instance, $\alpha_{0l'}$ equal to $\alpha_{l'0}$. Further examination of the table will show that two receptances of a beam are sometimes identical for a different reason—namely that of symmetry. Thus $\alpha_{00} = \alpha_{ll}$ and $\alpha_{0'0'} = \alpha_{l'l'}$ for the free-free beam; the physical meaning of these results should be evident to the reader.

It will be seen also from Table 7.1 (c) that, for the free-free beam,

$$\alpha_{0l'} = -\alpha_{l0'} \tag{7.2.31}$$

and it is important that the reader should appreciate the reason for the negative sign in this equation. It should be clear from the physical meaning of the receptances that the magnitudes of the two sides of the equation will be the same. The

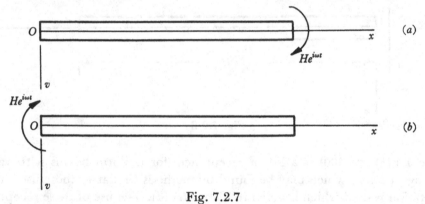

Fig. 7.2.7

minus sign arises because of our choice of slope and deflexion as co-ordinates. This may be seen by considering first the deflexions produced by a harmonic couple which is applied at the section $x = l$ of a free-free beam and secondly that due to an equal harmonic couple applied at the section $x = 0$; the two arrangements are shown in fig. 7.2.7 (a) and 7.2.7 (b). It will be seen that, if the couples deflect the beam equally then there will be a difference in sign of the deflexion in the two cases. If, for instance, the deflexion is positive at the section $x = 0$ in fig. 7.2.7 (a) at some specified instant, the deflexion at the section $x = l$ in fig. 7.2.7 (b) will be negative at this same instant if the two harmonic torques are in phase with one another. The reader may apply similar arguments to justify the presence of the negative sign in the relation

$$\alpha_{00'} = -\alpha_{ll'} \tag{7.2.32}$$

which holds for a free-free beam.

The change in sign is sometimes necessary when the end-conditions of a beam are changed over. Fig. 7.2.8 (a) shows a clamped-free beam, the clamp being on the left and the axis of x being drawn positive to the right. In fig. 7.2.8 (*b*), on the other hand, the beam has its right-hand end clamped while the direction of the

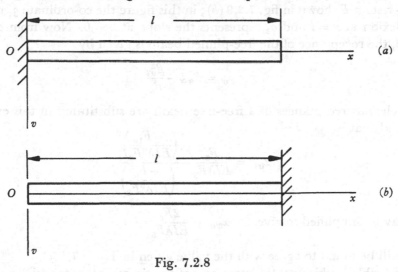

Fig. 7.2.8

x axis is the same as before. The cross-receptance α_{rl} for the beam of fig. 7.2.8 (*a*) is given by

$$\alpha_{rl} = \frac{F_1}{EI\lambda^2 F_4} \tag{7.2.33}$$

from Table 7.1 (*c*). The equivalent receptance for the beam of fig. 7.2.8 (*b*), namely $\alpha_{0'0}$ is given by

$$\alpha_{0'0} = \frac{-F_1}{EI\lambda^2 F_4}. \tag{7.2.34}$$

This result may be obtained by applying suitable boundary conditions to equation (7.2.4) as well as by direct argument.

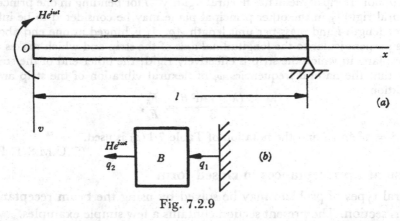

Fig. 7.2.9

Once the tip receptances of a free-free beam have been found (these being the functions in the first row of Table 7.1 (*c*)) the tip receptances for other types of beams may be deduced from them. This may be illustrated by the derivation of the receptance $\alpha_{0'0'}$ for a free-pinned beam. The beam is shown in fig. 7.2.9 (*a*) and it may be treated as a free-free beam with one of its co-ordinates (namely the deflexion at the section $x = l$) equal to zero. The free-free beam is represented by the sub-system B shown in fig. 7.2.9 (b); in this figure the co-ordinate q_1 represents the deflexion at $x = l$ and q_2 represents the slope at $x = 0$. Now from entry 7 of Table 2, the receptance of the free-pinned beam is given by

$$\alpha_{0'0'} = \beta_{0'0'} - \frac{\beta_{0'l}^2}{\beta_{ll}}. \qquad (7.2.35)$$

If the relevant receptances of a free-free beam are substituted in this expression, it is found that

$$\alpha_{0'0'} = \frac{F_6}{EI\lambda F_3} - \frac{\left(\dfrac{-F_{10}}{EI\lambda^2 F_3}\right)^2}{\left(\dfrac{-F_5}{EI\lambda^3 F_3}\right)}. \qquad (7.2.36)$$

This may be simplified to give $\qquad \alpha_{0'0'} = \dfrac{2F_2}{EI\lambda F_5} \qquad (7.2.37)$

which will be found to agree with the value given in Table 7.1 (c).

It is possible in this way to obtain any of the tip receptances which are given in Table 7.1 (c) from those which are given for the free-free beam. In doing this it is necessary to consider the free-free beam as a sub-system for which one, two or three co-ordinates may be fixed. It may be mentioned that the formulae became more complicated as more co-ordinates are fixed.

EXAMPLES 7.2

1. Deduce the tip receptance $\alpha_{0'0'}$ for a sliding-free beam using the receptances given for a free-free beam in Table 7.1 (c).

2. A uniform strip of metal has flexural rigidity EI for bending in one principal plane; the flexural rigidity in the other principal plane may be considered to be infinite. The strip is of length l and mass per unit length $A\rho$. It is hinged at one end about an axis which is perpendicular to the longitudinal axis of the strip and which makes an angle θ with the plane in which the strip is effectively rigid; the other end of the strip is free.

Show that the natural frequencies ω_r of flexural vibration of the strip are given by the equation

$$\frac{(\lambda l)^3 \tan^2 \theta}{3} = \frac{F_5}{F_4},$$

where $\lambda^4 = \omega^2 A\rho/EI$ and the notation of Table 7.1 (a) is used.

(C.U.M.S.T. Pt II, 1950)

7.3 Use of tip receptances in closed form

Several types of problem may be solved by using the beam receptances of the previous section. The present section contains a few simple examples.

Fig. 7.3.1 (a) shows a clamped-free beam to the free end of which is attached a disk. The mass of the disk is M and its moment of inertia for rotation about its axis is J. The axis of the disk passes through the end of the beam. The whole system may be regarded as being composed of the two sub-systems shown in fig. 7.3.1 (b).

The sub-systems B and C are linked at two co-ordinates if the attachment of the disk to the beam is rigid so that the former has the same rotational displacement as the tip of the cantilever. Let the linking co-ordinates be denoted by q_1 and q_2, where

$$q_1 = \text{deflexion at the end of the beam,}$$

$$q_2 = \text{slope at the end of the beam.}$$

The co-ordinates are indicated in fig. 7.3.1 (b) and the system is represented in the form of a block diagram in fig. 7.3.1 (c).

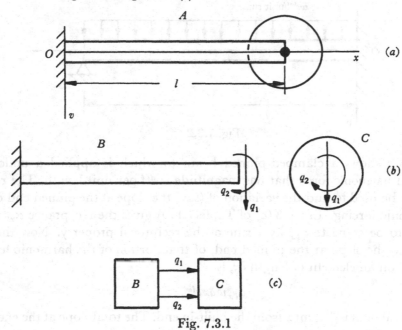

Fig. 7.3.1

Entry 9 in Table 3 gives the frequency equation, namely,

$$(\beta_{11} + \gamma_{11})(\beta_{22} + \gamma_{22}) - (\beta_{12} + \gamma_{12})^2 = 0. \qquad (7.3.1)$$

Now the receptances of B are found from Table 7.1 (c) to be

$$\beta_{11} = \frac{-F_5}{EI\lambda^3 F_4}, \quad \beta_{12} = \frac{F_1}{EI\lambda^2 F_4}, \quad \beta_{22} = \frac{F_6}{EI\lambda F_4}. \qquad (7.3.2)$$

The receptances of C are found by the method of §1.5 (see entry 1 of Table 1); they are

$$\gamma_{11} = -\frac{1}{M\omega^2}, \quad \gamma_{22} = -\frac{1}{J\omega^2}, \quad \gamma_{12} = 0. \qquad (7.3.3)$$

On substituting these values for the receptances β and γ in equation (7.3.1), it is found that

$$F_4^2 + \left(\frac{M\omega^2}{EI\lambda^3}\right) F_4 F_5 - \left(\frac{J\omega^2}{EI\lambda}\right) F_4 F_6 - \left(\frac{MJ\omega^4}{E^2 I^2 \lambda^4}\right)(F_1^2 + F_5 F_6) = 0 \qquad (7.3.4)$$

which is the frequency equation for the system.[†] It will be seen that if the constants of the system are specified numerically then the only unknown in equation (7.3.4) is ω so that the roots of the equation may be found arithmetically.

It is left to the reader to deduce the frequency equation for the system A if it were modified by arranging the disk so that it could turn freely about its axis without bending the beam. The two sub-systems would then be connected by one co-ordinate only. The reader may check his result by comparing it with the previous result when the latter is modified by putting $J = 0$.

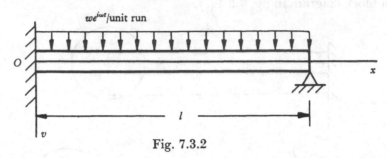

Fig. 7.3.2

Fig. 7.3.2 shows a clamped-pinned beam to which is applied a uniformly-distributed harmonic force that has magnitude $w\, e^{i\omega t}$ per unit length. The receptances may be used to find the variation of (say) the slope at the pinned end due to the harmonic forcing. Entry 3 (a) of Table 7.1 (b) gives the receptance $\alpha_{xl'}$ which we know to be equal to $\alpha_{l'x}$ by virtue of the reciprocal property. Now the contribution to the slope at the pinned end, of that portion of the harmonic loading which acts on an element of length δx, is

$$\alpha_{l'x}\, w\, \delta x\, e^{i\omega t},$$

where the element is distant x from the built-in end. The total slope at the end must therefore be

$$\left(\frac{\partial v}{\partial x}\right)_{x=l} = \frac{w\, e^{i\omega t}}{2EI\lambda^2 F_5}\int_0^l \left[F_8(\cos\lambda x - \cosh\lambda x) - F_{10}(\sin\lambda x - \sinh\lambda x)\right] dx. \quad (7.3.5)$$

This may be reduced to the result

$$\left(\frac{\partial v}{\partial x}\right)_{x=l} = -\frac{w(F_1 + F_{10})}{EI\lambda^3 F_5}\, e^{i\omega t}. \quad (7.3.6)$$

It has been shown by Duncan that free and forced vibrations of continuous beams may be analysed by the use of receptances.[‡] The following two simple examples illustrate a method of approach.

Fig. 7.3.3 (a) shows a uniform beam PQR which rests upon three supports, the two spans into which it is thereby divided being of unequal length. A receptance

† See W. J. Duncan, *Phil. Mag.* [7], vol. 32 (1941), p. 401.

‡ W. J. Duncan, *Phil. Mag.* [7], vol. 34 (1943), p. 49; see also *R. and M. 2000* (1947), p. 111. Duncan's treatment of continuous beams is based upon an extension of the 'equation of three moments' of structural theory and differs somewhat from the method outlined here.

method may be used to find the frequency equation for the beam. If the beam is divided at the support Q, it forms two sub-systems B and C which are the separate pinned-pinned beams PQ and QR. These sub-systems are linked by a single co-ordinate q_1 (see fig. 7.3.3 (b)), which is the slope at Q. The frequency equation for two systems linked in this way is

$$\beta_{11} + \gamma_{11} = 0 \tag{7.3.7}$$

and the two receptances are given in line 6 of Table 7.1 (c); they are

$$\left.\begin{aligned}
\beta_{11} &= \frac{-F_5}{2EI\lambda F_1} = \frac{\sin \lambda l_b \cosh \lambda l_b - \cos \lambda l_b \sinh \lambda l_b}{2EI\lambda \sin \lambda l_b \sinh \lambda l_b}, \\
\gamma_{11} &= \frac{-F_5}{2EI\lambda F_1} = \frac{\sin \lambda l_c \cosh \lambda l_c - \cos \lambda l_c \sinh \lambda l_c}{2EI\lambda \sin \lambda l_c \sinh \lambda l_c}.
\end{aligned}\right\} \tag{7.3.8}$$

The frequency equation in λ (which is equivalent to an equation in ω) may be obtained from these expressions and equation (7.3.7). It should be noted that, had the original beam been divided at an unsupported point, sub-systems would have been obtained which were linked by two co-ordinates instead of one.

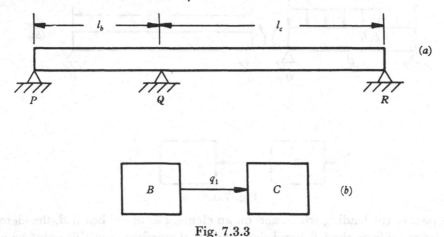

Fig. 7.3.3

Consider the particular case in which the spans of the system shown in fig. 7.3.3 (a) are of equal length l so that the system is symmetrical about the junction of the sub-systems. Free vibration of the system would then be possible without any bending moment occurring at the junction at any instant. This would mean that there would be no force between the sub-systems. If the expressions for β_{11} and γ_{11} are substituted in equation (7.3.7), the frequency equation is found to reduce to

$$F_5 = \cos \lambda l \sinh \lambda l - \sin \lambda l \cosh \lambda l = 0. \tag{7.3.9}$$

If this result is now compared with the entries in Table 7.1 (c), it will be found that it is the frequency equation of a clamped-pinned beam. The equation therefore will give only those natural frequencies of the system for which the modal shape is symmetrical; it will not give the frequencies corresponding to anti-symmetric modes in which the bending moment at the centre is zero. The missing frequencies are given by the equation

$$F_1 = \sin \lambda l \sinh \lambda l = 0 \tag{7.3.10}$$

because one-half of the beam may then be treated as a single system which has no force acting on it. This is an instance of theory which was discussed in § 1.9.

Fig. 7.3.4 (*a*) shows a continuous beam *PQR* which executes forced harmonic motion due to a distributed harmonic loading on one span. The motion may be found by first finding the bending moment which will be present at the central support. It is again convenient to divide the system at the intermediate support so that the sub-systems are linked by a single co-ordinate (see fig. 7.3.4 (*b*)); this co-ordinate is the slope at *Q* which may be denoted by q_2. Consider now the effect

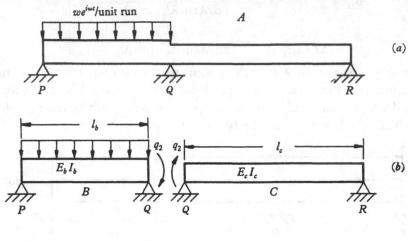

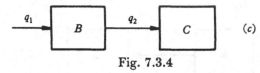

Fig. 7.3.4

of that part of the loading which acts on an element δx of the beam *B*, the element being distant *x* from the left-hand side, and let the deflexion at this point be q_1 so that the system may be represented as that of fig. 7.3.4 (*c*). Entry 1 of Table 2 (p. 40) shows that

$$\alpha_{12} = \beta_{12} - \frac{\beta_{12}\beta_{22}}{\beta_{22} + \gamma_{22}} = \frac{\beta_{12}\gamma_{22}}{\beta_{22} + \gamma_{22}}, \tag{7.3.11}$$

where

$$\beta_{22} = \frac{-F_{5b}}{2E_b I_b \lambda_b F_{1b}}, \quad \gamma_{22} = \frac{-F_{5c}}{2E_c I_c \lambda_c F_{1c}},$$

$$\left.\beta_{12} = \beta_{xl'} = \frac{(F_{7b} - F_{8b})\sin \lambda_b x - (F_{7b} + F_{8b})\sinh \lambda_b x}{-4E_b I_b \lambda_b^2 F_{1b}}.\right\} \tag{7.3.12}$$

In this expression the subscripts *b* and *c* distinguish quantities relating to the different spans.

The first two of these receptances are taken from line 6 of Table 7.1 (*c*) and the last is taken from entry 6 (*c*) of Table 7.1 (*b*). The contribution of the load $w\,\delta x\,e^{i\omega t}$ to q_2 is

$$\delta q_2 = \frac{\beta_{12}\gamma_{22}}{\beta_{22} + \gamma_{22}} w\,\delta x\,e^{i\omega t}. \tag{7.3.13}$$

This expression can now be integrated with respect to x giving

$$q_2 = \frac{w\gamma_{22}e^{i\omega t}}{\beta_{22}+\gamma_{22}}\int_0^{l_b}\beta_{12}\,dx = \left\{\frac{w\gamma_{22}(F_{6b}-F_{7b})}{2E_bI_b\lambda_b^3F_{1b}(\beta_{22}+\gamma_{22})}\right\}e^{i\omega t}. \qquad (7.3.14)$$

By considering system C alone it will be seen that the bending moment at the support Q is $M_c = q_2/\gamma_{22}$ so that†

$$M_c = \left\{\frac{w(F_{6b}-F_{7b})}{2E_bI_b\lambda_b^3F_{1b}(\beta_{22}+\gamma_{22})}\right\}e^{i\omega t}. \qquad (7.3.15)$$

Most of the receptances in closed form which we have used have been tip receptances; even in Table 7.1 (b), one of the two co-ordinates to which each

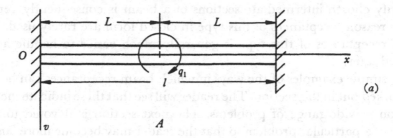

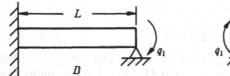

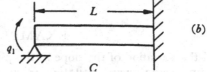

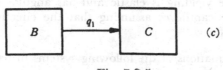

Fig. 7.3.5

receptance is referred is always measured at an extremity. This is not an essential restriction and receptances can be calculated at any section of a beam. This may be illustrated very simply by the determination of the direct receptance $\alpha_{L'L'}$ at the centre of a clamped-clamped beam of length $l = 2L$ (see fig. 7.3.5 (a)). The system may be divided into two similar sub-systems B and C which are clamped-pinned beams of length L as shown in fig. 7.3.5 (b). These are linked by a single co-ordinate q_1 which is the slope at the mid-point of the clamped-clamped beam. Entry 1 of Table 2 shows that

$$\alpha_{L'L'} \equiv \alpha_{11} = \frac{\beta_{11}\gamma_{11}}{\beta_{11}+\gamma_{11}} \qquad (7.3.16)$$

† This result will be found in Duncan's paper (1943), but his method of derivation differs from that given above.

and from line 3 of Table 7.1 (c), it is seen that

$$\beta_{11} = \gamma_{11} = \frac{F_3}{EI\lambda F_5}. \tag{7.3.17}$$

The required receptance is therefore

$$\alpha_{L'L'} = \frac{\beta_{11}}{2} = \frac{\cos \lambda L \cosh \lambda L - 1}{2EI\lambda(\cos \lambda L \sinh \lambda L - \sin \lambda L \cosh \lambda L)}. \tag{7.3.18}$$

The above problem is particularly simple because of the symmetry which ensures that the deflexion at the centre of the beam must always be zero. In general such symmetry will not be present and the calculation of a receptance between two arbitrarily chosen intermediate sections of a beam is consequently very tedious. For this reason receptances of this type in closed form are rarely used. The series form of receptances of this type is generally much easier to handle and will be described later.

Some simple examples of the ways in which beam receptances can be used have been worked out in this section. The reader will see that these indicate the method of attack on a wide range of problems. The next section is devoted to a detailed solution of a particular problem so that the reader may become more familiar with the numerical procedure for a real system.

EXAMPLES 7.3

1. Find the variation of the slope at the pinned end of the clamped-pinned beam of fig. 7.3.2 under the given excitation using equation (7.1.7) which contains w in the equation of motion. (Use a trial solution in the manner of §7.2.)

2. The clamped end of a clamped-free beam yields to allow rotation but not displacement of the root. The yielding is elastic and has angular stiffness k. Find the frequency equation for the cantilever assuming that the effective inertia associated with the yield is negligible.†

3. Find the frequency equations of the following systems in terms of the tabulated receptances of appropriate sub-systems:

   (a) a uniform continuous beam PQR resting on three simple (pinned) supports which divide the beam into two unequal spans PQ and QR, the supports P and R being rigid and Q being capable of deflexion with stiffness k;

   (b) a uniform beam which is clamped at each end, both clamps being capable of small elastic rotations in the plane of the beam with angular stiffness k but incapable of deflecting laterally; neglect inertia effects which are associated with rotation of the clamps;

   (c) an overhung uniform beam PQR which is pinned at P and Q and free at R and which has a connexion at Q which permits small local bending with angular stiffness k.

   Explain the notation used and describe in detail how the natural frequencies might be obtained from any one of these frequency equations.

(University College, London, 1958)

† This problem is discussed by W. J. Duncan, *R. and M. 2000* (1947), p. 110.

7.4 The numerical solution of beam problems

The examples which were discussed in §7.3 show how the tip receptances of beams can be used in analysis. Now the receptances are known in terms of the functions $F_1, F_2, ..., F_{10}$ which are defined in Table 7.1 (a). In numerical work it will be necessary to have tables of the functions available unless each of them is to be worked out, as required, from the appropriate trigonometric and hyperbolic functions. This procedure becomes very arduous and tabulated values of the functions F are therefore given in Table 7.1 (d) (p. 364). The range of values of the argument λl will be found sufficient to treat the vibration of any of the types of beam considered at frequencies up to and sometimes exceeding the third natural frequency. This range is found to be sufficient for most purposes.† The following example illustrates how the receptances, together with these tabulated values, may be used in numerical work.

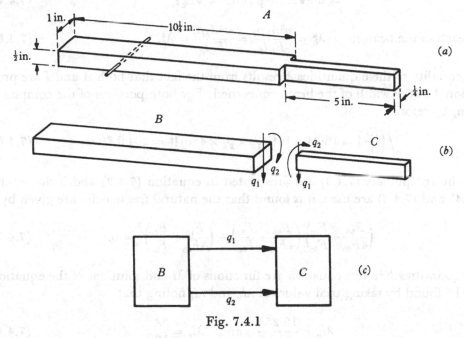

Fig. 7.4.1

Fig. 7.4.1 (a) shows a stepped beam which has free ends. Suppose that it is required to calculate the lowest natural frequency of flexural vibration in a plane parallel to those sides which have a depth of half an inch, assuming the material of the beam to be steel for which

$$E = 30 \times 10^6 \, \text{psi}, \quad \rho = 480 \, \text{lb./ft.}^3 \tag{7.4.1}$$

The beam may be split into the sub-systems B and C as shown in fig. 7.4.1 (b) so that the frequency equation is given by entry 9 of Table 3; it is

$$(\beta_{11} + \gamma_{11})(\beta_{22} + \gamma_{22}) - (\beta_{12} + \gamma_{12})^2 = 0. \tag{7.4.2}$$

† The authors are grateful to Miss J. Elliott who voluntarily undertook the computation of Table 7.1 (d).

The various receptances in this equation can be taken from the first line in Table 7·1 (c) and they are

$$\beta_{11} = \frac{-F_{5b}}{E_b I_b \lambda_b^3 F_{3b}}, \quad \beta_{22} = \frac{F_{6b}}{E_b I_b \lambda_b F_{3b}}, \quad \beta_{12} = \frac{F_{1b}}{E_b I_b \lambda_b^2 F_{3b}};$$

$$\gamma_{11} = \frac{-F_{5c}}{E_c I_c \lambda_c^3 F_{3c}}, \quad \gamma_{22} = \frac{F_{6c}}{E_c I_c \lambda_c F_{3c}}, \quad \gamma_{12} = \frac{-F_{1c}}{E_c I_c \lambda_c^2 F_{3c}}. \tag{7.4.3}$$

In these expressions, subscripts b and c have been added to the various symbols to distinguish between the two sub-systems. The values of the flexural rigidities of the beams are

$$E_b I_b = (30 \times 10^6 \times 144 \times 32) \frac{1}{12 \times 2^3 \times 144^2}$$

$$= 6 \cdot 94 \times 10^4 \, \text{pdl.ft.}^2 = 4 E_c I_c. \tag{7.4.4}$$

For each of the beams, $\lambda_b^4 = \dfrac{\omega^2 A_b \rho_b}{E_b I_b} = \dfrac{\omega^2 A_c \rho_c}{E_c I_c} = \lambda_c^4. \tag{7.4.5}$

The equality of these quantities λ results from the fact that both A and I are proportional to the width of the beam concerned. For both portions of the composite beam, therefore

$$\sqrt{\left(\frac{EI}{A\rho}\right)} = [6 \cdot 94 \times 10^4 / (\tfrac{1}{12} \times \tfrac{1}{24} \times 480)]^{\frac{1}{4}} = 204 \, \text{ft.}^2/\text{sec.} \tag{7.4.6}$$

If the receptances (7.4.3) are substituted in equation (7.4.2) and if the results (7.4.4) and (7.4.5) are used, it is found that the natural frequencies are given by

$$\left(\frac{F_{5b}}{4F_{3b}} + \frac{F_{5c}}{F_{3c}}\right)\left(\frac{F_{6b}}{4F_{3b}} + \frac{F_{6c}}{F_{3c}}\right) + \left(\frac{F_{1b}}{4F_{3b}} - \frac{F_{1c}}{F_{3c}}\right)^2 = 0. \tag{7.4.7}$$

The quantities F in this equation are functions of λl and solutions of the equation may be found by taking trial values of λl_b and λl_c noting that

$$\lambda l_b = \frac{10 \cdot 25 \lambda}{12} \quad \text{and} \quad \lambda l_c = \frac{5\lambda}{12}. \tag{7.4.8}$$

The subscript of λ has been discarded here in view of equation (7.4.5).

Consider the trial solution $\lambda l_b = 3 \cdot 50$. Table 7.1 (d) shows that

$$F_{1b} = -5 \cdot 8029, \quad F_{3b} = -16 \cdot 52,$$
$$F_{5b} = -9 \cdot 68, \quad F_{6b} = -21 \cdot 30. \tag{7.4.9}$$

The corresponding value of λl_c is

$$\lambda l_c = \frac{5}{10 \cdot 25} \lambda l_b = 1 \cdot 7073. \tag{7.4.10}$$

By interpolating in the table, it is found that

$$F_{1c} = \quad 2\cdot6414, \quad F_{3c} = -1\cdot3883,$$
$$F_{5c} = -3\cdot1848, \quad F_{6c} = \quad 2\cdot4574.$$

(7.4.11)

These values of the functions F make the left-hand side of equation (7.4.7) equal to

$$y = 0\cdot5135 \tag{7.4.12}$$

and this may be plotted against λ where

$$\lambda = \frac{3\cdot50 \times 12}{10\cdot25} = 4\cdot10\,\text{rad./ft.} \tag{7.4.13}$$

A curve may now be plotted for increasing values of λ and the first intersection of this curve with the axis of λ gives the first natural frequency of the composite system. In this problem the first intersection will be found to be at $\lambda = \lambda_1 = 3\cdot97$ rad./ft. This gives for the first natural frequency

$$\omega_1 = \lambda_1^2 \sqrt{\left(\frac{EI}{A\rho}\right)} = 3\cdot97^2 \times 204 = 3215\,\text{rad./sec.} \tag{7.4.14}$$

This corresponds to 511 cyc./sec. In a similar way the second intersection of the curve with the axis may be found, giving λ_2 and hence ω_2, and the process may be continued for the higher frequencies.

The calculation is straightforward and may be done by an inexperienced computer. It is desirable that a suitable calculating machine should be used.

The problem calculated above concerns an actual beam which was milled out of a single piece of metal and for which experimental results were available. They were obtained by inserting small pins in the beam as indicated in fig. 7.4.1 (a). The pins were placed at a point whose position was that of the calculated node of the system when oscillating in its first mode; the method by which this position was calculated will be discussed later. The bar was hung from the pins and the vibration was excited electromagnetically. The first natural frequency was found to be 495 cyc./sec. so that the calculated value is 3 % too high. The error is to be expected to be in this direction because of the sharp change in cross-section at the junction; this has the effect of interfering with the stress distribution which would be present at the end of the wide portion if the bending moment were applied to it over the whole of its end surface. To distribute the bending moment in this way, when it is being applied through the narrow cross-section of the other part of the beam, would require the introduction of additional constraint at the junction in order to keep the exposed face plane in accordance with simple bending theory. This constraint would raise the natural frequency so that it would become closer to the calculated value.

We will now find the modal shape of the system for its first frequency, using for this the calculated value rather than the experimental one. During free motion of the whole system in a principal mode, each sub-system executes forced vibration; by finding the deflexion shapes of the latter, the principal mode may be determined.

The receptances at the junction are given in line 1 of Table 7·1 (c) and their numerical values may be calculated for the frequency of 511 cyc./sec. (corresponding to $\lambda_b = \lambda_c = \lambda_1 = 3\cdot97$ rad./ft.). They are:

$$\left.\begin{aligned}
\beta_{11} &= -\ 0\cdot159 \times 10^{-6}\,\text{ft./pdl.}, \\
\beta_{12} &= +\ 0\cdot217 \times 10^{-6}\,\text{ft./pdl.ft.}, \\
\beta_{22} &= +\ 4\cdot25\ \times 10^{-6}\,\text{rad./pdl.ft.}, \\
\gamma_{12} &= +\ 7\cdot49\ \times 10^{-6}\,\text{ft./pdl.ft.}, \\
\gamma_{22} &= -29\cdot5\ \ \times 10^{-6}\,\text{rad./pdl.ft}
\end{aligned}\right\} \qquad (7.4.15)$$

Now the response of the sub-system B will be given by

$$\left.\begin{aligned}
q_1 &= \Psi_1 e^{i\omega_1 t} = [\beta_{11}\Phi_{1b} + \beta_{12}\Phi_{2b}]\,e^{i\omega_1 t}, \\
q_2 &= \Psi_2 e^{i\omega_1 t} = [\beta_{12}\Phi_{1b} + \beta_{22}\Phi_{2b}]\,e^{i\omega_1 t},
\end{aligned}\right\} \qquad (7.4.16)$$

where Φ_{1b} and Φ_{2b} are the amplitudes of the generalized forces corresponding to q_1 and q_2 respectively. The quantities Ψ_1 and Ψ_2 are the amplitudes of the co-ordinates q_1 and q_2.

In determining the modal shape, the amplitude Ψ_1 may be selected arbitrarily, and a convenient value here is $\Psi_1 = 10^{-6}$ ft. so that

$$\left.\begin{aligned}
\beta_{11}\Phi_{1b} + \beta_{12}\Phi_{2b} &= 10^{-6}, \\
\beta_{12}\Phi_{1b} + \beta_{22}\Phi_{2b} &= \Psi_2.
\end{aligned}\right\} \qquad (7.4.17)$$

The shape of the distortion of system B can now be determined if the force amplitudes Φ_{1b} and Φ_{2b} can be found. To do this it is necessary to consider also the sub-system C and so to find another expression for Ψ_2; in this way Ψ_2 can be eliminated from equation (7.4.17). For sub-system C,

$$\Psi_2 = \gamma_{12}\Phi_{1c} + \gamma_{22}\Phi_{2c} \qquad (7.4.18)$$

and in this expression the force amplitudes for C are related to those for B by the equilibrium conditions

$$\left.\begin{aligned}
\Phi_{1b} + \Phi_{1c} &= 0, \\
\Phi_{2b} + \Phi_{2c} &= 0.
\end{aligned}\right\} \qquad (7.4.19)$$

If Ψ_2 is now eliminated from equations (7.4.17) and (7.4.18) and equations (7.4.19) are used to get rid of Φ_{1c} and Φ_{2c}, it is found that

$$\left.\begin{aligned}
\beta_{11}\Phi_{1b} + \beta_{12}\Phi_{2b} &= 10^{-6}, \\
(\beta_{12} + \gamma_{12})\,\Phi_{1b} + (\beta_{22} + \gamma_{22})\,\Phi_{2b} &= 0.
\end{aligned}\right\} \qquad (7.4.20)$$

In this way, Φ_{1b} and Φ_{2b} may be calculated. They are

$$\left.\begin{aligned}
\Phi_{1b} &= -10\cdot7\ \ \text{pdl.}, \\
\Phi_{2b} &= -\ 3\cdot27\,\text{pdl.ft.}
\end{aligned}\right\} \qquad (7.4.21)$$

These values of the shearing force and bending couple at the junction of the sub-systems are those which correspond to free vibration of the complete system at the frequency of 511 cyc./sec. when the amplitude of deflexion at the step is 10^{-6} ft. During such motion therefore, the deflexion at any point of B is given by

$$v = [\beta_{xl}\Phi_{1b} + \beta_{xl'}\Phi_{2b}]e^{i\omega_1 t}, \tag{7.4.22}$$

where the axis along which x is measured is shown in fig. 7.4.2 and where the receptances are given in Table 7.1 (b). Substitution of the relevant expressions gives the amplitudes at any section of B as

$$v_{\text{max.}} = \left[\frac{F_8(\cos \lambda x + \cosh \lambda x) - F_{10}(\sin \lambda x + \sinh \lambda x)}{2EI\lambda^3 F_3}\right]\Phi_{1b}$$

$$+ \left[\frac{F_{10}(\cos \lambda x + \cosh \lambda x) + F_7(\sin \lambda x + \sinh \lambda x)}{2EI\lambda^2 F_3}\right]\Phi_{2b}, \tag{7.4.23}$$

where $\lambda = \lambda_1 = 3\cdot97$ rad./ft. The equation may be simplified by writing it in terms of the functions F. It then becomes

$$v_{\text{max.}} = \left[\frac{F_8(\lambda l) \cdot F_9(\lambda x) - F_{10}(\lambda l) \cdot F_7(\lambda x)}{2EI\lambda^3 \cdot F_3(\lambda l)}\right]\Phi_{1b}$$

$$+ \left[\frac{F_{10}(\lambda l) \cdot F_9(\lambda x) + F_7(\lambda l) \cdot F_7(\lambda x)}{2EI\lambda^2 \cdot F_3(\lambda l)}\right]\Phi_{2b}. \tag{7.4.24}$$

In this expression the arguments of the various functions have been written after them in parenthesis.

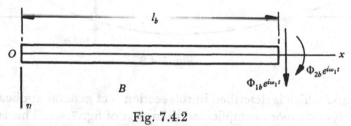

Fig. 7.4.2

Consider the amplitude at a point for which $x = 2$ in. or $\frac{1}{6}$ ft.—a point, that is, which is 8·25 in. from the step. The appropriate values of the arguments of the functions F are

$$\lambda l = 3\cdot97 \times \frac{10\cdot25}{12} = 3\cdot391,$$

$$\lambda x = 3\cdot97 \times \tfrac{2}{12} = 0\cdot662, \tag{7.4.25}$$

so that, at the section referred to,

$$v_{\text{max.}} = \left[\frac{(-15\cdot07)(2\cdot016) - (-15\cdot84)(1\cdot325)}{(2)(6\cdot94 \times 10^4)(3\cdot97)^3(-15\cdot40)}\right](-10\cdot7)$$

$$+ \left[\frac{(-15\cdot84)(2\cdot016) + (14\cdot58)(1\cdot325)}{(2)(6\cdot94 \times 10^4)(3\cdot97)^2(-15\cdot40)}\right](-3\cdot27)$$

$$= -1\cdot98 \times 10^{-6} \text{ ft.} \tag{7.4.26}$$

When making a calculation of this sort for a number of points along a beam, it is best to do the computation in tabular form. In this way, a curve of amplitude of displacement may be found and plotted for the beam B.

Equally, a curve showing the amplitude of displacement at points along C may be found since C is excited at its junction with B by (i) a shearing force of amplitude $\Phi_{1c} = +10.7$ pdl. and (ii) a bending couple of amplitude $\Phi_{2c} = +3.27$ pdl.ft. The frequency is, of course, the natural frequency of 511 cyc./sec.

The curves of deflexion for B and C may be joined together, the ordinate $\Psi'_1 = 10^{-6}$ ft. being common to both. This has been done for the stepped beam of fig. 7.4.1 and the curve is shown in fig. 7.4.3; the curve represents the first principal mode of the beam. As we stated earlier, small pins were inserted into opposite sides of the beam at a node of deflexion in order that the beam might be suspended at that point and the first natural frequency measured. The position of these pins is shown in fig. 7.4.3.

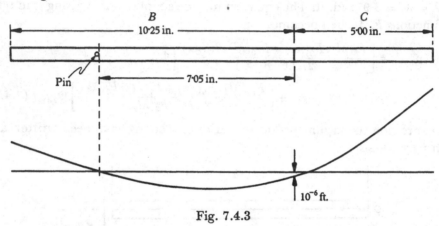

Fig. 7.4.3

The method which is described in this section is of general application and may be used for systems more complicated than that of fig. 7.4.1. This is illustrated in the following section where we shall take up the problem of frame vibration. The difficulties of computation become greater when more complicated systems are dealt with and, in particular, the calculation of modes quickly becomes tedious. The method of calculating natural frequencies, however, remains useful for a wide range of problems and can often be used by an inexperienced computer using a 'desk-calculator'.

EXAMPLES 7.4

1. A clamped-clamped beam of length 2 ft. is of rectangular cross-section, having depth $\frac{1}{4}$ in., and width 1 in., (cf. sub-system B of fig. 7.4.1). The material is steel for which

$$E = 30 \times 10^6 \text{ psi}, \quad \rho = 480 \text{ lb./ft.}^3$$

Calculate the cross-receptance between the deflexion and slope at a section distant 8 in. from the end $x = 0$ when the driving frequency is $\omega = 3600$ rad./sec.

2. A steel shaft 4 ft. long is carried at each end in bearings which effectively provide pinned support. The middle 1 ft. of its length has a diameter of 2 in. and, for a length of 1 ft. 6 in. at each end the diameter is 1 in. The shaft is of steel for which

$$E = 30 \times 10^6 \text{ psi}, \quad \rho = 480 \text{ lb./ft.}^3$$

Show that the frequency equation for symmetrical modes of flexural vibration can be written in the form

$$\left(\frac{2F_{1b}}{F_{5b}} - \frac{0 \cdot 3535 F_{2c}}{F_{6c}}\right)\left(\frac{2F_{2b}}{F_{5b}} + \frac{0 \cdot 1767 F_{1c}}{F_{6c}}\right) - \left(\frac{F_{6b}}{F_{5b}} + \frac{F_{5c}}{8F_{6c}}\right)^2 = 0,$$

where the subscript b refers to a shaft 1 ft. 6 in. long with diameter 1 in. and c refers to a shaft 6 in. long with diameter 2 in. Show further that the natural frequencies of antisymmetrical modes are given by the frequency equation

$$\left(\frac{2F_{1b}}{F_{5b}} + \frac{0 \cdot 3535 F_{1c}}{F_{5c}}\right)\left(\frac{2F_{2b}}{F_{5b}} + \frac{0 \cdot 1767 F_{2c}}{F_{5c}}\right) - \left(\frac{F_{6b}}{F_{5b}} - \frac{F_{6c}}{8F_{5c}}\right)^2 = 0.$$

Calculate the first three natural frequencies of flexural vibration of the shaft.

3. A steel shaft, whose span is 3 ft., is supported in long bearings that fully restrain the direction of its ends. The central length of 1 ft. of the shaft has diameter $\frac{3}{8}$ in., while, at each end, there is a length of 1 ft. having $\frac{1}{4}$ in. diameter.

The lowest natural frequency of flexural vibration is known to be less than 100 cyc./sec. and it is required to determine what natural frequencies this shaft possesses in the range 100–200 cyc./sec. Describe in detail a procedure for finding this information, using the data given to illustrate the method as far as possible. Actual calculation of these possible frequencies is *not* required.

For steel, Young's modulus = 30×10^6 psi, density = 480 lb./ft.³ (C.U.M.S.T. Pt II, 1957)

7.5 Frame vibration

It has now been shown how the vibration characteristics of a beam, which is composed of two or more uniform segments, may be found by using the receptances of the separate segments. There is no need, however, to restrict this technique to those cases in which the segments are joined to form a single straight member. Two of them are shown joined at right angles in fig. 7.5.1 (*a*), the two portions forming between them, a bent cantilever.

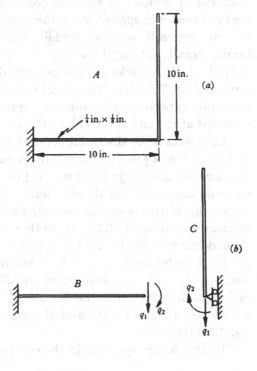

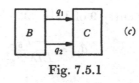

Fig. 7.5.1

Before discussing the calculations for this system, we note that the arrangement shown in fig. 7.5.1 (*a*) represents the first step in building up other types of system from straight members, and that there is no limit to the complexity of such systems.

It will become clear, on the other hand, that there are severe practical limitations on the extent to which our methods of analysis may be applied to systems which have a large number of members. It should be noted, in particular, that as the number of members increases, and as the number of ways in which they may be oriented relative to each other thereby increase, new terms have to be included in the equations of motion. Thus in going from the straight beam to the bent cantilever, the additional effect must be included of longitudinal motion of the member that is remote from the clamp, even if the discussion is confined to vibration in the plane of the cantilever. If the vibrations out of that plane are to be examined, then the torsional characteristics of the clamped member must be taken into account. Further, in analysing the motions of, say, a cantilever which is built up from three uniform beams, all the modes may, in general, involve torsion as well as bending if the bars are not coplanar. If, finally, systems are examined in which the uniform members are sufficient in number to be arranged to form closed polygons, as in a two-storey portal frame or in space frames, then torsional and flexural motion of all the members has generally to be allowed for. The process of setting up the equations is not fundamentally altered, however.

With these remarks on the general theory, we now return to the particular system of fig. 7.5.1 (a). Numerical values are given for the constants of the system and some experimental results for a system having the dimensions in question are included at the end of the calculations.

The system to be discussed is made from a steel strip 20 in. long and having a section of $\frac{1}{4}$ in. $\times \frac{1}{8}$ in.; a sharp bend is made in the strip at its mid-point, the bend being about an axis parallel to the $\frac{1}{4}$ in. sides. Only the vibrations of the strip in its own plane will be dealt with here.

For the purposes of calculation, the strip may be divided into the two subsystems shown in fig. 7.5.1 (b), so that B is a clamped-free beam and C is a free-pinned beam which is capable of longitudinal motion. This latter may be treated as 'rigid body' motion; that is to say it is assumed that the natural frequencies with which we shall be concerned are all much lower than the natural frequencies of longitudinal vibration of the bar C. The linking co-ordinates, q_1 and q_2, are indicated in fig. 7.5.1 (b) and the complete system is shown in block form in fig. 7.5.1 (c).

The frequency equation for the composite system is

$$(\beta_{11} + \gamma_{11})(\beta_{22} + \gamma_{22}) - (\beta_{12} + \gamma_{12})^2 = 0 \qquad (7.5.1)$$

this being entry 9 of Table 3. The relevant receptances are

$$\left. \begin{aligned} \beta_{11} &= \frac{-F_5}{EI\lambda^3 F_4}, \quad \beta_{22} = \frac{F_6}{EI\lambda F_4}, \quad \beta_{12} = \frac{F_1}{EI\lambda^2 F_4}; \\ \gamma_{11} &= \frac{-1}{A\rho l\omega^2}, \quad \gamma_{22} = \frac{F_4}{EI\lambda F_5}, \quad \gamma_{12} = 0. \end{aligned} \right\} \qquad (7.5.2)$$

In writing these expressions, no distinguishing suffixes have been added to $A, E, I, l,$ λ and ρ because all these quantities are the same for the two beams. Further, since the argument λl of the functions F is the same for each beam, these too carry no letter subscripts.

When the receptances have been substituted in equation (7.5.1) the resulting equation may be simplified by multiplying throughout by $EI\lambda^4$. This gives

$$\left(\frac{F_5}{F_4} + \frac{1}{\lambda l}\right)\left(\frac{F_6}{F_4} + \frac{F_4}{F_5}\right) + \left(\frac{F_1}{F_4}\right)^2 = 0 \qquad (7.5.3)$$

which is the frequency equation. This may be solved by taking trial values of λl and plotting the expression on the left-hand side against λl. In this way it is found that the first four roots are

$$\lambda_1 l = 1\cdot08, \quad \lambda_2 l = 1\cdot78, \quad \lambda_3 l = 3\cdot96, \quad \lambda_4 l = 4\cdot80. \qquad (7.5.4)$$

Instead of calculating $EI/A\rho$ directly from the values of the various constants for the material, this quantity was determined experimentally from a vibration test on a straight cantilever of the same section as the strip. In this way it was found that

$$\sqrt{\left(\frac{EI}{A\rho}\right)} = 7130 \,\text{in.}^2/\text{sec.} \qquad (7.5.5)$$

The first four natural frequencies of a bent cantilever made of the material used in this test should therefore be

$$\omega_1 = \frac{(\lambda_1 l)^2}{l^2}\sqrt{\left(\frac{EI}{A\rho}\right)} = \frac{1\cdot08^2 \times 7130}{10^2} = 83\cdot2\,\text{rad./sec. or } 13\cdot2\,\text{cyc./sec.,}$$

$$\omega_2 = \frac{1\cdot78^2 \times 7130}{10^2} = 226\,\text{rad./sec. or } 36\,\text{cyc./sec.,}$$

$$\omega_3 = \frac{3\cdot96^2 \times 7130}{10^2} = 1118\,\text{rad./sec. or } 178\,\text{cyc./sec.,}$$

$$\omega_4 = \frac{4\cdot80^2 \times 7130}{10^2} = 1643\,\text{rad./sec. or } 267\,\text{cyc./sec.}$$

$$(7.5.6)$$

The corresponding experimental results were

$$13\cdot2, \quad 36, \quad 177, \quad 264\,\text{cyc./sec.} \qquad (7.5.7)$$

so that errors in the calculated figures, which might arise from lack of exact correspondence between the real and assumed end conditions, are extremely small.

The principal modes may now be found. The process is slightly shorter than was the case with the stepped beam which was used in the previous section; the simplification arises because the cross-receptance γ_{12} happens to be zero. Thus one equation of motion for C is simply

$$\Psi_1 = \gamma_{11}\Phi_{1c}. \qquad (7.5.8)$$

If Ψ_1 is now given the value 1 ft. and the expression for γ_{11} which is given in equation (7.5.2) is used, it is found that

$$\Phi_{1c} = -A\rho l\omega^2 = -EI\lambda^4 l. \qquad (7.5.9)$$

Since there is no external force acting on the composite system,

$$\Phi_{1b} + \Phi_{1c} = 0, \qquad (7.5.10)$$

whence

$$\Phi_{1b} = EI\lambda^4 l. \qquad (7.5.11)$$

The equation of motion of B provides another expression for Ψ'_1, namely

$$\Psi'_1 = \beta_{11}\Phi_{1b} + \beta_{12}\Phi_{2b}. \qquad (7.5.12)$$

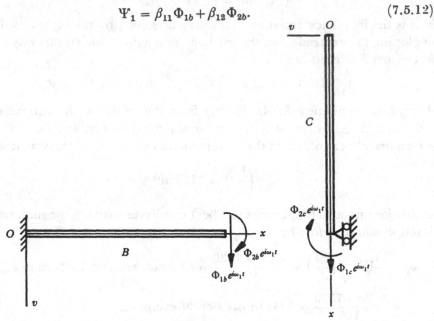

Fig. 7.5.2

From this, the value of Φ_{2b} can be found, since Ψ'_1 is unity and the receptances β_{11} and β_{12} are given in equation (7.5.2); thus

$$\Phi_{2b} = \frac{EI\lambda^2}{F_1}(F_4 + F_5\lambda l). \qquad (7.5.13)$$

Again,

$$\Phi_{2c} = -\Phi_{2b}. \qquad (7.5.14)$$

Equations (7.5.10), (7.5.11), (7.5.13) and (7.5.14) give the amplitudes of all the forces which act on the sub-systems through their junction when the composite system vibrates freely with unit amplitude of displacement at the right-angle bend. The numerical values of the forces may be found, for any particular mode, by using the appropriate value of λ. Thus in the first mode (for which $\lambda l = 1\cdot08$), with

$$EI = (30 \times 10^6 \times 32 \times 144) \times \frac{0\cdot25 \times 0\cdot125^3}{12^5} = 271 \text{ pdl.ft.}^2 \qquad (7.5.15)$$

it is found that

$$\left.\begin{array}{l} \Phi_{1b} = -\Phi_{1c} = 637 \text{ pdl.}, \\ \Phi_{2b} = -\Phi_{2c} = 345 \text{ pdl.ft.} \end{array}\right\} \qquad (7.5.16)$$

From these forces the principal mode of the composite system is found in a straightforward way, as follows. The amplitude of displacement at any section of B, distant x from the fixed end, is

$$v_{\text{max.}} = \beta_{xl}\Phi_{1b} + \beta_{xl'}\Phi_{2b} \tag{7.5.17}$$

(see fig. 7.5.2). This becomes, on substitution of the expressions for the receptances

$$v_{\text{max.}} = \left[\frac{F_7(\lambda l) \cdot F_{10}(\lambda x) - F_9(\lambda l) \cdot F_8(\lambda x)}{-2EI\lambda^3 F_4(\lambda l)}\right]\Phi_{1b}$$

$$+ \left[\frac{F_9(\lambda l) \cdot F_{10}(\lambda x) + F_8(\lambda l) \cdot F_8(\lambda x)}{-2EI\lambda^2 F_4(\lambda l)}\right]\Phi_{2b} \tag{7.5.18}$$

in which the value $\lambda l = 1\cdot08$ must be used in this case. Similarly the amplitude of displacement perpendicular to the axis of the beam C, of a point distant x from the end as indicated in fig. 7.5.2, will be

$$v_{\text{max.}} = \gamma_{xl'}\Phi_{2c}, \tag{7.5.19}$$

where $\gamma_{xl'}$ is a receptance of a free-pinned beam. On substituting the expression for the receptance, it is found that this relation becomes

$$v_{\text{max.}}$$
$$= \left[\frac{F_7(\lambda l) \cdot F_9(\lambda x) - F_9(\lambda l) \cdot F_7(\lambda x)}{-2EI\lambda^2 F_5(\lambda l)}\right]\Phi_{2c} \tag{7.5.20}$$

in which, again, $\lambda l = 1\cdot08$. The whole beam C, in addition to these transverse displacements, is subject to a longitudinal displacement equal to the transverse displacement of the end of the beam B; this is the displacement Ψ_1 which, for the purpose of calculation was arbitrarily taken as 1 ft. The first and second principal modes of the system A, calculated in this way, are shown in fig. 7.5.3 (a) and (b) respectively.

As a second example on frames, we shall examine the vibrations of a symmetrical clamped portal frame within its own plane. In general, the three members of a portal frame have to be treated as three double-linked sub-systems; but the symmetrical portal can be treated as two sub-systems only, by forming the equations for the symmetrical and anti-symmetrical vibrations separately. The symmetrical modes will be those in which there is no sideways

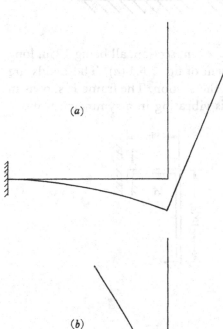

(a)

(b)

Fig. 7.5.3

motion of the tops of the stanchions and in which the centre of the top member remains horizontal. The anti-symmetrical or 'sway' vibrations, are those in which there is no vertical displacement, and no bending moment, at the centre of the top member. Sketches of the first symmetrical and first anti-symmetrical modes are shown in figs. 7.5.4 (*a*) and (*b*) respectively.

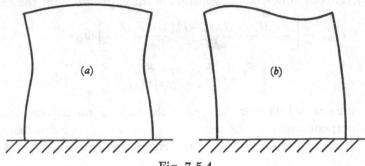

Fig. 7.5.4

The symmetrical frame has members of $\frac{1}{4}$ in. $\times \frac{1}{8}$ in. section, all being 10 in. long and of the same steel as was used for the system of fig. 7.5.1 (*a*). The bends are made about axes parallel to the longer sides of the section. The frame is shown in fig. 7.5.5 (*a*) and it will be seen that when it is vibrating in a symmetrical mode

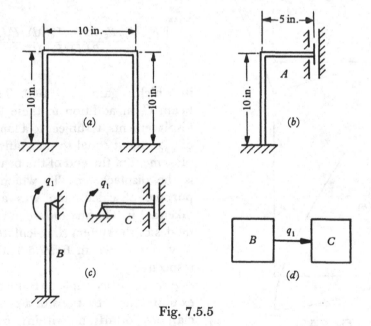

Fig. 7.5.5

each half of it will have the same modal form as the isolated half-frame shown in fig. 7.5.5 (*b*). In the latter, the constraint which would be provided by the other half of the frame is supplied instead by a guide. The half-frame can be divided into the sub-systems *B* and *C* shown in fig. 7.5.5 (*c*) one being a clamped-pinned beam, 10 in. long, and the other a pinned-sliding beam of length 5 in. *B* and *C* are connected by a single co-ordinate q_1 only, this being the slope of the beams at their junction. The other linking co-ordinate can be disregarded because it is the displacement which is zero for both sub-systems and is looked after by specifying pinned ends as above.

The frequency equation for the system, whose block representation is shown in fig. 7.5.5 (d), is

$$\beta_{11} + \gamma_{11} = 0. \qquad (7.5.21)$$

The receptances β_{11} and γ_{11} are given in Table 7.1 (c) and are

$$\beta_{11} = \frac{F_{3b}}{E_b I_b \lambda_b F_{5b}}, \quad \gamma_{11} = \frac{F_{6c}}{2E_c I_c \lambda_c F_{2c}}. \qquad (7.5.22)$$

In these expressions $E_b I_b$ and $E_c I_c$ are identical, as are also λ_b and λ_c. Equation (7.5.21) therefore reduces to

$$\frac{F_{3b}}{F_{5b}} + \frac{F_{6c}}{2F_{2c}} = 0. \qquad (7.5.23)$$

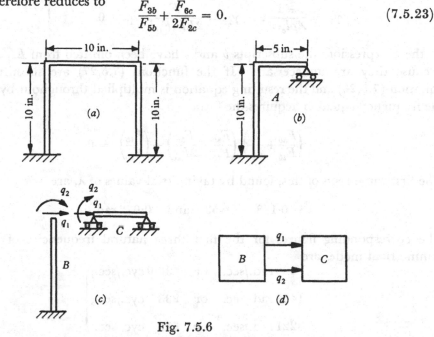

Fig. 7.5.6

By taking trial values of λ, and hence calculating the arguments λl_b and λl_c of the functions F, the left-hand side of equation (7.5.23) may be plotted and the roots of the equation found. In this way the first three roots

$$\lambda = 0\cdot357, \quad 0\cdot473 \quad \text{and} \quad 0\cdot744 \,\text{rad./in.} \qquad (7.5.24)$$

have been obtained. The corresponding natural frequencies are

$$0\cdot357^2 \sqrt{\left(\frac{EI}{A\rho}\right)} = 0\cdot357^2 \times 7130 = \left. \begin{array}{l} 909 \,\text{rad./sec.} \ \text{or} \ 145 \,\text{cyc./sec.,} \\ 0\cdot473^2 \times 7130 = 1595 \,\text{rad./sec.} \ \text{or} \ 256 \,\text{cyc./sec.,} \\ 0\cdot744^2 \times 7130 = 3948 \,\text{rad./sec.} \ \text{or} \ 628 \,\text{cyc./sec.} \end{array} \right\} \qquad (7.5.25)$$

When the portal frame is vibrating in one of its anti-symmetrical modes, each half of it can be treated as if it is an isolated half-frame having the form shown in fig. 7.5.6 (b). This half-frame may be divided into two sub-systems B and C

as in fig. 7.5.6 (c) and these are double-linked as shown in the block diagram (fig 7.5.6 (d)). One of the sub-systems is a clamped-free beam of length 10 in. and the other is a pinned-pinned beam of length 5 in., which is free to move longitudinally. The natural frequencies of the composite system are given by

$$(\beta_{11} + \gamma_{11})(\beta_{22} + \gamma_{22}) - (\beta_{12} + \gamma_{12})^2 = 0 \qquad (7.5.26)$$

and the appropriate receptances are

$$\beta_{11} = \frac{-F_{5b}}{EI\lambda^3 F_{4b}}, \quad \beta_{22} = \frac{F_{6b}}{EI\lambda F_{4b}}, \quad \beta_{12} = \frac{F_{1b}}{EI\lambda^2 F_{4b}};$$

$$\gamma_{11} = \frac{-1}{A\rho l_c \omega^2}, \quad \gamma_{22} = \frac{-F_{5c}}{2EI\lambda F_{1c}}, \quad \gamma_{12} = 0. \qquad (7.5.27)$$

In these expressions the subscripts b and c have been omitted from EI, $A\rho$ and λ because they are unnecessary. If the functions (7.5.27) are substituted into equation (7.5.26) and the resulting equation is multiplied throughout by $(EI)^2 \lambda^4$, the frequency equation acquires the form

$$\left(\frac{F_{5b}}{F_{4b}} + \frac{1}{\lambda l_c}\right)\left(\frac{F_{6b}}{F_{4b}} - \frac{F_{5c}}{2F_{1c}}\right) + \left(\frac{F_{1b}}{F_{4b}}\right)^2 = 0. \qquad (7.5.28)$$

The first three roots of this, found by taking trial values of λ, are

$$\lambda = 0\cdot178, \quad 0\cdot453 \quad \text{and} \quad 0\cdot672 \text{ rad./in.} \qquad (7.5.29)$$

The corresponding figures for the first three natural frequencies of the anti-symmetrical modes are

$$
\left.
\begin{array}{llll}
226 \text{ rad./sec.} & \text{or} & 36\cdot0 \text{ cyc./sec.,} \\
1463 \text{ rad./sec.} & \text{or} & 233 \quad \text{cyc./sec.,} \\
3221 \text{ rad./sec.} & \text{or} & 513 \quad \text{cyc./sec.}
\end{array}
\right\} \qquad (7.5.30)
$$

These results, together with those for the symmetrical modes, are given in the accompanying table by the side of the measured frequencies. It will be seen that the differences between the two columns are small.

NATURAL FREQUENCIES OF A SYMMETRICAL PORTAL FRAME

Calculated (cyc./sec.)	Measured (cyc./sec.)
36	36
145	145
233	233
256	254
513	517
628	626

We now turn to an unsymmetrical portal frame, for which the frequency calculation is somewhat longer. The frame which will be discussed is shown in fig. 7.5.7 (a). It is of mild steel of cross-section $\frac{1}{4}$ in. $\times \frac{1}{8}$ in., this material being from the same stock as before, and the lengths of the three members are 8, 10 and 12 in., arranged as shown in the figure. The bends are made about axes parallel to the longer sides of the section.

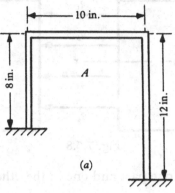

(a)

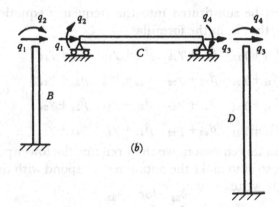

(b)

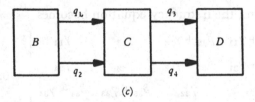

(c)

Fig. 7.5.7

The frame may be split up into three sub-systems B, C and D as shown in fig. 7.5.7 (b), these sub-systems being double-linked (as indicated in the block diagram) by the co-ordinates q_1, q_2, q_3 and q_4 (fig. 7.5.7 (c)). The frequency equation for a set of three sub-systems, linked in this way may be deduced as a special case of two sub-systems which are linked by four co-ordinates. This is illustrated in fig. 7.5.8 where the sub-system B has been divided by a line xx, into an upper and a lower portion. This line is to indicate that B comprises two systems which are not directly linked at all. The upper part is linked to C by q_1 and q_2 and the lower part is linked to C by q_3 and q_4. The division of B in this way requires that all cross-receptances

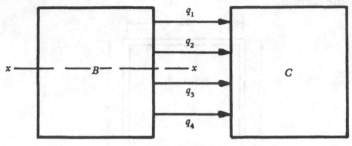

Fig. 7.5.8

between one co-ordinate in one part and one in the other shall be zero; that is to say, it is necessary that

$$\beta_{13} = \beta_{14} = \beta_{23} = \beta_{24} = 0. \tag{7.5.31}$$

These relations must be substituted into the frequency equation of entry 10 in Table 3 when $j = 4$, that is, in the formula

$$\begin{vmatrix} \beta_{11}+\gamma_{11} & \beta_{12}+\gamma_{12} & \beta_{13}+\gamma_{13} & \beta_{14}+\gamma_{14} \\ \beta_{21}+\gamma_{21} & \beta_{22}+\gamma_{22} & \beta_{23}+\gamma_{23} & \beta_{24}+\gamma_{24} \\ \beta_{31}+\gamma_{31} & \beta_{32}+\gamma_{32} & \beta_{33}+\gamma_{33} & \beta_{34}+\gamma_{34} \\ \beta_{41}+\gamma_{41} & \beta_{42}+\gamma_{42} & \beta_{43}+\gamma_{43} & \beta_{44}+\gamma_{44} \end{vmatrix} = 0. \tag{7.5.32}$$

When substituting in this equation, we shall rename the lower portion of system B as the sub-system D, so as to make the notation correspond with that of fig. 7.5.7 (c). We shall accordingly write

$$\delta_{33} \quad \text{for} \quad \beta_{33},$$
$$\delta_{34} \quad \text{for} \quad \beta_{34},$$
$$\delta_{44} \quad \text{for} \quad \beta_{44}.$$

With these modifications, the frequency equation becomes

$$\begin{vmatrix} \beta_{11}+\gamma_{11} & \beta_{12}+\gamma_{12} & \gamma_{13} & \gamma_{14} \\ \beta_{21}+\gamma_{21} & \beta_{22}+\gamma_{22} & \gamma_{23} & \gamma_{24} \\ \gamma_{31} & \gamma_{32} & \delta_{33}+\gamma_{33} & \delta_{34}+\gamma_{34} \\ \gamma_{41} & \gamma_{42} & \delta_{43}+\gamma_{43} & \delta_{44}+\gamma_{44} \end{vmatrix} = 0. \tag{7.5.33}$$

In this problem the determinant (7.5.33) simplifies slightly because the system C, which represents the horizontal member, has no coupling between its longitudinal and its flexural motion. Consequently

$$\gamma_{12} = \gamma_{14} = \gamma_{23} = \gamma_{34} = 0. \tag{7.5.34}$$

Those receptances which relate to flexural motion can be found in Table 7.1 (c); they are

$$\left.\begin{array}{lll}
\beta_{11} = \dfrac{-F_{5b}}{EI\lambda^3 F_{4b}}, & \beta_{22} = \dfrac{F_{6b}}{EI\lambda F_{4b}}, & \beta_{12} = \dfrac{F_{1b}}{EI\lambda^2 F_{4b}}; \\[2mm]
\gamma_{22} = \gamma_{44} = \dfrac{-F_{5c}}{2EI\lambda F_{1c}}, & \gamma_{24} = \dfrac{F_{8c}}{2EI\lambda F_{1c}}; & \\[2mm]
\delta_{33} = \dfrac{-F_{5d}}{EI\lambda^3 F_{4d}}, & \delta_{44} = \dfrac{F_{6d}}{EI\lambda F_{4d}}, & \delta_{34} = \dfrac{F_{1d}}{EI\lambda^2 F_{4d}}.
\end{array}\right\} \quad (7.5.35)$$

The receptances β and δ are those for clamped-free beams and the receptances γ are for a pinned-pinned beam. The receptances which relate to longitudinal motion of C are

$$\gamma_{11} = \gamma_{33} = \gamma_{13} = \frac{-1}{A\rho l_c \omega^2}. \quad (7.5.36)$$

When all the receptances are substituted in the determinant, the resulting expression can be simplified by multiplying each row by EI. The frequency equation then becomes

$$\begin{vmatrix}
\left(\dfrac{-F_{5b}}{\lambda^3 F_{4b}} - \dfrac{1}{\lambda^4 l_c}\right) & \dfrac{F_{1b}}{\lambda^2 F_{4b}} & -\dfrac{1}{\lambda^4 l_c} & 0 \\[3mm]
\dfrac{F_{1b}}{\lambda^2 F_{4b}} & \left(\dfrac{F_{6b}}{\lambda F_{4b}} - \dfrac{F_{5c}}{2\lambda F_{1c}}\right) & 0 & \dfrac{F_{8c}}{2\lambda F_{1c}} \\[3mm]
-\dfrac{1}{\lambda^4 l_c} & 0 & \left(\dfrac{-1}{\lambda^4 l_c} - \dfrac{F_{5d}}{\lambda^3 F_{4d}}\right) & \dfrac{F_{1d}}{\lambda^2 F_{4d}} \\[3mm]
0 & \dfrac{F_{8c}}{2\lambda F_{1c}} & \dfrac{F_{1d}}{\lambda^2 F_{4d}} & \left(\dfrac{-F_{5c}}{2\lambda F_{1c}} + \dfrac{F_{6d}}{\lambda F_{4d}}\right)
\end{vmatrix} = 0. \quad (7.5.37)$$

This equation can now be solved by taking trial values for λ, and it is convenient to leave it in the determinantal form during this process. In this way, the first six natural frequencies of the unsymmetrical frame of fig. 7.5.7 (a) have been calculated, using the value (7.5.5) for the elastic constants. The values of these frequencies, together with those which were found experimentally, are shown below.

NATURAL FREQUENCIES OF AN UNSYMMETRICAL PORTAL FRAME

Calculated (cyc./sec).	Measured (cyc./sec.)
38	39
134	135
194	197
346	346
439	436
606	598

EXAMPLE 7.5

1. A rectangular portal frame $ABCD$ is clamped at A and D. The stanchions AB and DC each have flexural rigidity 0.625×10^6 lb.wt.in.^{2} and the top member BC is rigid. AB and DC are both 30 in. long and BC has a mass of 40 lb. Show that, if each stanchion has a uniformly distributed mass of 9 lb. and if the effect of axial load on the stiffness of the stanchions can be neglected, then the frequency equation for vibration of this frame in its own plane is

$$\frac{F_3}{F_6} + \frac{3}{200\lambda} = 0,$$

where λ is in rad./in. and the argument of F_3 and F_6 is 30λ. Find the first two natural frequencies.

7.6 Natural frequencies; principal modes; characteristic functions; orthogonality

The modal shapes of uniform beams which have various types of end-fixings will now be dealt with. The types to be discussed are those for which the receptances are given in Tables 7.1 (b) and (c). The principal modes of the clamped-clamped beam will also be included in the treatment although this type of beam has no tip receptances so that it does not appear in Tables 7.1 (b) or (c).

7.6.1 Pinned-pinned beam

The form of the principal modes, and the values of the natural frequencies, are particularly simple to find for this type of beam. All the receptances of Tables 7.1 (b) and (c) which refer to a pinned-pinned beam become infinite if

$$F_1 = \sin \lambda l \sinh \lambda l = 0. \tag{7.6.1}$$

This is the frequency equation and its roots can be seen to be $\lambda l = \pi, 2\pi, 3\pi, \dots$ so that the rth root is given by

$$\lambda_r l = r\pi. \tag{7.6.2}$$

Now the natural frequencies are related to these roots by equation (7.2.5), namely

$$\lambda^4 = \frac{\omega^2 A \rho}{EI}. \tag{7.6.3}$$

Thus for all uniform beams $$\omega_r^2 = \frac{EI(\lambda_r l)^4}{A \rho l^4} \tag{7.6.4}$$

and for the pinned-pinned beam in particular

$$\omega_r^2 = \frac{r^4 \pi^4 EI}{A \rho l^4}. \tag{7.6.5}$$

The principal mode corresponding to a particular natural frequency ω_r can now be found by using the method of §4.3. In that method, one particular form of excitation is chosen, which in this problem can conveniently be a couple $H e^{i\omega t}$

applied at the end of the beam where $x = 0$, as shown in fig. 7.6.1. For this excitation, the deflexion at any section is given by entry 6 (a) of Table 7.1 (b); it is

$$v = \alpha_{x0'} H e^{i\omega t}$$

$$= \left\{ \frac{2F_1(\cos \lambda x - \cosh \lambda x) - (F_5 + F_6) \sin \lambda x - (F_5 - F_6) \sinh \lambda x}{4EI\lambda^2 F_1} \right\} H e^{i\omega t}. \quad (7.6.6)$$

If ω is now made to approach a natural frequency (ω_r say) and at the same time H is diminished indefinitely, so that the steady-state amplitude remains finite, then equation (7.6.6) will give the modal shape for free vibration at the relevant natural frequency. In this way it is found that free motion in the rth principal mode is of the form

$$v_r = \text{(constant)} \sin \lambda_r x \cdot e^{i\omega_r t} = \text{(constant)} \sin \frac{r\pi x}{l} \cdot e^{i\omega_r t}. \quad (7.6.7)$$

This result is obtained from equation (7.6.6) by writing out the expressions for F_1, F_5 and F_6 in full (see Table 7.1 (a)) and then letting $\lambda l \to \lambda_r l$ where λ_r is given by equation (7.6.2). The subscript r of v in this result is used, as usual, to signify a distortion in the rth principal mode.

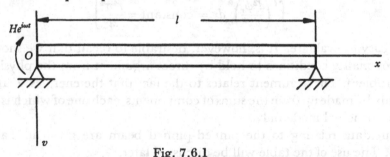

Fig. 7.6.1

It will be seen from equation (7.6.7) that the principal modes are sine curves. It is convenient to define unit value of the rth principal co-ordinate p_r as that which gives unit amplitude to the sine wave. Thus the distortion in the rth mode is

$$v_r(x, t) = p_r(t) \cdot \phi_r(x) \quad (7.6.8)$$

in which the characteristic function is

$$\phi_r(x) = \sin \frac{r\pi x}{l}. \quad (7.6.9)$$

Since any deflexion of the beam may be expressed as the sum of deflexions in the various principal modes, it may be expressed in the series form

$$v = \sum_{r=1}^{\infty} v_r = \sum_{r=1}^{\infty} p_r(t) \cdot \phi_r(x). \quad (7.6.10)$$

If this series is substituted into equations (7.1.9) and (7.1.11) and if the integrations are performed over the full length of the beam, then energy expressions are obtained having the form

$$2T = a_1 \dot{p}_1^2 + a_2 \dot{p}_2^2 + \ldots, \quad (7.6.11)$$

$$2V = c_1 p_1^2 + c_2 p_2^2 + \ldots. \quad (7.6.12)$$

The coefficients in these series are found to be

$$a_r = \frac{A\rho l}{2}, \quad c_r = \frac{r^4 \pi^4 EI}{2l^3}. \tag{7.6.13}$$

The expressions are related, as usual, to ω_r^2 by the relation $a_r \omega_r^2 = c_r$.

The absence of product terms in the expressions (7.6.11) and (7.6.12) shows that the orthogonality condition is met by the principal modes of a pinned-pinned beam. The reason for this absence of cross-terms is that

$$\int_0^l \phi_r(x) \cdot \phi_s(x)\, dx = \int_0^l \left(\frac{d^2\phi_r}{dx^2}\right)\left(\frac{d^2\phi_s}{dx^2}\right) dx = 0 \quad (r \neq s). \tag{7.6.14}$$

These integrals are simple to handle when the characteristic functions are sine functions. Similarly the integrals which it is required to evaluate in determining the coefficients a_r and c_r, namely

$$\left.\begin{array}{l} \displaystyle\int_0^l [\phi_r(x)]^2\, dx = \text{constant} = \frac{l}{2}, \\[3mm] \displaystyle\int_0^l \left(\frac{d^2\phi_r}{dx^2}\right)^2 dx = \text{constant} = \frac{r^4 \pi^4}{2l^3} \end{array}\right\} \tag{7.6.15}$$

are also easy to evaluate. It is, however, desirable to recall that we should expect the orthogonality conditions to hold by direct argument from the physical nature of the problem. The argument relates to the fact that the energies T and V must necessarily be made up from the sums of components, each one of which is associated with one principal mode only.

Certain data relating to the pinned-pinned beam are given in Table 7.2 (a) (p. 374). The use of the table will be discussed later.

7.6.2 Clamped-clamped beam

An expression for a receptance at any section of a clamped-clamped beam has not been found previously in this book. It will therefore be necessary to obtain the principal modes of such a beam by a slightly different method from that which was used for the pinned-pinned beam.

If the clamped-clamped beam is vibrating freely in its rth principal mode, the deflexion at any point may be written in the form

$$v_r = X_r(x)\, e^{i\omega_r t}. \tag{7.6.16}$$

This expression must satisfy the differential equation for free vibration. If it is substituted into that equation, it yields an ordinary differential equation governing the function $X_r(x)$, namely

$$\frac{d^4 X_r}{dx^4} - \lambda_r^4 X_r = 0. \tag{7.6.17}$$

It follows that the function $X_r(x)$ must be of the form

$$X_r = A \cos \lambda_r x + B \sin \lambda_r x + C \cosh \lambda_r x + D \sinh \lambda_r x \tag{7.6.18}$$

and the constants for this expression may be found from the boundary conditions

$$v_r = \frac{\partial v_r}{\partial x} = 0 \quad \text{at} \quad x = 0 \quad \text{and} \quad x = l. \tag{7.6.19}$$

When these are applied, they produce the equations

$$\left.\begin{array}{l} A(\cos \lambda_r l - \cosh \lambda_r l) + B(\sin \lambda_r l - \sinh \lambda_r l) = 0, \\ - A(\sin \lambda_r l + \sinh \lambda_r l) + B(\cos \lambda_r l - \cosh \lambda_r l) = 0, \end{array}\right\} \tag{7.6.20}$$

where $\qquad\qquad\qquad A = -C \quad \text{and} \quad B = -D. \tag{7.6.21}$

Equation (7.6.20) will have a non-zero solution for A and B, only if the determinant of the coefficients vanishes. That is to say, for A and B to be non-zero,

$$\begin{vmatrix} \cos \lambda_r l - \cosh \lambda_r l & \sin \lambda_r l - \sinh \lambda_r l \\ -(\sin \lambda_r l + \sinh \lambda_r l) & \cos \lambda_r l - \cosh \lambda_r l \end{vmatrix} = 0. \tag{7.6.22}$$

This determinantal equation simplifies to the form

$$F_3 = \cos \lambda_r l \cosh \lambda_r l - 1 = 0 \tag{7.6.23}$$

this being the frequency equation for the clamped-clamped beam.

This frequency equation may be compared with equation (7.6.1). Unlike that equation its roots cannot be found by inspection but only by calculation. The first three roots are, in fact,

$$\lambda_1 l = 4 \cdot 730, \quad \lambda_2 l = 7 \cdot 853, \quad \lambda_3 l = 10 \cdot 996. \tag{7.6.24}$$

When these roots have been found, the corresponding natural frequencies in any particular case may be calculated by the use of equation (7.6.4).

Equations (7.6.20) may now be used again in order to determine the principal modes. The first of the equations may be written

$$B = -A \frac{(\cosh \lambda_r l - \cos \lambda_r l)}{(\sinh \lambda_r l - \sin \lambda_r l)} \tag{7.6.25}$$

and the value of the ratio B/A, given by this expression, together with the results (7.6.21) may be substituted in equation (7.6.18). This gives

$$X_r = C \left\{ \cosh \lambda_r x - \cos \lambda_r x - \left[\frac{\cosh \lambda_r l - \cos \lambda_r l}{\sinh \lambda_r l - \sin \lambda_r l} \right] (\sinh \lambda_r x - \sin \lambda_r x) \right\}. \tag{7.6.26}$$

When this function is substituted back into equation (7.6.16), it gives an expression for the deflexion v_r during free vibration in the rth mode; it is

$$v_r = C \{ \cosh \lambda_r x - \cos \lambda_r x - \sigma_r (\sinh \lambda_r x - \sin \lambda_r x) \} e^{i\omega_r t}, \tag{7.6.27}$$

where C is an arbitrary constant and σ_r is the quantity

$$\sigma_r = \frac{\cosh \lambda_r l - \cos \lambda_r l}{\sinh \lambda_r l - \sin \lambda_r l}. \tag{7.6.28}$$

When a clamped-clamped beam is vibrating in its rth principal mode, the *shape* of the deflexion at any instant is given by the contents of the curly brackets of equation (7.6.27). This function can therefore be taken as the rth characteristic function so that

$$\phi_r(x) = \cosh \lambda_r x - \cos \lambda_r x - \sigma_r(\sinh \lambda_r x - \sin \lambda_r x). \qquad (7.6.29)$$

The *intensity* of the distortion in the rth principal mode will be defined by the rth principal co-ordinate p_r, and this may be expressed in the usual way as

$$v_r(x, t) = p_r(t) \cdot \phi_r(x). \qquad (7.6.30)$$

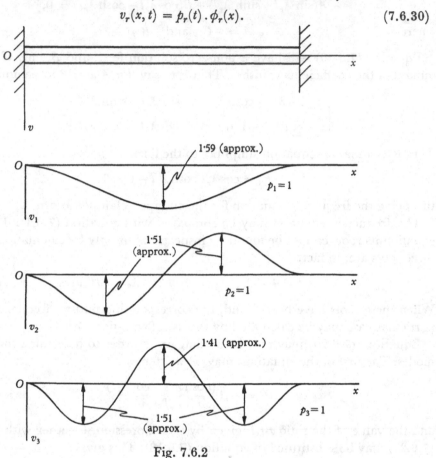

Fig. 7.6.2

This selection of the principal co-ordinate differs slightly from that which was made for the pinned-pinned beam. The latter beam had sinusoidal modes, and unit value of any principal co-ordinate was chosen so as to correspond to unit amplitude of the sine wave in question. Now if the modal shapes for the clamped-clamped beam are examined, it will be found that unit value of a principal co-ordinate does not correspond to unit deflexion of the point on the beam where the displacement is greatest. Nor are the maximum deflexions in different modes the same when the appropriate principal co-ordinates have unit value. The shapes of the first three modes are shown in fig. 7.6.2, the maximum deflexion for unit value of the principal co-ordinate being marked in each case.

We shall thus now drop the condition by which the statement $p_r = 1$ implies unit maximum deflexion in the rth mode for many types of beam. Its use for the pinned-pinned beam happened to be convenient.

The relations (7.6.14) and (7.6.15) have counterparts for the clamped-clamped beam. The more important ones are

$$\int_0^l \phi_r(x) \cdot \phi_s(x)\, dx = 0 \quad (r \neq s),$$ (7.6.31)

$$\int_0^l [\phi_r(x)]^2\, dx = l,$$ (7.6.32)

and a simple method of evaluating these definite integrals will be given at the end of this section. The results will, for the present, be assumed.

Since principal co-ordinates have been chosen, the corresponding coefficients in the energy functions may be formed. It will be found, by the use of equation (7.6.31) and (7.6.32) and the expression (7.1.9), that

$$a_r = A\rho l.$$ (7.6.33)

The coefficients c are most conveniently derived from the coefficients a by the relation

$$c_r = a_r \omega_r^2.$$ (7.6.34)

The functions $\phi_r(x)$ together with their derivatives, are tabulated in Table 7·2 (b) (p. 375).† They have several uses, some of which will be described in the next section. In setting out the tables, use is made of a convention which we shall employ extensively; this is to write

$$\phi_r' = \frac{1}{\lambda_r}\frac{d\phi_r}{dx}, \quad \phi_r'' = \frac{1}{\lambda_r^2}\frac{d^2\phi_r}{dx^2}, \quad \phi_r''' = \frac{1}{\lambda_r^3}\frac{d^3\phi_r}{dx^3}, \quad \text{etc.}$$ (7.6.35)

Each characteristic function, together with its first three derivatives, is tabulated. Higher derivatives may be obtained directly from the relations

$$\phi_r^{\mathrm{iv}} = \frac{1}{\lambda_r^4}\frac{d^4\phi_r}{dx^4} = \phi_r, \quad \phi_r^{\mathrm{v}} = \frac{1}{\lambda_r^5}\frac{d^5\phi_r}{dx^5} = \phi_r', \quad \text{etc.}$$ (7.6.36)

These relations will be seen to hold for a uniform beam, independently of its end fixings (see equation (7.2.4)).

7.6.3 Free-free beam

The characteristic functions for the free-free beam are related to those for the clamped-clamped beam for the following reason. At the ends of a clamped-clamped beam,

$$v = \frac{\partial v}{\partial x} = 0,$$ (7.6.37)

† Tables 7.2(b), (c) and (d) are taken, without any substantial alteration, from a University of Texas Publication (no. 4913, July 1949) entitled *Tables of Characteristic Functions Representing Normal Modes of Vibration of a Beam*, by Professor Dana Young and Dr R. P. Felgar. We are very grateful to these gentlemen, and to the Bureau of Engineering Research of Texas University, for permission to reproduce the tables here.

whereas at the ends of a free-free beam

$$\frac{\partial^2 v}{\partial x^2} = \frac{\partial^3 v}{\partial x^3} = 0. \tag{7.6.38}$$

Now if a characteristic function ϕ_r satisfies equations (7.6.37), then the function v given by its second derivative $d^2\phi_r/dx^2$ will satisfy (7.6.38). For the second and third derivatives required in the latter conditions correspond to the fourth and fifth derivatives of ϕ_r, which are related to the original characteristic function as in equations (7.6.36). Hence each characteristic function for the clamped-clamped beam gives rise to one for the free-free beam, and for the same reason the frequency equation is the same. These conclusions can be arrived at directly by using the same procedure as was used for either the pinned-pinned or the clamped-clamped beam.

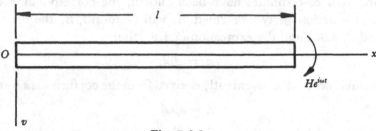

Fig. 7.6.3

Fig. 7.6.3 shows a free-free beam to which a harmonic exciting couple $H e^{i\omega t}$ is applied at the end $x = l$. The deflexion at any section is

$$v = \alpha_{xl'} H e^{i\omega t}$$
$$= \left\{ \frac{F_{10}(\cos \lambda x + \cosh \lambda x) + F_7(\sin \lambda x + \sinh \lambda x)}{2EI\lambda^2 F_3} \right\} H e^{i\omega t} \tag{7.6.39}$$

the receptance being entry 1 (g) of Table 7.1 (b). Evidently, then, the frequency equation of a free-free beam is

$$F_3 = \cos \lambda l \cosh \lambda l - 1 = 0. \tag{7.6.40}$$

This is the condition for all the receptances to become infinite, as may be checked by inspection of the receptances in Tables 7.1 (b) and (c). It is, however, evident from the nature of the problem, and may also be shown from the receptances, that the free-free beam has two zero frequencies in addition to the non-zero frequencies which it shares with the clamped-clamped beam.

If F_3 tends to zero, so that $\lambda \to \lambda_r$ in conformity with equation (7.6.40) and at the same time $H \to 0$ then, if F_7 and F_{10} are written out in full, equation (7.6.39) gives the result

$$v_r = (\text{constant}) \{(\cos \lambda_r l - \cosh \lambda_r l)(\cos \lambda_r x + \cosh \lambda_r x)$$
$$+ (\sin \lambda_r l + \sinh \lambda_r l)(\sin \lambda_r x + \sinh \lambda_r x)\} e^{i\omega_r t}. \tag{7.6.41}$$

The function within the curly brackets could be used as the rth characteristic function of the free-free beam. The slightly different form which we have used was chosen for convenience only and is obtained by re-arranging equation (7.6.41). This re-arrangement is accomplished by first rewriting (7.6.41) in the form

$$v_r = C(\cos\lambda_r l - \cosh\lambda_r l)\left\{\cosh\lambda_r x + \cos\lambda_r x\right.$$
$$\left. - \frac{(\sinh\lambda_r l + \sin\lambda_r l)}{(\cosh\lambda_r l - \cos\lambda_r l)}(\sinh\lambda_r x + \sin\lambda_r x)\right\}e^{i\omega_r t} \quad (7.6.42)$$

and noting that $C(\cos\lambda_r l - \cosh\lambda_r l)$ can be treated as a single constant. It can further be shown that

$$\frac{\sinh\lambda_r l + \sin\lambda_r l_r}{\cosh\lambda_r l - \cos\lambda_r l} = \sigma_r = \frac{\cosh\lambda_r l - \cos\lambda_r l}{\sinh\lambda_r l - \sin\lambda_r l} \quad (7.6.43)$$

This may be simply demonstrated by multiplying the numerator and denominator of the first fraction by $(\cosh\lambda_r l - \cos\lambda_r l)$ and simplifying the result by means of equation (7.6.40). The relationship having been established, the identity of the two forms of characteristic function becomes clear.

We have now, for the characteristic function of the free-free beam

$$\phi_r(x) = \cosh\lambda_r x + \cos\lambda_r x - \sigma_r(\sinh\lambda_r x + \sin\lambda_r x), \quad (7.6.44)$$

where σ_r has the same value as previously (equation (7.6.28)). It will be seen that this function is the second derivative of that for a clamped-clamped beam. Further, the functions ϕ_r', ϕ_r'' and ϕ_r''' for the free-free beam may be derived from the functions for a clamped-clamped beam. The relations are:

$$\left.\begin{aligned}
\phi_r \text{ (free-free)} &= \phi_r'' \text{ (clamped-clamped)},\\
\phi_r' \text{ (free-free)} &= \phi_r''' \text{ (clamped-clamped)},\\
\phi_r'' \text{ (free-free)} &= \phi_r \text{ (clamped-clamped)},\\
\phi_r''' \text{ (free-free)} &= \phi_r' \text{ (clamped-clamped)},
\end{aligned}\right\} (r = 1, 2, 3, \ldots). \quad (7.6.45)$$

It is therefore unnecessary to tabulate the functions for a free-free beam separately.

Table 7.2 (b) (p. 375) contains data for free-free beams. The derivation of the distortion modes follows that which we have described previously so that further elaboration is not necessary. Included in the table are the expressions

$$\phi_{-1}(x) = \frac{2\sqrt{3}}{l}\left(x - \frac{l}{2}\right), \quad \phi_0(x) = 1$$

representing the two necessary rigid body modes. These satisfy the equations of motion and the boundary conditions while the corresponding natural frequencies are $\omega_{-1} = \omega_0 = 0$, as may readily be verified. They are so scaled that they preserve the orthogonality relationships which follow, for $r, s = -1, 0, 1, 2, \ldots$.

7.6.4 Clamped-free, clamped-pinned, free-pinned, sliding-pinned, sliding-sliding, clamped-sliding and free-sliding beams

The remaining types of beam are covered in Table 7.2 (pp. 382–402). The methods of derivation are all similar to those discussed previously. Wherever they are relevant, expressions are given for rigid body modes which are so scaled as to preserve the orthogonality relations.

7.6.5 Orthogonality

The results (7.6.31) and (7.6.32) have been quoted for the clamped-clamped beam; their counterparts for all the remaining types of beam will be found in Tables 7.2. As mentioned above, we finish this section by showing how the definite integrals in question may be evaluated, because if an unsuitable method is used the working may become very tedious.

The characteristic functions for distortion of all uniform beams are known to satisfy the equation

$$\frac{d^4\phi_r}{dx^4} = \lambda_r^4\phi_r, \tag{7.6.46}$$

and it follows that two characteristic functions of any one beam satisfy the relation

$$\lambda_r^4\phi_r\phi_s - \lambda_s^4\phi_r\phi_s = \phi_s\frac{d^4\phi_r}{dx^4} - \phi_r\frac{d^4\phi_s}{dx^4} \tag{7.6.47}$$

which may be written as

$$\phi_r\phi_s = \frac{1}{\lambda_r^4 - \lambda_s^4}\left[\phi_s\frac{d^4\phi_r}{dx^4} - \phi_r\frac{d^4\phi_s}{dx^4}\right]. \tag{7.6.48}$$

Both sides of this equation may be integrated to give

$$\int_0^l \phi_r\phi_s\,dx = \frac{1}{\lambda_r^4 - \lambda_s^4}\int_0^l\left[\phi_s\frac{d^4\phi_r}{dx^4} - \phi_r\frac{d^4\phi_s}{dx^4}\right]dx. \tag{7.6.49}$$

The right-hand side of (7.6.49) may now be integrated by parts giving

$$\int_0^l \phi_r\phi_s\,dx = \frac{1}{\lambda_r^4 - \lambda_s^4}\left\{\left|\phi_s\frac{d^3\phi_r}{dx^3} - \phi_r\frac{d^3\phi_s}{dx^3}\right|_0^l - \int_0^l\left(\frac{d\phi_s}{dx}\frac{d^3\phi_r}{dx^3} - \frac{d\phi_r}{dx}\frac{d^3\phi_s}{dx^3}\right)dx\right\} \tag{7.6.50}$$

and the integral remaining on the right-hand side may be treated by parts in its turn. This leads to the result

$$\int_0^l \phi_r\phi_s\,dx = \frac{1}{\lambda_r^4 - \lambda_s^4}\left|\phi_s\frac{d^3\phi_r}{dx^3} - \phi_r\frac{d^3\phi_s}{dx^3} - \frac{d\phi_s}{dx}\frac{d^2\phi_r}{dx^2} + \frac{d\phi_r}{dx}\frac{d^2\phi_s}{dx^2}\right|_0^l. \tag{7.6.51}$$

Each of the four product terms on the right-hand side of this equation becomes zero at both limits. This can be seen for the first two terms by virtue of the fact that either the deflexion is zero, because the end is clamped or pinned, or else the shear force is zero because there is no clamp or pin. Similarly the second pair of terms vanishes either because the slope is zero or else because the bending moment is zero in the absence of a fixing to constrain the slope. It follows that

$$\int_0^l \phi_r\phi_s\,dx = 0 \quad (r \neq s) \tag{7.6.52}$$

for all uniform beams.

Let the other integral be denoted by Z; thus

$$Z = \int_0^l \phi_r^2\,dx. \tag{7.6.53}$$

In order to evaluate this quantity, we first write it in the form

$$Z = \int_0^l \phi_r\phi_r^{\text{iv}}\,dx. \tag{7.6.54}$$

This is permissible because ϕ_r satisfies equation (7.6.46). It is now possible to integrate by parts and, by the same argument as above, it is found that the integrated portion vanishes. This leaves

$$Z = -\int_0^l \phi_r' \phi_r''' \, dx, \tag{7.6.55}$$

and if the process is repeated, this becomes

$$Z = \int_0^l [\phi_r'']^2 \, dx. \tag{7.6.56}$$

By combining equations (7.6.53), (7.6.55) and (7.6.56) it is now found that

$$4Z = \int_0^l \{\phi_r^2 + [\phi_r'']^2 - 2\phi_r' \phi_r'''\} \, dx. \tag{7.6.57}$$

Now this expression is easy to integrate. For if ϕ_r has the form

$$\phi_r(x) = A \cos \lambda_r x + B \sin \lambda_r x + C \cosh \lambda_r x + D \sinh \lambda_r x, \tag{7.6.58}$$

where A, B, C and D are constants, then substitution into equation (7.6.57) shows that the bracketed quantity reduces to

$$2(A^2 + B^2 + C^2 - D^2) \tag{7.6.59}$$

so that

$$Z = \frac{(A^2 + B^2 + C^2 - D^2)\,l}{2}. \tag{7.6.60}$$

Consider the clamped-clamped beam as an example. The relevant constants are

$$A = -1, \quad B = \sigma_r, \quad C = 1, \quad D = -\sigma_r, \tag{7.6.61}$$

where σ_r has the meaning given in Table 7.2 (b). Substitution of these values into equation (7.6.60) shows that

$$Z = l \tag{7.6.62}$$

as was assumed in equation (7.6.32). The procedure is similar for other types of beam, and if the results are collected together they give

$$Z = \int_0^l \phi_r^2 \, dx = \begin{cases} \tfrac{1}{2}l & \text{(pinned-pinned, sliding-pinned, sliding-sliding),} \\ l & \text{(all other types of beam dealt with in this section).} \end{cases} \tag{7.6.63}$$

It is left as an exercise for the reader to verify that the results (7.6.52) and (7.6.63) are satisfied by the 0-modes quoted in Tables 7.2 (d), (f) and (g) and by the −1 and 0-modes in Table 7.2 (b).

Owing to the rather inconvenient form of the results for uniform beams it is sometimes helpful to have a simple approximate method of finding the frequencies and the positions of the nodes. Such a method has been given by Yates.†

† H. G. Yates, 'Vibration Diagnosis in Marine Geared Turbines', *Trans. N.E. Coast Inst. Engineers and Shipbuilders*, vol. 65, (1949), p. 260.

EXAMPLES 7.6

1. Show that the rth characteristic function of a clamped-sliding beam may be expressed in the form

$$\phi_r(x) = \cosh \lambda_r x - \cos \lambda_r x - \sigma_r(\sinh \lambda_r x - \sin \lambda_r x),$$

where

$$\sigma_r = \tanh \lambda_r l,$$

and where

$$\tan \lambda_r l = -\tanh \lambda_r l.$$

Find the values to which σ_r and $\lambda_r l$ tend when r is large.

2. Explain clearly why, and in what way, the characteristic functions for distortion of a free-sliding beam are related to those of a clamped-sliding beam. Do this without deriving the functions directly.

3. Show that the integral $\quad \displaystyle\int_0^l \left(\frac{d^2\phi_r}{dx^2}\right)\left(\frac{d^2\phi_s}{dx^2}\right) dx$

vanishes for all the types of beam, in accordance with equation (7.6.14).

Show that evaluation of the integral

$$\int_0^l \left(\frac{d^2\phi_r}{dx^2}\right)^2 dx,$$

in accordance with equation (7.6.15), leads to the general result

$$c_r = EI\lambda_r^4 Z.$$

7.7 Series representation of static deflexions of beams

It was shown in §4.7 that any static deflexion of a taut string could be expressed as a series of the relevant characteristic functions. Later, in §5.2, it was explained how useful approximations can be obtained by curtailing series of this type and thereby making them simpler to handle. In this section, these same ideas will be applied to beam problems.

Let any static force distribution $w(x)$ be applied to a beam, thereby producing deflexions in any or all of its principal modes. Consider one component $w_r(x)$ of the applied load, which produces deflexion in the rth principal mode only. This component is given by

$$w_r(x) = \left\{\frac{\displaystyle\int_0^l w(x)\,\phi_r(x)\,dx}{\displaystyle\int_0^l [\phi_r(x)]^2\,dx}\right\} \phi_r(x) = \left\{\frac{1}{Z}\int_0^l w(x)\,\phi_r(x)\,dx\right\}\phi_r(x)$$

$$= B_r \phi_r(x) \quad \text{(say)}, \tag{7.7.1}$$

where $\phi_r(x)$ is the rth characteristic function of the particular beam in question and Z is the appropriate constant given in equation (7.6.53).

The deflexion $v_r(x)$ in this mode may be found from equation (7.1.7), remembering that the time-derivative term is now zero. Thus $v_r(x)$ is obtained by integrating the equation

$$\frac{d^4 v_r}{dx^4} = \frac{B_r}{EI}\phi_r(x). \tag{7.7.2}$$

If that property of the characteristic functions which was pointed out in equation (7.6.36) is now used, this relation is seen to lead to the result

$$v_r(x) = \frac{B_r}{EI\lambda_r^4}\,\phi_r(x). \tag{7.7.3}$$

This shows that a static deflexion in a principal mode is produced by a load distribution which has the same form as the mode in question; that is to say, both w_r and v_r are proportional to ϕ_r. It follows from equation (7.7.3) that the rth principal co-ordinate, which in a static problem will evidently be constant, is

$$p_r = \frac{B_r}{EI\lambda_r^4}. \tag{7.7.4}$$

This value of p_r may be used in the expansion

$$v(x) = \sum_{r=1}^{\infty} v_r(x) = \sum_{r=1}^{\infty} p_r\phi_r(x). \tag{7.7.5}$$

The presence of the factor λ_r^4 in the denominator of the expression for p_r ensures that the series (7.7.5) converges, and further that useful approximate expressions may be obtained by curtailing it. It may be shown, by the method of §5.2 that the curtailment will lead automatically to the best possible approximation which is obtainable from a series of the selected number of terms.

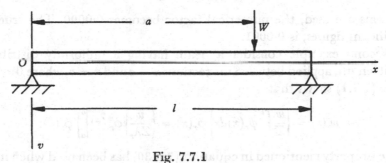

Fig. 7.7.1

Let a concentrated force of magnitude W be applied to a beam at the section $x = a$. From equation (7.7.1) the coefficient B_r will be

$$B_r = \frac{1}{Z}\int_0^l w(x)\,\phi_r(x)\,dx = \frac{W}{Z}\phi_r(a), \tag{7.7.6}$$

where $\phi_r(x)$ is the appropriate characteristic function. By using equation (7.7.4), p_r can be found and thereby a series of expressions is obtained for the beam deflexion which the force produces.

It may be noted here, in passing, that calculations which involve the assumption that a force is concentrated at a single point are necessarily approximate, as are also those assuming point application of couples. This matter will be discussed further in §7.9, but is mentioned here to emphasize that the approximations involved in the curtailment of series are not the only ones present.

Fig. 7.7.1 shows a pinned-pinned beam which supports a concentrated load

W at the section $x = a$. On using the expressions given in Table 7.2 (a) to substitute into equation (7.7.6) it is found that

$$B_r = \frac{2W}{l} \sin \frac{r\pi a}{l}. \tag{7.7.7}$$

From this it follows, by equation (7.7.4), that

$$p_r = \frac{2Wl^3}{EIr^4\pi^4} \sin \frac{r\pi a}{l}. \tag{7.7.8}$$

The deflexion at any point can therefore be written in the form

$$v = \frac{2Wl^3}{EI\pi^4} \sum_{r=1}^{\infty} \frac{1}{r^4} \sin \frac{r\pi a}{l} \sin \frac{r\pi x}{l}. \tag{7.7.9}$$

Now let $a = \frac{3}{4}l$ and suppose that the deflexion is required at the section $x = \frac{1}{4}l$; the series gives

$$v(\tfrac{1}{4}l) = \frac{2Wl^3}{EI\pi^4} \sum_{r=1}^{\infty} \frac{1}{r^4} \sin \frac{3\pi r}{4} \sin \frac{\pi r}{4}$$

$$= \frac{2Wl^3}{EI\pi^4} [\tfrac{1}{2} - \tfrac{1}{16} + \tfrac{1}{162} + \ldots], \tag{7.7.10}$$

$$\doteqdot 0 \cdot 0103 \frac{Wl^3}{EI} \quad \text{(from one term only).} \tag{7.7.11}$$

If two terms are used, the numerical factor becomes $0 \cdot 0090$. The true value, to two significant figures, is $0 \cdot 0091$.

As a second example consider a beam having a uniformly distributed load w per unit length, applied between the sections $x = a$ and $x = b$, where $0 \leqslant a \leqslant b \leqslant l$. Equation (7.7.1) shows that

$$w_r(x) = \left\{ \frac{w}{Z} \int_0^l \phi_r(x) dx \right\} \phi_r(x) = \left\{ \frac{w}{Z\lambda_r} \left| \phi_r'''(x) \right|_a^b \right\} \phi_r(x), \tag{7.7.12}$$

where the property mentioned in equation (7.6.36) has been used when integrating. The equation shows that

$$B_r = \frac{w}{Z\lambda_r} [\phi_r'''(b) - \phi_r'''(a)] \tag{7.7.13}$$

and from this the deflexion in the rth principal mode may be obtained.

The expression within the brackets in equation (7.7.13) takes on a simple form if the load extends along the full length of the beam (although the actual expression differs between different types of beam). For instance, a clamped-free beam will be found to give

$$[\phi_r'''(l) - \phi_r'''(0)] = 2\sigma_r. \tag{7.7.14}$$

Fig. 7.7.2 shows a clamped-free beam which supports a uniformly distributed load over the portion between $x = 0$ and $x = 0 \cdot 6l$; we shall use the above method to calculate the deflexion at the free end. From equations (7.7.4) and (7.7.13) it is found that

$$v_r(x) = \left\{ \frac{w[\phi_r'''(0 \cdot 6l) - \phi_r'''(0)]}{EI\lambda_r^5 l} \right\} \phi_r(x) \tag{7.7.15}$$

and the data which are required for evaluating this expression are to be found in Table 7.2 (c). The deflexion at the tip may be obtained by putting $x = l$, and the first two terms of the series for $v(l)$ are then:

$$
\left.
\begin{aligned}
v_1(l) &= \frac{wl^4[(-1 \cdot 090\,70) - (-1 \cdot 468\,19)]}{EI(12 \cdot 362 \times 1 \cdot 875)} \times 2 = 0 \cdot 0326\frac{wl^4}{EI}, \\
v_2(l) &= \frac{wl^4[(0 \cdot 430\,94) - (-2 \cdot 036\,93)]}{EI(485 \cdot 52 \times 4 \cdot 6941)} \times (-2) = -0 \cdot 0022\frac{wl^4}{EI}.
\end{aligned}
\right\}
\tag{7.7.16}
$$

These two terms give
$$
v(l) \doteqdot 0 \cdot 0304\frac{wl^4}{EI},
\tag{7.7.17}
$$

whereas the true value, to the same degree of accuracy, has a numerical factor of 0·0306.

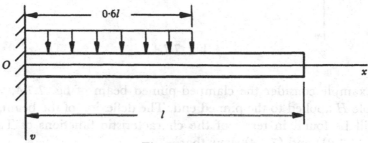

Fig. 7.7.2

It is possible also to find the static deflexion of a uniform beam when a couple is applied to it at a specified section. The problem may be approached as the limiting case of two loads W and $-W$ which act at the section $x = a$ and $x = a + \delta$, respectively, as shown in fig. 7.7.3. According to equation (7.7.1), the component of the loading which will produce the deflexion in the rth mode is

$$
w_r(x) = \frac{W}{Z}[\phi_r(a + \delta) - \phi_r(a)]\,\phi_r(x).
\tag{7.7.18}
$$

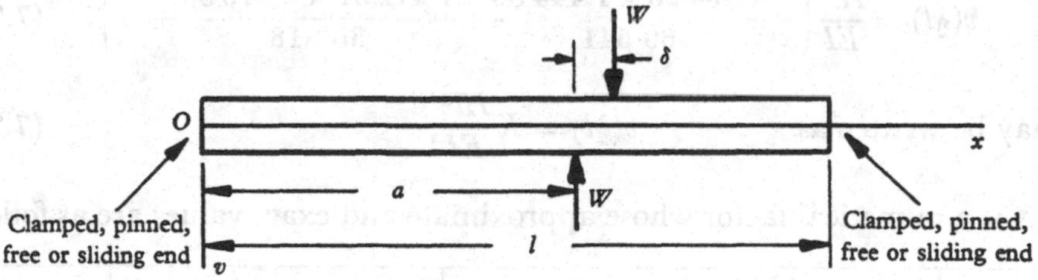

Fig. 7.7.3

The expression for B_r may now be written in the form

$$
B_r = \frac{W\delta}{Z}\left[\frac{\phi_r(a + \delta) - \phi_r(a)}{\delta}\right]
\tag{7.7.19}
$$

and it will be seen that the magnitude of the couple, $W\delta$, appears as a factor. This factor, which will be denoted by H, may be kept constant while δ becomes indefinitely small. Thus
$$
\lim_{\delta \to 0} W\delta = H.
\tag{7.7.20}
$$

The approach of δ to zero means that the quantity within the brackets in equation (7.7.19) becomes the derivative of ϕ_r with respect to x so that this limiting process leads to the result

$$B_r = \frac{H}{Z}\left(\frac{d\phi_r}{dx}\right)_{x=a} = \frac{H\lambda_r}{Z}\phi_r'(a). \qquad (7.7.21)$$

By using equation (7.7.4), the deflexion in the rth principal mode may now be found.

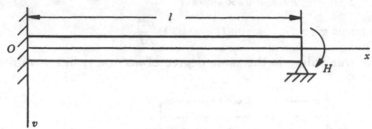

Fig. 7.7.4

As an example consider the clamped-pinned beam of fig. 7.7.4, which has a static couple H applied to the pinned end. The deflexion of the beam, due to this couple, will be found in terms of the characteristic functions of Table 7.2 (d). Equations (7.7.21) and (7.7.4) show that

$$v_r = \frac{H\phi_r'(l)}{EI\lambda_r^3 l}\phi_r(x), \qquad (7.7.22)$$

whence

$$v(x) = \frac{Hl^2}{EI}\sum_{r=1}^{\infty}\frac{\phi_r'(l)}{(\lambda_r l)^3}\phi_r(x). \qquad (7.7.23)$$

Using this series, the deflexion at the section $x = \frac{1}{2}l$ will be found to be

$$v(\tfrac{1}{2}l) = \frac{Hl^2}{EI}\left\{-\frac{1\cdot454\,20 \times 1\cdot444\,86}{60\cdot541} + \frac{1\cdot412\,51 \times 0\cdot570\,35}{353\cdot18} + \ldots\right\}. \qquad (7.7.24)$$

This may be written as

$$v(\tfrac{1}{2}l) = K\frac{Hl^2}{EI}, \qquad (7.7.25)$$

where K is a numerical factor whose approximate and exact values are as follows:

No. of terms	1	2	3	4	5	∞
Value of K	−0·0347	−0·0324	−0·0307	−0·0310	−0·0314	−0·03125

Just as the deflexion at any point of a beam may be found as a series, so also may the slope. Thus the slope of the clamped-free beam of fig. 7.7.2 can be found by differentiating the terms of the series whose typical term is given in equation (7.7.15); that is

$$\frac{dv}{dx} = \frac{w}{EIl}\sum_{r=1}^{\infty}\frac{[\phi_r'''(0\cdot6l) - \phi_r'''(0)]}{\lambda_r^4}\phi_r'(x). \qquad (7.7.26)$$

The slope at the section $x = 0.6l$ will be found to be

$$\left(\frac{dv}{dx}\right)_{0.6l} = \frac{wl^3}{EI}\left\{\frac{(-1.090\,70 + 1.468\,19)}{12.362} \times 1.346\,85\right.$$
$$\left. - \frac{(0.430\,94 + 2.036\,93)}{485.52} \times 0.860\,40 + \ldots\right\}$$

$$\doteq 0.0360\,\frac{wl^3}{EI}. \tag{7.7.27}$$

The coefficient is correct to three significant figures although three terms only have been used in finding it. The reader may check that the slope of the beam of fig. 7.7.2 within the segment $0.6l \leqslant x \leqslant l$ is given, correctly, as a constant by the series method.

Just as the deflexion series may be differentiated once to give the slope it may also be differentiated twice to give the bending moment (see equation (7.1.5)) or three times to give the shear force (equation (7.1.6)). In doing this, however, a watch must be kept on the convergence and this matter must now be discussed.

The convergence of the series (7.7.5) for a static deflexion $v(x)$ is ensured by the presence of the factor λ_r which occurs, raised to some power, in the coefficient p_r of the rth characteristic function. The power to which it is raised depends upon the power of λ_r which occurs in B_r as may be seen from equation (7.7.4) and this in its turn depends upon the nature of the loading $w(x)$. The powers of λ_r which are associated with the three types of loading which were considered above are shown in the accompanying table.

Type of loading	Power of λ_r in denominator	
	B_r	p_r
Point couple	-1	3
Point force	0	4
U.D. load	1	5

Now the methods of this section are only convenient if the series converge reasonably rapidly and it is usually necessary for λ_r to appear, raised to the third power or higher, in the denominator of p_r. The table shows that this is so for all three types of loading when it is the *deflexion* series which is required, though it will not always be so for the slope and bending moment series. It will hold for none of the types of loading when series are formed for the shear force. It is evident that, since the deflexion and its derivatives are real physical quantities which are finite at every point (except occasionally in the immediate vicinity of the load) the series which give these quantities must converge. It is therefore always permissible, mathematically, to sum the series. The difficulty which arises with the more slowly converging series is simply the practical one of having to take a large number of terms into account.

Each differentiation of a series reduces the power of λ_r in the denominator by one. In the following table, the powers of λ_r which occur in the series for different quantities are shown for various types of loading. The table is divided by heavy lines which indicate that the series to which the items below the line refer are not suitable for summing. By examining the table it will be found that the series converge most rapidly when the form of loading is 'smoothest'. The term 'smooth' here implies that a system of loading $w_1(x)$ is smoother than a system $w_2(x)$ if fewer differentiations are required of $w_1(x)$ to make it zero than are required of $w_2(x)$.

Series	Power of λ_r in denominator of rth coefficient		
	Point couple	Point force	U.D. loading
Deflexion	3	4	5
Slope	2	3	4
Bending moment	1	2	3
Shear force	—	1	2

This discussion on convergence is intended to be nothing more than a rough guide. The reader who wishes to pursue the matter further is referred to a work by Sokolnikoff.[†]

Finally, the reader may well ask why a set of functions which were developed in connexion with the dynamic problem of vibration should now be used for the solution of a purely statical problem. It could be argued that this introduces quite unnecessary complication into the problem. The answer to these doubts is that a set of deflexion functions, such as is provided by the ϕ_r's, is a simple tool to use once the functions have been tabulated. Now any such set of functions must satisfy the geometrical end conditions for the beam and must be mutually orthogonal and there are an infinite number of possible sets which have these properties. The particular set which has been developed through the vibration theory are worth tabulating for use with that theory and they thereby become available for the statical problem also. Some further points about their use are discussed by Felgar,[‡] by von Kármán[§] and by Bishop.[||]

The idea of using the characteristic functions in statical problems is not restricted to beams. It may be applied also, for instance, to plate and shell problems and is useful both in simple deflexion cases and when questions of elastic instability are involved.

EXAMPLES 7.7

1. A clamped-free beam carries a uniformly distributed load w per unit length along its whole span. Using a two-term approximation, estimate the deflexion and the slope at the free end.

[†] I. S. Sokolnikoff, *Advanced Calculus* (McGraw-Hill, London, 1st ed. 1939), §108.
[‡] R. P. Felgar, 'Formulas for Integrals containing Characteristic Functions of a Vibrating Beam', *Circular no. 14, Bur. of Eng. Res.*, Texas University (1950).
[§] Th. von Kármán, 'Use of Orthogonal Functions in Structural Problems', *Timoshenko Anniversary Volume* (Macmillan, London, 1938), pp. 114–24.
[||] R. E. D. Bishop, 'The Normal Functions of Beam Vibration in Series Solutions of Static Problems', *J. Roy. Aero. Soc.* vol. 57 (Aug. 1953), pp. 527–9.

2. Find the deflexion mid-way between the supports of the 'overhung' beam which is shown in the diagram.

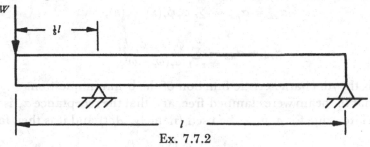

Ex. 7.7.2

3. Show that rigid body modes may be ignored in the series representation of static distortions of uniform beams.

7.8 The series form of beam receptances

We mentioned, at the end of § 7.3, that the closed form of beam receptances is often cumbersome and we shall now show that the series form can be more convenient.

The receptance at the rth principal co-ordinate of a beam can be written in the form

$$\alpha_r = \frac{1}{a_r(\omega_r^2 - \omega^2)} \tag{7.8.1}$$

(cf. equation (4.6.12)). Further, any other receptance α_{zy} can be expressed as a series of terms, each one of which contains one of these principal co-ordinate receptances. As in equation (4.6.19), then,

$$\alpha_{zy} = \alpha_{yz} = \sum_{r = -1,\, \text{or } 0,\, \text{or } 1}^{m} \alpha_r \frac{\partial y}{\partial p_r} \frac{\partial z}{\partial p_r}. \tag{7.8.2}$$

In this expression, α_{zy} is the receptance which gives the generalized displacement at a co-ordinate z when a harmonic force $Fe^{i\omega t}$ acts at a generalized co-ordinate y so that

$$z = \alpha_{zy} Fe^{i\omega t}. \tag{7.8.3}$$

In beam problems, y and z may be either displacements or slopes.

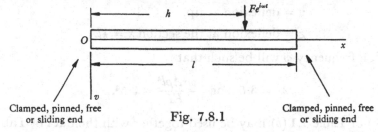

Fig. 7.8.1

Let a uniform beam have a force $Fe^{i\omega t}$ applied to it at the section $x = h$ as shown in fig. 7.8.1. Assuming that the beam has no rigid body modes—and this is only to reduce the cumbersome entry below the summation sign—deflexion at any section will be of the form (7.6.10) so that, in particular

$$v(h, t) = \sum_{r=1}^{\infty} p_r(t) \cdot \phi_r(h). \tag{7.8.4}$$

Now if the co-ordinates y and z of equation (7.8.2) are identified with $v(x,t)$ and $v(h,t)$ respectively, then

$$\alpha_{xh} = \alpha_{hx} = \sum_{r=1}^{\infty} \alpha_r \phi_r(x) \cdot \phi_r(h), \qquad (7.8.5)$$

or

$$\alpha_{xh} = \alpha_{hx} = \sum_{r=1}^{\infty} \frac{\phi_r(x) \cdot \phi_r(h)}{a_r(\omega_r^2 - \omega^2)}. \qquad (7.8.6)$$

Here, $\phi_r(x)$ is the rth characteristic function of the beam in question.

Suppose that the beam were clamped-free, and that the receptance α_{xl} is required. The appropriate value for a_r must be used (namely, $A\rho l$) and it is then found that

$$\alpha_{xl} = \frac{1}{A\rho l} \sum_{r=1}^{\infty} \frac{\phi_r(x) \cdot \phi_r(l)}{\omega_r^2 - \omega^2}, \qquad (7.8.7)$$

where the function ϕ_r must be taken from Table 7.2 (c). The expression (7.8.7) is an alternative form of entry 2 (a) of Table 7.1 (b). It could be found by treating the latter expression, which is a function of x, as if it were a static deflexion as in § 7.7. This method is usually, however, unnecessarily tedious.

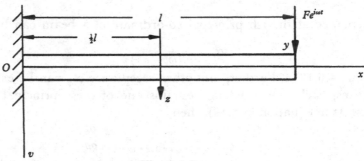

Fig. 7.8.2

The two forms of this receptance will now be calculated numerically for the particular problem of the clamped-free beam shown in fig. 7.8.2. The receptance α_{zy} will be found with the following notation used temporarily:

$$y = \text{deflexion at tip},$$
$$z = \text{deflexion at the section } x = \tfrac{1}{2}l.$$

The excitation frequency ω will be such that

$$\lambda l = 3 \cdot 5 \quad \text{or} \quad \frac{\omega^2 A\rho l^4}{EI} = 3 \cdot 5^4. \qquad (7.8.8)$$

Entry 2 (a) of Table 7.1 (b) may be used together with the data of Table 7.1 (d) to find the receptance directly in the closed form. Since, here,

$$\cos \lambda x - \cosh \lambda x = \cos \frac{\lambda l}{2} - \cosh \frac{\lambda l}{2} \qquad (7.8.9)$$

this quantity is given by F_{10} for the argument $1\cdot75$; its value is $-3\cdot1424$. Similarly the value of the term

$$\sin \lambda x - \sinh \lambda x$$

may be obtained directly. In this way, it is found that

$$\alpha_{zv} = \left(\frac{-16\cdot19 \times 3\cdot1424 + 16\cdot64 \times 1\cdot8064}{2 \times 3\cdot5^4 \times 14\cdot52}\right)\frac{l^3}{EI} = -0\cdot017\frac{l^3}{EI}. \quad (7.8.10)$$

When this receptance is to be evaluated from the series (7.8.7) it must first be noted that the forcing frequency is such that λl falls between $\lambda_1 l$ and $\lambda_2 l$ for a cantilever—see Table 7.2 (c). That is,

$$\omega_1 < \omega < \omega_2 \qquad (7.8.11)$$

and further ω is not very close either to ω_1 or to ω_2. It may be deduced that the first two terms of the series will be the most important and we shall therefore calculate these only. The values of $(\lambda_1 l)^4$ and $(\lambda_2 l)^4$ are given in Table 7.2 (c) and lead to the result

$$\alpha_{zv} \doteqdot \frac{l^3}{A\rho}\left\{\frac{0\cdot679\,05 \times 2}{\dfrac{EI}{A\rho}(12\cdot36 - 3\cdot5^4)} + \frac{1\cdot427\,33 \times (-2)}{\dfrac{EI}{A\rho}(485\cdot5 - 3\cdot5^4)}\right\} = -0\cdot018\frac{l^3}{EI}. \quad (7.8.12)$$

This value is close to the 'exact' figure above, and was found after a comparable amount of numerical work. If, however, we had been concerned with a receptance for which neither of the co-ordinates was measured at a tip then the closed-form calculation would have been very much more tedious. The series form, on the other hand, would not have involved any more work than in the example above.

The slope at some point of the beam of fig. 7.8.2 may be obtained by differentiating the series (7.6.10); thus, since there is no -1 or 0-mode,

$$\frac{\partial v}{\partial x} = \sum_{r=1}^{\infty} p_r(t)\frac{d\phi_r(x)}{dx} = \sum_{r=1}^{\infty} p_r(t)\lambda_r\phi_r'(x). \qquad (7.8.13)$$

If z of equation (7.8.3) is associated with this slope, and if y again represents the deflexion at the section $x = h$, then

$$\alpha_{x'h} = \alpha_{hx'} = \sum_{r=1}^{\infty} \alpha_r\lambda_r\phi_r'(x)\,\phi_r(h) = \sum_{r=1}^{\infty}\frac{\lambda_r\phi_r'(x)\,\phi_r(h)}{a_r(\omega_r^2 - \omega^2)}. \qquad (7.8.14)$$

This series may be used in the same way as the deflexion series. Similarly, the series for $\alpha_{xh'}$ will be found to be

$$\alpha_{xh'} = \alpha_{h'x} = \sum_{r=1}^{\infty} \alpha_r\lambda_r\phi_r(x)\,\phi_r'(h) = \sum_{r=1}^{\infty}\frac{\lambda_r\phi_r(x)\,\phi_r'(h)}{a_r(\omega_r^2 - \omega^2)}. \qquad (7.8.15)$$

The receptances (7.8.14) and (7.8.15) have the quantity λ_r in the numerator of each term in the series so that the rate of convergence is reduced. In general, then, slope receptances require more terms to be used when calculating their numerical values. This fact is not as troublesome as might at first be expected because in real problems there is a reduction in the amplitude of the higher principal modes because of the inevitable spreading of so-called concentrated forces. We shall deal with this matter in the next section.

Fig. 7.8.3 shows the clamped-free beam of the previous example, the co-ordinate z now representing the slope at the section where $x = \frac{1}{2}l$. Again we calculate the receptance α_{zy} by the two methods, using the same frequency as before. Entry 2 (b) of Table 7.1 (b) gives this receptance in closed form and the numerical values of Table 7.1 (d) lead to the result

$$\alpha_{zy} = \left(\frac{15 \cdot 64 \times (-3 \cdot 1424) + 16 \cdot 19 \times 3 \cdot 7744}{2 \times 3 \cdot 5^2 \times (-14 \cdot 52)} \right) \frac{l^2}{EI} = -0 \cdot 034 \frac{l^2}{EI}. \quad (7.8.16)$$

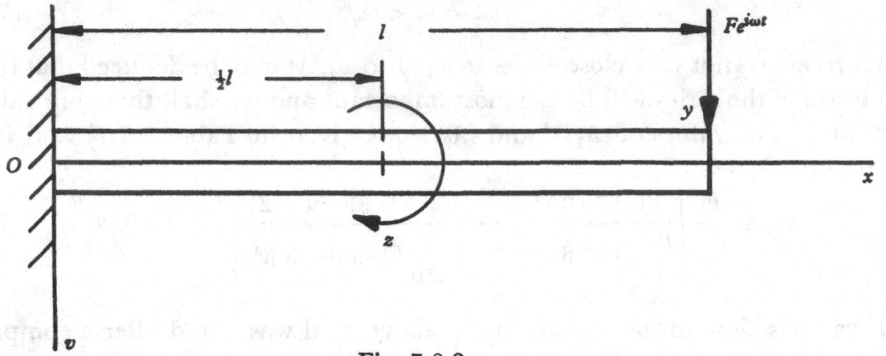

Fig. 7.8.3

Equation (7.8.14) gives, for this receptance,

$$\alpha_{zy} = \frac{l^2}{A\rho} \left\{ \frac{1 \cdot 8751 \times 1 \cdot 240\,52 \times 2}{\dfrac{EI}{A\rho}(12 \cdot 362 - 3 \cdot 5^4)} + \frac{4 \cdot 6941 \times (-0 \cdot 193\,07) \times (-2)}{\dfrac{EI}{A\rho}(485 \cdot 52 - 3 \cdot 5^4)} + \dots \right\}. \quad (7.8.17)$$

By curtailing this series, it is found that

$$
\left.
\begin{aligned}
\alpha_{zy} &\doteqdot -0 \cdot 034 \frac{l^2}{EI} \quad \text{(from 1 term),†} \\[4pt]
&\doteqdot -0 \cdot 028 \frac{l^2}{EI} \quad \text{(from 2 terms),} \\[4pt]
&\doteqdot -0 \cdot 034 \frac{l^2}{EI} \quad \text{(from 3 terms).}
\end{aligned}
\right\} \quad (7.8.18)
$$

Again, the co-ordinate y in equation (7.8.2) may be identified with the slope at a particular section $x = h$ and z with the slope at any section. This leads to the receptance series

$$\alpha_{zy} = \alpha_{x'h'} = \alpha_{h'x'} = \sum_{r=1}^{\infty} \alpha_r \lambda_r^2 \phi_r'(x)\,\phi_r'(h)$$

$$= \sum_{r=1}^{\infty} \frac{\lambda_r^2 \phi_r'(x)\,\phi_r'(h)}{a_r(\omega_r^2 - \omega^2)} \quad (7.8.19)$$

by equation (7.8.13). In this series the convergence is reduced still more by the presence of λ_r^2 in the numerators. As in the previous case the inaccuracies which are involved when a few terms only are used are less serious than might at first be

† This agreement with the correct result is fortuitous and cannot be foretold by inspection of the series.

expected. The difference between numerical answers derived from expressions in the closed form and from a few terms of the series form may now, however, be considerable.

The beam of fig. 7.8.4 may be used as an example. If the same value of λl is used as before and y and z represent the slopes at $x = l$ and $x = \frac{1}{2}l$ respectively, the following results will be found:

$$
\left.
\begin{aligned}
\alpha_{zy} &= -0 \cdot 059 \, \frac{l}{EI} \quad \text{(from the closed form),} \\[2mm]
&\doteq -0 \cdot 046 \, \frac{l}{EI} \quad \text{(1 term of series),} \\[2mm]
&\doteq -0 \cdot 021 \, \frac{l}{EI} \quad \text{(2 terms of series),} \\[2mm]
&\doteq -0 \cdot 068 \, \frac{l}{EI} \quad \text{(3 terms of series).}
\end{aligned}
\right\} \qquad (7.8.20)
$$

The slowness of the convergence is evident.

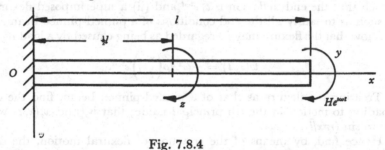

Fig. 7.8.4

The series receptances, which have been developed in this section, may be used in the same type of calculations as may receptances in the closed form. There is, however, one matter which must be noted. In the calculation of natural frequencies of—for instance—frames, the predominant terms of the series will not be known until a trial frequency has been selected. It may therefore be necessary to find the coefficients for more terms than would otherwise be necessary.

EXAMPLES 7.8

1. A uniform clamped-clamped beam is of length $l = 2L$ (see fig. 7.3.5 (a)) and it is excited with a frequency ω such that $\lambda l = 2 \cdot 00$.

   (a) Find the value of the direct receptance α_{LL}, (i) from the expression in closed form, and (ii) from two (finite) terms of the series form.

   (b) Repeat (a) for the direct receptance $\alpha_{L'L'}$.

2. A railway bridge consists of a uniform span simply supported at its ends, so that its principal modes of vibration in a vertical plane are sinusoidal. Find an expression for the direct receptance for vertical displacement at a point distant x from one end of the span. The length of the span is l, the relevant flexural rigidity is EI, and the mass per unit length is $A\rho$.

   If such a bridge has a span of 300 ft., a lowest natural frequency of vibration of 5 cyc./sec., and a total weight of 600 tons, estimate its first two natural frequencies if a locomotive weighing 150 tons is resting on the bridge 100 ft. from one end. The springs of the locomotive may be considered rigid for the purpose of the calculation.

(C.U.M.S.T. Pt II, 1952)

[NOTE. The answers given (p. 582) are 'exact' and they were calculated by the method of §7·5; the calculation is somewhat tedious. A very close approximation indeed may be found quickly by taking only two terms of the direct receptance in series form of the pinned-pinned beam.]

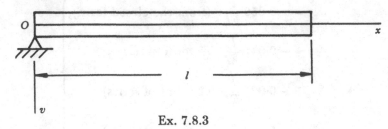

Ex. 7.8.3

3. The free end of the pinned-free beam shown experiences an imposed deflexion $v_l e^{i\omega t}$. Note that motion of the beam may be regarded as (i) a rigid body rotation about the pin such that the end deflexion is $v_l e^{i\omega t}$ and (ii) a superimposed flexural vibration which is such as to satisfy all the end conditions of a pinned-pinned beam.

(a) Show that the flexure may be regarded as being caused by a loading distribution

$$w(x, t) = -\omega^2 A\rho\left(\frac{x}{l}v_l\right)e^{i\omega t}.$$

(b) Treating the flexure as that of a pinned-pinned beam, find the contribution of this loading to motion in the rth principal mode; that is, find $w_r(x, t)$ which is proportional to $\sin(r\pi x/l)$.

(c) Hence find, by means of the equation of flexural motion, the displacement *in flexure* of any point as a series of the characteristic functions of a pinned-pinned beam.

(d) First adding the motion as a rigid body to that of flexure (found in (c)), find the *total* acceleration of any point of the beam.

(e) Find the harmonic force which must be applied at the free end by taking moments about the pin and hence find an expression for the direct receptance α_{ll}.

4. A uniform beam of length 4 ft. is clamped at one end and pinned at the other. Its lowest natural frequency of flexural vibration is 65 cyc./sec. and its total mass is 15 lb.

A concentrated mass of 4 lb. is attached to the beam at the section 1 ft. from the pinned end. Estimate the first two natural frequencies of this system.

(C.U.M.S.T. Pt II, 1957, part of question only)

7.9 The distribution of load over a small portion of a beam

The idea of a force or a couple which is concentrated at a single point is only an idealization and is not attainable in a real system. In this section we shall discuss the differences which arise when a force is distributed over a finite length instead of being concentrated at a point and, while the discussion will concern vibration problems, the approach is also applicable to static loading. The main purpose of the section is to indicate some of the limitations of the theory of beam flexure which has been developed above.

Let a beam be subjected to a force $W e^{i\omega t}$ which is uniformly distributed over a short segment 2ϵ, the centre of the segment being at the section of the beam where $x = h$. The arrangement is shown in fig. 7.9.1 and the loading may be expressed mathematically as follows:

$$w = \begin{cases} 0 & 0 \leqslant x < h - \epsilon, \\ w_0 e^{i\omega t} & h - \epsilon \leqslant x \leqslant h + \epsilon, \\ 0 & h + \epsilon < x \leqslant l, \end{cases} \tag{7.9.1}$$

where

$$w_0 = \frac{W}{2\epsilon}. \tag{7.9.2}$$

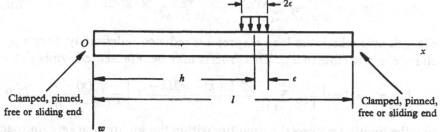

Fig. 7.9.1

Clamped, pinned, free or sliding end

Clamped, pinned, free or sliding end

It was shown in §4.8 that that component of the loading which produces motion in the rth principal mode may be found by treating the loading w as if it were a deflexion so that

$$w_r(x, t) = \frac{1}{Z}\left\{\int_0^l w(x, t)\, \phi_r(x)\, dx\right\} \phi_r(x). \tag{7.9.3}$$

Here, then,

$$w_r(x, t) = \frac{w_0 e^{i\omega t}}{Z}\left\{\int_{h-\epsilon}^{h+\epsilon} \phi_r(x)\, dx\right\} \phi_r(x), \tag{7.9.4}$$

where Z and $\phi_r(x)$ refer to the particular type of beam under consideration. This definite integral occurred earlier, in equation (7.7.12), and on evaluation it yields

$$w_r(x, t) = \frac{w_0 e^{i\omega t}}{Z\lambda_r}\left[\phi_r'''(h + \epsilon) - \phi_r'''(h - \epsilon)\right] \phi_r(x). \tag{7.9.5}$$

The response $v_r(x, t)$ of the beam to this loading may be found from equation (7.1.7) by seeking a solution of the form

$$v_r = (\text{constant})\, e^{i\omega t}\, \phi_r(x). \tag{7.9.6}$$

By this means it is found that

$$v_r = \frac{w_r}{A\rho(\omega_r^2 - \omega^2)}. \tag{7.9.7}$$

The full expression for the deflexion of the beam in its rth mode, due to the given loading, is therefore

$$v_r(x, t) = \frac{w_0 e^{i\omega t}\left[\phi_r'''(h + \epsilon) - \phi_r'''(h - \epsilon)\right]}{Z\lambda_r A\rho(\omega_r^2 - \omega^2)} \phi_r(x). \tag{7.9.8}$$

The total response, which is the sum of the responses in all the modes may now be written down. If this is done and the expression for w_0 is substituted from equation (7.9.2), it is found that

$$v(x, t) = \left\{\sum_{r=1}^{\infty} \frac{\left[\phi_r'''(h + \epsilon) - \phi_r'''(h - \epsilon)\right] \phi_r(x)}{2\epsilon\lambda_r A\rho Z(\omega_r^2 - \omega^2)}\right\} W e^{i\omega t}, \tag{7.9.9}$$

where it has again been assumed for simplicity of notation that there is no rigid body mode.

The function within the curly brackets is a receptance, though it differs slightly from the types of receptance which have been used up till now. It relates the displacement at a point, to a load which is distributed over a short length; such a load is not applied 'at a particular co-ordinate' in the sense in which this phrase has been used previously. The receptance will be denoted by α'_{xh} so that the suffix h refers to the section of the beam which is at the centre of the applied load. That is to say

$$\alpha'_{xh} = \frac{1}{2\epsilon A\rho Z}\sum_{r=1}^{\infty} \frac{[\phi_r'''(h+\epsilon) - \phi_r'''(h-\epsilon)]\,\phi_r(x)}{\lambda_r(\omega_r^2 - \omega^2)}. \qquad (7.9.10)$$

If the length over which the load is spread is reduced indefinitely, then α'_{xh} will approach indefinitely close to α_{xh}. This process may be represented symbolically as

$$\lim_{\epsilon\to 0}\alpha'_{xh} = \lim_{\epsilon\to 0}\left\{\frac{1}{A\rho Z}\sum_{r=1}^{\infty}\left\{\left[\frac{\phi_r'''(h+\epsilon) - \phi_r'''(h-\epsilon)}{2\epsilon}\right]\frac{\phi_r(x)}{\lambda_r(\omega_r^2 - \omega^2)}\right\}\right\} \qquad (7.9.11)$$

and during the limiting process the quantity within the square brackets approaches the value

$$\left(\frac{d\phi_r'''}{dx}\right)_{x=h} = \frac{1}{\lambda_r^3}\left(\frac{d^4\phi_r}{dx^4}\right)_{x=h} = \lambda_r\phi_r^{iv}(h) = \lambda_r\phi_r(h) \qquad (7.9.12)$$

by virtue of equation (7.6.36). Therefore

$$\lim_{\epsilon\to 0}\alpha'_{xh} = \frac{1}{A\rho Z}\sum_{r=1}^{\infty}\frac{\phi_r(h)\,\phi_r(x)}{\omega_r^2 - \omega^2} \qquad (7.9.13)$$

and this is identical with the series for α_{xh}, as would be expected from the above reasoning.

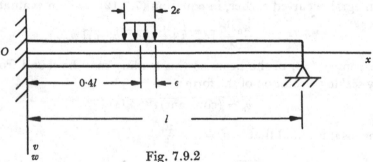

Fig. 7.9.2

When a real beam is subjected to a single alternating 'concentrated' force, that force is distributed over a small but finite length of the span. It is not likely that the distribution will be uniform but nevertheless the assumption of uniformity will probably be nearer to the truth than the assumption that the load is concentrated at a point. We shall examine a numerical case of loading of this kind in order to compare it with point loading. The method which will be used was chosen for its clarity rather than for its mathematical elegance.

Consider the clamped-pinned beam of fig. 7.9.2 which is loaded over a short length of which the mid-point is at $x = h = 0.4l$. We examine first the case when $2\epsilon = 0.12l$. From equation (7.9.9) it is seen that

$$v(x, t) = \left[\frac{-0.905\,66 + 1.470\,82}{0.12 \times 3.9266} \frac{\phi_1(x)}{\omega_1^2 - \omega^2} + \frac{1.064\,96 + 0.023\,81}{0.12 \times 7.0686} \frac{\phi_2(x)}{\omega_2^2 - \omega^2} + \dots \right] \frac{W e^{i\omega t}}{A\rho}. \tag{7.9.14}$$

The numerical values in this series are found in Table 7.2 (d). The expression reduces to

$$v(x, t) = \left[1.20\left(\frac{\phi_1(x)}{\omega_1^2 - \omega^2}\right) + 1.28\left(\frac{\phi_2(x)}{\omega_2^2 - \omega^2}\right) - 0.19\left(\frac{\phi_3(x)}{\omega_3^2 - \omega^2}\right) \right.$$
$$\left. - 1.25\left(\frac{\phi_4(x)}{\omega_4^2 - \omega^2}\right) - 0.54\left(\frac{\phi_5(x)}{\omega_5^2 - \omega^2}\right) + \dots \right] \frac{W e^{i\omega t}}{A\rho}. \tag{7.9.15}$$

Similar series may be found for other values of $2\epsilon/l$, the values of the coefficients in the series being dependent on this quantity. The accompanying table gives the coefficients for values of $2\epsilon/l$ between 0.12 and zero, the coefficients for the zero value being taken from equation (7.9.13).

$2\epsilon/l$	Coefficients of $\phi_r(x)/(\omega_r^2 - \omega^2)$				
	$r = 1$	$r = 2$	$r = 3$	$r = 4$	$r = 5$
0.12	1.20	1.28	−0.19	−1.25	−0.54
0.08	1.20	1.30	−0.20	−1.35	−0.60
0.04	1.21	1.31	−0.20	−1.38	−0.63
0.00	1.21	1.32	−0.20	−1.39	−0.64

It will be seen from the table that, as the length over which the load is spread is increased, the amplitudes of the lower modes is scarcely affected while those of the higher modes are progressively reduced. We shall show that the effect becomes steadily more pronounced, the higher the mode.

These characteristics apply to beams with all kinds of end-conditions. They lead to the following considerations which are pertinent whenever an applied load is slightly distributed, rather than concentrated:

(a) the amplitudes of the lower modes are given closely and simply by the first terms in the series receptance for a concentrated load.

(b) the amplitudes of the higher modes are over-estimated by the later terms in the series receptance for the concentrated load so that it is pointless to strive after great accuracy by taking into account a very large number of these terms.

(c) the receptance in the closed form for a concentrated load allows for all the higher modes in full and will also therefore introduce inaccuracies.

In order to show that these ideas are general, and do not apply merely to the particular beam which was examined above, consider the form of the various characteristic functions. Now any principal mode of a uniform beam will contain a definite number of nodes. For instance, Table 7.2 (b) shows that a clamped-clamped beam, vibrating in its fourth mode has nodes of deflexion at

$$x = 0, \quad 0.28l, \quad 0.50l, \quad 0.72l, \quad l.$$

Further, the number of nodes in the next higher mode is one greater. Thus the fifth mode has nodes at

$$x = 0, \quad 0{\cdot}23l, \quad 0{\cdot}41l, \quad 0{\cdot}59l, \quad 0{\cdot}77l, \quad l.$$

This shows that the number of nodes in a principal mode is, at least approximately, proportional to the number r of that mode provided that r is not small. Similar results will be found to hold for beams with other end conditions. In what follows, we shall be concerned more with the segments of beam between adjacent nodes rather than with the nodes themselves. The number of these segments, then, will increase with r. It will also be found, if the roots of the various frequency equations are examined, that for high values of r the approximate relation

$$\lambda_r l \propto r \tag{7.9.16}$$

holds. The approximation improves continuously as r is increased.

Now let equation (7.9.10) be written in the form

$$\alpha'_{xh} = \frac{1}{A\rho Z} \sum_{r=1}^{\infty} \frac{[\phi_r'''(h+\epsilon) - \phi_r'''(h-\epsilon)]\,\phi_r(x)}{\psi_r(\omega_r^2 - \omega^2)}, \tag{7.9.17}$$

where ψ_r is given by

$$\psi_r = \frac{2\epsilon}{l}\lambda_r l. \tag{7.9.18}$$

From the relation (7.9.16), and also from the relations between r and the number of segments, it follows that

$$\psi_r \propto 2\epsilon \frac{\text{(number of segments)}}{l} \tag{7.9.19}$$

for large values of r.

It may further be shown that the lengths of the segments become approximately equal to one another as r is increased and equation (7.9.19) may therefore be written

$$\psi_r \propto \frac{2\epsilon}{\text{(length of segment)}}. \tag{7.9.20}$$

The quantity ψ_r is thus proportional to the ratio of the length over which the load is distributed to the length of beam between adjacent nodes. This ratio will increase with r for a given load.

It is now necessary to examine the effect of these relations upon the various terms of the series (7.9.17). Let 2ϵ be much smaller than l. For the early terms of the series, ψ_r will be small because the denominator in equation (7.9.20) will be large compared with the numerator. The factor

$$[\phi_r'''(h+\epsilon) - \phi_r'''(h-\epsilon)]$$

will also be small for these terms because the length ϵ represents a small fraction of a segment, that is to say it is a small portion of the length in which ϕ_r changes from zero through a maximum to zero again. The first few terms of the series are thus the quotients of small quantities and will be found to be approximately equal to the limiting values which hold when $\epsilon \to 0$.

The later terms of the series are fractions for which the numerators, though not necessarily small, are always finite. This follows from the fact that $\phi_r'''(x)$ is a function which remains finite throughout the range $0 \leqslant x \leqslant l$; it will be found, for instance, that in a clamped-free beam the factor

$$[\phi_r'''(h+\epsilon) - \phi_r'''(h-\epsilon)]$$

cannot exceed 4. The denominators of the series, on the other hand, increase with r as is shown by equation (7.9.20) and this introduces a converging effect which is additional to that which is provided by the ω_r^2 in the denominators.

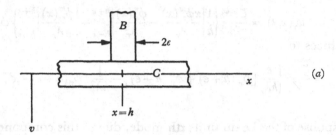

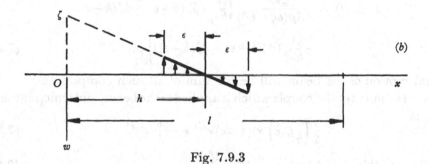

Fig. 7.9.3

Just as a single force which is applied to a beam is spread in reality over a finite length, so also is an applied couple. For instance if B and C of fig. 7.9.3 (a) represent portions of two beams which are rigidly attached to each other at right angles then, during vibration, the beam B will exert a couple on beam C, the couple being in fact spread over a length of C which is equal to the depth of B. An approximation toward the true load distribution which B exerts on C may be obtained by assuming that the stress distribution in B immediately next to C, is that due to simple bending. This load distribution is represented in fig. 7.9.3 (b) where 2ϵ has been written for the total depth of B and it may be expressed mathematically in the following form:

$$w(x, t) = \begin{cases} 0 & 0 \leqslant x < h-\epsilon, \\ \zeta\left(\dfrac{x}{h}-1\right) e^{i\omega t} & h-\epsilon \leqslant x \leqslant h+\epsilon, \\ 0 & h+\epsilon < x \leqslant l. \end{cases} \qquad (7.9.21)$$

The meaning of the factor ζ will be seen from the figure.

Now that component of the loading which produces motion of the beam in its rth principal mode is

$$w_r(x, t) = \frac{\zeta e^{i\omega t}}{Z} \left\{ \int_0^l \left(\frac{x}{h} - 1 \right) \phi_r(x) \, dx \right\} \phi_r(x)$$

$$= \frac{\zeta e^{i\omega t}}{Z} \left\{ \frac{1}{h} \int_{h-\epsilon}^{h+\epsilon} x \phi_r(x) \, dx - \int_{h-\epsilon}^{h+\epsilon} \phi_r(x) \, dx \right\} \phi_r(x). \quad (7.9.22)$$

The first of these integrals may be evaluated by parts and the second has been dealt with in equation (7.9.5). By these means it is found that

$$w_r(x, t) = \frac{\zeta e^{i\omega t}}{Z} \left\{ \frac{1}{h} \left| \frac{x \phi_r'''(x)}{\lambda_r} - \frac{\phi_r''(x)}{\lambda_r^2} \right|_{h-\epsilon}^{h+\epsilon} - \left| \frac{\phi_r'''(x)}{\lambda_r} \right|_{h-\epsilon}^{h+\epsilon} \right\} \phi_r(x) \quad (7.9.23)$$

which reduces to

$$w_r(x, t) = \frac{\zeta e^{i\omega t}}{Z} \left\{ \frac{\epsilon}{h\lambda_r} [\phi_r'''(h+\epsilon) + \phi_r'''(h-\epsilon)] - \frac{1}{h\lambda_r^2} [\phi_r''(h+\epsilon) - \phi_r''(h-\epsilon)] \right\} \phi_r(x).$$
$$(7.9.24)$$

The response of the beam in its rth mode, due to this component of the loading, will be

$$v_r(x, t) = \frac{\zeta e^{i\omega t}}{ZA\rho(\omega_r^2 - \omega^2)} \left\{ \frac{\epsilon}{h\lambda_r} [\phi_r'''(h+\epsilon) + \phi_r'''(h-\epsilon)] \right.$$

$$\left. - \frac{1}{h\lambda_r^2} [\phi_r''(h+\epsilon) - \phi_r''(h-\epsilon)] \right\} \phi_r(x). \quad (7.9.25)$$

The total motion of the beam will be the sum of all such components.

The total moment of the couple which is applied to the beam, and which causes this motion, is

$$\frac{1}{2} \left(\frac{\epsilon}{h} \zeta \cdot \epsilon \right) \times (\tfrac{2}{3} \cdot 2\epsilon) e^{i\omega t} = \frac{2\zeta \epsilon^3}{3h} e^{i\omega t}. \quad (7.9.26)$$

We shall write this as $He^{i\omega t}$ where $\quad H = \dfrac{2\zeta \epsilon^3}{3h}. \quad (7.9.27)$

If ζ is eliminated between equations (7.9.25) and (7.9.27) and the motions in all the modes are summed together, then a force-displacement relation is found which is, in the absence of rigid body modes,

$$v(x, t) = \left\{ \sum_{r=1}^{\infty} \frac{3\{\epsilon\lambda_r[\phi_r'''(h+\epsilon) + \phi_r'''(h-\epsilon)] - [\phi_r''(h+\epsilon) - \phi_r''(h-\epsilon)]\} \phi_r(x)}{2\epsilon^3 \lambda_r^2 ZA\rho(\omega_r^2 - \omega^2)} \right\} He^{i\omega t}.$$
$$(7.9.28)$$

Again we introduce a receptance notation so that equation (7.9.28) may be written in the form

$$v(x, t) = \alpha'_{xh'} He^{i\omega t}. \quad (7.9.29)$$

The behaviour of the function $\alpha'_{xh'}$, as ϵ is varied, may be studied in a similar way to that of α'_{xh}, but the treatment is more involved. We shall not, therefore, do this here, but take up a numerical example instead.

Let the clamped-pinned beam, shown in fig. 7.9.4, be acted on by a harmonic couple $He^{i\omega t}$ at the section $x = 0.4l$, the force distribution being centred around that section as in the figure. If $\epsilon \neq 0$, the deflexion may be found from equation

(7.9.28) whereas if the couple is concentrated (so that $\epsilon = 0$) equation (7.8.15) may be used. In each case, a series is found having the form

$$v(x, t) = \frac{H e^{i\omega t}}{A\rho l^2} \sum_{r=1}^{\infty} C_r \frac{\phi_r(x)}{\omega_r^2 - \omega^2},$$ (7.9.30)

where the values of the numerical coefficients C_r depend upon ϵ. Their values for various values of ϵ/l are shown in the accompanying table.

$2\epsilon/l$	Values of C_r				
	$r = 1$	$r = 2$	$r = 3$	$r = 4$	$r = 5$
0·12	3·08	−4·88	−13·91	−2·84	18·79
0·08	3·10†	−4·95	−14·20	−2·94	19·87
0·04	3·04†	−4·95	−14·37	−3·00	20·53
0·00	3·10	−4·96	−14·43	−3·02	20·76

† The values of those entries in this table which are marked with a dagger are as accurate as it is possible to make them when using Table 7·2 (d); owing, however, to the analytical form of the expression for $\alpha_{xh'}$, which involves the small difference between two much larger numbers, they are subject to some error.

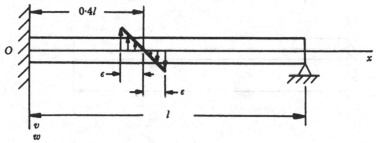

Fig. 7.9.4

The table shows that the early terms of $\alpha'_{xh'}$ are very nearly equal to those of the series for $\alpha_{xh'}$ but that the discrepancy between the two becomes appreciable at $r = 5$. This relation between the functions is similar to that which was found to relate α_{xh} and α'_{xh}; similar considerations about the use of the former still therefore apply.

The above computations are merely examples which illustrate the danger of using closed-form receptances, or large numbers of terms of series receptances, when treating frame and other beam problems. Similar considerations apply for other types of beam and types of loading and must always be remembered, particularly if the frequencies involved are higher than the fundamental frequencies of the beams involved. In particular, if one member of a frame has natural frequencies which are appreciably lower than those of the other members then it may merit special treatment.

It may also be noted that there is a second source of error when connexions between beams (such as are shown in fig. 7.9.3 (a)) are treated as points on the centre-lines of the connected beams. This second source arises from the stiffening effect which a welded connexion (say) of B will exert over the contiguous length of C. It is likely that this local stiffening will not greatly affect those modes of vibration for which the distance between the nodes greatly exceeds the length in question, because such modes will not involve appreciable flexure over that length. Thus the same considerations apply to the inaccuracies due to the stiffening as to the effects of distributing forces slightly.

From what has been said in this section, the reader may conclude that accurate calculation of frame vibrations is not always possible, even with all the receptance data which is available. It may sometimes be necessary therefore to resort to experiment at an early stage in an investigation.

7.10 Rayleigh's Principle

In this section, we shall discuss the application of Rayleigh's Principle, which is commonly used for calculating the lowest natural frequencies of beams. While it might be supposed that such a simple procedure could have been described earlier, it has been left till this point because our aim, throughout this book, has been to explain principles before practice. The experienced calculator would doubtless go direct to this simple method before embarking on more sophisticated processes.

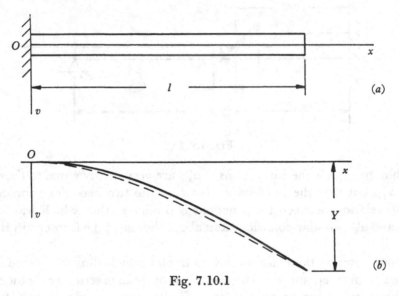

Fig. 7.10.1

It is, however, desirable that the reader should understand the approximations which are involved lest he should be tempted to use them where they are not justified. The nature of the approximations will be better appreciated in the light of the more complete theory which has gone before.

The use of the method of this section is often desirable as a first attack on a problem. The results so found may form a useful check on later results which, because of their mathematical complexity, are more liable to arithmetical errors.

Fig. 7.10.1 (a) shows a clamped-free beam and a suitable constrained mode, which may be used for the Rayleigh process, is shown by the full-line curve in fig. 7.10.1 (b). The shape of this mode is

$$v = Y\left(1 - \cos\frac{\pi x}{2l}\right), \tag{7.10.1}$$

where Y is a constant (the true first mode being shown dotted in the figure). The deflexion, when the beam is vibrating, may thus be written

$$v = p_1\left(1 - \cos\frac{\pi x}{2l}\right), \tag{7.10.2}$$

where p_1 is a principal (and, for this constrained motion, the only) co-ordinate. If this expression is substituted into the energy expressions (7.1.9) and (7.1.11), it gives

$$2T = A\rho \int_0^l \left(\frac{\partial v}{\partial t}\right)^2 dx = A\rho \dot{p}_1^2 \int_0^l \left(1 - \cos\frac{\pi x}{2l}\right)^2 dx$$

$$= A\rho l \dot{p}_1^2 \left(\frac{3\pi - 8}{2\pi}\right) = a_1 \dot{p}_1^2, \tag{7.10.3}$$

$$2V = EI \int_0^l \left(\frac{\partial^2 v}{\partial x^2}\right)^2 dx = EI p_1^2 \left(\frac{\pi^2}{4l^2}\right)^2 \int_0^l \cos^2\frac{\pi x}{2l} dx$$

$$= EI p_1^2 \frac{\pi^4}{32l^3} = c_1 p_1^2. \tag{7.10.4}$$

Since the motion is harmonic, it follows that

$$\omega_1^2 \doteq \omega_c^2 = \frac{c_1}{a_1} = \frac{\pi^5}{16(3\pi - 8)}\frac{EI}{A\rho l^4} = 13 \cdot 45 \frac{EI}{A\rho l^4}. \tag{7.10.5}$$

Remembering that $(\lambda_1 l)^4 = \omega_1^2 A\rho l^4/EI$, it will be seen from Table 7.2 (c) that the numerical factor should be 12·36. The error in ω_1 is 4·3 %.

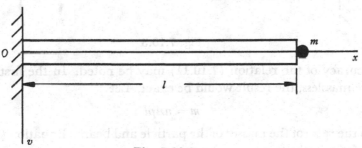

Fig. 7.10.2

It was shown in §5.3 that, when the natural frequencies of certain systems embodying taut strings were estimated, it was convenient to use the static deflexion curve due to gravity as an approximation to the shape of the first mode. This idea is also applicable to beams. Consider the clamped-free beam shown in fig. 7.10.2 which carries a concentrated mass m at its free end. The static deflexion curve of the beam, under the action of the particle's weight only, is

$$v = a_{xl}F = \frac{F}{6EI}(3lx^2 - x^3), \tag{7.10.6}$$

where a_{zl} is a 'flexibility' whose value is given in Table 7.3 (a).† If this is used as an approximate modal shape, and if the corresponding principal co-ordinate is taken as the tip deflexion so that

$$p_1 = v_l = \frac{Fl^3}{3EI} \tag{7.10.7}$$

and

$$v = p_1 \left(\frac{3lx^2 - x^3}{2l^3} \right), \tag{7.10.8}$$

then the energy expressions are

$$2V = p_1 F = \frac{3EI}{l^3} p_1^2 = c_1 p_1^2 \tag{7.10.9}$$

and

$$2T = p_1^2 A\rho \int_0^l \left(\frac{3lx^2 - x^3}{2l^3} \right)^2 dx + m\dot{p}_1^2$$

$$= \left[\left(\frac{33A\rho l}{140} \right) + m \right] \dot{p}_1^2 = a_1 \dot{p}_1^2. \tag{7.10.10}$$

These give the fundamental frequency as

$$\omega_1 \doteq \omega_c = \sqrt{\left(\frac{c_1}{a_1} \right)} = \sqrt{\left\{ \frac{3EI}{l^3 \left[\left(\frac{33A\rho l}{140} \right) + m \right]} \right\}}. \tag{7.10.11}$$

In this calculation the deflexion curve was used which is due to the weight of the concentrated mass only, instead of that due to both mass and beam. This is done because the function is simpler and for no other reason.

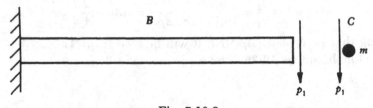

Fig. 7.10.3

The accuracy of the relation (7.10.11) may be noted. In the first place, if the beam were massless, the result would be exact. Let

$$m = nA\rho l \tag{7.10.12}$$

so that n is the ratio of the masses of the particle and beam. Equation (7.10.11) may then be written in the form

$$\omega_1 \doteq \sqrt{\left[\frac{3}{\left(\frac{33}{140} + n \right)} \left(\frac{EI}{A\rho l^4} \right) \right]}. \tag{7.10.13}$$

This may be compared with the exact result, which may be found by splitting the system into two sub-systems as shown in fig. 7.10.3. The frequency equation

$$\beta_{11} + \gamma_{11} = 0 \tag{7.10.14}$$

---

† The convention used for the subscripts of the flexibilities in Tables 7.3 (a), (b) and (c) is the same as that of the beam receptances. The flexibilities are the limiting values to which the receptances tend as the forcing frequency ω is diminished to zero. They are found by the methods of elementary Strength of Materials.

may be used, in which

$$\beta_{11} = \frac{-F_5}{EI\lambda^3 F_4}, \quad \gamma_{11} = \frac{-1}{m\omega^2}, \tag{7.10.15}$$

the former being taken from Table 7.1 (c). Since

$$m = nA\rho l \quad \text{and} \quad \lambda^4 = \frac{\omega^2 A\rho}{EI} \tag{7.10.16}$$

the frequency equation reduces to $\quad \lambda l = \frac{-F_4}{nF_5}$. $\tag{7.10.17}$

The first root of this equation may be obtained from Table 7.1 (d) for any particular value of n. It will be found that the error in ω_1, if it is calculated by the approximate method, is 1·45 % when $n = 0$, 0·05 % if $n = 1$ and smaller than the normal errors of calculation if $n = 2$. Even in the extreme case therefore, when there is no particle attached to the cantilever, the method gives a good estimate.

It may be mentioned at this point that the natural frequencies and modes of a beam are altered if it has an axial load applied to it. If the load is tensile the natural frequencies are raised because, superimposed upon the restoring force due to flexural stiffness, there is also a 'taut string' effect; it is sometimes possible, then, to apply Southwell's Theorem in the manner of §5.6. If the load is compressive the frequencies are lowered, finally becoming zero at the critical buckling load of the beam as a strut. The subject has been discussed in terms of receptances by Duncan.†

EXAMPLES 7.10

1. Estimate the lowest natural frequency of flexural vibration for a uniform clamped-clamped beam of length l, cross-sectional area A, density ρ and flexural rigidity EI. Compare the estimate with the exact result given in Table 7.2 (b).

2. A steel shaft 4 ft. long is carried at each end in bearings which do not constrain it in direction. The middle 1 ft. of its length has a diameter of 2 in. and for a length of 1 ft. 6 in. at each end the diameter is 1 in.
Using an approximate method, estimate the lowest natural frequency of the shaft.
Density: 480 lb./ft.³; Young's modulus: 30 × 10⁶ psi.

(C.U.M.S.T. Pt ɪɪ, 1948)

3. A $\frac{1}{4}$ in × $\frac{1}{8}$ in. steel strip of length 20 in. is clamped at one end and free at the other. A sharp right angle bend is made in this strip at the point 10 in. from each end, about an axis parallel to the sides of length $\frac{1}{4}$ in. Estimate the lowest natural frequency of flexural vibration of the bent cantilever in its own plane.
(For steel: Young's modulus = 30 × 10⁶ psi; density = 0·278 lb./in.³)

(C.U.M.S.T. Pt ɪɪ, 1955, part of question only)

4. A straight shaft of circular section is supported in end bearings which do not constrain the direction of the shaft. The mass per unit length of the shaft is m, and its flexural rigidity is EI.
Show from first principles that, if the shaft is subjected to an axial pull P which is small in relation to the Euler critical load for the shaft, each of the natural frequencies of flexural vibration will be raised by approximately $P/\sqrt{(4EIm)}$ rad./sec.

(C.U.M.S.T. Pt ɪɪ, 1954)

---

† W. J. Duncan, 'Mechanical Admittances and their Applications to Oscillation Problems', *R. and M. 2000* (1947), p. 113.

5. A support is made in the form of a rectangular portal frame *ABCD* which is clamped at *A* and *D*. The uprights *AB* and *DC* each have flexural rigidity 0.625×10^6 lb.in.^{2} and the top member *BC* is effectively rigid. *AB* and *DC* are both of length 30 in. Each upright has a uniformly distributed mass of 9 lb. and the top member has a mass of 40 lb. Estimate the lowest natural frequency of the frame in its own plane. The effect of axial load on the stiffness of the uprights may be neglected.

Give reasons why it is to be expected that the second natural frequency of the frame in its own plane is likely to be much higher than the first and describe briefly how this could be verified by calculation.

(C.U.M.S.T. Pt II, 1957)

7.11 Steady wave motions in uniform beams

We concluded Chapter 6 with a short discussion of steady wave motions in uniform shafts and bars and we shall now end the present chapter by doing the same for uniform beams in bending. It will be shown that the problem of waves in a beam is more complicated than that of waves in shafts or in bars. Certain useful results can easily be arrived at, however, although their technical significance will become more apparent when the effects of dissipation are taken up.

The wave expression which is similar to that of equations (6.9.1) and (6.9.5) is

$$v = (\text{constant}) \, (\cos \lambda x + i \sin \lambda x) \, e^{i\omega t}. \qquad (7.11.1)$$

This satisfies the equation of motion

$$\frac{\partial^2 v}{\partial t^2} + \frac{EI}{A\rho} \frac{\partial^4 v}{\partial x^4} = 0 \qquad (7.11.2)$$

provided that, as usual, $\qquad \lambda^4 = \dfrac{A\rho\omega^2}{EI}. \qquad (7.11.3)$

The solution (7.11.1) represents a travelling wave in the same way as before and the speed of the wave is found by writing equation (7.11.1) in the form

$$v = (\text{constant}) \, e^{i(\omega t + \lambda x)}. \qquad (7.11.4)$$

Thus the wave propagation speed is equal to ω/λ which may be written as

$$\omega^{\frac{1}{2}} \left[\frac{EI}{A\rho} \right]^{\frac{1}{4}}. \qquad (7.11.5)$$

This expression for the wave speed contains $\sqrt{\omega}$ so that there is a fundamental difference between this problem and that of the torsional motion of a shaft. In the latter, the wave speed was independent of the frequency. Here, the higher the frequency, the greater is the wave speed; very high-frequency oscillations are associated with very high wave-speeds. This latter prediction is not, in fact, correct as the theory does not apply to these high-frequency oscillations. This is because the assumptions made in deriving equation (7.11.2) become untenable for motions with very high frequencies.† The theory underlying equation (7.11.2) is, however, quite adequate for most practical purposes.

† See, for example, S. P. Timoshenko, *Vibration Problems in Engineering* (Van Nostrand, New York, 3rd ed. 1955), §§51 and 52.

There is a further complicating factor which must be allowed for before we treat the flexural wave motions in the same way as the torsional. It has already been shown that solutions of equation (7.11.2) may contain hyperbolic functions, these being necessary to satisfy a given set of end conditions. This necessity of satisfying end conditions still exists in the present problem as may be seen by considering the wave motion of a semi-infinite bar. Consider the beam shown in fig. 7.11.1 which is acted on by a bending couple of magnitude $H e^{i\omega t}$ at $x = 0$ and let the shearing force at this point be zero. If the bar extends to infinity in the direction of positive x, then waves may be expected which are of the form

$$v = (\text{constant}) (\cos \lambda x - i \sin \lambda x) e^{i\omega t}. \tag{7.11.6}$$

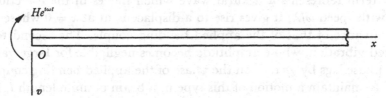

Fig. 7.11.1

Now the bending moment present at any section of the bar will be, under these circumstances,

$$M = EI\frac{\partial^2 v}{\partial x^2} = -(\text{constant}).EI\lambda^2(\cos \lambda x - i \sin \lambda x) e^{i\omega t}, \tag{7.11.7}$$

while the shear force will be

$$S = -EI\frac{\partial^3 v}{\partial x^3} = (\text{constant}).EI\lambda^3(\sin \lambda x + i \cos \lambda x) e^{i\omega t}. \tag{7.11.8}$$

Evidently if the constant is chosen to make the expression (7.11.7) equal to $-H e^{i\omega t}$ when $x = 0$ then it will not make the expression (7.11.8) equal to zero.

To overcome this difficulty, consider the solution

$$v = C(\cosh \lambda x - \sinh \lambda x) e^{i\omega t} = C e^{-\lambda x} e^{i\omega t}, \tag{7.11.9}$$

where C is a constant. This represents a steady motion which becomes negligibly small for large values of x, but which gives a finite bending moment and shear force at the origin. If this solution is combined with that of equation (7.11.6) so as to give a general expression

$$v = [B(\cos \lambda x - i \sin \lambda x) + C e^{-\lambda x}] e^{i\omega t} \tag{7.11.10}$$

then the two constants B and C may be chosen to satisfy the two end conditions at the origin. The relevant equations are

$$\left.\begin{aligned} H e^{i\omega t} &= -EI\left(\frac{\partial^2 v}{\partial x^2}\right)_{x=0} = EI\lambda^2(B-C) e^{i\omega t}, \\ 0 &= EI\left(\frac{\partial^3 v}{\partial x^3}\right)_{x=0} = EI\lambda^3(iB-C) e^{i\omega t}. \end{aligned}\right\} \tag{7.11.11}$$

From these relations, it is found that

$$B = \frac{H(1+i)}{2EI\lambda^2},$$
$$C = \frac{H(i-1)}{2EI\lambda^2}.$$

$$(7.11.12)$$

If these values are now inserted in equation (7.11.10) a solution is found which may be written in the form

$$v = \frac{H}{EI\lambda^2 \sqrt{2}} [e^{i(\omega t - \lambda x + \frac{1}{4}\pi)} - e^{-\lambda x} e^{i(\omega t - \frac{1}{4}\pi)}].$$

$$(7.11.13)$$

The first term represents a flexural wave which moves in the direction of increasing x with speed ω/λ; it gives rise to a displacement at $x = 0$ whose phase is $\frac{1}{4}\pi$ rad. in advance of that of the applied bending couple. The second term is a steady forced vibration, whose amplitude becomes negligible for large values of x and whose phase lags by $\frac{1}{4}\pi$ rad. on the phase of the applied bending couple.

In order to maintain a motion of this type in a beam of finite length l, it would be necessary to apply the bending couple $He^{i\omega t}$ at the section $x = 0$ together with the appropriate harmonic bending couple and lateral force at the other end, where $x = l$. These are the bending moment and shearing force which would exist at the section $x = l$ of the infinite beam; they are given by

$$EI\frac{\partial^2 v}{\partial x^2} \quad \text{and} \quad -EI\frac{\partial^3 v}{\partial x^3},$$

respectively. As will be found by forming the derivatives of v, the exciting couple and force are neither in phase with each other, nor with the couple $He^{i\omega t}$ which is applied at the end $x = 0$.

EXAMPLES 7.11

1. A semi-infinite beam has a pinned support at the end $x = 0$ and extends to infinity in the positive direction of x. A generalized force $He^{i\omega t}$, corresponding to the slope $\partial v/\partial x$, is applied to the beam at the pin. Show that the resulting deflexion at any section x is equal to

$$\frac{H}{2EI\lambda^2} [e^{i(\omega t - \lambda x)} - e^{-\lambda x} e^{i\omega t}].$$

If the beam were severed at the section $x = l$, what generalized forces would have to be applied at the cut to maintain the motion?

2. Show that the work done per cycle on the semi-infinite beam of the preceding example, by the applied couple $He^{i\omega t}$, is equal to

$$\frac{H^2\pi}{2EI\lambda}.$$

Show further that energy is transmitted at this rate across every section of the beam.

NOTATION

Supplementary List of Symbols used in Chapter 7

A	Area of cross-section of beam.
B_r	Coefficient of $\phi_r(x)$ in series representation of $w(x)$ (see equation (7.7.1)).
F	Amplitude of applied lateral force (see equation (7.2.7)).
$F_1, F_2, ..., F_{10}$	Abbreviations whose meanings are given in Table 7.1 (*a*); a letter subscript indicates a sub-system.
H	Amplitude of applied couple (see equations (7.2.8) and (7.9.27)).
M	Bending moment.
S	Shearing force.
W	Concentrated static force; see also equation (7.9.2).
w_0	See equation (7.9.1).
X	See equation (7.2.2).
X_r	See equation (7.6.16).
Z	See equation (7.6.63).
α	For the convention followed in assigning subscripts to the receptances, see the text following equation (7.2.16).
α'	See equations (7.9.10) and (7.9.29).
ϵ	See figs 7.9.1–7.9.4.
λ	$= \left[\dfrac{\omega^2 A\rho}{EI}\right]^{\frac{1}{4}}.$
λ_r	$= \left[\dfrac{\omega_r^2 A\rho}{EI}\right]^{\frac{1}{4}}.$
σ_r	Abbreviation used in expressions for the characteristic functions $\phi_r(x)$; see Tables 7.1 (*b*), (*c*), (*d*) and (*g*).
$\phi_r', \phi_r'',$ etc.	See equations (7.6.35).
ψ_r	See equation (7.9.18).

TABLE 7

FLEXURAL VIBRATION OF UNIFORM BEAMS

Notation

A	Area of cross-section.
a_{xh}, etc.	Flexibility between the sections at x and at h. If a static force F, acting in the direction of v is applied to a beam at the section where $x = h$, then the static deflexion and the slope for any x are

$$v = a_{xh}F, \quad \frac{dv}{dx} = a_{x'h}F.$$

If a static couple H, acting in the positive direction of dv/dx, is applied to the beam at any section $x = h$, then the resulting deflexion and slope at any section x are

$$v = a_{xh'}H, \quad \frac{dv}{dx} = a_{x'h'}H.$$

The flexibilities listed in Table 7.3 may be used as the receptances of the beams concerned when, for the purpose of vibration analysis, those beams may be considered as being devoid of mass. In accordance with Maxwell's Reciprocal Theorem, the subscripts of any flexibility may be reversed without changing its value; thus

$$a_{x'h} = a_{hx'}.$$

EI	Flexural rigidity of beam.
$F_1, F_2, ..., F_{10}$	Functions defined in Table 7·1 (*a*).
h	Particular value of x defining a section of a beam distant h from the origin $(0 \leqslant h \leqslant l)$.
l	Length of beam.
r, s	Subscripts used to indicate the rth and sth principal modes ($= 1, 2, 3, ...$ where there are no rigid body modes, $= 0, 1, 2, ...$ where there is one rigid body mode, $= -1, 0, 1, 2, ...$ where there are two rigid body modes).
v	Deflexion of beam.
x	Distance along the beam $(0 \leqslant x \leqslant l)$.
α_{xh}, etc.	Receptance between the sections x and h. If a harmonic lateral force $Fe^{i\omega t}$ is applied to the beam at the point $x = h$ (it being positive in the positive direction of v), then the response at any section x is given by

$$v = \alpha_{xh}Fe^{i\omega t}, \quad \frac{\partial v}{\partial x} = a_{x'h}Fe^{i\omega t}.$$

$\alpha_{xh'}$, etc. If a harmonic couple $He^{i\omega t}$ is applied to the beam at the point $x = h$ (it being positive in the positive direction of $\partial v/\partial x$), then the response at any section x is given by

$$v = \alpha_{xh'}He^{i\omega t}, \quad \frac{\partial v}{\partial x} = \alpha_{x'h'}He^{i\omega t}.$$

The order of the subscripts of α may be reversed without altering the value of the receptance; thus

$$\alpha_{x'h} = \alpha_{hx'}.$$

$\lambda \qquad = \left[\dfrac{\omega^2 A\rho}{EI}\right]^{\frac{1}{4}}.$

$\lambda_r \qquad = \left[\dfrac{\omega_r^2 A\rho}{EI}\right]^{\frac{1}{4}}.$

ρ \qquad Mass density.

σ_r \qquad Constant occurring in the expressions for the characteristic functions of the various types of beam.

Equation of motion: $\quad \dfrac{\partial^2 v}{\partial t^2} + \dfrac{EI}{A\rho}\dfrac{\partial^4 v}{\partial x^4} = 0.$

Boundary conditions: $\quad v = \dfrac{\partial v}{\partial x} = 0$ at a clamped end,

$$v = \frac{\partial^2 v}{\partial x^2} = 0 \text{ at a pinned end,}$$

$$\frac{\partial v}{\partial x} = \frac{\partial^3 v}{\partial x^3} = 0 \text{ at a sliding end,}$$

$$\frac{\partial^2 v}{\partial x^2} = \frac{\partial^3 v}{\partial x^3} = 0 \text{ at a free end.}$$

Energy expressions:

$$T = \frac{A\rho}{2}\int_0^l \left(\frac{\partial v}{\partial t}\right)^2 dx, \quad V = \frac{EI}{2}\int_0^l \left(\frac{\partial^2 v}{\partial x^2}\right)^2 dx.$$

Index of tables for flexural vibration of beams

7.1. Receptances:

(a) Notation used in Tables 7.1 (b), (c) and (d).

(b) Receptances of beams excited at an end.

(c) Tip receptances.

(d) Tables of the functions $F_1, F_2, ..., F_{10}$.

7.2. Characteristic functions and other data:
  (a) Pinned-pinned beam.
  (b) Clamped-clamped and free-free beams.
  (c) Clamped-free beam.
  (d) Clamped-pinned and free-pinned beams.
  (e) Sliding-pinned beam.
  (f) Sliding-sliding beam.
  (g) Clamped-sliding and free-sliding beams.

7.3. Flexibilities,:
  (a) Flexibilities of beams loaded at an end.
  (b) Tip flexibilities.
  (d) Flexibilities of clamped-clamped beams.

Note

It is a property of all the characteristic functions for distortion listed in Table 7.2 that

$$\phi_r^{iv}(x) = \phi_r(x), \quad \phi_r^{v}(x) = \phi_r'(x), \quad \phi_r^{vi}(x) = \phi_r''(x), \quad \text{etc.}$$

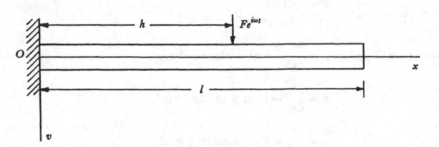

Clamped end   Free end

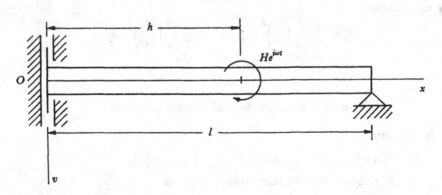

Sliding end   Pinned end

TABLE 7·1 (a)

NOTATION USED IN TABLES 7·1 (b), (c) AND (d)

$$F_1 = \sin \lambda l \sinh \lambda l.$$
$$F_2 = \cos \lambda l \cosh \lambda l.$$
$$F_3 = \cos \lambda l \cosh \lambda l - 1.$$
$$F_4 = \cos \lambda l \cosh \lambda l + 1.$$
$$F_5 = \cos \lambda l \sinh \lambda l - \sin \lambda l \cosh \lambda l.$$
$$F_6 = \cos \lambda l \sinh \lambda l + \sin \lambda l \cosh \lambda l.$$
$$F_7 = \sin \lambda l + \sinh \lambda l.$$
$$F_8 = \sin \lambda l - \sinh \lambda l.$$
$$F_9 = \cos \lambda l + \cosh \lambda l.$$
$$F_{10} = \cos \lambda l - \cosh \lambda l.$$

TABLE 7·1 (b)

RECEPTANCES OF BEAMS WHICH ARE EXCITED AT AN END

1	Free-free		
(a)	$\alpha_{x0} = \left\{ \dfrac{F_5(\cos \lambda x + \cosh \lambda x) + (F_1 + F_3)\sin \lambda x + (F_1 - F_3)\sinh \lambda x}{-2EI\lambda^3 F_3} \right\}$		
(b)	$\alpha_{x'0} = \left\{ \dfrac{(F_1 + F_3)\cos \lambda x + (F_1 - F_3)\cosh \lambda x - F_5(\sin \lambda x - \sinh \lambda x)}{-2EI\lambda^2 F_3} \right\}$		
(c)	$\alpha_{xl} = \left\{ \dfrac{F_8(\cos \lambda x + \cosh \lambda x) - F_{10}(\sin \lambda x + \sinh \lambda x)}{2EI\lambda^3 F_3} \right\}$		
(d)	$\alpha_{x'l} = \left\{ \dfrac{F_{10}(\cos \lambda x + \cosh \lambda x) + F_8(\sin \lambda x - \sinh \lambda x)}{-2EI\lambda^2 F_3} \right\}$		
(e)	$\alpha_{x0'} = \left\{ \dfrac{(F_1 - F_3)\cos \lambda x + (F_1 + F_3)\cosh \lambda x - F_6(\sin \lambda x + \sinh \lambda x)}{-2EI\lambda^2 F_3} \right\}$		
(f)	$\alpha_{x'0'} = \left\{ \dfrac{F_6(\cos \lambda x + \cosh \lambda x) + (F_1 - F_3)\sin \lambda x - (F_1 + F_3)\sinh \lambda x}{2EI\lambda F_3} \right\}$		
(g)	$\alpha_{xl'} = \left\{ \dfrac{F_{10}(\cos \lambda x + \cosh \lambda x) + F_7(\sin \lambda x + \sinh \lambda x)}{2EI\lambda^2 F_3} \right\}$		
(h)	$\alpha_{x'l'} = \left\{ \dfrac{F_7(\cos \lambda x + \cosh \lambda x) - F_{10}(\sin \lambda x - \sinh \lambda x)}{2EI\lambda F_3} \right\}$		

2	Clamped-free		
(a)	$\alpha_{xl} = \left\{ \dfrac{F_7(\cos \lambda x - \cosh \lambda x) - F_9(\sin \lambda x - \sinh \lambda x)}{-2EI\lambda^3 F_4} \right\}$		
(b)	$\alpha_{x'l} = \left\{ \dfrac{F_9(\cos \lambda x - \cosh \lambda x) + F_7(\sin \lambda x + \sinh \lambda x)}{2EI\lambda^2 F_4} \right\}$		
(c)	$\alpha_{xl'} = \left\{ \dfrac{F_9(\cos \lambda x - \cosh \lambda x) + F_8(\sin \lambda x - \sinh \lambda x)}{-2EI\lambda^2 F_4} \right\}$		
(d)	$\alpha_{x'l'} = \left\{ \dfrac{F_8(\cos \lambda x - \cosh \lambda x) - F_9(\sin \lambda x + \sinh \lambda x)}{-2EI\lambda F_4} \right\}$		

3	Clamped-pinned		
(a)	$\alpha_{xl'} = \left\{ \dfrac{F_8(\cos \lambda x - \cosh \lambda x) - F_{10}(\sin \lambda x - \sinh \lambda x)}{2EI\lambda^2 F_5} \right\}$		
(b)	$\alpha_{x'l'} = \left\{ \dfrac{F_{10}(\cos \lambda x - \cosh \lambda x) + F_8(\sin \lambda x + \sinh \lambda x)}{-2EI\lambda F_5} \right\}$		

TABLE 7·1 (b) (cont.)

4	Free-pinned
(a)	$\alpha_{x0} = \left\{ \dfrac{2F_1(\cos \lambda x + \cosh \lambda x) - (F_5 + F_6) \sin \lambda x + (F_5 - F_6) \sinh \lambda x}{2EI\lambda^3 F_5} \right\}$
(b)	$\alpha_{x'0} = \left\{ \dfrac{(F_5 + F_6) \cos \lambda x - (F_5 - F_6) \cosh \lambda x + 2F_1(\sin \lambda x - \sinh \lambda x)}{-2EI\lambda^2 F_5} \right\}$
(c)	$\alpha_{x0'} = \left\{ \dfrac{(F_5 - F_6) \cos \lambda x - (F_5 + F_6) \cosh \lambda x + 2F_2(\sin \lambda x + \sinh \lambda x)}{2EI\lambda^2 F_5} \right\}$
(d)	$\alpha_{x'0'} = \left\{ \dfrac{2F_2(\cos \lambda x + \cosh \lambda x) - (F_5 - F_6) \sin \lambda x - (F_5 + F_6) \sinh \lambda x}{2EI\lambda F_5} \right\}$
(e)	$\alpha_{xl'} = \left\{ \dfrac{F_7(\cos \lambda x + \cosh \lambda x) - F_9(\sin \lambda x + \sinh \lambda x)}{-2EI\lambda^2 F_5} \right\}$
(f)	$\alpha_{x'l'} = \left\{ \dfrac{F_9(\cos \lambda x + \cosh \lambda x) + F_7(\sin \lambda x - \sinh \lambda x)}{2EI\lambda F_5} \right\}$

5	Free-sliding
(a)	$\alpha_{x0} = \left\{ \dfrac{2F_2(\cos \lambda x + \cosh \lambda x) - (F_5 - F_6) \sin \lambda x - (F_5 + F_6) \sinh \lambda x}{-2EI\lambda^3 F_6} \right\}$
(b)	$\alpha_{x'0} = \left\{ \dfrac{(F_5 - F_6) \cos \lambda x + (F_5 + F_6) \cosh \lambda x + 2F_2(\sin \lambda x - \sinh \lambda x)}{2EI\lambda^2 F_6} \right\}$
(c)	$\alpha_{x0'} = \left\{ \dfrac{(F_5 + F_6) \cos \lambda x + (F_5 - F_6) \cosh \lambda x + 2F_1(\sin \lambda x + \sinh \lambda x)}{2EI\lambda^2 F_6} \right\}$
(d)	$\alpha_{x'0'} = \left\{ \dfrac{2F_1(\cos \lambda x + \cosh \lambda x) - (F_5 + F_6) \sin \lambda x + (F_5 - F_6) \sinh \lambda x}{2EI\lambda F_6} \right\}$
(e)	$\alpha_{xl} = \left\{ \dfrac{F_9(\cos \lambda x + \cosh \lambda x) + F_8(\sin \lambda x + \sinh \lambda x)}{-2EI\lambda^3 F_6} \right\}$
(f)	$\alpha_{x'l} = \left\{ \dfrac{F_8(\cos \lambda x + \cosh \lambda x) - F_9(\sin \lambda x - \sinh \lambda x)}{-2EI\lambda^2 F_6} \right\}$

6	Pinned-pinned
(a)	$\alpha_{x0'} = \left\{ \dfrac{2F_1(\cos \lambda x - \cosh \lambda x) - (F_5 + F_6) \sin \lambda x - (F_5 - F_6) \sinh \lambda x}{4EI\lambda^2 F_1} \right\}$
(b)	$\alpha_{x'0'} = \left\{ \dfrac{(F_5 + F_6) \cos \lambda x + (F_5 - F_6) \cosh \lambda x + 2F_1(\sin \lambda x + \sinh \lambda x)}{-4EI\lambda F_1} \right\}$
(c)	$\alpha_{xl'} = \left\{ \dfrac{(F_7 - F_8) \sin \lambda x - (F_7 + F_8) \sinh \lambda x}{-4EI\lambda^2 F_1} \right\}$
(d)	$\alpha_{x'l'} = \left\{ \dfrac{(F_7 - F_8) \cos \lambda x - (F_7 + F_8) \cosh \lambda x}{-4EI\lambda F_1} \right\}$

TABLE 7·1 (b) (cont.)

7	Sliding-pinned
(a)	$\alpha_{x0} = \left\{ \dfrac{(F_5 - F_6)\cos \lambda x + (F_5 + F_6)\cosh \lambda x + 2F_2(\sin \lambda x - \sinh \lambda x)}{-4EI\lambda^3 F_2} \right\}$
(b)	$\alpha_{x'0} = \left\{ \dfrac{2F_2(\cos \lambda x - \cosh \lambda x) - (F_5 - F_6)\sin \lambda x + (F_5 + F_6)\sinh \lambda x}{-4EI\lambda^2 F_2} \right\}$
(c)	$\alpha_{xl'} = \left\{ \dfrac{(F_9 - F_{10})\cos \lambda x - (F_9 + F_{10})\cosh \lambda x}{-4EI\lambda^2 F_2} \right\}$
(d)	$\alpha_{x'l'} = \left\{ \dfrac{(F_9 - F_{10})\sin \lambda x + (F_9 + F_{10})\sinh \lambda x}{4EI\lambda F_2} \right\}$

8	Clamped-sliding
(a)	$\alpha_{xl} = \left\{ \dfrac{F_{10}(\cos \lambda x - \cosh \lambda x) + F_7(\sin \lambda x - \sinh \lambda x)}{2EI\lambda^3 F_6} \right\}$
(b)	$\alpha_{x'l} = \left\{ \dfrac{F_7(\cos \lambda x - \cosh \lambda x) - F_{10}(\sin \lambda x + \sinh \lambda x)}{2EI\lambda^2 F_6} \right\}$

9	Sliding-sliding
(a)	$\alpha_{x0} = \left\{ \dfrac{(F_5 + F_6)\cos \lambda x - (F_5 - F_6)\cosh \lambda x + 2F_1(\sin \lambda x - \sinh \lambda x)}{-4EI\lambda^3 F_1} \right\}$
(b)	$\alpha_{x'0} = \left\{ \dfrac{2F_1(\cos \lambda x - \cosh \lambda x) - (F_5 + F_6)\sin \lambda x - (F_5 - F_6)\sinh \lambda x}{-4EI\lambda^2 F_1} \right\}$
(c)	$\alpha_{xl} = \left\{ \dfrac{(F_7 - F_8)\cos \lambda x + (F_7 + F_8)\cosh \lambda x}{-4EI\lambda^3 F_1} \right\}$
(d)	$\alpha_{x'l} = \left\{ \dfrac{(F_7 - F_8)\sin \lambda x - (F_7 + F_8)\sinh \lambda x}{4EI\lambda^2 F_1} \right\}$

TABLE 7·1(c)

TIP RECEPTANCES

#	Nature of support	End	α_{00}	$\alpha_{0'0}$ / $\alpha_{00'}$	α_{10} / α_{0l}	$\alpha_{1'0}$ / $\alpha_{0l'}$	$\alpha_{0'0'}$	$\alpha_{10'}$ / $\alpha_{0'l}$	$\alpha_{1'0'}$ / $\alpha_{0'l'}$	α_{ll}	$\alpha_{1'l}$ / $\alpha_{ll'}$	$\alpha_{l'l'}$	#
1	Free / Free	$x=0$ / $x=l$	$\dfrac{-F_5}{EI\lambda^3 F_3}$	$\dfrac{-F_1}{EI\lambda^2 F_3}$	$\dfrac{F_8}{EI\lambda^3 F_3}$	$\dfrac{F_{10}}{EI\lambda^2 F_3}$	$\dfrac{F_6}{EI\lambda F_3}$	$\dfrac{-F_{10}}{EI\lambda^2 F_3}$	$\dfrac{F_7}{EI\lambda F_3}$	$\dfrac{-F_5}{EI\lambda^3 F_3}$	$\dfrac{F_1}{EI\lambda^2 F_3}$	$\dfrac{F_6}{EI\lambda F_3}$	1
2	Clamped / Free	$x=0$ / $x=l$								$\dfrac{-F_5}{EI\lambda^3 F_4}$	$\dfrac{F_1}{EI\lambda^2 F_4}$	$\dfrac{F_6}{EI\lambda F_4}$	2
3	Clamped / Pinned	$x=0$ / $x=l$										$\dfrac{F_3}{EI\lambda F_5}$	3
4	Free / Pinned	$x=0$ / $x=l$	$\dfrac{2F_1}{EI\lambda^3 F_5}$	$\dfrac{-F_6}{EI\lambda^2 F_5}$		$\dfrac{-F_7}{EI\lambda^2 F_5}$	$\dfrac{2F_2}{EI\lambda F_5}$		$\dfrac{F_9}{EI\lambda F_5}$			$\dfrac{F_4}{EI\lambda F_5}$	4
5	Free / Sliding	$x=0$ / $x=l$	$\dfrac{-2F_2}{EI\lambda^3 F_6}$	$\dfrac{F_5}{EI\lambda^2 F_6}$	$\dfrac{-F_9}{EI\lambda^3 F_6}$		$\dfrac{2F_1}{EI\lambda F_6}$	$\dfrac{-F_8}{EI\lambda^2 F_6}$		$\dfrac{-F_4}{EI\lambda^3 F_6}$			5
6	Pinned / Pinned	$x=0$ / $x=l$					$\dfrac{-F_5}{2EI\lambda F_1}$		$\dfrac{F_8}{2EI\lambda F_1}$			$\dfrac{-F_5}{2EI\lambda F_1}$	6
7	Sliding / Pinned	$x=0$ / $x=l$	$\dfrac{-F_5}{2EI\lambda^3 F_2}$			$\dfrac{F_{10}}{2EI\lambda^2 F_2}$						$\dfrac{F_6}{2EI\lambda F_2}$	7
8	Clamped / Sliding	$x=0$ / $x=l$								$\dfrac{-F_3}{EI\lambda^3 F_6}$			8
9	Sliding / Sliding	$x=0$ / $x=l$	$\dfrac{-F_6}{2EI\lambda^3 F_1}$		$\dfrac{-F_7}{2EI\lambda^3 F_1}$					$\dfrac{-F_6}{2EI\lambda^3 F_1}$			9

TABLE 7·1 (d)

VALUES OF THE FUNCTIONS $F_1, F_2, ..., F_{10}$

λl	F_1	F_2	F_3	F_4	F_5
0·00	0·000 00	1·000 00	0·000 00	2·000 0	0·000 00
0·05	0·002 50	1·000 00	0·000 00	2·000 0	− 0·000 08
0·10	0·010 00	0·999 98	− 0·000 02	2·000 0	− 0·000 67
0·15	0·022 50	0·999 92	− 0·000 08	1·999 9	− 0·002 25
0·20	0·040 00	0·999 73	− 0·000 27	1·999 7	− 0·005 33
0·25	0·062 50	0·999 35	− 0·000 65	1·999 3	− 0·010 42
0·30	0·089 99	0·998 65	− 0·001 35	1·998 7	− 0·018 00
0·35	0·122 48	0·997 50	− 0·002 50	1·997 5	− 0·028 58
0·40	0·159 95	0·995 73	− 0·004 27	1·995 7	− 0·042 66
0·45	0·202 41	0·993 17	− 0·006 83	1·993 2	− 0·060 74
0·50	0·249 83	0·989 58	− 0·010 42	1·989 6	− 0·083 31
0·55	0·302 19	0·984 75	− 0·015 25	1·984 8	− 0·110 87
0·60	0·359 48	0·978 41	− 0·021 59	1·978 4	− 0·143 91
0·65	0·421 66	0·970 26	− 0·029 74	1·970 3	− 0·182 93
0·70	0·488 69	0·960 01	− 0·039 99	1·960 0	− 0·228 41
0·75	0·560 52	0·947 31	− 0·052 69	1·947 3	− 0·280 83
0·80	0·637 09	0·931 80	− 0·068 20	1·931 8	− 0·340 67
0·85	0·718 31	0·913 11	− 0·086 89	1·913 1	− 0·408 40
0·90	0·804 10	0·890 82	− 0·109 18	1·890 8	− 0·484 48
0·95	0·894 34	0·864 51	− 0·135 49	1·864 5	− 0·569 37
1·00	0·988 90	0·833 73	− 0·166 27	1·833 7	− 0·663 49
1·05	1·087 6	0·798 00	− 0·202 00	1·798 0	− 0·767 29
1·10	1·190 3	0·756 83	− 0·243 17	1·756 8	− 0·881 15
1·15	1·296 8	0·709 71	− 0·290 29	1·709 7	− 1·005 5
1·20	1·406 9	0·656 11	− 0·343 89	1·656 1	− 1·140 6
1·25	1·520 2	0·595 46	− 0·404 54	1·595 5	− 1·287 0
1·30	1·636 5	0·527 22	− 0·472 78	1·527 2	− 1·444 8
1·35	1·755 4	0·450 79	− 0·549 21	1·450 8	− 1·614 4
1·40	1·876 6	0·365 58	− 0·634 42	1·365 6	− 1·795 9
1·45	1·999 6	0·270 99	− 0·729 01	1·271 0	− 1·989 7
1·50	2·123 9	0·166 40	− 0·833 60	1·166 4	− 2·195 9
1·55	2·249 1	0·051 19	− 0·948 81	1·051 2	− 2·414 5
1·60	2·374 6	− 0·075 26	− 1·075 3	0·924 7	− 2·645 7
1·65	2·499 6	− 0·213 59	− 1·213 6	0·786 4	− 2·889 4
1·70	2·623 6	− 0·364 41	− 1·364 4	0·635 6	− 3·145 6
1·75	2·745 7	− 0·528 35	− 1·528 4	0·471 7	− 3·414 1
1·80	2·865 2	− 0·706 02	− 1·706 0	0·294 0	− 3·694 7
1·85	2·981 2	− 0·898 02	− 1·898 0	0·102 0	− 3·987 0
1·90	3·092 7	− 1·104 9	− 2·104 9	− 0·104 9	− 4·290 8
1·95	3·198 6	− 1·327 3	− 2·327 3	− 0·327 3	− 4·605 4
2·00	3·297 9	− 1·565 6			− 4·930 3
2·05	3·389 4	− 1·820 5			− 5·264 7
2·10	3·471 7	− 2·092 2	$F_2 - 1$	$F_2 + 1$	− 5·607 8
2·15	3·543 6	− 2·381 4			− 5·958 7
2·20	3·603 6	− 2·688 2			− 6·316 1
2·25	3·650 1	− 3·013 1			− 6·678 9

TABLE 7·1 (d) (cont.)

λl	F_1	F_2	F_3	F_4	F_5
2·25	3·650 1	−3·013 1			−6·678 9
2·30	3·681 5	−3·356 2			−7·045 7
2·35	3·696 2	−3·717 7	F_2-1	F_2+1	−7·414 7
2·40	3·692 2	−4·097 7			−7·784 3
2·45	3·667 8	−4·496 1			−8·152 5
2·50	3·620 9	−4·912 8			−8·517 1
2·55	3·549 4	−5·347 7			−8·875 8
2·60	3·451 1	−5·800 3			−9·226 1
2·65	3·323 9	−6·270 1			−9·565 1
2·70	3·165 3	−6·756 6			−9·889 8
2·75	2·972 9	−7·258 8			−10·20
2·80	2·744 2	−7·775 9			−10·48
2·85	2·476 6	−8·306 7			−10·74
2·90	2·167 5	−8·849 9			−10·98
2·95	1·814 1	−9·403 9			−11·18
3·00	1·413 7	−9·966 9			−11·34
3·05	0·963 5	−10·54			−11·46
3·10	0·460 6	−11·11			−11·53
3·15	−0·097 9	−11·69			−11·55
3·20	−0·714 8	−12·27			−11·51
3·25	−1·393 1	−12·84			−11·40
3·30	−2·135 5	−13·40			−11·23
3·35	−2·945 0	−13·96			−10·97
3·40	−3·824 3	−14·50			−10·64
3·45	−4·776 0	−15·02			−10·21
3·50	−5·802 9	−15·52			−9·68
3·55	−6·907 3	−15·99			−9·04
3·60	−8·091 7	−16·42			−8·29
3·65	−9·358 2	−16·82			−7·42
3·70	−10·71	−17·16			−6·42
3·75	−12·14	−17·46			−5·28
3·80	−13·67	−17·69			−3·99
3·85	−15·28	−17·85			−2·54
3·90	−16·98	−17·94			−0·93
3·95	−18·77	−17·94			0·86
4·00	−20·65	−17·85			2·83
4·05	−22·62	−17·66			4·99
4·10	−24·68	−17·35			7·36
4·15	−26·83	−16·92			9·93
4·20	−29·05	−16·35			12·72
4·25	−31·37	−15·64			15·74
4·30	−33·75	−14·77			19·00
4·35	−36·22	−13·74			22·50
4·40	−38·75	−12·52			26·25
4·45	−41·34	−11·11			30·25
4·50	−43·99	−9·49			34·52

TABLE 7·1(*d*) (*cont.*)

λl	F_1	F_2	F_3	F_4	F_5
4·50	−43·99	−9·49			34·52
4·55	−46·69	−7·65			39·05
4·60	−49·42	−5·58	$F_2 - 1$	$F_2 + 1$	43·86
4·65	−52·19	−3·26			48·94
4·70	−54·96	−0·68			54·29
4·75	−57·75	2·17			59·93
4·80	−60·52	5·32			65·84
4·85	−63·26	8·76			72·03
4·90	−65·96	12·52			78·49
4·95	−68·60	16·62			85·22
5·00	−71·16	21·05			92·21
5·05	−73·60	25·84			99·45
5·10	−75·92	31·00			106·9
5·15	−78·09	36·54			114·6
5·20	−80·07	42·47			122·5
5·25	−81·84	48·79			130·6
5·30	−83·36	55·53			138·9
5·35	−84·61	62·69			147·3
5·40	−85·55	70·26			155·8
5·45	−86·13	78·27			164·4
5·50	−86·32	86·70			173·0
5·55	−86·08	95·57			181·6
5·60	−85·35	104·9			190·2
5·65	−84·11	114·6			198·7
5·70	−82·29	124·7			207·0
5·75	−79·85	135·3			215·1
5·80	−76·73	146·2			223·0
5·85	−72·88	157·6			230·5
5·90	−68·24	169·3			237·5
5·95	−62·75	181·3			244·1
6·00	−56·36	193·7			250·0
6·05	−49·00	206·3			255·3
6·10	−40·61	219·2			259·8
6·15	−31·12	232·3			263·4
6·20	−20·47	245·5			266·0
6·25	−8·59	258·9			267·5
6·30	4·58	272·2			267·7
6·35	19·11	285·6			266·5
6·40	35·07	298·9			263·8
6·45	52·53	312·0			259·4
6·50	71·54	324·8			253·2
6·55	92·18	337·3			245·1
6·60	114·5	349·3			234·7
6·65	138·6	360·7			222·1
6·70	164·5	371·4			207·0
6·75	192·2	381·3			189·2

TABLE 7·1 (*d*) (*cont.*)

λl	F_1	F_2	F_3	F_4	F_5
6·75	192·2	381·3			189·2
6·80	221·8	390·3			168·5
6·85	253·4	398·1	F_2-1	F_2+1	144·7
6·90	287·0	404·7			117·7
6·95	322·6	409·9			87·3
7·00	360·2	413·4			53·1
7·05	400·0	415·1			15·1
7·10	441·7	414·8			−26·9
7·15	485·6	412·3			−73·3
7·20	531·5	407·4			−124·1
7·25	579·5	399·8			−179·6
7·30	629·5	389·4			−240·1
7·35	681·4	375·8			−305·6
7·40	735·1	358·7			−376·4
7·45	790·7	338·0			−452·7
7·50	848·0	313·4			−534·6
7·55	906·8	284·5			−622·3
7·60	967·0	251·0			−715·0
7·65	1 029	212·8			−815·8
7·70	1 091	169·4			−921·8
7·75	1 155	120·5			−1 034
7·80	1 219	65·8			−1 153
7·85	1 283	5·1			−1 278
7·90	1 347	−62·0			−1 409
7·95	1 411	−135·9			−1 547
8·00	1 475	−216·9			−1 691
8·05	1 537	−305·2			−1 842
8·10	1 598	−401·2			−1 999
8·15	1 656	−505·2			−2 162
8·20	1 713	−617·4			−2 330
8·25	1 766	−738·2			−2 504
8·30	1 815	−867·9			−2 683
8·35	1 860	−1 007			−2 867
8·40	1 900	−1 155			−3 055
8·45	1 934	−1 312			−3 247
8·50	1 962	−1 479			−3 442
8·55	1 982	−1 656			−3 639
8·60	1 994	−1 843			−3 838
8·65	1 997	−2 040			−4 037
8·70	1 990	−2 247			−4 237
8·75	1 971	−2 464			−4 435
8·80	1 940	−2 690			−4 631
8·85	1 896	−2 927			−4 823
8·90	1 837	−3 173			−5 009
8·95	1 762	−3 428			−5 189
9·00	1 670	−3 691			−5 361

TABLE 7·1 (d) (cont.)

λl	F_1	F_2	F_3	F_4	F_5
9·00	1 670	− 3 691			− 5 361
9·05	1 559	− 3 964			− 5 523
9·10	1 429	− 4 244	$F_2 - 1$	$F_2 + 1$	− 5 672
9·15	1 277	− 4 531			− 5 808
9·20	1 103	− 4 824			− 5 927
9·25	905	− 5 123			− 6 028
9·30	681	− 5 426			− 6 107
9·35	430	− 5 733			− 6 163
9·40	150	− 6 042			− 6 192
9·45	− 160	− 6 352			− 6 192
9·50	− 502	− 6 661			− 6 159
9·55	− 877	− 6 967			− 6 090
9·60	− 1 287	− 7 269			− 5 982
9·65	− 1 733	− 7 565			− 5 832
9·70	− 2 217	− 7 852			− 5 635
9·75	− 2 741	− 8 128			− 5 387
9·80	− 3 304	− 8 390			− 5 085
9·85	− 3 910	− 8 635			− 4 725
9·90	− 4 559	− 8 861			− 4 302
9·95	− 5 253	− 9 064			− 3 811
10·00	− 5 991	− 9 241			− 3 249
10·05	− 6 776	− 9 388			− 2 611
10·10	− 7 608	− 9 501			− 1 893
10·15	− 8 487	− 9 576			− 1 088
10·20	− 9 414	− 9 608			− 194
10·25	− 10 390	− 9 593			796
10·30	− 11 413	− 9 527			1 886
10 35	− 12 483	− 9 403			3 080
10·40	− 13 601	− 9 217			4 384
10·45	− 14 765	− 8 963			5 802
10·50	− 15 973	− 8 635			7 339
10·55	− 17 225	− 8 227			8 998
10·60	− 18 518	− 7 733			10 785
10·65	− 19 849	− 7 146			12 703
10·70	− 21 216	− 6 460			14 756
10·75	− 22 616	− 5 668			16 947
10·80	− 24 043	− 4 763			19 280
10·85	− 25 495	− 3 738			21 757
10·90	− 26 965	− 2 585			24 380
10·95	− 28 447	− 1 297			27 150
11·00	− 29 937	133			30 069

TABLE 7·1(d) (cont.)

λl	F_6	F_7	F_8	F_9	F_{10}
0·00	0·000 00	0·000 00	0·000 00	2·000 0	0·000 00
0·05	0·100 00	0·100 00	−0·000 04	2·000 0	−0·002 50
0·10	0·200 00	0·200 00	−0·000 33	2·000 0	−0·010 00
0·15	0·299 99	0·300 00	−0·001 13	2·000 0	−0·022 50
0·20	0·399 98	0·400 01	−0·002 67	2·000 1	−0·040 00
0·25	0·499 93	0·500 02	−0·005 21	2·000 3	−0·062 50
0·30	0·599 84	0·600 04	−0·009 00	2·000 7	−0·090 00
0·35	0·699 65	0·700 09	−0·014 29	2·001 3	−0·122 51
0·40	0·799 32	0·800 17	−0·021 33	2·002 1	−0·160 01
0·45	0·898 77	0·900 31	−0·030 38	2·003 4	−0·202 52
0·50	0·997 92	1·000 5	−0·041 67	2·005 2	−0·250 04
0·55	1·096 6	1·100 8	−0·055 46	2·007 6	−0·302 58
0·60	1·194 8	1·201 3	−0·072 01	2·010 8	−0·360 13
0·65	1·292 3	1·301 9	−0·091 56	2·014 9	−0·422 71
0·70	1·388 8	1·402 8	−0·114 37	2·020 0	−0·490 33
0·75	1·484 2	1·504 0	−0·140 68	2·026 4	−0·562 99
0·80	1·578 2	1·605 5	−0·170 75	2·034 1	−0·640 73
0·85	1·670 4	1·707 4	−0·204 84	2·043 5	−0·723 55
0·90	1·760 7	1·809 8	−0·243 19	2·054 7	−0·811 48
0·95	1·848 5	1·912 9	−0·286 07	2·067 9	−0·904 54
1·00	1·933 4	2·016 7	−0·333 73	2·083 4	−1·002 8
1·05	2·015 1	2·121 3	−0·386 43	2·101 4	−1·106 2
1·10	2·092 8	2·226 9	−0·444 44	2·122 1	−1·214 9
1·15	2·166 2	2·333 5	−0·508 01	2·145 9	−1·328 9
1·20	2·234 6	2·441 5	−0·577 42	2·173 0	−1·448 3
1·25	2·297 2	2·550 9	−0·652 93	2·203 7	−1·573 1
1·30	2·353 4	2·661 9	−0·734 82	2·238 4	−1·703 4
1·35	2·402 4	2·774 8	−0·823 37	2·277 3	−1·839 3
1·40	2·443 3	2·889 8	−0·918 85	2·320 9	−1·980 9
1·45	2·475 2	3·007 0	−1·021 6	2·369 3	−2·128 3
1·50	2·497 1	3·126 8	−1·131 8	2·423 1	−2·281 7
1·55	2·508 1	3·249 4	−1·249 8	2·482 7	−2·441 1
1·60	2·507 0	3·375 1	−1·376 0	2·548 3	−2·606 7
1·65	2·492 7	3·504 3	−1·510 6	2·620 4	−2·778 6
1·70	2·463 9	3·637 3	−1·654 0	2·699 5	−2·957 2
1·75	2·419 3	3·774 4	−1·806 4	2·785 9	−3·142 4
1·80	2·357 7	3·916 0	−1·968 3	2·880 3	−3·334 7
1·85	2·277 7	4·062 6	−2·140 0	2·982 9	−3·534 1
1·90	2·177 6	4·214 5	−2·321 9	3·094 4	−3·741 0
1·95	2·056 2	4·372 2	−2·514 2	3·215 3	−3·955 7
2·00	1·911 6	4·536 2	−2·717 6	3·346 0	−4·178 3
2·05	1·742 5	4·706 9	−2·932 2	3·487 2	−4·409 4
2·10	1·547 0	4·885 1	−3·158 6	3·639 5	−4·649 2
2·15	1·323 5	5·071 1	−3·397 3	3·803 3	−4·898 0
2·20	1·070 1	5·265 6	−3·648 6	3·979 4	−5·156 4
2·25	0·785 2	5·469 2	−3·913 1	4·168 4	−5·424 7

TABLE 7·1(d) (cont.)

λl	F_6	F_7	F_8	F_9	F_{10}
2·25	0·785 2	5·469 2	−3·913 1	4·168 4	−5·424 7
2·30	0·466 9	5·682 7	−4·191 3	4·370 9	−5·703 5
2·35	0·113 4	5·906 6	−4·483 6	4·587 8	−5·993 2
2·40	−0·277 3	6·141 7	−4·790 8	4·819 6	−6·294 3
2·45	−0·706 8	6·388 8	−5·113 3	5·067 1	−6·607 6
2·50	−1·177 1	6·648 7	−5·451 7	5·331 1	−6·933 4
2·55	−1·690 0	6·922 2	−5·806 8	5·612 5	−7·272 6
2·60	−2·247 2	7·210 2	−6·179 2	5·912 1	−7·625 9
2·65	−2·850 6	7·513 7	−6·569 7	6·230 8	−7·993 9
2·70	−3·501 8	7·833 6	−6·978 9	6·569 4	−8·377 5
2·75	−4·202 4	8·171 0	−7·407 7	6·929 0	−8·777 6
2·80	−4·954 0	8·526 9	−7·856 9	7·310 5	−9·195 0
2·85	−5·758 1	8·902 4	−8·327 5	7·715 0	−9·630 6
2·90	−6·615 8	9·298 8	−8·820 3	8·143 6	−10·09
2·95	−7·528 4	9·717 2	−9·336 4	8·597 4	−10·56
3·00	−8·496 9	10·16	−9·876 8	9·077 7	−11·06
3·05	−9·522 0	10·63	−10·44	9·585 5	−11·58
3·10	−10·60	11·12	−11·03	10·12	−12·12
3·15	−11·74	11·64	−11·66	10·69	−12·69
3·20	−12·94	12·19	−12·30	11·29	−13·28
3·25	−14·20	12·77	−12·98	11·92	−13·91
3·30	−15·51	13·38	−13·70	12·59	−14·56
3·35	−16·88	14·03	−14·44	13·29	−15·25
3·40	−18·30	14·71	−15·22	14·03	−15·97
3·45	−19·78	15·43	−16·04	14·81	−16·72
3·50	−21·30	16·19	−16·89	15·64	−17·51
3·55	−22·88	17·00	−17·79	16·50	−18·34
3·60	−24·50	17·84	−18·73	17·42	−19·21
3·65	−26·16	18·74	−19·71	18·38	−20·12
3·70	−27·86	19·68	−20·74	19·39	−21·08
3·75	−29·59	20·68	−21·82	20·45	−22·09
3·80	−31·35	21·73	−22·95	21·57	−23·15
3·85	−33·13	22·84	−24·14	22·75	−24·27
3·90	−34·92	24·00	−25·38	23·99	−25·44
3·95	−36·71	25·23	−26·68	25·29	−26·67
4·00	−38·50	26·53	−28·05	26·65	−27·96
4·05	−40·28	27·90	−29·48	28·09	−29·32
4·10	−42·03	29·34	−30·98	29·60	−30·75
4·15	−43·75	30·86	−32·56	31·19	−32·26
4·20	−45·41	32·46	−34·21	32·86	−33·84
4·25	−47·01	34·15	−35·94	34·61	−35·51
4·30	−48·53	35·93	−37·76	36·46	−37·26
4·35	−49·96	37·80	−39·67	38·39	−39·10
4·40	−51·27	39·77	−41·67	40·42	−41·04
4·45	−52·46	41·84	−43·77	42·56	−43·08
4·50	−53·49	44·03	−45·98	44·80	−45·22

TABLE 7·1(*d*) (*cont.*)

λl	F_6	F_7	F_8	F_9	F_{10}
4·50	−53·49	44·03	−45·98	44·80	−45·22
4·55	−54·35	46·32	−48·30	47·16	−47·48
4·60	−55·01	48·74	−50·73	49·64	−49·86
4·65	−55·46	51·29	−53·29	52·23	−52·36
4·70	−55·65	53·97	−55·97	54·97	−54·99
4·75	−55·58	56·79	−58·79	57·83	−57·76
4·80	−55·21	59·75	−61·75	60·85	−60·67
4·85	−54·51	62·88	−64·86	64·01	−63·74
4·90	−53·45	66·16	−68·12	67·34	−66·96
4·95	−51·99	69·61	−71·56	70·83	−70·36
5·00	−50·11	73·24	−75·16	74·49	−73·93
5·05	−47·77	77·06	−78·95	78·35	−77·68
5·10	−44·93	81·08	−82·93	82·39	−81·64
5·15	−41·56	85·31	−87·12	86·64	−85·79
5·20	−37·61	89·75	−91·52	91·11	−90·17
5·25	−33·05	94·42	−96·14	95·80	−94·77
5·30	−27·84	99·33	−101·0	100·7	−99·62
5·35	−21·93	104·5	−106·1	105·9	−104·7
5·40	−15·29	109·9	−111·5	111·3	−110·1
5·45	−7·87	115·6	−117·1	117·1	−115·7
5·50	0·38	121·6	−123·0	123·1	−121·6
5·55	9·49	127·9	−129·3	129·4	−127·9
5·60	19·51	134·6	−135·8	136·0	−134·4
5·65	30·48	141·6	−142·7	143·0	−141·3
5·70	42·44	148·9	−150·0	150·3	−148·6
5·75	55·44	156·6	−157·6	158·0	−156·2
5·80	69·51	164·7	−165·6	166·0	−164·3
5·85	84·70	173·2	−174·0	174·5	−172·7
5·90	101·0	182·1	−182·9	183·4	−181·6
5·95	118·6	191·5	−192·2	192·8	−190·9
6·00	137·3	201·4	−202·0	202·7	−200·8
6·05	157·3	211·8	−212·3	213·0	−211·1
6·10	178·6	222·7	−223·1	223·9	−221·9
6·15	201·2	234·2	−234·5	235·4	−233·4
6·20	225·1	246·3	−246·5	247·4	−245·4
6·25	250·3	259·0	−259·0	260·0	−258·0
6·30	276·8	272·3	−272·3	273·3	−271·3
6·35	304·7	286·3	−286·2	287·2	−285·2
6·40	333·9	301·0	−300·8	301·9	−299·9
6·45	364·5	316·5	−316·2	317·3	−315·4
6·50	396·3	332·8	−332·4	333·5	−331·6
6·55	429·4	349·9	−349·4	350·6	−348·7
6·60	463·8	367·9	−367·2	368·5	−366·6
6·65	499·3	386·8	−386·0	387·3	−385·5
6·70	535·9	406·6	−405·8	407·1	−405·3
6·75	573·5	427·5	−426·6	427·9	−426·1

TABLE 7·1(*d*) (*cont.*)

λl	F_6	F_7	F_8	F_9	F_{10}
6·75	573·5	427·5	−426·6	427·9	−426·1
6·80	612·1	449·4	−448·4	449·8	−448·1
6·85	651·5	472·5	−471·4	472·8	−471·1
6·90	691·7	496·7	−495·6	497·0	−495·3
6·95	732·4	522·2	−521·0	522·4	−520·8
7·00	773·6	549·0	−547·7	549·1	−547·6
7·05	815·1	577·1	−575·7	577·1	−575·7
7·10	856·6	606·7	−605·3	606·7	−605·3
7·15	897·9	637·8	−636·3	637·7	−636·4
7·20	939·0	670·5	−668·9	670·3	−669·1
7·25	979·3	704·9	−703·2	704·6	−703·5
7·30	1 019	741·0	−739·3	740·7	−739·6
7·35	1 057	779·0	−777·2	778·6	−777·6
7·40	1 094	818·9	−817·1	818·4	−817·6
7·45	1 129	860·9	−859·0	860·3	−859·5
7·50	1 161	905·0	−903·1	904·4	−903·7
7·55	1 191	951·3	−949·4	950·7	−950·1
7·60	1 218	1 000	−998·1	999·3	−998·8
7·65	1 241	1 051	−1 049	1 051	−1 050
7·70	1 260	1 105	−1 103	1 104	−1 104
7·75	1 275	1 162	−1 160	1 161	−1 161
7·80	1 284	1 221	−1 219	1 220	−1 220
7·85	1 288	1 284	−1 282	1 283	−1 283
7·90	1 285	1 350	−1 348	1 349	−1 349
7·95	1 275	1 419	−1 417	1 418	−1 418
8·00	1 258	1 491	−1 489	1 490	−1 491
8·05	1 232	1 568	−1 566	1 567	−1 567
8·10	1 196	1 648	−1 646	1 647	−1 647
8·15	1 151	1 733	−1 731	1 731	−1 732
8·20	1 095	1 821	−1 820	1 820	−1 821
8·25	1 027	1 915	−1 913	1 913	−1 914
8·30	947	2 013	−2 011	2 012	−2 012
8·35	854	2 116	−2 114	2 115	−2 116
8·40	746	2 224	−2 223	2 223	−2 224
8·45	622	2 338	−2 337	2 337	−2 338
8·50	483	2 458			
8·55	326	2 584			
8·60	151	2 717	$= -F_7$	$= F_7$	$= -F_7$
8·65	−43	2 856			
8·70	−257	3 002	approximately	approximately	approximately
8·75	−493	3 156			
8·80	−750	3 318			
8·85	−1 031	3 488			
8·90	−1 336	3 666			
8·95	−1 666	3 854			
9·00	−2 022	4 052			

TABLE 7·1 (d) (cont.)

λl	F_6	F_7	F_8	F_9	F_{10}
9·00	−2 022	4 052			
9·05	−2 404	4 260			
9·10	−2 815	4 478	$= -F_7$	$= F_7$	$= -F_7$
9·15	−3 253	4 707			
9·20	−3 721	4 949	approximately	approximately	approximately
9·25	−4 218	5 202			
9·30	−4 746	5 469			
9·35	−5 304	5 749			
9·40	−5 893	6 044			
9·45	−6 512	6 354			
9·50	−7 163	6 680			
9·55	−7 844	7 022			
9·60	−8 556	7 382			
9·65	−9 298	7 761			
9·70	−10 069	8 159			
9·75	−10 868	8 577			
9·80	−11 694	9 017			
9·85	−12 545	9 479			
9·90	−13 420	9 965			
9·95	−14 317	10 476			
10·00	−15 232	11 013			
10·05	−16 164	11 577			
10·10	−17 109	12 171			
10·15	−18 063	12 795			
10·20	−19 022	13 451			
10·25	−19 983	14 141			
10·30	−20 939	14 866			
10·35	−21 886	15 628			
10·40	−22 818	16 429			
10·45	−23 727	17 271			
10·50	−24 608	18 157			
10·55	−25 452	19 088			
10·60	−26 250	20 066			
10·65	−26 995	21 095			
10·70	−27 676	22 177			
10·75	−28 284	23 314			
10·80	−28 806	24 509			
10·85	−29 232	25 766			
10·90	−29 550	27 087			
10·95	−29 745	28 476			
11·00	−29 804	29 936			

TABLE 7·2 (a)

DATA FOR PINNED-PINNED BEAM

$$\phi_r(x) = \sin\frac{r\pi x}{l} = \sin\lambda_r x \quad (r = 1, 2, 3, \ldots).$$

$$a_r = \frac{A\rho l}{2}.$$

$$c_r = \frac{r^4\pi^4 EI}{2l^3}.$$

$$\omega_r = \sqrt{\left(\frac{c_r}{a_r}\right)} = \frac{r^2\pi^2}{l^2}\sqrt{\left(\frac{EI}{A\rho}\right)}.$$

Orthogonality

$$\int_0^l \phi_r(x)\,\phi_s(x)\,dx = \int_0^l \phi_r''(x)\,\phi_s''(x)\,dx = \begin{cases} 0 & (r \neq s), \\ \tfrac{1}{2}l & (r = s), \end{cases}$$

where
$$\phi_r''(x) = \frac{1}{\lambda_r^2}\frac{d^2\phi_r}{dx^2}, \text{ etc.}$$

Receptances in series form

$$\alpha_{xh} = \alpha_{hx} = \sum_{r=1}^{\infty} \frac{\phi_r(x)\,(\phi_r h)}{a_r(\omega_r^2 - \omega^2)}$$

other receptances by differentiation.

TABLE 7·2 (b)

DATA FOR CLAMPED-CLAMPED BEAM AND FREE-FREE BEAM

A. CLAMPED-CLAMPED BEAM

Characteristic function and its derivatives

$$\phi_r = \cosh \lambda_r x - \cos \lambda_r x - \sigma_r(\sinh \lambda_r x - \sin \lambda_r x), \quad (r = 1, 2, 3, \ldots),$$

$$\frac{1}{\lambda_r}\frac{d\phi_r}{dx} = \phi_r' = \sinh \lambda_r x + \sin \lambda_r x - \sigma_r(\cosh \lambda_r x - \cos \lambda_r x),$$

$$\frac{1}{\lambda_r^2}\frac{d^2\phi_r}{dx^2} = \phi_r'' = \cosh \lambda_r x + \cos \lambda_r x - \sigma_r(\sinh \lambda_r x + \sin \lambda_r x),$$

$$\frac{1}{\lambda_r^3}\frac{d^3\phi_r}{dx^3} = \phi_r''' = \sinh \lambda_r x - \sin \lambda_r x - \sigma_r(\cosh \lambda_r x + \cos \lambda_r x).$$

Boundary values

$$\phi_r(0) = \phi_r'(0) = 0,$$
$$\phi_r(l) = \phi_r'(l) = 0.$$

Orthogonality

$$\int_0^l \phi_r(x)\,\phi_s(x)\,dx = \int_0^l \phi_r''(x)\,\phi_s''(x)\,dx = \begin{cases} 0 & (r \neq s), \\ l & (r = s), \end{cases}$$

for $r, s = 1, 2, 3, \ldots$.

Receptances in series form

$$\alpha_{xh} = \alpha_{hx} = \sum_{r=1}^{\infty} \frac{\phi_r(x)\,\phi_r(h)}{a_r(\omega_r^2 - \omega^2)}, \quad \text{where} \quad a_r = A\rho l;$$

other receptances by differentiation.

B. FREE-FREE BEAM

Characteristic function

$$\phi_{-1} = \frac{2\sqrt{3}}{l}\left(x - \frac{l}{2}\right), \quad \phi_0 = 1,$$

$$\phi_r = \cosh \lambda_r x + \cos \lambda_r x - \sigma_r(\sinh \lambda_r x + \sin \lambda_r x), \quad (r = 1, 2, 3, \ldots),$$

where σ_r and λ_r are the same as for a clamped-clamped beam.

The characteristic function for distortion of a free-free beam is the same as the second derivative of that of a clamped-clamped beam; that is, for $r = 1, 2, 3, \ldots$,

$$\phi_r \text{ (free-free)} = \phi_r'' \text{ (clamped-clamped)},$$
$$\phi_r' \text{ (free-free)} = \phi_r''' \text{ (clamped-clamped)},$$
$$\phi_r'' \text{ (free-free)} = \phi_r \text{ (clamped-clamped)},$$
$$\phi_r''' \text{ (free-free)} = \phi_r' \text{ (clamped-clamped)}.$$

Boundary values

$$\phi_r''(0) = \phi_r'''(0) = 0,$$
$$\phi_r''(l) = \phi_r'''(l) = 0.$$

Orthogonality

$$\int_0^l \phi_r(x)\,\phi_s(x)\,dx = \begin{cases} 0 & (r \neq s), \\ l & (r = s), \end{cases}$$

for $r, s = -1, 0, 1, 2, \ldots$; and

$$\int_0^l \phi_r''(x)\,\phi_s''(x)\,dx = \begin{cases} 0 & (r \neq s), \\ l & (r = s), \end{cases}$$

for $r, s = 1, 2, 3, \ldots$.

Receptances in series form

$$\alpha_{xh} = \alpha_{hx} = \sum_{r=-1}^{\infty} \frac{\phi_r(x)\,\phi_r(h)}{a_r(\omega_r^2 - \omega^2)}, \quad \text{where} \quad a_r = A\rho l;$$

other receptances by differentiation.

GENERAL DATA (for A and B)

$$\left.\begin{array}{l} \cos\lambda_r l \cosh\lambda_r l - 1 = 0, \\[2mm] \sigma_r = \dfrac{\cosh\lambda_r l - \cos\lambda_r l}{\sinh\lambda_r l - \sin\lambda_r l}, \end{array}\right\} \quad (r = 1, 2, 3, \ldots)$$

TABLE 7·2 (*b*) (*cont.*)

Values of σ_r and $\lambda_r l$ and various powers

r	$\lambda_r l$	$(\lambda_r l)^2$	$(\lambda_r l)^3$	$(\lambda_r l)^4$
1	4·730 04	22·373 3	105·827	500·564
2	7·853 20	61·672 8	484·329	3 803·54
3	10·995 6	120·903	1 329·41	14 617·6
4	14·137 2	199·859	2 825·45	39 943·8
5	17·278 8	298·556	5 158·67	89 135·4

r	σ_r	$\dfrac{\omega_r}{\omega_1} = \dfrac{\lambda_r^2}{\lambda_1^2}$	$\dfrac{\omega_r^2}{\omega_1^2} = \dfrac{\lambda_r^4}{\lambda_1^4}$
1	0·982 502 2	1·0	1·0
2	1·000 777 3	2·756 54	7·598 50
3	0·999 966 5	5·403 92	29·202 3
4	1·000 001 5	8·932 95	79·797 6
5	0·999 999 9	13·344 3	178·070

For $r > 5$
$$\lambda_r l \fallingdotseq (2r+1)\,\pi/2$$
$$\sigma_r \fallingdotseq 1·0$$

TABLE 7·2 (b) (cont.)

CHARACTERISTIC FUNCTIONS AND DERIVATIVES
CLAMPED-CLAMPED BEAM

First mode

$\dfrac{x}{l}$	ϕ_1	$\phi_1' = \dfrac{1}{\lambda_1}\dfrac{d\phi_1}{dx}$	$\phi_1'' = \dfrac{1}{\lambda_1^2}\dfrac{d^2\phi_1}{dx^2}$	$\phi_1''' = \dfrac{1}{\lambda_1^3}\dfrac{d^3\phi_1}{dx^3}$
0·00	0·000 00	0·000 00	2·000 00	− 1·965 00
0·02	0·008 67	0·180 41	1·814 12	− 1·964 73
0·04	0·033 58	0·343 24	1·628 32	− 1·962 85
0·06	0·073 06	0·488 50	1·442 84	− 1·957 92
0·08	0·125 45	0·616 24	1·258 02	− 1·948 62
0·10	0·189 10	0·726 55	1·074 33	− 1·933 83
0·12	0·262 37	0·819 56	0·892 34	− 1·912 54
0·14	0·343 63	0·895 46	0·712 70	− 1·883 93
0·16	0·431 26	0·954 51	0·536 15	− 1·847 32
0·18	0·523 70	0·997 02	0·363 46	− 1·802 19
0·20	0·619 39	1·023 42	0·195 45	− 1·748 14
0·22	0·716 84	1·034 18	0·033 00	− 1·684 94
0·24	0·814 59	1·029 86	− 0·123 05	− 1·612 50
0·26	0·911 24	1·011 13	− 0·271 80	− 1·530 85
0·28	1·005 46	0·978 70	− 0·412 40	− 1·440 17
0·30	1·096 00	0·933 38	− 0·544 01	− 1·340 74
0·32	1·181 68	0·876 08	− 0·665 81	− 1·232 96
0·34	1·261 41	0·807 74	− 0·777 04	− 1·117 35
0·36	1·334 19	0·729 92	− 0·876 99	− 0·994 52
0·38	1·399 13	0·642 19	− 0·965 00	− 0·865 16
0·40	1·455 45	0·547 23	− 1·040 50	− 0·730 07
0·42	1·502 47	0·445 74	− 1·102 97	− 0·590 08
0·44	1·539 62	0·338 97	− 1·152 02	− 0·446 11
0·46	1·566 47	0·228 21	− 1·187 28	− 0·299 11
0·48	1·582 71	0·114 78	− 1·208 54	− 0·150 07
0·50	1·588 15	0·000 00	− 1·215 65	0·000 00
0·52	1·582 71	− 0·114 78	− 1·208 54	0·150 07
0·54	1·566 47	− 0·228 21	− 1·187 28	0·299 11
0·56	1·539 62	− 0·338 97	− 1·152 02	0·446 11
0·58	1·502 47	− 0·445 74	− 1·102 97	0·590 08
0·60	1·455 45	− 0·547 23	− 1·040 50	0·730 07
0·62	1·399 13	− 0·642 19	− 0·965 00	0·865 16
0·64	1·334 19	− 0·729 92	− 0·876 99	0·994 52
0·66	1·261 41	− 0·807 74	− 0·777 04	1·117 35
0·68	1·181 68	− 0·876 08	− 0·665 81	1·232 96
0·70	1·096 00	− 0·933 38	− 0·544 01	1·340 74
0·72	1·005 46	− 0·978 70	− 0·412 40	1·440 17
0·74	0·911 24	− 1·011 13	− 0·271 80	1·530 85
0·76	0·814 59	− 1·029 86	− 0·123 05	1·612 50
0·78	0·716 84	− 1·034 18	0·033 00	1·684 94
0·80	0·619 39	− 1·023 42	0·195 45	1·748 14
0·82	0·523 70	− 0·997 02	0·363 46	1·802 19
0·84	0·431 26	− 0·954 51	0·536 15	1·847 32
0·86	0·343 63	− 0·895 46	0·712 70	1·883 93
0·88	0·262 37	− 0·819 56	0·892 34	1·912 54
0·90	0·189 10	− 0·726 55	1·074 33	1·933 83
0·92	0·125 45	− 0·616 24	1·258 02	1·948 62
0·94	0·073 06	− 0·488 50	1·442 84	1·957 92
0·96	0·033 58	− 0·343 24	1·628 32	1·962 85
0·98	0·008 67	− 0·180 41	1·814 12	1·964 73
1·00	0·000 00	0·000 00	2·000 00	1·965 00

TABLE 7·2 (b) (cont.)

Second mode

$\dfrac{x}{l}$	ϕ_2	$\phi_2' = \dfrac{\mathrm{I}}{\lambda_2}\dfrac{d\phi_2}{dx}$	$\phi_2'' = \dfrac{\mathrm{I}}{\lambda_2^2}\dfrac{d^2\phi_2}{dx^2}$	$\phi_2''' = \dfrac{\mathrm{I}}{\lambda_2^3}\dfrac{d^3\phi_2}{dx^3}$
0·00	0·000 00	0·000 00	2·000 00	− 2·001 55
0·02	0·023 38	0·289 44	1·685 68	− 2·000 31
0·04	0·088 34	0·529 55	1·372 02	− 1·992 05
0·06	0·187 15	0·720 55	1·060 61	− 1·970 80
0·08	0·312 14	0·862 96	0·753 86	− 1·931 86
0·10	0·455 73	0·957 76	0·454 86	− 1·871 76
0·12	0·610 58	1·006 44	0·167 13	− 1·788 13
0·14	0·769 58	1·011 05	− 0·105 54	− 1·679 75
0·16	0·926 02	0·974 27	− 0·359 23	− 1·546 52
0·18	1·073 63	0·899 40	− 0·590 10	− 1·389 33
0·20	1·206 74	0·790 30	− 0·794 50	− 1·210 02
0·22	1·320 32	0·651 38	− 0·969 18	− 1·011 27
0·24	1·410 05	0·487 55	− 1·111 33	− 0·796 51
0·26	1·472 45	0·304 10	− 1·218 76	− 0·569 77
0·28	1·504 85	0·106 60	− 1·289 91	− 0·335 55
0·30	1·505 50	− 0·099 16	− 1·324 02	− 0·098 72
0·32	1·473 57	− 0·307 36	− 1·321 06	0·135 66
0·34	1·409 14	− 0·512 24	− 1·281 81	0·362 46
0·36	1·313 14	− 0·708 19	− 1·207 86	0·576 65
0·38	1·187 40	− 0·889 97	− 1·101 57	0·773 40
0·40	1·034 57	− 1·052 71	− 0·966 05	0·948 23
0·42	0·857 94	− 1·192 09	− 0·805 07	1·097 14
0·44	0·661 50	− 1·304 48	− 0·622 96	1·216 70
0·46	0·449 73	− 1·386 93	− 0·424 56	1·304 14
0·48	0·227 51	− 1·437 28	− 0·215 08	1·357 44
0·50	0·000 00	− 1·454 20	0·000 00	1·375 32
0·52	− 0·227 51	− 1·437 28	0·215 08	1·357 44
0·54	− 0·449 73	− 1·386 93	0·424 56	1·304 14
0·56	− 0·661 50	− 1·304 48	0·622 96	1·216 70
0·58	− 0·857 94	− 1·192 09	0·805 07	1·097 14
0·60	− 1·034 57	− 1·052 71	0·966 05	0·948 23
0·62	− 1·187 40	− 0·889 97	1·101 57	0·773 40
0·64	− 1·313 14	− 0·708 19	1·207 86	0·576 65
0·66	− 1·409 14	− 0·512 24	1·281 81	0·362 46
0·68	− 1·473 57	− 0·307 36	1·321 06	0·135 66
0·70	− 1·505 50	− 0·099 16	1·324 02	− 0·098 72
0·72	− 1·504 85	0·106 60	1·289 91	− 0·335 55
0·74	− 1·472 45	0·304 10	1·218 76	− 0·569 77
0·76	− 1·410 05	0·487 55	1·111 33	− 0·796 51
0·78	− 1·320 32	0·651 38	0·969 18	− 1·011 27
0·80	− 1·206 74	0·790 30	0·794 50	− 1·210 02
0·82	− 1·073 63	0·899 40	0·590 10	− 1·389 33
0·84	− 0·926 02	0·974 27	0·359 23	− 1·546 52
0·86	− 0·769 58	1·011 05	0·105 54	− 1·679 75
0·88	− 0·610 58	1·006 44	− 0·167 13	− 1·788 13
0·90	− 0·455 73	0·957 76	− 0·454 86	− 1·871 76
0·92	− 0·312 14	0·862 96	− 0·753 86	− 1·931 86
0·94	− 0·187 15	0·720 55	− 1·060 61	− 1·970 80
0·96	− 0·088 34	0·529 55	− 1·372 02	− 1·992 05
0·98	− 0·023 38	0·289 44	− 1·685 68	− 2·000 31
1·00	0·000 00	0·000 00	− 2·000 00	− 2·001 55

TABLE 7·2 (b) (cont.)

Third mode

$\dfrac{x}{l}$	ϕ_3	$\phi_3' = \dfrac{1}{\lambda_3}\dfrac{d\phi_3}{dx}$	$\phi_3'' = \dfrac{1}{\lambda_3^2}\dfrac{d^2\phi_3}{dx^2}$	$\phi_3''' = \dfrac{1}{\lambda_3^3}\dfrac{d^3\phi_3}{dx^3}$
0·00	0·000 00	0·000 00	2·000 00	− 1·999 93
0·02	0·044 81	0·391 47	1·560 38	− 1·996 58
0·04	0·165 10	0·686 46	1·123 23	− 1·974 69
0·06	0·339 75	0·886 09	0·694 28	− 1·919 98
0·08	0·548 04	0·993 03	0·281 89	− 1·822 80
0·10	0·770 05	1·012 02	− 0·103 93	− 1·677 95
0·12	0·987 20	0·950 06	− 0·452 52	− 1·484 47
0·14	1·182 65	0·816 49	− 0·753 48	− 1·245 35
0·16	1·341 90	0·622 85	− 0·997 38	− 0·966 98
0·18	1·453 17	0·382 56	− 1·176 57	− 0·658 67
0·20	1·507 82	0·110 50	− 1·285 72	− 0·331 99
0·22	1·500 59	− 0·177 59	− 1·322 20	− 0·000 05
0·24	1·429 71	− 0·465 73	− 1·286 37	0·323 33
0·26	1·296 90	− 0·738 33	− 1·181 65	0·624 25
0·28	1·107 19	− 0·980 87	− 1·014 43	0·889 56
0·30	0·868 64	− 1·180 57	− 0·793 87	1·107 62
0·32	0·591 86	− 1·326 94	− 0·531 45	1·268 80
0·34	0·289 49	− 1·412 22	− 0·240 51	1·366 06
0·36	− 0·024 45	− 1·431 71	0·064 38	1·395 29
0·38	− 0·335 28	− 1·383 99	0·368 11	1·355 54
0·40	− 0·628 37	− 1·270 99	0·655 69	1·249 12
0·42	− 0·889 87	− 1·097 82	0·913 01	1·081 48
0·44	− 1·107 39	− 0·872 57	1·127 47	0·860 96
0·46	− 1·270 60	− 0·605 86	1·288 60	0·598 42
0·48	− 1·371 74	− 0·310 31	1·388 52	0·306 69
0·50	− 1·406 00	0·000 00	1·422 38	0·000 00
0·52	− 1·371 74	0·310 31	1·388 52	− 0·306 69
0·54	− 1·270 60	0·605 86	1·288 60	− 0·598 42
0·56	− 1·107 39	0·872 57	1·127 47	− 0·860 96
0·58	− 0·889 87	1·097 82	0·913 01	− 1·081 48
0·60	− 0·628 37	1·270 99	0·655 69	− 1·249 12
0·62	− 0·335 28	1·383 99	0·368 11	− 1·355 54
0·64	− 0·024 45	1·431 71	0·064 38	− 1·395 29
0·66	0·289 49	1·412 22	− 0·240 51	− 1·366 06
0·68	0·591 86	1·326 94	− 0·531 45	− 1·268 80
0·70	0·868 64	1·180 57	− 0·793 87	− 1·107 62
0·72	1·107 19	0·980 87	− 1·014 43	− 0·889 56
0·74	1·296 90	0·738 33	− 1·181 65	− 0·624 25
0·76	1·429 71	0·465 73	− 1·286 37	− 0·323 33
0·78	1·500 59	0·177 59	− 1·322 20	0·000 05
0·80	1·507 82	− 0·110 50	− 1·285 72	0·331 99
0·82	1·453 17	− 0·382 56	− 1·176 57	0·658 67
0·84	1·341 90	− 0·622 85	− 0·997 38	0·966 98
0·86	1·182 65	− 0·816 49	− 0·753 48	1·245 35
0·88	0·987 20	− 0·950 06	− 0·452 52	1·484 47
0·90	0·770 05	− 1·012 02	− 0·103 93	1·677 95
0·92	0·548 04	− 0·993 03	0·281 89	1·822 80
0·94	0·339 75	− 0·886 09	0·694 28	1·919 98
0·96	0·165 10	− 0·686 46	1·123 23	1·974 69
0·98	0·044 81	− 0·391 47	1·560 38	1·996 58
1·00	0·000 00	0·000 00	2·000 00	1·999 93

TABLE 7·2 (b) (cont.)

Fourth mode

$\dfrac{x}{l}$	ϕ_4	$\phi_4' = \dfrac{1}{\lambda_4}\dfrac{d\phi_4}{dx}$	$\phi_4'' = \dfrac{1}{\lambda_4^2}\dfrac{d^2\phi_4}{dx^2}$	$\phi_4''' = \dfrac{1}{\lambda_4^3}\dfrac{d^3\phi_4}{dx^3}$
0·00	0·000 00	0·000 00	2·000 00	− 2·000 00
0·02	0·072 41	0·485 57	1·435 02	− 1·993 00
0·04	0·259 58	0·812 07	0·876 58	− 1·948 24
0·06	0·516 97	0·983 25	0·339 37	− 1·839 60
0·08	0·801 77	1·007 89	− 0·156 33	− 1·653 33
0·10	1·074 49	0·900 88	− 0·588 02	− 1·387 36
0·12	1·300 78	0·683 45	− 0·934 12	− 1·050 12
0·14	1·453 08	0·382 42	− 1·176 73	− 0·658 79
0·16	1·512 08	0·028 94	− 1·303 80	− 0·237 25
0·18	1·467 65	− 0·343 51	− 1·310 68	0·186 49
0·20	1·319 23	− 0·701 22	− 1·200 92	0·582 86
0·22	1·075 50	− 1·012 71	− 0·986 34	0·923 49
0·24	0·753 48	− 1·250 91	− 0·686 30	1·183 64
0·26	0·377 00	− 1·395 15	− 0·326 40	1·344 42
0·28	− 0·025 37	− 1·432 65	0·063 48	1·394 39
0·30	− 0·422 68	− 1·359 44	0·451 36	1·330 56
0·32	− 0·784 13	− 1·180 58	0·805 69	1·158 76
0·34	− 1·081 59	− 0·909 72	1·097 76	0·893 19
0·36	− 1·291 86	− 0·567 93	1·303 95	0·555 37
0·38	− 1·398 58	− 0·182 05	1·407 55	0·172 45
0·40	− 1·393 51	0·217 53	1·400 10	− 0·224 94
0·42	− 1·277 26	0·599 23	1·281 98	− 0·605 06
0·44	− 1·059 20	0·932 89	1·062 44	− 0·937 59
0·46	− 0·756 76	1·192 08	0·758 79	− 1·196 04
0·48	− 0·394 07	1·356 29	0·395 04	− 1·359 83
0·50	0·000 00	1·412 51	0·000 00	− 1·415 92
0·52	0·394 07	1·356 29	− 0·395 04	− 1·359 83
0·54	0·756 76	1·192 08	− 0·758 79	− 1·196 04
0·56	1·059 20	0·932 89	− 1·062 44	− 0·937 59
0·58	1·277 26	0·599 23	− 1·281 98	− 0·605 06
0·60	1·393 51	0·217 53	− 1·400 10	− 0·224 94
0·62	1·398 58	− 0·182 05	− 1·407 55	0·172 45
0·64	1·291 86	− 0·567 93	− 1·303 95	0·555 37
0·66	1·081 59	− 0·909 72	− 1·097 76	0·893 19
0·68	0·784 13	− 1·180 58	− 0·805 69	1·158 76
0·70	0·422 68	− 1·359 44	− 0·451 36	1·330 56
0·72	0·025 37	− 1·432 65	− 0·063 48	1·394 39
0·74	− 0·377 00	− 1·395 15	0·326 40	1·344 42
0·76	− 0·753 48	− 1·250 91	0·686 30	1·183 64
0·78	− 1·075 50	− 1·012 71	0·986 34	0·923 49
0·80	− 1·319 23	− 0·701 22	1·200 92	0·582 86
0·82	− 1·467 65	− 0·343 51	1·310 68	0·186 49
0·84	− 1·512 08	0·028 94	1·303 80	− 0·237 25
0·86	− 1·453 08	0·382 42	1·176 73	− 0·658 79
0·88	− 1·300 78	0·683 45	0·934 12	− 1·050 12
0·90	− 1·074 49	0·900 88	0·588 02	− 1·387 36
0·92	− 0·801 77	1·007 89	0·156 33	− 1·653 33
0·94	− 0·516 97	0·983 25	− 0·339 37	− 1·839 60
0·96	− 0·259 58	0·812 07	− 0·876 58	− 1·948 24
0·98	− 0·072 41	0·485 57	− 1·435 02	− 1·993 00
1·00	0·000 00	0·000 00	− 2·000 00	− 2·000 00

TABLE 7·2 (b) (cont.)

Fifth mode

$\dfrac{x}{l}$	ϕ_5	$\phi_5'=\dfrac{1}{\lambda_5}\dfrac{d\phi_5}{dx}$	$\phi_5''=\dfrac{1}{\lambda_5^2}\dfrac{d^2\phi_5}{dx^2}$	$\phi_5'''=\dfrac{1}{\lambda_5^3}\dfrac{d^3\phi_5}{dx^3}$
0·00	0·000 00	0·000 00	2·000 00	− 2·000 00
0·02	0·105 67	0·571 81	1·309 96	− 1·987 43
0·04	0·367 91	0·906 94	0·634 09	− 1·908 94
0·06	0·706 32	1·015 17	0·002 91	− 1·724 40
0·08	1·045 91	0·918 67	− 0·543 91	− 1·420 67
0·10	1·321 78	0·653 59	− 0·966 46	− 1·008 91
0·12	1·483 81	0·268 80	− 1·232 31	− 0·520 30
0·14	1·500 43	− 0·177 81	− 1·322 42	− 0·000 21
0·16	1·360 90	− 0·624 65	− 1·234 90	0·498 65
0·18	1·075 51	− 1·012 69	− 0·986 32	0·923 51
0·20	0·673 60	− 1·291 64	− 0·610 48	1·228 51
0·22	0·199 59	− 1·425 40	− 0·154 91	1·380 72
0·24	− 0·292 69	− 1·395 97	0·324 32	1·364 34
0·26	− 0·746 58	− 1·205 25	0·768 97	1·182 87
0·28	− 1·109 52	− 0·874 70	1·125 38	0·858 86
0·30	− 1·339 38	− 0·442 62	1·350 61	0·431 41
0·32	− 1·409 54	0·040 46	1·417 49	− 0·048 38
0·34	− 1·312 08	0·517 81	1·317 72	− 0·523 41
0·36	− 1·058 81	0·933 26	1·062 82	− 0·937 21
0·38	− 0·679 87	1·237 90	0·682 73	− 1·240 67
0·40	− 0·220 21	1·395 84	0·222 26	− 1·397 77
0·42	0·265 75	1·388 50	− 0·264 25	− 1·389 83
0·44	0·720 46	1·216 84	− 0·719 33	− 1·217 71
0·46	1·090 11	0·901 19	− 1·089 23	− 0·901 72
0·48	1·330 98	0·478 92	− 1·330 23	− 0·479 17
0·50	1·414 57	0·000 00	− 1·413 86	0·000 00
0·52	1·330 98	− 0·478 92	− 1·330 23	0·479 17
0·54	1·090 11	− 0·901 19	− 1·089 23	0·901 72
0·56	0·720 46	− 1·216 84	− 0·719 33	1·217 71
0·58	0·265 75	− 1·388 50	− 0·264 25	1·389 83
0·60	− 0·220 21	− 1·395 84	0·222 26	1·397 77
0·62	− 0·679 87	− 1·237 90	0·682 73	1·240 67
0·64	− 1·058 81	− 0·933 26	1·062 82	0·937 21
0·66	− 1·312 08	− 0·517 81	1·317 72	0·523 41
0·68	− 1·409 54	− 0·040 46	1·417 49	0·048 38
0·70	− 1·339 38	0·442 62	1·350 61	− 0·431 41
0·72	− 1·109 52	0·874 70	1·125 38	− 0·858 86
0·74	− 0·746 58	1·205 25	0·768 97	− 1·182 87
0·76	− 0·292 69	1·395 97	0·324 32	− 1·364 34
0·78	0·199 59	1·425 40	− 0·154 91	− 1·380 72
0·80	0·673 60	1·291 64	− 0·610 48	− 1·228 51
0·82	1·075 51	1·012 69	− 0·986 32	− 0·923 51
0·84	1·360 90	0·624 65	− 1·234 90	− 0·498 65
0·86	1·500 43	0·177 81	− 1·322 42	0·000 21
0·88	1·483 81	− 0·268 80	− 1·232 31	0·520 30
0·90	1·321 78	− 0·653 59	− 0·966 46	1·008 91
0·92	1·045 91	− 0·918 67	− 0·543 91	1·420 67
0·94	0·706 32	− 1·015 17	0·002 91	1·724 40
0·96	0·367 91	− 0·906 94	0·634 09	1·908 94
0·98	0·105 67	− 0·571 81	1·309 96	1·987 43
1·00	0·000 00	0·000 00	2·000 00	2·000 00

TABLE 7·2(c)

DATA FOR CLAMPED-FREE BEAM

Characteristic function and its derivatives

$$\phi_r = \cosh \lambda_r x - \cos \lambda_r x - \sigma_r(\sinh \lambda_r x - \sin \lambda_r x), \ (r = 1, 2, 3, \ldots),$$

$$\frac{1}{\lambda_r}\frac{d\phi_r}{dx} = \phi_r' = \sinh \lambda_r x + \sin \lambda_r x - \sigma_r(\cosh \lambda_r x - \cos \lambda_r x),$$

$$\frac{1}{\lambda_r^2}\frac{d^2\phi_r}{dx^2} = \phi_r'' = \cosh \lambda_r x + \cos \lambda_r x - \sigma_r(\sinh \lambda_r x + \sin \lambda_r x),$$

$$\frac{1}{\lambda_r^3}\frac{d^3\phi_r}{dx^3} = \phi_r''' = \sinh \lambda_r x - \sin \lambda_r x - \sigma_r(\cosh \lambda_r x + \cos \lambda_r x).$$

Boundary values

$$\phi_r(0) = \phi_r'(0) = 0,$$

$$\phi_r''(l) = \phi_r'''(l) = 0.$$

Orthogonality

$$\int_0^l \phi_r(x)\,\phi_s(x)\,dx = \int_0^l \phi_r''(x)\,\phi_s''(x)\,dx = \begin{cases} 0 & (r \neq s), \\ l & (r = s), \end{cases}$$

for $r, s = 1, 2, 3, \ldots$.

Receptances in series form

$$\alpha_{xh} = \alpha_{hx} = \sum_{r=1}^{\infty} \frac{\phi_r(x)\,\phi_r(h)}{a_r(\omega_r^2 - \omega^2)}, \quad \text{where} \quad a_r = A\rho l;$$

other receptances by differentiation.

General data

$$\cos \lambda_r l \cosh \lambda_r l + 1 = 0, \quad \sigma_r = \frac{\sinh \lambda_r l - \sin \lambda_r l}{\cosh \lambda_r l + \cos \lambda_r l}, \quad (r = 1, 2, 3, \ldots),$$

Values of σ_r and $\lambda_r l$ and various powers

r	$\lambda_r l$	$(\lambda_r l)^2$	$(\lambda_r l)^3$	$(\lambda_r l)^4$
1	1·875 10	3·516 02	6·592 90	12·362 4
2	4·694 09	22·034 5	103·432	485·519
3	7·854 76	61·697 2	484·617	3 806·55
4	10·995 5	120·902	1 329·38	14 617·3
5	14·137 2	199·860	2 825·45	39 943·8

r	σ_r	$\dfrac{\omega_r}{\omega_1} = \dfrac{\lambda_r^2}{\lambda_1^2}$	$\dfrac{\omega_r^2}{\omega_1^2} = \dfrac{\lambda_r^4}{\lambda_1^4}$	
1	0·734 095 5	1·0	1·0	**For $r > 5$**
2	1·018 466 4	6·266 89	39·273 9	$\lambda_r l \eqsim (2r-1)\,\pi/2$
3	0·999 224 5	17·547 5	307·914	$\sigma_r \eqsim 1\cdot0$
4	1·000 033 6	34·386 1	1 182·40	
5	0·999 998 6	56·842 6	3 231·08	

TABLE 7·2 (c) (cont.)

CHARACTERISTIC FUNCTIONS AND DERIVATIVES
CLAMPED-FREE BEAM

First mode

$\dfrac{x}{l}$	ϕ_1	$\phi_1' = \dfrac{1}{\lambda_1}\dfrac{d\phi_1}{dx}$	$\phi_1'' = \dfrac{1}{\lambda_1^2}\dfrac{d^2\phi_1}{dx^2}$	$\phi_1''' = \dfrac{1}{\lambda_1^3}\dfrac{d^3\phi_1}{dx^3}$
0·00	0·000 00	0·000 00	2·000 00	− 1·468 19
0·02	0·001 39	0·073 97	1·944 94	− 1·468 17
0·04	0·005 52	0·145 88	1·889 88	− 1·468 05
0·06	0·012 31	0·215 72	1·834 83	− 1·467 73
0·08	0·021 68	0·283 50	1·779 80	− 1·467 10
0·10	0·033 55	0·349 21	1·724 80	− 1·466 07
0·12	0·047 84	0·412 86	1·669 85	− 1·464 55
0·14	0·064 49	0·474 46	1·614 96	− 1·462 45
0·16	0·083 40	0·534 00	1·560 16	− 1·459 68
0·18	0·104 52	0·591 48	1·505 49	− 1·456 17
0·20	0·127 74	0·646 92	1·450 96	− 1·451 82
0·22	0·153 01	0·700 31	1·396 60	− 1·446 56
0·24	0·180 24	0·751 67	1·342 47	− 1·440 32
0·26	0·209 36	0·801 00	1·288 59	− 1·433 02
0·28	0·240 30	0·848 32	1·235 00	− 1·424 59
0·30	0·272 97	0·893 64	1·181 75	− 1·414 97
0·32	0·307 30	0·936 96	1·128 89	− 1·404 10
0·34	0·343 22	0·978 31	1·076 46	− 1·391 91
0·36	0·380 65	1·017 71	1·024 51	− 1·378 34
0·38	0·419 52	1·055 16	0·973 09	− 1·363 34
0·40	0·459 77	1·090 70	0·922 27	− 1·346 85
0·42	0·501 31	1·124 35	0·872 09	− 1·328 84
0·44	0·544 08	1·156 12	0·822 62	− 1·309 24
0·46	0·588 00	1·186 06	0·773 92	− 1·288 01
0·48	0·633 01	1·214 18	0·726 03	− 1·265 12
0·50	0·679 05	1·240 52	0·679 05	− 1·240 52
0·52	0·726 03	1·265 12	0·633 01	− 1·214 18
0·54	0·773 92	1·288 01	0·588 00	− 1·186 06
0·56	0·822 62	1·309 24	0·544 08	− 1·156 12
0·58	0·872 09	1·328 84	0·501 31	− 1·124 35
0·60	0·922 27	1·346 85	0·459 77	− 1·090 70
0·62	0·973 09	1·363 34	0·419 52	− 1·055 16
0·64	1·024 51	1·378 34	0·380 65	− 1·017 71
0·66	1·076 46	1·391 91	0·343 22	− 0·978 31
0·68	1·128 89	1·404 10	0·307 30	− 0·936 96
0·70	1·181 75	1·414 97	0·272 97	− 0·893 64
0·72	1·235 00	1·424 59	0·240 30	− 0·848 32
0·74	1·288 59	1·433 02	0·209 36	− 0·801 00
0·76	1·342 47	1·440 32	0·180 24	− 0·751 67
0·78	1·396 60	1·446 56	0·153 01	− 0·700 31
0·80	1·450 96	1·451 82	0·127 74	− 0·646 92
0·82	1·505 49	1·456 17	0·104 52	− 0·591 48
0·84	1·560 16	1·459 68	0·083 40	− 0·534 00
0·86	1·614 96	1·462 45	0·064 49	− 0·474 46
0·88	1·669 85	1·464 55	0·047 84	− 0·412 86
0·90	1·724 80	1·466 07	0·033 55	− 0·349 21
0·92	1·779 80	1·467 10	0·021 68	− 0·283 50
0·94	1·834 83	1·467 73	0·012 31	− 0·215 72
0·96	1·889 88	1·468 05	0·005 52	− 0·145 88
0·98	1·944 94	1·468 17	0·001 39	− 0·073 97
1·00	2·000 00	1·468 19	0·000 00	0·000 00

TABLE 7·2 (c) (cont.)

Second mode

$\dfrac{x}{l}$	ϕ_2	$\phi_2'=\dfrac{1}{\lambda_2}\dfrac{d\phi_2}{dx}$	$\phi_2''=\dfrac{1}{\lambda_2^2}\dfrac{d^2\phi_2}{dx^2}$	$\phi_2'''=\dfrac{1}{\lambda_2^3}\dfrac{d^3\phi_2}{dx^3}$
0·00	0·000 00	0·000 00	2·000 00	−2·036 93
0·02	0·008 53	0·178 79	1·808 77	−2·036 66
0·04	0·033 01	0·339 62	1·617 64	−2·034 83
0·06	0·071 74	0·482 53	1·426 80	−2·030 02
0·08	0·123 05	0·607 54	1·236 60	−2·020 97
0·10	0·185 26	0·714 75	1·047 50	−2·006 58
0·12	0·256 70	0·804 28	0·860 04	−1·985 90
0·14	0·335 73	0·876 31	0·674 84	−1·958 14
0·16	0·420 70	0·931 08	0·492 61	−1·922 67
0·18	0·510 02	0·968 92	0·314 09	−1·879 01
0·20	0·602 11	0·990 20	0·140 07	−1·826 82
0·22	0·695 44	0·995 39	−0·028 65	−1·765 92
0·24	0·788 52	0·985 02	−0·191 23	−1·696 25
0·26	0·879 92	0·959 70	−0·346 87	−1·617 91
0·28	0·968 27	0·920 13	−0·494 75	−1·531 13
0·30	1·052 27	0·867 07	−0·634 10	−1·436 24
0·32	1·130 68	0·801 36	−0·764 19	−1·333 73
0·34	1·202 36	0·723 89	−0·884 31	−1·224 16
0·36	1·266 26	0·635 65	−0·993 84	−1·108 21
0·38	1·321 41	0·537 64	−1·092 22	−0·986 67
0·40	1·366 94	0·430 94	−1·178 95	−0·860 40
0·42	1·402 09	0·316 65	−1·253 65	−0·730 34
0·44	1·426 19	0·195 93	−1·316 00	−0·597 48
0·46	1·438 71	0·069 95	−1·365 78	−0·462 91
0·48	1·439 20	−0·060 12	−1·402 89	−0·327 72
0·50	1·427 33	−0·193 07	−1·427 33	−0·193 07
0·52	1·402 89	−0·327 72	−1·439 20	−0·060 12
0·54	1·365 78	−0·462 91	−1·438 71	0·069 95
0·56	1·316 00	−0·597 48	−1·426 19	0·195 93
0·58	1·253 65	−0·730 34	−1·402 09	0·316 65
0·60	1·178 95	−0·860 40	−1·366 94	0·430 94
0·62	1·092 22	−0·986 67	−1·321 41	0·537 64
0·64	0·993 84	−1·108 21	−1·266 26	0·635 65
0·66	0·884 31	−1·224 16	−1·202 36	0·723 89
0·68	0·764 19	−1·333 73	−1·130 68	0·801 36
0·70	0·634 10	−1·436 24	−1·052 27	0·867 07
0·72	0·494 75	−1·531 13	−0·968 27	0·920 13
0·74	0·346 87	−1·617 91	−0·879 92	0·959 70
0·76	0·191 23	−1·696 25	−0·788 52	0·985 02
0·78	0·028 65	−1·765 92	−0·695 44	0·995 39
0·80	−0·140 07	−1·826 82	−0·602 11	0·990 20
0·82	−0·314 09	−1·879 01	−0·510 02	0·968 92
0·84	−0·492 61	−1·922 67	−0·420 70	0·931 08
0·86	−0·674 84	−1·958 14	−0·335 73	0·876 31
0·88	−0·860 04	−1·985 90	−0·256 70	0·804 28
0·90	−1·047 50	−2·006 58	−0·185 26	0·714 75
0·92	−1·236 60	−2·020 97	−0·123 05	0·607 54
0·94	−1·426 80	−2·030 02	−0·071 74	0·482 53
0·96	−1·617 64	−2·034 83	−0·033 01	0·339 62
0·98	−1·808 77	−2·036 66	−0·008 53	0·178 79
1·00	−2·000 00	−2·036 93	0·000 00	0·000 00

TABLE 7·2 (c) (cont.)

Third mode

$\dfrac{x}{l}$	ϕ_3	$\phi_3' = \dfrac{1}{\lambda_3}\dfrac{d\phi_3}{dx}$	$\phi_3'' = \dfrac{1}{\lambda_3^2}\dfrac{d^2\phi_3}{dx^2}$	$\phi_3''' = \dfrac{1}{\lambda_3^3}\dfrac{d^3\phi_3}{dx^3}$
0·00	0·000 00	0·000 00	2·000 00	− 1·998 45
0·02	0·023 39	0·289 53	1·686 10	− 1·997 21
0·04	0·088 39	0·529 79	1·372 87	− 1·988 92
0·06	0·187 27	0·720 99	1·061 89	− 1·967 66
0·08	0·312 38	0·863 67	0·755 58	− 1·928 71
0·10	0·456 14	0·958 79	0·457 02	− 1·868 54
0·12	0·611 20	1·007 85	0·169 74	− 1·784 80
0·14	0·770 49	1·012 91	− 0·102 45	− 1·676 29
0·16	0·927 28	0·976 65	− 0·355 63	− 1·542 86
0·18	1·075 35	0·902 37	− 0·585 94	− 1·385 40
0·20	1·209 01	0·793 94	− 0·789 75	− 1·205 75
0·22	1·323 24	0·655 80	− 0·963 75	− 1·006 56
0·24	1·413 76	0·492 85	− 1·105 15	− 0·791 24
0·26	1·477 07	0·310 40	− 1·211 72	− 0·563 80
0·28	1·510 56	0·114 05	− 1·281 89	− 0·328 72
0·30	1·512 48	− 0·090 41	− 1·314 85	− 0·090 85
0·32	1·482 03	− 0·297 11	− 1·310 55	0·144 79
0·34	1·419 31	− 0·500 26	− 1·269 74	0·373 10
0·36	1·325 34	− 0·694 22	− 1·193 98	0·589 08
0·38	1·201 96	− 0·873 68	− 1·085 56	0·787 97
0·40	1·051 85	− 1·033 74	− 0·947 53	0·965 33
0·42	0·878 41	− 1·170 03	− 0·783 59	1·117 23
0·44	0·685 68	− 1·278 81	− 0·598 02	1·240 30
0·46	0·478 22	− 1·357 04	− 0·395 55	1·331 88
0·48	0·261 03	− 1·402 47	− 0·181 30	1·390 04
0·50	0·039 37	− 1·413 66	0·039 37	1·413 66
0·52	− 0·181 30	− 1·390 04	0·261 03	1·402 47
0·54	− 0·395 55	− 1·331 88	0·478 22	1·357 04
0·56	− 0·598 02	− 1·240 30	0·685 68	1·278 81
0·58	− 0·783 59	− 1·117 23	0·878 41	1·170 03
0·60	− 0·947 53	− 0·965 33	1·051 85	1·033 74
0·62	− 1·085 56	− 0·787 97	1·201 96	0·873 68
0·64	− 1·193 98	− 0·589 08	1·325 34	0·694 22
0·66	− 1·269 74	− 0·373 10	1·419 31	0·500 26
0·68	− 1·310 55	− 0·144 79	1·482 03	0·297 11
0·70	− 1·314 85	0·090 85	1·512 48	0·090 41
0·72	− 1·281 89	0·328 72	1·510 56	− 0·114 05
0·74	− 1·211 72	0·563 80	1·477 07	− 0·310 40
0·76	− 1·105 15	0·791 24	1·413 76	− 0·492 85
0·78	− 0·963 75	1·006 56	1·323 24	− 0·655 80
0·80	− 0·789 75	1·205 75	1·209 01	− 0·793 94
0·82	− 0·585 94	1·385 40	1·075 35	− 0·902 37
0·84	− 0·355 63	1·542 86	0·927 28	− 0·976 65
0·86	− 0·102 45	1·676 29	0·770 49	− 1·012 91
0·88	0·169 74	1·784 80	0·611 20	− 1·007 85
0·90	0·457 02	1·868 54	0·456 14	− 0·958 79
0·92	0·755 58	1·928 71	0·312 38	− 0·863 67
0·94	1·061 89	1·967 66	0·187 27	− 0·720 99
0·96	1·372 87	1·988 92	0·088 29	− 0·529 79
0·98	1·686 10	1·997 21	0·023 39	− 0·289 53
1·00	2·000 00	1·998 45	0·000 00	0·000 00

TABLE 7·2 (c) (cont.)

Fourth mode

$\dfrac{x}{l}$	ϕ_4	$\phi_4' = \dfrac{1}{\lambda_4}\dfrac{d\phi_4}{dx}$	$\phi_4'' = \dfrac{1}{\lambda_4^2}\dfrac{d^2\phi_4}{dx^2}$	$\phi_4''' = \dfrac{1}{\lambda_4^3}\dfrac{d^3\phi_4}{dx^3}$
0·00	0·000 00	0·000 00	2·000 00	−2·000 07
0·02	0·044 82	0·391 47	1·560 35	−1·996 72
0·04	0·165 10	0·686 45	1·123 17	−1·974 82
0·06	0·339 74	0·886 06	0·694 20	−1·920 12
0·08	0·548 01	0·992 98	0·281 79	−1·822 94
0·10	0·770 02	1·011 94	−0·104 07	−1·678 09
0·12	0·987 14	0·949 94	−0·452 70	−1·484 63
0·14	1·182 56	0·816 33	−0·753 68	−1·245 52
0·16	1·341 77	0·622 64	−0·997 62	−0·967 17
0·18	1·452 99	0·382 30	−1·176 87	−0·658 91
0·20	1·507 58	0·110 17	−1·286 08	−0·332 28
0·22	1·500 27	−0·178 01	−1·322 62	−0·000 38
0·24	1·429 28	−0·466 24	−1·286 88	0·322 90
0·26	1·296 34	−0·738 95	−1·182 26	0·623 70
0·28	1·106 48	−0·981 64	−1·015 18	0·888 88
0·30	0·867 74	−1·181 54	−0·794 78	1·106 76
0·32	0·590 73	−1·328 13	−0·532 58	1·267 72
0·34	0·288 08	−1·413 68	−0·241 91	1·364 69
0·36	−0·026 21	−1·433 51	0·062 64	1·393 57
0·38	−0·337 48	−1·386 22	0·365 94	1·353 39
0·40	−0·631 12	−1·273 76	0·652 99	1·246 43
0·42	−0·893 30	−1·101 26	0·909 64	1·078 12
0·44	−1·111 66	−0·876 83	1·123 27	0·856 75
0·46	−1·275 92	−0·611 15	1·283 36	0·593 15
0·48	−1·378 36	−0·316 90	1·381 99	0·300 11
0·50	−1·414 24	−0·008 19	1·414 24	−0·008 19
0·52	−1·381 99	0·300 12	1·378 36	−0·316 90
0·54	−1·283 36	0·593 16	1·275 92	−0·611 15
0·56	−1·123 27	0·856 75	1·111 66	−0·876 84
0·58	−0·909 64	1·078 12	0·893 30	−1·101 26
0·60	−0·652 99	1·246 43	0·631 12	−1·273 76
0·62	−0·365 94	1·353 39	0·337 48	−1·386 22
0·64	−0·062 64	1·393 57	0·026 21	−1·433 51
0·66	0·241 91	1·364 69	−0·288 08	−1·413 68
0·68	0·532 58	1·267 72	−0·590 73	−1·328 13
0·70	0·794 78	1·106 76	−0·867 74	−1·181 53
0·72	1·015 18	0·888 88	−1·106 48	−0·981 64
0·74	1·182 26	0·623 70	−1·296 34	−0·738 95
0·76	1·286 88	0·322 90	−1·429 28	−0·466 24
0·78	1·322 62	−0·000 39	−1·500 27	−0·178 01
0·80	1·286 08	−0·332 28	−1·507 58	0·110 17
0·82	1·176 87	−0·658 90	−1·452 99	0·382 30
0·84	0·997 62	−0·967 17	−1·341 77	0·622 64
0·86	0·753 68	−1·245 52	−1·182 56	0·816 33
0·88	0·452 70	−1·484 63	−0·987 14	0·949 94
0·90	0·104 07	−1·678 09	−0·770 02	1·011 94
0·92	−0·281 79	−1·822 94	−0·548 01	0·992 98
0·94	−0·694 20	−1·920 12	−0·339 74	0·886 06
0·96	−1·123 17	−1·974 82	−0·165 10	0·686 45
0·98	−1·560 35	−1·996 72	−0·044 82	0·391 47
1·00	−2·000 00	−2·000 07	0·000 00	0·000 00

TABLE 7·2 (c) (cont.)

Fifth mode

$\dfrac{x}{l}$	ϕ_5	$\phi_5' = \dfrac{1}{\lambda_5}\dfrac{d\phi_5}{dx}$	$\phi_5'' = \dfrac{1}{\lambda_5^2}\dfrac{d^2\phi_5}{dx^2}$	$\phi_5''' = \dfrac{1}{\lambda_5^3}\dfrac{d^3\phi_5}{dx^3}$
0·00	0·000 00	0·000 00	2·000 00	− 2·000 00
0·02	0·072 41	0·485 57	1·435 02	− 1·993 00
0·04	0·259 58	0·812 07	0·876 58	− 1·948 24
0·06	0·516 97	0·983 25	0·339 37	− 1·839 59
0·08	0·801 77	1·007 89	− 0·156 33	− 1·653 32
0·10	1·074 49	0·900 89	− 0·588 01	− 1·387 36
0·12	1·300 78	0·683 46	− 0·934 11	− 1·050 11
0·14	1·453 09	0·382 43	− 1·176 72	− 0·658 78
0·16	1·512 09	0·028 95	− 1·303 78	− 0·237 23
0·18	1·467 67	− 0·343 48	− 1·310 66	0·186 51
0·20	1·319 25	− 0·701 19	− 1·200 90	0·582 89
0·22	1·075 53	− 1·012 67	− 0·986 31	0·923 52
0·24	0·753 53	− 1·250 86	− 0·686 26	1·183 68
0·26	0·377 06	− 1·395 09	− 0·326 34	1·344 48
0·28	− 0·025 29	− 1·432 57	0·063 55	1·394 46
0·30	− 0·422 57	− 1·359 34	0·451 46	1·330 65
0·32	− 0·783 99	− 1·180 45	0·805 82	1·158 89
0·34	− 1·081 40	− 0·909 54	1·097 93	0·893 37
0·36	− 1·291 62	− 0·567 70	1·304 18	0·555 61
0·38	− 1·398 26	− 0·181 74	1·407 86	0·172 76
0·40	− 1·393 10	0·217 94	1·400 51	− 0·224 52
0·42	− 1·276 70	0·599 78	1·289 53	− 0·604 50
0·44	− 1·058 46	0·933 61	1·063 17	− 0·936 86
0·46	− 0·755 79	1·193 04	0·759 76	− 1·195 08
0·48	− 0·392 78	1·357 57	0·396 32	− 1·358 55
0·50	0·001 70	1·414 21	0·001 70	− 1·414 21
0·52	0·396 32	1·358 55	− 0·392 78	− 1·357 57
0·54	0·759 76	1·195 08	− 0·755 79	− 1·193 04
0·56	1·063 17	0·936 86	− 1·050 46	− 0·933 61
0·58	1·282 53	0·604 50	− 1·276 70	− 0·599 78
0·60	1·400 51	0·224 52	− 1·393 10	− 0·217 94
0·62	1·407 86	− 0·172 76	− 1·398 26	0·181 74
0·64	1·304 18	− 0·555 61	− 1·291 62	0·567 70
0·66	1·097 93	− 0·893 37	− 1·081 40	0·909 54
0·68	0·805 82	− 1·158 89	− 0·783 99	1·180 45
0·70	0·451 46	− 1·330 65	− 0·422 57	1·359 34
0·72	0·063 55	− 1·394 46	− 0·025 29	1·432 57
0·74	− 0·326 34	− 1·344 48	0·377 06	1·395 09
0·76	− 0·686 26	− 1·183 68	0·753 53	1·250 86
0·78	− 0·986 31	− 0·923 52	1·075 53	1·012 67
0·80	− 1·200 90	− 0·582 89	1·319 25	0·701 19
0·82	− 1·310 66	− 0·186 51	1·467 67	0·343 48
0·84	− 1·303 78	0·237 23	1·512 09	− 0·028 95
0·86	− 1·176 72	0·658 78	1·453 09	− 0·382 43
0·88	− 0·934 11	1·050 11	1·300 78	− 0·683 46
0·90	− 0·588 01	1·387 36	1·074 49	− 0·900 89
0·92	− 0·156 33	1·653 32	0·801 77	− 1·007 89
0·94	0·339 37	1·839 59	0·516 97	− 0·983 25
0·96	0·876 58	1·948 24	0·259 58	− 0·812 07
0·98	1·435 02	1·993 00	0·072 41	− 0·485 57
1·00	2·000 00	2·000 00	0·000 00	0·000 00

TABLE 7·2 (d)

DATA FOR CLAMPED-PINNED BEAM AND FREE-PINNED BEAM

A. Clamped-pinned beam

Characteristic function and its derivatives

$$\phi_r = \cosh \lambda_r x - \cos \lambda_r x - \sigma_r(\sinh \lambda_r x - \sin \lambda_r x), \quad (r = 1, 2, 3, \ldots).$$

$$\frac{1}{\lambda_r}\frac{d\phi_r}{dx} = \phi_r' = \sinh \lambda_r x + \sin \lambda_r x - \sigma_r(\cosh \lambda_r x - \cos \lambda_r x),$$

$$\frac{1}{\lambda_r^2}\frac{d^2\phi_r}{dx^2} = \phi_r'' = \cosh \lambda_r x + \cos \lambda_r x - \sigma_r(\sinh \lambda_r x + \sin \lambda_r x),$$

$$\frac{1}{\lambda_r^3}\frac{d^3\phi_r}{dx^3} = \phi_r''' = \sinh \lambda_r x - \sin \lambda_r x - \sigma_r(\cosh \lambda_r x + \cos \lambda_r x).$$

Boundary values

$$\phi_r(0) = \phi_r'(0) = 0,$$

$$\phi_r(l) = \phi_r''(l) = 0.$$

Orthogonality

$$\int_0^l \phi_r(x)\,\phi_s(x)\,dx = \int_0^l \phi_r''(x)\,\phi_s''(x)\,dx = \begin{cases} 0 & (r \neq s), \\ l & (r = s), \end{cases}$$

for $r, s = 1, 2, 3, \ldots$.

Receptances in series form

$$\alpha_{xh} = \alpha_{hx} = \sum_{r=1}^{\infty} \frac{\phi_r(x)\,\phi_r(h)}{a_r(\omega_r^2 - \omega^2)}, \quad \text{where} \quad a_r = A\rho l;$$

other receptances by differentiation.

B. Free-pinned beam

Characteristic function

$$\phi_0 = \frac{\sqrt{3}}{l}(l - x),$$

$$\phi_r = \cosh \lambda_r x + \cos \lambda_r x - \sigma_r(\sinh \lambda_r x + \sin \lambda_r x), \quad (r = 1, 2, 3, \ldots),$$

where σ_r and λ_r are the same as for a clamped-pinned beam.

The characteristic function for distortion of a free-pinned beam is the same as the second derivative of that of a clamped-pinned beam; that is, for $r = 1, 2, 3, \ldots,$

$$\phi_r \text{ (free-pinned)} = \phi_r'' \text{ (clamped-pinned)},$$

$$\phi_r' \text{ (free-pinned)} = \phi_r''' \text{ (clamped-pinned)},$$

$$\phi_r'' \text{ (free-pinned)} = \phi_r \text{ (clamped-pinned)},$$

$$\phi_r''' \text{ (free-pinned)} = \phi_r' \text{ (clamped-pinned)}.$$

Boundary values

$$\phi_r''(0) = \phi_r'''(0) = 0,$$
$$\phi_r(l) = \phi_r''(l) = 0.$$

Orthogonality

$$\int_0^l \phi_r(x)\,\phi_s(x)\,dx = \begin{cases} 0 & (r \neq s), \\ l & (r = s), \end{cases}$$

for $r, s = 0, 1, 2, \ldots$; and

$$\int_0^l \phi_r''(x)\,\phi_s''(x)\,dx = \begin{cases} 0 & (r \neq s), \\ l & (r = s), \end{cases}$$

for $r, s = 1, 2, 3, \ldots$.

Receptances in series form

$$\alpha_{xh} = \alpha_{hx} = \sum_{r=0}^{\infty} \frac{\phi_r(x)\,\phi_r(h)}{a_r(\omega_r^2 - \omega^2)}, \quad \text{where} \quad a_r = A\rho l;$$

other receptances by differentiation.

TABLE 7·2 (d) (cont.)

GENERAL DATA (for A and B)

$$\left.\begin{array}{l} \tan \lambda_r l = \tanh \lambda_r l, \\ \sigma_r = \cot \lambda_r l = \coth \lambda_r l, \end{array}\right\} (r = 1, 2, 3, \ldots).$$

Values of σ_r and $\lambda_r l$ and various powers

r	$\lambda_r l$	$(\lambda_r l)^2$	$(\lambda_r l)^3$	$(\lambda_r l)^4$
1	3·926 60	15·418 2	60·541 2	237·721
2	7·068 58	49·964 9	353·181	2 496·49
3	10·210 2	104·248	1 064·39	10 867·6
4	13·351 8	178·270	2 380·22	31 780·1
5	16·493 4	272·031	4 486·71	74 000·8

r	σ_r	$\dfrac{\omega_r}{\omega_1} = \dfrac{\lambda_r^2}{\lambda_1^2}$	$\dfrac{\omega_r^2}{\omega_1^2} = \dfrac{\lambda_r^4}{\lambda_1^4}$
1	1·000 777 3	1·0	1·0
2	1·000 001 4	3·240 64	10·501 7
3	1·000 000 0	6·761 34	45·715 7
4	1·000 000 0	11·562 3	133·686
5	1·000 000 0	17·643 5	311·293

For $r > 5$
$$\lambda_r l \doteqdot (4r+1)\,\pi/4$$
$$\sigma_r \doteqdot 1\cdot0$$

TABLE 7·2 (d) (cont.)
CHARACTERISTIC FUNCTIONS AND DERIVATIVES
CLAMPED-PINNED BEAM

First mode

$\dfrac{x}{l}$	ϕ_1	$\phi_1' = \dfrac{1}{\lambda_1}\dfrac{d\phi_1}{dx}$	$\phi_1'' = \dfrac{1}{\lambda_1^2}\dfrac{d^2\phi_1}{dx^2}$	$\phi_1''' = \dfrac{1}{\lambda_1^3}\dfrac{d^3\phi_1}{dx^3}$
0·00	0·000 00	0·000 00	2·000 00	− 2·001 55
0·02	0·006 00	0·150 89	1·842 82	− 2·001 40
0·04	0·023 38	0·289 44	1·685 68	− 2·000 31
0·06	0·051 14	0·415 66	1·528 69	− 1·997 45
0·08	0·088 34	0·529 55	1·372 02	− 1·992 03
0·10	0·134 00	0·631 16	1·215 90	− 1·983 36
0·12	0·187 15	0·720 55	1·060 60	− 1·970 79
0·14	0·246 85	0·797 78	0·906 47	− 1·953 79
0·16	0·312 14	0·862 96	0·753 86	− 1·931 87
0·18	0·382 08	0·916 23	0·603 18	− 1·904 64
0·20	0·455 74	0·957 76	0·454 86	− 1·871 77
0·22	0·532 21	0·987 75	0·309 35	− 1·832 99
0·24	0·610 58	1·006 43	0·167 12	− 1·788 12
0·26	0·689 99	1·014 10	0·028 66	− 1·737 06
0·28	0·769 58	1·011 05	− 0·105 54	− 1·679 75
0·30	0·848 52	0·997 64	− 0·235 00	− 1·616 20
0·32	0·926 01	0·974 27	− 0·359 23	− 1·546 52
0·34	1·001 29	0·941 37	− 0·477 75	− 1·470 82
0·36	1·073 63	0·899 40	− 0·590 09	− 1·389 32
0·38	1·142 33	0·848 86	− 0·695 82	− 1·302 29
0·40	1·206 75	0·790 29	− 0·794 50	− 1·210 02
0·42	1·266 26	0·724 27	− 0·885 74	− 1·112 88
0·44	1·320 32	0·651 38	− 0·969 18	− 1·011 28
0·46	1·368 41	0·572 26	− 1·044 47	− 0·905 66
0·48	1·410 06	0·487 55	− 1·111 33	− 0·796 52
0·50	1·444 86	0·397 94	− 1·169 50	− 0·684 37
0·52	1·472 45	0·304 10	− 1·218 75	− 0·569 77
0·54	1·492 53	0·206 75	− 1·258 94	− 0·453 30
0·56	1·504 85	0·106 61	− 1·289 92	− 0·335 55
0·58	1·509 22	0·004 40	− 1·311 62	− 0·217 15
0·60	1·505 50	− 0·099 16	− 1·324 02	− 0·098 72
0·62	1·493 63	− 0·203 32	− 1·327 14	0·019 10
0·64	1·473 57	− 0·307 36	− 1·321 06	0·135 66
0·66	1·445 37	− 0·410 57	− 1·305 88	0·250 33
0·68	1·409 13	− 0·512 24	− 1·281 80	0·362 47
0·70	1·364 98	− 0·611 67	− 1·249 04	0·471 45
0·72	1·313 13	− 0·708 20	− 1·207 86	0·576 66
0·74	1·253 84	− 0·801 17	− 1·158 58	0·677 50
0·76	1·187 41	− 0·889 96	− 1·101 57	0·773 40
0·78	1·114 18	− 0·974 00	− 1·037 25	0·863 82
0·80	1·034 57	− 1·052 70	− 0·966 06	0·948 23
0·82	0·948 99	− 1·125 56	− 0·888 49	1·026 16
0·84	0·857 95	− 1·192 10	− 0·805 07	1·097 14
0·86	0·761 94	− 1·251 87	− 0·716 36	1·160 78
0·88	0·661 51	− 1·304 48	− 0·622 95	1·216 70
0·90	0·557 24	− 1·349 60	− 0·525 47	1·264 58
0·92	0·449 74	− 1·386 93	− 0·424 55	1·304 14
0·94	0·339 62	− 1·416 21	− 0·320 86	1·335 15
0·96	0·227 52	− 1·437 27	− 0·215 07	1·357 43
0·98	0·114 10	− 1·449 96	− 0·107 89	1·370 85
1·00	0·000 00	− 1·454 20	0·000 00	1·375 33

TABLE 7·2 (d) (cont.)

Second mode

$\dfrac{x}{l}$	ϕ_2	$\phi_2' = \dfrac{1}{\lambda_2}\dfrac{d\phi_2}{dx}$	$\phi_2'' = \dfrac{1}{\lambda_2^2}\dfrac{d^2\phi_2}{dx^2}$	$\phi_2''' = \dfrac{1}{\lambda_2^3}\dfrac{d^3\phi_2}{dx^3}$
0·00	0·000 00	0·000 00	2·000 00	−2·000 00
0·02	0·019 04	0·262 76	1·717 29	−1·999 10
0·04	0·072 41	0·485 57	1·435 02	−1·993 00
0·06	0·154 46	0·668 57	1·154 24	−1·977 27
0·08	0·259 58	0·812 07	0·876 58	−1·948 24
0·10	0·382 23	0·916 66	0·604 15	−1·903 05
0·12	0·516 97	0·983 25	0·339 37	−1·839 60
0·14	0·658 51	1·013 10	0·084 94	−1·756 56
0·16	0·801 76	1·007 89	−0·156 33	−1·653 33
0·18	0·941 92	0·969 66	−0·381 58	−1·530 01
0·20	1·074 49	0·900 88	−0·588 02	−1·387 36
0·22	1·195 34	0·804 41	−0·773 00	−1·226 76
0·24	1·300 78	0·683 45	−0·934 12	−1·050 12
0·26	1·387 59	0·541 52	−1·069 27	−0·859 85
0·28	1·453 08	0·382 42	−1·176 73	−0·658 79
0·30	1·495 10	0·210 17	−1·255 18	−0·450 11
0·32	1·512 08	0·028 94	−1·303 80	−0·237 24
0·34	1·503 05	−0·157 04	−1·322 24	−0·023 81
0·36	1·467 65	−0·343 50	−1·310 68	0·186 49
0·38	1·406 11	−0·526 25	−1·269 83	0·389 93
0·40	1·319 23	−0·701 22	−1·200 92	0·582 86
0·42	1·208 39	−0·864 56	−1·105 69	0·761 80
0·44	1·075 50	−1·012 70	−0·986 34	0·923 49
0·46	0·922 92	−1·142 43	−0·845 53	1·064 96
0·48	0·753 48	−1·250 90	−0·686 31	1·183 64
0·50	0·570 35	−1·335 77	−0·512 04	1·277 36
0·52	0·377 00	−1·395 15	−0·326 40	1·344 42
0·54	0·177 15	−1·427 70	−0·133 23	1·383 65
0·56	−0·025 36	−1·432 65	0·063 48	1·394 38
0·58	−0·226 61	−1·409 78	0·259 68	1·376 54
0·60	−0·422 68	−1·359 44	0·451 36	1·330 56
0·62	−0·609 73	−1·282 56	0·634 60	1·257 45
0·64	−0·784 13	−1·180 58	0·805 69	1·158 76
0·66	−0·942 44	−1·055 49	0·961 12	1·036 50
0·68	−1·081 58	−0·909 72	1·097 76	0·893 19
0·70	−1·198 82	−0·746 12	1·212 81	0·731 72
0·72	−1·291 86	−0·567 93	1·303 95	0·555 37
0·74	−1·358 88	−0·378 66	1·369 30	0·367 69
0·76	−1·398 58	−0·182 05	1·407 55	0·172 45
0·78	−1·410 19	0·018 00	1·417 89	−0·026 43
0·80	−1·393 51	0·217 52	1·400 10	−0·224 94
0·82	−1·348 90	0·412 56	1·354 50	−0·419 12
0·84	−1·277 26	0·599 23	1·281 98	−0·605 06
0·86	−1·180 04	0·773 83	1·183 99	−0·779 04
0·88	−1·059 19	0·932 88	1·062 44	−0·937 59
0·90	−0·917 15	1·073 23	0·919 76	−1·077 52
0·92	−0·756 76	1·192 08	0·758 79	−1·196 04
0·94	−0·581 22	1·287 06	0·582 71	−1·290 78
0·96	−0·394 06	1·356 29	0·395 04	−1·359 83
0·98	−0·199 02	1·398 39	0·199 51	−1·401 83
1·00	0·000 00	1·412 51	0·000 00	−1·415 92

TABLE 7·2 (d) (cont.)

Third mode

$\dfrac{x}{l}$	ϕ_3	$\phi_3' = \dfrac{1}{\lambda_3}\dfrac{d\phi_3}{dx}$	$\phi_3'' = \dfrac{1}{\lambda_3^2}\dfrac{d^2\phi_3}{dx^2}$	$\phi_3''' = \dfrac{1}{\lambda_3^3}\dfrac{d^3\phi_3}{dx^3}$
0·00	0·000 00	0·000 00	2·000 00	−2·000 00
0·02	0·038 86	0·366 72	1·591 73	−1·997 31
0·04	0·144 10	0·650 20	1·185 32	−1·979 61
0·06	0·298 79	0·851 22	0·785 08	−1·935 09
0·08	0·486 26	0·971 68	0·397 42	−1·855 35
0·10	0·690 37	1·014 91	0·030 09	−1·735 37
0·12	0·895 84	0·985 93	−0·308 45	−1·573 31
0·14	1·088 57	0·891 48	−0·609 68	−1·370 37
0·16	1·256 04	0·740 02	−0·865 60	−1·130 46
0·18	1·387 59	0·541 52	−1·069 27	−0·859 85
0·20	1·474 76	0·307 25	−1·215 23	−0·566 78
0·22	1·511 47	0·049 39	−1·299 88	−0·260 98
0·24	1·494 19	−0·219 34	−1·321 68	0·046 83
0·26	1·422 02	−0·486 16	−1·281 37	0·345 51
0·28	1·296 62	−0·738 64	−1·181 95	0·623 97
0·30	1·122 12	−0·965 20	−1·028 63	0·871 71
0·32	0·904 89	−1·155 56	−0·828 67	1·079 34
0·34	0·653 24	−1·301 07	−0·591 10	1·238 93
0·36	0·377 03	−1·395 12	−0·326 37	1·344 45
0·38	0·087 27	−1·433 30	−0·045 96	1·391 99
0·40	−0·204 39	−1·413 64	0·238 07	1·379 96
0·42	−0·486 16	−1·336 65	0·513 62	1·309 19
0·44	−0·746 58	−1·205 25	0·768 97	1·182 87
0·46	−0·975 04	−1·024 71	0·993 30	1·006 46
0·48	−1·162 23	−0·802 34	1·177 11	0·787 46
0·50	−1·300 50	−0·547 26	1·312 63	0·535 13
0·52	−1·384 22	−0·269 94	1·394 11	0·260 05
0·54	−1·410 01	0·018 18	1·418 07	−0·026 24
0·56	−1·376 87	0·305 22	1·383 44	−0·311 79
0·58	−1·286 24	0·579 29	1·291 60	−0·584 65
0·60	−1·141 94	0·829 07	1·146 31	−0·833 44
0·62	−0·950 00	1·044 22	0·953 56	−1·047 78
0·64	−0·718 44	1·215 82	0·721 34	−1·218 73
0·66	−0·456 91	1·336 78	0·459 27	−1·339 15
0·68	−0·176 28	1·402 10	0·178 21	−1·404 03
0·70	0·111 74	1·409 06	−0·110 17	−1·410 64
0·72	0·395 19	1·357 42	−0·393 91	−1·358 70
0·74	0·662 27	1·249 31	−0·661 23	−1·250 36
0·76	0·901 88	1·089 24	−0·901 03	−1·090 10
0·78	1·104 04	0·883 87	−1·103 35	−0·884 58
0·80	1·260 35	0·641 75	−1·259 80	−0·642 33
0·82	1·364 32	0·372 94	−1·363 86	−0·373 41
0·84	1·411 60	0·088 60	−1·411 24	−0·089 00
0·86	1·400 25	−0·199 43	−1·399 96	0·199 10
0·88	1·330 72	−0·479 18	−1·330 49	0·478 91
0·90	1·205 90	−0·739 04	−1·205 73	0·738 81
0·92	1·030 98	−0·968 20	−1·030 85	0·968 00
0·94	0·813 23	−1·157 13	−0·813 13	1·156 95
0·96	0·561 68	−1·297 98	−0·561 62	1·297 82
0·98	0·286 80	−1·384 90	−0·286 77	1·384 76
1·00	0·000 00	−1·414 29	0·000 00	1·414 14

TABLE 7·2 (d) (cont.)

Fourth mode

$\dfrac{x}{l}$	ϕ_4	$\phi_4'=\dfrac{1}{\lambda_4}\dfrac{d\phi_4}{dx}$	$\phi_4''=\dfrac{1}{\lambda_4^2}\dfrac{d^2\phi_4}{dx^2}$	$\phi_4'''=\dfrac{1}{\lambda_4^3}\dfrac{d^3\phi_4}{dx^3}$
0·00	0·000 00	0·000 00	2·000 00	− 2·000 00
0·02	0·064 96	0·462 78	1·466 33	− 1·994 08
0·04	0·234 51	0·783 57	0·937 92	− 1·956 00
0·06	0·471 04	0·965 21	0·426 62	− 1·862 87
0·08	0·738 20	1·014 41	− 0·050 91	− 1·701 71
0·10	1·002 04	0·942 70	− 0·475 81	− 1·468 93
0·12	1·232 37	0·766 64	− 0·829 47	− 1·169 55
0·14	1·404 07	0·507 51	− 1·095 59	− 0·815 99
0·16	1·498 25	0·190 41	− 1·262 06	− 0·426 60
0·18	1·503 06	− 0·157 04	− 1·322 23	− 0·023 80
0·20	1·414 22	− 0·506 24	− 1·275 77	0·367 79
0·22	1·235 02	− 0·829 44	− 1·129 01	0·723 43
0·24	0·975 82	− 1·101 40	− 0·894 66	1·020 24
0·26	0·653 24	− 1·301 07	− 0·591 10	1·238 93
0·28	0·288 79	− 1·412 95	− 0·241 21	1·365 37
0·30	− 0·092 74	− 1·428 07	0·129 17	1·391 64
0·32	− 0·465 10	− 1·344 55	0·492 99	1·316 66
0·34	− 0·802 50	− 1·167 72	0·823 86	1·146 36
0·36	− 1·081 50	− 0·909 63	1·097 85	0·893 28
0·38	− 1·282 66	− 0·588 23	1·295 18	0·575 71
0·40	− 1·392 01	− 0·226 02	1·401 60	0·216 44
0·42	− 1·402 00	0·151 52	1·409 34	− 0·158 86
0·44	− 1·312 09	0·517 80	1·317 71	− 0·523 42
0·46	− 1·128 77	0·846 97	1·133 08	− 0·851 27
0·48	− 0·865 13	1·115 80	0·868 43	− 1·119 10
0·50	− 0·539 94	1·305 30	0·542 46	− 1·307 82
0·52	− 0·176 28	1·402 10	0·178 21	− 1·404 03
0·54	0·200 00	1·399 37	− 0·198 53	− 1·400 84
0·56	0·562 22	1·297 34	− 0·561 09	− 1·298 47
0·58	0·884 66	1·103 26	− 0·883 79	− 1·104 13
0·60	1·144 45	0·830 92	− 1·143 79	− 0·831 59
0·62	1·323 17	0·499 63	− 1·322 66	− 0·500 14
0·64	1·408 13	0·132 89	− 1·407 74	− 0·133 28
0·66	1·393 30	− 0·243 29	− 1·393 01	0·242 99
0·68	1·279 73	− 0·602 26	− 1·279 50	0·602 03
0·70	1·075 46	− 0·918 54	− 1·075 29	0·918 37
0·72	0·794 97	− 1·169 74	− 0·794 84	1·169 60
0·74	0·458 14	− 1·338 02	− 0·458 04	1·337 92
0·76	0·088 84	− 1·411 46	− 0·088 76	1·411 38
0·78	− 0·286 76	− 1·384 86	0·286 82	1·384 80
0·80	− 0·642 02	− 1·260 10	0·642 06	1·260 05
0·82	− 0·951 76	− 1·046 02	0·951 80	1·045 98
0·84	− 1·194 05	− 0·757 79	1·194 07	0·757 76
0·86	− 1·351 68	− 0·415 85	1·351 70	0·415 83
0·88	− 1·413 51	− 0·044 43	1·413 52	0·044 41
0·90	− 1·375 13	0·330 14	1·375 14	− 0·330 15
0·92	− 1·239 28	0·681 30	1·239 29	− 0·681 31
0·94	− 1·015 58	0·984 16	1·015 59	− 0·984 18
0·96	− 0·719 89	1·217 27	0·719 90	− 1·217 28
0·98	− 0·373 17	1·364 09	0·373 18	− 1·364 09
1·00	0·000 00	1·414 21	0·000 00	− 1·414 22

TABLE 7·2 (d) (cont.)

Fifth mode

$\dfrac{x}{l}$	ϕ_5	$\phi_5' = \dfrac{1}{\lambda_5}\dfrac{d\phi_5}{dx}$	$\phi_5'' = \dfrac{1}{\lambda_5^2}\dfrac{d^2\phi_5}{dx^2}$	$\phi_5''' = \dfrac{1}{\lambda_5^3}\dfrac{d^3\phi_5}{dx^3}$
0·00	0·000 00	0·000 00	2·000 00	− 2·000 00
0·02	0·096 85	0·550 98	1·341 19	− 1·989 02
0·04	0·339 74	0·886 07	0·694 24	− 1·920 05
0·06	0·658 51	1·013 11	0·084 94	− 1·756 56
0·08	0·987 17	0·950 00	− 0·452 62	− 1·484 55
0·10	1·267 55	0·726 28	− 0·883 20	− 1·110 64
0·12	1·453 08	0·382 43	− 1·176 72	− 0·658 79
0·14	1·512 00	− 0·032 74	− 1·313 29	− 0·165 97
0·16	1·429 50	− 0·465 99	− 1·286 62	0·323 12
0·18	1·208 40	− 0·864 54	− 1·105 67	0·761 82
0·20	0·868 19	− 1·181 05	− 0·794 32	1·107 19
0·22	0·442 39	− 1·378 25	− 0·389 28	1·325 14
0·24	− 0·025 33	− 1·432 61	0·063 52	1·394 42
0·26	− 0·486 16	− 1·336 65	0·513 62	1·309 19
0·28	− 0·891 58	− 1·099 54	0·911 32	1·079 80
0·30	− 1·198 72	− 0·746 02	1·212 91	0·731 83
0·32	− 1·375 05	− 0·313 60	1·385 26	0·303 40
0·34	− 1·402 00	0·151 52	1·409 34	− 0·158 86
0·36	− 1·276 98	0·599 50	1·282 26	− 0·604 78
0·38	− 1·013 69	0·982 27	1·017 48	− 0·986 07
0·40	− 0·640 67	1·258 71	0·643 40	− 1·261 44
0·42	− 0·098 28	1·399 12	0·200 24	− 1·401 09
0·44	0·265 70	1·388 46	− 0·264 29	− 1·389 87
0·46	0·701 19	1·227 92	− 0·700 18	− 1·228 94
0·48	1·061 18	0·934 87	− 1·060 45	− 0·935 60
0·50	1·306 82	0·540 93	− 1·306 30	− 0·541 46
0·52	1·411 61	0·088 61	− 1·411 24	− 0·088 99
0·54	1·364 23	− 0·373 31	− 1·363 95	0·373 04
0·56	1·169 77	− 0·795 00	− 1·169 57	0·794 81
0·58	0·849 19	− 1·131 00	− 0·849 05	1·130 86
0·60	0·437 06	− 1·345 05	− 0·436 96	1·344 95
0·62	− 0·022 18	− 1·414 08	0·022 25	1·414 00
0·64	− 0·479 02	− 1·330 63	0·479 07	1·330 58
0·66	− 0·884 21	− 1·103 71	0·884 25	1·103 68
0·68	− 1·194 05	− 0·757 79	1·194 07	0·757 76
0·70	− 1·375 13	− 0·330 15	1·375 15	0·330 13
0·72	− 1·407 93	0·133 08	1·407 94	− 0·133 10
0·74	− 1·288 92	0·581 96	1·288 92	− 0·581 97
0·76	− 1·030 91	0·968 09	1·030 92	− 0·968 10
0·78	− 0·661 75	1·249 83	0·661 76	− 1·249 84
0·80	− 0·221 23	1·396 80	0·221 23	− 1·396 80
0·82	0·243 14	1·393 15	− 0·243 14	− 1·393 16
0·84	0·681 30	1·239 28	− 0·681 30	− 1·239 29
0·86	1·046 00	0·951 78	− 1·046 00	− 0·951 78
0·88	1·297 90	0·561 65	− 1·297 90	− 0·561 65
0·90	1·409 85	0·110 96	− 1·409 85	− 0·110 96
0·92	1·369 78	− 0·351 70	− 1·369 78	0·351 70
0·94	1·182 01	− 0·776 44	− 1·182 01	0·776 44
0·96	0·866 78	− 1·117 45	− 0·866 78	1·117 45
0·98	0·458 09	− 1·337 97	− 0·458 09	1·337 97
1·00	0·000 00	− 1·414 21	0·000 00	1·414 21

TABLE 7·2 (e)

DATA FOR SLIDING-PINNED BEAM

$$\phi_r(x) = \cos \frac{(2r-1)\,\pi x}{2l} = \cos \lambda_r x \quad (r = 1, 2, 3, \ldots).$$

$$a_r = \frac{A\rho l}{2}, \quad c_r = \frac{(2r-1)^4 \pi^4 EI}{32l^3}.$$

$$\omega_r = \sqrt{\left(\frac{c_r}{a_r}\right)} = \frac{(2r-1)^2 \pi^2}{4l^2} \sqrt{\left(\frac{EI}{A\rho}\right)}.$$

Orthogonality

$$\int_0^l \phi_r(x)\,\phi_s(x)\,dx = \int_0^l \phi_r''(x)\,\phi_s''(x)\,dx = \begin{cases} 0 & (r \neq s), \\ \tfrac{1}{2}l & (r = s), \end{cases}$$

where
$$\phi_r''(x) = \frac{1}{\lambda_r^2}\frac{d^2\phi_r}{dx^2}, \quad \text{etc.}$$

Receptances in series form

$$\alpha_{xh} = \alpha_{hx} = \sum_{r=1}^{\infty} \frac{\phi_r(x)\,\phi_r(h)}{a_r(\omega_r^2 - \omega^2)},$$

other receptances by differentiation.

TABLE 7·2 (f)

DATA FOR SLIDING-SLIDING BEAM

$$\phi_0(x) = \frac{1}{\sqrt{2}},$$

$$\psi_r(x) = \cos \frac{r\pi x}{l} = \cos \lambda_r x \quad (r = 1, 2, 3, \ldots).$$

$$a_r = \frac{A\rho l}{2}, \quad c_r = \frac{r^4 \pi^4 EI}{2l^3} \quad (r = 0, 1, 2, \ldots).$$

$$\omega_r = \sqrt{\left(\frac{c_r}{a_r}\right)} = \frac{r^2 \pi^2}{l^2} \sqrt{\left(\frac{EI}{A\rho}\right)}.$$

Orthogonality

$$\int_0^l \phi_r(x)\,\phi_s(x)\,dx = \int_0^l \phi_r''(x)\,\phi_s''(x)\,dx = \begin{cases} 0 & (r \neq s), \\ \tfrac{1}{2}l & (r = s). \end{cases}$$

where
$$\phi_r''(x) = \frac{1}{\lambda_r^2}\frac{d^2\phi_r}{dx^2}, \quad \text{etc.}$$

Receptances in series form

$$\alpha_{xh} = \alpha_{hx} = \sum_{r=0}^{\infty} \frac{\phi_r(x)\,\phi_r(h)}{a_r(\omega_r^2 - \omega^2)},$$

other receptances by differentiation.

TABLE 7·2 (g)

DATA FOR CLAMPED-SLIDING BEAM AND FREE-SLIDING BEAM

A. Clamped-sliding beam

Characteristic function and its derivatives

$$\phi_r = \cosh \lambda_r x - \cos \lambda_r x - \sigma_r (\sinh \lambda_r x - \sin \lambda_r x) \quad (r = 1, 2, 3, \ldots),$$

$$\frac{1}{\lambda_r} \frac{d\phi_r}{dx} = \phi_r' = \sinh \lambda_r x + \sin \lambda_r x - \sigma_r (\cosh \lambda_r x - \cos \lambda_r x),$$

$$\frac{1}{\lambda_r^2} \frac{d^2\phi_r}{dx^2} = \phi_r'' = \cosh \lambda_r x + \cos \lambda_r x - \sigma_r (\sinh \lambda_r x + \sin \lambda_r x),$$

$$\frac{1}{\lambda_r^3} \frac{d^3\phi_r}{dx^3} = \phi_r''' = \sinh \lambda_r x - \sin \lambda_r x - \sigma_r (\cosh \lambda_r x + \cos \lambda_r x).$$

Boundary values

$$\phi_r(0) = \phi_r'(0) = 0,$$
$$\phi_r'(l) = \phi_r'''(l) = 0.$$

Orthogonality

$$\int_0^l \phi_r(x)\, \phi_s(x)\, dx = \int_0^l \phi_r''(x)\, \phi_s''(x)\, dx = \begin{cases} 0 & (r \neq s), \\ l & (r = s), \end{cases}$$

for $r, s = 1, 2, 3, \ldots$.

Receptances in series form

$$\alpha_{xh} = \alpha_{hx} = \sum_{r=1}^{\infty} \frac{\phi_r(x)\, \phi_r(h)}{a_r(\omega_r^2 - \omega^2)}, \quad \text{where} \quad a_r = A\rho l;$$

other receptances by differentiation.

B. Free-sliding beam

Characteristic function

$$\phi_0 = 1,$$

$$\phi_r = \cosh \lambda_r x + \cos \lambda_r x - \sigma_r (\sinh \lambda_r x + \sin \lambda_r x) \quad (r = 1, 2, 3, \ldots),$$

where σ_r and λ_r are the same as for a clamped-sliding beam.

The characteristic function for distortion of a free-sliding beam is the same as the second derivative of that of a clamped-sliding beam; that is, for $r = 1, 2, 3, \ldots$

$$\phi_r \text{ (free-sliding)} = \phi_r'' \text{ (clamped-sliding)},$$
$$\phi_r' \text{ (free-sliding)} = \phi_r''' \text{ (clamped-sliding)},$$
$$\phi_r'' \text{ (free-sliding)} = \phi_r \text{ (clamped-sliding)},$$
$$\phi_r''' \text{ (free-sliding)} = \phi_r' \text{ (clamped-sliding)}.$$

Boundary values

$$\phi_r''(0) = \phi_r'''(0) = 0,$$
$$\phi_r'(l) = \phi_r'''(l) = 0.$$

Orthogonality

$$\int_0^l \phi_r(x)\,\phi_s(x)\,dx = \begin{cases} 0 & (r \neq s), \\ l & (r = s), \end{cases}$$

for $r, s = 0, 1, 2, \ldots$; and

$$\int_0^l \phi_r''(x)\,\phi_s''(x)\,dx = \begin{cases} 0 & (r \neq s), \\ l & (r = s), \end{cases}$$

for $r, s = 1, 2, 3, \ldots$.

Receptances in series form

$$\alpha_{xh} = \alpha_{hx} = \sum_{r=0}^{\infty} \frac{\phi_r(x)\,\phi_r(h)}{a_r(\omega_r^2 - \omega^2)}, \quad \text{where} \quad a_r = A\rho l;$$

other receptances by differentiation.

TABLE 7·2 (g) (cont.)
GENERAL DATA (for A and B)

$$\left.\begin{aligned} \tan \lambda_r l + \tanh \lambda_r l &= 0, \\ \sigma_r &= \tanh \lambda_r l, \end{aligned}\right\} \quad (r = 1, 2, 3, \ldots).$$

Values of σ_r and $\lambda_r l$ and various powers

r	$\lambda_r l$	$(\lambda_r l)^2$	$(\lambda_r l)^3$	$(\lambda_r l)^4$
1	2·365 02	5·593 32	13·228 3	31·285 2
2	5·497 80	30·225 8	166·176	913·602
3	8·639 38	74·638 9	644·834	5 570·963
4	11·781 0	138·791	1 635·10	19 263·0
5	14·922 6	222·683	3 323·00	49 587·7

r	σ_r	$\dfrac{\omega_r}{\omega_1} = \dfrac{\lambda_r^2}{\lambda_1^2}$	$\dfrac{\omega_r^2}{\omega_1^2} = \dfrac{\lambda_r^4}{\lambda_1^4}$
1	0·982 502 2	1·0	1·0
2	0·999 966 4	5·403 92	29·202 3
3	0·999 999 9	13·344 3	178·070
4	1·000 000 0	24·813 8	615·722
5	1·000 000 0	39·812 3	1 585·02

For $r > 5$

$$\lambda_r l \doteqdot (4r - 1)\,\pi/4,$$
$$\sigma_r \doteqdot 1\cdot0.$$

TABLE 7·2 (g) (cont.)

CHARACTERISTIC FUNCTIONS AND DERIVATIVES
CLAMPED-SLIDING BEAM

First mode

$\dfrac{x}{l}$	ϕ_1	$\phi_1' = \dfrac{1}{\lambda_1}\dfrac{d\phi_1}{dx}$	$\phi_1'' = \dfrac{1}{\lambda_1^2}\dfrac{d^2\phi_1}{dx^2}$	$\phi_1''' = \dfrac{1}{\lambda_1^3}\dfrac{d^3\phi_1}{dx^3}$
0·00	0·000 00	0·000 00	2·000 00	− 1·965 00
0·02	0·002 20	0·092 40	1·907 05	− 1·964 97
0·04	0·008 67	0·180 41	1·814 12	− 1·964 73
0·06	0·019 20	0·264 02	1·721 20	− 1·964 09
0·08	0·033 58	0·343 24	1·628 32	− 1·962 85
0·10	0·051 60	0·418 06	1·535 52	− 1·960 85
0·12	0·073 06	0·488 50	1·442 84	− 1·957 92
0·14	0·097 74	0·554 56	1·350 32	− 1·953 89
0·16	0·125 45	0·616 24	1·258 02	− 1·948 62
0·18	0·155 97	0·673 57	1·166 00	− 1·941 98
0·20	0·189 10	0·726 55	1·074 33	− 1·933 83
0·22	0·224 64	0·775 21	0·983 08	− 1·924 05
0·24	0·262 37	0·819 56	0·892 34	− 1·912 54
0·26	0·302 10	0·859 64	0·802 18	− 1·899 20
0·28	0·343 63	0·895 46	0·712 70	− 1·883 93
0·30	0·386 75	0·927 07	0·623 99	− 1·866 66
0·32	0·431 26	0·954 50	0·536 15	− 1·847 32
0·34	0·476 98	0·977 81	0·449 27	− 1·825 85
0·36	0·523 70	0·997 02	0·363 46	− 1·802 19
0·38	0·571 23	1·012 21	0·278 82	− 1·776 29
0·40	0·619 39	1·023 42	0·195 45	− 1·748 14
0·42	0·667 99	1·030 72	0·113 48	− 1·717 69
0·44	0·716 84	1·034 18	0·032 99	− 1·684 94
0·46	0·765 76	1·033 87	− 0·045 88	− 1·649 88
0·48	0·814 58	1·029 86	− 0·123 05	− 1·612 50
0·50	0·863 13	1·022 25	− 0·198 39	− 1·572 82
0·52	0·911 24	1·011 13	− 0·271 80	− 1·530 85
0·54	0·958 73	0·996 57	− 0·343 18	− 1·486 63
0·56	1·005 46	0·978 70	− 0·412 40	− 1·440 17
0·58	1·051 27	0·957 60	− 0·479 38	− 1·391 52
0·60	1·096 00	0·933 38	− 0·544 01	− 1·340 74
0·62	1·139 52	0·906 17	− 0·606 18	− 1·287 86
0·64	1·181 68	0·876 08	− 0·665 81	− 1·232 96
0·66	1·222 35	0·843 23	− 0·722 79	− 1·176 10
0·68	1·261 41	0·807 74	− 0·777 04	− 1·117 35
0·70	1·298 73	0·769 76	− 0·828 47	− 1·056 79
0·72	1·334 19	0·729 42	− 0·876 99	− 0·994 52
0·74	1·367 69	0·686 85	− 0·922 52	− 0·930 61
0·76	1·399 13	0·642 19	− 0·965 00	− 0·865 16
0·78	1·428 41	0·595 61	− 1·004 34	− 0·798 28
0·80	1·455 45	0·547 23	− 1·040 49	− 0·730 07
0·82	1·480 16	0·497 22	− 1·073 39	− 0·660 63
0·84	1·502 46	0·445 74	− 1·102 97	− 0·590 08
0·86	1·522 30	0·392 94	− 1·129 20	− 0·518 54
0·88	1·539 62	0·338 97	− 1·152 01	− 0·446 11
0·90	1·554 36	0·284 01	− 1·171 39	− 0·372 93
0·92	1·566 47	0·228 21	− 1·187 28	− 0·299 11
0·94	1·575 93	0·171 75	− 1·199 68	− 0·224 78
0·96	1·582 71	0·114 78	− 1·208 54	− 0·150 07
0·98	1·586 79	0·057 47	− 1·213 87	− 0·075 10
1·00	1·588 15	0·000 00	− 1·215 64	0·000 00

TABLE 7·2 (g) (cont.)

Second mode

$\dfrac{x}{l}$	ϕ_2	$\phi_2' = \dfrac{1}{\lambda_2}\dfrac{d\phi_2}{dx}$	$\phi_2'' = \dfrac{1}{\lambda_2^2}\dfrac{d^2\phi_2}{dx^2}$	$\phi_2''' = \dfrac{1}{\lambda_2^3}\dfrac{d^3\phi_2}{dx^3}$
0·00	0·000 00	0·000 00	2·000 00	−1·999 93
0·02	0·011 65	0·207 82	1·780 11	−1·999 50
0·04	0·044 82	0·391 47	1·560 38	−1·996 58
0·06	0·096 85	0·550 99	1·341 21	−1·988 95
0·08	0·165 10	0·686 46	1·123 23	−1·974 69
0·10	0·246 94	0·798 07	0·907 25	−1·952 15
0·12	0·339 75	0·886 09	0·694 28	−1·919 98
0·14	0·440 95	0·950 90	0·485 42	−1·877 13
0·16	0·548 03	0·993 03	0·281 89	−1·822 80
0·18	0·658 52	1·013 14	0·085 00	−1·756 48
0·20	0·770 05	1·012 02	−0·103 93	−1·677 94
0·22	0·880 34	0·990 62	−0·283 55	−1·587 19
0·24	0·987 20	0·950 05	−0·452 53	−1·484 47
0·26	1·088 61	0·891 54	−0·609 58	−1·370 29
0·28	1·182 65	0·816 48	−0·753 48	−1·245 34
0·30	1·267 61	0·726 37	−0·883 09	−1·110 54
0·32	1·341 90	0·622 84	−0·997 38	−0·966 97
0·34	1·404 15	0·507 63	−1·095 46	−0·815 88
0·36	1·453 17	0·382 56	−1·176 58	−0·658 67
0·38	1·487 99	0·249 53	−1·240 14	−0·496 83
0·40	1·507 83	0·110 49	−1·285 73	−0·331 99
0·42	1·512 14	0·032 55	−1·313 10	−0·165 81
0·44	1·500 59	−0·177 60	−1·322 21	−0·000 03
0·46	1·473 08	−0·322 66	−1·313 19	0·163 60
0·48	1·429 71	−0·465 74	−1·286 37	0·323 33
0·50	1·370 80	−0·604 91	−1·242 29	0·477 44
0·52	1·296 90	−0·738 32	−1·181 64	0·624 24
0·54	1·208 72	−0·864 19	−1·105 34	0·762 12
0·56	1·107 19	−0·980 86	−1·014 43	0·889 56
0·58	0·993 41	−1·086 79	−0·910 14	1·005 16
0·60	0·868 63	−1·180 57	−0·793 86	1·107 62
0·62	0·734 27	−1·260 98	−0·667 09	1·195 83
0·64	0·591 86	−1·326 95	−0·531 44	1·268 80
0·66	0·443 02	−1·377 59	−0·388 65	1·325 75
0·68	0·289 49	−1·412 22	−0·240 51	1·366 06
0·70	0·133 06	−1·430 35	−0·088 86	1·389 30
0·72	−0·024 44	−1·431 71	0·064 39	1·395 28
0·74	−0·181 17	−1·416 21	0·217 34	1·383 96
0·76	−0·335 27	−1·383 99	0·368 11	1·355 53
0·78	−0·484 93	−1·335 40	0·514 83	1·310 39
0·80	−0·628 36	−1·270 99	0·655 69	1·249 12
0·82	−0·763 88	−1·191 49	0·788 96	1·172 50
0·84	−0·889 87	−1·097 83	0·913 01	1·081 48
0·86	−1·004 82	−0·991 10	1·026 30	0·977 21
0·88	−1·107 39	−0·872 57	1·127 47	0·860 96
0·90	−1·196 33	−0·743 65	1·215 25	0·734 18
0·92	−1·270 60	−0·605 85	1·288 59	0·598 41
0·94	−1·329 30	−0·460 83	1·346 58	0·455 33
0·96	−1·371 74	−0·310 31	1·388 52	0·306 68
0·98	−1·397 41	−0·156 09	1·413 89	0·154 28
1·00	−1·406 00	−0·000 00	1·422 38	−0·000 00

TABLE 7·2 (g) (cont.)

Third mode

$\dfrac{x}{l}$	ϕ_3	$\phi_3' = \dfrac{1}{\lambda_3}\dfrac{d\phi_3}{dx}$	$\phi_3'' = \dfrac{1}{\lambda_3^2}\dfrac{d^2\phi_3}{dx^2}$	$\phi_3''' = \dfrac{1}{\lambda_3^3}\dfrac{d^3\phi_3}{dx^3}$
0·00	0·000 00	0·000 00	2·000 00	− 2·000 00
0·02	0·028 14	0·315 72	1·654 50	− 1·998 35
0·04	0·105 67	0·571 81	1·309 96	− 1·987 43
0·06	0·222 32	0·768 60	0·968 67	− 1·959 58
0·08	0·367 91	0·906 94	0·634 09	− 1·908 94
0·10	0·532 46	0·988 36	0·310 54	− 1·831 35
0·12	0·706 31	1·015 17	0·002 91	− 1·724 40
0·14	0·880 31	0·990 58	− 0·283 63	− 1·587 26
0·16	1·045 91	0·918 67	− 0·543 91	− 1·420 67
0·18	1·195 34	0·804 41	− 0·773 00	− 1·226 75
0·20	1·321 78	0·653 59	− 0·966 46	− 1·008 92
0·22	1·419 47	0·472 70	− 1·120 53	− 0·771 64
0·24	1·483 81	0·268 80	− 1·232 31	− 0·520 30
0·26	1·511 47	0·049 39	− 1·299 88	− 0·260 98
0·28	1·500 43	− 0·177 81	− 1·322 41	− 0·000 21
0·30	1·450 02	− 0·405 03	− 1·300 25	0·255 26
0·32	1·360 90	− 0·624 65	− 1·234 90	0·498 65
0·34	1·235 01	− 0·829 44	− 1·129 01	0·723 43
0·36	1·075 51	− 1·012 69	− 0·986 32	0·923 50
0·38	0·886 64	− 1·168 44	− 0·811 60	1·093 41
0·40	0·673 60	− 1·291 64	− 0·610 47	1·228 51
0·42	0·442 39	− 1·378 25	− 0·389 28	1·325 14
0·44	0·199 59	− 1·425 40	− 0·154 91	1·380 72
0·46	− 0·047 82	− 1·431 44	0·085 42	1·393 85
0·48	− 0·292 69	− 1·395 97	0·324 32	1·364 34
0·50	− 0·527 89	− 1·319 86	0·554 50	1·293 26
0·52	− 0·746 58	− 1·205 25	0·768 97	1·182 87
0·54	− 0·942 36	− 1·055 41	0·961 20	1·036 58
0·56	− 1·109 52	− 0·874 70	1·125 37	0·858 86
0·58	− 1·243 16	− 0·668 41	1·256 50	0·655 09
0·60	− 1·339 38	− 0·442 62	1·350 61	0·431 41
0·62	− 1·395 38	− 0·203 98	1·404 83	0·194 55
0·64	− 1·409 54	0·040 46	1·417 49	− 0·048 38
0·66	− 1·381 48	0·283 45	1·388 18	− 0·290 11
0·68	− 1·312 08	0·517 81	1·317 72	− 0·523 41
0·70	− 1·203 44	0·736 57	1·208 19	− 0·741 28
0·72	− 1·058 81	0·933 26	1·062 82	− 0·937 21
0·74	− 0·882 54	1·102 04	0·885 92	− 1·105 35
0·76	− 0·679 87	1·237 90	0·682 73	− 1·240 67
0·78	− 0·456 88	1·336 81	0·459 30	− 1·339 12
0·80	− 0·220 20	1·395 84	0·222 26	− 1·397 77
0·82	0·023 09	1·413 24	− 0·021 34	− 1·414 84
0·84	0·265 75	1·388 50	− 0·264 25	− 1·389 83
0·86	0·500 53	1·322 38	− 0·499 24	− 1·323 46
0·88	0·720 45	1·216 84	− 0·719 33	− 1·217 71
0·90	0·918 95	1·075 03	− 0·917 96	− 1·075 72
0·92	1·090 11	0·901 19	− 1·089 23	− 0·901 72
0·94	1·228 83	0·700 49	− 1·228 03	− 0·700 88
0·96	1·330 98	0·478 92	− 1·330 23	− 0·479 17
0·98	1·393 51	0·243 08	− 1·392 80	− 0·243 21
1·00	1·414 56	− 0·000 00	− 1·413 86	− 0·000 00

TABLE 7·2 (g) (cont.)

Fourth mode

$\dfrac{x}{l}$	ϕ_4	$\phi_4' = \dfrac{1}{\lambda_4}\dfrac{d\phi_4}{dx}$	$\phi_4'' = \dfrac{1}{\lambda_4^2}\dfrac{d^2\phi_4}{dx^2}$	$\phi_4''' = \dfrac{1}{\lambda_4^3}\dfrac{d^3\phi_4}{dx^3}$
0·00	0·000 00	0·000 00	2·000 00	− 2·000 00
0·02	0·051 16	0·415 73	1·529 01	− 1·995 90
0·04	0·187 21	0·720 77	1·061 24	− 1·969 23
0·06	0·382 23	0·916 66	0·604 15	− 1·903 05
0·08	0·610 89	1·007 14	0·168 43	− 1·786 46
0·10	0·849 06	0·998 70	− 0·233 33	− 1·614 43
0·12	1·074 49	0·900 89	− 0·588 02	− 1·387 36
0·14	1·267 55	0·726 28	− 0·883 20	− 1·110 64
0·16	1·411 91	0·490 20	− 1·108 24	− 0·793 88
0·18	1·495 10	0·210 18	− 1·255 18	− 0·450 10
0·20	1·508 99	− 0·094 78	− 1·319 43	− 0·094 78
0·22	1·450 02	− 0·405 03	− 1·300 25	0·255 26
0·24	1·319 24	− 0·701 20	− 1·200 91	0·582 87
0·26	1·122 12	− 0·965 20	− 1·028 63	0·871 71
0·28	0·868 19	− 1·181 05	− 0·794 32	1·107 19
0·30	0·570 38	− 1·335 74	− 0·512 02	1·277 38
0·32	0·244 29	− 1·419 86	− 0·198 18	1·373 75
0·34	− 0·092 74	− 1·428 07	0·129 17	1·391 64
0·36	− 0·422 62	− 1·359 39	0·451 41	1·330 61
0·38	− 0·727 55	− 1·217 19	0·750 29	1·194 45
0·40	− 0·991 02	− 1·008 98	1·008 98	0·991 02
0·42	− 1·198 72	− 0·746 02	1·212 91	0·731 83
0·44	− 1·339 39	− 0·442 62	1·350 60	0·431 41
0·46	− 1·405 42	− 0·115 39	1·414 28	0·106 53
0·48	− 1·393 30	0·217 73	1·400 30	− 0·224 73
0·50	− 1·303 80	0·538 43	1·309 33	− 0·543 96
0·52	− 1·141 94	0·829 07	1·146 31	− 0·833 44
0·54	− 0·916 73	1·073 65	0·920 18	− 1·077 10
0·56	− 0·640 68	1·258 71	0·643 40	− 1·261 44
0·58	− 0·329 06	1·374 06	0·331 22	− 1·376 22
0·60	0·000 85	1·413 36	0·000 85	− 1·415 07
0·62	0·330 81	1·374 47	− 0·329 47	− 1·375 81
0·64	0·642 57	1·259 54	− 0·641 51	− 1·260 61
0·66	0·918 88	1·074 96	− 0·918 04	− 1·075 80
0·68	1·144 45	0·830 92	− 1·143 79	− 0·831 59
0·70	1·306 83	0·540 93	− 1·306 30	− 0·541 46
0·72	1·397 01	0·221 02	− 1·396 60	− 0·221 44
0·74	1·410 02	− 0·111 12	− 1·409 69	0·110 79
0·76	1·345 13	− 0·437 15	− 1·344 87	0·436 89
0·78	1·205 92	− 0·739 03	− 1·205 71	0·738 82
0·80	1·000 08	− 1·000 08	− 0·999 92	0·999 92
0·82	0·738 99	− 1·205 88	− 0·738 86	1·205 75
0·84	0·437 07	− 1·345 05	− 0·436 97	1·344 95
0·86	0·111 00	− 1·409 89	− 0·110 92	1·409 81
0·88	− 0·221 20	− 1·396 83	0·221 26	1·396 77
0·90	− 0·541 17	− 1·306 59	0·541 22	1·306 54
0·92	− 0·831 23	− 1·144 14	0·831 27	1·144 10
0·94	− 1·075 36	− 0·918 47	1·075 39	0·918 44
0·96	− 1·260 06	− 0·642 05	1·260 09	0·642 03
0·98	− 1·375 13	− 0·330 15	1·375 15	0·330 13
1·00	− 1·414 21	− 0·000 00	1·414 22	− 0·000 00

TABLE 7·2 (g) (cont.)

Fifth mode

$\dfrac{x}{l}$	ϕ_5	$\phi_5' = \dfrac{1}{\lambda_5}\dfrac{d\phi_5}{dx}$	$\phi_5'' = \dfrac{1}{\lambda_5^2}\dfrac{d^2\phi_5}{dx^2}$	$\phi_5''' = \dfrac{1}{\lambda_5^3}\dfrac{d^3\phi_5}{dx^3}$
0·00	0·000 00	0·000 00	2·000 00	− 2·000 00
0·02	0·080 21	0·507 87	1·403 72	− 1·991 80
0·04	0·285 52	0·838 65	0·815 51	− 1·939 68
0·06	0·563 65	0·997 21	0·253 28	− 1·814 14
0·08	0·864 72	0·994 84	− 0·258 59	− 1·600 97
0·10	1·143 32	0·850 51	− 0·693 59	− 1·300 24
0·12	1·360 90	0·590 93	− 1·027 22	− 0·924 62
0·14	1·487 88	0·249 38	− 1·240 30	− 0·496 96
0·16	1·505 36	− 0·136 27	− 1·321 67	− 0·047 43
0·18	1·406 12	− 0·526 24	− 1·269 82	0·389 94
0·20	1·194 69	− 0·881 82	− 1·093 56	0·780 69
0·22	0·886 64	− 1·168 44	− 0·811 61	1·093 41
0·24	0·506 88	− 1·358 44	− 0·451 21	1·302 77
0·26	0·087 27	− 1·433 30	− 0·045 97	1·391 99
0·28	− 0·336 38	− 1·385 11	0·367 02	1·354 46
0·30	− 0·727 55	− 1·217 19	0·750 29	1·194 45
0·32	− 1·052 38	− 0·943 67	1·069 25	0·926 80
0·34	− 1·282 66	− 0·588 23	1·295 18	0·575 71
0·36	− 1·398 42	− 0·181 89	1·407 71	0·172 60
0·38	− 1·389 71	0·239 70	1·396 60	− 0·246 59
0·40	− 1·257 52	0·639 48	1·262 63	− 0·644 60
0·42	− 1·013 69	0·982 27	1·017 48	− 0·986 07
0·44	− 0·679 90	1·237 88	0·682 71	− 1·240 69
0·46	− 0·285 74	1·383 79	0·287 83	− 1·385 87
0·48	0·133 86	1·407 16	− 0·132 31	− 1·408 71
0·50	0·541 77	1·305 99	− 0·540 62	− 1·307 14
0·52	0·901 88	1·089 24	− 0·901 03	− 1·090 10
0·54	1·182 33	0·776 12	− 1·181 69	− 0·776 75
0·56	1·358 30	0·394 32	− 1·357 83	− 0·394 79
0·58	1·414 21	− 0·022 39	− 1·413 86	0·022 04
0·60	1·345 13	− 0·437 15	− 1·344 87	0·436 89
0·62	1·157 13	− 0·813 28	− 1·156 94	0·813 08
0·64	0·866 85	− 1·117 52	− 0·866 71	1·117 38
0·66	0·499 94	− 1·322 97	− 0·499 84	1·322 86
0·68	0·088 84	− 1·411 46	− 0·088 76	1·411 38
0·70	− 0·330 11	− 1·375 17	0·330 17	1·375 11
0·72	− 0·719 87	− 1·217 29	0·719 91	1·217 25
0·74	− 1·045 98	− 0·951 80	1·046 01	0·951 77
0·76	− 1·279 61	− 0·602 15	1·279 63	0·602 13
0·78	− 1·400 10	− 0·199 27	1·400 11	0·199 26
0·80	− 1·396 80	0·221 23	1·396 81	− 0·221 24
0·82	− 1·270 00	0·622 16	1·270 01	− 0·622 17
0·84	− 1·030 91	0·968 09	1·030 92	− 0·968 10
0·86	− 0·700 68	1·228 43	0·700 69	− 1·228 43
0·88	− 0·308 50	1·380 15	0·308 50	− 1·380 16
0·90	0·110 96	1·409 85	− 0·110 96	− 1·409 86
0·92	0·520 61	1·314 90	− 0·520 61	− 1·314 90
0·94	0·880 23	1·103 69	− 0·884 23	− 1·103 70
0·96	1·169 67	0·794 91	− 1·169 67	− 0·794 91
0·98	1·351 70	0·415 84	− 1·351 69	− 0·415 84
1·00	1·414 21	− 0·000 00	− 1·414 21	− 0·000 00

TABLE 7.3 (a)

FLEXIBILITIES OF BEAMS LOADED AT AN END

1. *Clamped-free*

    (a) $\quad a_{xl} = \dfrac{x^2(3l - x)}{6EI}$, $\qquad$ (c) $\quad a_{xl'} = \dfrac{x^2}{2EI}$,

    (b) $\quad a_{x'l} = \dfrac{x(2l - x)}{2EI}$, $\qquad$ (d) $\quad a_{x'l'} = \dfrac{x}{EI}$.

2. *Pinned-pinned*

    (a) $\quad a_{x0'} = \dfrac{x(l - x)\,(2l - x)}{6lEI}$, $\qquad$ (c) $\quad a_{xl'} = -\dfrac{x(l^2 - x^2)}{6lEI}$,

    (b) $\quad a_{x'0'} = \dfrac{2l^2 - 6lx + 3x^2}{6lEI}$, $\qquad$ (d) $\quad a_{x'l'} = -\dfrac{(l^2 - 3x^2)}{6lEI}$.

3. *Sliding-pinned*

    (a) $\quad a_{x0} = \dfrac{2l^3 - 3lx^2 + x^3}{6EI}$, $\qquad$ (c) $\quad a_{xl'} = -\dfrac{(l^2 - x^2)}{2EI}$,

    (b) $\quad a_{x'0} = -\dfrac{x(2l - x)}{2EI}$, $\qquad$ (d) $\quad a_{x'l'} = \dfrac{x}{EI}$.

4. *Clamped-pinned*

    (a) $\quad u_{xl'} = -\dfrac{x^2(l - x)}{4lEI}$, $\qquad$ (b) $\quad a_{x'l'} = -\dfrac{x(2l - 3x)}{4lEI}$.

5. *Clamped-sliding*

    (a) $\quad a_{xl} = \dfrac{x^2(3l - 2x)}{12EI}$, $\qquad$ (b) $\quad a_{x'l} = \dfrac{x(l - x)}{2EI}$.

TABLE 7·3(b)

TIP FLEXIBILITIES

Nature of support	End	a_{00}	$a_{l'0}$ $a_{0l'}$	$a_{0'0'}$	$a_{0'l'}$ $a_{l'0'}$	a_{ll}	$a_{l'l}$ $a_{ll'}$	$a_{l'l'}$	
Clamped Free	$x=0$ $x=l$					$\dfrac{l^3}{3EI}$	$\dfrac{l^2}{2EI}$	$\dfrac{l}{EI}$	1
Pinned Pinned	$x=0$ $x=l$			$\dfrac{l}{3EI}$	$-\dfrac{l}{6EI}$			$\dfrac{l}{3EI}$	2
Sliding Pinned	$x=0$ $x=l$	$\dfrac{l^3}{3EI}$	$-\dfrac{l^2}{2EI}$					$\dfrac{l}{EI}$	3
Clamped Pinned	$x=0$ $x=l$							$\dfrac{l}{4EI}$	4
Clamped Sliding	$x=0$ $x=l$					$\dfrac{l^3}{12EI}$			5

TABLE 7·3(c)

FLEXIBILITIES OF CLAMPED-CLAMPED BEAMS

$$a_{xx} = \frac{x^3(l-x)^3}{3l^3EI},$$

$$a_{x'x} = -\frac{x^2(l-x)^2(2x-l)}{2l^3EI} = a_{xx'},$$

$$a_{x'x'} = \frac{x(l-x)(l^2-3lx+3x^2)}{l^3EI}.$$

TABLE 8(a)

CHARACTERISTIC FUNCTIONS AND DERIVATIVES FOR:

LATERAL VIBRATION OF UNIFORM FIXED-FIXED OR FREE-FREE TAUT STRING.

TORSIONAL VIBRATION OF UNIFORM CIRCULAR CLAMPED-CLAMPED OR FREE-FREE SHAFT.

LONGITUDINAL VIBRATION OF UNIFORM CLAMPED-CLAMPED OR FREE-FREE BAR.

FLEXURAL VIBRATION OF UNIFORM PINNED-PINNED OR SLIDING-SLIDING BEAM.

$\dfrac{x}{l}$	$\sin\dfrac{\pi x}{l}$	$\cos\dfrac{\pi x}{l}$	$\dfrac{x}{l}$	$\sin\dfrac{\pi x}{l}$	$\cos\dfrac{\pi x}{l}$
0·00	0·000 00	1·000 00	0·50	1·000 00	0·000 00
0·02	0·062 79	0·998 03	0·52	0·998 03	−0·062 79
0·04	0·125 33	0·992 11	0·54	0·992 11	−0·125 33
0·06	0·187 38	0·982 29	0·56	0·982 29	−0·187 38
0·08	0·248 69	0·968 58	0·58	0·968 58	−0·248 69
0·10	0·309 02	0·951 06	0·60	0·951 06	−0·309 02
0·12	0·368 12	0·929 78	0·62	0·929 78	−0·368 12
0·14	0·425 78	0·904 83	0·64	0·904 83	−0·425 78
0·16	0·481 75	0·876 31	0·66	0·876 31	−0·481 75
0·18	0·535 83	0·844 33	0·68	0·844 33	−0·535 83
0·20	0·587 79	0·809 02	0·70	0·809 02	−0·587 79
0·22	0·637 42	0·770 51	0·72	0·770 51	−0·637 42
0·24	0·684 55	0·728 97	0·74	0·728 97	−0·684 55
0·26	0·728 97	0·684 55	0·76	0·684 55	−0·728 97
0·28	0·770 51	0·637 42	0·78	0·637 42	−0·770 51
0·30	0·809 02	0·587 79	0·80	0·587 79	−0·809 02
0·32	0·844 33	0·535 83	0·82	0·535 83	−0·844 33
0·34	0·876 31	0·481 75	0·84	0·481 75	−0·876 31
0·36	0·904 83	0·425 78	0·86	0·425 78	−0·904 83
0·38	0·929 78	0·368 12	0·88	0·368 12	−0·929 78
0·40	0·951 06	0·309 02	0·90	0·309 02	−0·951 06
0·42	0·968 58	0·248 69	0·92	0·248 69	−0·968 58
0·44	0·982 29	0·187 38	0·94	0·187 38	−0·982 29
0·46	0·992 11	0·125 33	0·96	0·125 33	−0·992 11
0·48	0·998 03	0·062 79	0·98	0·062 79	−0·998 03
0·50	1·000 00	0·000 00	1·00	0·000 00	−1·000 00

TABLE 8(*b*)

CHARACTERISTIC FUNCTIONS AND DERIVATIVES FOR:

LATERAL VIBRATION OF UNIFORM FIXED-FREE TAUT STRING.

TORSIONAL VIBRATION OF UNIFORM CIRCULAR CLAMPED-FREE SHAFT.

LONGITUDINAL VIBRATION OF UNIFORM CLAMPED-FREE BAR.

FLEXURAL VIBRATION OF UNIFORM SLIDING-PINNED BEAM

$\dfrac{x}{l}$	$\sin\dfrac{\pi x}{2l}$	$\cos\dfrac{\pi x}{2l}$	$\dfrac{x}{l}$	$\sin\dfrac{\pi x}{2l}$	$\cos\dfrac{\pi x}{2l}$
0·00	0·000 00	1·000 00	0·50	0·707 11	0·707 11
0·02	0·031 41	0·999 51	0·52	0·728 97	0·684 55
0·04	0·062 79	0·998 03	0·54	0·750 11	0·661 31
0·06	0·094 11	0·995 56	0·56	0·770 51	0·637 42
0·08	0·125 33	0·992 11	0·58	0·790 16	0·612 91
0·10	0·156 43	0·987 69	0·60	0·809 02	0·587 79
0·12	0·187 38	0·982 29	0·62	0·827 08	0·562 08
0·14	0·218 14	0·975 92	0·64	0·844 33	0·535 83
0·16	0·248 69	0·968 58	0·66	0·860 74	0·509 04
0·18	0·278 99	0·960 29	0·68	0·876 31	0·481 75
0·20	0·309 02	0·951 06	0·70	0·891 01	0·453 99
0·22	0·338 74	0·940 88	0·72	0·904 83	0·425 78
0·24	0·368 12	0·929 78	0·74	0·917 75	0·397 15
0·26	0·397 15	0·917 75	0·76	0·929 78	0·368 12
0·28	0·425 78	0·904 83	0·78	0·940 88	0·338 74
0·30	0·453 99	0·891 01	0·80	0·951 06	0·309 02
0·32	0·481 75	0·876 31	0·82	0·960 29	0·278 99
0·34	0·509 04	0·860 74	0·84	0·968 58	0·248 69
0·36	0·535 83	0·844 33	0·86	0·975 92	0·218 14
0·38	0·562 08	0·827 08	0·88	0·982 29	0·187 38
0·40	0·587 79	0·809 02	0·90	0·987 69	0·156 43
0·42	0·612 91	0·790 16	0·92	0·992 11	0·125 33
0·44	0·637 42	0·770 51	0·94	0·995 56	0·094 11
0·46	0·661 31	0·750 11	0·96	0·998 03	0·062 79
0·48	0·684 55	0·728 97	0·98	0·999 51	0·031 41
0·50	0·707 11	0·707 11	1·00	1·000 00	0·000 00

CHAPTER 8

VISCOUS DAMPING

We may avail ourselves of an artifice due to Rayleigh, and assume that the deviation of any particle of the fluid from the state of uniform flow is resisted by a force proportional to the *relative* velocity.

This law of resistance does not profess to be altogether a natural one, but it serves to represent in a rough way the effect of small dissipative forces.

HORACE LAMB, *Hydrodynamics* (1916)

The mechanical systems which have been discussed in this book so far have had the property of being able to vibrate freely with constant amplitude for an indefinite period. Real systems do not do this because their movement involves the dissipation of energy, and the energy has to be drawn from the energy of vibration. As a result, the amplitude is progressively reduced. This dissipation or *damping* effect prevents the direct receptances of a real system from ever being infinite, since a finite force is always required to maintain an energy input if the motion is to be steady.

The term dissipation is used here to denote the transfer of energy from those forms which are associated with the motion of vibration to other forms. Commonly the energy is turned to heat, but a fraction of it may be radiated in the form of sound waves in the surrounding air or of stress waves through the ground on which the system is supported.

Fortunately the effects of damping may be treated by making small modifications to the foregoing theory; it is not necessary to rewrite that theory from the start. We are free to regard a damped system as an ideal conservative system which is acted on by a set of external damping forces in addition to the exciting forces. If the matter is looked at in this way it should not be surprising that the principal modes and frequencies of the undamped system are still important. Indeed, if the special case of excitation at a natural frequency is considered, it will be seen that the effects of the exciting forces must just cancel those of the damping forces and that the undamped system is left vibrating freely with no net excitation.

It will also be shown later that, if the excitation frequency is not close to a natural frequency, then the motion is often very little altered by the damping forces.

In real problems it is rare for the magnitudes of the damping forces to be known with anything like the same accuracy as the elastic and inertia forces. It is therefore pointless to attempt to construct a detailed mathematical treatment of the damping effects. The treatment to be used should rather be selected for its convenience and simplicity, provided that it gives as good an estimate as possible using such knowledge of the damping forces as may exist in each case. Now the mathematics may be made tractable if the equations of motion are linear, as are all the foregoing equations for undamped motion. Strictly, the form of linearity that we require will only be preserved if, during harmonic motion, the damping forces themselves vary harmonically. If the forces are due to oil viscosity in bearings this may be so; but if they arise from dry friction they are likely to be far from sinusoidal. Nevertheless, useful numerical treatment of problems involving dry friction can be obtained by using linear equations.

There are two useful systems for setting up linear damping equations. Either it may be assumed that all the damping forces are proportional to the instantaneous velocity or that the damping forces are harmonic and in quadrature with the displacement, their magnitudes being proportional to the displacement. The first system is known as *viscous* damping and will be described in this chapter, while the second is *hysteretic* damping which is treated in Chapter 9.

The analysis will be developed in terms of systems in which the damping forces are specified as acting at particular points, just as the masses and stiffnesses of finite-freedom systems have been idealized. In practice, the locations of the damping effects in real systems are distributed and, as we mentioned earlier, their magnitudes are often unknown. The assessment of damping is a subject in which research is needed.

8.1 The complex receptance of a system with a single degree of freedom

Fig. 8.1.1 shows a simple oscillator of the type discussed in Chapter 1, here modified by the addition of a massless damping device or dashpot between the mass M and the abutment. The dashpot is supposed to exert a force $-b\dot{x}$ which is positive in the positive direction of x, so that b is a 'damping coefficient'. The equation for forced harmonic motion is

$$M\ddot{x} + b\dot{x} + kx = F e^{i\omega t}. \tag{8.1.1}$$

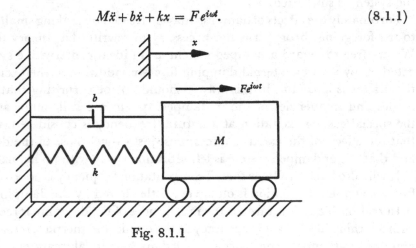

Fig. 8.1.1

As before, a solution may be sought in which x varies at the forcing frequency ω. Using the trial solution

$$x = X e^{i\omega t} \tag{8.1.2}$$

in which X is a constant, it is found that

$$x = X e^{i\omega t} = \frac{F e^{i\omega t}}{(k - M\omega^2) + ib\omega}. \tag{8.1.3}$$

The receptance α, at x, is therefore given by

$$\alpha = \frac{1}{(k - M\omega^2) + ib\omega} \tag{8.1.4}$$

and it is a complex quantity.

In order to examine the implications of this complexity more fully, we divide the receptance into its real and imaginary parts; it becomes

$$\alpha = \left[\frac{k - M\omega^2}{(k - M\omega^2)^2 + b^2\omega^2}\right] - i\left[\frac{b\omega}{(k - M\omega^2)^2 + b^2\omega^2}\right]. \tag{8.1.5}$$

This shows that the displacement x has one component

$$\frac{(k - M\omega^2)\, F\, e^{i\omega t}}{(k - M\omega^2)^2 + b^2\omega^2} \tag{8.1.6}$$

which is in phase with the applied force and another component

$$\frac{b\omega F\, e^{i\omega t}}{(k - M\omega^2)^2 + b^2\omega^2} \tag{8.1.7}$$

which has a phase-lag of $\frac{1}{2}\pi$ rad. behind the applied force. This component is said to be *in quadrature* with the excitation.

The in-phase and quadrature components may be plotted separately. They are represented in fig. 8.1.2 (a) and (b) by $\mathscr{R}(\alpha)$ and $\mathscr{I}(\alpha)$ respectively, being shown plotted against ω in the manner of §3.3. It will be seen that the curves of $\mathscr{R}(\alpha)$ possess two 'horns' and it may be shown by differentiation that these correspond to the values

$$\omega = \omega_1 \sqrt{(1 \pm 2\nu)}, \tag{8.1.8}$$

where ν is a dimensionless measure of the damping given by

$$\nu = \frac{b}{2\sqrt{(kM)}} = \frac{b}{2M\omega_1}. \tag{8.1.9}$$

The physical significance of this quantity will be discussed later. As the damping is decreased, the horns become closer together.

The values of $\mathscr{R}(\alpha)$ at the two peaks are

$$\left.\begin{array}{ll} \dfrac{1}{4k\nu(1 - \nu)} & \text{when} \quad \omega = \omega_1\sqrt{(1 - 2\nu)}, \\[2mm] \dfrac{-1}{4k\nu(1 + \nu)} & \text{when} \quad \omega = \omega_1\sqrt{(1 + 2\nu)}. \end{array}\right\} \tag{8.1.10}$$

These values increase and the peaks become more pointed as ν is diminished. In the limit, when $b = 0$, we already know that the curve of fig. 8.1.2 (a) has an asymptote at $\omega = \omega_1$.

It may be mentioned that the curves shown correspond to

$$k = 8\,\text{lb.wt./ft.} \quad \text{or} \quad 256\,\text{pdl./ft.,}$$
$$M = 1\,\text{lb.,}$$
$$\nu = 0\cdot01,\ 0\cdot1\ \text{and}\ 0\cdot3.$$

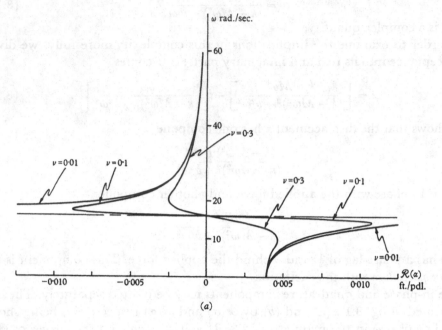

(a)

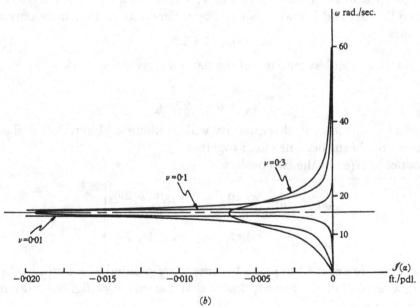

(b)

Fig. 8.1.2

It is sometimes more convenient to express complex receptances in polar form. To do this, the denominator of equation (8.1.4) is written as

$$\sqrt{[(k - M\omega^2)^2 + b^2\omega^2]} \, e^{i\zeta},$$

where
$$\zeta = \tan^{-1}\left(\frac{b\omega}{k - M\omega^2}\right). \tag{8.1.11}$$

The receptance will now be seen to be

$$\alpha = \frac{e^{-i\zeta}}{\sqrt{[(k - M\omega^2)^2 + b^2\omega^2]}}. \tag{8.1.12}$$

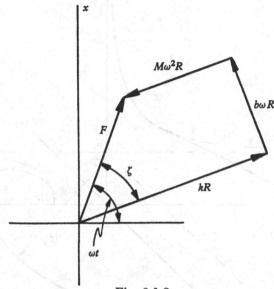

Fig. 8.1.3

This quantity may be interpreted as shown in § 1.3. The factor $e^{-i\zeta}$ means that the force vector must be rotated backwards—clockwise on the conventional Argand diagram—through an angle ζ to give the direction of the displacement vector. The length of the displacement vector is found by multiplying that of the force vector by the factor

$$\frac{1}{\sqrt{[(k - M\omega^2)^2 + b^2\omega^2]}}.$$

The result may also be obtained by the direct use of rotating vectors as described in § 1.2. Fig. 8.1.3 illustrates the procedure. The solution (8.1.2) is represented by a line of length R which rotates with angular velocity ω, R being the amplitude of x so that, in the original notation,
$$R = |X|. \tag{8.1.13}$$

The spring force kx is thus represented by a parallel line of length kR and this is shown in the diagram. The dashpot force $b\dot{x}$ and the reversed inertia force $M\ddot{x}$ are represented by lines of length $b\omega R$ and $M\omega^2 R$ respectively, these being $\frac{1}{2}\pi$ and π rad. respectively in advance of the displacement vector. By adding the three force vectors, kR, $b\omega R$ and $M\omega^2 R$, the resultant force vector is obtained, as may be seen from equation (8.1.1). From the geometry of the vector polygon we have

$$F^2 = [(k - M\omega^2)^2 + b^2\omega^2] R^2 \tag{8.1.14}$$

or

$$R = \frac{F}{\sqrt{[(k - M\omega^2)^2 + b^2\omega^2]}}. \tag{8.1.15}$$

as shown previously. The expression for the phase angle ζ, given in equation (8.1.11), may be deduced by inspection of the vector diagram.

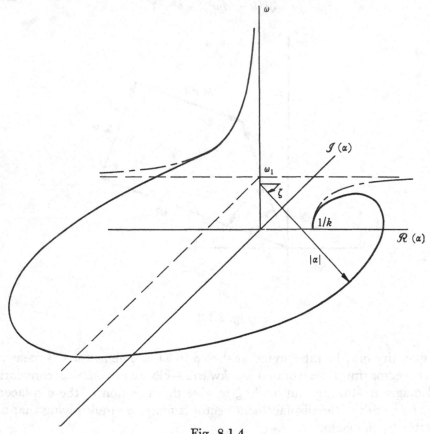

Fig. 8.1.4

In numerical work, it is usually required to find R and ζ for a given F rather than the other way. If the vector diagram of fig. 8.1.3 is used, it is then necessary to draw the polygon for an arbitrary value of R, thereby obtaining a corresponding value of F. This is then compared with the specified F and the arbitrary value of R can be modified to give the correct value of F.

The curves of the real and imaginary parts of the receptance α can be combined into a single three-dimensional curve. This is sometimes useful when the effects of coupling one damped system to another are to be investigated. A three-dimensional curve of this sort is sketched in fig. 8.1.4. In this figure, the broken line represents the curve of fig. 3.4.5 (b) for which $b = 0$ and which therefore lies wholly in the $(\omega, \mathscr{R}(\alpha))$ plane. This curve lies close to the full line curve for values of ω which are remote from the value $\omega = \omega_1 = \sqrt{(k/M)}$.

The curves of fig. 8.1.2 are projections of three-dimensional curves, like that of fig. 8.1.4, on the $(\omega, \mathscr{R}(\alpha))$ and the $(\omega, \mathscr{I}(\alpha))$ planes. The third projection, on the $(\mathscr{R}(\alpha), \mathscr{I}(\alpha))$ plane, is sometimes useful. It is shown in fig. 8.1.5, plotted for the same system having the same three values of damping which were used for fig. 8.1.2. Each curve represents the locus, on the Argand diagram, of the end of the line representing the complex receptance for a particular intensity of damping as the frequency is varied. A few values of the frequencies are marked on the curves.

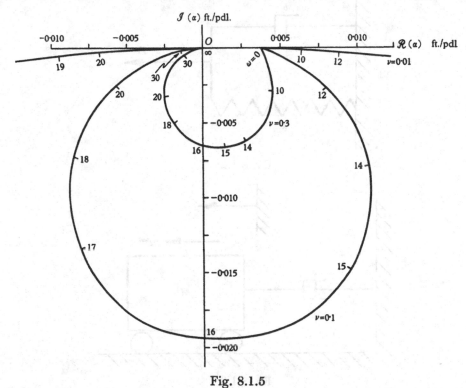

Fig. 8.1.5

EXAMPLES 8.1

1. Evaluate the receptance α for the damped oscillator of fig. 8.1.1, where

$$k = 7680 \text{ pdl./ft.}, \quad M = 40 \text{ lb.}, \quad b = 96 \text{ pdl.sec./ft.}$$

for $\omega = 10$ rad./sec. Express the result in Cartesian and in polar form.

2. The system shown (see page 414) is that of fig. 8.1.1 with $M = 0$.

(a) If k and ω have some given values, what are the limiting values to which the receptance α tends as b varies from a very small value to a very large one?

(b) What is the corresponding range of the phase angle ζ?

(c) Calculate α in the Cartesian and polar forms for

$$k = 7680 \text{ pdl./ft.}, \quad b = 96 \text{ pdl.sec./ft.}, \quad \omega = 10 \text{ rad./sec.}$$

3. The system is that of the simple damped oscillator with zero spring stiffness.

(*a*) If *b* and ω have given values, what are the limiting values to which the receptance α tends as *M* varies from a very small magnitude to a very large one?

(*b*) What is the corresponding range of the phase angle ζ?

(*c*) Calculate α in the Cartesian and polar forms for

$$b = 96 \text{ pdl.sec./ft.,} \quad M = 40 \text{ lb.,} \quad \omega = 10 \text{ rad./sec.}$$

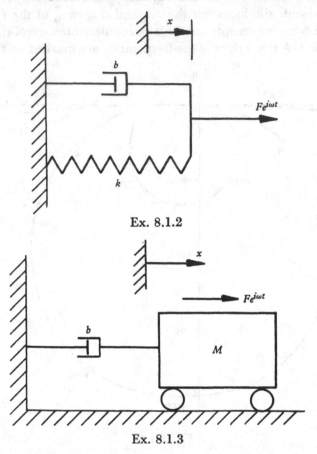

Ex. 8.1.2

Ex. 8.1.3

4. Sketch the three-dimensional curve of $1/\alpha$ against ω for the simple oscillator of fig. 8.1.1 with $\mathscr{R}(1/\alpha)$ and $\mathscr{I}(1/\alpha)$ plotted horizontally and ω vertically. Show by a dotted line the curve corresponding to $b = 0$ and note that, if $b > 0$, the curve has no intersection with the ω axis.

8.2 Specification of the damping in a system having one degree of freedom; energy considerations

We shall still restrict attention, in this section, to systems having one degree of freedom. The reason for treating such systems at length is that damped multi-degree-of-freedom systems can often be analysed over a limited frequency range as if they had but one degree of freedom. Results for the simple system are therefore of wide application.

The amount of damping in a system depends upon the coefficient b. It was defined non-dimensionally in § 8.1 by the ratio

$$\nu = \frac{b}{2\sqrt{(kM)}} = \frac{b}{2M\omega_1}. \tag{8.2.1}$$

This ratio ν is equal to the actual value of b divided by the value which b would have if the free motion of the system were just non-oscillatory. This will be explained more fully when free damped motion is discussed in Chapter 10. It will be sufficient for the present to regard ν as the ratio of the damping constant b to a certain standard or 'critical' damping constant. It may be noted that if vibration is troublesome in a particular mechanical system, then ν must be considerably less than unity. It is possible for values of ν of the order of 0·001 to be found in mechanical systems.

The equation of steady forced motion may be written in the form

$$\ddot{x} + 2\nu\omega_1\dot{x} + \omega_1^2 x = \omega_1^2\left(\frac{F}{k}\right)e^{i\omega t}. \tag{8.2.2}$$

The steady-state solution to this equation is

$$x = \left[\frac{1}{\sqrt{\left[\left(1 - \frac{\omega^2}{\omega_1^2}\right)^2 + 4\nu^2\frac{\omega^2}{\omega_1^2}\right]}}\right]\frac{F}{k}e^{i(\omega t - \zeta)}, \tag{8.2.3}$$

where

$$\zeta = \tan^{-1}\left[\frac{2\nu\frac{\omega}{\omega_1}}{1 - \frac{\omega^2}{\omega_1^2}}\right]. \tag{8.2.4}$$

The factor F/k in equation (8.2.3) is the extension which would be produced in the spring by the force F alone. The factor within the large brackets is thus a dimensionless multiplier or 'magnification factor' and we shall denote it by N. It will be seen that N and ζ are both functions of the dimensionless quantities ν and ω/ω_1. They may be plotted, as shown in fig. 8.2.1 against a base of ω/ω_1. The figure contains the curves for $\nu = 0$ which are reproduced from fig. 1.2.2.

The advantage of using the ratio ν, rather than the quantity b, can now be seen. The ratio enables curves of N and ζ to be drawn which are applicable to any system having one degree of freedom. We shall show later that ν, together with ω_1, are quantities which may be measured directly for a real system; they may then be used to obtain N and ζ from the curves of fig. 8.2.1.

It will be found from the curves of fig. 8.2.1 that if the damping is small—if for instance ν is less than about 0·01—then the amplitude and phase are almost independent of the value of ν, except when ω/ω_1 is close to unity. But if the system is nearly in resonance, then the damping is important in limiting the oscillation amplitude. It will be found that the maximum amplitude occurs when ω/ω_1 is slightly less than unity, and is given in fact by the value

$$\frac{\omega}{\omega_1} = \sqrt{(1 - 2\nu^2)}. \tag{8.2.5}$$

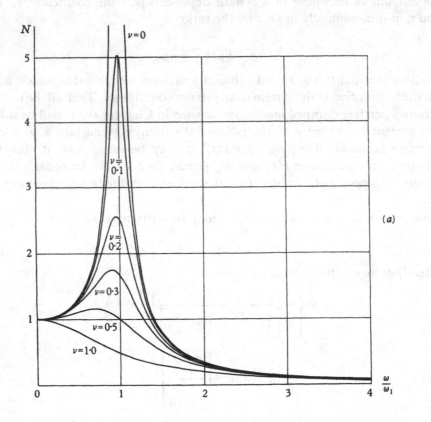

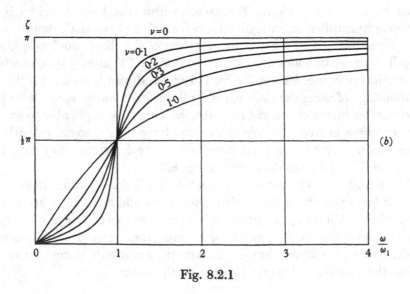

Fig. 8.2.1

The corresponding value of N is

$$\frac{1}{2\nu\sqrt{(1-\nu^2)}}. \tag{8.2.6}$$

If $\nu \ll 1$, the maximum value of N may be taken to occur when $\omega = \omega_1$ and to be given by

$$N \doteqdot \frac{1}{2\nu}. \tag{8.2.7}$$

The state of affairs which obtains when $\omega = \omega_1$ is often referred to as the *resonant condition*; the corresponding vector diagram, of the type shown in fig. 8.1.3, has a rectangular form.

There is another dimensionless quantity which is often used to indicate the amount of damping in a system. This is the so-called *Q-factor*. It may conveniently be defined in terms of energy dissipation and this idea is therefore introduced at this stage.

The dashpot of fig. 8.1.1 absorbs energy which has to be fed to the system by the exciting force. The instantaneous rate of dissipation is $b\dot{x} \cdot \dot{x} = b\dot{x}^2$ so that the total energy dissipated in one cycle is

$$E = \int_0^{2\pi/\omega} b\dot{x}^2 \, dt. \tag{8.2.8}$$

For steady sinusoidal motion in which $x = R\sin\omega t$, the integral gives

$$E = bR^2\pi\omega. \tag{8.2.9}$$

It is preferable at this stage to return to the trigonometrical representation in place of the complex exponential, because the latter does not lend itself to the representation of the work done in a cycle, this not being itself a sinusoidally varying quantity. We therefore write $F\sin\omega t$ for the applied force.

The rate of working of the applied force is $F\sin\omega t \cdot \dot{x}$. The work done per cycle is, therefore,

$$w = \int_0^{2\pi/\omega} F\sin\omega t \cdot \dot{x} \cdot dt \tag{8.2.10}$$

and if x is written $R\sin(\omega t - \zeta)$, then the integral becomes

$$w = FR\omega \int_0^{2\pi/\omega} \sin\omega t \cos(\omega t - \zeta) \, dt = FR\pi\sin\zeta. \tag{8.2.11}$$

This expression may be interpreted to mean that the work is done entirely by the time-component $F\sin\zeta$ of the applied force, this component being in phase with the velocity and in quadrature with the displacement (see fig. 8.1.3). Thus we may write

$$F\sin\omega t \equiv F[\sin(\omega t - \zeta)\cos\zeta + \cos(\omega t - \zeta)\sin\zeta] \tag{8.2.12}$$

in which $F\sin\omega t$ appears as the sum of the component $F\cos\zeta$ which is in phase with x and $F\sin\zeta$ which is in phase with $\dot{x}$.

Although this discussion concerns single-degree-of-freedom systems, it should be noted at this point that the transfer of energy, by means of a sinusoidal force which is undergoing a sinusoidal displacement in quadrature with it, occurs in multi-degree-of-freedom systems. Let B and C of fig. 8.2.2 represent coupled sub-systems with a common co-ordinate q_2. Let a sinusoidal force be applied to B at q_1 and suppose that energy is dissipated within C. Energy must be fed through the connexion at the co-ordinate q_2 and this can occur only if the generalized force Q_{c2} acting on C, has a component in quadrature with q_2. Though there may be no energy dissipated in B there must be an energy input at q_1 and therefore the force Q_1 must also have a component in quadrature with q_1. Thus the presence of energy dissipation at any point in a system will affect the phase relationships between forces and displacements throughout the system.

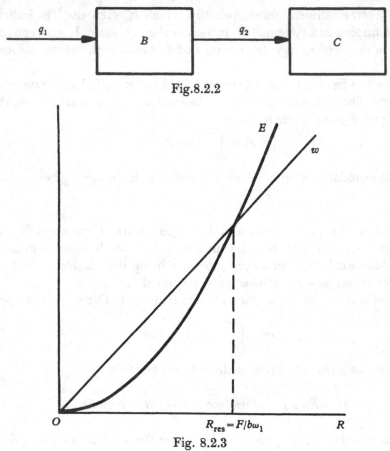

Fig.8.2.2

Fig. 8.2.3

Returning now to the system having one degree of freedom, we see that it must be possible to equate the energy input per cycle and the energy loss per cycle. Thus $w = E$ so that

$$R = \frac{F \sin \zeta}{b\omega}. \tag{8.2.13}$$

It will be found that this agrees with the results found previously.

The equality of w and E, during steady motion, can be represented graphically. Let the exciting frequency be made equal to ω_1 so that $\zeta = \frac{1}{2}\pi$ and $w = FR\pi$. Evidently w may be plotted against the amplitude of displacement R, giving a straight line as shown in fig. 8.2.3. On the other hand if E is plotted against R it gives a parabola in accordance with equation (8.2.9). The intersection of the two curves gives the resonant amplitude R_{res}. The motion evidently remains at this amplitude because if it were to increase the dissipation would exceed the input and so reduce it again. Conversely if the motion were to decrease the input would exceed the dissipation.

Consider now the energy which is present within the system. When harmonic motion occurs at the resonant frequency with amplitude R, the maximum kinetic energy is equal to $\frac{1}{2}MR^2\omega_1^2$. The energy dissipated per cycle in the resonant condition may be expressed in terms of this; thus,

$$\frac{\text{Maximum kinetic energy}}{\text{Energy dissipated per cycle}} = \frac{MR^2\omega_1^2}{2bR^2\pi\omega_1} = \frac{1}{2\pi}\frac{M\omega_1}{b}. \qquad (8.2.14)$$

It will be seen that this ratio is independent of the amplitude. The Q-factor, mentioned above, may now be defined as

$$Q = \frac{M\omega_1}{b} \qquad (8.2.15)$$

so that the right-hand side of equation (8.2.14) may be written $(1/2\pi)\,Q$. If the damping in a system is small, then Q will be large and the system is said to be one possessing 'high Q'. Similarly a 'low Q' system has heavy damping.

The expression for the stored energy, which is used above, is that for the maximum kinetic energy. The maximum potential energy may be used equally well, and is equal to $\frac{1}{2}kR^2$. The ratio (8.2.14) then becomes

$$\frac{kR^2}{2bR^2\pi\omega_1} = \frac{1}{2\pi}\frac{k}{b\omega_1} \qquad (8.2.16)$$

and from this it is found that $\qquad Q = \dfrac{k}{b\omega_1} \qquad (8.2.17)$

by the same definition. The two expressions for Q are related to the quantity ν which was used earlier; thus

$$Q = \frac{M\omega_1}{b} = \frac{k}{b\omega_1} = \frac{1}{2\nu}. \qquad (8.2.18)$$

It will now be seen from equation (8.2.7) that the Q-factor is equal to the magnification factor when the system is in resonance.

The term 'Q-factor' was first used in electric circuit theory and its application was subsequently extended to mechanical vibration problems. The letter Q was originally used to stand for 'quality'. This was because the electric circuits for which the factor was defined were designed to have low damping; they performed better if their Q values were high. In mechanical systems this is by no means always the case; nevertheless the notation is now well established so that it is used without implication of the desirability of high or low damping.

The term 'Q-factor' has been used with slightly different meanings and this has sometimes led to confusion. Although the definitions that have already been given will be used in this book, it may be helpful if we digress briefly to explain this matter. In electrical practice Q may not always be defined for a *resonant* frequency, and the mechanical equivalents of the definitions of Q then become

$$Q = \frac{M\omega}{b} \quad \text{and} \quad Q = \frac{k}{b\omega},$$

where $\omega \neq \omega_1$. It happens that the electrical equivalent of the first of these (namely $L\omega/r$, where L is the inductance and r the resistance of a coil) is a useful quantity because r is found to be frequency-dependent and $L\omega/r$ to be almost constant over a wide frequency range. The Q-factor may therefore be specified for a particular coil, without reference to the frequency.

This latter use of the term 'Q-factor' has a counterpart in mechanical vibration theory. This is because the source of energy loss in a mechanical system is often largely within that part of the mechanism which provides the flexibility. It may be due for instance to a lack of perfect elasticity in the material, this being particularly common if the material is non-metallic. In these cases, the energy loss per cycle in the flexible component is often found to be nearly independent of ω so that $b\omega$ is constant over a range of frequency (see equation (8.2.9)). It follows that, for this member, the quantity

$$Q = \frac{k}{b\omega}$$

is constant. It is therefore convenient to define Q for the spring alone (and to specify it by a constant) rather than for the whole system. As we mentioned previously, however, this use of the term 'Q-factor' will not be employed in this book; we shall use it in the form of equations (8.2.18).

The intensity of the damping may be found, for a system having one degree of freedom, by means of a free vibration test. It is then necessary to observe the rate at which the motion decays, as will be explained in Chapter 10. We shall now show how the intensity of damping can also be estimated by means of a forced vibration test, using a response curve in which the amplitude R of the response is plotted against driving frequency ω.

Consider the particular values of ω, one above and one below ω_1, for which

$$b\omega = \pm (k - M\omega^2) = \pm M(\omega_1^2 - \omega^2). \tag{8.2.19}$$

It will be found on substitution in equation (8.2.3) that, for these values of ω, the amplitude of the response x has $1/\sqrt{2}$ of its peak value. That is to say N has $1/\sqrt{2}$ of its maximum value. Now the difference between these two frequencies and the natural frequency is small and they are approximately equidistant from the natural frequency. Let them be denoted by

$$\omega_1 - \epsilon_1 \quad \text{and} \quad \omega_1 + \epsilon_2.$$

It follows that
$$\left.\begin{aligned}
b(\omega_1 - \epsilon_1) &= M[\omega_1^2 - (\omega_1 - \epsilon_1)^2] \doteq 2M\epsilon_1\omega_1, \\
b(\omega_1 + \epsilon_2) &= -M[\omega_1^2 - (\omega_1 + \epsilon_2)^2] \doteq 2M\epsilon_2\omega_1,
\end{aligned}\right\} \tag{8.2.20}$$

where the approximation is obtained by neglecting ϵ^2 compared with $\omega_1\epsilon$.

If equations (8.2.20) are added together, they give

$$2b\omega_1 + b(\epsilon_2 - \epsilon_1) \doteq 2M\omega_1(\epsilon_1 + \epsilon_2) \qquad (8.2.21)$$

which may be simplified to $\qquad b \doteq M(\epsilon_1 + \epsilon_2) \qquad (8.2.22)$

if $(\epsilon_2 - \epsilon_1)$ is neglected. It follows therefore that b can be found from M, together with $(\epsilon_1 + \epsilon_2)$. The latter quantity can be measured directly off a response curve since it is the width of the peak at $1/\sqrt{2}$ of the height of the maximum.

EXAMPLES 8.2

1. An oscillatory system, which represents a vibration-measuring instrument, is shown in its rest position in fig. (a). Motion $\phi(t)$ of the abutment A disturbs the system as shown in fig. (b). Using the notation of this section, find:

    (i) the differential equation governing the absolute displacement $x(t)$, and

    (ii) that governing the relative displacement $y(t)$.

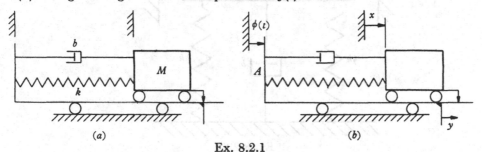

        (a)   (b)

Ex. 8.2.1

2. If the disturbance in Ex. 8.2.1 is harmonic of the form

$$\phi = \Phi e^{i\omega t},$$

find the steady-state solution for the absolute displacement x in the form

$$x = R e^{i(\omega t - \alpha)}.$$

3. Repeat Ex. 8.2.2 using the rotating-vector method of solution. Examine the following three cases:

    (i) $\omega/\omega_1 < 1$.

    (ii) $\omega/\omega_1 = 1$.

    (iii) $\omega/\omega_1 > 1$.

4. If the disturbance in Ex. 8.2.1 is

$$\phi = \Phi e^{i\omega t}$$

find the steady-state solution for the relative displacement y in the form

$$y = S e^{i(\omega t - \beta)}.$$

Verify that the solution (found analytically) can be obtained by vector subtraction from the solution of Ex. 8.2.3 noting that

$$y = x - \phi.$$

5. The first solution given for Ex. 8.2.4 (see p. 584) is expressed in terms of a quantity L which is a non-dimensional 'magnification factor for relative displacement'. Sketch curves of L plotted (vertically) against the ratio ω/ω_1 for $\nu = 0$, $\nu = 0.3$ and $\nu = 1$.

6. The figure represents an unbalanced machine of total mass M that vibrates due to the arm of length e which carries a mass m and which rotates with the constant speed ω rad./sec. Find the steady-state displacement in the form

$$x = R\, e^{i(\omega t - \gamma)}.$$

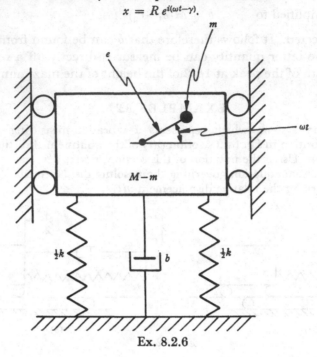

Ex. 8.2.6

7. Show that the response of the system of fig. 8.1.1 can be found by first finding the phase angle ζ from the formula

$$\cot \zeta = Q\left(\frac{\omega_1}{\omega} - \frac{\omega}{\omega_1}\right)$$

and then using the relation

$$\frac{R}{R_{\text{res.}}} = \frac{\omega_1}{\omega} \sin \zeta,$$

where $R_{\text{res.}}$ is the amplitude of the displacement in the resonance state (when $\omega = \omega_1$).[†]

8.3 Systems having several degrees of freedom; analysis and synthesis of damped systems

We have shown that the direct receptance at x of the simple damped oscillator of fig. 8.1.1 is a complex quantity. This is a general property of the receptances of damped systems. In this section, we shall first show how receptances may be found directly from the equations of motion; the treatment is similar to that which was given for undamped systems in § 1.7.

Consider the system having two degrees of freedom which is shown in fig. 8.3.1, the co-ordinates being as indicated. The equations of motion are

$$\left.\begin{aligned}
M_1\ddot{x}_1 + b\dot{x}_1 + kx_1 - b\dot{x}_2 - kx_2 &= 0, \\
-b\dot{x}_1 - kx_1 + M_2\ddot{x}_2 + b\dot{x}_2 + kx_2 &= F_2\, e^{i\omega t}.
\end{aligned}\right\} \tag{8.3.1}$$

[†] This type of result is suggested in a paper by H. G. Yates 'A Universal Resonance Chart', *The Engineer* (1 Oct. 1943), p. 268.

Into these, substitute the trial solution

$$x_1 = X_1 e^{i\omega t}, \quad x_2 = X_2 e^{i\omega t}, \tag{8.3.2}$$

which gives
$$(k - M_1 \omega^2 + ib\omega) X_1 - (k + ib\omega) X_2 = 0, \atop - (k + ib\omega) X_1 + (k - M_2 \omega^2 + ib\omega) X_2 = F_2. \left.\right\} \tag{8.3.3}$$

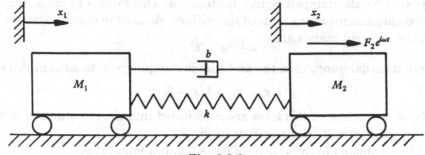

Fig. 8.3.1

If these equations are solved for X_1 and X_2 they give

$$X_1 e^{i\omega t} = x_1 = \frac{(k + ib\omega) F_2 e^{i\omega t}}{(k - M_1 \omega^2 + ib\omega)(k - M_2 \omega^2 + ib\omega) - (k + ib\omega)^2} = \alpha_{12} F_2 e^{i\omega t}, \atop X_2 e^{i\omega t} = x_2 = \frac{(k - M_1 \omega^2 + ib\omega) F_2 e^{i\omega t}}{(k - M_1 \omega^2 + ib\omega)(k - M_2 \omega^2 + ib\omega) - (k + ib\omega)^2} = \alpha_{22} F_2 e^{i\omega t}. \left.\right\} \tag{8.3.4}$$

It will be seen that α_{12} is unchanged if M_1 and M_2 are interchanged. If this is done, however, it is equivalent to replacing α_{12} by α_{21}. It follows that the reciprocal relation
$$\alpha_{12} = \alpha_{21} \tag{8.3.5}$$

holds in this problem. It will be shown later than it remains true in the general case.

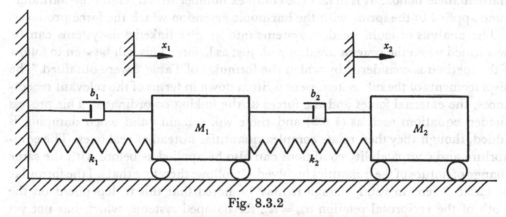

Fig. 8.3.2

Although it is possible to find the receptances of more complicated systems by this direct method, it soon becomes very tedious. We shall leave the matter at this stage and subsequently show how a new set of energy coefficients may be introduced to make the work easier. The matter will be discussed in § 8.5.

For the present it will be sufficient to show how many of the receptances for undamped systems may be modified to allow for damping. Consider for example the system shown in fig. 8.3.2. This is the translational counterpart of system 8 of Table 1, modified by the addition of dashpots across the springs. Now if, in the receptances for the undamped system, each of the stiffness coefficients k is replaced by a 'complex stiffness'

$$K = k + ib\omega \qquad (8.3.6)$$

the receptance for the damped system is obtained. This follows because, if x_a and x_b are the displacements at the ends of the stiffness element in question, so that the forces on the adjacent masses are $\pm k(x_a - x_b)$

when there is no damping, then the addition of damping will transform this into

$$\pm [k(x_a - x_b) + b(\dot{x}_a - \dot{x}_b)].$$

Evidently, if expressions of this form are substituted into the equations of motion, then the receptances will be changed accordingly.

If the idea is applied to the system of fig. 8.3.2, it is found that

$$\alpha_{11} = \frac{(k_2 + ib_2\omega) - M_2\omega^2}{\{M_1 M_2 \omega^4 - [(k_1 + ib_1\omega) M_2 + (k_2 + ib_2\omega)(M_1 + M_2)]\omega^2 + (k_1 + ib_1\omega)(k_2 + ib_2\omega)\}}.$$

$$(8.3.7)$$

This expression may now be simplified and, if desired, thrown into the polar form.

It may be noted that if the damping effect associated with a dashpot whose coefficient is b_r is due to the physical properties of a spring whose stiffness is k_r, as discussed in the previous section, then the 'complex stiffness'

$$K_r = k_r + ib_r\omega \qquad (8.3.8)$$

may be regarded as a quantity of some physical significance and not merely a mathematical fiction. It is in fact the complex number which relates the harmonic force applied to the spring with the harmonic extension which the force produces.

The analysis of multi-freedom systems into simpler linked sub-systems can be performed when the systems are damped, just as before. This will be seen to follow if the method is considered by which the formulae of Table 2 were obtained. The displacements of the sub-system were written down in terms of the relevant receptances, the external forces and the forces at the linking co-ordinates. This process yielded equations such as (1.8.2) and these will remain valid when damping is added, though they then relate complex quantities instead of real ones. The equilibrium and compatability conditions can also be applied as before with the same change of nature of the quantities involved. It follows therefore that all the formulae of Table 2 remain valid, though it may be noted that this is dependent on the truth of the reciprocal relation $\alpha_{rs} = \alpha_{sr}$ for damped systems, which has not yet been proved here.

The receptances of damped systems cannot be used, as were those of undamped systems, for finding frequency equations. The previous technique was based on the idea of frequencies at which the ratio of applied force to displacement become infinitely small. When damping is present this does not occur because if motion is steady some finite force must be applied in order to provide the energy input. Thus the very concept of a natural frequency needs some modification before it is applied to damped systems.

The applicability of Table 2 to damped systems may be illustrated by deriving from it the receptance of the system of fig. 8.1.1. Consider first the receptance for an isolated damper, one end of which is anchored; the arrangement is shown in fig. 8.3.3. The applied force $Fe^{i\omega t}$ and displacement x are related by the equation

$$b\dot{x} = Fe^{i\omega t}. \qquad (8.3.9)$$

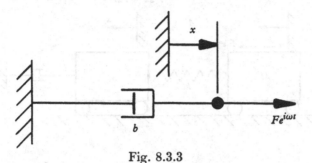

Fig. 8.3.3

The usual form of trial solution (cf. equation (8.1.2)) gives for x, the expression $\alpha Fe^{i\omega t}$, where

$$\alpha = \frac{1}{ib\omega} = \frac{e^{-i\pi/2}}{b\omega}. \qquad (8.3.10)$$

This receptance will be seen to be purely imaginary.

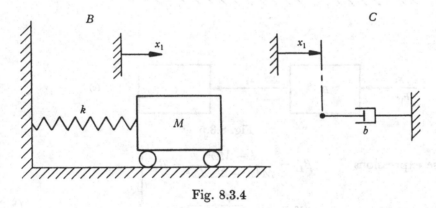

Fig. 8.3.4

Let the system of fig. 8.1.1 be divided into the sub-systems shown in fig. 8.3.4. The expression given in entry 1 of Table 2 for α_{22} may here be used in the form

$$\frac{1}{\alpha_{11}} = \frac{1}{\beta_{11}} + \frac{1}{\gamma_{11}}, \qquad (8.3.11)$$

where
$$\beta_{11} = \frac{1}{k - M\omega^2}, \quad \gamma_{11} = \frac{1}{ib\omega}. \qquad (8.3.12)$$

The expression (8.1.4) for the required receptance is now obtained by substitution.

As a second illustration of the use of Table 2 with damped systems, consider the arrangement shown in fig. 8.3.5 (a). If the system is divided into sub-systems B and C as shown at (b) and (c), then entry 1 of the table shows that

$$\left.\begin{aligned} \alpha_{11} &= \beta_{11} - \frac{\beta_{12}^2}{\beta_{22} + \gamma_{22}}, \\ \alpha_{12} &= \beta_{12} - \frac{\beta_{12}\beta_{22}}{\beta_{22} + \gamma_{22}}. \end{aligned}\right\} \tag{8.3.13}$$

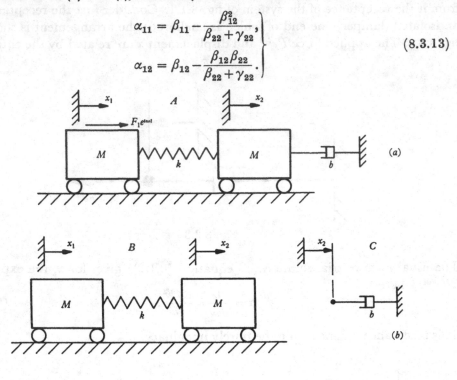

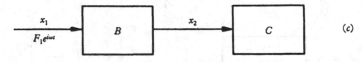

Fig. 8.3.5

In these expressions

$$\left.\begin{aligned} \beta_{11} &= \frac{k - M\omega^2}{M^2\omega^4 - 2kM\omega^2} = \beta_{22}, \\ \beta_{12} &= \frac{k}{M^2\omega^4 - 2kM\omega^2}, \\ \gamma_{22} &= \frac{1}{ib\omega}. \end{aligned}\right\} \tag{8.3.14}$$

After simplification, the following expressions are found for the system of fig. 8.3.5 (a);

$$\left.\begin{aligned} \alpha_{11} &= \frac{(k - M\omega^2) + ib\omega}{M^2\omega^4 - 2kM\omega^2 + ib\omega(k - M\omega^2)}, \\ \alpha_{12} &= \frac{k}{M^2\omega^4 - 2kM\omega^2 + ib\omega(k - M\omega^2)}. \end{aligned}\right\} \tag{8.3.15}$$

These results illustrate a matter which has been mentioned already, namely that in a damped system there are phase differences between the motions of different points. This follows because, whereas the denominators of α_{11} and α_{12} are the same, the numerator of α_{11} is complex while that of α_{12} is real. It is of interest now to examine how the amplitude and phase of the motion vary along the spring. Before doing this, however, we shall give particular values to the damping and to the frequency in order to simplify the algebra, choosing values for which the results have some practical interest. First, let the forcing frequency be equal to the natural frequency of the undamped system, so that

$$\omega^2 = \frac{2k}{M}. \tag{8.3.16}$$

Secondly let the damping be such that

$$\frac{2k}{b\omega} = 50. \tag{8.3.17}$$

When these values are inserted in the expressions for the receptances, they give

$$\alpha_{11} = -\frac{1+25i}{k}, \quad \alpha_{12} = \frac{25i}{k}. \tag{8.3.18}$$

Now consider the displacement x at some point on the spring distant y from x_1 and $(l-y)$ from x_2. If the spring is uniform the displacement at this point will be

$$x_1 + \frac{y}{l}(x_2 - x_1) \quad \text{or} \quad \left(\frac{l-y}{l}\right)x_1 + \left(\frac{y}{l}\right)x_2. \tag{8.3.19}$$

It follows that $$x = \left[\left(\frac{l-y}{l}\right)\alpha_{11} + \left(\frac{y}{l}\right)\alpha_{12}\right]F_1 e^{i\omega t}. \tag{8.3.20}$$

If the values of α_{11} and α_{12}, given in equations (8.3.18), are now introduced, they give

$$x = \left[-\left(1-\frac{y}{l}\right) - 25i\left(1-\frac{2y}{l}\right)\right]\frac{F}{k}e^{i\omega t}. \tag{8.3.21}$$

This expression shows that the displacement at any point of the spring is composed of one component which is in antiphase with the force and whose magnitude varies linearly with y, being zero when $y = l$, and one component which is in quadrature with the force and which also varies linearly but is zero at the mid-point of the spring. The two components are plotted separately in fig. 8.3.6 as curves (a) and (b) respectively. The amplitude of the displacement x at any point will evidently be

$$\frac{F}{k}\sqrt{\left[\left(1-\frac{y}{l}\right)^2 + 625\left(1-\frac{2y}{l}\right)^2\right]}. \tag{8.3.22}$$

This quantity is nearly numerically equal to the amplitude of the quadrature component except near the mid-point of the spring where the quadrature component is small. The magnitude given in equation (8.3.22) is plotted in fig. 8.3.6 as curve (c).

The phase-advance of the displacement, relative to the exciting force, is given by the angle ξ, where

$$\tan \xi = \frac{-25\left(1-\dfrac{2y}{l}\right)}{-\left(1-\dfrac{y}{l}\right)}.$$

The variation of ξ with y is given in fig. 8.3.7.

These results show that when a long elastic member, to which masses are attached at the ends, is in resonance, and when energy is being supplied at one end of the

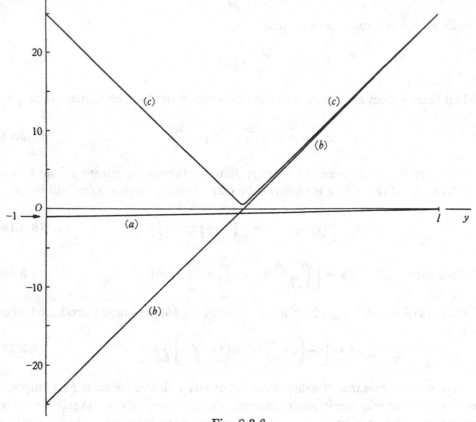

Fig. 8.3.6

member and absorbed at the other, then there is no point on the member where the displacement is zero. This is to be expected because energy could not be transmitted through the member without finite displacement at every point. Thus the node, which would be present in an undamped system, is replaced by a point of minimum amplitude, and the phase of the motion changes rapidly through nearly 180° in the vicinity of this point.

This example illustrates that, as with undamped systems, the chief use of Table 2 is in numerical work rather than in the derivation of analytical expressions; but with damped systems the arithmetic involves complex numbers. Further, however, when Table 2 is used with undamped systems, it is usually in order to build up a receptance expression from which a frequency equation is to be derived. With damped systems on the other hand, frequency equations cannot be found in this way and the table is used only to find specific receptances.

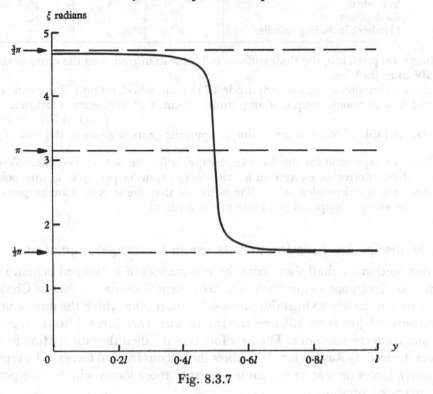

Fig. 8.3.7

EXAMPLES 8.3

1. The figure shows a light shaft with an epicyclic gear at one end and a light viscous damping device at the other. There are six planets carried on spindles attached to a flywheel.

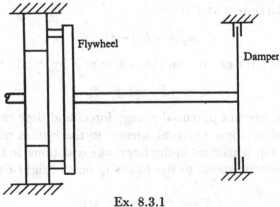

Ex. 8.3.1

The constants relating to the system are as follows:

	Mass (lb.)	Radius of gyration (in.)	Number of teeth
Sun wheel	4	2	48
Each planet	1	1	24
Flywheel, including spindles	10	3	—

The diametral pitch is 9, the shaft stiffness is 3×10^4 lb.in./rad. and the damper constant is 3×10^2 lb.in./rad./sec.

Find, as a function of ω, the amplitude of the sun wheel motion if the sun wheel is subjected to a harmonic torque of amplitude F lb.in. and frequency ω rad./sec.

(C.U.M.S.T. Pt II, 1954)

[Note. A table of receptances of simple epicyclic gears is given at the end of §3.2.]

2. Find an expression for the force in the spring for the last worked example of this section. Hence derive an expression for the energy transfer per cycle at any point and show that this is independent of y. Show further that the energy transfer per cycle is equal to the energy dissipated per cycle at the dashpot.

8.4 The dissipation function and its use in Lagrange's equations

In this section we shall show how the receptances of a damped system may be obtained by Lagrange's equations. The treatment is similar to that of Chapter 3; but it is now necessary to introduce a new function from which the damping forces may be derived, just as the stiffness and inertia forces are derived from the potential and kinetic energy functions. The new function is called the 'dissipation function' and was devised by Rayleigh.† It enables the viscous friction forces to be separated from other forces present in the same way that those forces which have potential were separated previously.

Consider a system with generalized co-ordinates $q_1, q_2, ..., q_n$. Suppose that the system is undamped except for the presence of a single dashpot which is attached between two points Y and Z of the system. Let the displacements of Y and Z, along the direction YZ, be y and z and let the dashpot have a constant b so that the force which it exerts on the system at Y is

$$R_y = -b(\dot{y} - \dot{z}). \tag{8.4.1}$$

In the same way, the force exerted on the system at Z by the dashpot is

$$R_z = -R_y = b(\dot{y} - \dot{z}). \tag{8.4.2}$$

The forces R_y and R_z are not potential energy forces and they can be regarded as if they were applied by some external agency to the system minus its damper. They would be taken into account in the Lagrange equations, in the form (3.1.12), by including their contributions to the forces Q on the right-hand side of those equations.

† *Theory of Sound* (1894), §81.

Since y and z are functions of the generalized co-ordinates, it follows that

$$\left.\begin{aligned} \dot{y} &= \frac{\partial y}{\partial q_1}\dot{q}_1 + \frac{\partial y}{\partial q_2}\dot{q}_2 + \ldots + \frac{\partial y}{\partial q_n}\dot{q}_n, \\ \dot{z} &= \frac{\partial z}{\partial q_1}\dot{q}_1 + \frac{\partial z}{\partial q_2}\dot{q}_2 + \ldots + \frac{\partial z}{\partial q_n}\dot{q}_n. \end{aligned}\right\} \tag{8.4.3}$$

The contribution of R_y and R_z to the generalized force Q_1 is obtained by considering the work which they do in a virtual displacement δq_1. This is

$$\delta W_1 = R_y\,\delta y + R_z\,\delta z = R_y(\delta y - \delta z), \tag{8.4.4}$$

where δy and δz may be written

$$\delta y = \frac{\partial y}{\partial q_1}\delta q_1, \quad \delta z = \frac{\partial z}{\partial q_1}\delta q_1. \tag{8.4.5}$$

The component $(Q_1)_D$ of Q_1, which is due to the damper, is therefore

$$(Q_1)_D = \frac{\delta W_1}{\delta q_1} = R_y\left(\frac{\partial y}{\partial q_1} - \frac{\partial z}{\partial q_1}\right). \tag{8.4.6}$$

This expression, in conjunction with equations (8.4.1) and (8.4.3) gives

$$(Q_1)_D = \left[-b\left(\frac{\partial y}{\partial q_1} - \frac{\partial z}{\partial q_1}\right)^2\right]\dot{q}_1 + \left[-b\left(\frac{\partial y}{\partial q_1} - \frac{\partial z}{\partial q_1}\right)\left(\frac{\partial y}{\partial q_2} - \frac{\partial z}{\partial q_2}\right)\right]\dot{q}_2 + \ldots$$
$$+ \left[-b\left(\frac{\partial y}{\partial q_1} - \frac{\partial z}{\partial q_1}\right)\left(\frac{\partial y}{\partial q_n} - \frac{\partial z}{\partial q_n}\right)\right]\dot{q}_n. \tag{8.4.7}$$

Similar expressions may be found for the forces $(Q_2)_D$, $(Q_3)_D$, etc.

The expressions within the square brackets have a constant value for small displacements, the restriction being similar to that which was imposed previously in order that the coefficients a and c might be treated as constants. If the quantities in question are denoted by $-b_{11}$, $-b_{12}$, etc., then the set of equations of which (8.4.7) is a member takes the form

$$\left.\begin{aligned} (Q_1)_D &= -(b_{11}\dot{q}_1 + b_{12}\dot{q}_2 + \ldots + b_{1n}\dot{q}_n), \\ (Q_2)_D &= -(b_{21}\dot{q}_1 + b_{22}\dot{q}_2 + \ldots + b_{2n}\dot{q}_n), \\ &\ldots\ldots\ldots\ldots\ldots\ldots\ldots\ldots\ldots\ldots\ldots\ldots\ldots\ldots\ldots\ldots \\ (Q_n)_D &= -(b_{n1}\dot{q}_1 + b_{n2}\dot{q}_2 + \ldots + b_{nn}\dot{q}_n). \end{aligned}\right\} \tag{8.4.8}$$

It will be seen from the form of equation (8.4.7) that the constants b are such that, in equations (8.4.8)

$$b_{rs} = b_{sr}. \tag{8.4.9}$$

Equations (8.4.8) have been derived for a system containing a single viscous damper. Evidently, if a number of such dampers were present, then similar sets of equations would hold for each. The Lagrange equations for the whole system could then be formed by adding all the separate contributions to the forces $(Q)_D$. If this were done the final coefficients b would still obey equation (8.4.9) since they would be the sums of terms, all of which obey that relation.

The system of fig. 8.4.1 illustrates these ideas. It has two generalized co-ordinates q_1 and q_2 as indicated and it contains two damping devices, with constants b_1 and b_2. The damper whose constant is b_1 exerts a torque of magnitude $-b_1\dot{q}_2$ in the direction of the angular displacement q_2. The other damper exerts a force $b_2(\dot{q}_1 + r\dot{q}_2)$ which opposes closure of the gap AB. For a displacement δq_1, during which q_2 remains fixed, the virtual work done by the damper forces is

$$\delta W_1 = -b_2(\dot{q}_1 + r\dot{q}_2)\,\delta q_1 \qquad (8.4.10)$$

so that
$$(Q_1)_D = -b_2(\dot{q}_1 + r\dot{q}_2). \qquad (8.4.11)$$

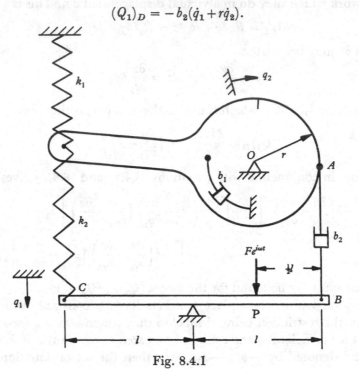

Fig. 8.4.1

The coefficients of this expression are such that

$$b_{11} = b_2, \quad b_{12} = rb_2. \qquad (8.4.12)$$

Again, for a displacement δq_2, in which q_1 remains constant, the virtual work of the damping forces is
$$\delta W_2 = -b_1\dot{q}_2\delta q_2 - b_2(\dot{q}_1 + r\dot{q}_2)\, r\,\delta q_2 \qquad (8.4.13)$$

so that
$$(Q_2)_D = -rb_2\dot{q}_1 - (b_1 + r^2 b_2)\,\dot{q}_2. \qquad (8.4.14)$$

The coefficients of this are
$$b_{21} = rb_2 = b_{12}, \quad b_{22} = b_1 + r^2 b_2. \qquad (8.4.15)$$

We shall now introduce, into the discussion of the general case, the 'dissipation function'; it will be denoted by the symbol D. This function is of the second degree in the generalized velocities and it is equal to half the rate at which energy is dissipated in the dampers.

If one damper only is present in a system, the dissipation function is given by

$$-2D = R_y\dot{y} + R_z\dot{z} = R_y(\dot{y} - \dot{z}) \qquad (8.4.16)$$

and, by equation (8.4.1), this becomes

$$-2D = -b(\dot{y}-\dot{z})^2. \tag{8.4.17}$$

Now if the expressions (8.4.3) are substituted for $\dot{y}$ and $\dot{z}$, it is found that

$$-2D = -b\left[\left(\frac{\partial y}{\partial q_1}-\frac{\partial z}{\partial q_1}\right)^2 \dot{q}_1^2 + \left(\frac{\partial y}{\partial q_2}-\frac{\partial z}{\partial q_2}\right)^2 \dot{q}_2^2 + \dots + \left(\frac{\partial y}{\partial q_n}-\frac{\partial z}{\partial q_n}\right)^2 \dot{q}_n^2\right.$$
$$\left.+ 2\left(\frac{\partial y}{\partial q_1}-\frac{\partial z}{\partial q_1}\right)\left(\frac{\partial y}{\partial q_2}-\frac{\partial z}{\partial q_2}\right)\dot{q}_1\dot{q}_2 + \dots\right]. \tag{8.4.18}$$

If the coefficients of the various terms of this expression are compared with those in equation (8.4.7), it is seen that the present function may be written

$$2D = b_{11}\dot{q}_1^2 + b_{22}\dot{q}_2^2 + \dots + b_{nn}\dot{q}_n^2 + 2b_{12}\dot{q}_1\dot{q}_2 + \dots. \tag{8.4.19}$$

Further, as in the previous case, the form of this can be seen to be unaltered if a large number of dampers are present instead of a single one.

Now the rth generalized damping force $(Q_r)_D$ may be formed from D in a similar way to that in which the generalized stiffness forces are formed from V; thus

$$(Q_r)_D = -\frac{\partial D}{\partial \dot{q}_r}. \tag{8.4.20}$$

In the example discussed above,

$$2D = b_1\dot{q}_2^2 + b_2(\dot{q}_1 + r\dot{q}_2)^2 \tag{8.4.21}$$

and by applying equation (8.4.20) it is deduced that

$$(Q_1)_D = -b_2(\dot{q}_1 + r\dot{q}_2), \quad (Q_2)_D = -rb_2\dot{q}_1 - (b_1 + r^2 b_2)\dot{q}_2. \tag{8.4.22}$$

These will be seen to agree with the previous expressions.

The generalized force of equation (8.4.20) can be included in Lagrange's equations. That is to say, for a system with viscous damping,

$$\frac{d}{dt}\left(\frac{\partial T}{\partial \dot{q}_r}\right) + \frac{\partial D}{\partial \dot{q}_r} + \frac{\partial V}{\partial q_r} = Q_r \quad (r = 1, 2, \dots, n), \tag{8.4.23}$$

where Q_r now contains only those forces which are not derivable either from a potential function, or from the dissipation function.

The equations of motion for the system of fig. 8.4.1 can now be formed by using equations (8.4.23). If the mass of the thin rigid rod CB is M and if the moment of inertia of the upper body, about its support O, is I then the three energy functions are

$$\left.\begin{aligned}
2T &= \frac{Ml^2}{3}\left(\frac{\dot{q}_1}{l}\right)^2 + I\dot{q}_2^2 = \frac{M}{3}\dot{q}_1^2 + I\dot{q}_2^2, \\
2D &= b_2\dot{q}_1^2 + (b_1 + r^2 b_2)\dot{q}_2^2 + 2rb_2\dot{q}_1\dot{q}_2, \\
2V &= k_1[(2l-r)q_2]^2 + k_2[q_1 + (2l-r)q_2]^2 \\
&= k_2q_1^2 + (k_1 + k_2)(2l-r)^2 q_2^2 + 2k_2(2l-r)q_1q_2.
\end{aligned}\right\} \tag{8.4.24}$$

If, further, a harmonic force $Fe^{i\omega t}$ acts on the system at P then

$$\left.\begin{aligned}
Q_1 &= (Fe^{i\omega t})\frac{\partial}{\partial q_1}\left(-\frac{q_1}{2}\right) = -\frac{F}{2}e^{i\omega t},\\
Q_2 &= (Fe^{i\omega t})\frac{\partial}{\partial q_2}\left(-\frac{q_1}{2}\right) = 0.
\end{aligned}\right\} \quad (8.4.25)$$

Using all these expressions, we finally obtain

$$\left.\begin{aligned}
\frac{M}{3}\ddot{q}_1 + b_2\dot{q}_1 + k_2 q_1 + rb_2\dot{q}_2 + k_2(2l-r)q_2 &= -\frac{F}{2}e^{i\omega t},\\
rb_2\dot{q}_1 + k_2(2l-r)q_1 + I\ddot{q}_2 + (b_1+r^2 b_2)\dot{q}_2 + (k_1+k_2)(2l-r)^2 q_2 &= 0.
\end{aligned}\right\} \quad (8.4.26)$$

EXAMPLE 8.4

1. Find the equations of motion of the system of fig. 8.3.1 using Lagrange's equations with generalized co-ordinates q_1 and q_2 such that

$$q_1 = x_1 + 2x_2,$$
$$q_2 = x_1 - x_2,$$

x_1 and x_2 being the co-ordinates shown in the figure.

8.5 The derivation of receptances using the Lagrangian equations

Let a viscously damped system have generalized co-ordinates $q_1, q_2, \ldots, q_n$ and let the energy functions for the system be

$$\left.\begin{aligned}
2T &= a_{11}\dot{q}_1^2 + a_{22}\dot{q}_2^2 + \ldots + a_{nn}\dot{q}_n^2 + 2a_{12}\dot{q}_1\dot{q}_2 + \ldots,\\
2D &= b_{11}\dot{q}_1^2 + b_{22}\dot{q}_2^2 + \ldots + b_{nn}\dot{q}_n^2 + 2b_{12}\dot{q}_1\dot{q}_2 + \ldots,\\
2V &= c_{11}q_1^2 + c_{22}q_2^2 + \ldots + c_{nn}q_n^2 + 2c_{12}q_1 q_2 + \ldots.
\end{aligned}\right\} \quad (8.5.1)$$

Suppose that the system is acted on by forces $Q_1, Q_2, \ldots, Q_n$ which vary harmonically and are given by

$$Q_1 = \Phi_1 e^{i\omega t}, \quad Q_2 = \Phi_2 e^{i\omega t}, \quad \ldots, \quad Q_n = \Phi_n e^{i\omega t}. \quad (8.5.2)$$

If the Lagrangian method is now used to set up the equations of motion for the system, it is found that

$$\left.\begin{aligned}
(a_{11}\ddot{q}_1 + b_{11}\dot{q}_1 + c_{11}q_1) &+ (a_{12}\ddot{q}_2 + b_{12}\dot{q}_2 + c_{12}q_2) + \ldots\\
&+ (a_{1n}\ddot{q}_n + b_{1n}\dot{q}_n + c_{1n}q_n) = \Phi_1 e^{i\omega t},\\
(a_{21}\ddot{q}_1 + b_{21}\dot{q}_1 + c_{21}q_1) &+ (a_{22}\ddot{q}_2 + b_{22}\dot{q}_2 + c_{22}q_2) + \ldots\\
&+ (a_{2n}\ddot{q}_n + b_{2n}\dot{q}_n + c_{2n}q_n) = \Phi_2 e^{i\omega t},\\
&\cdots\cdots\cdots\cdots\cdots\cdots\cdots\cdots\cdots\cdots\cdots\cdots\cdots\cdots\cdots\\
(a_{n1}\ddot{q}_1 + b_{n1}\dot{q}_1 + c_{n1}q_1) &+ (a_{n2}\ddot{q}_2 + b_{n2}\dot{q}_2 + c_{n2}q_2) + \ldots\\
&+ (a_{nn}\ddot{q}_n + b_{nn}\dot{q}_n + c_{nn}q_n) = \Phi_n e^{i\omega t}.
\end{aligned}\right\} \quad (8.5.3)$$

The trial solution for steady motion, namely

$$q_1 = \Psi_1 e^{i\omega t}, \quad q_2 = \Psi_2 e^{i\omega t}, \quad \ldots, \quad q_n = \Psi_n e^{i\omega t} \quad (8.5.4)$$

may now be substituted into these equations. In this way the amplitudes Ψ may be found from the n algebraic equations

$$
\begin{aligned}
&(c_{11} - \omega^2 a_{11} + i\omega b_{11})\,\Psi_1 + (c_{12} - \omega^2 a_{12} + i\omega b_{12})\,\Psi_2 + \ldots \\
&\qquad\qquad + (c_{1n} - \omega^2 a_{1n} + i\omega b_{1n})\,\Psi_n = \Phi_1, \\
&(c_{21} - \omega^2 a_{21} + i\omega b_{21})\,\Psi_1 + (c_{22} - \omega^2 a_{22} + i\omega b_{22})\,\Psi_2 + \ldots \\
&\qquad\qquad + (c_{2n} - \omega^2 a_{2n} + i\omega b_{2n})\,\Psi_n = \Phi_2, \\
&\cdots\cdots\cdots\cdots\cdots\cdots\cdots\cdots\cdots\cdots\cdots\cdots\cdots\cdots\cdots\cdots\cdots\cdots\cdots \\
&(c_{n1} - \omega^2 a_{n1} + i\omega b_{n1})\,\Psi_1 + (c_{n2} - \omega^2 a_{n2} + i\omega b_{n2})\,\Psi_2 + \ldots \\
&\qquad\qquad + (c_{nn} - \omega^2 a_{nn} + i\omega b_{nn})\,\Psi_n = \Phi_n.
\end{aligned} \quad (8.5.5)
$$

These equations are comparable with equations (3.2.2). But the presence of the damping renders the coefficients of the amplitudes Ψ complex and introduces multiples of ω whereas only terms in ω^2 were present previously.

The equations may be solved by the same method as before, the full solution being obtained by adding a set of special solutions, each one of which corresponds to the vanishing of all but one of the quantities Φ on the right-hand side. Thus if

$$
\begin{aligned}
\Phi_1 = \Phi_2 = \ldots = \Phi_{s-1} = \Phi_{s+1} = \ldots = \Phi_n = 0, \\
\Phi_s \neq 0,
\end{aligned} \quad (8.5.6)
$$

then the solution of equations (8.5.5) for Ψ_r is

$$
\begin{aligned}
\Psi_r = \alpha_{rs}\Phi_s, \\
q_r = \alpha_{rs}\Phi_s e^{i\omega t},
\end{aligned} \quad (8.5.7)
$$

or

where the receptance α_{rs} is given by

$$\alpha_{rs} = (-1)^{r+s}\frac{\Delta_{rs}}{\Delta}. \quad (8.5.8)$$

In this expression, Δ represents the determinant of the coefficients; that is

$$\Delta = \begin{vmatrix} c_{11} - \omega^2 a_{11} + i\omega b_{11} & c_{12} - \omega^2 a_{12} + i\omega b_{12} & \ldots & c_{1n} - \omega^2 a_{1n} + i\omega b_{1n} \\ c_{21} - \omega^2 a_{21} + i\omega b_{21} & c_{22} - \omega^2 a_{22} + i\omega b_{22} & \ldots & c_{2n} - \omega^2 a_{2n} + i\omega b_{2n} \\ \cdots\cdots\cdots\cdots\cdots\cdots\cdots\cdots\cdots\cdots\cdots\cdots\cdots\cdots\cdots\cdots \\ c_{n1} - \omega^2 a_{n1} + i\omega b_{n1} & c_{n2} - \omega^2 a_{n2} + i\omega b_{n2} & \ldots & c_{nn} - \omega^2 a_{nn} + i\omega b_{nn} \end{vmatrix} \quad (8.5.9)$$

and Δ_{rs} is the determinant which may be derived from Δ by omitting the rth column and sth row.

As with undamped systems, corresponding rows and columns of the determinant Δ are identical. It follows that the reciprocal relation

$$\alpha_{rs} = \alpha_{sr}. \quad (8.5.10)$$

still holds. The relation is an equality of complex numbers and implies that the real and imaginary parts are separately equal.

If now the solutions for Φ_1, Φ_2, etc., are found in the same way as that for Φ_s, then the total motion due to the presence of all the forcing terms is

$$\left.\begin{aligned}
q_1 &= [\alpha_{11}\Phi_1 + \alpha_{12}\Phi_2 + \ldots + \alpha_{1n}\Phi_n]\, e^{i\omega t}, \\
q_2 &= [\alpha_{21}\Phi_1 + \alpha_{22}\Phi_2 + \ldots + \alpha_{2n}\Phi_n]\, e^{i\omega t}, \\
&\cdots\cdots\cdots\cdots\cdots\cdots\cdots\cdots\cdots\cdots\cdots\cdots \\
q_n &= [\alpha_{n1}\Phi_1 + \alpha_{n2}\Phi_2 + \ldots + \alpha_{nn}\Phi_n]\, e^{i\omega t}.
\end{aligned}\right\} \quad (8.5.11)$$

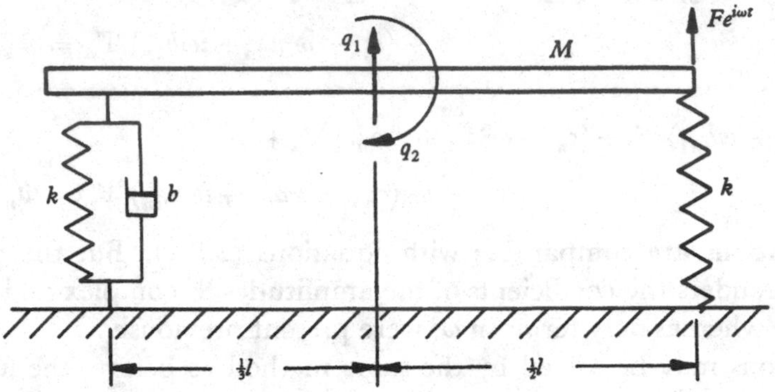

Fig. 8.5.1

As an example of the calculation of receptances of damped systems consider the arrangement shown in fig. 8.5.1. The rigid uniform beam, which is of length l and mass M, is supported on two springs of stiffness k. The springs are distant $\tfrac{1}{2}l$ and $\tfrac{1}{3}l$ from the mid-point of the beam. The spring which is nearer to the centre has a viscous damper of constant b fixed across it but the other spring is undamped. A force $F e^{i\omega t}$ acts vertically above the undamped spring. Suitable generalized co-ordinates are q_1, the vertical displacement upward of the mid-point, and q_2, the rotation. The three energy functions are

$$\left.\begin{aligned}
2T &= M\dot{q}_1^2 + \frac{Ml^2}{12}\dot{q}_2^2, \\
2D &= b\left(\dot{q}_1 + \frac{l}{3}\dot{q}_2\right)^2, \\
2V &= k\left(q_1 + \frac{l}{3}q_2\right)^2 + k\left(q_1 - \frac{l}{2}q_2\right)^2.
\end{aligned}\right\} \quad (8.5.12)$$

The energy coefficients are therefore

$$\left.\begin{aligned}
a_{11} &= M, & a_{22} &= \frac{Ml^2}{12}, & a_{12} &= 0; \\
b_{11} &= b, & b_{22} &= \frac{bl^2}{9}, & b_{12} &= \frac{bl}{3}; \\
c_{11} &= 2k, & c_{22} &= \frac{13kl^2}{36}, & c_{12} &= -\frac{kl}{6}.
\end{aligned}\right\} \quad (8.5.13)$$

The generalized forces Q_1 and Q_2 are

$$\left.\begin{aligned}
Q_1 &= Fe^{i\omega t}\frac{\partial}{\partial q_1}\left(q_1 - \frac{l}{2}q_2\right) = Fe^{i\omega t},\\
Q_2 &= Fe^{i\omega t}\frac{\partial}{\partial q_2}\left(q_1 - \frac{l}{2}q_2\right) = -\frac{Fl}{2}e^{i\omega t}.
\end{aligned}\right\} \tag{8.5.14}$$

The determinant Δ of the coefficients may be written

$$\Delta = \begin{vmatrix} 2k - M\omega^2 + ib\omega & -\dfrac{kl}{6} + i\dfrac{bl}{3}\omega \\[2ex] -\dfrac{kl}{6} + i\dfrac{bl}{3}\omega & \dfrac{13kl^2}{36} - \dfrac{Ml^2}{12}\omega^2 + i\dfrac{bl^2}{9}\omega \end{vmatrix} \tag{8.5.15}$$

and from this the three receptances may be formed immediately. They are

$$\left.\begin{aligned}
\alpha_{11} &= \frac{\dfrac{13kl^2}{36} - \dfrac{Ml^2}{12}\omega^2 + i\dfrac{bl^2}{9}\omega}{\Delta},\\
\alpha_{22} &= \frac{2k - M\omega^2 + ib\omega}{\Delta},\\
\alpha_{12} = \alpha_{21} &= \frac{\dfrac{kl}{6} - i\dfrac{bl}{3}\omega}{\Delta}.
\end{aligned}\right\} \tag{8.5.16}$$

The total responses to the exciting force can now be written in the form

$$\left.\begin{aligned}
q_1 &= \alpha_{11}Q_1 + \alpha_{12}Q_2,\\
q_2 &= \alpha_{21}Q_1 + \alpha_{22}Q_2,
\end{aligned}\right\} \tag{8.5.17}$$

and, in fact, are given by

$$\left.\begin{aligned}
q_1 &= \left[\frac{10k - 3M\omega^2 + 10ib\omega}{36\Delta}\right]Fl^2 e^{i\omega t},\\
q_2 &= -\left[\frac{5k - 3M\omega^2 + 5ib\omega}{6\Delta}\right]Fl\, e^{i\omega t}.
\end{aligned}\right\} \tag{8.5.18}$$

Returning to the general expression (8.5.8) for α_{rs}, it will be seen that the denominator is of degree $2n$ in ω as in the undamped case, but that odd powers of ω are now present and these terms have imaginary coefficients. It follows that, in general, the equation

$$\Delta = 0 \tag{8.5.19}$$

will now have complex roots. If, for instance, the expression for Δ in the above example is expanded and equated to zero, it is found that

$$\left[\frac{M^2l^2}{12}\omega^4 - \frac{19kMl^2}{36}\omega^2 + \frac{25k^2l^2}{36}\right] - i\left[\frac{7Mbl^2}{36}\omega^3 - \frac{25kbl^2}{36}\omega\right] = 0. \tag{8.5.20}$$

It is usually only of interest to discuss forced vibration for real frequencies. Evidently, when damping is present, a real exciting frequency will not make Δ zero so that an infinite ratio of displacement to force is never achieved. Similarly Δ_{rr} will not vanish for a real frequency so that a direct receptance will never be zero.

EXAMPLES 8.5

1. Find the direct receptance at x_1 for the system shown, using the method of this section.

Hence show that, due to the presence of the dashpot, the amplitude of the driving point cannot be zero when the system is excited at the co-ordinate x_1 in the manner shown in the diagram.

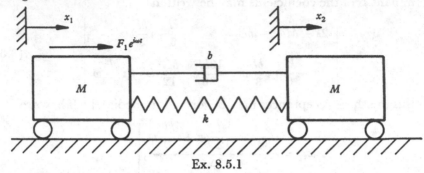

Ex. 8.5.1

2. Show that if the system of Ex. 8.5.1 has a high Q-factor (i.e. if $k/b\omega_2$ is very large) then

(i) if $\omega = \omega_2$, the response to the given forcing is large, and

(ii) if ω is equal to the anti-resonance frequency of the system in the absence of its damping, the response at x_1 is very small.

Find approximate values for the receptances at x_1 in these two cases.

8.6 The effects of damping on the principal modes

When a receptance α_{rs} has been found in closed form for an undamped system, it can be expanded as a set of partial fractions. We used this treatment in § 3.4 to introduce the idea of principal modes. When viscous damping is present, however, this type of analysis becomes much less strightforward although it can still be used when the damping is 'hysteretic' as will be described in the next chapter. For the present, we shall develop the theory differently.

The principal modes and principal co-ordinates of a damped system have not yet been defined here. A generalization of the previous definitions is possible but has not been found useful. It will be necessary still to refer to the principal modes, the principal co-ordinates and natural frequencies which a damped system would have if the damping were removed. These will still be referred to as the principal modes, the principal co-ordinates and the natural frequencies of the damped system.

Consider again a system in which a single viscous damper with coefficient b is attached between two points Y and Z. The previous discussion of this problem may be developed in terms of the principal co-ordinates $p_1, p_2, ..., p_n$ instead of the co-ordinates q. The displacements y and z of Y and Z will then be

$$y = \frac{\partial y}{\partial p_1} p_1 + \frac{\partial y}{\partial p_2} p_2 + ... + \frac{\partial y}{\partial p_n} p_n,$$
$$z = \frac{\partial z}{\partial p_1} p_1 + \frac{\partial z}{\partial p_2} p_2 + ... + \frac{\partial z}{\partial p_n} p_n. \qquad (8.6.1)$$

The force R_y which the damper exerts at Y will be

$$R_y = -b\left[\left(\frac{\partial y}{\partial p_1}-\frac{\partial z}{\partial p_1}\right)\dot{p}_1 + \left(\frac{\partial y}{\partial p_2}-\frac{\partial z}{\partial p_2}\right)\dot{p}_2 + \dots + \left(\frac{\partial y}{\partial p_n}-\frac{\partial z}{\partial p_n}\right)\dot{p}_n\right] \quad (8.6.2)$$

if y and z are both taken to be positive in the same direction. The force R_z which the damper exerts at Z will be equal to $-R_y$ and, if this force acting between Y and Z is included in the equations of motion, these become

$$\left.\begin{aligned} a_1\ddot{p}_1 + c_1 p_1 &= P_1 + R_y\left(\frac{\partial y}{\partial p_1}-\frac{\partial z}{\partial p_1}\right), \\ a_2\ddot{p}_2 + c_2 p_2 &= P_2 + R_y\left(\frac{\partial y}{\partial p_2}-\frac{\partial z}{\partial p_2}\right), \\ \dots\dots\dots\dots\dots\dots\dots\dots\dots\dots\dots\dots\dots \\ a_n\ddot{p}_n + c_n p_n &= P_n + R_y\left(\frac{\partial y}{\partial p_n}-\frac{\partial z}{\partial p_n}\right). \end{aligned}\right\} \quad (8.6.3)$$

In these equations $P_1, P_2, \dots, P_n$ represent the generalized forces which correspond to the principal co-ordinates. R_y may now be eliminated from equations (8.6.2) and (8.6.3) to give, for the rth equation,

$$a_r\ddot{p}_r + c_r p_r = P_r - b\left[\left(\frac{\partial y}{\partial p_1}-\frac{\partial z}{\partial p_1}\right)\left(\frac{\partial y}{\partial p_r}-\frac{\partial z}{\partial p_r}\right)\dot{p}_1 + \left(\frac{\partial y}{\partial p_2}-\frac{\partial z}{\partial p_2}\right)\left(\frac{\partial y}{\partial p_r}-\frac{\partial z}{\partial p_r}\right)\dot{p}_2 + \dots \right.$$
$$\left. + \left(\frac{\partial y}{\partial p_n}-\frac{\partial z}{\partial p_n}\right)\left(\frac{\partial y}{\partial p_r}-\frac{\partial z}{\partial p_r}\right)\dot{p}_n\right]. \quad (8.6.4)$$

If the system has a number of dampers, they can all be treated in this way; thus a set of damping coefficients can be formed by adding up the terms due to the various damping elements. We shall write, for the damping terms in the p_1 equation,

$$-[b_{11}\dot{p}_1 + b_{12}\dot{p}_2 + \dots + b_{1n}\dot{p}_n] \quad (8.6.5)$$

for those in the p_2 equation

$$-[b_{21}\dot{p}_1 + b_{22}\dot{p}_2 + \dots + b_{2n}\dot{p}_n] \quad (8.6.6)$$

and so on. By the same argument as before

$$b_{rs} = b_{sr}. \quad (8.6.7)$$

The full equations of motion may now be written in terms of the principal co-ordinates

$$\left.\begin{aligned} a_1\ddot{p}_1 + b_{11}\dot{p}_1 + c_1 p_1 + b_{12}\dot{p}_2 + \dots + b_{1n}\dot{p}_n &= P_1, \\ b_{21}\dot{p}_1 + a_2\ddot{p}_2 + b_{22}\dot{p}_2 + c_2 p_2 + \dots + b_{2n}\dot{p}_n &= P_2, \\ \dots\dots\dots\dots\dots\dots\dots\dots\dots\dots\dots\dots\dots\dots \\ b_{n1}\dot{p}_1 + b_{n2}\dot{p}_2 + \dots + a_n\ddot{p}_n + b_{nn}\dot{p}_n + c_n p_n &= P_n. \end{aligned}\right\} \quad (8.6.8)$$

It will be seen that, whereas for an undamped system the principal co-ordinates provide a set of equations which are all independent, so that they may be solved directly without further manipulation, these co-ordinates do not generally have this advantage for the damped system. The fact that all the co-ordinates occur in each equation means that, in general, if only one of the forces P is present, say P_r, then motion will take place at all the co-ordinates and not merely at p_r. That is to say, motion will occur in all the principal modes.

The above derivation of the p equations might have been dispensed with on the grounds that the co-ordinates p can be treated as a special set of co-ordinates q, so that the result could be quoted from the previous section. An outline of the derivation has, however, been included here because it indicates why the final equations contain the terms containing b_{12}, b_{13}, ..., etc.

Another aspect of the matter may be seen in terms of the discussion of § 3.6. It was there shown that the principal co-ordinates p were linear functions of the generalized co-ordinates q and that the choice of coefficients was just sufficient to enable all the product terms in the series for T and V to be made equal to zero. The introduction of the dissipation function does not bring with it any more choice of coefficients and it is therefore not possible to arrange for the vanishing of the product terms in the series for D. It is in fact possible to select a new set of co-ordinates for which there are no product terms in the T and D expansions but which will contain them in the V expansions; again yet another set may be chosen based on the D and V expansions. These other sets of co-ordinates are not known to be of engineering interest at present.

There are certain special cases when the product terms in all three expansions are zero so that the equations governing the principal co-ordinates remain independent of each other. This arises in particular if all the potential energy is stored in springs and if each spring has a damper connected across it whose constant is proportional to the spring stiffness. If this is so, the coefficients b in the dissipation function D are all proportional to the corresponding coefficients c of the potential energy function V and there are no product terms in either D or V if the principal co-ordinates are used. This state of affairs arises if the damping forces are due to imperfect elasticity of the material of which the springs are made, though it must be mentioned immediately that damping from this cause is unlikely to obey a viscous law with much precision.

Consider the problem of the vibration of a uniform beam which is simply supported at its ends. Provided there are no damping forces at the supports it is reasonable to suppose that the damping forces will be spread uniformly along the beam, whether they arise from air resistance or from energy loss within the material. If it is assumed that the damping force per unit length of beam is proportional to the velocity and is of magnitude b, then the force per unit length at a distance x along the beam is

$$b\left[p_1 \sin \frac{\pi x}{l} + p_2 \sin \frac{2\pi x}{l} + ... \right]. \tag{8.6.9}$$

The damping term in the rth equation is

$$\int_0^l \left\{ \sin \frac{r\pi x}{l} \cdot b\left[p_1 \sin \frac{\pi x}{l} + p_2 \sin \frac{2\pi x}{l} + ... \right] \right\} dx, \tag{8.6.10}$$

which reduces to $\frac{1}{2}bl\dot{p}_r$. The equations for the beam vibration are therefore

$$
\left.
\begin{aligned}
\frac{A\rho l}{2}\ddot{p}_1 + \frac{bl}{2}\dot{p}_1 + \frac{\pi^4 EI}{2l^3}p_1 &= P_1, \\
\frac{A\rho l}{2}\ddot{p}_2 + \frac{bl}{2}\dot{p}_2 + \frac{16\pi^4 EI}{2l^3}p_2 &= P_2, \\
\cdots\cdots\cdots\cdots\cdots\cdots\cdots\cdots\cdots
\end{aligned}
\right\}
\tag{8.6.11}
$$

Each of these equations contains one co-ordinate only and each may be solved independently as is the case when there is no damping.

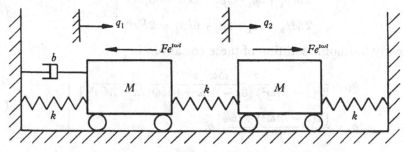

Fig. 8.6.1

Consider now an example in which the damping forces are not of this special kind. The system of fig. 8.6.1 would be symmetrical if it were not for the damper at the left-hand end. The principal modes of the undamped system consist of

(i) an in-phase, equal-amplitude motion of the two masses for which the frequency is given by $\omega_1^2 = k/M$, and

(ii) an antiphase equal-amplitude motion for which $\omega_2^2 = 3k/M$.

We shall take unit value of p_1 to correspond to unit displacement of both masses to the right and unit value of p_2 to correspond to unit displacement of each of the masses away from the centre. Thus the generalized co-ordinates q_1 and q_2, indicated in the figure, are related to the principal co-ordinates by

$$
\left.
\begin{aligned}
q_1 &= p_1 - p_2, \\
q_2 &= p_1 + p_2.
\end{aligned}
\right\}
\tag{8.6.12}
$$

The energy functions for the system are

$$
\left.
\begin{aligned}
2T &= M(\dot{q}_1^2 + \dot{q}_2^2) = 2M\dot{p}_1^2 + 2M\dot{p}_2^2, \\
2D &= b\dot{q}_1^2 = b\dot{p}_1^2 - 2b\dot{p}_1\dot{p}_2 + b\dot{p}_2^2, \\
2V &= kq_1^2 + k(q_2 - q_1)^2 + kq_2^2 = 2kp_1^2 + 6kp_2^2.
\end{aligned}
\right\}
\tag{8.6.13}
$$

The equations of motion may now be set up by means of Lagrange's equations. They are

$$
\left.
\begin{aligned}
2M\ddot{p}_1 + b\dot{p}_1 - b\dot{p}_2 + 2kp_1 &= P_1, \\
2M\ddot{p}_2 - b\dot{p}_1 + b\dot{p}_2 + 6kp_2 &= P_2.
\end{aligned}
\right\}
\tag{8.6.14}
$$

Let equal and opposite forces $Fe^{i\omega t}$ be applied to the two masses so that, if the damper were not present, motion would be produced in the second principal mode only. Thus the forces P_1 and P_2 are

$$P_1 = (-Fe^{i\omega t})\frac{\partial q_1}{\partial p_1} + (Fe^{i\omega t})\frac{\partial q_2}{\partial p_1} = 0, \left.\rule{0pt}{2.2em}\right\}$$

$$P_2 = (-Fe^{i\omega t})\frac{\partial q_1}{\partial p_2} + (Fe^{i\omega t})\frac{\partial q_2}{\partial p_2} = 2Fe^{i\omega t}. \quad (8.6.15)$$

When the damper is present the equations of motion are

$$2M\ddot{p}_1 + b\dot{p}_1 - b\dot{p}_2 + 2kp_1 = 0, \left.\rule{0pt}{1.6em}\right\}$$

$$2M\ddot{p}_2 - b\dot{p}_1 + b\dot{p}_2 + 6kp_2 = 2Fe^{i\omega t} \quad (8.6.16)$$

and the steady harmonic solution of these equations is

$$p_1 = \frac{ib\omega Fe^{i\omega t}}{2[(k-M\omega^2)(3k-M\omega^2) + ib\omega(2k-M\omega^2)]}, \left.\rule{0pt}{3em}\right\}$$

$$p_2 = \frac{[2(k-M\omega^2) + ib\omega]}{ib\omega}p_1. \quad (8.6.17)$$

The presence of the first-mode component in the motion arises from the lack of symmetry which is brought about by the damper.

8.7 The behaviour of systems in which the damping is small

It would be extremely tedious to undertake the arithmetical solution of a set of equations having the form (8.6.8), though it could be done by using, for example, a relaxation method. Fortunately it is possible to obtain, simply, approximate solutions to the equations if the damping forces are small compared with the stiffness and inertia forces. These conditions are met in practical problems because it is only when the damping is small that vibration is troublesome. An approximate solution is usually quite satisfactory because the damping constants themselves are rarely known with any accuracy and an exact solution would therefore be of little value. Indeed knowledge of the magnitude of damping in real systems is usually so incomplete that it is rarely possible to do more than estimate the order of magnitude of the vibration amplitude.

It will be sufficient to examine the algebraic problem when only one of the forces P is present; let this force be P_1. The equations of motion are thus

$$a_1\ddot{p}_1 + b_{11}\dot{p}_1 + c_1 p_1 + b_{12}\dot{p}_2 + \ldots + b_{1n}\dot{p}_n = P_1, \left.\rule{0pt}{1.6em}\right\}$$

$$b_{21}\dot{p}_1 + a_2\ddot{p}_2 + b_{22}\dot{p}_2 + c_2 p_2 + \ldots + b_{2n}\dot{p}_n = 0,$$

$$\cdots\cdots\cdots\cdots\cdots\cdots\cdots\cdots\cdots\cdots\cdots\cdots\cdots\cdots\cdots\cdots \quad (8.7.1)$$

$$b_{n1}\dot{p}_1 + b_{n2}\dot{p}_2 + \ldots + a_n\ddot{p}_n + b_{nn}\dot{p}_n + c_n p_n = 0.$$

If P_1 is harmonic, of the form $\Xi e^{i\omega t}$, and a steady solution is sought of the type

$$p_1 = \Pi_1 e^{i\omega t}, \quad p_2 = \Pi_2 e^{i\omega t}, \quad \ldots, \quad p_n = \Pi_n e^{i\omega t} \quad (8.7.2)$$

then the equations of motion give the following set of algebraic equations governing the amplitudes

$$(c_1 - \omega^2 a_1 + i\omega b_{11})\, \Pi_1 + i\omega[b_{12}\, \Pi_2 + b_{13}\, \Pi_3 + \ldots + b_{1n}\, \Pi_n] = \Xi,$$

$$i\omega b_{21}\, \Pi_1 + (c_2 - \omega^2 a_2 + i\omega b_{22})\, \Pi_2 + i\omega[b_{23}\, \Pi_3 + \ldots + b_{2n}\, \Pi_n] = 0,$$

$$\ldots\ldots\ldots\ldots\ldots\ldots\ldots\ldots\ldots\ldots\ldots\ldots\ldots\ldots\ldots\ldots\ldots\ldots$$

$$i\omega[b_{n1}\, \Pi_1 + b_{n2}\, \Pi_2 + \ldots + b_{n(n-1)}\, \Pi_{n-1}] + (c_n - \omega^2 a_n + i\omega b_{nn})\, \Pi_n = 0. \tag{8.7.3}$$

Consider now the behaviour of the system when it resonates so that $\omega^2 = c_1/a_1 = \omega_1^2$. If there were no damping, the response at p_1 would be infinite and all the other co-ordinates p would be zero. The presence of the damping must reduce the amplitude to a finite value though it may still be large compared with the value that it would have if ω were appreciably different from ω_1. The effects of the damping terms may be examined more fully by introducing first the coefficients $b_{11}, b_{22}, \ldots, b_{nn}$ only; when this is done the first equation becomes the same as that for a damped system having a single degree of freedom which we know to have an amplitude peak very close to the frequency $\omega = \omega_1$; the co-ordinates $p_2, p_3, \ldots, p_n$ in the other equations remain zero. The variation of p_1 is given by

$$p_1 = \Pi_1\, e^{i\omega t} = \frac{\Xi\, e^{i\omega t}}{i\omega_1 b_{11}}. \tag{8.7.4}$$

Now let the coefficients b_{12} and b_{21} be introduced. They couple the first and second principal co-ordinates so that, if p_1 is not zero, p_2 must have some finite value. In fact the ratio of the two co-ordinates is found to be

$$\frac{p_2}{p_1} = \frac{\Pi_2}{\Pi_1} = \frac{i\omega b_{12}}{c_2 - \omega_1^2 a_2 + i\omega_1 b_{22}}. \tag{8.7.5}$$

Now provided that the frequencies ω_1 and ω_2, corresponding to the first and second principal modes, are not very nearly equal, the term $(c_2 - \omega_1^2 a_2)$ in the denominator of equation (8.7.5) may be taken to be much greater than ωb_{12}; this is the condition that the damping is small. It follows that p_2/p_1 is itself small or, in other words, p_1 is very much larger than p_2. In the special case when $\omega_2 \doteqdot \omega_1$ (or $c_2 \doteqdot \omega_1^2 a_2$) however, p_1 and p_2 may be of the same order of magnitude although the damping is small.

The introduction of the terms $b_{13}, b_{14}, \ldots, b_{1n}$ will, similarly, produce small motions at the co-ordinates $p_3, p_4, \ldots, p_n$. But the presence of these motions does not affect the response at p_1 materially. Thus, when b_{12} and b_{21} only are present p_1 is given by

$$i\omega_1[b_{11} p_1 + b_{12} p_2] = \Xi\, e^{i\omega t} \tag{8.7.6}$$

in which $p_2 \ll p_1$. It follows that p_1 is still approximately equal to the value (8.7.4).

When the terms b_{23}, b_{24}, b_{34}, etc., are introduced there will be some change in the relative values of $p_2, p_3, \ldots, p_n$. These quantities, however, will always remain small compared with p_1 because of the relations of the form (8.7.5). Thus the lightly damped system at resonance may be treated approximately as a one-degree-of-freedom system by neglecting coupling terms; the values of the non-resonating modes may be estimated subsequently if necessary.

Now consider the motion when the exciting frequency is not resonant. If, again, all the damping terms are zero except $b_{11}, b_{22}, \ldots, b_{nn}$, then p_1 will be given by

$$p_1 = \Pi_1 e^{i\omega t} = \frac{\Xi\, e^{i\omega t}}{c_1 - \omega^2 a_1 + i\omega b_{11}}. \tag{8.7.7}$$

If the damping is small, the magnitude of $(c_1 - \omega^2 a_1)$ will be much larger than that of ωb_{11} so that

$$p_1 = \Pi_1 e^{i\omega t} \doteqdot \frac{\Xi\, e^{i\omega t}}{c_1 - \omega^2 a_1}. \tag{8.7.8}$$

This shows that the damping does not greatly affect the amplitude of p_1. If the terms b_{12} and b_{21} are now introduced, the ratio of p_2 to p_1 is again given by

$$\frac{p_2}{p_1} = \frac{\Pi_2}{\Pi_1} = \frac{i\omega b_{12}}{c_2 - \omega^2 a_2 + i\omega b_{22}}, \tag{8.7.9}$$

which is small except when $c_2 - \omega^2 a_2$ is very small. This would mean that the exciting frequency is very close to the second natural frequency; the second mode is not directly excited by the external forces but is excited through the coupling terms. This condition produces comparable amplitudes of p_2 and p_1. For the present we shall assume that

$$c_2 - \omega^2 a_2 \gg \omega b_{22} \tag{8.7.10}$$

so that p_2 is much smaller than p_1. The effects of introducing the other damping terms will be similar to that already discussed above for a resonant frequency.

Summing up the results of this discussion, we can say that when a single generalized force P_r acts on a lightly damped system the corresponding co-ordinate p_r will in general have a much larger amplitude than the others. If the exciting frequency is allowed to approach the natural frequency of the mode in question, the amplitude of that mode will become large and the amplitudes of the other modes will rise, maintaining the amplitude ratios approximately. The effects of a number of forces P_r, P_s etc. can be examined by superposition.

We have also seen that special cases arise (a) when a damped system has close natural frequencies and (b) when a generalized force P_r corresponding to one mode acts on a system with a frequency corresponding to another.

The fact that the amplitude of a mode at resonance is almost independent of the coupling with the other modes permits the definition of a Q-factor for each mode of a system. The maximum kinetic energy which is stored as a result of resonant motion in the rth mode is $\frac{1}{2} a_r \Pi_r^2 \omega_r^2$. The energy dissipated per cycle due to this mode only, neglecting contributions from the damping coupling terms, is

$$E = \int_0^{2\pi/\omega_r} b_{rr}\, \dot{p}_r^2\, dt = b_{rr} \pi \Pi_r^2 \omega_r. \tag{8.7.11}$$

If the Q-factor is now defined through the ratio

$$\frac{\text{Maximum kinetic energy stored}}{\text{Energy dissipated per cycle}} = \frac{\frac{1}{2} a_r \Pi_r^2 \omega_r^2}{b_{rr} \pi \Pi_r^2 \omega_r} = \frac{1}{2\pi}\left(\frac{\omega_r a_r}{b_{rr}}\right) \tag{8.7.12}$$

then, by analogy with the definition of § 8.2,

$$Q = \frac{\omega_r a_r}{b_{rr}} = \frac{c_r}{\omega_r b_{rr}}. \tag{8.7.13}$$

It will be realized that Q is not necessarily the same for all modes of a system and indeed in some practical cases the Q's vary greatly from mode to mode.

As an example of the general theory consider the system of fig. 8.7.1. It consists of three flywheels, each having a moment of inertia I mounted on a shaft as shown so that the stiffness of the portions of shaft between adjacent flywheels is k. Damping torques act on the two end flywheels, their magnitudes being

$$\frac{\sqrt{(Ik)}}{10} \times (\text{angular velocity}).$$

The angular displacements of the three disks are q_1, q_2, q_3 and a torque $F e^{i\omega t}$ acts at q_3.

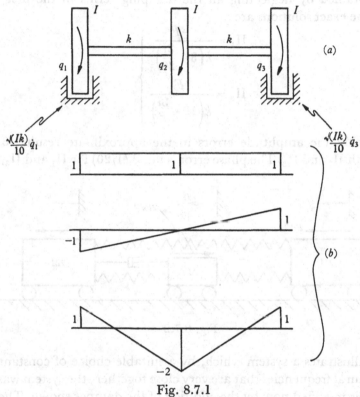

Fig. 8.7.1

The principal modes of the system are illustrated in the figure, at (b), and we shall take unit values of the principal co-ordinates p_1, p_2, p_3 to correspond to the displacements indicated. The equations of motion of the system are

$$3I\ddot{p}_1 + \frac{\sqrt{(Ik)}}{5} \dot{p}_1 + \frac{\sqrt{(Ik)}}{5} \dot{p}_3 = F e^{i\omega t},$$

$$2I\ddot{p}_2 + \frac{\sqrt{(Ik)}}{5} \dot{p}_2 + 2kp_2 = F e^{i\omega t},$$

$$\frac{\sqrt{(Ik)}}{5} \dot{p}_1 + 6I\ddot{p}_3 + \frac{\sqrt{(Ik)}}{5} \dot{p}_3 + 18p_3 = F e^{i\omega t}.$$

(8.7.14)

Trial solutions of the type of equations (8.7.2) give the usual form of algebraic equations. If ω is made equal to $\omega_2 = \sqrt{(k/I)}$, which is the lower of the two non-zero natural frequencies, then these equations may be solved to give

$$\Pi_2 = \frac{5F}{ik}. \tag{8.7.15}$$

This is an exact expression because of the absence of coupling terms in the second equation. The approximate amplitudes of p_1 and p_3 are

$$\Pi_1 \doteqdot -\frac{F}{3k}, \quad \Pi_3 \doteqdot \frac{F}{12k}. \tag{8.7.16}$$

These are obtained by neglecting all the damping terms in the first and third equations. The exact solutions are

$$\left.\begin{aligned}
\Pi_1 &= \frac{-F}{k\left(3 - \dfrac{3i}{20}\right)}, \\
\Pi_3 &= \frac{F}{k\left(12 - \dfrac{3i}{5}\right)}.
\end{aligned}\right\} \tag{8.7.17}$$

These show that the amplitude errors in the approximate results (8.7.16) are $0\cdot13\%$ for both Π_1 and Π_3. The phase error is $\tan^{-1}(1/20)$ for Π_1 and Π_3, this being less than $3°$.

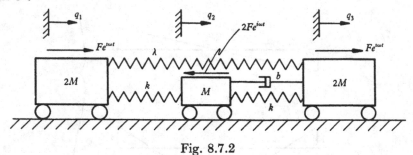

Fig. 8.7.2

Fig. 8.7.2 illustrates a system which, by a suitable choice of constants, may be given two natural frequencies that are very close together; the system was discussed in §3.10, but is modified now by the addition of the damper shown. The principal co-ordinates have been shown to be such that we can write

$$\left.\begin{aligned}
q_1 &= p_1 - p_2 + p_3, \\
q_2 &= p_1 - 4p_3, \\
q_3 &= p_1 + p_2 + p_3.
\end{aligned}\right\} \tag{8.7.18}$$

Further, with the applied force as shown, the corresponding forces P are

$$P_1 = P_2 = 0, \quad P_3 = 10Fe^{i\omega t}, \tag{8.7.19}$$

so that if it were not for the coupling terms due to the damper, only the third principal mode would be excited.

The energy functions are

$$2T = 2M\dot{q}_1^2 + M\dot{q}_2^2 + 2M\dot{q}_3^2 = 5M\dot{p}_1^2 + 4M\dot{p}_2^2 + 20M\dot{p}_3^2,$$

$$2D = b(\dot{q}_2 - \dot{q}_3)^2 = b\dot{p}_2^2 + 25b\dot{p}_3^2 + 10b\dot{p}_2\dot{p}_3,$$

$$2V = k(q_2 - q_1)^2 + k(q_3 - q_2)^2 + \lambda(q_3 - q_1)^2 = (4\lambda + 2k)\,p_2^2 + 50kp_3^2. \tag{8.7.20}$$

From these, the following equations of motion may be found.

$$5M\ddot{p}_1 = 0,$$

$$4M\ddot{p}_2 + b\dot{p}_2 + 5b\dot{p}_3 + (4\lambda + 2k)\,p_2 = 0,$$

$$20M\ddot{p}_3 + 5b\dot{p}_2 + 25b\dot{p}_3 + 50kp_3 = 10Fe^{i\omega t}. \tag{8.7.21}$$

The algebraic equations, which are derived from these by substituting trial harmonic solutions, are

$$-5M\omega^2\Pi_1 = 0,$$

$$(4\lambda + 2k - 4M\omega^2 + ib\omega)\,\Pi_2 + 5ib\omega\Pi_3 = 0,$$

$$5ib\omega\Pi_2 + (50k - 20M\omega^2 + 25ib\omega)\,\Pi_3 = 10F. \tag{8.7.22}$$

From these equations exact values for the amplitudes Π may be found. In particular, if $\omega^2 = \omega_3^2 = 5k/2M$ they yield results which may be written in the form

$$\Pi_1 = 0,$$

$$\Pi_2 = \left(\frac{-5ib\omega_3}{4(\lambda - 2k) + ib\omega_3}\right)\Pi_3,$$

$$\Pi_3 = \left(\frac{4(\lambda - 2k) + ib\omega_3}{10ib\omega_3(\lambda - 2k)}\right)F. \tag{8.7.23}$$

As we should expect, there is a large response at p_3 and a much smaller one at p_2 provided that the damping is small and $(\lambda - 2k)$ is not small.

An estimate of these results may be made by the approximate method; the procedure is to neglect the coupling terms in equations (8.7.22). By this means it is found that

$$\Pi_1 = \Pi_2 = 0, \quad \Pi_3 = \frac{10F}{25ib\omega_3}, \tag{8.7.24}$$

which will be seen to be a close approximation to the correct results (8.7.23).

Now it was shown in § 3.10 that, if λ approaches the value $2k$, two natural frequencies of the system of fig. 8.7.2 become equal. Under these conditions equations (8.7.23) yield

$$\Pi_1 = 0,$$

$$\Pi_2 \to -5\Pi_3,$$

$$\Pi_3 \to \infty. \tag{8.7.25}$$

These values conflict with those of equations (8.7.24) and illustrate the fact that the approximations which were used, and which are normally valid for lightly damped systems, are valueless when the equal-frequency condition is approached.

The particular equal-frequency system that is discussed above was chosen for its convenience as an example, and is not of direct interest. Equal-frequency systems in general are fairly common in engineering practice, however. They arise, in particular, in connexion with the vibration of solids of revolution. As a simple example of this consider the first mode of flexural vibration of a cantilever of circular cross-section. The cantilever is free to vibrate in any plane, and motions in any two perpendicular planes constitute orthogonal modes having equal frequencies. But in practice small departures from true symmetry make the frequencies in different planes slightly different, and introduce coupling between the planes. Consequently the two planes in which vibrations occur in principal modes are not arbitrary but are determined by the asymmetry of the cantilever; they remain perpendicular in order to satisfy the orthogonality condition.

EXAMPLES 8.7

1. Using the principal co-ordinates which are employed in the text (and whose unity values are indicated in fig. 8.7.1 (b)), find the response of the system of fig. 8.7.1 (a) when it resonates at the frequency $\omega_3 = \sqrt{(3k/I)}$. Find (a) the approximate solutions by neglecting the coupling of the modes and (b) the exact solutions.
What are the Q-factors of the system?

2. If the driving frequency ω of the system of fig. 8.7.1 has the value $\sqrt{(2k/I)}$, then the conditions are not near resonance. Calculate the responses at the co-ordinates p_1, p_2 and p_3 (as they are defined in the text) (a) by neglecting the damping completely, and (b) by solving the equations of motion exactly.
Calculate the percentage errors in the amplitudes and the phase errors of the approximate solutions.

NOTATION

Supplementary List of Symbols used in Chapter 8

E	Energy dissipated per cycle.
K	Complex stiffness [see equations (8.3.6) and (8.3.8)].
Q	Q-factor; a dimensionless measure of freedom from damping [see equations (8.2.15) and (8.7.13)].
$(Q_r)_D$	Contribution to rth generalized force of damping forces.
R	Amplitude of x [$= |X|$].
R_y, R_z	Generalized forces corresponding to y, z exerted on a system by a damping element (a dashpot).
w	Work done per cycle.
X	Complex amplitude of displacement x [see equation (8.1.2)].
Y, Z	Points of a system connected by a damping element (a dashpot).
y, z	Generalized displacements of Y and Z measured in the direction YZ.
ζ	Phase angle [see equation (8.1.11)].
ν	Damping factor [see equation (8.1.9)].
Ξ	Amplitude of generalized harmonic force P_1.

CHAPTER 9

HYSTERETIC DAMPING

In problems relating to vibrations, nature has provided us with a range of mysteries which for their elucidation require the exercise of a certain amount of mathematical dexterity.

C. E. INGLIS, The Fiftieth James Forrest Lecture delivered at the
Institution of Civil Engineers (1944)

The previous chapter was devoted to a theory of damping in which it is not suggested that real damping forces are represented with any accuracy. The theory provides qualitative information about the effects of damping and it can sometimes be used quantitatively if data are available. An alternative treatment is introduced in the present chapter which may be regarded as a modification of the previous one. It leads to slightly simpler algebra and may in some cases give a closer approximation to the real process of damping. It is for the former, rather than for the latter reason that it is discussed here. Another advantage of the new treatment is that it allows many of the receptance formulae to be applied directly to damped systems.

The equations of damped motion still remain linear in the new method. It follows that, according to this theory, different motions can occur simultaneously without interaction and also that a harmonic excitation produces a harmonic response.

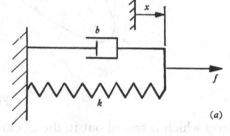

9.1 Comparison of viscous and hysteretic damping in an anchored spring

Fig. 9.1.1 (a) represents a linear spring which has a dashpot, or source of viscous damping, connected across it. If x denotes the extension of the spring then the force, f, which must be applied to the end is

$$f = kx + b\dot{x}, \qquad (9.1.1)$$

where k is the spring constant and b the damping coefficient. If the extension varies harmonically so that

$$x = R \sin \omega t \qquad (9.1.2)$$

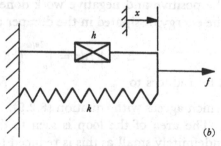

Fig. 9.1.1

then the force is $\qquad f = kR \sin \omega t + bR\omega \cos \omega t. \qquad (9.1.3)$

This equation and the preceding one express f and x as functions of time. The time t

may be eliminated between the equations so that f is given as a function of x. If this is done it is found that

$$f = kx \pm b\omega \sqrt{(R^2 - x^2)}, \qquad (9.1.4)$$

where the positive sign is used when $\cos \omega t$ is positive.

By means of this equation, f can be plotted against x just as it might be for the undamped spring; the latter case would be represented by a straight-line relation. For the present problem, the curve of f against x is shown in fig. 9.1.2. It is a closed

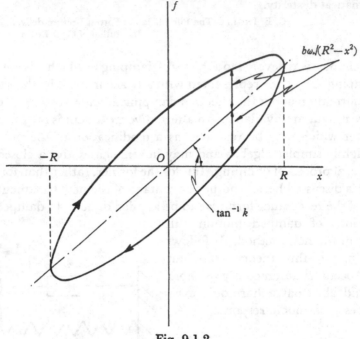

Fig. 9.1.2

loop which is traced out in the clockwise direction as t increases. The area below any arc of the curve represents the work done by the force on the system during the relevant change of x and the total area of the loop represents the difference between the positive and negative work done in one cycle of the motion. This is equal to the energy dissipated in the damper and is given by

$$E = \oint f \, dx = \int_0^{2\pi/\omega} (kR \sin \omega t + bR\omega \cos \omega t) \, R\omega \cos \omega t \, dt. \qquad (9.1.5)$$

This reduces to
$$E = bR^2 \pi \omega \qquad (9.1.6)$$

which agrees with equation (8.2.9).

The area of the loop is seen to depend upon the frequency ω and to become indefinitely small as this is reduced to zero. But if the dashpot is regarded merely as a means of representing the energy dissipation within a real spring, then this linear dependence of E on ω does not accord with physical behaviour. In fact that part of the energy loss per cycle which is due to imperfect elasticity of the material of the spring is usually independent of frequency. Now it is possible to modify the mathematical treatment so that the area of the loop remains constant with frequency. That is to say, that loop may be chosen which corresponds to some particular frequency and then the assumption made that it gives the correct relation of f to x at all frequencies.

A convenient way to introduce this idea is to rederive the above theory, but to begin by taking the damping coefficient as being inversely proportional to frequency. Thus the damping force is taken as $-h\dot{x}/\omega$ instead of $-b\dot{x}$, where h is a constant. A damper with this characteristic will be called here a 'hysteretic' damper, and h will be called the 'hysteretic damping coefficient'; the damper will be indicated diagrammatically as shown in fig. 9.1.1 (b).

With hysteretic damping, the area of the (f, x) loop is obtained by writing h/ω for b so that

$$E = hR^2\pi. \tag{9.1.7}$$

It was shown in § 8.3 that when a spring has a viscous damper connected across it, and when it is subjected to a simple harmonic extension, then the force-displacement relation may be regarded as being specified by a complex stiffness, $k + i\omega b$. If a spring is hysteretically damped, then again it may be treated as having a complex stiffness

$$K = k + ih. \tag{9.1.8}$$

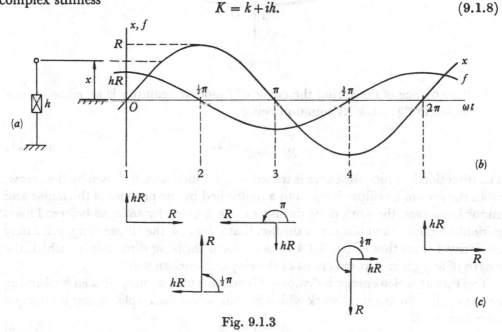

Fig. 9.1.3

Thus if the force is f and the extension is x as above,

$$f = Kx = (k + ih)x = k(1 + i\mu)x, \tag{9.1.9}$$

where $\mu = h/k$. The constant μ is a dimensionless measure of the damping.

The force f and displacement x across an isolated damper may be plotted against time or, more conveniently, against the parameter ωt. This is done in fig. 9.1.3 (b) for a hysteretic damper without an associated spring, as shown at (a) in the same figure. The origin is chosen when the displacement x is zero so that the two curves are given by

$$\left.\begin{array}{l} x = R \sin \omega t, \\ f = hR \cos \omega t. \end{array}\right\} \tag{9.1.10}$$

The ordinates of the curves may be obtained as the projections on a vertical line, of rotating vectors. The instantaneous positions of the vectors, for four particular values of ωt, are shown in fig. 9.1.3 (c).

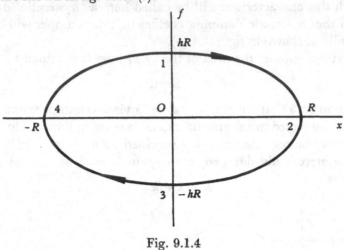

Fig. 9.1.4

In the absence of the spring the curve of f plotted against x is an ellipse of the form shown in fig. 9.1.4, its equation being

$$\frac{x^2}{R^2} + \frac{f^2}{h^2 R^2} = 1. \tag{9.1.11}$$

The direction in which the curve is traced out, as t increases, is shown by the arrow. Since the area of an ellipse is equal to π multiplied by the product of the major and minor semi-axes, the work done during a cycle is given by $hR^2\pi$, as before. This is evident physically and illustrates the fact that if $b\omega = h$, the ellipse of fig. 9.1.2 may be derived from that of fig. 9.1.4 by a uniform shearing distortion in which the length of any given vertical chord of the ellipse is unchanged.

The rate at which energy is dissipated is not constant, as may be seen by forming an expression for the total work which is done when the displacement is changed from 0 to x_1. If

$$x = R \sin \omega t \tag{9.1.12}$$

this is

$$W = \int_0^{x_1} f\,dx = \int_0^{\omega t_1} (hR \cos \omega t)\,(R \cos \omega t)\,d(\omega t) \tag{9.1.13}$$

which reduces to

$$W = \frac{hR^2 \omega t_1}{2} + \frac{hR^2}{4} \sin 2\omega t_1. \tag{9.1.14}$$

If W is plotted against ωt_1, it gives the curve shown in fig. 9.1.5 (a). This is the sum of two components, one of which increases linearly while the other fluctuates. The corresponding curve of x_1 against ωt_1 is shown in fig. 9.1.5 (b). The slope of the curve in fig. 9.1.5 (a) is never negative, as would be expected on physical grounds; its value at any moment is

$$\frac{dW}{d(\omega t_1)} = \frac{hR^2}{2}(1 + \cos 2\omega t_1). \tag{9.1.15}$$

The curve of W against x_1 may also be plotted from the equation

$$W = \frac{hR^2}{2}\left\{\sin^{-1}\left(\frac{x_1}{R}\right) \pm \left(\frac{x_1}{R}\right)\sqrt{\left[1 - \left(\frac{x_1}{R}\right)^2\right]}\right\}. \tag{9.1.16}$$

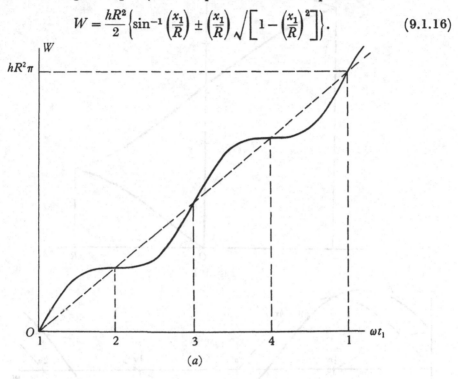

(a)

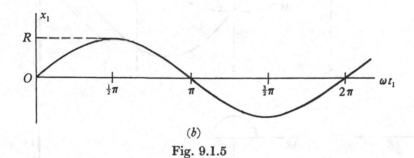

(b)

Fig. 9.1.5

This result is found by eliminating ωt_1 in equation (9.1.14) by substitution from equation (9.1.12) and it is shown plotted in fig. 9.1.6. The numbers on the curve are placed at points corresponding to those marked on figs. 9.1.3, 9.1.4 and 9.1.5.

Similar curves of f and x against ωt, and of f against x, may be plotted for a hysteretic damper and spring in parallel, as indicated in fig. 9.1.1 (b). If the instantaneous value of x is given by

$$x = R\sin\omega t \tag{9.1.17}$$

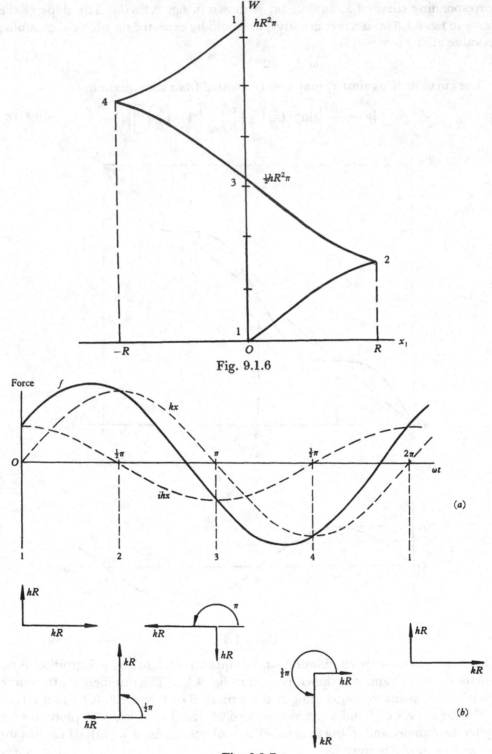

Fig. 9.1.6

Fig. 9.1.7

as before, then the value of the force is

$$f = kR \sin \omega t + hR \cos \omega t. \tag{9.1.18}$$

The curves of f against ωt are shown in fig. 9.1.7 (a), where the full line represents the total force and the broken lines represent the spring force and the damper force separately. The appropriate vector diagrams are shown at (b).

Again the parameter ωt may be eliminated to give the (f, x) relation. This has the form

$$f = kx \pm h \sqrt{(R^2 - x^2)} \tag{9.1.19}$$

which corresponds to the viscous damping equation (9.1.4). If this is plotted, it gives the ellipse of fig. 9.1.8 which is obtainable from that of fig. 9.1.4 by a shearing distortion as mentioned previously. It is the term kx in equation (9.1.19) which produces this shear.

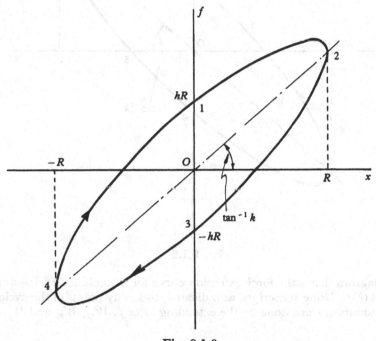

Fig. 9.1.8

When a spring and damper are coupled in parallel in this way the instantaneous value of the rate of doing work on the system will contain a term due to the energy storage in the spring as well as a term due to energy dissipation. It follows that the curves of fig. 9.1.5 and 9.1.6 must be appropriately modified.

The theory of hysteretic damping will be developed further in the following sections. But it may be emphasized here that it does not represent, any more than does viscous damping, the true damping laws which exist in practice. In certain problems, however, notably where dissipation is due to imperfect elasticity, hysteretic damping is a better approximation than is viscous damping.

EXAMPLES 9.1

1. Show that the curve of fig. 9.1.2 is an ellipse whose major axis is inclined at an angle

$$\tfrac{1}{2}\tan^{-1}\left(\frac{2k}{1-k^2-b^2\omega^2}\right)$$

to the x-axis.

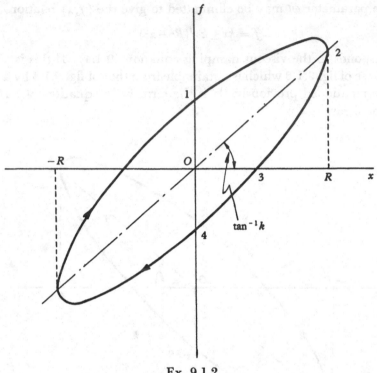

Ex. 9.1.2

2. The diagram shows the force-extension curve for a combined spring-and-damper (see fig. 9.1.1 (b)). Using subscripts, as indicated, to signify points of the cycle, find the following amounts of work done by the extending force f: W_{12}, W_{23} and W_{34}.

9.2 Systems having one degree of freedom

Consider the spring-mass system of fig. 9.2.1 which has a hysteretic damper of constant h connected across the spring and which is acted on by a force $F e^{i\omega t}$. The displacement x of the mass is governed by the equation

$$M\ddot{x}+\frac{h}{\omega}\dot{x}+kx = F e^{i\omega t}. \tag{9.2.1}$$

When considering solutions of this equation, it must be remembered that hysteretic damping has been defined for steady harmonic motion only so that other types of solution are meaningless. The appropriate trial solution is

$$x = X e^{i\omega t} = \alpha F e^{i\omega t} \tag{9.2.2}$$

and this leads to the result

$$[-M\omega^2 + (k+ih)]\,Xe^{i\omega t} = Fe^{i\omega t}. \tag{9.2.3}$$

In this equation the complex stiffness $(k+ih)$ may also be written in its polar form so that the equation becomes

$$[-M\omega^2 + k\sqrt{(1+\mu^2)}\,e^{i\tan^{-1}\mu}]\,Xe^{i\omega t} = Fe^{i\omega t}. \tag{9.2.4}$$

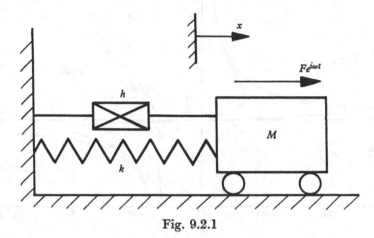

Fig. 9.2.1

From equation (9.2.3), the receptance α is found to be

$$\alpha = \left[\frac{k-M\omega^2}{(k-M\omega^2)^2 + h^2}\right] - i\left[\frac{h}{(k-M\omega^2)^2 + h^2}\right]. \tag{9.2.5}$$

This result may also be written in the polar form

$$\alpha = \frac{e^{-i\eta}}{\sqrt{[(k-M\omega^2)^2 + h^2]}}, \tag{9.2.6}$$

where the phase angle η is given by

$$\tan\eta = \frac{h}{k-M\omega^2}. \tag{9.2.7}$$

The real and imaginary parts of α are shown plotted against ω in fig. 9.2.2. The curves have been drawn for the particular values $M = 1\,\text{lb.}$, $k = 8\,\text{lb.wt./ft.}$ (i.e. $256\,\text{pdl./ft.}$), and $\mu = h/k = 0.02$, 0.2 and 0.6. In fig. 9.2.3 the real and imaginary parts of α are plotted on an Argand diagram for variation of the parameter ω. Fig. 9.2.4 shows a three-dimensional graph in which the real and imaginary parts of α are measured along two of the axes while ω is measured along the third. These graphs are constructed for the same values of M and k as above but with μ equal to 0.2 and 0.6 only. The curves of fig. 9.2.2 and 9.2.3 are projections on the relevant planes of the three-dimensional curve of fig. 9.2.4.

The curves are generally similar to the corresponding curves for viscous damping but there are some minor differences which will be mentioned later.

The force-displacement relation for steady harmonic motion may be obtained

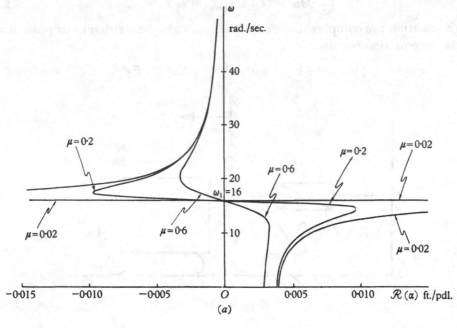

(a)

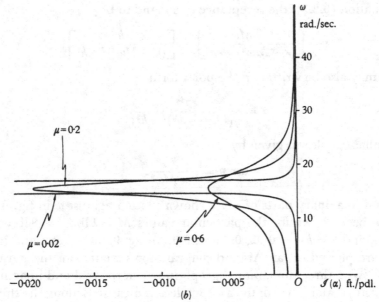

(b)

Fig. 9.2.2

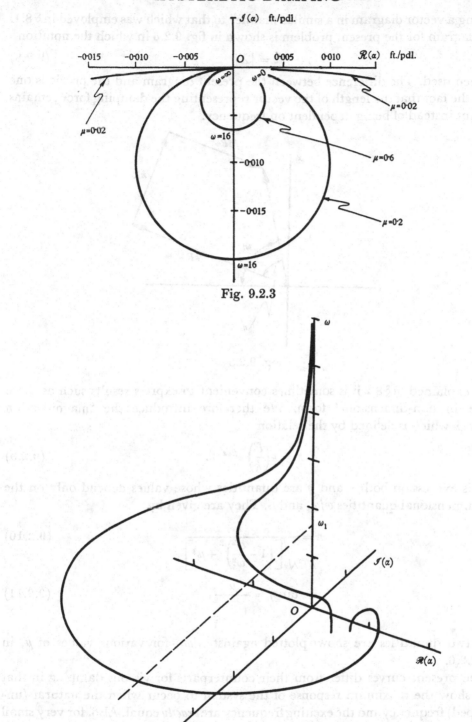

Fig. 9.2.3

Fig. 9.2.4

by using a vector diagram in a similar manner to that which was employed in § 8.1. The diagram for the present problem is shown in fig. 9.2.5 in which the notation

$$R = |X| \tag{9.2.8}$$

has been used. The difference between the present diagram and the previous one lies in the fact that the length of the vector representing the damping force remains constant instead of being dependent on frequency.

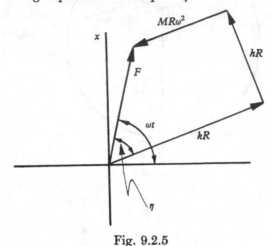

Fig. 9.2.5

As explained in § 8.2 it is sometimes convenient to express results such as those above in non-dimensional form. We therefore introduce the 'magnification factor' n which is defined by the relation

$$x = n\left(\frac{F}{k}\right) e^{i(\omega t - \eta)}. \tag{9.2.9}$$

In this expression both n and η are quantities whose values depend only on the non-dimensional quantities ω/ω_1 and μ. They are given by

$$n = \frac{1}{\sqrt{\left[\left(1 - \dfrac{\omega^2}{\omega_1^2}\right)^2 + \mu^2\right]}} \tag{9.2.10}$$

and

$$\tan \eta = \frac{\mu}{1 - \dfrac{\omega^2}{\omega_1^2}}. \tag{9.2.11}$$

The two quantities are shown plotted against ω/ω_1, for various values of μ, in fig. 9.2.6.

The present curves differ from their counterparts for viscous damping in that they show the maximum response of the system to occur when the natural (undamped) frequency and the exciting frequency are *exactly* equal. Also, for very small values of ω/ω_1, the phase angle in the present case tends to the value $\tan^{-1}\mu$, whereas for viscous damping it tends to zero.

It is possible to use a Q-factor for hysteretic damping. It may be defined as a multiple of the ratio of vibrational energy stored at resonance to the energy dissipated per cycle. Thus

$$\frac{1}{2\pi}Q = \frac{T_{max.}}{E} = \frac{\frac{1}{2}MR^2\omega_1^2}{hR^2\pi} = \frac{\frac{1}{2}kR^2}{hR^2\pi} = \frac{V_{max.}}{E} \qquad (9.2.12)$$

which gives

$$Q = \frac{k}{h} = \frac{1}{\mu}. \qquad (9.2.13)$$

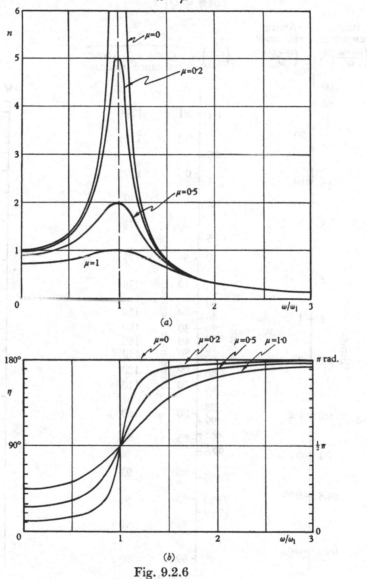

Fig. 9.2.6

It will be seen that the amplitude at resonance is

$$R_{res.} = \frac{F}{h} = Q\frac{F}{k}. \qquad (9.2.14)$$

This shows that the magnification factor at resonance is equal to Q.

Equations (9.2.10) and (9.2.11) give the magnification factor and phase angle for any specified values of ω/ω_1 and μ. It is sometimes convenient to rearrange the latter equation, introducing the Q factor. From equation (9.2.7), it is seen that

$$\cot\eta = \frac{k - M\omega^2}{h} = Q\left[1 - \frac{M\omega^2}{Qh}\right] \qquad (9.2.15)$$

or

$$\cot\eta = Q\left[1 - \frac{\omega^2}{\omega_1^2}\right]. \qquad (9.2.16)$$

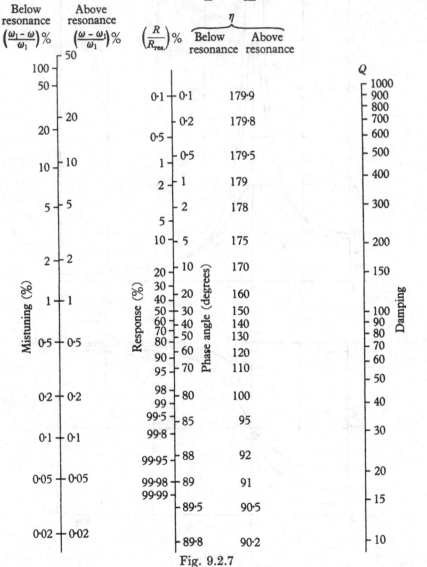

Fig. 9.2.7

This relationship, between η, Q and ω/ω_1, can be represented in the form of a nomogram as shown in fig. 9.2.7. In constructing the chart, the percentage mistuning,

$$\frac{\omega_1 - \omega}{\omega_1} \times 100$$

has been used instead of the frequency ratio ω/ω_1. This makes the scale more convenient for most purposes.

Instead of considering the ratio of the amplitude to the static deflexion F/k, as we do when using the magnification factor, it is convenient to consider the ratio of the amplitude at a given frequency to the amplitude at resonance. Thus if the amplitude is R,

$$\frac{R}{R_{\text{res.}}} = \left(\frac{F}{\sqrt{[(k - M\omega^2)^2 + h^2]}}\right)\Big/\left(\frac{F}{h}\right) = \frac{h}{\sqrt{[(k - M\omega^2)^2 + h^2]}}. \tag{9.2.17}$$

Now in view of equation (9.2.7) this may be written

$$\frac{R}{R_{\text{res.}}} = \sin \eta. \tag{9.2.18}$$

This relation makes it possible to read $R/R_{\text{res.}}$ from the nomogram of fig. 9.2.7 simply by adding a suitable scale to the η-line. On the chart, this ratio has been expressed as a percentage.

The nomogram is used by joining with a straight edge the point giving the relevant degree of mistuning on the left-hand line with the required Q-factor on the right-hand line. The intersection on the middle line then gives both the phase angle η and the amplitude ratio $R/R_{\text{res.}}$ of the response.

EXAMPLES 9.2

1. Show that, if α is the complex receptance at x of the system of fig. 9.2.1, then
   (a) the maximum values of $\mathscr{R}(\alpha)$ occur at the frequencies
$$\omega = \omega_1 \sqrt{(1 \pm \mu)},$$
   where $\mu = h/k$.
   (b) the curve of α on an Argand diagram, in which ω is taken as a parameter (cf. fig. 9.2.3), is a circle of radius $\frac{1}{2h}$ whose centre lies on the negative imaginary axis and which passes through the origin.

2. Consider the system of fig. 9.2.1 in which the hysteretic damping force is replaced by a damping force which is proportional to the *acceleration* of the mass; let the damping force (which is in phase with velocity) be of magnitude $-f\ddot{x}$.
   (a) Find the complex receptance at x.
   (b) Find the magnification factor N_a and the phase angle ξ (between x and the driving force) and sketch curves of these quantities for a few intensities of damping.
   (c) Sketch the rotating-vector representation of the motion.
   (d) Derive the Q-factor of the system.

3. The length of a simple pendulum is 1 ft. and the mass of the small bob is 1 lb. If the Q-factor for small swinging oscillations is 10, find the coefficient of the hysteretic damper which may be thought of as damping rotational motion of the pendulum arm.

4. (a) The response at x in the system of fig. 9.2.1 can be expressed in the form
$$x = [ne^{-i\eta}]\frac{F}{k}e^{i\omega t},$$
where the quantity in brackets is a complex dimensionless magnification factor (see equation (9.2.10)). Show that this quantity can be plotted in the manner of fig. 9.2.4

and that curves like those of figs. 9.2.2 and 9.2.3 may be obtained by projection. What form does the curve of $1/[n\,e^{-i\eta}]$ against ω/ω_1 take?

(*b*) Examine also the case where the harmonic applied force $F e^{i\omega t}$ is replaced by the excitation produced by a small unbalanced mass m rotating at speed ω with eccentricity e.

[NOTE. These types of curve—or rather their projections of the form shown in fig. 9.2.3—are used by Kennedy and Pancu in an interesting treatment of aircraft vibration testing.† Their work, which extends to multi-degree-of-freedom systems, contains the implicit assumption that all coupling between principal modes can be ignored.]

9.3 Hysteretic damping of systems having several degrees of freedom; analysis and synthesis of damped systems

It was shown in § 8.3 that systems which are subject to viscous damping may be split up into sub-systems, just as conservative systems could; similar arguments apply to hysteretically damped systems. Here again, the receptances will be complex quantities, although the variation of the receptances with frequency will be governed by slightly different laws from those which would hold with viscous damping.

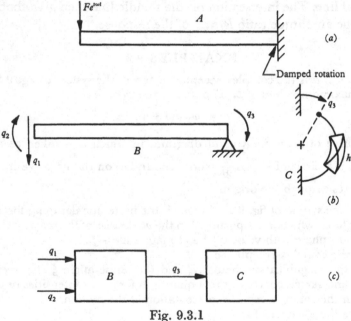

Fig. 9.3.1

Consider the beam shown in fig. 9.3.1 (*a*). One end is hinged to a support, the hinge having some friction in it which is assumed to obey the hysteretic law. The other end of the beam has no support, but is acted on by a force $F e^{i\omega t}$. Let the deflexion and slope at this latter end be denoted by q_1 and q_2 respectively, and suppose that it is required to find α_{12} in order to obtain the amplitude at q_2 due to the applied force. The co-ordinate q_3 is used to denote the slope at the hinged end.

† C. C. Kennedy and C. D. P. Pancu, 'Use of Vectors in Vibration Measurement and Analysis', *J. Aero. Sciences*, vol. 14, no. 11 (Nov. 1947), pp. 603–25.

The system may be split into the two parts shown in fig. 9.3.1 (*b*), one being a free-pinned beam and the other a hysteretic damper only; these are sub-systems *B* and *C* respectively. The block diagram of fig. 9.3.1 (*c*) shows the way in which the sub-systems are connected. The required receptance α_{12} is given by

$$\alpha_{12} = \beta_{12} - \frac{\beta_{23}\beta_{13}}{\beta_{33}+\gamma_{33}} \tag{9.3.1}$$

which result may be found by the method of §1.8. The receptances for the beam may be taken from Table 7.1 (*c*) using the notation previously established; they are

$$\beta_{12} = \frac{-F_6}{EI\lambda^2 F_5}, \quad \beta_{13} = \frac{-F_7}{EI\lambda^2 F_5}, \\ \beta_{23} = \frac{F_9}{EI\lambda F_5}, \quad \beta_{33} = \frac{F_4}{EI\lambda F_5}. \tag{9.3.2}$$

The receptance γ_{33} is that of the hysteretic damper only. If the damper is acted on by a torque $\Phi_3 e^{i\omega t}$, then the displacement produced is given by

$$ihq_3 = \Phi_3 e^{i\omega t}. \tag{9.3.3}$$

If the value of q_3 is now sought in the usual way, this gives

$$q_3 = \Psi_3 e^{i\omega t} = \left(\frac{1}{ih}\right)\Phi_3 e^{i\omega t} \tag{9.3.4}$$

so that the receptance is
$$\gamma_{33} = \frac{1}{ih}. \tag{9.3.5}$$

This expression, together with those of equations (9.3.2), enables α_{12} to be found from equation (9.3.1).

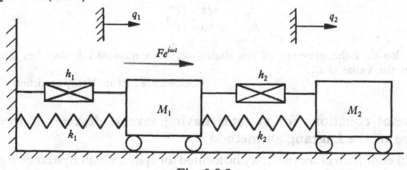

Fig. 9.3.2

The previous discussion has treated hysteretic dampers which have one side anchored. It is also possible to have such dampers with both their sides connected to movable points, the force in the damper depending upon the relative displacement. Fig. 9.3.2 illustrates a system in which this is so. The equations of motion may be written down directly by Newton's laws; they are

$$M_1\ddot{q}_1 = Fe^{i\omega t} + k_2(q_2-q_1) + ih_2(q_2-q_1) - k_1 q_1 - ih_1 q_1, \\ M_2\ddot{q}_2 = -k_2(q_2-q_1) - ih_2(q_2-q_1). \tag{9.3.6}$$

By assuming harmonic solutions for q_1 and q_2 in the usual way, the following receptances may be derived from these equations:

$$\alpha_{11} = \frac{(k_2 - M_2\omega^2) + ih_2}{\Delta}, \quad \alpha_{21} = \frac{k_2 + ih_2}{\Delta}, \quad (9.3.7)$$

where

$$\Delta = \{M_1 M_2 \omega^4 - [k_1 M_2 + k_2(M_1 + M_2)]\,\omega^2 + k_1 k_2 - h_1 h_2\}$$
$$- i\{[h_1 M_2 + h_2(M_1 + M_2)]\,\omega^2 - (k_1 h_2 + k_2 h_1)\}. \quad (9.3.8)$$

These receptances are complex showing that, as might be expected, the displacements are not in phase with the applied force. It may also be noted that the ratio of the real to the imaginary part is not the same for the two receptances; this shows that the displacements at q_1 and q_2 are not in phase with each other.

The above receptances may be obtained directly, from entry 8 of Table 1. It is only necessary to replace the spring constants in the table by the complex stiffnesses $(k_1 + ih_1)$ and $(k_2 + ih_2)$ since these give the force-displacement relations across the springs. This method of deriving receptances may be used in conjunction with any of the formulae of Table 1.

EXAMPLE 9.3

1. Two masses, $4M$ and M, are joined by a spring of stiffness $3k$ and a hysteretic damper, exerting a force hv/ω, where v is the relative velocity of the masses and ω is the frequency. The larger mass is suspended from a fixed support by a spring of stiffness k so that the mass M hangs below it.

If a disturbing force $Fe^{i\omega t}$, where $\omega^2 = k/M$, be applied to the mass $4M$, show that the amplitudes R_1 and R_2 of the motions of the masses $4M$ and M will be in the ratio

$$\frac{R_1}{R_2} = \sqrt{\left(\frac{4 + H^2}{9 + H^2}\right)},$$

where $H = h/k$.

Verify also that the presence of the damping, however small h may be, causes an increase in the value of R_1.

(C.U.M.S.T. Pt II, 1948, modified slightly)

9.4 General equations for systems having several degrees of freedom; use of the Lagrangian method

General equations of motion may be formed for systems with hysteretic damping just as they may when the damping is viscous. But, with hysteretic damping, the equations must be restricted to those governing steady harmonic motion only. Certain results can be obtained more readily from the general equations for hysteretic damping than from those for viscous damping.

Consider a system having generalized co-ordinates $q_1, q_2, \ldots, q_n$ and suppose that a hysteretic damper is connected between two points Y and Z of the system. Let y and z denote the displacements of Y and Z along the direction YZ when the system is distorted. The extension of the damper during this distortion may be denoted by x, so that

$$x = y - z. \quad (9.4.1)$$

During harmonic oscillation, the damper exerts a force of magnitude hx at the points Y and Z. If P_y denotes the force which is exerted on the system at Y in the direction YZ, and P_z denotes the force exerted at Z in the same direction, then

$$P_y = -ihx, \quad P_z = -P_y = ihx, \tag{9.4.2}$$

where the multiplier i may be used because x is a harmonically-varying quantity.

It was shown in Chapter 8 that if a single viscous damper of constant b were joined between points Y and Z of a system, as is the hysteretic damper in the present case, then the contribution of the damper force to the generalized force Q_r corresponding to the rth co-ordinate q_r is

$$(Q_r)_D = -b\left[\left(\frac{\partial y}{\partial q_1} - \frac{\partial z}{\partial q_1}\right)\left(\frac{\partial y}{\partial q_r} - \frac{\partial z}{\partial q_r}\right)\dot{q}_1 + \left(\frac{\partial y}{\partial q_2} - \frac{\partial z}{\partial q_2}\right)\left(\frac{\partial y}{\partial q_r} - \frac{\partial z}{\partial q_r}\right)\dot{q}_2 + \cdots \right.$$
$$\left. + \left(\frac{\partial y}{\partial q_n} - \frac{\partial z}{\partial q_n}\right)\left(\frac{\partial y}{\partial q_r} - \frac{\partial z}{\partial q_r}\right)\dot{q}_n\right]. \tag{9.4.3}$$

The result follows from that of equation (8.4.7) and this expression may be adapted to the problem of hysteretic damping. Since we are concerned with harmonic motion only, we can write

$$q_1 = \Psi_1 e^{i\omega t}, \quad q_2 = \Psi_2 e^{i\omega t}, \quad \cdots, \quad q_n = \Psi_n e^{i\omega t}, \tag{9.4.4}$$

where the amplitudes Ψ will be complex if the motions at the different co-ordinates are not in phase. The factors

$$\left(\frac{\partial y}{\partial q_1} - \frac{\partial z}{\partial q_1}\right), \quad \text{etc.,}$$

in equation (9.4.3) relate to geometrical properties of the system and are not affected by the time-variation of the co-ordinates; these terms are constants. If the coefficient b of viscous damping is replaced by the constant h/ω for a hysteretic damper, then the rth generalized force due to the hysteretic damper can be written

$$(Q_r)_H = -ih\left[\left(\frac{\partial y}{\partial q_1} - \frac{\partial z}{\partial q_1}\right)\left(\frac{\partial y}{\partial q_r} - \frac{\partial z}{\partial q_r}\right)\Psi_1 e^{i\omega t}\right.$$
$$+ \left(\frac{\partial y}{\partial q_2} - \frac{\partial z}{\partial q_2}\right)\left(\frac{\partial y}{\partial q_r} - \frac{\partial z}{\partial q_r}\right)\Psi_2 e^{i\omega t} + \cdots$$
$$\left. + \left(\frac{\partial y}{\partial q_n} - \frac{\partial z}{\partial q_n}\right)\left(\frac{\partial y}{\partial q_r} - \frac{\partial z}{\partial q_r}\right)\Psi_n e^{i\omega t}\right]. \tag{9.4.5}$$

This may be abbreviated to

$$(Q_r)_H = -i[d_{r1}q_1 + d_{r2}q_2 + \cdots + d_{rn}q_n], \tag{9.4.6}$$

where it will be seen, by examining equation (9.4.5), that

$$d_{rs} = d_{sr} = h\frac{\partial x}{\partial q_r}\frac{\partial x}{\partial q_s}. \tag{9.4.7}$$

If now similar calculations are made for hysteretic dampers at other points, they will yield similar expressions. Corresponding coefficients in these expressions can be added together to give a single expression for $(Q_r)_H$. This will have the same form as equation (9.4.6) and equation (9.4.7) will still be valid.

It was shown in the preceding chapter that the viscous damping terms in the equations of motion could be derived by differentiation of a certain dissipation function. This could be done because the coefficients b_{rs} conformed to a reciprocal relation similar to that of equation (9.4.7). In the same way, therefore, the hysteretic damping terms can be derived by differentiating a suitable quadratic function of the co-ordinates q. This function will be denoted by S and is defined by the equation

$$2S = i[d_{11}q_1^2 + d_{22}q_2^2 + \ldots + d_{nn}q_n^2 + 2d_{12}q_1q_2 + \ldots]. \qquad (9.4.8)$$

It will be seen that the generalized force of equation (9.4.6) can be written

$$(Q_r)_H = -\frac{\partial S}{\partial q_r}. \qquad (9.4.9)$$

The presence of the quantity i and the restricted meaning of the hysteretic damping coefficients prevent any physical meaning being given to the function S which would be comparable with that given to the function D. Nevertheless it is possible, purely as a matter of convenience and without suggesting anything other than an algebraic interpretation, to include the derivatives of S within the Lagrangian equations so that they become

$$\frac{d}{dt}\left(\frac{\partial T}{\partial \dot{q}_r}\right) + \frac{\partial V}{\partial q_r} + \frac{\partial S}{\partial q_r} = Q_r \quad (r = 1, 2, \ldots, n). \qquad (9.4.10)$$

In these equations, Q_r now represents generalized forces other than those derivable from a potential function V or a hysteretic damping function S.

Consider the form of the function S when it refers to a single damper. If the expressions (9.4.7) for the constants d_{rs} are substituted in equation (9.4.8), it is found that

$$2S = ih\left[\left(\frac{\partial x}{\partial q_1}\right)^2 q_1^2 + \left(\frac{\partial x}{\partial q_2}\right)^2 q_2^2 + \ldots + \left(\frac{\partial x}{\partial q_n}\right)^2 q_n^2 + 2\frac{\partial x}{\partial q_1}\frac{\partial x}{\partial q_2} q_1 q_2 + \ldots\right]. \qquad (9.4.11)$$

Now the displacement x between the ends of the damper is of the form

$$x = \frac{\partial x}{\partial q_1} q_1 + \frac{\partial x}{\partial q_2} q_2 + \ldots + \frac{\partial x}{\partial q_n} q_n \qquad (9.4.12)$$

so that the quantity in square brackets in equation (9.4.11) is simply x^2. Hence the function S for a single source of hysteretic damping is given by

$$2S = ihx^2. \qquad (9.4.13)$$

The use of the function S may be illustrated in an elementary manner by reference to the system of fig. 9.2.1. If q_1 is identified with the displacement x, then the various functions are

$$2T = M\dot{q}_1^2, \quad 2V = kq_1^2, \quad 2S = ihq_1^2, \quad Q_1 = Fe^{i\omega t}. \qquad (9.4.14)$$

The reader may use these to construct the equation of motion.

As a slightly more complicated example consider the system of fig. 9.3.2. This has two degrees of freedom and the functions for it are

$$\left.\begin{aligned}
2T &= M_1 \dot{q}_1^2 + M_2 \dot{q}_2^2, \\
2V &= k_1 q_1^2 + k_2 (q_2 - q_1)^2, \\
2S &= ih_1 q_1^2 + ih_2 (q_2 - q_1)^2, \\
Q_1 &= F e^{i\omega t}, \quad Q_2 = 0.
\end{aligned}\right\} \tag{9.4.15}$$

From these, equations (9.3.6) may be obtained by using the Lagrangian method.

The function S has the same general algebraic form as the energy functions T and V. Thus, when the equations of motion are formed in the usual way it is found that they have the form

$$(a_{1r} \ddot{q}_1 + a_{2r} \ddot{q}_2 + \ldots + a_{nr} \ddot{q}_n)$$

$$+ [(c_{1r} + id_{1r}) q_1 + (c_{2r} + id_{2r}) q_2 + \ldots + (c_{nr} + id_{nr}) q_n] = \Phi_r e^{i\omega t} \quad (r = 1, 2, \ldots, n),$$

$$\tag{9.4.16}$$

where $\Phi_r e^{i\omega t}$ is an external generalized force corresponding to the co-ordinate q_r. The hysteretic damping has the effect, therefore, of replacing the real stability coefficients c by complex quantities without otherwise affecting the equations.

This last fact suggests the possibility of regarding the complex stability coefficients as being derived from a single function which contains complex coefficients. The new function is merely the sum of the functions V and S and we shall denote it by U so that

$$U = V + S. \tag{0.1.17}$$

With this function, the equations of motion may be written as

$$\frac{d}{dt}\left(\frac{\partial T}{\partial \dot{q}_r}\right) + \frac{\partial U}{\partial q_r} = \Phi_r e^{i\omega t} \quad (r = 1, 2, \ldots, n) \tag{9.4.18}$$

provided that the restricted meaning of the equations is borne in mind.

The advantage of this procedure is that the function U may be formed as a single unit if the complex stiffnesses of the system are used in exactly the same way as are the stiffnesses in undamped systems when the function V is formed. Thus if the extension of a particular spring within the system is x, if its stiffness is k and if a hysteretic damper of constant h is connected across it, then the total contribution to the function U will be

$$U = \frac{(k + ih) x^2}{2}. \tag{9.4.19}$$

Contributions to V which arise from forces other than elasticity—that is to say gravity, etc.—must of course all be included.

For the system of fig. 9.3.2,

$$2U = (k_1 + ih_1) q_1^2 + (k_2 + ih_2) (q_2 - q_1)^2. \tag{9.4.20}$$

From this expression, equations (9.3.6) may be formed by using Lagrange's equations in the form (9.4.18).

As a second example on the function U, consider the arrangement shown in fig. 9.4.1. For this,

$$2T = M\dot{q}_1^2 + m(\dot{q}_1 + l\dot{q}_2)^2 \\ = (M+m)\,\dot{q}_1^2 + ml^2\dot{q}_2^2 + 2ml\dot{q}_1\dot{q}_2, \\ 2U = (k_1 + ih_1)\,q_1^2 + ih_2\,q_2^2 + mglq_2^2. \tag{9.4.21}$$

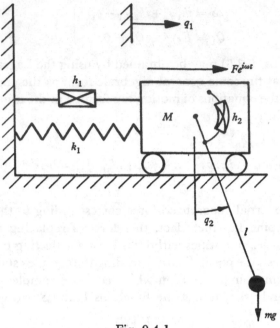

Fig. 9.4.1

The generalized forces Q are

$$Q_1 = Fe^{i\omega t}, \quad Q_2 = 0, \tag{9.4.22}$$

and from these the equations of motion are found to be

$$(M+m)\,\ddot{q}_1 + ml\ddot{q}_2 + (k_1 + ih_1)\,q_1 = Fe^{i\omega t}, \\ ml\ddot{q}_1 + ml^2\ddot{q}_2 + (mgl + ih_2)\,q_2 = 0. \tag{9.4.23}$$

EXAMPLE 9.4

1. The figure (p. 471) shows two pendula which are coupled by a spring that dissipates energy; this is allowed for by the damper. Find expressions for the energy functions S and U and write down the equations of motion for the case where a force $Fe^{i\omega t}$ acts as shown.

9.5 The closed and series forms of receptances

The equations of motion for systems with hysteretic damping may be solved to give receptances in the same way as may those for undamped systems. We now examine the forms which the receptances take.

Consider a system which has generalized co-ordinates $q_1, q_2, ..., q_n$, and let the system execute steady harmonic motion in which

$$q_1 = \Psi_1 e^{i\omega t}, \quad q_2 = \Psi_2 e^{i\omega t}, \quad ..., \quad q_n = \Psi_n e^{i\omega t}. \tag{9.5.1}$$

The equations of motion may be written in the form

$$\left.\begin{array}{l}
\xi_{11}\Psi_1 + \xi_{12}\Psi_2 + \ldots + \xi_{1n}\Psi_n = \Phi_1, \\
\xi_{21}\Psi_1 + \xi_{22}\Psi_2 + \ldots + \xi_{2n}\Psi_n = \Phi_2, \\
\cdots\cdots\cdots\cdots\cdots\cdots\cdots\cdots\cdots\cdots\cdots\cdots\cdots \\
\xi_{n1}\Psi_1 + \xi_{n2}\Psi_2 + \ldots + \xi_{nn}\Psi_n = \Phi_n,
\end{array}\right\} \tag{9.5.2}$$

where
$$\xi_{uv} = \xi_{vu} = c_{uv} - \omega^2 a_{uv} + i d_{uv}. \tag{9.5.3}$$

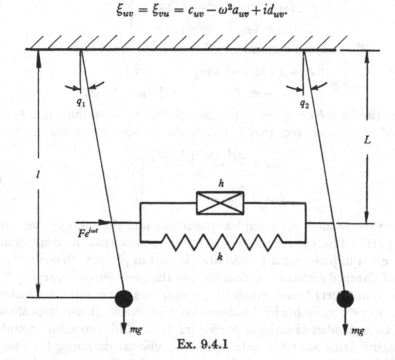

Ex. 9.4.1

The set of equations (9.5.2) may be solved to give the amplitudes Ψ as linear functions of the Φ's; thus

$$\Psi_r = \alpha_{r1}\Phi_1 + \alpha_{r2}\Phi_2 + \ldots + \alpha_{rn}\Phi_n \quad (r = 1, 2, \ldots, n). \tag{9.5.4}$$

In this expression the receptances α are functions of the coefficients ξ and are therefore complex quantities. If Δ denotes the determinant of the coefficients, thus

$$\Delta = \begin{vmatrix}
\xi_{11} & \xi_{12} & \cdots & \xi_{1n} \\
\xi_{21} & \xi_{22} & \cdots & \xi_{2n} \\
\cdots & \cdots & \cdots & \cdots \\
\xi_{n1} & \xi_{n2} & \cdots & \xi_{nn}
\end{vmatrix} \tag{9.5.5}$$

and if Δ_{rs} is the determinant formed from Δ by omitting the rth column and sth row, then by the usual rule we have

$$\alpha_{rs} = (-1)^{r+s}\frac{\Delta_{rs}}{\Delta}. \tag{9.5.6}$$

Once again, because ξ_{uv} and ξ_{vu} are the same, the reciprocal relation

$$\alpha_{rs} = \alpha_{sr} \tag{9.5.7}$$

holds good.

For an illustration of the process of forming receptances consider the system of fig. 9.4.1. From equations (9.4.21), it is seen that

$$\left.\begin{array}{l} \xi_{11} = k_1 - \omega^2(M+m) + ih_1, \\ \xi_{22} = mgl - ml^2\omega^2 + ih_2, \\ \xi_{12} = \xi_{21} = -ml\omega^2. \end{array}\right\} \tag{9.5.8}$$

From these coefficients, the determinant Δ may be formed; thus

$$\Delta = \begin{vmatrix} k_1 - \omega^2(M+m) + ih_1 & -ml\omega^2 \\ -ml\omega^2 & mgl - ml^2\omega^2 + ih_2 \end{vmatrix}. \tag{9.5.9}$$

In the figure the force $Fe^{i\omega t}$ is the only source of excitation, so that $Q_2 = 0$. Accordingly α_{11} and α_{21} only are required to specify the motion. These are

$$\left.\begin{array}{l} \alpha_{11} = \dfrac{mgl - ml^2\omega^2 + ih_2}{\Delta}, \\[2ex] \alpha_{21} = \dfrac{ml\omega^2}{\Delta}. \end{array}\right\} \tag{9.5.10}$$

Both of these receptances are complex quantities, and the ratio of the real to the imaginary part is different for the two. They thus show that the displacements at q_1 and q_2 are not in phase with Q_1 and are also not in phase with each other. This behaviour of damped systems was described in the preceding chapter.

It will be remembered that, when the receptances of conservative systems had been shown to be obtainable in the closed form of (9.5.6), it was then shown that they could be expanded as series of partial fractions. This procedure could not be followed in the same way for systems having viscous damping because of the presence of odd-degree terms in ω in the expansion of Δ. For systems with hysteretic damping however, the odd-degree terms are not present and it is therefore possible to expand the receptances as in Chapter 3.

Let the n roots of the equation $\qquad \Delta = 0 \tag{9.5.11}$

be denoted by $\zeta_1, \zeta_2, ..., \zeta_n$. It follows that

$$\Delta = (\text{constant}) \, (\zeta_1 - \omega^2) \, (\zeta_2 - \omega^2) \, ... \, (\zeta_n - \omega^2), \tag{9.5.12}$$

where it must be remembered that the quantities ζ are complex. The expansion of the determinant Δ_{rs} will give a polynomial in ω^2 of degree $(n-1)$, and any receptance may therefore be written in the form

$$\alpha_{rs} = \frac{f(\omega^2)}{(\zeta_1 - \omega^2) \, (\zeta_2 - \omega^2) \, ... \, (\zeta_n - \omega^2)}, \tag{9.5.13}$$

where $f(\omega^2)$ is a function of degree $(n-1)$. An expression such as this can be expanded to give

$$\alpha_{rs} = \frac{{}_1B_{rs}}{\zeta_1 - \omega^2} + \frac{{}_2B_{rs}}{\zeta_2 - \omega^2} + ... + \frac{{}_nB_{rs}}{\zeta_n - \omega^2}. \tag{9.5.14}$$

In this expansion the coefficients B are complex constants.

The quantities $\zeta_1, \zeta_2, ..., \zeta_n$ may be referred to loosely as the 'squares of the natural frequencies of the damped system'. This phraseology, however, has no physical significance because only real values of ω^2 give the equations of motion any meaning. It may, however, be noted that the ζ's may be found by solving 'frequency equations' which have been formed from receptances of sub-systems, using the formulae of Table 3, just as with undamped systems.

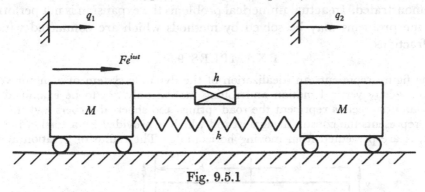

Fig. 9.5.1

The system of fig. 9.5.1 may be used to illustrate the foregoing theory. The system consists of two equal masses M which are joined by a spring of complex stiffness $K = k + ih$. Using the co-ordinates indicated, the equations of motion are

$$M\ddot{q}_1 = K(q_2 - q_1) + Fe^{i\omega t}, \left.\begin{array}{l} \\ \\ \end{array}\right\}$$
$$M\ddot{q}_2 = K(q_1 - q_2). \qquad (9.5.15)$$

Since these refer to harmonic motion they may be written in the form

$$(K - M\omega^2)\,\Psi_1 - K\Psi_2 = F, \left.\begin{array}{l} \\ \\ \end{array}\right\}$$
$$-K\Psi_1 + (K - M\omega^2)\,\Psi_2 = 0. \qquad (9.5.16)$$

These may be solved for the amplitudes Ψ giving

$$\Psi_1 = \left(\frac{K - M\omega^2}{\Delta}\right) F, \quad \Psi_2 = \left(\frac{K}{\Delta}\right) F, \qquad (9.5.17)$$

where
$$\Delta = \begin{vmatrix} K - M\omega^2 & -K \\ -K & K - M\omega^2 \end{vmatrix}. \qquad (9.5.18)$$

Now this determinant may be factorized, giving

$$\Delta = M^2\omega^2\left(\omega^2 - \frac{2K}{M}\right). \qquad (9.5.19)$$

In the previous notation, therefore,

$$\zeta_1 = 0, \quad \zeta_2 = \frac{2(k + ih)}{M}. \qquad (9.5.20)$$

The receptances of equations (9.5.17) may now be expanded into partial fractions. They become

$$\alpha_{11} = \frac{K - M\omega^2}{\Delta} = -\frac{1}{2M\omega^2} + \frac{1}{4(k + ih) - 2M\omega^2}, \left.\begin{array}{l} \\ \\ \\ \end{array}\right\}$$
$$\alpha_{21} = \frac{K}{\Delta} = -\frac{1}{2M\omega^2} - \frac{1}{4(k + ih) - 2M\omega^2}. \qquad (9.5.21)$$

These results can, alternatively, be obtained directly from the corresponding results for the conservative system by writing $(k+ih)$ in place of k.

The above example is extremely simple and more complicated problems would demand very arduous algebraic work during the process of expansion. This, however, is not important because, as in the development of the theory for undamped systems, it is only the possibility of the expansion into partial fractions which needs to be demonstrated. In actual numerical problems the expansion is not performed, though the problems may be solved by methods which are connected with the partial fractions.

EXAMPLES 9.5

1. The figure represents an idealization of the dynamic system of a motor vehicle suspension, where vertical motion and rolling motion only are to be examined. The springs k and dampers h represent the road springs and shock absorbers and the spiral spring λ represents the constraint against roll which is provided by a torsion bar. The points A, A are prevented from moving horizontally. The radius of gyration about G is R.

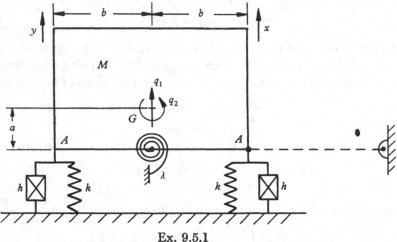

Ex. 9.5.1

Show that the displacements q_1 and q_2, indicated in the figure, are principal co-ordinates for the undamped system.

Find the direct receptance at the displacement x.

2. The arrangement of Ex. 9.5.1 is modified by deleting the damper on the left-hand side. Obtain the receptance at x by treating the system as a combination of an undamped system and of a single damper.

3. Again for the system with a damper at the right-hand side only, obtain the receptance at y, using receptance formulae of Table 2. Use the result to find the receptance at y when the damper at that side is replaced and check that the result agrees with the solution of Ex. 9.5.1.

9.6 Properties of receptances; motion in modes

In this section the general expressions (9.5.13) and (9.5.14) will be examined more closely. It may be mentioned that here, as elsewhere in this chapter, the algebraic manipulations will be the same as those used previously for real functions although we are now concerned with complex quantities. Thus the expression (9.5.6) for a receptance is the same as that of equation (3.2.10) although the latter referred specifically to undamped systems. Now the properties of sets of linear equations like (9.5.2) are discussed in works on algebra† and it is shown that the essential properties of the equations are the same if the coefficients are real or if they are complex; mathematically, the previous theory for undamped systems is a special case of that now used.

Consider first the general expressions for a direct receptance. The direct receptance α_{rr} has been shown to be given by

$$\alpha_{rr} = \frac{\Delta_{rr}}{\Delta}. \tag{9.6.1}$$

Here, Δ is a determinant which, when expanded, gives a polynomial of degree n in ω^2. Further, Δ_{rr} gives a polynomial of degree $(n-1)$ in ω^2. It may be noted that the coefficient of the highest-degree term in either determinant is dependent upon the inertia coefficients only and is therefore unaffected by the introduction of the hysteretic damping terms. The degrees of the polynomials are therefore the same as they would be if the system were undamped. In particular,

$$\Delta_{rr} = 0 \tag{9.6.2}$$

must have $(n-1)$ roots, although the determinants Δ_{rs} $(r \neq s)$ may be of lower degree in ω^2. This matter was mentioned in § 3.3. Let the $(n-1)$ roots of equation (9.6.2) be denoted by $\eta_1, \eta_2, ..., \eta_{n-1}$. The receptance α_{rr} may then be written

$$\alpha_{rr} = C \frac{(\eta_1 - \omega^2)(\eta_2 - \omega^2) \cdots (\eta_{n-1} - \omega^2)}{(\zeta_1 - \omega^2)(\zeta_2 - \omega^2) \cdots (\zeta_n - \omega^2)}, \tag{9.6.3}$$

where C is a constant.

It was shown in § 3.3 that, in the absence of damping, the values of the antiresonance frequencies which are given by the roots of equation (9.6.2) occur alternately with the natural frequencies given by the roots of the equation

$$\Delta = 0. \tag{9.6.4}$$

Thus the value of α_{rr} was shown to alternate between zero and infinity as ω^2 was varied. When hysteretic damping is present the roots η and ζ are complex whereas ω^2 is, by definition of this type of damping, necessarily real. Neither zero nor infinite values of α_{rr} are therefore possible and this is to be expected from the energy considerations which were mentioned earlier. Maximum and minimum values of α_{rr} will occur however and, for small damping, they will do so when ω^2 has almost those values at which infinite and zero receptances are obtained for the system with its damping deleted. This is evident from the fact that, if the damping constants are diminished indefinitely the real and imaginary parts of the η's and ζ's tend as follows:

$$\left. \begin{array}{ll} \mathscr{R}(\eta_i) \to \Omega_i^2, & \mathscr{I}(\eta_i) \to 0 \quad (i = 1, 2, ..., (n-1)), \\ \mathscr{R}(\zeta_j) \to \omega_j^2, & \mathscr{I}(\zeta_j) \to 0 \quad (j = 1, 2, ..., n). \end{array} \right\} \tag{9.6.5}$$

† See, for example, G. Birkhoff and S. MacLane, *A Survey of Modern Algebra* (Macmillan, London, 1941), or W. L. Ferrar, *Algebra* (ed. 2, Oxford University Press 1957).

A fuller mathematical treatment of the values of the receptances for small damping is given in the next section.

The possibility of writing a direct receptance in the form of equation (9.6.3) is illustrated by α_{11} of equation (9.5.21) which may be written

$$\alpha_{11} = \frac{\left(\dfrac{k+ih}{M} - \omega^2\right)}{(0-\omega^2)\left[\dfrac{2(k+ih)}{M} - \omega^2\right]}. \tag{9.6.6}$$

From this it will be seen that the minimum amplitude of displacement q_1 for a given amplitude of the force Q_1—that is to say the minimum value of $|\alpha_{11}|$—occurs when

$$\omega^2 \doteqdot \frac{k}{M}. \tag{9.6.7}$$

This is the value of ω^2 which corresponds to anti-resonance in the absence of damping. It will further be seen that the value

$$\omega^2 = \frac{2k}{M} \tag{9.6.8}$$

which is a resonant frequency of the undamped system, is close to the value which makes $|\alpha_{11}|$ a maximum. In each case the difference between the frequency which makes $|\alpha_{11}|$ stationary and the corresponding frequency of the undamped system, is due to the presence of the other brackets within the expression for α_{11}.

We return now to consider the cross-receptances as well as the direct ones and, in doing so, we revert to the partial fractions of equation (9.5.14). The responses at the co-ordinates $q_1, q_2, ..., q_n$, due to a single exciting force $\Phi_s e^{i\omega t}$, can be written down using the partial fraction form for all the receptances; thus

$$\left.\begin{aligned}
q_1 &= \left[\frac{{}_1B_{1s}}{\zeta_1-\omega^2} + \frac{{}_2B_{1s}}{\zeta_2-\omega^2} + ... + \frac{{}_nB_{1s}}{\zeta_n-\omega^2}\right]\Phi_s e^{i\omega t}, \\
q_2 &= \left[\frac{{}_1B_{2s}}{\zeta_1-\omega^2} + \frac{{}_2B_{2s}}{\zeta_2-\omega^2} + ... + \frac{{}_nB_{2s}}{\zeta_n-\omega^2}\right]\Phi_s e^{i\omega t}, \\
&\cdots\cdots\cdots\cdots\cdots\cdots\cdots\cdots\cdots\cdots\cdots\cdots \\
q_n &= \left[\frac{{}_1B_{ns}}{\zeta_1-\omega^2} + \frac{{}_2B_{ns}}{\zeta_2-\omega^2} + ... + \frac{{}_nB_{ns}}{\zeta_n-\omega^2}\right]\Phi_s e^{i\omega t}.
\end{aligned}\right\} \tag{9.6.9}$$

Before examining these equations in their general form, we shall take up a simple illustrative example. Consider the system of fig. 9.4.1, taking the following numerical values:

$$\left.\begin{aligned}
l = 1\,\text{ft.}, \quad M = 2\,\text{lb.}, \quad m = 1\,\text{lb.}, \quad k_1 = 256\,\text{pdl./ft.}, \\
h_1 = 50\,\text{pdl./ft.}, \quad h_2 = 10\,\text{ft.pdl./rad.}
\end{aligned}\right\}$$

Equation (9.5.9) shows that the determinant Δ for the equations of motion of the system is

$$\Delta = \begin{vmatrix} 256 - 3\omega^2 + 50i & -\omega^2 \\ -\omega^2 & 32 - \omega^2 + 10i \end{vmatrix}. \tag{9.6.10}$$

This factorizes to
$$\Delta = 2(27 \cdot 66 + 8 \cdot 07i - \omega^2)\,(148 \cdot 34 + 31 \cdot 93i - \omega^2) \qquad (9.6.11)$$
so that the roots ζ_1 and ζ_2 are
$$\zeta_1 = 27 \cdot 66 + 8 \cdot 07i, \quad \zeta_2 = 148 \cdot 34 + 31 \cdot 93i, \qquad (9.6.12)$$
and the receptances α_{11} and α_{12} are
$$\left.\begin{aligned}
\alpha_{11} &= \frac{32 - \omega^2 + 10i}{2(\zeta_1 - \omega^2)\,(\zeta_2 - \omega^2)}, \\[2mm]
\alpha_{12} &= \alpha_{21} = \frac{\omega^2}{2(\zeta_1 - \omega^2)\,(\zeta_2 - \omega^2)}.
\end{aligned}\right\} \qquad (9.6.13)$$

These expressions may be split into partial fractions so that, if a force $F e^{i\omega t}$ pdl. acts at q_1, the displacements are
$$\left.\begin{aligned}
q_1 &= \alpha_{11} F e^{i\omega t} = \left[\frac{(0 \cdot 019 + 0 \cdot 004i)}{(27 \cdot 66 + 8 \cdot 07i - \omega^2)} + \frac{(0 \cdot 481 - 0 \cdot 004i)}{(148 \cdot 34 + 31 \cdot 93i - \omega^2)}\right] F e^{i\omega t} \text{ (ft.)}, \\[2mm]
q_2 &= \alpha_{21} F e^{i\omega t} = \left[\frac{(0 \cdot 117 + 0 \cdot 010i)}{(27 \cdot 66 + 8 \cdot 07i - \omega^2)} + \frac{(-0 \cdot 617 - 0 \cdot 010i)}{(148 \cdot 34 + 31 \cdot 93i - \omega^2)}\right] F e^{i\omega t} \text{ (rad.)}.
\end{aligned}\right\}$$
$$(9.6.14)$$

We again emphasize that numerical calculations of this type are not required in practice. They are made here in order to help the reader to visualize the meaning of the general equation (9.6.9).

Returning now to these general equations, consider the set of numerators in one column of the fractions. For an undamped system, these numerators defined a modal shape. Thus, taking the mth column, the modal shape corresponding to excitation at q_s would be
$$_m q_1 : _m q_2 : \dots : _m q_n : : _m B_{1s} : _m B_{2s} : \dots : _m B_{ns} \qquad (9.6.15)$$
although, for undamped systems, we used the symbol A instead of B. Further it was shown that if the motion were produced by excitation at some co-ordinate q_r, other than q_s, then the modal shape, thereby found to be associated with the mth frequency, was identical with that found for excitation at q_s. This was because in each case the motion found was free, being obtained as a limiting case of forced motion as the natural frequency ω_m was approached.

Now in the present case this latter argument, based on physical reasoning, cannot be applied in quite the same way because the equations refer to steady forced motion only, in which ω^2 has a real value. For such motion the denominators never vanish and a free vibration in which the modal shape is given by (9.6.15) is not possible. But while the physical argument is no longer applicable it is still possible to follow the argument through algebraically and thereby to show that the ratios of the set of numerators
$$_m B_{1s} : _m B_{2s} : \dots : _m B_{ns}$$
are equal to the ratios $\qquad _m B_{1r} : _m B_{2r} : \dots : _m B_{nr}.$

We first note that, if the n relations of equations (9.5.2) are written with the right-hand side zero, thus

$$\left.\begin{aligned}
\xi_{11}\Psi_1' + \xi_{12}\Psi_2' + \ldots + \xi_{1n}\Psi_n' &= 0, \\
\xi_{21}\Psi_1' + \xi_{22}\Psi_2' + \ldots + \xi_{2n}\Psi_n' &= 0, \\
&\cdots\cdots\cdots\cdots\cdots\cdots\cdots\cdots\cdots \\
\xi_{n1}\Psi_1' + \xi_{n2}\Psi_2' + \ldots + \xi_{nn}\Psi_n' &= 0,
\end{aligned}\right\} \tag{9.6.16}$$

then they form n relations between the $(n-1)$ quantities

$$\frac{\Psi_2'}{\Psi_1'}, \quad \frac{\Psi_3'}{\Psi_1'}, \quad \ldots, \quad \frac{\Psi_n'}{\Psi_1'}. \tag{9.6.17}$$

This is evident if the equations (9.6.16) are divided throughout by Ψ_1' and if the coefficients $\xi_{11}, \xi_{21}, \ldots, \xi_{n1}$ are then moved over to the right-hand side. Now a set of values of the $(n-1)$ quantities Ψ_2'/Ψ_1', etc., which satisfy all n equations, can only be found if a certain relation exists between the coefficients ξ_{rs}. This relation is in fact that the determinant Δ shall vanish. This occurs if ω^2 has one of the values $\zeta_1, \zeta_2, \ldots, \zeta_n$ so that, for each of these roots, a set of values of the quantities (9.6.17) may be found. Such a set of values constitutes a mode of the damped system, since it is a set of ratios of the amplitudes of the co-ordinates which holds during vibration 'with zero excitation'. We shall here call modes of this sort 'damped principal modes', the adjective 'damped' being used to distinguish these modes from the principal modes which the system possesses in the absence of its damping.

This approach to the existence of characteristic modes and frequencies is often used for undamped systems for which the physical meaning of the equations for free vibrations is clear. But, as we have said, when it is applied to systems with hysteretic damping the physical meaning is lost because the equations have been formed on the understanding that they refer only to steady forced motion, that is to say, to real values of ω^2. There is, however, no objection to defining the damped principal modes of the system as those sets of ratios of the co-ordinates q which are obtained by solving equations (9.6.16) for each of the roots $\zeta_1, \zeta_2, \ldots, \zeta_n$ of equation (9.6.4). Noting the existence of these modes, it can now be seen that they must be identical with the various sets of ratios of the numerators that were mentioned earlier. Thus the first mode will be given by the ratios

$$_1q_1 : {}_1q_2 : \ldots : {}_1q_n :: {}_1B_{1s} : {}_1B_{2s} : \ldots : {}_1B_{ns} \tag{9.6.18}$$

whatever the number s may be. This can be seen from the fact that if ω^2 is set equal to ζ_1 in equations (9.6.9) and, at the same time, Φ_s is allowed to tend to zero, then the ratios of the q's thereby obtained will be the same as those obtained by solving equations (9.6.16) with $\omega^2 = \zeta_1$. For equations (9.6.9) are then merely solutions of equations (9.6.16).

Having defined the damped principal modes of a system it is evidently possible to use a set of corresponding co-ordinates which will represent the magnitudes of the distortions in these modes. These co-ordinates will be linear functions of the co-ordinates q, as are the principal co-ordinates of undamped systems. In the present case, however, the coefficients in the linear functions will be complex.

If the motion in one damped principal mode only is considered, and the appropriate co-ordinate corresponding to this mode is denoted by p'_m then the variation of p'_m with ω, for a given exciting force, follows the same law as does the co-ordinate of a system with one degree of freedom. This is shown by the form of the terms of the mth column in equation (9.6.9) which are such that

$$p'_m = \frac{(\text{constant}) \times e^{i\omega t}}{\zeta_m - \omega^2}. \tag{9.6.19}$$

It is now possible to define a Q-factor for a co-ordinate such as this by treating the motion in the corresponding mode as if it were that of a system with one degree of freedom. For such a system, consisting of a mass M and a hysteretically damped anchoring spring of complex stiffness $K = k + ih$ (as in fig. 9.2.1), the displacement x is given by

$$x = \frac{F e^{i\omega t}}{k + ih - M\omega^2} \tag{9.6.20}$$

and the Q-factor is

$$Q = \frac{k}{h}. \tag{9.6.21}$$

Accordingly the Q-factor for the mth damped principal mode of a multi-freedom system can be defined as

$$Q = \frac{\mathscr{R}(\zeta_m)}{\mathscr{I}(\zeta_m)}. \tag{9.6.22}$$

The relation between this definition, which has been obtained by algebraic manipulation of the equations of motion, and the definition of Q by the ratio of energy stored to energy lost per cycle, will be seen later, when the special question of light damping is discussed.

The Q-factors, defined as above, may now be found for the system of fig. 9.4.1, using equation (9.6.14). For the first mode,

$$Q = \frac{27\cdot66}{8\cdot07} = 3\cdot43 \tag{9.6.23}$$

and for the second

$$Q = \frac{148\cdot34}{31\cdot93} = 4\cdot65. \tag{9.6.24}$$

These show that the damping is heavy and severe vibration would not normally be met with in a system having such low Q-factors.

The independence of the damped principal modes, and the possibility of finding an appropriate Q-factor for each, means that the nomogram of fig. 9.2.7 can be used for systems with more than one degree of freedom as well as for single-degree-of-freedom systems.

In most numerical work concerning damped systems, the analysis is simplified by various approximations. These are justifiable provided that the systems do not possess nearly-equal natural frequencies and that the damping terms are not large; they may usually be used for Q-factors of 10 or more. The theory is discussed in the next section.

EXAMPLES 9.6

1. Show that the damped principal modes of the system shown are identical with its principal modes and that the value of the anti-resonance frequency is unaltered by the incorporation of the dampers.

Find the Q-factor for the motion in its two modes.

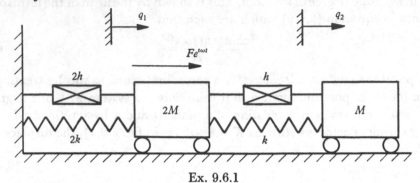

Ex. 9.6.1

2. Show that the displacements q_1 and q_2 of the system of Ex. 9.6.1 are out of phase with each other by the angle

$$\tan^{-1}\left(\frac{h}{k - M\omega^2}\right) - \tan^{-1}\left(\frac{h}{k}\right).$$

3. Use the following numerical values in the system of Ex. 9.5.1:

$$a = 1 \text{ ft.}$$
$$b = 2 \text{ ft.}$$
$$R = 1 \text{ ft.}$$
$$M = 2000 \text{ lb.}$$
$$k = 50000 \text{ pdl./ft.}$$
$$\lambda = 400000 \text{ ft.pdl./rad.}$$
$$h = 20000 \text{ pdl./ft.}$$

Using these values, and assuming that one damper only is present, obtain approximate values for ζ_1 and ζ_2.

4. Find the modal shapes which correspond to the solutions of Ex. 9.6.3.

9.7 Lightly damped systems

When the damping within a system is light, so that the various Q-factors are greater than (say) 10, the following assumptions are usually made in vibration analysis.

(a) That the resonant frequencies, that is to say the frequencies at which the response is a maximum, are the same for the damped system as they would be if it were undamped.

(b) That, similarly, the anti-resonance frequencies are not altered by the damping effects.

(c) That the modal shapes for the damped system, which are given by the columns of numerators in equations (9.6.9), are identical with those for the undamped system. It will be seen that this assumption involves neglecting any phase differences within one mode.

In this section we shall examine the validity of these assumptions.

An idea of the degree of approximation which is involved in the assumptions may be obtained by applying them to the system of fig. 9.4.1. If l, M, m and k, have the values previously quoted, but if h_1 and h_2 are zero, then it is found that

$$\left.\begin{aligned} q_1 = \alpha_{11} F e^{i\omega t} = \left[\frac{0 \cdot 018}{27 \cdot 60 - \omega^2} + \frac{0 \cdot 482}{148 \cdot 40 - \omega^2}\right] F e^{i\omega t} \text{ (ft.),} \\ q_2 = \alpha_{21} F e^{i\omega t} = \left[\frac{0 \cdot 114}{27 \cdot 60 - \omega^2} - \frac{0 \cdot 614}{148 \cdot 40 - \omega^2}\right] F e^{i\omega t} \text{ (rad.).} \end{aligned}\right\} \quad (9.7.1)$$

These expressions are comparable with those of equations (9.6.14) which refer to the damped system. The values of $27 \cdot 60$ and $148 \cdot 40$ for ω_1^2 and ω_2^2 may be compared with $27 \cdot 66$ and $148 \cdot 34$ which correspond to the minimum values of the denominators in equations (9.6.14). Again the modal shapes given by equations (9.7.1) are

$$\left.\begin{aligned} {}_1q_1 : {}_1q_2 :: 0 \cdot 018 : 0 \cdot 114, \\ {}_2q_1 : {}_2q_2 :: 0 \cdot 482 : -0 \cdot 614. \end{aligned}\right\} \quad (9.7.2)$$

in units of feet and radians whereas the corresponding ratios between the q's, for the damped system, are

$$\left.\begin{aligned} {}_1q_1 : {}_1q_2 :: (0 \cdot 019 + 0 \cdot 004i) : (0 \cdot 117 + 0 \cdot 010i), \\ {}_2q_1 : {}_2q_2 :: (0 \cdot 481 - 0 \cdot 004i) : (-0 \cdot 617 - 0 \cdot 010i). \end{aligned}\right\} \quad (9.7.3)$$

Finally the anti-resonance frequency of the undamped system can be shown to be

$$\Omega = 4\sqrt{2} \text{ rad./sec.} \quad (9.7.4)$$

and this value is not altered by the damping.

These results show that, in this problem, where damping is comparatively heavy, the simplifying assumptions mentioned do not introduce any appreciable error. We shall now modify the general theory of damped systems in order to justify the use of the assumptions.

Consider a system whose stiffness coefficients are $c_{uv} + i d_{uv}$, where d_{uv} is small compared with c_{uv}. The quantities $\zeta_1, \zeta_2, ..., \zeta_n$, which are the roots of the equation

$$\Delta = 0 \quad (9.7.5)$$

will be functions of the quantites $(c_{uv} + i d_{uv})$. It will be shown next, by Taylor's theorem, that any root ζ_r may be put in the form

$$\zeta_r = \omega_r^2 + i[\text{real function of order } d_{uv}] + [\text{functions of higher orders of } d_{uv}], \quad (9.7.6)$$

where ω_r is the rth natural frequency for the system when the damping is deleted.

Equation (9.7.5) is satisfied by substituting one of the roots ζ_r for ω^2. The expression which is thereby obtained has the form

$$f(\zeta_r, z_{11}, z_{22}, \dots, z_{nn}, z_{12}, \dots) = 0, \qquad (9.7.7)$$

where
$$z_{uv} = c_{uv} + id_{uv}. \qquad (9.7.8)$$

Consider now the way in which ζ_r varies as the coefficients d are varied, the coefficients c being held constant. For the particular case when all the d's are zero ζ_r has the value ω_r^2. If the coefficients d are not zero then, according to Taylor's theorem,

$$\zeta_r = \omega_r^2 + \sum_{u=1}^{n} \sum_{v=1}^{n} \left[\frac{\partial \zeta_r}{\partial z_{uv}} (id_{uv}) \right] + \dots, \qquad (9.7.9)$$

where the derivatives are all evaluated at the point where all the d's are zero.

To obtain the derivatives we have to differentiate the implicit function (9.7.7) of ζ_r and the quantities z. This may be done by using a theorem in the theory of the complex variable which states that, provided the function is analytic and that equation (9.7.5) has no repeated roots, then

$$\left(\frac{\partial \zeta_r}{\partial z_{uv}} \right)_{\zeta_r = \omega_r^2} = - \frac{\left(\dfrac{\partial f}{\partial z_{uv}} \right)_{\zeta_r = \omega_r^2}}{\left(\dfrac{\partial f}{\partial \zeta_r} \right)_{\zeta_r = \omega_r^2}}. \qquad (9.7.10)$$

The two derivatives on the right-hand side of this equation are real because they both refer to the point where all the d's vanish. It follows that their quotient is also real and hence all the terms contained within

$$\sum_{u=1}^{n} \sum_{v=1}^{n} \left[\frac{\partial \zeta_r}{\partial z_{uv}} (id_{uv}) \right]$$

of equation (9.7.9) are imaginary.

From this it can be seen that, provided

$$d_{uv} \left(\frac{\partial \zeta_r}{\partial z_{uv}} \right)_{\zeta_r = \omega_r^2} \ll \omega_r^2 \qquad (9.7.11)$$

for all values of u, v and r and that the higher-order terms are negligible, then the damping of the system may be treated as light. Under these conditions any receptance (9.5.13) may be written as

$$\alpha_{rs} \doteqdot \frac{f(\omega^2)}{(\omega_1^2 + iO[d] - \omega^2)(\omega_2^2 + iO[d] - \omega^2) \dots (\omega_n^2 + iO[d] - \omega^2)}, \qquad (9.7.12)$$

where $O[d]$ signifies a real quantity of the first order in the coefficients d. The properties of receptances of this form, and subsequently the modal shapes which may be derived from them, can now be examined.

Consider first a direct receptance α_{rr}. The roots of

$$\Delta_{rr} = 0 \qquad (9.7.13)$$

will be modified, in a similar way to those of equation (9.7.5) by the presence of the coefficients d. Thus the rth anti-resonance frequency, Ω_r, of the undamped system is related to the rth root of equation (9.7.13) by the approximate relation

$$\eta_r \doteqdot \Omega_r^2 + iO[d] \quad (m = 1, 2, ..., n-1), \tag{9.7.14}$$

where the function of order d is real.

Direct receptances may therefore be written in the general form

$$\alpha_{rr} \doteqdot K \frac{(\Omega_1^2 + iO[d] - \omega^2)\,(\Omega_2^2 + iO[d] - \omega^2) \dots (\Omega_{n-1}^2 + iO[d] - \omega^2)}{(\omega_1^2 + iO[d] - \omega^2)\,(\omega_2^2 + iO[d] - \omega^2) \dots (\omega_n^2 + iO[d] - \omega^2)}, \tag{9.7.15}$$

where K is a constant. This expression shows that the frequencies at which minimum and maximum responses occur differ only slightly from the anti-resonance and resonance frequencies respectively. Further, the minimum values of α_{rr} are small, being of order of d, while the maxima are large, being of order $1/d$.

To examine the modal shapes for the case where the damping is small it is necessary to obtain the series form of the general expression for a receptance. In doing this, we shall first show that the numerators $_mB_{rs}$ of the partial fractions of equations (9.5.14) have the form

$$_mB_{rs} \doteqdot {_mA_{rs}} + iO[d] \quad (m = 1, 2, ..., n), \tag{9.7.16}$$

where $_mA_{rs}$ is the corresponding numerator when the damping is absent (see equations (3.4.3)) and where again the function of order d is real.

The constants $_mB_{rs}$ of equation (9.5.14) may be found explicitly as follows. If both sides of the equations

$$\alpha_{rs} = (-1)^{r+s}\frac{\Delta_{rs}}{\Delta} = \sum_{m=1}^{n} \frac{_mB_{rs}}{\zeta_m - \omega^2} \tag{9.7.17}$$

are multiplied by Δ and then ω^2 is put equal to ζ_m, it is found that

$$_mB_{rs} = \frac{(-1)^{r+s}\,(\Delta_{rs})_{\omega^2 = \zeta_m}}{\left(\dfrac{\Delta}{\zeta_m - \omega^2}\right)_{\omega^2 = \zeta_m}} \tag{9.7.18}$$

provided that equation (9.7.5) has no repeated roots. This defines the quantities $_mB_{rs}$ as analytic functions of the coefficients z of equation (9.7.8) and of the ζ's. The latter are themselves analytic functions of the coefficients z so that the quantities $_mB_{rs}$ are analytic functions of the z's.

Taylor's theorem may now be used as before to show how the quantities $_mB_{rs}$ vary with the hysteretic damping constants, d_{uv}. When all the latter are zero the $_mB_{rs}$ numerators are identical with the $_mA_{rs}$ numerators of the undamped system. When the system is lightly damped the $_mB_{rs}$'s are given by

$$_mB_{rs} = {_mA_{rs}} + \sum_{u=1}^{n} \sum_{v=1}^{n} \left[\frac{\partial\, _mB_{rs}}{\partial z_{uv}}\,(id_{uv}) \right] + \dots. \tag{9.7.19}$$

Since, as before, all the derivatives in this series are formed for a point where all the z's are real, the $_mB_{rs}$ terms differ from the $_mA_{rs}$'s by containing imaginary components of the first degree in the coefficients d_{uv}.

If the system is excited by a force Q_s at q_s, then the total motion is given by

$$
\left.
\begin{aligned}
q_1 &= \left[\frac{{}_1A_{1s}+iO[d]}{\omega_1^2+iO[d]-\omega^2} + \frac{{}_2A_{1s}+iO[d]}{\omega_2^2+iO[d]-\omega^2} + \cdots + \frac{{}_nA_{1s}+iO[d]}{\omega_n^2+iO[d]-\omega^2} \right] \Phi_s e^{i\omega t}, \\
q_2 &= \left[\frac{{}_1A_{2s}+iO[d]}{\omega_1^2+iO[d]-\omega^2} + \frac{{}_2A_{2s}+iO[d]}{\omega_2^2+iO[d]-\omega^2} + \cdots + \frac{{}_nA_{2s}+iO[d]}{\omega_n^2+iO[d]-\omega^2} \right] \Phi_s e^{i\omega t}, \\
&\cdots\cdots\cdots\cdots\cdots\cdots\cdots\cdots\cdots\cdots\cdots\cdots \\
q_n &= \left[\frac{{}_1A_{ns}+iO[d]}{\omega_1^2+iO[d]-\omega^2} + \frac{{}_2A_{ns}+iO[d]}{\omega_2^2+iO[d]-\omega^2} + \cdots + \frac{{}_nA_{ns}+iO[d]}{\omega_n^2+iO[d]-\omega^2} \right] \Phi_s e^{i\omega t}.
\end{aligned}
\right\} \quad (9.7.20)
$$

The modes given by these equations are complex but, because the complex terms in the numerators are only of order d in magnitude, the phase differences within any one mode are small. The modal shapes are only slightly different from those of the undamped system and if ω is made equal to one of the natural frequencies then the distortion in the corresponding mode is much larger than that in any other.

The effects of the imaginary terms in equations (9.7.20) may be illustrated by referring once more to the system of fig. 9.4.1. Let the same values of l, M, m and k as before be retained but let the damping constants be reduced, h_1 from 50 pdl./ft. to 30 and h_2 from 10 ft.pdl./rad. to 5. With these changes the results corresponding to those of equations (9.6.14) are

$$
\left.
\begin{aligned}
q_1 &= \alpha_{11} F e^{i\omega t} = \left[\frac{0{\cdot}018+0{\cdot}001i}{27{\cdot}61+4{\cdot}13i-\omega^2} + \frac{0{\cdot}482-0{\cdot}001i}{148{\cdot}39+18{\cdot}31i-\omega^2} \right] F e^{i\omega t} \text{ (ft.)}, \\
q_2 &= \alpha_{21} F e^{i\omega t} = \left[\frac{0{\cdot}115+0{\cdot}004i}{27{\cdot}61+4{\cdot}13i-\omega^2} + \frac{-0{\cdot}615-0{\cdot}004i}{148{\cdot}39+18{\cdot}31i-\omega^2} \right] F e^{i\omega t} \text{ (rad.)}.
\end{aligned}
\right\} \quad (9.7.21)
$$

These results will be seen to approximate very closely to those of equations (9.7.1) although, even with the reduced value of the damping, the Q-factors are low, being 6·69 for the first mode and 8·10 for the second. The receptances α_{11} and α_{12} of equations (9.7.21) are represented in figs. 9.7.1 and 9.7.2 respectively.

The nature of the approximation involved when the resonant motion in one mode is assumed to account for the whole of a system's motion can be seen from equations (9.7.20). In particular it will be seen that a system having two or more natural frequencies which are nearly equal demands special consideration. For if the frequency of excitation is nearly equal to these frequencies in such a system, then the motion will, in general, be large in more than one mode.

The definition given in equation (9.6.22) of the Q-factors for a multi-freedom, hysteretically damped system was based on the ratio of the real and imaginary parts of the relevant roots ζ, and not on the idea of stored- and dissipated-energies. The reason for this was that, in the general case, there are phase differences within each mode and this prevents a simple expression for either the total kinetic or potential energy from being written down. Also, although it would be possible (as we showed) to use a set of co-ordinates which would be the equivalent of principal co-ordinates in an undamped system, it would not be possible to attribute simple physical meanings to the energy coefficients a and c which would thereby be

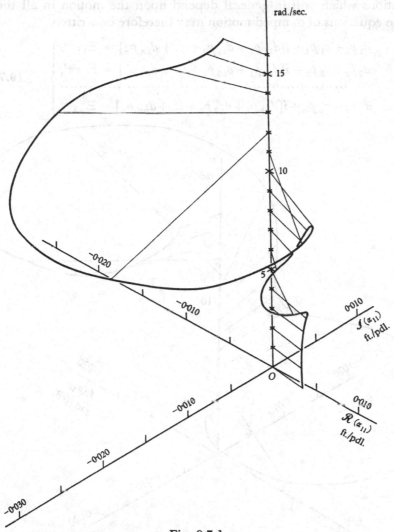

Fig. 9.7.1

introduced. When the damping is light, however, it is possible to define the Q-factors in terms of the ratios of stored and dissipated energies.

To do this, consider first the equations of motion of the damped system written in terms of the principal co-ordinates of the undamped system. If the damping were effaced, the equations would be

$$
\left.
\begin{aligned}
a_1\ddot{p}_1 + c_1 p_1 &= \Xi_1 e^{i\omega t}, \\
a_2\ddot{p}_2 + c_2 p_2 &= \Xi_2 e^{i\omega t}, \\
&\cdots\cdots\cdots\cdots\cdots \\
a_n\ddot{p}_n + c_n p_n &= \Xi_n e^{i\omega t},
\end{aligned}
\right\}
\tag{9.7.22}
$$

where the constants Ξ are the amplitudes of the harmonic applied generalized forces P. When the damping forces are present there are quadrature terms in all these equations which will in general depend upon the motion in all the other modes. The equations of damped motion may therefore be written

$$a_1 \ddot{p}_1 + c_1 p_1 + i[d_{11} p_1 + d_{12} p_2 + \ldots + d_{1n} p_n] = \Xi_1 e^{i\omega t},$$
$$a_2 \ddot{p}_2 + c_2 p_2 + i[d_{21} p_1 + d_{22} p_2 + \ldots + d_{2n} p_n] = \Xi_2 e^{i\omega t}, \qquad (9.7.23)$$
$$\cdots\cdots\cdots\cdots\cdots\cdots\cdots\cdots\cdots\cdots\cdots\cdots\cdots\cdots\cdots$$
$$a_n \ddot{p}_n + c_n p_n + i[d_{n1} p_1 + d_{n2} p_2 + \ldots + d_{nn} p_n] = \Xi_n e^{i\omega t},$$

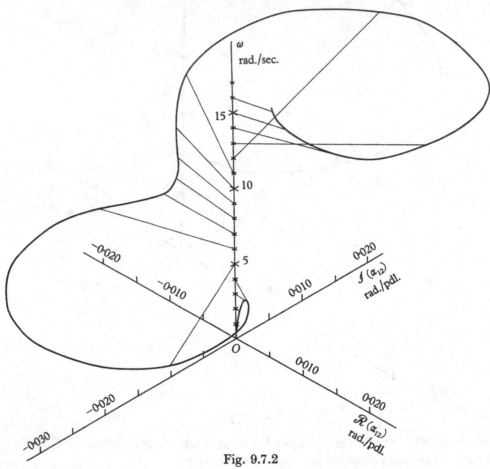

Fig. 9.7.2

where the coefficients d are, of course, not the same as those previously used in the q-equations. From these equations it may be shown, by arguments similar to those used in § 8.7, that, if the damping is light and if there are no other natural frequencies close to ω_r, then for resonance in the rth mode

$$p_r \doteqdot \frac{\Xi_r e^{i\omega_r t}}{i d_{rr}}, \qquad p_s \ll p_r \quad (s \neq r). \qquad (9.7.24)$$

Under these conditions the stored energy may be taken as that due to p_r only and the Q-factor may be defined in terms of this provided that the rate of dissipation can be found.

Consider first the work which is done per cycle on a single damping element whose constant is h. This work is

$$E = \pi h R^2, \tag{9.7.25}$$

where R represents the amplitude of the relevant relative displacement, x. Following the argument of §9.4 we can write the quantity hx^2 in the form

$$hx^2 = h\left[\frac{\partial x}{\partial p_1}p_1 + \frac{\partial x}{\partial p_2}p_2 + \ldots + \frac{\partial x}{\partial p_n}p_n\right]^2 \tag{9.7.26}$$

and this may be expanded to give

$$hx^2 = d_{11}p_1^2 + d_{22}p_2^2 + \ldots + d_{nn}p_n^2 + 2d_{12}p_1p_2 + \ldots, \tag{9.7.27}$$

where the coefficients d refer to the system with a single damper only. This second form of hx^2 holds by virtue of the definition of the d's which was adopted originally.

Now if only the motion at p_r is sufficiently large to be considered, then the energy (9.7.25) will be

$$E = \pi d_{rr} \Pi_r^2, \tag{9.7.28}$$

where Π_r is the amplitude of p_r. If, rather than the single damper, a number of dampers are present, then equation (9.7.28) will still be valid, but the value of d_{rr} will be the sum of a number of contributions from the various damping sources.

The maximum potential energy of the system during this resonant motion will be $\frac{1}{2}c_r \Pi_r^2$. The Q-factor may be defined therefore by the ratio

$$\frac{V_{\text{max.}}}{E} = \frac{\frac{1}{2}c_r \Pi_r^2}{\pi d_{rr} \Pi_r^2} = \frac{1}{2\pi} Q \tag{9.7.29}$$

so that

$$Q = \frac{c_r}{d_{rr}} = \frac{a_r \omega_r^2}{d_{rr}}. \tag{9.7.30}$$

This fits in with the previous definition, although the present discussion is limited to small damping whereas the previous one was general.

The agreement may be expected because if the small imaginary terms in the numerators in equations (9.7.20) are neglected then the equations show that the various columns correspond to motion in the various principal modes. This being so, then close to resonance the relevant co-ordinate p_r is proportional to

$$\frac{1}{\omega_r^2 + iO[d] - \omega^2} \quad \text{or} \quad \frac{1}{c_r + ia_r.O[d] - a_r \omega^2}$$

so that the quantity d_{rr} must be identified with $a_r.O[d]$. Now the ratio of this quantity to the constant c_r was used in the earlier definition of Q so that for small damping the two definitions are equivalent.

EXAMPLE 9.7

1. (a) Sketch the general form of the (three-dimensional) curve of a direct receptance for a multi-freedom system (cf. fig. 9.7.1). Examine the effects on the shape of the curve of diminishing the damping within the system to zero.

   (b) Do the same, taking a cross-receptance.

9.8 Infinite-freedom systems

The analysis of infinite-freedom systems which experience damping forces has not yet been discussed explicitly in this book. This does not mean, however, that the foregoing theory of damped vibration has no connexion with continuous systems. For the assumption is made that, as far as their general dynamical behaviour is concerned, they may be regarded as having large but finite freedom.

When the damping of an infinite-freedom system may be regarded as concentrated at discrete points, the analysis can be handled directly by the methods which have already been described. If, for instance, a uniform heavy shaft is excited into torsional vibration, and if it is affected by damping forces which act within one or more short bearings, then equations of motion may be derived by using the complex receptances at the bearings together with the real receptances of the undamped shaft.

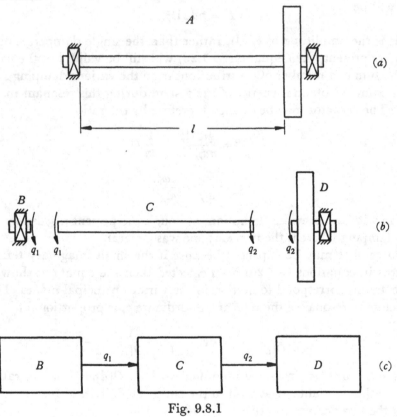

Fig. 9.8.1

Consider as an example a uniform shaft of length l which is supported in bearings at its ends and which carries a flywheel of inertia I as shown in fig. 9.8.1 (a). Let the second polar moment of area of the shaft be J, its density ρ and shear modulus G, and let hysteretic damping torques, of coefficient h, act at each of the bearings. The receptance δ_{22} of the bearing and flywheel together, which may be regarded as a sub-system D to which the shaft is attached at a co-ordinate q_2 (fig. 9.8.1 (b)), is given by

$$\frac{1}{\delta_{22}} = -I\omega^2 + ih. \tag{9.8.1}$$

If now a torque $F e^{i\omega t}$ acts at the end of the shaft remote from the flywheel, and if the receptances of the shaft alone are denoted by γ's then, calling q_1 the rotation at the point of excitation and neglecting for the moment the damping at that point, the direct receptance, α_{11}' say, at q_1 of C and D together is given by the formula

$$\alpha_{11}' = \gamma_{11} - \frac{\gamma_{12}^2}{\gamma_{22} + \delta_{22}} \tag{9.8.2}$$

(see entry 1 in Table 2). Here, δ_{22} is taken from equation (9.8.1) and the γ's may be taken from Table 5. Finally the damping at the bearing which comprises sub-system B may be taken into account by using the relation

$$\frac{1}{\alpha_{11}} = \frac{1}{\alpha_{11}'} + ih \tag{9.8.3}$$

in which α_{11} is the receptance of the whole system which comprises both bearings, flywheel and shaft. This result could equally well be obtained by means of entry 2 in Table 2 using the sub-systems shown in fig. 9.8.1 (c).

Another problem of this sort, in which the damping forces are localized, was discussed in § 9.3.

This type of problem does not need further discussion and we now go on to consider some simple problems in which damping forces are uniformly distributed in continuous bodies as they would be for instance if they were due to elastic hysteresis. It is possible to examine them by assuming either viscous† or hysteretic damping; we shall discuss only the latter as it leads to slightly simpler calculations.

The sources of hysteretic damping within a uniform shaft may be thought of as being in parallel with the elastic constraint which is present between any two sections. It may thus be taken into account by replacing the shear modulus G by a complex quantity, the equation of motion being otherwise unaltered. The differential equation of motion derived in Chapter 6 is

$$G \frac{\partial^2 \theta}{\partial x^2} = \rho \frac{\partial^2 \theta}{\partial t^2}. \tag{9.8.4}$$

This is now modified, to allow for hysteretic damping, by writing it

$$G(1 + i\mu) \frac{\partial^2 \theta}{\partial x^2} = \rho \frac{\partial^2 \theta}{\partial t^2} \tag{9.8.5}$$

remembering that the restriction to harmonic motion must now be observed.

It would be possible to write the complex form of G into the closed-form receptances of Table 5 and thereby to obtain the motion.‡ In fact, however, this introduces algebraic complication. It is more practicable sometimes to insert the damping

---

† See, for example, *Theory of Sound* (1894), §148.
‡ See N. O. Myklestad, 'The Concept of Complex Damping', *J. Appl. Mech.* vol. 19 (1952), p. 284.

constants into the series form of the receptances and then to curtail the series in the usual way. It is not often necessary to make this type of calculation and the matter will not be pursued here. Instead we shall examine the effect of the damping on the wave motion which was described in § 6.9, because under certain conditions the damped wave motion gives a simple solution to problems which would otherwise be troublesome.

Let the symbol G, implicit in equation (6.9.7), be replaced by the complex stiffness $G(1 + i\mu)$ so that that equation becomes

$$\theta = A \exp\left[i\omega\left(t - x \sqrt{\left(\frac{\rho}{G(1 + i\mu)} \right)} \right) \right]. \qquad (9.8.6)$$

Consider now the quantity $\sqrt{\left(\dfrac{\rho}{G(1 + i\mu)} \right)}$.

If a denotes the wave velocity $\sqrt{(G/\rho)}$, the expression may be written as

$$\frac{1}{a}\left(\frac{1}{1 + i\mu} \right)^{\frac{1}{2}}$$

which may be thrown into the alternative form

$$\frac{e^{-\frac{1}{2}i\tan^{-1}\mu}}{a(1 + \mu^2)^{\frac{1}{4}}},$$

which will be required later. If the damping is small so that μ is much less than unity, then μ^2 may be neglected in comparison with unity; the first result then becomes

$$\sqrt{\left(\frac{\rho}{G(1 + i\mu)} \right)} \doteqdot \frac{1}{a}\left[1 - i\left(\frac{\mu}{2} \right) \right]. \qquad (9.8.7)$$

If this is now substituted into equation (9.8.6), it is found that (for small damping)

$$\theta \doteqdot A\, e^{-x\mu\omega/2a}\, e^{i\omega(t - x/a)} \qquad (9.8.8)$$

This differs from equation (6.9.7) by containing the quantity $e^{-x\mu\omega/2a}$ as a factor. The solution still represents a wave which moves in the direction of increasing x, but the exponential factor makes the wave amplitude diminish continuously as the distance travelled increases. It is possible to show in a similar fashion that a wave which moves in the direction of decreasing x also diminishes with distance travelled.

These results may be expected from energy considerations for it has already been shown in § 6.9 that a travelling wave is associated with a transfer of energy along the shaft in the direction of the wave motion. Evidently if energy is absorbed at all points along the shaft then the energy transferred must continuously decrease and consequently the amplitude of displacement decreases also.

The steady forced motion of a hysteretically damped shaft, when one end is acted on by an exciting torque $Fe^{i\omega t}$ can now be shown to reduce to a particularly simple form provided that the shaft is sufficiently long. For purposes of argument we may consider a semi-infinite shaft ($0 \leqslant x < \infty$) to which the torque is applied at the section where $x = 0$. Energy is fed to the shaft at this point only, and the waves can only travel therefore in the direction of increasing x. Now at a point along the shaft where $x\mu\omega/2a = 4$ the amplitude of the wave will have fallen to about 2 % of its initial value. It follows that if the shaft is clamped at some point beyond this, then the clamp will have very little effect on the motion. The shaft may therefore still be treated approximately by regarding it as semi-infinite for the purposes of analysis, if it is a free-clamped shaft of length l, where

$$\frac{l\mu\omega}{2a} > 4. \tag{9.8.9}$$

Again if the shaft is free-free, and of similar length, then the deletion of the portion of shaft beyond $x = l$ will not affect the motion of the portion $0 \leqslant x \leqslant l$ appreciably because the torque at the section $x = l$ will be very small compared with that at $x = 0$; again therefore the shaft may be treated as semi-infinite for the purposes of analysis.

In those problems in which equation (9.8.9) is not satisfied it is necessary to take into account wave motions moving both forward and backwards in the shaft, and to find those motions which will satisfy the end conditions. This process is usually tedious and it will usually, therefore, be better to use a curtailed form of the series receptances as mentioned above.

Consider now the problem of hysteretically damped flexural motions of a uniform beam. It was shown in Chapter 7 that the differential equation governing the flexural motion of uniform beams is

$$\frac{\partial^2 v}{\partial t^2} + \frac{EI}{A\rho} \frac{\partial^4 v}{\partial x^4} = 0. \tag{9.8.10}$$

If dissipation effects are uniformly distributed along the beam they may be taken into account by assuming Young's modulus to be complex.[†] Thus we shall now discuss briefly approximate harmonic solutions of the equation

$$\frac{\partial^2 v}{\partial t^2} + \frac{E(1+i\mu) I}{A\rho} \frac{\partial^4 v}{\partial x^4} = 0. \tag{9.8.11}$$

The problem of undamped flexural waves was introduced in §7.11. It was shown, in particular, that the motion of the semi-infinite beam of fig. 7.11.1, which is excited by a couple $He^{i\omega t}$ at the end $x = 0$, is given by

$$v = \frac{H}{EI\lambda^2 \sqrt{2}} \left[e^{i(\omega t - \lambda x + \frac{1}{4}\pi)} - e^{-\lambda x} e^{i(\omega t - \frac{1}{4}\pi)} \right] \tag{9.8.12}$$

(see equation (7.11.13)). Now this solution satisfies the equation of damped motion (9.8.11) and the boundary conditions at the point of excitation if E is replaced by $E(1+i\mu)$ and λ is replaced by

$$\lambda' = \left[\frac{A\rho\omega^2}{E(1+i\mu) I} \right]^{\frac{1}{4}} = \frac{\lambda}{(1+i\mu)^{\frac{1}{4}}}. \tag{9.8.13}$$

† See N. O. Myklestad, *loc. cit.*

This solution for damped wave motion assumes a simple form when the damping is small, so that $\mu \ll 1$. When this is the case

$$\lambda' \doteq \lambda(1 - \tfrac{1}{4}i\mu). \tag{9.8.14}$$

If this value is used in equation (9.8.12) and E is replaced by the complex modulus, it is found that, to the first power of μ,

$$v = \frac{H(1 - \tfrac{1}{2}i\mu)}{EI\lambda^2 \sqrt{2}} \left[e^{-\tfrac{1}{4}\lambda\mu x} e^{i(\omega t - \lambda x + \tfrac{1}{4}\pi)} - e^{-\lambda x} e^{i(\omega t + \tfrac{1}{4}\lambda\mu x - \tfrac{1}{4}\pi)} \right]. \tag{9.8.15}$$

The first term still represents a wave which travels in the direction of increasing x, but the wave amplitude decreases exponentially by virtue of the $e^{-\tfrac{1}{4}\lambda\mu x}$ factor. This motion diminishes to about 2 % of its initial value when $\tfrac{1}{4}\lambda\mu x$ is equal to 4. The second term, which previously represented a steady vibration, now represents a vibration whose phase varies along the beam. This could be interpreted as a wave, moving in the direction of decreasing x with speed $4\omega/\lambda\mu$, whose amplitude increases exponentially as it moves. This interpretation is mathematically valid but is misleading as a description of the phenomena, particularly when the value of μ is small compared with unity, as is here assumed. In such cases the exponential factor $e^{-\lambda x}$ reduces the amplitude to a negligible value before the quantity $\tfrac{1}{4}\lambda\mu x$ in the harmonic factor has changed the phase of that factor very much. Thus the motion is better regarded as a modification of the steady forced motion which was found in the undamped beam.

Since both terms of equation (9.8.15) diminish exponentially with positive values of x, the total motion becomes negligible for sufficiently large values of x. Thus if the beam is long enough there will be no appreciable motion at the far end and the constraints which act at that end need not be considered in finding the motion. If $\mu \ll 1$, which may be expected in practical cases, then the minimum length for which the above assumptions hold will be determined by the first term of equation (9.8.15) and it will be given approximately by

$$\frac{\lambda\mu l}{4} > 4. \tag{9.8.16}$$

If the damping is too light for this condition to hold, then the problem will be better dealt with by using receptances in the series form and inserting the appropriate damping factors in each term.

EXAMPLES 9.8

1. A non-metallic bar has Young's modulus 2×10^6 psi and the material has internal damping which may be represented by writing E in the complex form

$$2 \times 10^6 (1 + 0 \cdot 2i) \text{ psi}.$$

The density is $0 \cdot 06$ lb./in.^{3} One end of the bar is acted on by a harmonically varying axial force with a frequency of 5000 cyc./sec. If the other is free, how long must the bar be for the amplitude at the other end to be less than 4% of that at the driven end?

2. A steel shaft of 1 in. diameter carries a series of flywheels, each of which has moment of inertia 70 lb.in.² The flywheels are keyed to the shaft at intervals of 2·0 in., this being the distance between the central planes of adjacent wheels, and it may be assumed that the wheels do not affect the torsional stiffness of the shaft.

The shaft is supported in bearings placed between each pair of wheels and each of these exerts a damping torque against torsional vibration. When the frequency of the vibration is 100 cyc./sec. the amplitude of this torque is 10000 lb.in./rad. The entire system may be regarded as if it were a uniform shaft with hysteretic damping.

Show that, if a harmonic torque of frequency 100 cyc./sec. is applied to one end, then it would be necessary for the shaft to be about 360 in. long for the motion at the other end to be 2% of that at the excited end.

NOTATION

Supplementary List of Symbols used in Chapter 9

$_kB_{rs}$ Numerator of kth partial fraction in series representation of α_{rs} [see equation (9.5.14)].

E Energy dissipated per cycle.

f Force applied to spring-damper combination (see fig. 9.1.1).

K Complex stiffness [see equation (9.1.8)].

P_y, P_z Generalized forces corresponding to y, z exerted on a system by a damping element (a hysteretic damper).

Q Q-factor; a dimensionless measure of freedom from damping [see equation (9.2.12)].

$(Q_r)_H$ Contribution to the rth generalized force of hysteretic damping forces.

W Work done per cycle.

x $= y - z$.

Y, Z Points of a system connected by a damping element.

y, z Generalized displacements of points Y, Z measured in the direction YZ.

ζ_r rth root of equation (9.5.11).

η Phase angle [see equation (9.2.7)].

η_r rth root of equation (9.6.2).

μ $= h/k$.

ξ_{rs} $= c_{rs} - \omega^2 a_{rs} + i d_{rs}$.

CHAPTER 10

FREE VIBRATION

The complete independence of the normal coordinates leads to an interesting theorem concerning the relation of the subsequent motion to the initial disturbance. For if the forces which act upon the system be of such a character that they do no work on the displacement indicated by $\delta\phi_1$ then $\Phi_1 = 0$. No such forces, however long continued, can produce any effect on the motion ϕ_1. If it exist, they cannot destroy it; if it do not exist, they cannot generate it.

LORD RAYLEIGH, *Theory of Sound* (1894)

In our discussion of undamped systems, free vibration has so far been regarded as a limiting case of forced vibration when the frequency approaches a natural frequency. This conception is convenient in developing the theory, but is not sufficient for all purposes. It cannot be extended to damped systems which perform free vibration. In the present chapter we shall discuss free vibration more fully, and show that motion is established if a system is given some specified configuration and velocities and is then left to vibrate freely.

10.1 Free vibration of a damped system with one degree of freedom

It is convenient to devote some space to the single-degree-of-freedom system first. The theory will be developed for viscous damping because the equations will then be linear. We have already explained that the alternative form of linear damping, namely hysteretic, is only defined for steady forced motion.

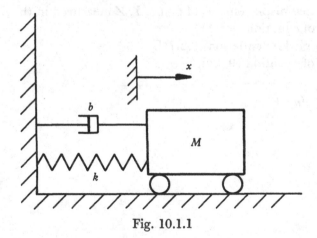

Fig. 10.1.1

The system to be discussed is shown in fig. 10.1.1. The displacement from the equilibrium configuration is x, and the equation for free motion is

$$M\ddot{x} + b\dot{x} + kx = 0, \tag{10.1.1}$$

where k is the spring stiffness and b is the viscous damping coefficient. We shall adopt the idea of a damping factor ν as in Chapter 8, this being defined by

$$\nu = \frac{b}{2\sqrt{(kM)}}. \tag{10.1.2}$$

This allows the equation of motion to be written in the form

$$\ddot{x} + 2\nu\omega_1\dot{x} + \omega_1^2 x = 0, \tag{10.1.3}$$

where ω_1 is the natural frequency when the damping is removed; that is to say

$$\omega_1^2 = \frac{k}{M}. \tag{10.1.4}$$

The dissipation of energy in the damper will make a free sinusoidal motion impossible, and a trial solution having $e^{i\omega t}$ as a factor is therefore inappropriate. Consider, instead, the trial solution

$$x = A e^{\lambda t}, \tag{10.1.5}$$

where real or complex values of λ are to be regarded as possible. Substitution leads to the auxiliary equation

$$\lambda^2 + 2\nu\omega_1\lambda + \omega_1^2 = 0 \tag{10.1.6}$$

which has the roots $\qquad \lambda = [-\nu \pm \sqrt{(\nu^2 - 1)}]\,\omega_1. \tag{10.1.7}$

The general solution of the equation of motion is accordingly

$$x = A e^{[-\nu + \sqrt{(\nu^2 - 1)}]\omega_1 t} + B e^{[-\nu - \sqrt{(\nu^2 - 1)}]\omega_1 t}, \tag{10.1.8}$$

where A and B are constants which must be chosen to suit the values of x and $\dot{x}$ which correspond to the beginning of the motion, when $t = 0$.

The character of the free motion depends upon the value of ν. If ν were zero, the result (10.1.8) would reduce to a steady harmonic motion. We shall show that if ν is not zero, but is much smaller than unity, then the motion is a slowly decaying vibration. As ν is increased further the vibration decays more rapidly and when ν equals or exceeds one there is no oscillation but only a gradual approach to the equilibrium position.

Consider first the case when $\nu > 1$. The two values of λ are then real quantities, and are negative since ν is always positive. The motion therefore consists of two exponential components, and may be expressed in the form

$$x = A e^{\lambda_1 t} + B e^{\lambda_2 t}. \tag{10.1.9}$$

The general shape of the corresponding curve of displacement against time is shown in fig. 10.1.2. Suppose that x_0 and $\dot{x}_0$ are the initial values of x and $\dot{x}$ respectively, so that

$$\left.\begin{array}{l} x_0 = A + B, \\ \dot{x}_0 = \lambda_1 A + \lambda_2 B. \end{array}\right\} \tag{10.1.10}$$

These equations may be solved to give

$$A = \frac{\dot{x}_0 - x_0\lambda_2}{\lambda_1 - \lambda_2}, \quad B = \frac{x_0\lambda_1 - \dot{x}_0}{\lambda_1 - \lambda_2}, \tag{10.1.11}$$

whence
$$x = \frac{1}{\lambda_1 - \lambda_2} [(\dot{x}_0 - x_0 \lambda_2) e^{\lambda_1 t} + (x_0 \lambda_1 - \dot{x}_0) e^{\lambda_2 t}]. \tag{10.1.12}$$

For the particular case when $\nu = 1$ the two roots of (10.1.6) are equal. The solution (10.1.12) cannot then be used directly because of the term $(\lambda_1 - \lambda_2)$ in the denominator. This difficulty may be avoided by writing

$$\lambda_2 = \lambda_1 + \epsilon \tag{10.1.13}$$

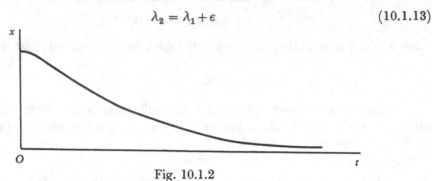

Fig. 10.1.2

and then considering the limiting case as ϵ tends to zero. If $\lambda_1 + \epsilon$ is substituted for λ_2 in equation (10.1.12) it is found that

$$x = \frac{e^{\lambda_1 t}}{-\epsilon} [(\dot{x}_0 - x_0 \lambda_1) (1 - e^{\epsilon t}) - x_0 \epsilon]. \tag{10.1.14}$$

Now
$$\lim_{\epsilon \to 0} \left(\frac{1 - e^{\epsilon t}}{-\epsilon} \right) = t$$

so that the result (10.1.14) tends to the limit

$$x = e^{\lambda_1 t} [(\dot{x}_0 - x_0 \lambda_1) t + x_0] \tag{10.1.15}$$

as λ_1 and λ_2 tend to equality. When this special case arises the damping is said to be 'critical'; it has the value which is just sufficient to prevent oscillation during free motion. From the definition of ν it follows that, for critical damping,

$$b = 2 \sqrt{(kM)} = 2M\omega_1 = b_{\text{crit}}. \tag{10.1.16}$$

The critical value of damping is of importance in the theory of control and in the design of instruments which are required to respond to suddenly changing values of the quantities which they measure.

In numerical work, the form (10.1.15) of the expression for free motion may be more convenient than that of equation (10.1.12) if the roots of the auxiliary equation are close, as well as if they are equal. The reason for this is that if the expression (10.1.14) is worked out numerically for small values of ϵ, then it will involve the determination of the quantity $(1 - e^{\epsilon t})$ where $e^{\epsilon t}$ is close to unity. The inaccuracies resulting from this process may well be greater than those introduced by assuming at the start that

$$\left(\frac{1 - e^{\epsilon t}}{-\epsilon} \right)$$

is equal to t.

Consider now the motion when $\nu < 1$. The solution may now be written in the form

$$x = e^{-\nu\omega_1 t}\{A\,e^{i[\omega_1\sqrt{(1-\nu^2)}]t} + B\,e^{-i[\omega_1\sqrt{(1-\nu^2)}]t}\} \qquad (10.1.17)$$

and this may be expressed in the alternative form

$$x = e^{-\nu\omega_1 t}[G\cos\Omega t + H\sin\Omega t], \qquad (10.1.18)$$

where

$$\Omega = \omega_1\sqrt{(1-\nu^2)} \qquad (10.1.19)$$

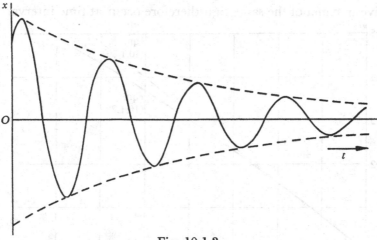

Fig. 10.1.3

and where G and H are new arbitrary constants which are to be determined by the initial conditions. The form (10.1.18) shows that the motion is oscillatory and is continuously diminished by the term outside the bracket. A curve of displacement plotted against time now has the general form shown in fig. 10.1.3.

The frequency Ω of the oscillation is less than ω_1, that of the system without damping, and is related to it by the equation

$$\frac{\Omega^2}{\omega_1^2} + \nu^2 = 1. \qquad (10.1.20)$$

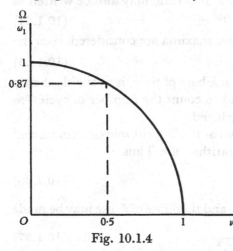

Fig. 10.1.4

This shows that if Ω/ω_1 is plotted against ν then the curve is a circular arc as shown in fig. 10.1.4. Now in most engineering structures in which vibration is troublesome, the damping factor ν is small compared with unity, and may well be of the order of 0·001 or less. For such values of ν the difference between Ω and ω_1 is negligible and, even if ν is as great as 0·5, Ω is 87 % of ω_1. It follows that ω_1 can usually be obtained with good accuracy by measuring the frequency of free vibrations. We shall show later that this important fact also holds good for systems with more than one degree of freedom.

The damping factor ν can often be measured experimentally by recording the rate of decay of a free vibration. Equation (10.1.18) may be written

$$x = R e^{-\nu\omega_1 t} \sin (\Omega t + \psi), \qquad (10.1.21)$$

where R and ψ are new constants which may be found in terms of G and H. Now the maxima and minima of x occur whenever $\dot{x} = 0$, that is whenever

$$\Omega \cos (\Omega t + \psi) - \nu\omega_1 \sin (\Omega t + \psi) = 0 \qquad (10.1.22)$$

and successive maxima of the same sign therefore occur at time intervals of $2\pi/\Omega$.

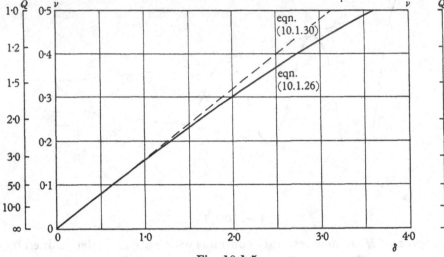

Fig. 10.1.5

The ratio of the first of two successive positive maximum values to the second is thus

$$r = \frac{R e^{-\nu\omega_1(t_1 - 2\pi/\Omega)} \sin [\Omega(t_1 - 2\pi/\Omega) + \psi]}{R e^{-\nu\omega_1 t_1} \sin (\Omega t_1 + \psi)} = e^{2\pi\nu\omega_1/\Omega}, \qquad (10.1.23)$$

where t_1 is the value of t at the second of the two. This ratio may also be written as

$$r = e^{2\pi\nu/\sqrt{(1-\nu^2)}} \qquad (10.1.24)$$

and it follows directly that, if n successive positive maxima are considered, then the ratio of the first to the last is

$$r^n = e^{2\pi\nu n/\sqrt{(1-\nu^2)}}. \qquad (10.1.25)$$

If a continuous record of the motion, plotted on a base of time, is available then it is easy to measure the heights of the peaks and to count the number of cycles, so that r can be calculated and ν subsequently deduced.

The power of e in equation (10.1.24) is known as the logarithmic decrement and is denoted by δ; it is equal to the Naperian logarithm of r. Thus

$$\delta = \frac{2\pi\nu}{\sqrt{(1-\nu^2)}}. \qquad (10.1.26)$$

When calculating ν from a measured value of r, and therefore of δ, use may be made of the relation

$$\nu = \frac{1}{\sqrt{\left[1 + \left(\dfrac{2\pi}{\delta}\right)^2\right]}} \qquad (10.1.27)$$

This function is shown plotted in fig. 10.1.5.

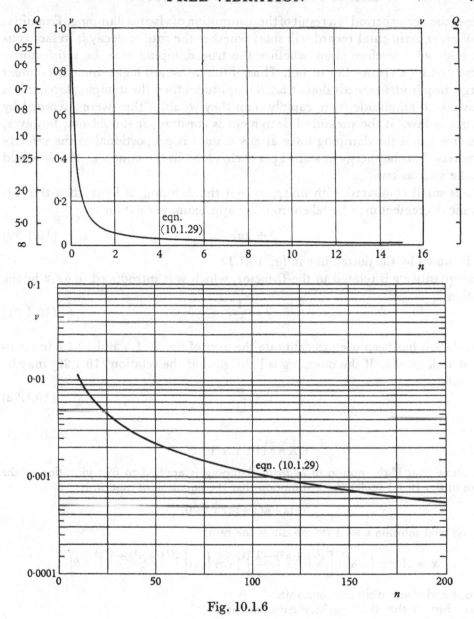

Fig. 10.1.6

If, on an experimental record, the height of the peaks is found to be halved after some number n swings, then

$$n\delta = \log_e 2 = 0.693. \tag{10.1.28}$$

Equation (10.1.27) now gives

$$\nu = \cfrac{1}{\sqrt{\left[1 + \left(\cfrac{2\pi n}{0.693}\right)^2\right]}} \tag{10.1.29}$$

and this relation, between n and ν is shown plotted in fig. 10.1.6.

The fact that the ratio of the heights of adjacent peaks is constant during the whole of the decay period is a result of the assumption of viscous damping. Examination of an experimental record will show whether the rate of decay is in fact constant and will therefore show whether the true damping may be satisfactorily represented by a viscous law or not. Thus, if the measured logarithmic decrement at large amplitudes exceeds that at small amplitudes then the damping forces must increase with amplitude more rapidly than they would if they were governed by the viscous law. If the measured decrement is constant, it should not, however, be assumed that the damping force at any instant is proportional to the velocity but merely that the energy absorbed per cycle varies in the same way that it would with the viscous law.

If ν is small compared with unity, so that the damping is light, then the logarithmic decrement may be taken from the approximate relation

$$\delta \doteqdot 2\pi\nu. \qquad (10.1.30)$$

This is shown by the dotted line in fig. 10.1.15.

The quantity ν is related to the Q-factor, which was introduced in § 8.2 by the equation

$$Q = \frac{1}{2\nu}. \qquad (10.1.31)$$

This relation has been used to calibrate the axes of fig. 10.1.5 and 10.1.6 in terms of Q as well as of ν. If the damping is light, so that the relation (10.1.30) may be used, then

$$\delta \doteqdot \frac{\pi}{Q}. \qquad (10.1.32)$$

EXAMPLES 10.1

1. Show that if the notion of *hysteretic* damping is applied to free vibration of the system of fig. 10.1.1 under the assumption that the equation of motion is

$$M\ddot{x} + k(1+i\mu)\,x = 0$$

then the trial solution $x = X\,e^{i\omega t}$ produces the result

$$x = A \exp\left\{-\omega_1 t\left[\frac{\sqrt{(1+\mu^2)}-1}{2}\right]^{\frac{1}{2}}\right\} \cos\left\{\omega_1 t\left[\frac{\sqrt{(1+\mu^2)}+1}{2}\right]^{\frac{1}{2}}+\theta\right\},$$

where A and θ are arbitrary constants.

Show further that the logarithmic decrement is

$$\frac{2\pi\mu}{1+\sqrt{(1+\mu^2)}}.$$

2. The expressions

$$M\ddot{x} + (k+ih)\,x \quad \text{and} \quad M\ddot{x} + \frac{h}{\omega}\,\dot{x} + kx$$

have been shown to be equivalent for harmonic oscillations of frequency ω; this equivalence breaks down for other motions. Instead of using the first (as in Ex. 10.1.1), use the second as an extension of the meaning of hysteretic damping to cover free vibration.

Show that the trial solution $\qquad x = X\,e^{(p+i\omega)t}$

in the equation $\qquad\qquad\qquad M\ddot{x} + \dfrac{h}{\omega}\,\dot{x} + kx = 0,$

where p and ω are unknown, leads to the result

$$x = A\exp\left\{-\omega_1 t\left[\frac{1-\sqrt{(1-\mu^2)}}{2}\right]^{\frac{1}{2}}\right\}\cos\left\{\omega_1 t\left[\frac{1+\sqrt{(1-\mu^2)}}{2}\right]^{\frac{1}{2}}+\theta\right\},$$

where A and θ are arbitrary constants.

Show further that the appropriate logarithmic decrement is

$$\frac{2\pi\mu}{1+\sqrt{(1-\mu^2)}}.$$

3. Show that the logarithmic decrement of a single-degree-of-freedom system having small viscous damping is given by the approximate relation

$$\delta = \pi\left(\frac{\omega_a-\omega_b}{\omega_1}\right),$$

where ω_1 is the resonant frequency at which the amplitude is $R_{\max}$. and where ω_a and ω_b are the frequencies at which the amplitude is $R_{\max}./\sqrt{2}$ for the same excitation.

4. Assuming that the free vibration of a single-degree-of-freedom system with light damping is sensibly sinusoidal show that

$$\delta = \frac{E}{2V_{\max}},$$

where δ is the logarithmic decrement, E is the energy dissipated per cycle and V is the potential energy.

10.2 Free vibration of damped systems with several degrees of freedom having independent modes

General equations for the motion of damped multi-freedom systems have been derived in Chapter 8. If the external forces, which are represented by the terms on the right-hand side of equation (8.5.3) are put equal to zero, then the equations for free vibration are left. They are

$$\left.\begin{aligned}
(a_{11}\ddot{q}_1 + b_{11}\dot{q}_1 + c_{11}q_1) + (a_{12}\ddot{q}_2 + b_{12}\dot{q}_2 + c_{12}q_2) + \ldots + (a_{1n}\ddot{q}_n + b_{1n}\dot{q}_n + c_{1n}q_n) &= 0, \\
(a_{21}\ddot{q}_1 + b_{21}\dot{q}_1 + c_{21}q_1) + (a_{22}\ddot{q}_2 + b_{22}\dot{q}_2 + c_{22}q_2) + \ldots + (a_{2n}\ddot{q}_n + b_{2n}\dot{q}_n + c_{2n}q_n) &= 0, \\
\cdots\cdots\cdots\cdots\cdots\cdots\cdots\cdots\cdots\cdots\cdots\cdots\cdots\cdots\cdots\cdots\cdots\cdots\cdots\cdots& \\
(a_{n1}\ddot{q}_1 + b_{n1}\dot{q}_1 + c_{n1}q_1) + (a_{n2}\ddot{q}_2 + b_{n2}\dot{q}_2 + c_{n2}q_2) + \ldots + (a_{nn}\ddot{q}_n + b_{nn}\dot{q}_n + c_{nn}q_n) &= 0.
\end{aligned}\right\}$$

$$(10.2.1)$$

Equations of this form may be obtained for any set of n generalized co-ordinates of the system. In particular, the n principal co-ordinates may be used, these being defined for the system when its damping is deleted. If this is done, the terms containing coefficients a_{rs} and c_{rs} $(r \neq s)$ will vanish. Writing the co-ordinates $p_1, p_2, ..., p_n$ as before, we then have

$$
\left.
\begin{aligned}
&a_1 \ddot{p}_1 + b_{11} \dot{p}_1 + c_1 p_1 + b_{12} \dot{p}_2 + b_{13} \dot{p}_3 + ... + b_{1n} \dot{p}_n = 0, \\
&b_{21} \dot{p}_1 + a_2 \ddot{p}_2 + b_{22} \dot{p}_2 + c_2 p_2 + b_{23} \dot{p}_3 + . + b_{2n} \dot{p}_n = 0, \\
&\cdots\cdots\cdots\cdots\cdots\cdots\cdots\cdots\cdots\cdots\cdots\cdots\cdots\cdots\cdots\cdots\cdots\cdots\cdots \\
&b_{n1} \dot{p}_1 + b_{n2} \dot{p}_2 + b_{n3} \dot{p}_3 + ... + a_n \ddot{p}_n + b_{nn} \dot{p}_n + c_n p_n = 0,
\end{aligned}
\right\}
\quad (10.2.2)
$$

where the constants b_{rs} are different from those of equations (10.2.1). This is a set of n simultaneous equations governing the co-ordinates p; but if all the coefficients b_{rs} $(r \neq s)$ are zero, just as the corresponding coefficients a and c are, then equations (10.2.2) reduce to a set of n equations, each containing one co-ordinate only. This special set of equations may be very easily solved. Now this condition arises in some physical problems and this section is therefore devoted to a discussion of this particular form of the equations.

The problems to which this special form of the equations of motion apply are those in which the damping forces are proportional to the spring forces throughout. This state of affairs prevails if all the damping is due to elastic hysteresis in the material and if the damping force thus generated is proportional to the elastic force. No real material behaves in quite this way; but the behaviour of a complete structure may sometimes, nevertheless, be represented approximately by equations of this type. Similar equations hold for the vibration of uniform beams and shafts if the source of damping is uniformly distributed along their length. Thus, a uniform beam on rigid supports, immersed in a viscous fluid to provide damping forces, will be governed by equations which approximate to the special form.

The equations which are to be discussed are

$$
\left.
\begin{aligned}
&a_1 \ddot{p}_1 + b_{11} \dot{p}_1 + c_1 p_1 = 0, \\
&a_2 \ddot{p}_2 + b_{22} \dot{p}_2 + c_2 p_2 = 0, \\
&\cdots\cdots\cdots\cdots\cdots\cdots\cdots\cdots \\
&a_n \ddot{p}_n + b_{nn} \dot{p}_n + c_n p_n = 0,
\end{aligned}
\right\}
\quad (10.2.3)
$$

and each is similar to that governing the motion of a system having one degree of freedom. The latter was discussed in the preceding section. If follows that the motion in each mode will decay exponentially and that for any mode, say the sth, a constant ν_s may be defined such that

$$
\nu_s = \frac{b_{ss}}{2\sqrt{(a_s c_s)}} = \frac{b_{ss}}{2 a_s \omega_s} \quad (s = 1, 2, ..., n). \quad (10.2.4)
$$

The corresponding motion can then be written

$$
p_s = A_s e^{[-\nu_s + \sqrt{(\nu_s^2 - 1)}]\omega_s t} + B_s e^{[-\nu_s - \sqrt{(\nu_s^2 - 1)}]\omega_s t}, \quad (10.2.5)
$$

where A_s and B_s are constants which correspond to A and B of equation (10.1.8) and which are determined by the initial conditions.

Again, the character of the motion in each mode, whether it is oscillatory or exponential, will depend upon whether ν_s is less than or greater than unity.

In general ν_s may vary widely between different modes. If this is so in a system which is thrown into free motion in all the modes simultaneously, then motion in some of the modes will die away more rapidly than that in others. If, on the other hand, the damping forces are related to the elastic forces in such a way that the energy dissipated per cycle in any mode is proportional to the energy stored in that mode, then the logarithmic decrement is the same for each mode. Under these conditions the rate of decay of motion in each mode with time will be proportional to its frequency, so that after an interval only the motions in the lower modes will persist.

If the initial displacements and velocities of the system are known then this information may be used to determine the $2n$ constants $A_1, A_2, ..., A_n$ and $B_1, B_2, ..., B_n$. If the displacements and velocities are known in terms of the principal co-ordinates p then this may be done directly, motion in each mode being treated just as the motion of the one-degree-of-freedom system is treated in the preceding section. If, however, the initial conditions are known in terms of some other co-ordinates $q_1, q_2, ..., q_n$, then the algebraic process of determining the A's and B's may be lengthy, even though the modal shapes are known as sets of ratios between the q's. This matter is discussed in § 10.7.

When all the damping in a multi-freedom system is light, it is usually possible to neglect the effects of the terms containing the coefficients b_{rs} $(r \neq s)$ when finding the free motion. This possibility was first pointed out by Rayleigh;[†] it reduces the general case to the special one dealt with above and results in a great saving of labour.

EXAMPLE 10.2

1. Show that the equations governing a pair of principal co-ordinates of the given system are independent, provided that

$$b_1 = 2b_2.$$

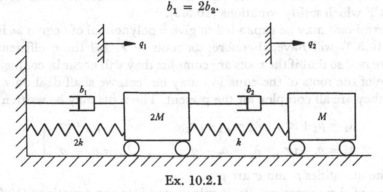

Ex. 10.2.1

10.3 General free vibration of damped systems having several degrees of freedom

We come now to discuss the general case of damped free vibration, where the damping coefficients b_{rs} $(r \neq s)$ do not vanish in equations (10.2.2). It is possible to use, for this discussion, either these equations or equations (10.2.1) which govern

† *Theory of Sound* (1894), §102.

the co-ordinates q. Algebraically there is little to choose between the two sets; but the physical interpretation of the results is rather easier if the q-equations are used. We shall therefore revert to the latter. The theory shows that certain 'modes' of free vibration exist which degenerate, when the damping is zero, to the principal modes. These 'modes' will be obtained as sets of ratios between the co-ordinates $q_1, q_2, ..., q_n$.

If the trial solution

$$q_1 = \Psi_1' e^{\lambda t}, \quad q_2 = \Psi_2' e^{\lambda t}, \quad ..., \quad q_n = \Psi_n' e^{\lambda t} \qquad (10.3.1)$$

is substituted into equations (10.2.1) then a set of n algebraic equations is obtained. These are

$$\left.\begin{aligned}
(a_{11}\lambda^2 + b_{11}\lambda + c_{11})\,\Psi_1' + (a_{12}\lambda^2 + b_{12}\lambda + c_{12})\,\Psi_2' + ... + (a_{1n}\lambda^2 + b_{1n}\lambda + c_{1n})\,\Psi_n' = 0, \\
(a_{21}\lambda^2 + b_{21}\lambda + c_{21})\,\Psi_1' + (a_{22}\lambda^2 + b_{22}\lambda + c_{22})\,\Psi_2' + ... + (a_{2n}\lambda^2 + b_{2n}\lambda + c_{2n})\,\Psi_n' = 0, \\
\cdots\cdots\cdots\cdots\cdots\cdots\cdots\cdots\cdots\cdots\cdots\cdots\cdots\cdots\cdots\cdots\cdots\cdots \\
(a_{n1}\lambda^2 + b_{n1}\lambda + c_{n1})\,\Psi_1' + (a_{n2}\lambda^2 + b_{n2}\lambda + c_{n2})\,\Psi_2' + ... + (a_{nn}\lambda^2 + b_{nn}\lambda + c_{nn})\,\Psi_n' = 0.
\end{aligned}\right\}$$
$$(10.3.2)$$

These equations can be satisfied simultaneously by a set of the amplitudes Ψ' only if the determinant Δ of the coefficients of the Ψ's is zero; that is to say,

$$\Delta = \begin{vmatrix}
a_{11}\lambda^2 + b_{11}\lambda + c_{11} & a_{12}\lambda^2 + b_{12}\lambda + c_{12} & ... & a_{1n}\lambda^2 + b_{1n}\lambda + c_{1n} \\
a_{21}\lambda^2 + b_{21}\lambda + c_{21} & a_{22}\lambda^2 + b_{22}\lambda + c_{22} & ... & a_{2n}\lambda^2 + b_{2n}\lambda + c_{2n} \\
\cdots\cdots\cdots\cdots\cdots & & & \\
a_{n1}\lambda^2 + b_{n1}\lambda + c_{n1} & a_{n2}\lambda^2 + b_{n2}\lambda + c_{n2} & ... & a_{nn}\lambda^2 + b_{nn}\lambda + c_{nn}
\end{vmatrix} = 0. \qquad (10.3.3)$$

Corresponding to each root of this equation there is a set of ratios between the amplitudes Ψ' which satisfy equations (10.3.2).

This determinant may be expanded to give a polynomial of degree $2n$ in λ. The equation (10.3.3) will have, therefore, $2n$ roots in λ. All the coefficients in the equation are real so that if the roots are complex they will occur in conjugate pairs. While some of the roots of the equation may be real, we shall deal only with the case when they are all complex for the present. The roots may be written as

$$\left.\begin{aligned}
\lambda_1 = \rho_1 + i\sigma_1, \quad \lambda_2 = \rho_2 + i\sigma_2, \quad ..., \quad \lambda_n = \rho_n + i\sigma_n, \\
\lambda_1' = \rho_1 - i\sigma_1, \quad \lambda_2' = \rho_2 - i\sigma_2, \quad ..., \quad \lambda_n' = \rho_n - i\sigma_n,
\end{aligned}\right\} \qquad (10.3.4)$$

where all the quantities ρ and σ are real.

If any one of these roots, say λ_s, is substituted into the equations (10.3.2) then they may be used to determine the $(n-1)$ ratios between the amplitudes Ψ' which determine the sth mode. Since λ_s is complex the equations will have complex coefficients and the ratios of the Ψ's will themselves be complex. These ratios may be written in the form

$$\frac{\Psi_1'}{{}_sC_1 e^{i_s\theta_1}} = \frac{\Psi_2'}{{}_sC_2 e^{i_s\theta_2}} = ... = \frac{\Psi_n'}{{}_sC_n e^{i_s\theta_n}} \quad (= S, \text{ say}), \qquad (10.3.5)$$

where S is an arbitrary complex number. It may further be shown that the quantities $_sC_1 e^{i_s\theta_1}$, etc., are the co-factors of any row of the determinant Δ when λ in that determinant is put equal to λ_s. The instantaneous values of the co-ordinates q, when the system is vibrating freely in the 'λ_s-mode', are given by

$$
\left.\begin{aligned}
q_1 &= S._sC_1 e^{\rho_s t} e^{i(\sigma_s t + _s\theta_1)}, \\
q_2 &= S._sC_2 e^{\rho_s t} e^{i(\sigma_s t + _s\theta_2)}, \\
&\cdots\cdots\cdots\cdots\cdots\cdots \\
q_n &= S._sC_n e^{\rho_s t} e^{i(\sigma_s t + _s\theta_n)}.
\end{aligned}\right\} \qquad (10.3.6)
$$

There are $2n$ of these sets, and the total motion may be compounded from them in any proportions. The apparent contradiction between the existence of $2n$ different motions in the damped system and the n modes of an undamped system arises because the phase of the undamped modes is not included when defining them. Now an undamped mode in any phase can be regarded as being the resultant of components of two similar modes which are in quadrature with one another; if each undamped mode is looked upon in this way then the undamped system, like the damped system, has $2n$ modes. It happens, however, to be more convenient to treat the damped and undamped systems differently and it is not suggested that the approach used in previous chapters needs altering.

Before discussing the nature of the free motion given by equations (10.3.6) it is convenient to combine the solutions in pairs, each pair corresponding to a pair of conjugate roots of Δ. When the root $(\rho_s + i\sigma_s)$ is substituted into the first of equations (10.3.2) it gives

$$
[a_{11}(\rho_s + i\sigma_s)^2 + b_{11}(\rho_s + i\sigma_s) + c_{11}]\,\Psi_1' + [a_{12}(\rho_s + i\sigma_s)^2 + b_{12}(\rho_s + i\sigma_s) + c_{12}]\,\Psi_2' + \ldots
$$
$$
+ [a_{1n}(\rho_s + i\sigma_s)^2 + b_{1n}(\rho_s + i\sigma_s) + c_{1n}]\,\Psi_n' = 0 \quad (10.3.7)
$$

and similarly for all the others. Now if the conjugate root $\rho_s - i\sigma_s$ is substituted into these same equations then all the terms containing an odd power of σ_s, that is to say all the imaginary terms, have their sign changed. It follows that all the imaginary terms in the solutions to the equations will also change sign so that the ratios of the Ψ's will be given by

$$
\frac{\Psi_1'}{_sC_1 e^{-i_s\theta_1}} = \frac{\Psi_2'}{_sC_2 e^{-i_s\theta_2}} = \ldots = \frac{\Psi_n'}{_sC_n e^{-i_s\theta_n}} \quad (= S', \text{ say}). \qquad (10.3.8)
$$

The corresponding instantaneous values of the co-ordinates q are thus

$$
\left.\begin{aligned}
q_1 &= S'._sC_1 e^{\rho_s t} e^{-i(\sigma_s t + _s\theta_1)}, \\
q_2 &= S'._sC_2 e^{\rho_s t} e^{-i(\sigma_s t + _s\theta_2)}, \\
&\cdots\cdots\cdots\cdots\cdots\cdots \\
q_n &= S'._sC_n e^{\rho_s t} e^{-i(\sigma_s t + _s\theta_n)}.
\end{aligned}\right\} \qquad (10.3.9)
$$

Suppose that this motion is now combined with that of equation (10.3.6). In doing this the complex exponential can be replaced by the trigonometrical form, and the constants S and $\bar{S}'$ can be replaced by others for simplicity. The process gives

$$\left.\begin{aligned}
{}_s q_1 &= {}_s C_1\, e^{\rho_s t}\,[G_s \cos(\sigma_s t + {}_s\theta_1) + H_s \sin(\sigma_s t + {}_s\theta_1)], \\
{}_s q_2 &= {}_s C_2\, e^{\rho_s t}\,[G_s \cos(\sigma_s t + {}_s\theta_2) + H_s \sin(\sigma_s t + {}_s\theta_2)], \\
&\cdots\cdots\cdots\cdots\cdots\cdots\cdots\cdots\cdots\cdots\cdots\cdots\cdots\cdots\cdots\cdots\cdots\cdots\cdots\cdots\cdots \\
{}_s q_n &= {}_s C_n\, e^{\rho_s t}\,[G_s \cos(\sigma_s t + {}_s\theta_n) + H_s \sin(\sigma_s t + {}_s\theta_n)].
\end{aligned}\right\} \tag{10.3.10}$$

The prefixed suffix has here been added to the co-ordinates q to indicate the modal number. It will be seen that there are n possible values for s so that equations (10.3.10) define a motion corresponding to one of the n principal modes of an undamped system; we shall refer to these motions as being in the 'damped modes'.

The distinction between a damped mode and a principal mode is a fundamental one and it is to this that we next turn our attention. There is one difference, however, which may be noted immediately. Owing to the phase differences within any one damped mode, a fixed distortion cannot be regarded as being made up of one *damped* mode alone. It is, in general, not possible to start a vibration in one damped mode only, by releasing the system from some particular static distortion.

Each modal number s has two constants, G_s and H_s, associated with it. There are thus $2n$ constants which may be chosen to satisfy any initial values of the n co-ordinates and of the rates of change of those co-ordinates.

It may be shown[†] that the factors ρ_s are all negative provided that the dissipation function D, given by the quadratic form

$$2D = b_{11}\dot{q}_1^2 + b_{22}\dot{q}_2^2 + \ldots + b_{nn}\dot{q}_n^2 + 2b_{12}\dot{q}_1\dot{q}_2 + \ldots \tag{10.3.11}$$

is positive definite as are the corresponding energy functions involving the coefficients a and c. We shall not present the argument here but leave the matter as being evident on physical grounds. The function (10.3.11) represents the rate at which energy is dissipated in the system and if this is necessarily positive, as it must be from the nature of the individual damping effects, then there must be a continuous decrease of stored energy with time. This could not be so if one or more of the factors ρ_s were positive.

It follows from this that any one damped mode decays exponentially, retaining its 'shape' as it does so. But it must be noted that this 'shape', which is defined by the ratios of the instantaneous values of the co-ordinates q, depends not only upon the quantities ${}_s C_1, {}_s C_2, \ldots, {}_s C_n$, but also upon the phase angles ${}_s\theta_1, {}_s\theta_2, \ldots, {}_s\theta_n$. These angles are all zero for the principal modes of a conservative system; their presence in the damped modes shows that the displacements at the various co-ordinates are not necessarily in phase.[‡]

Since each of the damped modes decays independently it is possible to define a damping factor ν and a corresponding logarithmic decrement for each, as with a one-degree-of-freedom system.

† See, for example, H. Lamb, *Higher Mechanics* (Cambridge University Press, 1920), §97.
‡ Strictly the term 'phase' cannot be applied because it refers to simple harmonic motion; it is used here to refer to the harmonically varying factor in the motion and not to the exponential factor.

As an example of a system of the type discussed above consider the arrangement shown in fig. 10.3.1 in which the system can vibrate in torsion, the flywheels being controlled by the shafts which have torsional stiffnesses of k and K, and by the damper of constant b. If the damping were absent the system would have the principal modes

$$\left.\begin{aligned} q_1:q_2&::1:1, \\ q_1:q_2&::1:-1, \end{aligned}\right\} \tag{10.3.12}$$

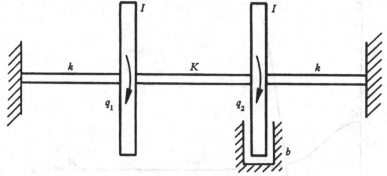

Fig. 10. 3. 1

and principal co-ordinates p_1 and p_2 could be defined such that

$$\left.\begin{aligned} q_1 &= p_1+p_2, \\ q_2 &= p_1-p_2, \end{aligned}\right\} \tag{10.3.13}$$

or, transposing this relation

$$\left.\begin{aligned} p_1 &= \tfrac{1}{2}(q_1+q_2), \\ p_2 &= \tfrac{1}{2}(q_1-q_2). \end{aligned}\right\} \tag{10.3.14}$$

When written in terms of the co-ordinates p, the equations for the damped system are

$$\left.\begin{aligned} 2I\ddot{p}_1+b\dot{p}_1+2kp_1-b\dot{p}_2 &= 0, \\ -b\dot{p}_1+2I\ddot{p}_2+b\dot{p}_2+2(k+2K)\,p_2 &= 0. \end{aligned}\right\} \tag{10.3.15}$$

The trial solutions

$$p_1 = \Pi_1 e^{\lambda t}, \quad p_2 = \Pi_2 e^{\lambda t}, \tag{10.3.16}$$

give the homogeneous algebraic equations

$$\left.\begin{aligned} \left[\lambda^2+\left(\frac{b}{2I}\right)\lambda+\omega_1^2\right]\Pi_1-\left(\frac{b}{2I}\right)\lambda\Pi_2 &= 0, \\ -\left(\frac{b}{2I}\right)\lambda\Pi_1+\left[\lambda^2+\left(\frac{b}{2I}\right)\lambda+\omega_2^2\right]\Pi_2 &= 0, \end{aligned}\right\} \tag{10.3.17}$$

where ω_1 and ω_2 are the natural frequencies of the system when the damper is omitted. That is to say

$$\omega_1^2 = \frac{k}{I}, \quad \omega_2^2 = \frac{k+2K}{I}. \tag{10.3.18}$$

We shall now examine the way in which the free motion of the system changes as the damping is increased from zero and, in order to simplify the algebra, we consider the special case for which $K = 2k$ so that

$$\omega_2^2 = 5\omega_1^2. \tag{10.3.19}$$

Further simplification is possible by introducing a dimensionless parameter α as a measure of the damping such that

$$\frac{b}{2I} = \alpha\omega_1. \tag{10.3.20}$$

These values may be substituted into equations (10.3.17) and the determinantal relation corresponding to equation (10.3.3) can be expanded to give

$$\lambda^4 + 2\alpha\omega_1\lambda^3 + 6\omega_1^2\lambda^2 + 6\alpha\omega_1^3\lambda + 5\omega_1^4 = 0. \tag{10.3.21}$$

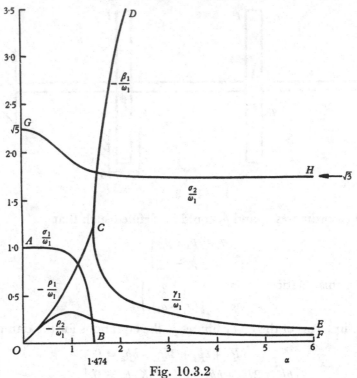

Fig. 10.3.2

This may be made non-dimensional by the substitution of

$$\Lambda = \frac{\lambda}{\omega_1} \tag{10.3.22}$$

which yields

$$\Lambda^4 + 2\alpha\Lambda^3 + 6\Lambda^2 + 6\alpha\Lambda + 5 = 0. \tag{10.3.23}$$

We require to know how the roots of this equation change as α is increased from zero. And, for any particular value of α, the roots may be substituted back in equation (10.3.17) to find the corresponding ratio of Π_1 to Π_2.

Fig. 10.3.2 shows the values of the real and imaginary parts of the roots of Λ plotted against α.† When $\alpha = 0$ the roots are

$$\Lambda = i, \quad -i, \quad i\sqrt{5}, \quad -i\sqrt{5} \tag{10.3.24}$$

and these are represented by the points A and G in fig. 10.3.2. As α is increased the magnitude of the imaginary parts falls, as shown by the drooping curves of that figure. At the same time the roots acquire real parts which increase with α as

† The authors gratefully acknowledge the assistance of Dr P. M. E. Percival who devised practical numerical methods for the calculations of figs. 10.3.2 and 10.3.3.

shown. The quantities plotted on the figure are ρ_1/ω_1, ρ_2/ω_1 and σ_1/ω_1, σ_2/ω_1 because of the dimensionless nature of the parameter Λ.

When $\alpha = 1{\cdot}474$ the pair of complex roots Λ_1, Λ_1' which correspond to the first damped mode, change to form two real roots. The nature of the free motion, when the roots are real, has not yet been dealt with and this portion of the curve will be referred to later when the theory has been covered. The other two roots of Λ remain complex for $\alpha > 1{\cdot}474$ and the motion to which they correspond remains similar to that already described. Before discussing this example further we return to examine the general case when some of the roots of Δ are real.

If the damping coefficients within a system are increased sufficiently some (or all) of the roots of equation (10.3.3) become real; this corresponds to non-oscillatory motion as already described for systems having one degree of freedom. Since the total number of roots and the number of complex roots are both even it follows that the number of real roots will be even also. They may therefore be divided into pairs. In selecting the pairs it is possible to pair those roots which appear together whenever a pair of complex roots are lost with increasing damping, as takes place for instance in the above example when α passes through the value $1{\cdot}474$.

Let the sth pair of roots λ_s and λ_s' be real, having the values

$$\lambda_s = \beta_s, \quad \lambda_s' = \gamma_s, \tag{10.3.25}$$

where, in general, β_s and γ_s are not equal. The roots may be substituted in equations (10.3.2) as before and they will give sets of ratios of the amplitudes Ψ which may be written in the form

$$\frac{\Psi_1'}{{}_sD_1} = \frac{\Psi_2'}{{}_sD_2} = \dots = \frac{\Psi_n'}{{}_sD_n} = A_s \tag{10.3.26}$$

for the root β_s, and

$$\frac{\Psi_1'}{{}_sE_1} = \frac{\Psi_2'}{{}_sE_2} = \dots = \frac{\Psi_n'}{{}_sE_n} = B_s \tag{10.3.27}$$

for the root γ_s. In these expressions A_s and B_s are arbitrary constants, and the denominators are the co-factors of the determinant Δ when β_s or γ_s, as appropriate, is substituted into it. The ratios defined in this way are all real. The 'sth damped mode', corresponding to the real roots is now

$$
\left.\begin{aligned}
{}_sq_1 &= A_s \cdot {}_sD_1\, e^{\beta_s t} + B_s \cdot {}_sE_1\, e^{\gamma_s t}, \\
{}_sq_2 &= A_s \cdot {}_sD_2\, e^{\beta_s t} + B_s \cdot {}_sE_2\, e^{\gamma_s t}, \\
&\cdots\cdots\cdots\cdots\cdots\cdots\cdots\cdots\cdots\cdots \\
{}_sq_n &= A_s \cdot {}_sD_n\, e^{\beta_s t} + B_s \cdot {}_sE_n\, e^{\gamma_s t}.
\end{aligned}\right\} \tag{10.3.28}
$$

It is evident, as before, on physical grounds, that β_s and γ_s must be negative for a stable system. It may be proved algebraically† that this is so if the quadratic functions formed from the a's, b's and c's are all positive definite.

The above treatment of the real roots of Δ, in which they are selected in pairs, emphasizes the essential similarity of the results whether the roots are pure imaginary, complex, or real. In each case n 'modes' are obtained and each of these has two constants associated with it which may be chosen to suit initial conditions of displacement and velocity. When the roots are imaginary, the constants define the

† Lamb, *loc. cit.*

amplitude and phase of the sinusoidal motion; when the roots are complex the constants define the amplitude and phase of the sinusoidal factor in the motion, the net motion being obtained by multiplying this factor by an 'exponential-decay' term; when the roots are real the constants define the magnitudes of two components which decay exponentially at different rates. It will be seen that these last two components could be treated as separate 'modes'; but this would involve a change in the total number of modes when roots are made real by increasing the damping, and hence some of the simplicity of the treatment would be lost.

It is possible to develop the theory further and to show for example that orthogonal relations between damped modes exist just as they do for undamped ones. We shall not do this here because the results so far obtained are sufficient for most engineering purposes.

Returning again to the example, it is found that there are two real roots and two complex roots for all values of α above $1 \cdot 474$. This result may at first seem surprising, in that it might be expected that all the roots would be made real if the damping were sufficiently great. The reason for this will be realized by considering how the system will behave if the damping is made very large. In this state free motion of the damped flywheel is negligible although the other can still vibrate by twisting its two adjacent shafts. In the extreme case, when the damping is infinite, the damped flywheel becomes locked and the undamped flywheel, with its two shafts, constitutes a conservative system with a single degree of freedom.

These results can be seen to follow algebraically because, if α becomes very large, the first, third and last terms of equation (10.3.23) become negligible compared with the second and fourth. With these last two terms only, the root $\pm i \sqrt{3}$ is found; therefore, as $\alpha \to \infty$

$$\left. \begin{aligned} \lambda_2 &\to i\sqrt{3}\,\omega_1 = i\sqrt{\left(\frac{3k}{I}\right)}, \\ \lambda_2' &\to -i\sqrt{3}\,\omega_1 = -i\sqrt{\left(\frac{3k}{I}\right)}. \end{aligned} \right\} \tag{10.3.29}$$

It can readily be seen that this is consistent with the above argument because the total stiffness controlling the motion of the single flywheel is $3k$. Further, as $\alpha \to \infty$, equations (10.3.17) show that Π_1 and Π_2 approach equality so that, from equations (10.3.13), q_2 tends to zero but q_1 does not.

The relative amplitudes and phase differences between the distortions in the principal modes, during vibration in a single damped mode, can be found from equations (10.3.17). They are R and Γ respectively, where

$$\frac{\Pi_2}{\Pi_1} = \frac{\Lambda^2 + \alpha\Lambda + 1}{\alpha\Lambda} = R\,e^{i\Gamma} \tag{10.3.30}$$

which are to be evaluated for the appropriate value of Λ for any given α. R and Γ, obtained in this way for the root

$$\Lambda = \frac{\rho_1}{\omega_1} + i\frac{\sigma_1}{\omega_1} \tag{10.3.31}$$

are plotted in fig. 10.3.3. The curve of R against α passes through the origin because $\alpha = 0$ corresponds to zero damping and therefore to motion taking place in the first principal mode only. The curve does not extend beyond $\alpha = 1\cdot474$ because the first damped mode ceases to be periodic beyond this value. The phase angle Γ is also plotted in fig. 10.3.3 and will be seen to vary between $\frac{1}{2}\pi$ and π.

Similar curves may be drawn for the second damped mode; but it is more convenient to calculate these from the expression

$$\frac{\Pi_1}{\Pi_2} = \frac{\Lambda^2 + \alpha\Lambda + 5}{\alpha\Lambda} = R'\,e^{i\Gamma'} \tag{10.3.32}$$

in which the values

$$\Lambda = \frac{\rho_2}{\omega_1} \pm i\frac{\sigma_2}{\omega_1} \tag{10.3.33}$$

are used. The curves of R' and Γ', for the positive sign in equation (10.3.33), are plotted in fig. 10.3.3. The curve of R' passes through the origin as expected and the

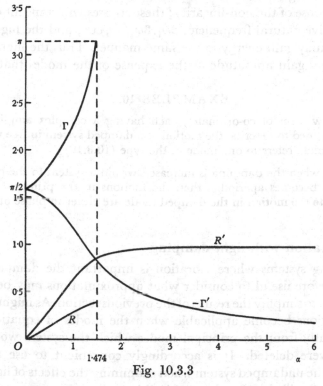

Fig. 10.3.3

curve of Γ' varies from $-\frac{1}{2}\pi$ to zero. The ratio Π_1/Π_2 tends to unity as α tends to infinity.

It can be seen from this example that the free motion of a heavily damped system is extremely complex even when the system is relatively simple. As a result it becomes difficult to examine the modal shapes of a heavily damped system by observing the free motion only. This fact has been found relevant in connexion with the observation of vibration of aircraft when they are supported on the ground by partially deflated tyres or air bags.

Although the theory developed above shows that any initial disturbance will start a motion in which each damped mode steadily decays there is nevertheless a class of real systems which do not, apparently, behave in this way. This class consists of certain musical instruments, in particular the pianoforte. Jeans† shows that if a single note on the keyboard is struck then all the harmonics produced do not decay continuously; some of them, on the contrary, increase in amplitude for some time after the initial disturbance. Indeed it is commonly accepted by musicians that certain harmonics do not reach their maximum amplitude until some time after a note is struck. Additional theory is evidently needed to explain this behaviour. It seems likely that it depends on the presence of two factors—first non-linearity in the stiffness or damping effects and secondly the existence of natural frequencies which are in simple ratios to each other. These factors interact.

Consider for instance a single string which has a natural frequency ω_1. If the string is set vibrating stresses will be produced in the frame at frequencies $2\omega_1$, $3\omega_1$, ..., etc., because of the non-linearity; these stresses will transmit energy to the strings which have natural frequencies $2\omega_1, 3\omega_1, ...$, etc., and the higher modes of the string itself may gain energy in the same manner. Thus the higher-frequency modes will slowly gain amplitude at the expense of the modes that are initially excited.

EXAMPLES 10.3

1. Explain how a set of co-ordinates, each having a complex amplitude and frequency, might be used to describe the motion of a damped system in free vibration such that each co-ordinate refers to one mode of the type (10.3.10).

2. Show that, when the damping is increased within a system to the point at which a damped mode becomes aperiodic, then the motions in the principal modes which together constitute the motion in the damped mode are either in phase or in anti-phase with each other.

10.4 Free vibration with light damping

In engineering systems where vibration is important the damping is usually light. It is therefore useful to consider what approximations may be used, under these conditions, to simplify the results of the previous section. As might be expected, the approximations become applicable when the modes of equations (10.3.10) differ only slightly from the principal modes which the system would possess if the damping were deleted. It is accordingly convenient to use the principal co-ordinates of the undamped system when examining the effects of light damping; equations (10.2.2) will therefore be used in this section.

It will be recalled that a solution to these equations was obtained by substituting

$$p_1 = \Pi_1 e^{\lambda t}, \quad p_2 = \Pi_2 e^{\lambda t}, \quad ..., \quad p_n = \Pi_n e^{\lambda t} \qquad (10.4.1)$$

into them. This gives

$$\left.\begin{aligned}
(a_1 \lambda^2 + b_{11} \lambda + c_1) \Pi_1 + b_{12} \lambda \Pi_2 + b_{13} \lambda \Pi_3 + ... + b_{1n} \lambda \Pi_n &= 0, \\
b_{21} \lambda \Pi_1 + (a_2 \lambda^2 + b_{22} \lambda + c_2) \Pi_2 + b_{23} \lambda \Pi_3 + ... + b_{2n} \lambda \Pi_n &= 0, \\
\cdots\cdots\cdots\cdots\cdots\cdots\cdots\cdots\cdots\cdots\cdots\cdots\cdots\cdots\cdots\cdots\cdots\cdots\cdots\cdots\cdots \\
b_{n1} \lambda \Pi_1 + b_{n2} \lambda \Pi_2 + b_{n3} \lambda \Pi_3 + ... + (a_n \lambda^2 + b_{nn} \lambda + c_n) \Pi_n &= 0.
\end{aligned}\right\} \qquad (10.4.2)$$

† Sir James Jeans, *Science and Music* (Cambridge University Press, 1937).

This set of equations has a solution, other than the trivial one

$$\Pi_1 = \Pi_2 = \ldots = \Pi_n = 0,$$

only if λ satisfies the equation $\Delta = 0$, (10.4.3)

where Δ is the determinant

$$\begin{vmatrix} a_1\lambda^2 + b_{11}\lambda + c_1 & b_{12}\lambda & \ldots & b_{1n}\lambda \\ b_{21}\lambda & a_2\lambda^2 + b_{22}\lambda + c_2 & \ldots & b_{2n}\lambda \\ \cdots\cdots\cdots\cdots\cdots\cdots\cdots\cdots\cdots\cdots\cdots\cdots\cdots\cdots\cdots \\ b_{n1}\lambda & b_{n2}\lambda & \ldots & a_n\lambda^2 + b_{nn}\lambda + c_n \end{vmatrix}. \quad (10.4.4)$$

The roots of equation (10.4.3) are identical with those of equation (10.3.3) because they are the complex natural frequencies of the system; these frequencies must be independent of the choice of co-ordinates.

Now let the particular root λ_s, which is equal to $\rho_s + i\sigma_s$, be substituted into the simultaneous equations (10.4.2). These then become

$$\left.\begin{aligned} [a_1(\rho_s + i\sigma_s)^2 + b_{11}(\rho_s + i\sigma_s) + c_1]\,\Pi_1 + b_{12}(\rho_s + i\sigma_s)\,\Pi_2 + \ldots + b_{1n}(\rho_s + i\sigma_s)\,\Pi_n &= 0, \\ b_{21}(\rho_s + i\sigma_s)\,\Pi_1 + [a_2(\rho_s + i\sigma_s)^2 + b_{22}(\rho_s + i\sigma_s) + c_2]\,\Pi_2 + \ldots + b_{2n}(\rho_s + i\sigma_s)\,\Pi_n &= 0, \\ \cdots\cdots\cdots\cdots\cdots\cdots\cdots\cdots\cdots\cdots\cdots\cdots\cdots\cdots\cdots\cdots\cdots\cdots\cdots\cdots \\ b_{n1}(\rho_s + i\sigma_s)\,\Pi_1 + b_{n2}(\rho_s + i\sigma_s)\,\Pi_2 + \ldots + [a_n(\rho_s + i\sigma_s)^2 + b_{nn}(\rho_s + i\sigma_s) + c_n]\,\Pi_n &= 0. \end{aligned}\right\}$$

(10.4.5)

From these it is known that the ratios between the amplitudes $\Pi_1, \Pi_2, \ldots, \Pi_n$ are given by

$$\frac{\Pi_1}{\alpha_1} = \frac{\Pi_2}{\alpha_2} = \ldots = \frac{\Pi_n}{\alpha_n} = S, \quad (10.4.6)$$

where S is an arbitrary constant. The α's are the co-factors of the terms of any one row of the determinant Δ when that determinant is evaluated for $\lambda = \lambda_s$ and they are the same as the quantities $_sC_1 e^{i_s\theta_1}$, etc., which we used in equation (10.3.5) except that they now refer specifically to the principal co-ordinates p.

This brief statement of the way in which the algebraic theory may be developed for the co-ordinates p follows the treatment which was used for the co-ordinates q. It gives, finally, results corresponding to those of equations (10.3.10) and it is unnecessary to repeat all the argument here. We shall consider immediately the way in which the theory can be modified when the damping is light.

The amount of damping present is determined by the magnitudes of the various damping coefficients b. In discussing the case of light damping in which these coefficients are small, we shall divide them into two classes, namely the direct-coefficients $b_{11}, b_{22}, \ldots, b_{nn}$ and the cross-coefficients b_{uv}, that is to say $b_{12}, b_{13}, \ldots$. We shall not, for the present, pay attention to the magnitudes of the coefficients b relative to the other energy coefficients. Instead we shall develop a theory which will hold good provided all the coefficients b are sufficiently small, reserving until later an examination of how large they can be before the treatment becomes invalid.

Now the determinant Δ may be expanded in the form

$$\Delta = (a_1\lambda^2 + b_{11}\lambda + c_1)(a_2\lambda^2 + b_{22}\lambda + c_2) \dots (a_n\lambda^2 + b_{nn}\lambda + c_n) + O[b^2, \lambda], \quad (10.4.7)$$

where the term $O[b^2, \lambda]$ denotes an expression in which all the terms are of the second degree or higher in the cross-coefficients b_{uv}, and which is also a function of λ. The roots of equation (10.4.3) are evidently dependent on, and therefore functions of, the coefficients b. Equation (10.4.7) shows that if only the coefficients $b_{11}, b_{22}, \dots, b_{nn}$ were present, the coefficients b_{uv} being zero, then the roots would be

$$\left.\begin{aligned}
\lambda_1 &= -\frac{b_{11}}{2a_1} + i\sqrt{\left[\frac{c_1}{a_1} - \frac{b_{11}^2}{4a_1^2}\right]}, & \lambda_1' &= -\frac{b_{11}}{2a_1} - i\sqrt{\left[\frac{c_1}{a_1} - \frac{b_{11}^2}{4a_1^2}\right]}, \\
\lambda_2 &= -\frac{b_{22}}{2a_2} + i\sqrt{\left[\frac{c_2}{a_2} - \frac{b_{22}^2}{4a_2^2}\right]}, & \lambda_2' &= -\frac{b_{22}}{2a_2} - i\sqrt{\left[\frac{c_2}{a_2} - \frac{b_{22}^2}{4a_2^2}\right]}, \\
&\cdots\cdots\cdots\cdots\cdots & &\cdots\cdots\cdots\cdots\cdots \\
\lambda_n &= -\frac{b_{nn}}{2a_n} + i\sqrt{\left[\frac{c_n}{a_n} - \frac{b_{nn}^2}{4a_n^2}\right]}, & \lambda_n' &= -\frac{b_{nn}}{2a_n} - i\sqrt{\left[\frac{c_n}{a_n} - \frac{b_{nn}^2}{4a_n^2}\right]}.
\end{aligned}\right\} \quad (10.4.8)$$

If the direct coefficients b are small, so that their second powers may be neglected, these roots may be written approximately, as usual for light damping,

$$\left.\begin{aligned}
\lambda_1 &= -\frac{b_{11}}{2a_1} + i\omega_1, & \lambda_1' &= -\frac{b_{11}}{2a_1} - i\omega_1, \\
\lambda_2 &= -\frac{b_{22}}{2a_2} + i\omega_2, & \lambda_2' &= -\frac{b_{22}}{2a_2} - i\omega_2, \\
&\cdots\cdots\cdots & &\cdots\cdots\cdots \\
\lambda_n &= -\frac{b_{nn}}{2a_n} + i\omega_n, & \lambda_n' &= -\frac{b_{nn}}{2a_n} - i\omega_n,
\end{aligned}\right\} \quad (10.4.9)$$

where $\omega_1, \omega_2, \dots, \omega_n$ are the natural frequencies of the undamped system.

We must now examine how these roots will be modified by the presence of the cross-coefficients b_{uv}. To do this consider a particular root λ_s and let it be expressed in ascending powers of the coefficients b_{uv}, using Taylor's theorem. We have

$$\lambda_s = -\frac{b_{ss}}{2a_s} + i\omega_s + \sum_{u=1}^{n}\sum_{v=1}^{n}\left[\left(\frac{\partial\lambda_s}{\partial b_{uv}}\right)b_{uv}\right] + \dots\text{(higher degree terms)} \quad (10.4.10)$$

and when evaluating this, the partial derivatives all relate to the condition where

$$\left.\begin{aligned}
b_{uv} &= 0 \quad (u \neq v), \\
\lambda_s &= -\frac{b_{ss}}{2a_s} + i\omega_s.
\end{aligned}\right\} \quad (10.4.11)$$

Now λ_s is not given as an explicit function of the coefficients b_{uv} and the differentiation cannot, therefore, be performed directly. Instead, the relation

$$\frac{\partial\lambda_s}{\partial b_{uv}} = -\left(\frac{\partial\Delta}{\partial b_{uv}}\right)\bigg/\left(\frac{\partial\Delta}{\partial\lambda_s}\right) \quad (u \neq v) \quad (10.4.12)$$

must be used, because the function Δ relates the b's and the root λ_s. But it has already been pointed out that, in equation (10.4.7) the b_{uv} terms in the expansion of Δ are all of the second degree or higher. When Δ is differentiated with respect to the b_{uv} therefore these terms all vanish if b_{uv} is put equal to zero. The numerator of equation (10.4.12) is therefore zero, whereas the denominator is not. The addition of cross-damping terms to the equations thus produces second-order changes only in the values of the roots of equation (10.4.3), if the coefficients b_{uv} are of the first order of smallness.

Any pairs of roots λ_s, λ'_s, as given by equations (10.4.9) may now be substituted into the equations (10.4.2) in order to find the time variation of the damped motion. During this motion the ratios of the displacements at the different co-ordinates is given by the co-factors of Δ, as in equation (10.4.6). And, in finding these ratios for the root λ_s it is convenient to use the co-factors of the sth row. We begin by finding the co-factor of the sth term of this row; this is the quantity α_s of equations (10.4.6).

When finding the co-factors of Δ it is necessary to write λ_s for λ. With this substitution it can be seen, from an examination of Δ, that α_s can be written in the form

$$\alpha_s = (a_1 \lambda_s^2 + b_{11} \lambda_s + c_1)(a_2 \lambda_s^2 + b_{22} \lambda_s + c_2) \dots (a_n \lambda_s^2 + b_{nn} \lambda_s + c_n)_{[s]} + O[b^2, \lambda_s], \quad (10.4.13)$$

where the suffix $_{[s]}$ after the product indicates that the term $(a_s \lambda_s^2 + b_{ss} \lambda_s + c_s)$ is missing and where $O[b^2, \lambda_s]$ denotes terms containing λ_s which are of the second degree or higher in the coefficients b_{uv}. If the b_{uv} terms are sufficiently small, this last expression $O[b^2, \lambda_s]$ will be negligible so that only the product is left. This then, with λ_s made equal to

$$-\frac{b_{ss}}{2a_s} + i\omega_s \quad (10.4.14)$$

gives the required co-factor α_s for use in equation (10.4.6). Similar treatment holds good for the conjugate root λ'_s and it will be seen to give the conjugate result for α_s.

We now require the other co-factors of the sth row of Δ. Consider the rth of these which has been denoted by α_r. This determinant can be expressed in the form

$$\alpha_r = (-1)^{r+s} \Delta_{rs}, \quad (10.4.15)$$

where

$$\Delta_{rs} = (-1)^{r+s} \begin{vmatrix} b_{rs}\lambda_s & b_{r1}\lambda_s & b_{r2}\lambda_s & \dots & b_{rn}\lambda_s \\ b_{1s}\lambda_s & (a_1\lambda_s^2 + b_{11}\lambda_s + c_1) & b_{12}\lambda_s & \dots & b_{1n}\lambda_s \\ b_{2s}\lambda_s & b_{21}\lambda_s & (a_2\lambda_s^2 + b_{22}\lambda_s + c_2) & \dots & b_{2n}\lambda_s \\ \dots\dots\dots\dots\dots\dots\dots\dots\dots\dots\dots\dots\dots\dots\dots\dots\dots\dots \\ b_{ns}\lambda_s & b_{n1}\lambda_s & b_{n2}\lambda_s & \dots & (a_n\lambda_s^2 + b_{nn}\lambda_s + c_n) \end{vmatrix} .$$

$$(10.4.16)$$

This arrangement of the determinant is obtained from Δ by moving the sth column to the first column and the rth row to the first row. This gives $b_{rs}\lambda_s$ as the first term of the first row, and there is no other $b_{rs}\lambda_s$ term because it is removed in forming the co-factor from Δ. It can now be seen that α_r can be put in the form

$$\alpha_r = b_{rs}\lambda_s[(a_1\lambda_s^2 + b_{11}\lambda_s + c_1)(a_2\lambda_s^2 + b_{22}\lambda_s + c_2) \dots$$
$$(a_n\lambda_s^2 + b_{nn}\lambda_s + c_n)]_{[rs]} + O[b^3, \lambda_s], \quad (10.4.17)$$

where the subscript $_{[rs]}$ indicates that the factors

$$(a_r\lambda_s^2 + b_{rr}\lambda_s + c_r) \quad \text{and} \quad (a_s\lambda_s^2 + b_{ss}\lambda_s + c_s)$$

are missing from the product and where $O[b^3, \lambda_s]$ denotes a function of λ_s whose terms are of the third degree or higher in the coefficients b_{uv}. If the latter coefficients are sufficiently small, this function can be neglected in comparison with the product function, though it should be noted that this is itself a quantity of the first order of smallness since it contains the factor $b_{rs}\lambda_s$.

Collecting these results together, then, we have

$$\left.\begin{array}{l}\alpha_s = (a_1\lambda_s^2 + b_{11}\lambda_s + c_1)(a_2\lambda_s^2 + b_{22}\lambda_s + c_2) \dots (a_n\lambda_s^2 + b_{nn}\lambda_s + c_n)_{[s]},\\[2mm]\alpha_r = b_{rs}\lambda_s[(a_1\lambda_s^2 + b_{11}\lambda_s + c_1)(a_2\lambda_s^2 + b_{22}\lambda_s + c_2) \dots (a_n\lambda_s^2 + b_{nn}\lambda_s + c_n)]_{[rs]}.\end{array}\right\} \quad (10.4.18)$$

These results correspond to the root

$$\lambda_s = -\frac{b_{ss}}{2a_s} + i\omega_s \quad (10.4.19)$$

of equation (10.4.3) and are only appropriate when the coefficients b are sufficiently small. If the conjugate root

$$\lambda_s' = -\frac{b_{ss}}{2a_s} - i\omega_s \quad (10.4.20)$$

is used, then the quantity λ_s in equations (10.4.18) must be replaced by λ_s'.

The motions at the principal co-ordinates p_r and p_s can now be written down by the use of equations (10.4.6). If the ratio Π_r/Π_s is formed, it is found that

$$\frac{\Pi_r}{\Pi_s} = \frac{\alpha_r}{\alpha_s} = \frac{b_{rs}\lambda_s}{a_r\lambda_s^2 + b_{rr}\lambda_s + c_r}. \quad (10.4.21)$$

When the value of λ_s is inserted from equation (10.4.19) and terms of higher degree than the first in b are neglected, this becomes,

$$\frac{\Pi_r}{\Pi_s} = \frac{ib_{rs}\omega_s}{c_r - a_r\omega_s^2}. \quad (10.4.22)$$

The ratio is thus pure imaginary and if λ_s' is used instead of λ_s, the conjugate result is arrived at. And when the results corresponding to λ_s and λ_s' are taken together in the manner of the preceding section, it is found that if

$$p_s = e^{-(b_{ss}t/2a_s)}[G_s\cos\omega_s t + H_s\sin\omega_s t] \quad (10.4.23)$$

then $\quad p_r = \dfrac{b_{rs}\omega_s}{c_r - a_r\omega_s^2}e^{-(b_{ss}t/2a_s)}\left[G_s\cos\left(\omega_s t + \dfrac{\pi}{2}\right) + H_s\sin\left(\omega_s t + \dfrac{\pi}{2}\right)\right]. \quad (10.4.24)$

The factor b_{rs} in the numerator of equation (10.4.22) shows that the fraction is small and therefore the amplitude of the rth principal co-ordinate is small compared with that of the sth principal co-ordinate when the system is vibrating in the sth damped mode—that is when the motion corresponds to the sth pair of roots, λ_s and λ'_s, of equation (10.4.3). In the analysis p_r is any principal co-ordinate other that the sth and the result shows therefore that all the principal co-ordinates are small compared with the sth. Further, all these co-ordinates are in quadrature with the sth since the ratio Π_r/Π_s is imaginary.

These results may be illustrated by reference to the system of fig. 10.3.1. The curves of σ_1/ω_1 and σ_2/ω_1 shown in fig. 10.3.2, have zero slope when α is zero so that for small damping, the frequencies σ_1 and σ_2 are very nearly equal to the natural frequencies ω_1 and ω_2. The curves OC and OF of that figure, on the other hand, have a negative finite slope at the origin so that, to the first order, a decaying motion is predicted; these two curves happen to have the *same* slope at $\alpha = 0$ because $b_{11} = b_{22}$. Fig. 10.3.3 shows that the entrainment of the principal modes is small when the damping is small and that the entrained principal mode is nearly in quadrature with the predominating principal mode in either damped mode.

It is now necessary to return to the question of how large the coefficients b may be, in comparison with the other energy coefficients, before these approximations cease to be of value. First it is necessary for the sth mode to be lightly damped; otherwise the approximation for λ_s in equation (10.4.9) breaks down. The approximation of that equation requires that

$$b_{ss} \ll \sqrt{(2a_s c_s)}. \tag{10.4.25}$$

Secondly the cross-coefficients b_{uv} must be sufficiently small for the second-degree terms in equation (10.4.10) to be negligible in comparison with $b_{ss}/2a_s$. In discussing this condition it is convenient to assume that the co-ordinates are defined in such a way that all the inertia coefficients a are equal. The condition will then hold provided that any $b_{uv}\lambda_s$ term in equations (10.4.2) is small compared with $a\lambda_s^2$. That is to say if

$$\frac{b_{uv}}{a} \ll \lambda_s \tag{10.4.26}$$

for all the cross-coefficients b_{uv}.

Other conditions must be fulfilled if the relation (10.4.22) is to be valid. So far no restriction has been placed on the magnitude of the coefficients b_{rr}; if one of these is large, however, so that b_{rr} is not much smaller than $a_r\omega_s$, then the factor α_s which contains it will contain an appreciable imaginary component and so p_r and p_s will not be in quadrature.

These conclusions can be expressed in another way. If all the inertia coefficients are equal to a then the damping may certainly be treated as 'small' if

$$\left|\frac{b}{a\lambda_1}\right| \quad \text{or} \quad \frac{b}{a\omega_1} \ll 1 \tag{10.4.27}$$

when b represents any damping coefficient, either direct or cross.

Finally a new condition is introduced by the denominator of equation (10.4.22). If ω_r is very close to ω_s then this denominator will itself be small and the amplitude of the rth co-ordinate p_r may then be comparable with that of the sth for motion in the sth damped mode. This condition may also interfere with the phase relations and it is discussed in the next section.

This deduction of the form of the free motion in a lightly damped system may also be made by a simple algebraic argument as follows, starting from equations (10.4.2). The damping is supposed to be light so that the coefficients b are small. We examine solutions which only differ from the solutions which are valid when the b's are all zero by having in any mode, say the sth, small components of $\Pi_1, \Pi_2, \ldots, \Pi_n$ in addition to a large Π_s component. These other small components can be neglected, in comparison with Π_s, in the sth equation; this then becomes

$$(a_s\lambda^2 + b_{ss}\lambda + c_s)\,\Pi_s = 0 \qquad (10.4.28)$$

and determines the two roots of λ as previously discussed. If one of these roots is now substituted into one of the other equations, say the rth, and if in this equation we neglect all the b terms except that containing Π_s then we have

$$(a_r\lambda_s^2 + c_r)\,\Pi_r + b_{rs}\lambda_s\,\Pi_s = 0 \qquad (10.4.29)$$

which gives
$$\frac{\Pi_r}{\Pi_s} = \frac{-b_{rs}\lambda_s}{a_r\lambda_s^2 + c_r}. \qquad (10.4.30)$$

This equation corresponds to the relation (10.4.22) and illustrates the quadrature phase difference between the Π_s and Π_r components. For the numerator $b_{rs}\lambda_s$ has a large imaginary part compared with its real part whereas the denominator is mainly real. This general argument was given by Rayleigh.† It reveals very simply how the relationships between the Π's come about, but it is not a guide for the limits of application of the approximate theory. The foregoing theory is more complete and provides this; and, as we shall show in the next section, it can be extended to the case of close natural frequencies.

It is also possible to explain the behaviour of the system by direct argument as follows. If the coefficients b_{rs} $(r \neq s)$ were absent, then the system could vibrate freely in, say, the sth mode with the appropriate frequency and damping factor. The addition of a single damping element which would introduce a small b_{rs} term would cause a force to be transmitted to the rth mode because of the motion in the sth. This force would be in phase with the rate of change of the displacement causing it, and the displacement which it would itself cause would be in phase—or anti-phase—with it provided that the rth frequency were not very close to the sth and the rth mode were not very heavily damped. The quadrature rth component thus introduced would not appreciably affect the sth frequency or damping provided b_{rs} is sufficiently small. The introduction of other cross-damping coefficients would have a similar effect, each new motion being almost unaffected by the others.

EXAMPLE 10.4

1. Verify that the result given in equation (10.1.18), which refers to the lightly damped vibration of a system having one degree of freedom, is in agreement with equations (10.4.9).

† *Theory of Sound*, vol. 1 (1894), §102.

10.5 Free vibration of lightly damped systems having two natural frequencies which are nearly equal

The conclusions of the preceding section are invalid if two natural frequencies of the undamped system are nearly equal. This is because the denominator of equation (10.4.22) may then become so small that it is comparable in value with the numerator. Thus if ω_r is very close to ω_s the amplitude of the rth principal co-ordinate may be comparable with that of the sth co-ordinate, when the system is vibrating in the sth damped mode. This result is not surprising since motion at the rth co-ordinate p_r is excited, through the damping forces, at a frequency which is close to resonance so that the amplitude may be large. The proximity of the rth and sth resonance peaks also suggests that the phase relations between the rth and sth modes will not be the same as they would be if ω_r and ω_s were not close to each other. We shall show that this may be so and shall also investigate more fully the amplitude relations. Finally we shall show, by examining a simplified case, that small changes in the frequency of free vibration are produced by the damping when two of the undamped frequencies are very close, the changes being such that the frequencies are drawn closer together.

We examine the motion in the sth mode, when the frequency ω_r is very close to ω_s, using the results of the previous section and noting how they are affected by this special condition. We assume that the damping of the rth and sth modes is small so that equation (10.4.9) is still valid when the cross-coefficients b_{uv} are zero. For the present we shall continue to assume that even when the b_{uv} coefficients are not zero, all the second-degree terms in the b's, in the expansion (10.4.10) of λ_s, can be neglected; but we shall return to this matter later.

Consider the expressions for α_r and α_s, for it is these expressions which determine the relative phase and the amplitude ratio of the rth and sth principal co-ordinates during motion in the sth damped mode. The expressions are those of equations (10.4.18) when

$$-\frac{b_{ss}}{2a_s} + i\omega_s$$

is substituted for λ_s. The propinquity of ω_r and ω_s affects the value of the factor

$$\left[a_r\left(-\frac{b_{ss}}{2a_s} + i\omega_s \right)^2 + b_{rr}\left(-\frac{b_{ss}}{2a_s} + i\omega_s \right) + c_r \right] \tag{10.5.1}$$

which occurs in α_s but not in α_r. This factor also occurs in the expressions for $\alpha_1, \alpha_2, ..., \alpha_{r-1}, \alpha_{r+1}$, etc. If second-degree terms in the b's are neglected, the factor becomes

$$(c_r - a_r\omega_s^2) + i\left(b_{rr} - \frac{a_r}{a_s}b_{ss} \right)\omega_s \tag{10.5.2}$$

from which it can be seen that the real part is not very much greater than the imaginary part if c_r/a_r, that is to say ω_r^2, is very close to ω_s^2.

Consider the extreme case when

$$(c_r - a_r\omega_s^2) \ll \left(b_{rr} - \frac{a_r}{a_s}b_{ss} \right)\omega_s \tag{10.5.3}$$

so that the expression (10.5.2) is almost a pure imaginary. In this event the result (10.4.22) may be replaced by

$$\left.\begin{aligned}
\frac{\Pi_r}{\Pi_s} &= \frac{\alpha_r}{\alpha_s} = \frac{b_{rs}}{b_{rr} - \dfrac{a_r}{a_s} b_{ss}}, \\
\frac{\Pi_u}{\Pi_s} &= \frac{\alpha_u}{\alpha_s} = \frac{i b_{us} \omega_s}{c_u - a_u \omega_s^2},
\end{aligned}\right\} \tag{10.5.4}$$

where p_u is any principal co-ordinate other than the rth or sth. Thus the amplitudes of the rth and sth co-ordinates are comparable and the corresponding distortions are in phase. The amplitude of any other co-ordinate is small compared with them and the corresponding distortions are in quadrature.

It will be seen that the quantity

$$b_{rr} - \frac{a_r}{a_s} b_{ss}$$

which occurs in the denominator of the expressions for Π_r/Π_s will be zero if

$$\frac{b_{ss}}{a_s} = \frac{b_{rr}}{a_r}. \tag{10.5.5}$$

This relation is likely to hold good if the two modes whose frequencies are close are similar in shape, and this is usually the case in real systems where close frequencies are met with. It applies for instance to straight circular bars which have flexural modes in two planes at right-angles. For this type of system the assumed inequality (10.5.3) is not a valid one and the nature of the motion must be examined further.

A full investigation of the free motions of a system with close natural frequencies cannot be made without allowing for the effects of the second-order terms which we previously neglected in the Taylor expansions. To attempt to introduce these terms into the general theory would involve us in an excessively complicated analysis. This is because we now have two criteria of 'smallness' to consider instead of one, namely the difference between ω_r and ω_s and also the amount of the damping; further, the introduction of specific second-order terms into the Taylor expansion leads to excessive complication. Instead of doing this, therefore, we now examine the theory for a lightly damped system with two degrees of freedom only; this will illustrate the main features of the motion of more complex systems.

Let the principal co-ordinates of the system be p_1 and p_2 so that the equations of motion are

$$\left.\begin{aligned}
a_1 \ddot{p}_1 + b_{11} \dot{p}_1 + c_1 p_1 + b_{12} \dot{p}_2 &= 0, \\
b_{21} \dot{p}_1 + a_2 \ddot{p}_2 + b_{22} \dot{p}_2 + c_2 p_2 &= 0.
\end{aligned}\right\} \tag{10.5.6}$$

The co-ordinates may be chosen in such a way that $a_2 = a_1$, as was done in dealing with the system of fig. 10.3.1. Let the stability coefficient c_2 be equal to $c_1 + \epsilon$ where ϵ/c_1 is small so that ω_2 is nearly equal to ω_1. Suppose further that the damping is small; that is to say

$$\left.\begin{aligned}
\frac{\epsilon}{c_1} &\ll 1, \\
\frac{b}{a_1 \omega_1} &\ll 1,
\end{aligned}\right\} \tag{10.5.7}$$

where b is the absolute value of b_{11}, b_{22} or b_{12}.

The determinant Δ for this system will have two conjugate pairs of roots and both pairs will be approximately equal to

$$-\frac{b_{11}}{2a_1} \pm i\omega_1$$

which represents the approximate pair of roots for the first damped mode when ω_2 is not nearly equal to ω_1. Taking the positive sign, we may now seek a solution of the determinantal equation

$$\Delta = \begin{vmatrix} a_1\lambda^2 + b_{11}\lambda + c_1 & b_{12}\lambda \\ b_{21}\lambda & a_1\lambda^2 + b_{22}\lambda + c_1 + \epsilon \end{vmatrix} = 0 \qquad (10.5.8)$$

which is of the form $\qquad \lambda = -\dfrac{b_{11}}{2a_1} + i\omega_1 + h = \lambda_1 + h \quad$ (say), $\qquad (10.5.9)$

where h is a real or complex quantity whose components are small compared with ω_1. If this trial solution is substituted into equation (10.5.8) and all terms of higher degree than the second in

$$\frac{b}{a_1\omega_1}, \quad \frac{h}{\omega_1} \quad \text{and} \quad \frac{\epsilon}{c_1}$$

are neglected, then it is found that

$$4\left(\frac{h}{\omega_1}\right)^2 + 2\left(\frac{h}{\omega_1}\right)\left[\frac{b_{22}-b_{11}}{a_1\omega_1} - i\frac{\epsilon}{c_1}\right] - \frac{b_{12}^2}{a_1^2\omega_1^2} = 0. \qquad (10.5.10)$$

This equation is a quadratic in h/ω_1 and it has the roots

$$\frac{h}{\omega_1} = \frac{-\left[\dfrac{b_{22}-b_{11}}{a_1\omega_1} - i\dfrac{\epsilon}{c_1}\right] \pm \sqrt{\left[\left(\dfrac{b_{22}-b_{11}}{a_1\omega_1} - i\dfrac{\epsilon}{c_1}\right)^2 + \dfrac{4b_{12}^2}{a_1^2\omega_1^2}\right]}}{4}. \qquad (10.5.11)$$

The roots, h_1 and h_2 say, are complex in general. The roots for the free motion of the complete system therefore differ in both frequency and damping factor from those which would hold for the first principal mode if it were entirely independent of the second. For both the real and imaginary parts of equation (10.5.9) contain contributions from h.

The modal shape for the frequency $\lambda_1 + h_1$ can be obtained from the relation

$$\frac{\Pi_1}{\Pi_2} = \frac{-b_{12}(\lambda_1+h_1)}{a_1(\lambda_1+h_1)^2 + b_{11}(\lambda_1+h_1) + c_1}$$

$$= \frac{a_1(\lambda_1+h_1)^2 + b_{22}(\lambda_1+h_1) + c_1 + \epsilon}{-b_{12}(\lambda_1+h_1)} \qquad (10.5.12)$$

and similarly for $\lambda_1 + h_2$. And as both the real and imaginary parts of h are small compared with the imaginary part of λ_1 this relation can be written approximately as

$$\frac{\Pi_1}{\Pi_2} \doteqdot \frac{-\dfrac{b_{12}}{a_1\omega_1}}{\dfrac{2h_1}{\omega_1}} \doteqdot \frac{-\left(\dfrac{b_{22}-b_{11}}{a_1\omega_1}\right) - \dfrac{2h_1}{\omega_1} + i\dfrac{\epsilon}{c_1}}{\dfrac{b_{12}}{a_1\omega_1}}. \qquad (10.5.13)$$

An idea of the form of the motion in particular cases can be obtained from this expression.

Before examining these results, however, we now go back to equation (10.5.8) and seek a solution of the remaining form

$$\lambda = -\frac{b_{11}}{2a_1} - i\omega_1 + h' = \lambda_1' + h' \quad \text{(say)}. \qquad (10.5.14)$$

The small quantity h' may now be evaluated just as before and, in fact, is governed by equation (10.5.10) with i replaced by $-i$. The roots h_1' and h_2' are thus the complex conjugates of h_1 and h_2. If $\lambda_1 + h_1$ is replaced by $\lambda_1' + h_1'$ in equation (10.5.12), then the conjugates of the previous expressions for Π_1/Π_2 are arrived at. The frequencies and amplitude ratios are therefore unchanged and only the phase relations are altered.

Consider first the case when $b_{11} = b_{22}$ which, as mentioned earlier, is common in practice. Equation (10.5.11) now reduces to

$$\frac{h}{\omega_1} = \frac{i\frac{\epsilon}{c_1} \pm \sqrt{\left[\frac{4b_{12}^2}{a_1^2\omega_1^2} - \frac{\epsilon^2}{c_1^2}\right]}}{4} \doteqdot \frac{i\epsilon}{4c_1}\left[1 \pm \sqrt{\left(1 - \frac{4b_{12}^2}{a_1^2\omega_1^2}\frac{c_1^2}{\epsilon^2}\right)}\right]. \qquad (10.5.15)$$

Suppose now that

$$1 \gg \frac{\epsilon}{c_1} \gg \frac{b_{12}}{a_1\omega_1}. \qquad (10.5.16)$$

In this event, equation (10.5.15) gives

$$\frac{h}{\omega_1} = i\frac{b_{12}^2}{2a_1^2\omega_1^2}\frac{c_1}{\epsilon} \quad \text{or} \quad \left(\frac{i\epsilon}{2c_1} - i\frac{b_{12}^2}{2a_1^2\omega_1^2}\frac{c_1}{\epsilon}\right). \qquad (10.5.17)$$

Substitution of the first of these values into equations (10.5.13) shows that

$$\frac{\Pi_1}{\Pi_2} = i\frac{\left(\frac{\epsilon}{c_1}\right)}{\left(\frac{b_{12}}{a_1\omega_1}\right)} = iR \quad \text{(say)}, \qquad (10.5.18)$$

where $R \gg 1$, while the second gives

$$\frac{\Pi_1}{\Pi_2} = i\frac{\left(\frac{b_{12}}{a_1\omega_1}\right)}{\left(\frac{\epsilon}{c_1}\right)} = \frac{i}{R}. \qquad (10.5.19)$$

Now the meaning of these results can best be seen when it is remembered that, if b_{12} is zero then the roots of equation (10.5.8) are

$$-\frac{b_{11}}{2a_1} + i\omega_1 \quad \text{and} \quad -\frac{b_{11}}{2a_1} + i\omega_1\left(1 + \frac{\epsilon}{2c_1}\right). \qquad (10.5.20)$$

The two imaginary values of h which we have just found for small b_{12} show that the natural frequencies differ from these. Thus the lower frequency is raised by

$$\frac{b_{12}^2}{2a_1^2\omega_1^2}\frac{c_1}{\epsilon}\omega_1$$

and the higher frequency is lowered by the same amount; the b_{12} factor has the effect, therefore, of drawing the two frequencies closer together. The expressions (10.5.18) and (10.5.19) show that the two principal modes are almost in quadratur during motion in either damped mode; at the lower frequency, amplitude of the first principal mode is much greater than that of the second whereas the opposite holds at the higher frequency. The motion in the second principal mode lags behind that in the first.

If the calculations are repeated, using the conjugate root λ_1' in place of λ_1 then the frequency and amplitude relations are unchanged but the phase relation alters from lagging to leading.

The motion of the system can be depicted as that of a particle moving in a plane. The two free motions of the particle, when b_{12} is zero, correspond to two motions along perpendicular directions. When b_{12} is introduced the motions become ellipses which diminish in size as the motion is damped out. The ratio of the axes of the ellipses is given by

$$\frac{\left(\dfrac{\epsilon}{c_1}\right)}{\left(\dfrac{b_{12}}{a_1\omega_1}\right)} \quad \text{or} \quad \frac{\epsilon}{\omega_1 b_{12}}.$$

If ϵ is very large compared to $(\omega_1 b_{12})$ then the ellipses are long and thin; they become progressively rounder as ϵ is reduced. If ϵ is made too small then the approximations in the above treatment become invalid and the results do not apply.

Dropping the assumption (10.5.16), we now examine the special case

$$1 \gg \frac{\epsilon}{c_1} = \frac{2b_{12}}{a_1\omega_1}. \tag{10.5.21}$$

For this value, the square root in equation (10.5.15) is so small that the roots are coincident. These roots are both

$$\frac{h}{\omega_1} = \frac{i\epsilon}{4c_1} \tag{10.5.22}$$

and the corresponding modal ratios (10.5.13) are both

$$\frac{\Pi_1}{\Pi_2} = i \tag{10.5.23}$$

showing that the amplitudes of the motion in the two principal modes are equal and in quadrature; the free motion of the particle therefore becomes a circle. The conjugate value of λ_1 changes the sign of the result (10.5.23) so that a second circular motion, in the opposite direction to the first, is also possible. The frequency for these motions is the imaginary part of λ, namely

$$\omega_1\left(1 + \frac{\epsilon}{4c_1}\right) \tag{10.5.24}$$

which is midway between the two natural frequencies which exist when b_{12} is zero.

A further special case can be examined when

$$1 \gg \frac{b_{12}}{a_1 \omega_1} \gg \frac{\epsilon}{c_1}. \tag{10.5.25}$$

This gives the roots of equation (10.5.15) as

$$\frac{h}{\omega_1} = \frac{i\epsilon}{4c_1} \pm \frac{b_{12}}{2a_1 \omega_1}. \tag{10.5.26}$$

Taking the positive sign the modal shape is

$$\frac{\Pi_1}{\Pi_2} = -1. \tag{10.5.27}$$

While the frequency is still $\qquad \omega_1\left(1 + \dfrac{\epsilon}{4a_1}\right) \qquad$ (10.5.28)

the damping component of λ_1 is reduced from $-b_{11}/2a_1$ to

$$-\left(\frac{b_{11} - b_{12}}{2a_1}\right).$$

With the negative sign in equation (10.5.25) the modal shape becomes

$$\frac{\Pi_1}{\Pi_2} = +1 \tag{10.5.29}$$

so that there are now two linear motions which are perpendicular. The frequency for the second motion is identical with that of the first but the damping component is raised from $-b_{11}/2a_1$ to

$$-\left(\frac{b_{11} + b_{12}}{2a_1}\right).$$

Summing up these results, using the particle system for illustration, we begin with b_{12} zero so that the particle has two linear perpendicular motions whose frequencies are in the ratio

$$\left(1 + \frac{\epsilon}{2c_1}\right) : 1.$$

Introduction of a small coefficient b_{12} draws the frequencies together slightly and transforms the motions into thin ellipses whose axes coincide with the original perpendicular motions. If b_{12} is increased further the frequencies are drawn closer until eventually they coincide. The motion has by then been transformed into a circle and motion in either direction round it is possible. If b_{12} is increased still further the motions become linear again but they are at 45° to the original linear motions. The damping of one of them is increased and of the other is reduced.

All these results will be modified if the coefficients b_{11} and b_{22} differ, and in any particular case the motion can be found by using the preceding methods. In general, elliptical motions (whose axes are neither coincident with, nor equally inclined to, the principal mode directions) are possible.

EXAMPLE 10.5

1. Consider the case where $b_{12} = 0$ in the two-degree-of-freedom system to which equation (10.5.10) refers. Show that its prediction of the frequencies of the damped modes is correct; that is to say, verify the results (10.5.20).

10.6 The complete solution of the equations of motion for harmonic excitation

In this section, we shall examine the problem of the motion which follows the application of a harmonic force, at some instant $t = 0$, to a mechanical system. As before it will be simplest to start with some consideration of a system having a single degree of freedom.

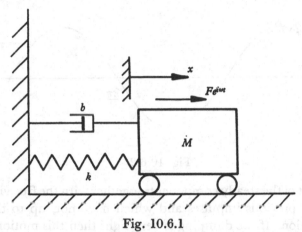

Fig. 10.6.1

The simple oscillator of fig. 10.6.1 moves in accordance with the equation

$$M\ddot{x} + b\dot{x} + kx = Fe^{i\omega t}. \tag{10.6.1}$$

The solution to this equation is the sum of (*a*) the complementary function which, when substituted for x, makes the left-hand side of equation (10.6.1) equal to zero and (*b*) the particular integral which is any solution that makes the left-hand side equal to the right-hand side. The general solution may therefore be written

$$x = e^{-\nu\omega_1 t}[G\cos\Omega t + H\sin\Omega t] + \alpha_{xx}Fe^{i\omega t}, \tag{10.6.2}$$

where α_{xx} is the complex direct receptance at x which is given by equation (8.1.5). The constants G and H must be chosen to fit the values of x and $\dot{x}$ which occur at $t = 0$ and, when finding these, it is convenient to replace the complex exponential by the appropriate trigonometric function.

Although the values of G and H affect the motion during the time immediately after the application of the force $Fe^{i\omega t}$, their influence becomes progressively diminished, as time goes on, by the presence of the factor $e^{-\nu\omega_1 t}$. In other words, the total motion consists of a steady-state component which remains constant throughout and a free vibration which dies away because of the damping. At the instant when $t = 0$ the combined motion due to both components fulfills the initial conditions. The presence of the initial free vibration and its subsequent decay are illustrated in fig. 10.6.2. If the system is very lightly damped, then the free component may persist for a long time, but the exponential term will always reduce it to negligible magnitude eventually.

The behaviour of real, lightly damped, systems is complicated by the fact that the exciting forces are rarely perfectly sinusoidal but consist of one or more sinusoidal components together with small random variations. If, for instance, the exciting forces arise from the firing strokes of an engine then the small differences between one stroke and another will produce the random variations on top of the steady components. These variations can be regarded as an irregular sequence of impulses, each of which starts off a free damped vibration; at any instant the total

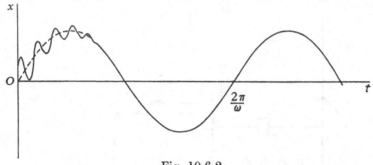

Fig. 10.6.2

motion will consist of the steady components together with the free vibrations which were initiated by previous impulses and which have not, up to that time, been dissipated by friction. If the damping is very light then this motion may be large because it will contain components generated by a large number of impulses. It consists of a motion having the frequency Ω whose amplitude fluctuates irregularly.

If the forcing frequency ω is very close to the free frequency Ω then 'beats' will occur between them if the two motions are present together; this is discussed further in §11.1. The phenomenon may be particularly noticeable if light damping and irregular forcing combine to produce a large component having the free vibration frequency Ω.

There are, however, many cases where reasonably regular forcing and moderate damping result in the forced component being very much larger and more important than the free component. Indeed it is for this reason that so much attention has been devoted to forced vibration in earlier chapters. It is perhaps surprising that in assessing the forced motion it is often permissible, as explained in Chapter 8, to neglect the damping terms in the equations.

The above argument is developed for a single-degree-of-freedom system but it is applicable also to multi-freedom systems. The complete solutions of the equations of motion (that is, of equations (8.5.3)) for such a system consist of the complementary functions, which are the sums of solutions such as those of equations (10.3.10) and particular integrals such as those discussed in §8.5 and which are given by complex receptances. As with the single-freedom case, the free motion due to an initial disturbance will die away; but free motion, in several modes simultaneously, can be maintained by irregularities in the excitation if the damping is light. The forced vibration can be found by the methods which have been described earlier and for this the damping may either be neglected or dealt with by assuming the viscous or the hysteretic law to hold good.

A particular case, which needs special treatment, arises when the forcing frequency and one of the frequencies of free vibration are the same, and when the damping is sufficiently small to be neglected. We shall discuss this by reference to

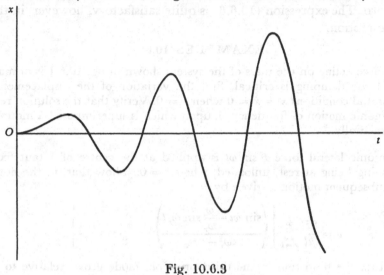

Fig. 10.6.3

the single-freedom system. Assuming that at $t = 0$ the mass is at rest in the equilibrium position, then the result (10.6.2) becomes

$$x = \frac{F}{M(\omega_1^2 - \omega^2)}\left[\sin \omega t - \frac{\omega}{\omega_1}\sin \omega_1 t\right] \qquad (10.6.3)$$

and this form of the expression is indeterminate when $\omega = \omega_1$. If, however, we write

$$\frac{\omega}{\omega_1} = 1 - \epsilon \qquad (10.6.4)$$

then the limit of the result (10.6.3) may be found, as ϵ tends to zero. The expression becomes

$$x \doteq \frac{F}{M}\frac{[\sin \omega_1(1-\epsilon)\,t - (1-\epsilon)\sin \omega_1 t]}{\omega_1^2[1-(1-2\epsilon)]} \qquad (10.6.5)$$

when $\epsilon \ll 1$. This may now be expanded and procedure to the limit $\epsilon \to 0$ then gives

$$x = \frac{F}{2M\omega_1^2}[\sin \omega_1 t - \omega_1 t \cos \omega_1 t]. \qquad (10.6.6)$$

It will be found that this solution satisfies both the equation of undamped motion and the initial conditions $x = \dot{x} = 0$ when $t = 0$.

Fig. 10.6.3 represents the graph of displacement against time. It shows that the motion is an oscillation whose amplitude increases linearly with time; such a result may be expected on the grounds that energy is being fed into the system continuously while none is being dissipated. The fact that the amplitude can be made to exceed any chosen value, no matter how large, by choosing a sufficiently large value for t, shows that the expression breaks down at these values because an infinite amplitude cannot occur in a real system. This breakdown is due to the neglect of damping which will in practice limit the amplitude eventually unless some part of the system fails before this happens. A more complete treatment, allowing for damping, would show that the rate of growth of amplitude eventually falls off to zero. The expression (10.6.6) is quite satisfactory, however, in the early stages of the motion.

EXAMPLES 10.6

1. If the force acting on the mass of the system shown in fig. 10.6.1 is of magnitude $F \sin \omega_1 t$ and the damping is critical, find the variation of the displacement which satisfies the initial conditions $x = \dot{x} = 0$ when $t = 0$. Verify that the solution represents a steady harmonic motion of frequency ω_1 upon which is superimposed a motion which decays exponentially.

2. A harmonic lateral force $F \sin \omega t$ is applied at the centre of a taut fixed-fixed string, the string being at rest, unloaded, when $t = 0$. Show that, in the notation of Chapter 4, subsequent motion is given by

$$v = \frac{2F}{\mu l} \sum_{r=1}^{\infty} \left[\left(\frac{\sin \omega t - \dfrac{\omega}{\omega_r} \sin \omega_r t}{\omega_r^2 - \omega^2} \right) \sin \frac{r\pi}{2} \sin \frac{r\pi x}{l} \right].$$

3. Show that the distortion in the third principal mode grows relative to those in the remaining modes if $\omega = \omega_3$ in Ex. 10.6.2.

10.7 Initial conditions in free vibration of systems with several degrees of freedom

The evaluation of the two relevant constants, which are determined by the initial displacement and velocity, when forming the expression for the free motion of a system having one degree of freedom does not involve algebraic difficulty. The same problem for a multi-freedom system, however, can be troublesome because of the large number of constants which are required and which are related by simultaneous equations. The difficulty can be resolved by use of the orthogonality relations and the process is described in this section. The argument is not essential for the understanding of other parts of this book (apart from Chapter 11) and may therefore be omitted if the reader has no need for it.

The treatment will be restricted to undamped systems. A similar method is applicable when damping is allowed for, but this complication is rarely necessary.

It was pointed out in Chapter 4 that there are two types of problem concerning distortion in principal modes. There is first the problem of determining the components in the various modes which are produced by a specified loading; the second problem is that of expressing a given distortion, which is known in terms of some set of co-ordinates that are not principal co-ordinates, as a set of components of the principal modes. It is the second type of problem that concerns us in this section.

Fig. 10.7.1, which is identical with fig. 3.7.1, represents a system which can be used to illustrate the nature of the problem. When this system was discussed in § 3.7 the following values were taken for the parameters:

$$m = 16\,\text{lb.}, \quad l = 2\,\text{ft.}, \quad h = 1\,\text{ft.},$$
$$k = 1024\,\text{pdl./ft.}, \quad \lambda = 512\,\text{pdl./ft.}$$

$$(10.7.1)$$

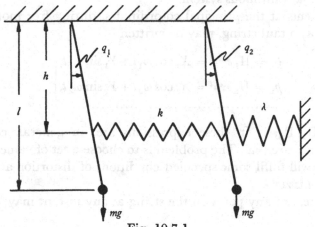

Fig. 10.7.1

and for the first principal mode the shape and frequency were found to be

$$\frac{q_1}{q_2} = 1{\cdot}28 \quad \text{with} \quad \omega_1 = \sqrt{19{\cdot}5} = 4{\cdot}41\,\text{rad./sec.} \tag{10.7.2}$$

while for the second mode

$$\frac{q_1}{q_2} = -0{\cdot}78 \quad \text{with} \quad \omega_2 = \sqrt{52{\cdot}5} = 7{\cdot}25\,\text{rad./sec.} \tag{10.7.3}$$

The complete solution for free motion therefore has the form

$$q_1 = 1{\cdot}28(A\cos\omega_1 t + B\sin\omega_1 t) - 0{\cdot}78(C\cos\omega_2 t + D\sin\omega_2 t),$$
$$q_2 = (A\cos\omega_1 t + B\sin\omega_1 t) + (C\cos\omega_2 t + D\sin\omega_2 t),$$

$$(10.7.4)$$

where A, B, C, D are constants which may be chosen to fit any given set of initial values of $q_1, q_2, \dot{q}_1$ and $\dot{q}_2$.

If the initial conditions, when $t = 0$, are

$$q_1(0) = 0{\cdot}04\,\text{rad.}, \qquad \dot{q}_1(0) = -0{\cdot}1\,\text{rad./sec.},$$
$$q_2(0) = 0, \qquad\qquad \dot{q}_2(0) = \;\;0{\cdot}2\,\text{rad./sec.},$$

$$(10.7.5)$$

for instance, then the constants may be found from the four equations

$$0{\cdot}04 = 1{\cdot}28A - 0{\cdot}78C,$$
$$0 = A + C,$$
$$-0{\cdot}1 = 1{\cdot}28 \times 4{\cdot}41B - 0{\cdot}78 \times 7{\cdot}25D,$$
$$0{\cdot}2 = 4{\cdot}41B + 7{\cdot}25D.$$

$$(10.7.6)$$

The solution of these equations is

$$A = 0 \cdot 019, \quad B = 0 \cdot 006, \quad C = -0 \cdot 019, \quad D = 0 \cdot 024. \qquad (10.7.7)$$

Since there are only four unknowns, the equations can be solved directly without excessive labour. If the number of degrees of freedom is great, however, the labour becomes prohibitive and the approach evidently fails completely if the system has infinite freedom. An alternative process will now be developed, firstly in connexion with a continuous system.

The displacements at the principal co-ordinates, during free motion of a continuous system, say a taut string, may be written

$$\begin{aligned} p_1 &= \Pi_1 e^{i\omega_1 t} = X_1 \cos \omega_1 t + Y_1 \sin \omega_1 t, \\ p_2 &= \Pi_2 e^{i\omega_2 t} = X_2 \cos \omega_2 t + Y_2 \sin \omega_2 t, \\ &\cdots\cdots\cdots\cdots\cdots\cdots\cdots\cdots\cdots\cdots\cdots \end{aligned} \qquad (10.7.8)$$

The constants $\Pi_1, \Pi_2, \dots$ in these expressions are, in general, complex while $X_1, X_2, \dots, Y_1, Y_2, \dots$ are real. The problem is to choose a set of values for these real constants which will fulfil some specified conditions of distortion and velocities of the system when t is zero.

The displacement of any point on the string at any instant may be found from the relation

$$v(x, t) = \sum_{r=1}^{\infty} p_r(t) \, \phi_r(x), \qquad (10.7.9)$$

where $\phi_1(x), \phi_2(x), \dots$ are the characteristic functions, so that, at any instant after the motion is started

$$v(x, t) = (X_1 \cos \omega_1 t + Y_1 \sin \omega_1 t) \, \phi_1(x)$$

$$+ (X_2 \cos \omega_2 t + Y_2 \sin \omega_2 t) \, \phi_2(x)$$

$$+ \cdots\cdots\cdots\cdots\cdots\cdots\cdots\cdots \qquad (10.7.10)$$

When $t = 0$ this gives

$$\begin{aligned} v &= v(x, 0) = X_1 \phi_1(x) + X_2 \phi_2(x) + \dots, \\ \frac{\partial v}{\partial t} &= \dot{v}(x, 0) = \omega_1 Y_1 \phi_1(x) + \omega_2 Y_2 \phi_2(x) + \dots. \end{aligned} \qquad (10.7.11)$$

If now the functions $v(x, 0)$ and $\dot{v}(x, 0)$ are specified analytically then the technique described in Chapter 4 may be used to find the constants X and Y. The process is explained by the following example.

A fixed-free taut string is set vibrating freely by being given an initial displacement

$$v(x, 0) = \frac{x^2}{l^2} E \qquad (10.7.12)$$

and an initial velocity $\qquad \dot{v}(x, 0) = \frac{x}{l} F, \qquad (10.7.13)$

where E and F are constants. Equation (4.7.6) now gives

$$X_r = \frac{\dfrac{E}{l^2}\displaystyle\int_0^l x^2 \sin\frac{(2r-1)\,\pi x}{2l}\,dx}{\displaystyle\int_0^l \sin^2\frac{(2r-1)\,\pi x}{2l}\,dx}$$

$$= \frac{16E}{(2r-1)^2\,\pi^2}\left[\sin\frac{(2r-1)\,\pi}{2} - \frac{2}{(2r-1)\,\pi}\right]. \tag{10.7.14}$$

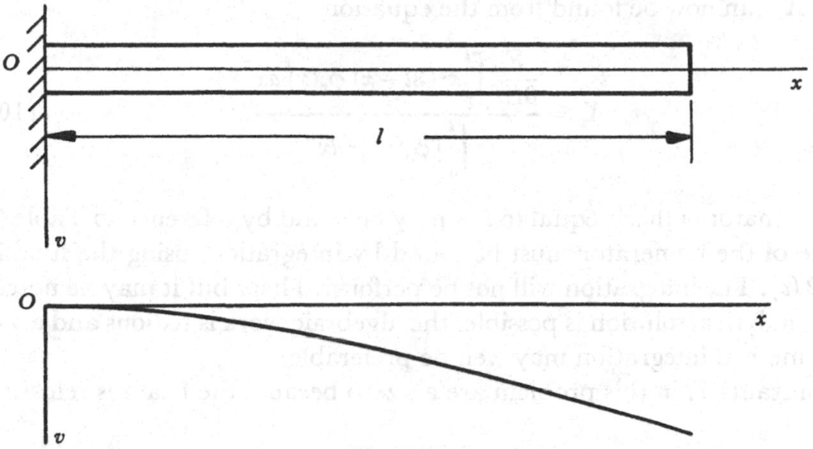

Fig. 10.7.2

The Y_r coefficients may be found as multiplies of F by a similar algebraic process. The result will be found to be

$$\omega_r Y_r = \frac{8F}{(2r-1)^2\,\pi^2}\sin\frac{(2r-1)\,\pi}{2}. \tag{10.7.15}$$

Using these results the complete solution for the free motion is

$$v(x,t) = \left[\frac{16E}{\pi^2}\left(1 - \frac{2}{\pi}\right)\cos\omega_1 t + \frac{8F}{\omega_1\pi^2}\sin\omega_1 t\right]\sin\frac{\pi x}{2l}$$

$$+ \left[-\frac{16E}{9\pi^2}\left(1 + \frac{2}{3\pi}\right)\cos\omega_2 t - \frac{8F}{9\omega_2\pi^2}\sin\omega_2 t\right]\sin\frac{3\pi x}{2l}$$

$$+ \dots\dots\dots\dots\dots\dots\dots\dots\dots\dots\dots\dots\dots\dots\dots\dots\dots, \tag{10.7.16}$$

where

$$\omega_1 = \frac{\pi}{2l}\sqrt{\left(\frac{S}{\mu}\right)}, \quad \omega_2 = \frac{3\pi}{2l}\sqrt{\left(\frac{S}{\mu}\right)}, \quad \dots \tag{10.7.17}$$

as shown in Table 4.

As a second example consider a clamped-free uniform beam, as shown in fig. 10.7.2. Let the beam be deflected by a static force F applied to its free end. The deflexion at any point is given by entry 1 of Table 7.3 (a); that is to say

$$v(x) = a_{xl}F = \frac{x^2(3l-x)F}{6EI} \tag{10.7.18}$$

which distortion has the form shown in the figure. If the force F is now suddenly removed, as would be the case if the beam represented a reed which is plucked, then the beam will be set vibrating in all its modes. The above method can now be used to find the magnitude of each of the components. The initial conditions may be written

$$v(x, 0) = \frac{x^2(3l-x)\,F}{6EI}, \quad \frac{\partial v}{\partial t} = \dot{v}(x, 0) = 0. \qquad (10.7.19)$$

Using the appropriate characteristic functions $\phi_r(x)$ for a clamped-free beam the constants X_r can now be found from the equation

$$X_r = \frac{\dfrac{F}{6EI}\displaystyle\int_0^l x^2(3l-x)\,\phi_r(x)\,dx}{\displaystyle\int_0^l [\phi_r(x)]^2\,dx}. \qquad (10.7.20)$$

The denominator of this is equal to l as may be found by reference to Table 7.2 (c). The value of the numerator must be found by integration, using the functions of Table 7.2 (c). The integration will not be performed here but it may be noted that, while an analytical solution is possible, the algebraic work is tedious and a step-by-step arithmetical integration may well be preferable.

The constants Y_r in this problem are all zero because the beam is released from rest.

The above technique, which is essential for dealing with infinite-freedom systems, may also be applied usefully to finite-freedom systems. Consider again the system of fig. 10.7.1. We introduce suitable principal co-ordinates p_1 and p_2 which are defined by the relations

$$\left.\begin{aligned} q_1 &= 1\!\cdot\!28p_1 - 0\!\cdot\!78p_2, \\ q_2 &= p_1 + p_2. \end{aligned}\right\} \qquad (10.7.21)$$

These equations may be solved so as to obtain the co-ordinates p in terms of the co-ordinates q; thus

$$\left.\begin{aligned} p_1 &= 0\!\cdot\!48q_1 + 0\!\cdot\!37q_2, \\ p_2 &= -0\!\cdot\!48q_1 + 0\!\cdot\!63q_2. \end{aligned}\right\} \qquad (10.7.22)$$

The initial values of p_1, $\dot{p}_1$, p_2 and $\dot{p}_2$ are, therefore,

$$\left.\begin{aligned} p_1(0) &= 0\!\cdot\!48 \times 0\!\cdot\!04 + 0\!\cdot\!37 \times 0 &&= 0\!\cdot\!019, \\ \dot{p}_1(0) &= 0\!\cdot\!48 \times (-0\!\cdot\!1) + 0\!\cdot\!37 \times 0\!\cdot\!2 &&= 0\!\cdot\!026, \\ p_2(0) &= -0\!\cdot\!48 \times 0\!\cdot\!04 + 0\!\cdot\!63 \times 0 &&= -0\!\cdot\!019, \\ \dot{p}_2(0) &= -0\!\cdot\!48 \times (-0\!\cdot\!1) + 0\!\cdot\!63 \times 0\!\cdot\!2 &&= 0\!\cdot\!174. \end{aligned}\right\} \qquad (10.7.23)$$

The values of the principal co-ordinates at any instant are given by

$$\left.\begin{aligned} p_1 &= X_1 \cos 4\!\cdot\!41t + Y_1 \sin 4\!\cdot\!41t, \\ p_2 &= X_2 \cos 7\!\cdot\!25t + Y_2 \sin 7\!\cdot\!25t, \end{aligned}\right\} \qquad (10.7.24)$$

and the set of constants which fit the initial conditions (10.7.23) are

$$X_1 = 0 \cdot 019, \qquad Y_1 = \frac{0 \cdot 026}{4 \cdot 41} = 0 \cdot 006,$$

$$X_2 = -0 \cdot 019, \qquad Y_2 = \frac{0 \cdot 174}{7 \cdot 25} = 0 \cdot 024. \tag{10.7.25}$$

These will be found to agree with the previous solution.

The above method suffers from the disadvantage that the simultaneous equations (10.7.21) have to be solved. This can be avoided by using the process which was employed for infinite-freedom systems; it was used previously in § 3.8.

We begin by writing down the general expression for the co-ordinates q in terms of the principal co-ordinates p when the system is vibrating freely (cf. equation (3.5.9)); they are

$$
\begin{aligned}
q_1 &= A_1[X_1 \cos \omega_1 t + Y_1 \sin \omega_1 t] + B_1[X_2 \cos \omega_2 t + Y_2 \sin \omega_2 t] + \ldots \\
&\qquad + N_1[X_n \cos \omega_n t + Y_n \sin \omega_n t], \\
q_2 &= A_2[X_1 \cos \omega_1 t + Y_1 \sin \omega_1 t] + B_2[X_2 \cos \omega_2 t + Y_2 \sin \omega_2 t] + \ldots \\
&\qquad + N_2[X_n \cos \omega_n t + Y_n \sin \omega_n t], \\
&\quad \cdots\cdots\cdots\cdots\cdots\cdots\cdots\cdots\cdots\cdots\cdots\cdots\cdots\cdots\cdots\cdots\cdots \\
q_n &= A_n[X_1 \cos \omega_1 t + Y_1 \sin \omega_1 t] + B_n[X_2 \cos \omega_2 t + Y_2 \sin \omega_2 t] + \ldots \\
&\qquad + N_n[X_n \cos \omega_n t + Y_n \sin \omega_n t].
\end{aligned} \tag{10.7.26}
$$

When $t = 0$, these reduce to

$$
\begin{aligned}
q_1(0) &= A_1 X_1 + B_1 X_2 + \ldots + N_1 X_n, \\
q_2(0) &= A_2 X_1 + B_2 X_2 + \ldots + N_2 X_n, \\
&\cdots\cdots\cdots\cdots\cdots\cdots\cdots\cdots\cdots\cdots\cdots \\
q_n(0) &= A_n X_1 + B_n X_2 + \ldots + N_n X_n,
\end{aligned} \tag{10.7.27}
$$

and

$$
\begin{aligned}
\dot{q}_1(0) &= A_1 \omega_1 Y_1 + B_1 \omega_2 Y_2 + \ldots + N_1 \omega_n Y_n, \\
\dot{q}_2(0) &= A_2 \omega_1 Y_1 + B_2 \omega_2 Y_2 + \ldots + N_2 \omega_n Y_n, \\
&\cdots\cdots\cdots\cdots\cdots\cdots\cdots\cdots\cdots\cdots\cdots \\
\dot{q}_n(0) &= A_n \omega_1 Y_1 + B_n \omega_2 Y_2 + \ldots + N_n \omega_n Y_n.
\end{aligned} \tag{10.7.28}
$$

It is now necessary to find the values of the constants $X_1, X_2, \ldots, X_n, Y_1, Y_2, \ldots, Y_n$ without going through the process of solving the set of simultaneous equations, which would be necessary if the processes of the above example were followed exactly. Compare the equations (10.7.27) with equations (3.5.9). The quantities on the left-hand side represent static deflexions at the co-ordinates q and the principal co-ordinates p of the previous equations must now be imagined to have the static values $X_1, X_2, \ldots, X_n$. These values can now be written down separately by using the method of § 3.8. The value, for example, of X_r will be

$$
X_r = \frac{\left\{ \begin{aligned} (c_{11} R_1 + c_{12} R_2 + \ldots + c_{1n} R_n)\, q_1(0) + (c_{21} R_1 + c_{22} R_2 + \ldots + c_{2n} R_n)\, q_2(0) + \ldots \\ + (c_{n1} R_1 + c_{n2} R_2 + \ldots + c_{nn} R_n)\, q_n(0) \end{aligned} \right\}}{[c_{11} R_1^2 + c_{22} R_2^2 + \ldots + c_{nn} R_n^2 + 2c_{12} R_1 R_2 + \ldots]}
\tag{10.7.29}
$$

by virtue of equation (3.8.26).

The constants Y may be found in a similar manner. That is, the quantities $\dot{q}_1(0), \dot{q}_2(0), ..., \dot{q}_n(0)$ may be used in the equations in the same way as the quantities $q_1(0), q_2(0), ..., q_n(0)$, and the constants $A_1\omega_1, A_2\omega_1, ..., B_1\omega_2, B_2\omega_2, ...$ are used like $A_1, A_2, ..., B_1, B_2, ...$ were previously. This gives

$$Y_r = \frac{\left\{\begin{array}{l}(c_{11}R_1 + c_{12}R_2 + ... + c_{1n}R_n)\,\dot{q}_1(0) + (c_{21}R_1 + c_{22}R_2 + ... + c_{2n}R_n)\,\dot{q}_2(0) + ... \\ \qquad\qquad\qquad + (c_{n1}R_1 + c_{n2}R_2 + ... + c_{nn}R_n)\,\dot{q}_n(0)\end{array}\right\}}{[c_{11}R_1^2 + c_{22}R_2^2 + ... + c_{nn}R_n^2 + 2c_{12}R_1R_2 + ...]\,\omega_r}.$$

(10.7.30)

This process may be illustrated by returning once more to the system of fig. 10.7.1. The potential energy of the system is given by

$$2V = (mgl + kh^2)\,q_1^2 + (mgl + kh^2 + \lambda h^2)\,q_2^2 - 2kh^2 q_1 q_2 \qquad (10.7.31)$$

and from this the stability coefficients can be calculated for the values (10.7.1) of the various parameters; they are

$$c_{11} = 2048, \quad c_{22} = 2560, \quad c_{12} = -1024. \qquad (10.7.32)$$

These values may be substituted into the expressions (10.7.29) and (10.7.30) with the initial conditions (10.7.5) and it is thus found that

$$\left.\begin{array}{l}X_1 = \dfrac{\left\{\begin{array}{l}(2048 \times 1{\cdot}28 - 1024 \times 1) \times 0{\cdot}04 \\ \quad + (-1024 \times 1{\cdot}28 + 2560 \times 1) \times 0\end{array}\right\}}{[2048 \times 1{\cdot}28^2 + 2560 \times 1^2 - 2 \times 1024 \times 1{\cdot}28 \times 1]} = 0{\cdot}019, \\[4ex] Y_1 = \dfrac{\left\{\begin{array}{l}(2048 \times 1{\cdot}28 - 1024 \times 1) \times (-0{\cdot}1) \\ \quad + (-1024 \times 1{\cdot}28 + 2560 \times 1) \times 0{\cdot}2\end{array}\right\}}{[2048 \times 1{\cdot}28^2 + 2560 \times 1^2 - 2 \times 1024 \times 1{\cdot}28 \times 1] \times 4{\cdot}41} = 0{\cdot}006, \\[4ex] X_2 = \dfrac{\left\{\begin{array}{l}(-2048 \times 0{\cdot}78 - 1024 \times 1) \times 0{\cdot}04 \\ \quad + (1024 \times 0{\cdot}78 + 2560 \times 1) \times 0\end{array}\right\}}{[2048 \times 0{\cdot}78^2 + 2560 \times 1^2 + 2 \times 1024 \times 0{\cdot}78 \times 1]} = -0{\cdot}019, \\[4ex] Y_2 = \dfrac{\left\{\begin{array}{l}(-2048 \times 0{\cdot}78 - 1024 \times 1) \times (-0{\cdot}1) \\ \quad + (1024 \times 0{\cdot}78 + 2560 \times 1) \times 0{\cdot}2\end{array}\right\}}{[2048 \times 0{\cdot}78^2 + 2560 \times 1^2 + 2 \times 1024 \times 0{\cdot}78 \times 1] \times 7{\cdot}25} = 0{\cdot}024.\end{array}\right\} \quad (10.7.33)$$

These results are the same as those obtained previously.

EXAMPLES 10.7

1. A free-free taut string of length l is released from rest at the instant $t = 0$ with the displacement

$$v(x, 0) = A \sin\frac{\pi x}{l},$$

where A is a constant. Show that the deflexion at any later time t is given by

$$v(x, t) = A\left[\frac{2}{\pi} - \frac{4}{3\pi}\cos\omega_2 t \cos\frac{2\pi x}{l} - \frac{4}{15\pi}\cos\omega_4 t \cos\frac{4\pi x}{l} - \frac{4}{35\pi}\cos\omega_6 t \cos\frac{6\pi x}{l} - ...\right],$$

where $\omega_2, \omega_4, ...$ are the 2nd, 4th, ... non-zero natural frequencies of the taut string.

2. The angular deflexion and velocity of a free-free shaft of length l, at the instant $t = 0$, are

$$\theta(x, 0) = \frac{xE}{l}, \quad \dot{\theta}(x, 0) = \frac{xF}{l},$$

where E and F are constants. Show that the deflexion at any later instant is

$$\theta(x, t) = \frac{E}{2} + \frac{Ft}{2} + \sum_{r=1}^{\infty} \left\{ \frac{2(\cos r\pi - 1)}{r^2\pi^2} \left[E \cos \omega_r t + \frac{F}{\omega_r} \sin \omega_r t \right] \cos \frac{r\pi x}{l} \right\},$$

where ω_r is the rth non-zero natural frequency of a free-free shaft.

3. A clamped-free bar is acted upon by a constant axial tensile force F which is applied at the free end as shown. What will be the subsequent longitudinal displacement u if the force is suddenly removed at the instant $t = 0$?

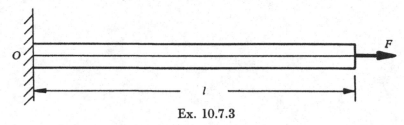

Ex. 10.7.3

NOTATION

Supplementary List of Symbols used in Chapter 10

A_s, B_s	Arbitrary constants associated with sth damped mode [see equation (10.3.28)].
$A_1, A_2, ..., B_1, B_2, ...,$ etc.	See equations (10.7.26).
$b_{\text{crit.}}$	$= 2M\omega_1$.
$_sC_r$	See equation (10.3.5).
$_sD_r, _sE_r$	See equations (10.3.26) and (10.3.27).
G_s, H_s	Arbitrary constants associated with sth damped mode [see equations (10.3.10)].
h, h'	See equations (10.5.9) and (10.5.14).
K	Torsional stiffness (see fig. 10.3.1).
R, Γ	Abbreviations used in worked example [see equation (10.3.30)].
R', Γ'	Abbreviations used in worked example [see equation (10.3.32)].
X_r, Y_r	See equations (10.7.8).
α	Dimensionless measure of damping used in worked example [see equation (10.3.20)].
α_s	See equation (10.4.6).
β_s, γ_s	Values of λ_s and λ'_s respectively when sth pair of roots of determinantal equation is real.
δ	Logarithmic decrement [see equation (10.1.26)].
ϵ	Value of $c_2 - c_1$ for system having close natural frequencies, a_2 being equal to a_1.
$_s\theta_r$	See equation (10.3.5).

Λ	Complex dimensionless frequency used in worked example ($= \lambda/\omega_1$).
λ	Complex frequency used in trial solutions [as in equations (10.1.5) and (10.3.1)].
λ_s, λ_s'	sth pair of roots of determinantal equation [see equations (10.3.4) and (10.3.25)].
ν	Damping factor [see equations (10.1.2)].
ρ_s	See equations (10.3.4).
σ_s	See equations (10.3.4).
Ω	Frequency of free damped vibration of single-freedom system [see equation (10.1.19)].

CHAPTER 11

NON-HARMONIC AND TRANSIENT VIBRATION

Fourier's Theorem....is not only one of the most beautiful results of modern analysis, but may be said to furnish an indispensible instrument in the treatment of nearly every recondite question in modern physics. To mention only sonorous vibrations, the propagation of electric signals along a telegraph wire, and the conduction of heat by the earth's crust, as subjects in their generality intractable without it, is to give but a feeble idea of its importance.

LORD KELVIN AND PETER GUTHRIE TAIT, *Treatise on Natural Philosophy* (1867)

Throughout all the previous chapters of this book, the mechanical systems which have been discussed have either been excited by steady harmonic forces or have moved freely with no external forces at all. Many problems arise, however, in which the external forces vary with time in some non-harmonic way and we shall show in this chapter that the resulting motion can still be analysed by using some of the theorems which have been developed for the free, and for the harmonically-forced, motion.

The first type of excitation to be considered is that in which the external forces are steadily fluctuating, but are non-sinusoidal. This arises in almost all reciprocating engines and is of great technical importance. Its analysis is straightforward and a brief discussion only is required.

The second type of excitation to be considered is that of transient forces; these forces vary with time but do not repeat their variation after regular intervals. The most usual variety of force in this category is that which lasts for a limited (and often brief) time only, and it is to this that the name 'transient' strictly applies; the term is also used, however, for a force which varies with time in an irregular manner over an indefinite period. When a motor vehicle is driven over a pot hole on an otherwise smooth road it is subjected to a transient excitation of the first kind. The other kind arises when the vehicle is being driven over a continuously rough road.

A matter which will not be discussed in this book, but of which the reader should be aware, is the possibility of analysing certain types of transient problem by the use of the travelling-wave concept. This is often a convenient alternative to vibration theory and it does not involve use of the principal modes and natural frequencies. If we wished to calculate the motion of the water in a pond after a stone has been dropped into it, then it would be possible, theoretically, to determine a large number of natural frequencies and modes, and then to express the motion in terms of these. Such an analysis would be extremely complicated and would not use the relatively simple physical picture of the set of circular ripples which emanates from the disturbance during the early stages of the motion. Now it is possible to analyse this early part of the motion by restricting the analysis to the spreading ripples only and this process is very much simpler than the other; indeed the one method is practicable and the other is not. One reason for the greater simplicity of the wave method is that the shape of the boundaries of the pond does not affect the analysis for the early stages of the motion, which it evidently should not on physical grounds. In the vibration method, on the other hand, the modes and frequencies will be dependent on the boundaries and the effects of the latter only cancel out in the final analysis; they therefore introduce quite unnecessary complexity in the previous stages. These arguments become less important if we are concerned with the motion after the ripples have crossed the pond several times, and this problem then submits best to the principal mode analysis.

There are a number of engineering problems to which these ideas apply. For instance, if a long metal bar is struck longitudinally at one end by a hammer, then a wave of stress passes along the bar; this is analogous to the ripple spreading on the pond. In this problem, the bar's distortion during the first stages of the motion is most easily examined by considering the stress wave rather than the motion of a large number of principal modes of the bar. In the later stages of the motion, when the wave has been reflected backward and forward along the bar several times, the modal analysis becomes useful.

Wave theory is an important complement to vibration theory and becomes useful when the duration of a transient force is very short compared with the lowest period of natural vibration of the system to which it is applied. The theory has been discussed by Timoshenko[†] and more fully by Kolsky.[‡]

11.1 Excitation by non-harmonic periodic forces

Any force which varies with time in such a way that its value at any instant t is equal to its value at a preceding instant $(t-\tau)$, where τ is a constant, may be expressed as a set of sinusoidal components with periods τ, $\frac{1}{2}\tau$, $\frac{1}{3}\tau$, etc., in conformity with Fourier's theorem, by using the methods of harmonic analysis.[§] The motion which is produced by each of these components can be found by the methods already described and the net motion will be the sum of all the component motions, by virtue of the linearity of the equations of motion.

Suppose that the exciting force acts at q_r, and that the motion is required at q_s. If the amplitude of the force component with the lowest frequency is F_1, then the corresponding displacement at q_s is given by the expression

$$\alpha_{sr}(\tau) \, F_1 \sin\left(\frac{2\pi t}{\tau} + \psi_1\right). \tag{11.1.1}$$

Here, ψ_1 is a phase angle which, like F_1, will depend on the form of the periodic force function, and the receptance has been written in the form $\alpha_{sr}(\tau)$ to emphasize

† S. P. Timoshenko, *Vibration Problems in Engineering* (Van Nostrand, New York, 3rd ed. 1955).

‡ H. Kolsky, *Stress Waves in Solids* (Oxford University Press, 1953).

§ See, for example, Th. von Kármán and M. A. Biot, *Mathematical Methods in Engineering* (McGraw-Hill, London, 1940), p. 335.

that it is a function of the period τ. The total motion will be the sum of a series of such terms; thus

$$q_s = \alpha_{sr}(\tau)\, F_1 \sin\left(\frac{2\pi t}{\tau} + \psi_1\right) + \alpha_{sr}(\tfrac{1}{2}\tau)\, F_2 \sin\left(\frac{4\pi t}{\tau} + \psi_2\right) + \alpha_{sr}(\tfrac{1}{3}\tau)\, F_3 \sin\left(\frac{6\pi t}{\tau} + \psi_3\right) + \ldots.$$

$$(11.1.2)$$

Now in general the various receptances in this series will not be equal to one another and, if the system is damped, they will be complex quantities whose arguments differ as well as their moduli. The instantaneous displacement at q_s therefore consists of Fourier components which differ in relative phase and in relative amplitude from the corresponding components of the force. These components combine to give a displacement-time curve whose shape may be quite unlike that of the curve representing the force which causes the motion, although it will have the same periodicity. In particular, it may happen that one of the displacement components of the series (11.1.2) is very much larger than the others because the corresponding force component has a frequency which is very close to a natural frequency of the system. The total displacement of the system will then be almost purely sinusoidal, the motion having this particular frequency.

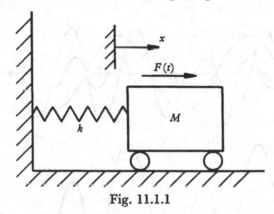

Fig. 11.1.1

As a very simple example, consider the motion of the oscillator of fig. 11.1.1 when it is acted on by the periodic force whose variation with time is plotted in fig. 11.1.2 (a). The force function $F(t)$ is in this case given by the series

$$F(t) = kE[\sin \omega t + \tfrac{1}{2} \sin 2\omega t + \tfrac{1}{4} \sin 3\omega t], \qquad (11.1.3)$$

where E is a constant, and the separate components are plotted in fig. 11.1.2 (b). If the natural frequency $\omega_1 = \sqrt{(k/M)}$ is such that

$$\omega_1 = \frac{5\omega}{2} \qquad (11.1.4)$$

then the corresponding displacement components are as plotted in fig. 11.1.2 (c). These components combine to give the curve of fig. 11.1.2 (d) which will be seen to differ greatly from the force curve at (a). In the displacement curve, the second harmonic is close to resonance and is therefore predominant; the third is in anti-phase to that of the force because its frequency is greater than ω_1.

In many engineering problems, the period τ of the fluctuating force is fixed by the running-speed of a machine. The designer needs to examine whether resonance may be produced at any of the natural frequencies of the system by any of the components of the force. He can use for this purpose a chart arranged as in fig. 11.1.3.

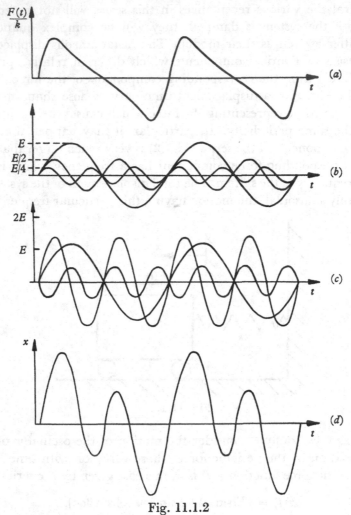

Fig. 11.1.2

Distances along the horizontal axis denote running speed (which is proportional to $1/\tau$) and those along the vertical axis denote frequency. Horizontal lines are drawn at $\omega_1, \omega_2, \ldots$, etc., these being the natural frequencies. Straight lines are drawn through the origin with slopes given by the multiples of the running speed which correspond to the excitation frequency. The intersection of one of these lines with a horizontal line indicates a state of resonance. And if the running speed lies within a range, as indicated by the two vertical dotted lines, then the resonances within this range can be readily seen. It will be observed that, as the speed falls, the resonances become closer together. Consequently, at low speeds, two resonances may be sufficiently close for the total response to contain large components due to each.

Suppose for instance that the natural frequency ω_1 lies between the forcing frequencies $10\pi/\tau$ and $12\pi/\tau$. The displacement will then contain large components at these frequencies, but components having other frequencies will be relatively

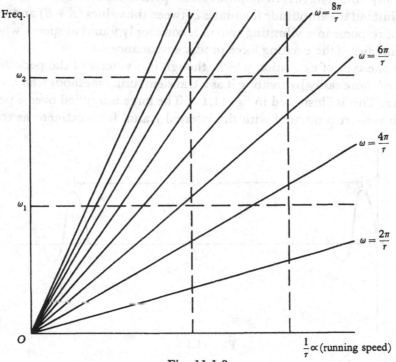

Fig. 11.1.3

small. The displacement measured at some co-ordinate q can therefore be written approximately as

$$q \doteq A \sin\left(\frac{10\pi t}{\tau} + \psi_5\right) + B \sin\left(\frac{12\pi t}{\tau} + \psi_6\right), \qquad (11.1.5)$$

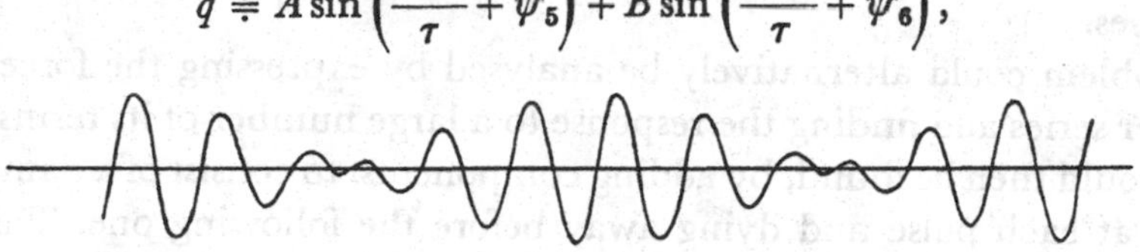

Fig. 11.1.4

where A and B are equal to the products of α and F terms as in equation (11.1.2). Now equation (11.1.5) may be thrown into the form

$$q \doteq (A+B) \sin\left(\frac{11\pi t}{\tau} + \frac{\psi_5 + \psi_6}{2}\right) \cos\left(\frac{\pi t}{\tau} + \frac{\psi_6 - \psi_5}{2}\right)$$

$$+ (B-A) \cos\left(\frac{11\pi t}{\tau} + \frac{\psi_5 + \psi_6}{2}\right) \sin\left(\frac{\pi t}{\tau} + \frac{\psi_6 - \psi_5}{2}\right). \qquad (11.1.6)$$

Each of these terms is the product of a sinusoid having the frequency $11\pi/\tau$ and of one with the frequency π/τ. They each correspond therefore to a 'beating motion' as illustrated in fig. 11.1.4 in which the amplitude of the high-frequency component is 'modulated' by the low-frequency component so that it varies between $A+B$ and zero. The two beating motions together form a single beating motion which does not at any instant have zero amplitude, except in the special case when A and B are equal; instead the amplitude fluctuates between the values $(A+B)$ and $(A-B)$. This type of response in a vibrating system is commonly found at speeds where the higher harmonics of the exciting force produce resonance.

There is one type of excitation which, though it is strictly of the periodic type, can be solved more easily by treating it as a transient, using methods which we shall discuss later. This is illustrated in fig. 11.1.5. The force is applied over a period of time which is short compared with the interval τ and, for treatment as transient

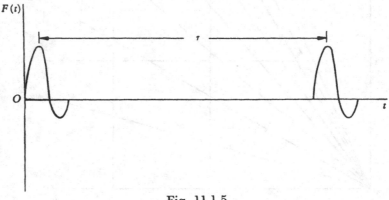

Fig. 11.1.5

excitation to be suitable, it is also necessary for a free vibration to die away to negligible amplitude within the interval τ. Under these conditions each pulse produces a motion which can be analysed without reference to the effect of preceding pulses.

This problem could alternatively be analysed by expressing the force function as a Fourier series and finding the response to a large number of its terms. The net response would then be found, by adding components, to consist of a damped wave beginning at each pulse and dying away before the following one. This type of analysis would probably be more tedious than a transient analysis using a single pulse; but as τ is reduced, so that the free motion has less time to decay, the Fourier method becomes appropriate.

EXAMPLE 11.1

1. A fixed-free taut string has, applied to its free end, a periodic lateral force of 'square' wave-form. The period of this excitation is τ and its maximum value is F, as shown in the figure. Show that the response at the rth principal co-ordinate p_r (chosen in the manner of Chapter 4) may be expressed in the form

$$p_r = \frac{8F}{\pi \mu l} \sin \frac{(2r-1)\,\pi}{2} \left[\frac{\sin \omega t}{\omega_r^2 - \omega^2} + \frac{\sin 3\omega t}{3(\omega_r^2 - 9\omega^2)} + \frac{\sin 5\omega t}{5(\omega_r^2 - 25\omega^2)} + \cdots \right],$$

where $\omega = 2\pi/\tau$ and the notation of Chapter 4 is used.

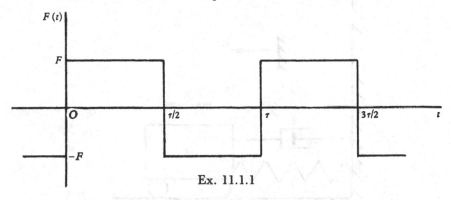

Ex. 11.1.1

11.2 Transient excitation of systems having a single degree of freedom; the Duhamel integral

In this section, we shall study the motion of a single-degree-of-freedom system which is at rest in its position of equilibrium at the instant $t = 0$ and which is then disturbed by a force $F(t)$ which begins to act at this instant and which varies with time in any prescribed manner. This covers all possible cases of transient excitation of such a system because, when the system it not at rest when $t = 0$, its initial motion can be added to the subsequent transient motion by virtue of the linearity of the equations.

For the purposes of the discussion, let the system be a damped oscillator, as shown in fig. 11.2.1, having a mass M, a spring stiffness k and a viscous damping constant b. Consider first the ensuing motion when a steady force of magnitude F_1 begins to act on the system at $t = 0$. After this instant the displacement x will be governed by the equation

$$M\ddot{x} + b\dot{x} + kx = F_1 \tag{11.2.1}$$

with the initial conditions $x = \dot{x} = 0$ when $t = 0$. The solution of this equation is

$$x = \frac{F_1}{k} \left[1 - e^{-\nu \omega_1 t} \left(\frac{\nu \omega_1}{\Omega} \sin \Omega t + \cos \Omega t \right) \right], \tag{11.2.2}$$

where

$$\Omega = \sqrt{\left[\frac{k}{M} - \frac{b^2}{4M^2} \right]} = \omega_1 \sqrt{(1 - \nu^2)}, \qquad \nu = \frac{b}{2M\omega_1}, \qquad \omega_1 = \sqrt{\left(\frac{k}{M} \right)}. \tag{11.2.3}$$

Consider now the change in this motion which is produced if the force F_1 ceases to act at some instant δt after it began. The subsequent motion can conveniently be found by adding to the result (11.2.2) a similar expression which gives the motion due to a force $-F_1$ which begins to act at the instant $t = \delta t$, the effect of these two forces together being that of a force F_1 acting during the interval $0 \leqslant t \leqslant \delta t$. The new motion, for $t \geqslant \delta t$, is

$$x = \frac{F_1}{k}\left[1 - e^{-\nu\omega_1 t}\left(\frac{\nu\omega_1}{\Omega}\sin\Omega t + \cos\Omega t\right)\right]$$
$$- \frac{F_1}{k}\left[1 - e^{-\nu\omega_1(t-\delta t)}\left(\frac{\nu\omega_1}{\Omega}\sin\Omega(t-\delta t) + \cos\Omega(t-\delta t)\right)\right] \quad (11.2.4)$$

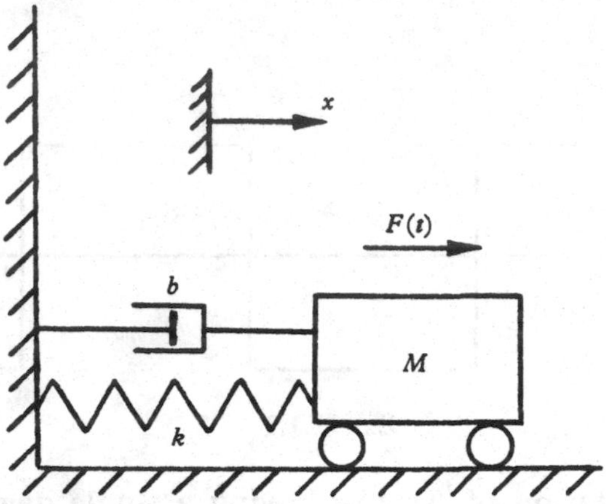

Fig. 11.2.1

and it is clear that this expression is of the functional form

$$x = f(t) - f(t - \delta t). \quad (11.2.5)$$

Suppose now that the interval δt is very small. In this event, equation (11.2.5) may be written in the alternative form

$$x = \frac{df}{dt}\delta t. \quad (11.2.6)$$

If the differentiation of equation (11.2.2) is carried out, then it is found that the expression (11.2.6) gives

$$x = \frac{F_1}{k}\left(\Omega + \frac{\nu^2\omega_1^2}{\Omega}\right)\delta t\, e^{-\nu\omega_1 t}\sin\Omega t \quad (11.2.7)$$

which, by virtue of equation (11.2.3), reduces to

$$x = \frac{F_1\delta t}{M\Omega}e^{-\nu\omega_1 t}\sin\Omega t. \quad (11.2.8)$$

This then is the motion at any instant t due to a force F_1 acting for a short interval δt when $t = 0$. The formal derivation which is given above makes it clear that similar expressions may be found for *any* system having one degree of freedom when it is excited in this way. The equation of motion will, in general, govern a generalized co-ordinate q rather than a simple displacement x, but it will be of the same mathematical form as (11.2.1) so that the same technique of solution must be applicable.

This general case will be taken up in a later section. For the present, however, it should be noted that equation (11.2.8) can be derived by considering the dynamics of the spring-mass system of fig. 11.2.1 directly. The system is acted on by an impulse of magnitude $F_1 . \delta t$ at the instant $t = 0$. This impulse gives the mass a velocity $(F_1/M) \delta t$ and the system is thus set oscillating in such a way that it has this velocity and zero displacement when $t = 0$. It can be seen that equation (11.2.8) fulfils these requirements. These arguments require some elaboration if they are to be applied to other types of mechanical system, rather than the simple oscillator, but it is not necessary to supply this because our original treatment establishes the solution.

The result (11.2.8) may now be used to solve the more general problem of finding the motion which is produced, in the same system, by a force $F(t)$ which varies with time. Such a force may be regarded as a sequence of impulses each of which contributes a term of the form (11.2.8) to the net motion. The impulse of magnitude $F(t) \delta t$ which occurs at time t will produce at the later instant t_1, the displacement

$$\frac{F(t) \delta t}{M\Omega} e^{-\nu\omega_1(t_1 - t)} \sin \Omega(t_1 - t), \tag{11.2.9}$$

where $t < t_1$. The total displacement x at the instant t_1 can now be found by integrating over the period of time between $t = 0$ and $t = t_1$. This gives

$$x = \frac{1}{M\Omega} \int_0^{t_1} F(t) e^{-\nu\omega_1(t_1 - t)} \sin \Omega(t_1 - t) \, dt \tag{11.2.10}$$

which is known as the Duhamel integral. It is left to the reader to show that this expression satisfies equation (11.2.1) when F_1 in the latter is replaced by $F(t)$ and also that it satisfies the initial conditions.

As an example of the use of equation (11.2.10), consider the motion of the simple oscillator when it is acted on by a force which increases linearly with time from $t = 0$ to $t = \tau$, the initial value of the force being zero and the final value P. In the interval $0 \leqslant t \leqslant \tau$ the force is, therefore,

$$F(t) = \frac{Pt}{\tau} \tag{11.2.11}$$

and for $t > \tau$ the force is zero. Before applying equation (11.2.10), it must be noted that two forms of solution are to be expected. One of these will give the displacement at any instant t_1 up to $t = \tau$, during which interval the force is acting, and the other will give the displacement at time t_1, subsequent to $t = \tau$. The first form of solution is obtained by integrating the equation

$$x = \frac{1}{M\Omega} \int_0^{t_1} \frac{Pt}{\tau} e^{-\nu\omega_1(t_1 - t)} \sin \Omega(t_1 - t) \, dt, \tag{11.2.12}$$

where $t_1 < \tau$ while, in the second, allowance must be made for the vanishing of the force after the time $t = \tau$ by taking τ instead of t_1 as a limit. Thus if $t_1 > \tau$, the motion is

$$x = \frac{1}{M\Omega} \int_0^{\tau} \frac{Pt}{\tau} e^{-\nu\omega_1(t_1 - t)} \sin \Omega(t_1 - t) \, dt. \tag{11.2.13}$$

In this particular instance the integration can be carried out analytically, it gives, for equation (11.2.12),

$$x = \frac{P}{k}\left\{\left(\frac{t_1}{\tau} - \frac{2\nu}{\omega_1\tau}\right) + e^{-\nu\omega_1 t_1}\left[\left(\frac{1}{\Omega\tau} - \frac{2\Omega}{\omega_1^2\tau}\right)\sin\Omega t_1 + \frac{2\nu}{\omega_1\tau}\cos\Omega t_1\right]\right\} \quad (11.2.14)$$

and for equation (11.2.13)

$$x = \frac{P}{k}\left\{e^{-\nu\omega_1(t_1-\tau)}\left[\left(\frac{\nu\omega_1}{\Omega} - \frac{1}{\Omega\tau} + 2\frac{\Omega}{\omega_1^2\tau}\right)\sin\Omega(t_1-\tau) + \left(1 - \frac{2\nu}{\omega_1\tau}\right)\cos\Omega(t_1-\tau)\right]\right.$$
$$\left. + e^{-\nu\omega_1 t_1}\left[\left(\frac{1}{\Omega\tau} - 2\frac{\Omega}{\omega_1^2\tau}\right)\sin\Omega t_1 + \frac{2\nu}{\omega_1\tau}\cos\Omega t_1\right]\right\}. \quad (11.2.15)$$

The first of these expressions contains a term that represents a steady deflexion which increases with t_1. This is a non-oscillatory component which is present only while the force continues to act. The other term is a damped oscillation which becomes negligible compared with the non-oscillatory term if t_1 is sufficiently great. In the second expression all the terms are oscillatory because the force has ceased to act. The whole expression becomes negligible for large values of t_1 since all the disturbance is damped out.

From the first expression it can be seen that if a force is applied to a system very gradually then the deflexion produced is the static deflexion alone with no oscillatory component. The condition is obtained analytically by making τ very large, so that the rate of application of the load, P/τ, becomes very small; if t_1 is then set equal to τ, the displacement when the loading is complete is obtained. This gives

$$x = \frac{P}{k^2\tau}(k\tau - b) \quad (11.2.16)$$

which is approximately P/k if $k\tau$ is very large compared with b. It will be found that to neglect b/k in comparison with τ is permissible only if the time of loading τ is so long that

$$\omega_1\tau \gg 2\nu. \quad (11.2.17)$$

In many problems it is not possible to evaluate the Duhamel integral analytically and then graphical or numerical methods must be used. Some idea of the nature of the motion can often be obtained by inspection. The process is illustrated in fig. 11.2.2 in which the lower curve represents the function

$$e^{-\nu\omega_1(t_1-t)}\sin\Omega(t_1-t) \quad (11.2.18)$$

between $t = 0$ and $t = t_1$, and the upper curve represents the function $F(t)$. The Duhamel integral is given by the area below that curve which is formed by multiplying the ordinates of the two given curves. In estimating this area the following points may be noted. First, if the interval is sufficiently long for the amplitude of the lower curve to be very small at the start then the portion of $F(t)$ which occurs early in the interval makes little contribution to the final motion. This is to be expected in view of the fact that motion due to disturbance in the early part of the interval becomes greatly reduced by damping towards the end.

Secondly, if we are concerned with the maximum displacement which a given $F(t)$ will produce, and not merely with the displacement at some particular instant, then it is necessary to examine the way in which the area of the product curve changes when the relative phase between the upper and lower curves is altered. This follows because, if the displacement at some later instant $t_1 + \Delta t$, is required, the lower curve must be moved to the right through a distance Δt while the upper curve is fixed. Now if the $F(t)$ curve consists of a single peak whose breadth is less or equal to a half wave of the lower curve then the greatest value of the integral evidently

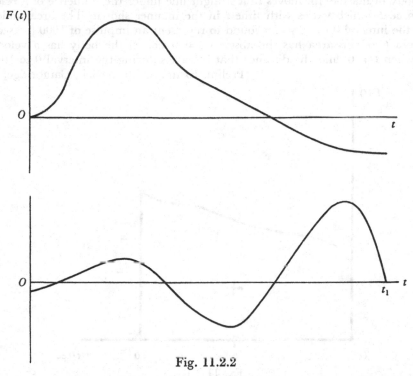

Fig. 11.2.2

occurs when the peak is opposite to a peak of the lower curve. If $F(t)$ consists of a peak which is broader than a half wave of the lower curve, then some of the areas covered by the integral will be negative, and it may be more difficult to estimate the position which gives the maximum displacement.

When a varying force $F(t)$ acts on the mass of the system shown in fig. 11.2.1 the equation of motion is

$$M\ddot{x} + b\dot{x} + kx = F(t) \tag{11.2.19}$$

and a mathematically identical equation will govern the transient motion of any other system having only one degree of freedom. The Duhamel integral offers one means of solving this equation and its underlying principle may be used in the solution of more complicated problems, as we shall show in this chapter. The reader should note, however, that other methods are sometimes more convenient for solving equation (11.2.19). It may, for instance, be possible to solve it analytically by the addition of a particular integral to the complementary function and there are also graphical and numerical methods which do not employ the notion of a series of impulses, as with the Duhamel integral.† These alternative methods, with their many refinements, may sometimes be convenient for multi-degree-of-freedom systems as well, since the use of principal co-ordinates may make it possible to obtain a separate equation of motion for each co-ordinate.

EXAMPLES 11.2

1. A body of mass 40 lb. moves in a straight line under the influence of a resultant applied force F which varies with time t in the manner shown. The area under the curve for the interval 0 to 10 sec. is found to represent an impulse of 1600 pdl.sec. and the centroid C of this area has the abscissa $t = 6$ sec. If the body has a velocity of 2 ft./sec. when $t = 0$, find the distance that it travels during the interval 0 to 10 sec.

(Prelim. Exam. in Mech. Sci., Cambridge, 1954)

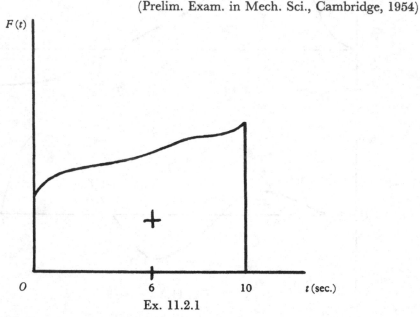

Ex. 11.2.1

2. A mass M is attached to a fixed abutment through a dashpot whose damping coefficient is b (cf. Ex. 8.1.3). The mass is at rest at time $t = 0$ when a transient force $F(t)$ is applied to it in the direction of its motion. Show from first principles that the velocity of the mass is given by

$$\frac{1}{M} e^{-bt/M} \int_0^t e^{b\tau/M} F(\tau) \, d\tau$$

(when $t > 0$) and verify that this solution satisfies the equation of motion and the initial condition.

3. A simple oscillator, like that of fig. 11.2.1 with its damper deleted, has a force

$$F(t) = F_0 \sin \omega t \quad (0 \leqslant t \leqslant \pi/\omega)$$

† See, for example, S. P. Timoshenko and D. H. Young, *Advanced Dynamics* (McGraw-Hill, London, 1947), ch. 1; see also R. E. D. Bishop, 'On the Graphical Solution of Transient Vibration Problems,' *Proc. Inst. Mech. Eng.* vol. 168 (1954), p. 229; or L. S. Jacobsen and R. S. Ayre, *Engineering Vibrations* (McGraw-Hill, London, 1958).

applied to it, where $\omega \neq \sqrt{(k/M)}$. If the mass starts from rest in its equilibrium position at time $t = 0$, find the displacement x during the interval $0 \leqslant t \leqslant \pi/\omega$

(a) by setting up and solving the equation of motion using a particular integral and complementary function;

(b) by means of the Duhamel integral.

4. An instrument for indicating the acceleration of a motor car consists of a graduated drum of moment of inertia 100 grm.cm.², mounted on a spindle. The spindle carries a gear of pitch circle diameter 1 cm., which meshes with a straight rack, the rack being free to slide in the direction of the car's motion and having a mass of 80 grm. The drum is prevented from rotating freely by a spring, which is such that an acceleration of g will produce one revolution of the drum when conditions are steady. The whole mechanism is immersed in oil so that there is a resistance to motion of the drum of 720 dyne-cm./rad./sec.

Initially, the car is travelling along a level road at steady speed and the drum is at rest on its spindle. The car is then accelerated at 5 ft./sec.² What is the reading of the indicator in ft./sec.² at an instant ⅜ sec. later? (C.U.M.S.T. Pt I, 1948)

11.3 Impulsive excitation of multi-degree-of-freedom systems

In § 11.2 we showed how the effect of any transient force, acting on a single-degree-of-freedom system, can be obtained by integration once the effect of a single impulse is known. The motion produced by an impulse is a free vibration which fits initial conditions of zero displacement and non-zero velocity. These ideas are directly applicable to multi-freedom systems and, in using them, we break down the problem of transient excitation into three stages. These are, first to find the initial motion produced by an impulse, secondly to find the free vibration which follows the initial conditions so produced and thirdly to integrate the effects of a sequence of impulses which form the transient exciting force. The first of these stages only will be dealt with in this section.

The motion following an impulse was obtained previously by finding the effect of a steady force acting during a time interval δt and then considering the limiting case as δt is reduced indefinitely. The problem of a system having many degrees of freedom may be dealt with in a similar fashion. Consider the Lagrangian equation

$$\frac{d}{dt}\left(\frac{\partial T}{\partial \dot{q}_r}\right) + \frac{\partial D}{\partial \dot{q}_r} + \frac{\partial V}{\partial q_r} = Q_r \quad (r = 1, 2, \dots, n) \quad (11.3.1)$$

which contains the dissipation function D (cf. equation (8.4.23)). As usual, solutions of this equation are limited to small values of the co-ordinates q, as discussed in Chapter 3. Now let the equation be integrated, with respect to time, over the interval t to $t+\epsilon$. During this interval it is not necessary to take Q_r as constant because it can be included within the integral. The process gives

$$\int_t^{t+\epsilon} \left[\frac{d}{dt}\left(\frac{\partial T}{\partial \dot{q}_r}\right)\right] dt = \int_t^{t+\epsilon} \left[Q_r - \frac{\partial D}{\partial \dot{q}_r} - \frac{\partial V}{\partial q_r}\right] dt. \quad (11.3.2)$$

As ϵ is made to approach zero the first term on the right-hand side, that is

$$\lim_{\epsilon \to 0}\left[\int_t^{t+\epsilon} Q_r dt\right] = I_r \quad \text{(say)}, \quad (11.3.3)$$

represents a quantity which may be thought of as an impulse. It should be noted, however that, in the application with which we shall be concerned, it is an indefinitely small impulse because Q_r is finite.

The second and third terms on the right-hand side of equation (11.3.2) may be expanded in the forms

$$\left. \begin{aligned} -\int_t^{t+\epsilon} (b_{r1}\dot{q}_1 + b_{r2}\dot{q}_2 + \ldots + b_{rn}\dot{q}_n)\, dt, \\ -\int_t^{t+\epsilon} (c_{r1}q_1 + c_{r2}q_2 + \ldots + c_{rn}q_n)\, dt, \end{aligned} \right\} \tag{11.3.4}$$

respectively. These will be of the second order of smallness if the system is at rest in the equilibrium position at the beginning of the interval ϵ.

The term on the left-hand side of equation (11.3.2) is of the first order of smallness because it may be written

$$\int_t^{t+\epsilon} (a_{r1}\ddot{q}_1 + a_{r2}\ddot{q}_2 + \ldots + a_{rn}\ddot{q}_n)\, dt \tag{11.3.5}$$

and $\ddot{q}_1, \ddot{q}_2, \ldots, \ddot{q}_n$ are not indefinitely small (as are the quantities q_r and $\dot{q}_r$). The expression (11.3.5) may be integrated to give

$$a_{r1}\dot{q}_1 + a_{r2}\dot{q}_2 + \ldots + a_{rn}\dot{q}_n \tag{11.3.6}$$

and the equations which determine the velocities $\dot{q}_s$ produced by the impulses I_r can therefore be written

$$\left. \begin{aligned} a_{11}\dot{q}_1 + a_{12}\dot{q}_2 + \ldots + a_{1n}\dot{q}_n &= I_1, \\ a_{21}\dot{q}_1 + a_{22}\dot{q}_2 + \ldots + a_{2n}\dot{q}_n &= I_2, \\ \cdots\cdots\cdots\cdots\cdots\cdots\cdots\cdots\cdots & \\ a_{n1}\dot{q}_1 + a_{n2}\dot{q}_2 + \ldots + a_{nn}\dot{q}_n &= I_n. \end{aligned} \right\} \tag{11.3.7}$$

These equations will be seen to contain the inertia coefficients but not the damping- nor the stability coefficients. The initial motion produced by an impulse is therefore dependent upon the mass distribution only.

A formal solution to the equations may be obtained by a similar method to that which was used in Chapter 3 for the equations of steady forced vibration. That is to say, the equations may be solved for one impulse at a time, say I_s, because the solutions for several impulses can be found by superposition. Let Δ' represent the determinant formed from the coefficients a and let Δ'_{is} be derived from Δ' by omitting the ith column and sth row. The velocity $\dot{q}_i$ produced by the impulse I_s will be given by

$$\dot{q}_i = (-1)^{i+s} \frac{\Delta'_{is}}{\Delta'} I_s. \tag{11.3.8}$$

For an example of this type of calculation, consider the system of fig. 11.3.1 when it is acted on by an impulse I_1 applied to the mass M as shown. The coefficients a are

$$a_{11} = M + m, \quad a_{22} = ml^2, \quad a_{12} = ml \tag{11.3.9}$$

so that the equations which determine the velocities are

$$(M+m)\,\dot{q}_1 + ml\dot{q}_2 = I_1,\\ ml\dot{q}_1 + ml^2\dot{q}_2 = 0.$$ (11.3.10)

The solutions of these are

$$\dot{q}_1 = \frac{I_1}{M},\\ \dot{q}_2 = -\frac{I_1}{Ml}.$$ (11.3.11)

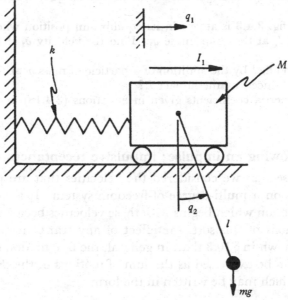

Fig. 11.3.1

Equations of the type (11.3.7) may evidently be formed for any set of generalized co-ordinates. If the principal co-ordinates are chosen then the equations are independent and may be written

$$a_1\,p_1 = I_1,\\ a_2\,p_2 = I_2,\\ \cdots\cdots\cdots\\ a_n\,p_n = I_n,$$ (11.3.12)

where the impulses I must now be those relating to the co-ordinates p. These equations may be solved immediately and may lead to simpler algebra in the analysis of the subsequent motion. There are, however, an infinite number of other sets of co-ordinates which give independent impulse equations but which are not principal co-ordinates. This is evident from the fact that for a given mass-distribution within a system there are an infinite number of possible spring systems; and for each spring system there will be a set of principal co-ordinates which will produce equations of the form (11.3.12). The selection of one of these sets, for the purpose of forming the impulse equations, may be easier than finding the true principal co-ordinates.

EXAMPLES 11.3

1. The quantities

$$(-1)^{i+s}\frac{\Delta'_{is}}{\Delta'}$$

of equation (11.3.8) are comparable with receptances in several respects (cf. equation (3.2.10)). Show that, like the latter, they obey a reciprocity relation of the type

$$\alpha_{rs} = \alpha_{sr}$$

and that they may be used in conjunction with the results quoted in Table 2.

2. The system of fig. 3.4.3 is at rest in its equilibrium position when it is subjected to a sudden impulse I_1 at the co-ordinate q_1. Find the velocity $\dot{q}_1$ that is produced in this way.

If the system is modified by the addition of a particle of mass m at the point B, what value of $\dot{q}_1$ will be produced by the impact I_1?

[NOTE. Use the inertia coefficients given in equations (3.4.18) and also entry 1 in Table 2.]

11.4 Motion following an impulse; impulsive receptances

In the previous section we considered the velocities that are produced by an impulse which acts on a multi-degree-of-freedom system. It is now necessary to examine the free motion which begins with these velocities because, by subsequent integration of motions of this sort, the effect of any transient excitation can be obtained. It was shown in § 10.3 that, in general, the free motion of a system with viscous damping can be expressed as the sum of motions in the different damped modes, the sth of which may be written in the form

$$\begin{aligned}
{}_sq_1 &= {}_sC_1 e^{\rho_s t}[G_s\cos(\sigma_s t + {}_s\theta_1) + H_s\sin(\sigma_s t + {}_s\theta_1)],\\
{}_sq_2 &= {}_sC_2 e^{\rho_s t}[G_s\cos(\sigma_s t + {}_s\theta_2) + H_s\sin(\sigma_s t + {}_s\theta_2)],\\
&\cdots\cdots\cdots\cdots\cdots\cdots\cdots\cdots\cdots\cdots\cdots\cdots\cdots\cdots\\
{}_sq_n &= {}_sC_n e^{\rho_s t}[G_s\cos(\sigma_s t + {}_s\theta_n) + H_s\sin(\sigma_s t + {}_s\theta_n)].
\end{aligned} \tag{11.4.1}$$

(see equation (10.3.10)). For a given set of initial values of the velocities $\dot{q}_1, \dot{q}_2, ..., \dot{q}_n$ arising from an impulse I_s and for zero initial values of the displacements $q_1, q_2, ..., q_n$ it is possible to select the appropriate set of values of the $2n$ constants $G_1, H_1, G_2, H_2, ..., G_n, H_n$. At least in theory, this will then enable the displacement at any co-ordinate, say q_i, to be found; it will be of the form

$$q_i = I_s[L_1 e^{\rho_1 t}\cos(\sigma_1 t + \phi_1) + L_2 e^{\rho_2 t}\cos(\sigma_2 t + \phi_2) + ... + L_n e^{\rho_n t}\cos(\sigma_n t + \phi_n)], \tag{11.4.2}$$

where the coefficients $L_1, L_2, ..., L_n$ and the phase angles $\phi_1, \phi_2, ..., \phi_n$ are dependent on the coefficients G and H.

The function in the square brackets in equation (11.4.2) gives the displacement at the co-ordinate q_i which is produced by a unit impulse at the co-ordinate q_s. It is therefore a function which serves a similar purpose to a receptance and we shall refer to it as the 'impulsive receptance', $\alpha_{is}(t)$. The bracketed t after the symbol α serves to distinguish these functions from the other kind of receptances.†

The impulsive receptances must not be assumed to obey all the rules which govern the other receptances. They do, however, fulfil one of the rules which have been deduced for the other ones; this is that they enable the motion due to a number of impulses to be written down as a set of expressions, each one of which gives the motion from one impulse only.

It would be extremely tedious to find an impulsive receptance in a numerical problem on a damped system, even if the number of co-ordinates were small. This is because the determination of the damped modes is in itself exceedingly laborious, and there is now the additional task of finding the coefficients G and H. Once more the approximations which are allowable when the damping is light are helpful because the damped modes may then be regarded as having the same shape as the principal modes of the undamped system; and the methods described in § 10.7 can be used to find the amplitudes of distortion in these modes which fit the initial conditions produced by the impulse. When the damping is zero, as it may sometimes be taken to be, the calculations are further simplified by the absence of the exponential terms.

If, in the case of light or zero damping, the principal co-ordinates are used from the start then there will be no cross-impulsive receptances. The direct impulsive receptances at other co-ordinates can be built up from those relating to the principal co-ordinates.

The remainder of this section is devoted to the calculation of some simple impulsive receptances. As a first example we take the system of fig. 11.3.1 for which we have already found, in the preceding section, the velocities that are produced by an impulse I_1. These are

$$\dot{q}_1 = \frac{I_1}{M}, \quad \dot{q}_2 = -\frac{I_1}{Ml} \qquad (11.4.3)$$

as given in equation (11.3.11). To find expressions for the free motion which arises from these velocities it is convenient to introduce principal co-ordinates which are defined by the equations

$$\left. \begin{aligned} q_1 &= \chi_1 p_1 + \chi_2 p_2, \\ q_2 &= p_1 + p_2, \end{aligned} \right\} \qquad (11.4.4)$$

where χ_1 and χ_2 are the constants of equation (1.4.8). The initial values of p_1 and p_2, which correspond to zero displacement and to the velocities of equation (11.4.3) are obtained from

$$\left. \begin{aligned} 0 &= \chi_1 X_1 + \chi_2 X_2, \\ 0 &= X_1 + X_2, \\ \frac{I_1}{Ml} &= \chi_1 \omega_1 Y_1 + \chi_2 \omega_2 Y_2, \\ -\frac{I_1}{Ml} &= \omega_1 Y_1 + \omega_2 Y_2, \end{aligned} \right\} \qquad (11.4.5)$$

† See W. J. Duncan, *R. and M.* **2000** (1947), p. 94.

where the notation is that of equations (10.7.8). The solutions of these equations are

$$X_1 = X_2 = 0,$$

$$Y_1 = \frac{l + \chi_2}{Ml\omega_1(\chi_1 - \chi_2)},$$

$$Y_2 = \frac{-(l + \chi_1)}{Ml\omega_2(\chi_1 - \chi_2)},$$

(11.4.6)

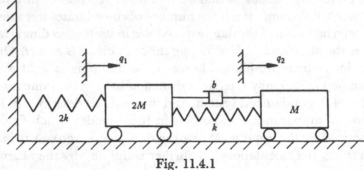

Fig. 11.4.1

and these permit expressions to be written down for the displacements q_1 and q_2 at any subsequent time t. These are

$$q_1 = \frac{\chi_1(l + \chi_2)\,I_1}{Ml\omega_1(\chi_1 - \chi_2)} \sin \omega_1 t - \frac{\chi_2(l + \chi_1)\,I_1}{Ml\omega_2(\chi_1 - \chi_2)} \sin \omega_2 t,$$

$$q_2 = \frac{(l + \chi_2)\,I_1}{Ml\omega_1(\chi_1 - \chi_2)} \sin \omega_1 t - \frac{(l + \chi_1)\,I_1}{Ml\omega_2(\chi_1 - \chi_2)} \sin \omega_2 t,$$

(11.4.7)

and they may be simplified by substituting into them the expressions for χ_1 and χ_2. The impulsive receptances $\alpha_{11}(t)$ and $\alpha_{21}(t)$ are thus found to be

$$\alpha_{11}(t) = \frac{1}{Ml(\omega_2^2 - \omega_1^2)} [\omega_1 \chi_1 \sin \omega_1 t - \omega_2 \chi_2 \sin \omega_2 t],$$

$$\alpha_{21}(t) = \frac{1}{Ml(\omega_2^2 - \omega_1^2)} [\omega_1 \sin \omega_1 t - \omega_2 \sin \omega_2 t].$$

(11.4.8)

As a second example, consider the system shown in fig. 11.4.1 in which the damping is light so that the usual approximations are valid. In view of this, it is permissible to begin by finding the modes and frequencies for the undamped system and to use these, modified by the introduction of an exponential decay term, to give the free motion of the damped system.

The natural frequencies of the undamped system are given by

$$\omega_1^2 = \frac{k}{2M}, \qquad \omega_2^2 = \frac{2k}{M}$$

(11.4.9)

and the corresponding modes are

$$q_1 : q_2 :: 1 : 2,$$

$$q_1 : q_2 :: 1 : -1.$$

(11.4.10)

Principal co-ordinates p_1 and p_2 may therefore be defined by the equations

$$\left. \begin{aligned} q_1 &= p_1 + p_2, \\ q_2 &= 2p_1 - p_2, \end{aligned} \right\} \tag{11.4.11}$$

and the full equations of free motion, in terms of the principal co-ordinates, are found to be

$$\left. \begin{aligned} 6M\ddot{p}_1 + \ b\dot{p}_1 + 3kp_1 - 2bp_2 &= 0, \\ -2b\dot{p}_1 + 3M\ddot{p}_2 + 4b\dot{p}_2 + 6kp_2 &= 0. \end{aligned} \right\} \tag{11.4.12}$$

Provided the damping is light the free motion of the system may now be written as

$$\left. \begin{aligned} p_1 &= e^{-bt/12M}[G_1 \cos \omega_1 t + H_1 \sin \omega_1 t], \\ p_2 &= e^{-2bt/3M}[G_2 \cos \omega_2 t + H_2 \sin \omega_2 t]. \end{aligned} \right\} \tag{11.4.13}$$

The initial values of the velocities, which are produced by an impulse, and for which the coefficients G and H must be chosen, can be written directly in terms of the co-ordinates q. Let an impulse I_1 act on the left-hand mass in the direction of q_1. The conditions of motion produced by this impulse are

$$\left. \begin{aligned} q_1 &= 0, \quad \dot{q}_1 = \frac{I_1}{2M}, \\ q_2 &= 0, \quad \dot{q}_2 = 0, \end{aligned} \right\} \tag{11.4.14}$$

when $t = 0$, and the corresponding constants are

$$\left. \begin{aligned} G_1 &= G_2 = 0, \\ H_1 &= H_2 = \frac{I_1}{3\sqrt{(2kM)}}. \end{aligned} \right\} \tag{11.4.15}$$

These values may be substituted into equations (11.4.13) and the resulting expressions may be used in equations (11.4.11). This gives the displacement at q_1 and q_2 due to the impulse I_1 and the impulsive receptances may be written down immediately from these relations. They are

$$\left. \begin{aligned} \alpha_{11}(t) &= \frac{1}{3\sqrt{(2kM)}} e^{-bt/12M} \sin\sqrt{\left(\frac{k}{2M}\right)} t + \frac{1}{3\sqrt{(2kM)}} e^{-2bt/3M} \sin\sqrt{\left(\frac{2k}{M}\right)} t, \\ \alpha_{21}(t) &= \frac{\sqrt{2}}{3\sqrt{(kM)}} e^{-bt/12M} \sin\sqrt{\left(\frac{k}{2M}\right)} t - \frac{1}{3\sqrt{(2kM)}} e^{-2bt/3M} \sin\sqrt{\left(\frac{2k}{M}\right)} t. \end{aligned} \right\} \tag{11.4.16}$$

For our final example, we take an infinite-freedom system although we shall assume it to be undamped for the sake of simplicity. The system is a uniform free-free bar of length l, cross-section A and density ρ and we shall find the impulsive receptance which gives the displacement u at a point distant x along the bar if an impulse I_1 is applied axially to the end $x = 0$. Let u_x and u_0 be the displacements at these two points. There are no inertia coefficients corresponding to these co-ordinates and the initial velocities $\dot{u}_x$ and $\dot{u}_0$ cannot therefore be found directly. Instead it is convenient to work immediately in terms of the principal co-ordinates, defined as in Table 6.

For the co-ordinate p_0 which represents a rigid-body motion in the x-direction, the impulse equation is

$$A\rho l \dot{p}_0 = I_1 \qquad (11.4.17)$$

and for the rth principal co-ordinate p_r ($r = 1, 2, \ldots$) the equation is

$$\frac{A\rho l}{2} \dot{p}_r = I_1. \qquad (11.4.18)$$

These results can be used to find the values of the principal co-ordinates at any subsequent instant t. They are

$$p_0 = \frac{I_1}{A\rho l} t, \quad p_r = \frac{2I_1}{A\rho l \omega_r} \sin \omega_r t. \qquad (11.4.19)$$

The displacements u_x and u_0 can now be formed and the impulsive receptances will then be found to be

$$\left. \begin{aligned} \alpha_{x0}(t) &= \frac{t}{A\rho l} + \sum_{r=1}^{\infty} \frac{2}{A\rho l \omega_r} \cos \frac{r\pi x}{l} \sin \omega_r t, \\ \alpha_{00}(t) &= \frac{t}{A\rho l} + \sum_{r=1}^{\infty} \frac{2}{A\rho l \omega_r} \sin \omega_r t. \end{aligned} \right\} \qquad (11.4.20)$$

EXAMPLES 11.4

1. Do impulsive receptances obey the results given in Table 2?

2. Is the reciprocal property of receptances (i.e. $\alpha_{rs} = \alpha_{sr}$) also true, in general, for impulsive receptances?

11.5 Transient excitation of multi-degree-of-freedom systems

Once the appropriate impulsive receptance has been found, the Duhamel integral may be used to establish the motion due to any transient force. Thus if the force $F(t)$ begins to act at time $t = 0$ at the co-ordinate q_r, then the displacement at q_s at a later instant t_1 will be

$$q_s = \int_0^{t_1} F(t) \, \alpha_{sr}(t_1 - t) \, dt. \qquad (11.5.1)$$

This expression can be obtained by an argument which is identical with that used in § 11.2. We shall now illustrate its use with the impulsive receptances found in the previous section.

Let a constant force F_1 begin to act, at the instant $t = 0$, at the co-ordinate q_1 of the system of fig. 11.3.1. The subsequent displacement at the co-ordinate q_2 is given in terms of the impulsive receptance $\alpha_{21}(t)$ of equation (11.4.8). By substituting this in equation (11.5.1), it is found that

$$q_2 = \int_0^t \frac{F_1}{Ml(\omega_2^2 - \omega_1^2)} [\omega_1 \sin \omega_1(t_1 - t) - \omega_2 \sin \omega_2(t_1 - t)] \, dt \qquad (11.5.2)$$

which gives

$$q_2 = \frac{F_1}{Ml(\omega_2^2 - \omega_1^2)} (\cos \omega_2 t_1 - \cos \omega_1 t_1). \qquad (11.5.3)$$

As a second example let the system of fig. 11.5.1 be acted on by a sinusoidal force which persists for one cycle only and which is applied at the co-ordinate q_1. Suppose that the forcing function is

$$F(t) = F_1 \sin \sqrt{\left(\frac{k}{M}\right)} t \quad (0 \leqslant t \leqslant 2\pi \sqrt{(M/k)}) \tag{11.5.4}$$

the frequency of this excitation being the geometric mean of the two natural frequencies which are given in equations (11.4.9). We shall find an expression for the motion at q_1 which remains after the force has ceased to act.

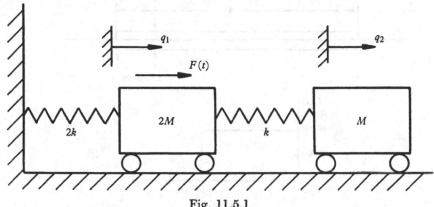

Fig. 11.5.1

The direct impulsive receptance is given by equation (11.4.16) when b is put equal to zero. This permits an expression to be written down for the displacement q_1 at any instant t_1, where $t_1 > 2\pi \sqrt{(M/k)}$; it is

$$q_1 = \int_0^{2\pi \sqrt{(M/k)}} \left\{ F_1 \left[\frac{1}{3\sqrt{(2kM)}} \sin \sqrt{\left(\frac{k}{2M}\right)}(t_1 - t) \right. \right.$$
$$\left. \left. + \frac{1}{3\sqrt{(2kM)}} \sin \sqrt{\left(\frac{2k}{M}\right)}(t_1 - t) \right] \sin \sqrt{\left(\frac{k}{M}\right)} t \right\} dt. \tag{11.5.5}$$

The integration may be performed by first converting the products of sines into cosine-differences and the expression finally reduces to

$$q_1 = \frac{F_1}{3k} \left\{ \sqrt{2} \left[\sin \omega_1 t_1 - \sin(\omega_1 t_1 - \sqrt{2} . \pi) \right] - \frac{1}{\sqrt{2}} \left[\sin \omega_2 t_1 - \sin(\omega_2 t_1 - 2\sqrt{2} . \pi) \right] \right\},$$
$$\tag{11.5.6}$$

where ω_1 and ω_2 are as given in equations (11.4.9).

The third impulsive receptance which was found in the previous article, and which we shall now use in a problem on transient forcing, is that for excitation at the end of a uniform free-free bar. Consider a transient force which rises instantaneously to a steady value F_1 and then falls instantaneously to zero after a certain time τ. In virtue of equation (11.4.20), the displacement u_x at any point x along the bar, at an instant t_1 where $t_1 > \tau$, will be given by

$$u_x = \int_0^\tau F_1 \left[\frac{t_1 - t}{A\rho l} + \sum_{r=1}^\infty \frac{2}{A\rho l \omega_r} \cos \frac{r\pi x}{l} \sin \omega_r(t_1 - t) \right] dt. \tag{11.5.7}$$

Integration of this yields

$$u_x = F_1 \left[\frac{2t_1\tau - \tau^2}{2A\rho l} + \sum_{r=1}^{\infty} \frac{2}{A\rho l \omega_r^2} \cos\frac{r\pi x}{l} \{\cos\omega_r(t_1-\tau) - \cos\omega_r t_1\} \right] \quad (t_1 > \tau). \quad (11.5.8)$$

Now it happens that this is a problem which may be treated by the wave method which was mentioned in the introductory portion of this chapter. The solution obtained by that method can be compared with the series solution (11.5.8) and we shall use a numerical example for the purpose of comparison.

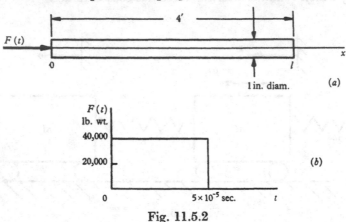

Fig. 11.5.2

Let the dimensions of the bar be as shown in fig. 11.5.2 (a) and let it be of steel for which

$$E = 13\cdot 8 \times 10^{10}\,\text{pdl./ft.}^2,$$

$$\rho = 480\,\text{lb./ft.}^3.$$

These figures give the frequencies $\omega_1 = 13\,320\,\text{rad./sec.}$, $\omega_2 = 26\,640\,\text{rad./sec.}, \ldots$, etc. Suppose that the force function is given by the rectangle shown in fig. 11.5.2 (b). A rectangular pulse of this kind can be obtained, approximately, by firing a lead rifle bullet, with a fairly bluff nose-profile, at the end of the bar. By substituting into equation (11.5.8) and taking the first two terms only of the cosine series, the displacement in feet is found to be

$$u_x = (6\cdot 11t_1 - 1\cdot 53 \times 10^{-4})$$

$$+ [-0\cdot 000\,280 \cos 13\,320t_1 + 0\cdot 000\,833 \sin 13\,320t_1] \cos\frac{\pi x}{4}$$

$$+ [-0\cdot 000\,272 \cos 26\,640t_1 + 0\cdot 000\,337 \sin 26\,640t_1] \cos\frac{2\pi x}{4} \ldots \text{ft.} \quad (11.5.9)$$

This function is shown plotted in fig. 11.5.3 for the value $t_1 = 8 \times 10^{-5}\,\text{sec.}$ The three heavy straight lines shown in this figure (curve A) represent the result which is found by the wave method and this is an 'exact' solution; that is to say there are no approximations within the calculation although the problem is somewhat idealized before the calculation can be started. The solution outlined above, on the other hand, is only approximate because of the neglect of the higher-frequency terms. The manner in which the approximation improves as extra terms are added is shown by the remaining curves in fig. 11.5.3. Curve B is that due to the non-sinusoidal term; it is the rigid body displacement which corresponds to the zero natural frequency. Curve C is obtained by adding the $\cos(\frac{1}{4}\pi x)$ term to B and curve D is found by adding the $\cos(\frac{1}{2}\pi x)$ term to C. It will be seen that curve D is becoming sufficiently close to the exact result for it to be a useful approximation.

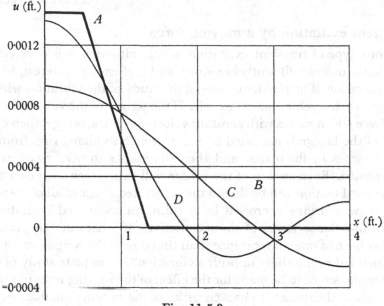

Fig. 11.5.3

The subject of impulsive receptances is, in all its generality, a very broad one and only a few of the properties of these functions have been touched on in this book. The reader who requires more information on the subject should consult the work of Duncan.†

EXAMPLES 11.5

1. Show, by the use of the appropriate impulsive receptance that the variation of q_1, in the system of fig. 11.3.1, which corresponds to the motion (11.5.3) at q_2 is

$$q_1 = \frac{F_1}{k} - \frac{F_1}{Ml(\omega_2^2 - \omega_1^2)}[\chi_1 \cos \omega_1 t - \chi_2 \cos \omega_2 t].$$

2. Two particles, whose masses are 32 lb. and 16 lb., are joined by a light spring of stiffness 100 lb.wt./ft. The 32 lb. mass is attached by a light spring of stiffness 200 lb.wt./ft. to a rigid abutment so that the system hangs freely. The system remains at rest in its position of stable equilibrium until the instant $t = 0$ when a downward

† W. J. Duncan, *Conferencias de Dinámica Técnica y Aeroelasticidad* (Madrid, Instituto Nacional de Técnica Aeronáutica Esteban Terradas, 1951), pp. 37–57 (in Spanish); W. J. Duncan, *The Principles of the Control and Stability of Aircraft* (Cambridge University Press, 1952), ch. 4; W. J. Duncan, 'Solution of Ordinary Linear Differential Equations with Variable Coefficients by Impulsive Admittances', *Q.J.M.A.M.* vol. 6, (1953), pp. 122–7; see also, W. J. Duncan, 'Indicial Admittances for Linear Systems with Variable Coefficients', *J. Roy. Aero. Soc.* vol. 61 (1957), pp. 46–7.

vertical force of magnitude $6t$ lb.wt. is applied to the upper particle. This force increases until the instant $t = \pi$ sec. and, for $t \geqslant \pi$ sec., it remains constant at 6π lb.wt.

Find equations governing the variations of the principal co-ordinates in the interval $0 \leqslant t \leqslant \pi$ sec., taking unit values of these co-ordinates to correspond to unit downward displacement of the upper mass in each principal mode. Hence, or otherwise, find the displacements of the two particles from their initial positions (*a*) for the interval $0 \leqslant t \leqslant \pi$ sec., and (*b*) for $t \geqslant \pi$ sec.

(C.U.M.S.T. Pt ii, 1955)

11.6 Transient excitation by a moving force

There is one type of transient excitation which, although it is covered by the treatment above, differs sufficiently in nature, and is of enough interest, to deserve special consideration. The problem arises in the study of the vibration which is set up in a bridge when a vehicle passes over it. If the weight of the vehicle is regarded as a steady force which moves with constant velocity over the bridge, then evidently any section of the bridge is subjected to a force which suddenly rises from zero as the vehicle reaches it, fluctuates, and then diminishes to zero again when the vehicle has passed. Each section of the bridge will experience such excitations in turn and the total motion will be due to the integrated effect of all of them.

When a railway bridge is crossed by a train, an additional excitation arises from the out-of-balance forces of the locomotive's mechanism. In practice this excitation has been of more importance than that due to the weight and it was this problem which led to the study of bridge vibration. A complete study of the problem requires allowance to be made for the effect of the moving mass of the vehicle because this alters the dynamic characteristics of the system; the mass is, further, mounted on springs whose effect must be allowed for. A complete treatment of these matters will not be attempted here because they belong to the class of non-linear vibration problems which has been excluded from this book.

It will be useful, however, to examine the problem of excitation by a moving concentrated pulsating force because, from this, some of the more complex treatments may be built up. It should be noted that, with some systems, problems of *steady* vibration may arise from a moving force. Examples of this are the excitation of turbine disks when they rotate and are acted on by a jet of steam from a single nozzle, and the excitation of rotating gear wheels which are acted on by pulsating forces at the meshing point. These problems may be treated by a technique which is similar to that which we shall develop here.

Certain problems of travelling loads yield separate equations of motion, each of the form of equation (11.2.19), when principal co-ordinates are used. The generalized forces at these co-ordinates may be found by the standard means and the solution of the equations presents no essentially new feature. Some problems of this type are given at the end of this section. It is not our purpose, however, to elaborate further on this aspect of transient vibration.

In fig. 11.6.1, AB represents a bridge or other structure across which a moving concentrated force passes. The force, which is a function of time, is denoted by $F(t)$ and h is the distance that it moves along the bridge in some interval τ, so that $F(\tau)$

is the value of the force when it is at the point indicated. The distance h and the interval τ are related by the velocity V so that

$$V = \frac{dh}{d\tau}. \tag{11.6.1}$$

We shall here restrict attention to those problems in which V is constant. Now the bridge has principal co-ordinates $p_1, p_2, \ldots$, etc., and $\alpha_{p_r h}(t)$ will be used to denote the impulsive receptance which gives the displacement in the rth principal mode due to an impulse at a point distance h along the bridge. We shall assume for simplicity that there is no dissipation although an allowance for viscous damping can be made readily, provided that the usual assumptions about light damping are used.

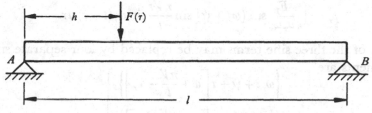

Fig. 11.6.1

During the interval $\delta\tau$, when the force is moving across the segment δh, the bridge experiences an impulse $F(\tau) . \delta\tau$. The displacement, at a subsequent instant t, in the rth principal mode, due to the impulse, will be

$$p_r = F(\tau) . \delta\tau . \alpha_{p_r h}(t - \tau). \tag{11.6.2}$$

Therefore, if the bridge is at rest in equilibrium when the force first reaches it, the total displacement in the rth principal mode is

$$p_r = \int_0^t F(\tau) . \alpha_{p_r h}(t - \tau) \, d\tau. \tag{11.6.3}$$

In evaluating this integral it must be remembered that h is a function of τ and that the impulsive receptance is therefore a function of τ as well as of $(t - \tau)$.

The integration of equation (11.6.3) can be performed, graphically if necessary, provided $F(\tau)$, $\alpha_{p_r h}(t)$ and V are known. Some ideas about the nature of the motion, and of the factors which affect it, can be obtained by assuming that $F(t)$ is a sine function and by expanding the characteristic function $\phi_r(x)$ of the bridge as a series of sine terms. To make the treatment simpler we assume that the characteristic function $\phi_r(x)$ is zero at the ends of the bridge so that it may be written as a half-range sine series; thus

$$\phi_r(x) = \sin\frac{\pi x}{l} + t_2 \sin\frac{2\pi x}{l} + t_3 \sin\frac{3\pi x}{l} + \ldots, \tag{11.6.4}$$

where $t_2, t_3, \ldots$ are constants. Further, let

$$F(t) = F_1 \sin(\omega t + \psi). \tag{11.6.5}$$

The impulsive receptance $\alpha_{p_r h}(t)$ may now be found by the method of §11.4; it is given by

$$\alpha_{p_r h}(t) = \frac{\dfrac{\partial v}{\partial p_r} \sin \omega_r t}{a_r \omega_r} = \frac{\phi_r(h) \sin \omega_r t}{a_r \omega_r}, \tag{11.6.6}$$

where v is the deflexion under the force at time τ.

Equation (11.6.3) now becomes, after substitution for the impulsive receptance,

$$p_r = \int_0^t \frac{F_1 \sin(\omega\tau + \psi) \cdot \phi_r(h) \sin \omega_r(t-\tau)}{a_r \omega_r} \, d\tau \tag{11.6.7}$$

and this integral may be evaluated for each term of $\phi_r(h)$ separately. With $h = V\tau$, the first term gives

$$p_r = \int_0^t \frac{F_1}{a_r \omega_r} \sin(\omega\tau + \psi) \sin \frac{\pi V\tau}{l} \sin \omega_r(t-\tau) \, d\tau. \tag{11.6.8}$$

The product of the three sine terms may be replaced by four separate sine terms whose arguments are

$$\left.\begin{aligned}
&\left\{\omega_r t + \psi + \tau\left[\omega + \frac{\pi V}{l} - \omega_r\right]\right\}, \\
&\left\{\omega_r t + \psi + \tau\left[\omega - \frac{\pi V}{l} - \omega_r\right]\right\}, \\
&\left\{-\omega_r t + \psi + \tau\left[\omega + \frac{\pi V}{l} + \omega_r\right]\right\}, \\
&\left\{-\omega_r t + \psi + \tau\left[\omega - \frac{\pi V}{l} + \omega_r\right]\right\},
\end{aligned}\right\} \tag{11.6.9}$$

and the quantities in the square brackets will appear in the denominators when the integration is carried out. Rather than attempt to evaluate the complete expression we merely note that for certain values of the parameters one or other of the denominators will be very small and the corresponding term will become very large. This indicates that severe vibration is possible in the rth principal mode for these values of the parameters.

The out-of-balance force from a locomotive has the frequency of revolution of the driving wheels, so that in the railway bridge problem the velocity V and the frequency ω are related by the driving wheel radius R through the equation

$$\omega = \frac{V}{R}. \tag{11.6.10}$$

Now in any bridge which is sufficiently long for vibration to be troublesome l will be very much greater than R; ω will therefore be greater than $\pi V/l$ so that only the first two of expressions (11.6.9) can have small values for the quantity within the square brackets. These values occur when

$$\omega \doteq \frac{\omega_r}{1 + \dfrac{\pi R}{l}} \quad \text{and} \quad \omega \doteq \frac{\omega_r}{1 - \dfrac{\pi R}{l}}, \tag{11.6.11}$$

respectively. This shows that the greatest amplitudes do not occur when ω is equal to ω_r, as would be the case if the locomotive were allowed to spin its wheels while it remained stationary on the bridge. Instead they occur at two values of ω, one slightly above and one slightly below the natural frequency ω_r.

It will be remembered that the above results have been deduced by considering one particular term of the series for $\phi_r(x)$ and one particular bridge frequency ω_r. It may appear therefore that a complete solution would lead to such a large number of possible dangerous values of ω that it would not be possible to avoid them all. In fact this is not so. Many bridges are sufficiently like uniform beams for their modal shapes to be effectively sinusoidal so that each of the series (11.6.4) for $\phi_1(x)$, $\phi_2(x)$, ..., etc., have one term only. Further, it is usually the case that only the lowest natural frequency is within the range in which the above-mentioned dangerous condition can occur. In these circumstances the only troublesome values of ω are

$$\omega \doteq \frac{\omega_1}{1 + \dfrac{\pi R}{l}} \quad \text{or} \quad \omega \doteq \frac{\omega_1}{1 - \dfrac{\pi R}{l}}. \tag{11.6.12}$$

A numerical example will give an idea of the type of condition possible. Let the bridge be 100 ft. long and the driving wheels be 6 ft. in diameter. Then $\pi R/l$ is approximately 0·094 so that the dangerous conditions occur about 10 % below and 10 % above the wheel speed at which resonance would occur with a stationary locomotive.

In some arch bridges the lowest mode has a shape which contains a large component $t_3 \sin (3\pi x/l)$. By using the analysis for this term it will be found that the relations (11.6.12) are replaced by

$$\frac{\omega_1}{1 + \dfrac{3\pi R}{l}} \quad \text{and} \quad \frac{\omega_1}{1 - \dfrac{3\pi R}{l}} \tag{11.6.13}$$

which shows that the difference between the two dangerous speeds becomes great and in the above example 10 % becomes approximately 30 %. With a bridge of this type, the other dangerous speeds will still exist so that there will be four in all.

The above examination of excitation by a moving force has been kept brief because its main purpose is to show that the methods developed in the earlier part of this chapter are equally applicable to this type of problem. A complete study of the bridge problem—but one which is restricted to bridges with sinusoidal modes —has been given by Inglis.[†]

EXAMPLES 11.6

1. The pinned-pinned beam shown in the figure, on p. 564, is subjected to a transient distributed load, as indicated. Show that the rth principal co-ordinate p_r is governed by the equation

$$\ddot{p}_r + \omega_r^2 p_r = \frac{2F(t)}{A\rho l} \int_0^l X(x) \sin \frac{r\pi x}{l} \, dx.$$

† C. E. Inglis, *A Mathematical Treatise on Vibrations in Railway Bridges* (Cambridge University Press, 1934).

2. The figure shows a pinned-pinned beam which is being crossed by a travelling force of constant intensity w_0/unit length, with constant speed V. Show that, if the beam is at rest in equilibrium before the load passes on to it, the variation of the rth principal co-ordinate p_r is given by

$$p_r = \frac{2w_0}{A\rho r\pi\omega_r^2} + \frac{2w_0}{A\rho r\pi(\omega_r^2 - \xi_r^2)}\left[\frac{\xi_r^2}{\omega_r^2}\cos\omega_r t - \cos\xi_r t\right],$$

where $\xi_r = r\pi V/l$.

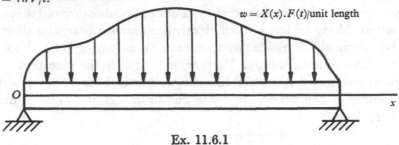

$w = X(x).F(t)$/unit length

Ex. 11.6.1

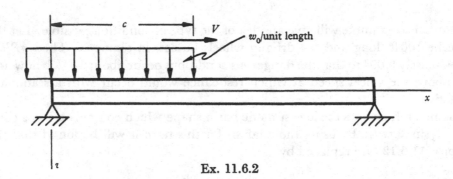

c

V

w_0/unit length

x

t

Ex. 11.6.2

3. A load like that considered in Ex. 11.6.2 crosses a fixed-fixed taut string which is initially at rest in its equilibrium position. The velocity V is equal to $\surd(S/\mu)$ in the notation of Chapter 4. Show that a resonant increase of amplitude occurs in all principal modes of the string, the rth principal co-ordinate being given by

$$p_r = \frac{w_0 l^2}{r^3\pi^3 S}[2 - \omega_r t\sin\omega_r t].$$

11.7 Transient vibration due to passage through resonance

A particular example of transient excitation occurs when the forcing frequency, to which a system is subjected, varies with time. This happens when the force arises from lack of balance in a rotating shaft, for instance, if the shaft undergoes steady angular acceleration or retardation; it happens therefore every time a machine is started up or shut down. Now in many machines there are certain 'critical' speeds at which troublesome resonances occur. When these are below the normal running speeds, in a given machine, they have to be passed through as the machine is accelerated from rest to its operating speed. Intuition suggests that if the acceleration is rapid, so that the time during which the machine is running near the resonance is small, the amplitude attained will be small. In this article we shall examine this problem and show that this conclusion is correct. A method will be developed by which the amplitude can be predicted.

The theory will be restricted to that for a system having only a single degree of freedom, like that shown in fig. 11.2.1. The equation for the forced motion will therefore have the form

$$M\ddot{x} + b\dot{x} + kx = Fe^{i[\omega(t)]t},$$
(11.7.1)

where $\omega(t)$ implies that the forcing frequency ω is not constant but depends on the time. If the acceleration of the exciting device is constant, then

$$\omega(t) = Kt,$$
(11.7.2)

where K is a constant. The equation for the forced motion will then be

$$M\ddot{x} + b\dot{x} + kx = Fe^{iKt^2}.$$
(11.7.3)

To find the displacement x at any instant, assuming that the system was at rest when excitation started and that the acceleration is uniform from the instant $t = 0$ to time t_1, equation (11.2.10) shows that we must evaluate the integral

$$x = \frac{F}{M\Omega} \int_0^{t_1} e^{iKt^2} e^{-\nu\omega_1(t_1-t)} \sin \Omega (t_1 - t) \, dt,$$
(11.7.4)

where Ω, ω_1 and ν are defined as in equation (11.2.3). The solution of this has been given by Lewis† who uses the method of contour integration. The process is fairly long, but Lewis gives his results in non-dimensional form so that they can be used directly for numerical problems. If the damping is neglected, then the solution is simpler and this case has been treated by Ellington and McCallion‡ who obtain the solution by the use of Fresnel's integrals. Rather than repeat either of these methods here we shall use instead an approximate process which, it is hoped, will illustrate the nature of the problem more effectively. The approximation is sufficiently close for the method to be suitable for practical use.

The function e^{iKt^2} is not sinusoidal because the frequency parameter Kt is constantly changing. It may nevertheless be replaced by a set of successive sine wave cycles, each one having a shorter period than the preceding one, without greatly changing the overall shape of the function. This replacement of the function by another becomes a better and better approximation as the rate of change of frequency is diminished. The approximation is very close if the percentage change of period per cycle is five or less, and it is with rates of change of this order that we are usually concerned in engineering problems.

A constant acceleration implies a uniform rate of change of frequency; nevertheless within a limited frequency range we may assume a uniform rate of change of period without altering the nature of the excitation significantly. Since we shall only be concerned with frequencies which are close to resonance the restriction to

† F. M. Lewis, 'Vibration during Acceleration through a Critical Speed', *Trans. A.S.M.E.* vol. 54 (1932), pp. (APM) 253–61.
‡ J. P. Ellington and H. McCallion, 'On Running a Machine through its Resonant Frequency', *J. Roy. Aero. Soc.* vol. 60 (1956), pp. 620–1.

a limited frequency range is not troublesome. It may further be remarked that in real problems the acceleration may well vary through the speed range and that to assume constant acceleration is therefore an approximation in any case.

Let the period of successive forcing cycles differ by α and suppose that there will be one cycle of excitation whose period is exactly equal to the period $2\pi/\omega_1$ for free vibration of the system in the absence of its damping. The cycles which follow this cycle, which we shall call the 'true resonance' cycle, will have periods

$$\left(\frac{2\pi}{\omega_1} - \alpha\right), \quad \left(\frac{2\pi}{\omega_1} - 2\alpha\right), \quad \dots, \quad \text{etc.}$$

and those which precede it will similarly have periods

$$\left(\frac{2\pi}{\omega_1} + \alpha\right), \quad \left(\frac{2\pi}{\omega_1} + 2\alpha\right), \quad \dots, \quad \text{etc.}$$

The reader will be able to deduce subsequently, or may alternatively regard as evident, that if there is no true resonance cycle so that the period changes from one value greater than $2\pi/\omega_1$ to a succeeding value which is less, the analysis remains valid.

Consider a single-degree-of-freedom system which is subjected to a series of single sinusoidal pulsations, all having the same amplitude but having periodic times of

$$\left(\frac{2\pi}{\omega_1} + m\alpha\right), \quad \left(\frac{2\pi}{\omega_1} + (m-1)\alpha\right), \quad \dots, \quad \left(\frac{2\pi}{\omega_1} + \alpha\right), \quad \left(\frac{2\pi}{\omega_1}\right), \quad \left(\frac{2\pi}{\omega_1} - \alpha\right),$$

$$\dots, \left(\frac{2\pi}{\omega_1} - (n-1)\alpha\right), \quad \left(\frac{2\pi}{\omega_1} - n\alpha\right). \quad (11.7.5)$$

Suppose that the rate of change of period is slow, such that

$$\alpha \ll \frac{2\pi}{\omega_1} \tag{11.7.6}$$

and assume also that it is only in the neighbourhood of resonance that the excitation need be considered so that $m\alpha$ and $n\alpha$ will not exceed, say, 20 % of $2\pi/\omega_1$. The motion produced by this excitation may be found by summing the motions due to each cycle separately and it is necessary first therefore to find an expression for the motion produced by a single cycle of this excitation. In doing this we shall neglect the damping entirely in the first place; this simplifies the algebra and it will be shown later that the final results, obtained by so doing, can be modified for light damping without repeating all the analysis.

The motion produced by the single cycle of the excitation $F \sin \omega t$, which commences at time $t = 0$ and finishes when $t = 2\pi/\omega$, must be found for the instant t_1. It is given by equation (11.2.10), being

$$x = \frac{F}{M\omega_1} \int_0^{2\pi/\omega} \sin \omega t \sin \omega_1(t_1 - t) \, dt. \tag{11.7.7}$$

This gives, for the instant when $t_1 = 2\pi/\omega$,

$$x = \frac{F}{M}\frac{\omega}{\omega_1}\frac{1}{(\omega^2 - \omega_1^2)}\sin\frac{2\pi\omega_1}{\omega}. \qquad (11.7.8)$$

By a similar calculation the velocity at this instant may be shown to be

$$\dot{x} = \frac{-F\omega}{M(\omega^2 - \omega_1^2)}\left(1 - \cos\frac{2\pi\omega_1}{\omega}\right). \qquad (11.7.9)$$

This displacement and this velocity determine the initial conditions of a free vibration whose frequency is ω_1.

Let the amplitude of this vibration be R and let the displacement be written

$$x = R\sin(\omega_1 t + \theta), \qquad (11.7.10)$$

where θ is a phase angle and where time is measured from the end of the single cycle of excitation which produced the motion. The initial conditions of the motion therefore give

$$R\sin\theta = \frac{F}{M}\frac{\omega}{\omega_1}\cdot\frac{1}{(\omega^2 - \omega_1^2)}\sin\frac{2\pi\omega_1}{\omega} \qquad (11.7.11)$$

and

$$R\omega_1\cos\theta = \frac{-F\omega}{M(\omega^2 - \omega_1^2)}\left(1 - \cos\frac{2\pi\omega_1}{\omega}\right). \qquad (11.7.12)$$

These may be solved to give

$$R = \frac{2F}{M}\frac{\omega}{\omega_1}\cdot\frac{1}{\omega^2 - \omega_1^2}\sin\frac{\pi\omega_1}{\omega} \qquad (11.7.13)$$

and

$$\theta = \frac{\pi\omega_1}{\omega} + \frac{\pi}{2}. \qquad (11.7.14)$$

We shall be concerned with the values of these quantities when ω is nearly equal to ω_1. Thus let

$$\frac{1}{\omega_1} - \frac{1}{\omega} = \frac{n\alpha}{2\pi} \qquad (11.7.15)$$

which gives the value of ω during the nth cycle after the true resonance. Noting that this equation gives

$$\frac{\omega_1}{\omega} = 1 - \frac{n\alpha\omega_1}{2\pi} \qquad (11.7.16)$$

and hence that

$$\sin\frac{\pi\omega_1}{\omega} = \sin\frac{n\alpha\omega_1}{2} \qquad (11.7.17)$$

we find, for the amplitude R,

$$R = \frac{2F}{M}\frac{\omega}{\omega_1}\cdot\frac{1}{\omega^2 - \omega_1^2}\sin\frac{n\alpha\omega_1}{2}. \qquad (11.7.18)$$

This reduces, in virtue of equation (11.7.16), to

$$R = \frac{F\pi}{M\omega_1^2} \qquad (11.7.19)$$

if $n\alpha\omega_1/2\pi$ is neglected in comparison with unity (cf. equation (11.7.6)). This expression for R is independent of n showing that, provided the exciting frequency is nearly equal to ω_1, the amplitude of the motion produced by each separate sinusoidal cycle of excitation is the same. The total motion is therefore the sum of a number of components all of which are of equal amplitude, and in order to find it we only require the relative phases of these components.

In considering the phases of the components of free motions which are produced by the various single forcing cycles we shall use a 'standard' free vibration $x = R \sin \omega_1 t$ for the purpose of reference. Thus it will be convenient to take the origin of time for each of these motions as the end of the true resonance cycle.

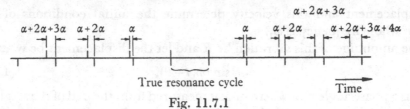

Fig. 11.7.1

Consider, first, the true resonance cycle itself. The forcing frequency is $\omega = \omega_1$ and the system is assumed to start with the conditions $x = \dot{x} = 0$ at the beginning of the cycle. It can readily be shown by the use of equation (10.6.6) that, at the end of this cycle, the initial conditions of the subsequent free motion (11.7.10) are such that $\theta = \frac{3}{2}\pi$. That is to say this particular component of the free motion leads the standard free vibration by a phase angle θ_0, where

$$\theta_0 = \theta = \frac{3\pi}{2}. \tag{11.7.20}$$

In fig. 11.7.1 the intervals occupied by each of the forcing cycles are marked off along the top of the time base while the periods of the standard free vibration $x = R \sin \omega_1 t$ are shown below it. The intervals indicated by the pairs of vertical lines show how soon, before the end of a period of the standard free vibration cycle, the end of a forcing cycle occurs. From these intervals it will be seen that the nth forcing cycle after the true resonance ends at a time $\alpha + 2\alpha + 3\alpha + \ldots + n\alpha$, or

$$\frac{n(n+1)\,\alpha}{2}$$

before the end of the corresponding free vibration cycle. The total phase angle θ_0 by which the free motion resulting from this cycle of forced motion will lead the standard free vibration is thus

$$\theta_0 = \theta + \frac{n(n+1)\,\alpha\omega_1}{2}, \tag{11.7.21}$$

where θ is given, in accordance with equation (11.7.14), by

$$\theta = \pi\omega_1 \left[\frac{\dfrac{2\pi}{\omega_1} - n\alpha}{2\pi} \right] + \frac{\pi}{2} = \frac{3\pi}{2} - \frac{n\alpha\omega_1}{2}. \tag{11.7.22}$$

That is to say the total phase angle is

$$\theta_0 = \frac{3\pi}{2} - \frac{n\alpha\omega_1}{2} + \frac{n(n+1)\,\alpha\omega_1}{2} = \frac{3\pi}{2} + \frac{n^2\alpha\omega_1}{2}. \tag{11.7.23}$$

It will be seen further that the mth cycle before the true resonance cycle ends at a time $\alpha + 2\alpha + 3\alpha + \ldots + (m-1)\alpha$, or

$$\frac{m(m-1)\alpha}{2}$$

before the end of the corresponding cycle of the standard free vibration. The total phase angle by which the free motion resulting from this cycle of forcing leads the standard free vibration is therefore

$$\theta_0 = \theta + \frac{m(m-1)\alpha\omega_1}{2}, \qquad (11.7.24)$$

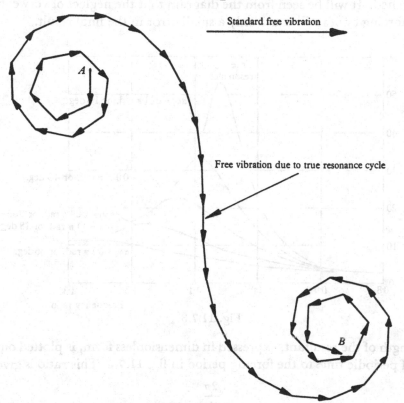

Standard free vibration

Free vibration due to true resonance cycle

Fig. 11.7.2

where θ is now given by

$$\theta = \pi\omega_1 \left[\frac{\dfrac{2\pi}{\omega_1} + m\alpha}{2\pi} \right] + \frac{\pi}{2} = \frac{3\pi}{2} + \frac{m\alpha\omega_1}{2}. \qquad (11.7.25)$$

That is to say $\qquad \theta_0 = \dfrac{3\pi}{2} + \dfrac{m\alpha\omega_1}{2} + \dfrac{m(m-1)\alpha\omega_1}{2} = \dfrac{3\pi}{2} + \dfrac{m^2\alpha\omega_1}{2}. \qquad (11.7.26)$

These phase angles and amplitudes allow a vector diagram to be drawn in which the components due to successive cycles are added in order. Such a diagram is shown in fig. 11.7.2 where the vectors begin at the point A and finish at B. It will be seen that, at the start of the diagram the angles between the vectors are large because of the rapid changes required by the expressions (11.7.26) as m is diminished; as a result the diagram curls round. As m is reduced the curl becomes less and once the resonant cycle is passed it reverses and then becomes rapid again in conformity with equation (11.7.23). The total amplitude at the end of any cycle is given by the length of the resultant vector drawn from the starting point A to the end of the relevant component vector. As the process goes on the resultant vector increases in length, the growth being small in the early stages because of the curl but reaching a maximum rate in the vicinity of the true resonance condition. The rate then falls off again and the amplitude becomes almost constant once the region of B is reached. It will be seen from the diagram that the neglect of very early and very late forcing cycles will cause only a small error in the final result.

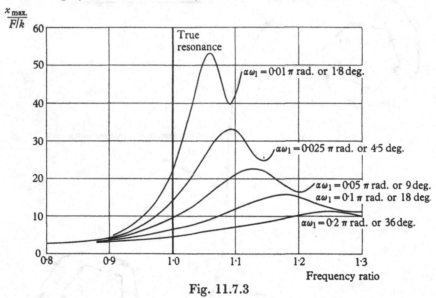

Fig. 11.7.3

The length of the resultant, expressed in dimensionless form, is plotted on a base of ratio of periodic time to the forcing period in fig. 11.7.3. This ratio is given by

$$\frac{\dfrac{2\pi}{\omega_1}}{\dfrac{2\pi}{\omega_1} - n\alpha} \qquad (11.7.27)$$

which is equal to

$$1 + \frac{n\alpha\omega_1}{2\pi} \qquad (11.7.28)$$

provided that

$$\frac{n\alpha\omega_1}{2\pi} \ll 1. \qquad (11.7.29)$$

Five different curves are drawn corresponding to five different rates of change of period, these being denoted by the magnitude of the product $\alpha\omega_1$ in degrees. This product divided by 360, gives the proportional change of frequency per cycle, and it will be seen that the values between 0·5 and 10% are plotted. The ordinates in fig. 11.7.3 represent the ratio of instantaneous amplitude to static deflexion. It will be noticed that the curves start rising again at the end; this is due to the curled end of the vector diagram and from inspection of this is it seen that the curve will oscillate but will finally settle down to a value about midway between the maximum and minimum.

The above process may easily be modified to allow for light damping. As a first approximation the effect of damping during any cycle of excitation may be neglected when estimating the motion produced by that cycle when it is complete.

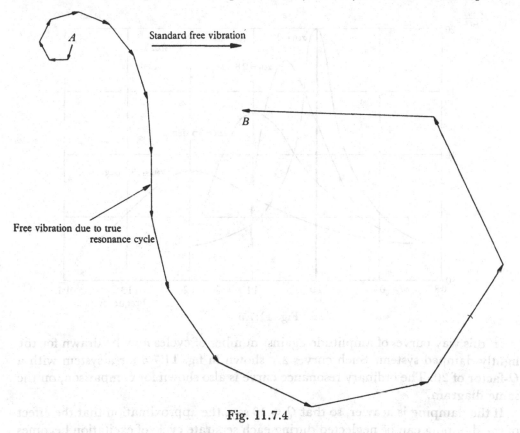

Fig. 11.7.4

After this instant the motion will decay exponentially so that at any subsequent instant the components due to the early excitation cycles will be less than those due to the later ones. The vector diagram giving the motion at the nth cycle after true resonance will now differ from that previously drawn by the fact that, whereas the component due to the nth cycle will be as before, that due to the $(n-1)$th will bear some ratio l to the previous one where l is the decrement; similarly the $(n-2)$ component will be l^2 times its original value and so on. The diagram will therefore consist of a set of vectors, the length of each one of which bears the ratio $1/l$ to that of the preceding one. Such a diagram is sketched in fig. 11.7.4. Evidently the diagrams for different cycles should differ in size, the final vector of each diagram having the same length. It is, however, unnecessary to draw more than one diagram because the ratios of the vectors would be the same for all the diagrams; this being so let a single diagram be drawn in which the final vector has the same length as it would have in the undamped case, the lengths of all the earlier vectors being reduced in the appropriate ratios. The final resultant vector may then be measured direct but the resultant vector for the end of the preceding cycle will have to be divided by the decrement after measurement because all the components which form it will be too small in this ratio. The resultant of the cycle before this must similarly be divided by the square of the decrements and so on.

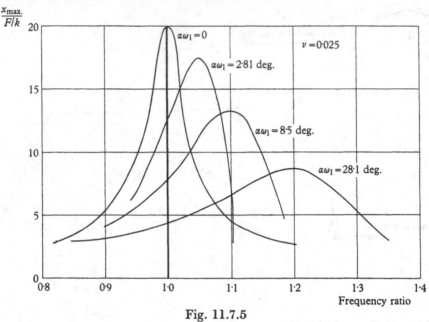

Fig. 11.7.5

In this way curves of amplitude against number of cycles may be drawn for the lightly damped system. Such curves are shown in fig. 11.7.5 for a system with a Q-factor of 20. The ordinary resonance curve is also shown for comparison, on the same diagram.

If the damping is heavier, so that $Q = 10$ say, the approximation that the effect of the damping can be neglected during each separate cycle of excitation becomes inadmissible. The damping during that cycle materially affects the phase angle, particularly in the cycles which are very close to resonance. An expression for the phase angle θ must be used, which is dependent both on the Q-factor and on the rate of passing through resonance. Details of the method are given by Hother-Lushington and Johnson.[†]

† S. Hother-Lushington and D. C. Johnson, 'The Acceleration of a Single-Degree-of-Freedom System through its Resonant Frequency,' *J. Roy. Aero. Soc.*, vol. 62 (1958), 752–7.

NOTATION

Supplementary List of Symbols used in Chapter 11

I_r	Generalized impulse applied at the co-ordinate q_r [see equation (11.3.3)].
K	Constant used in specifying a continuously varying frequency [see equation (11.7.2)].
R	See equation (11.7.10).
t_1	Time at which distortion is measured, the distortion being caused by an impulse applied at some earlier time.
V	Velocity of load crossing a beam (§ 11.6 only).
α	Difference between periods of successive forcing cycles (§ 11.7 only).
$\alpha_{rs}(\tau)$	Receptance of usual type; the bracketed symbol τ indicates that the appropriate frequency is $\omega = 2\pi/\tau$ (§ 11.1 only).
$\alpha_{rs}(t)$	Impulsive cross-receptance between q_r and q_s.
Δ', Δ'_{rs}	See equation (11.3.8).
θ	See equation (11.7.10).
ν	Damping factor [see equation (11.2.3)].
τ	Period of a non-harmonic periodic force.
ψ	Phase angle [see equation (11.1.1)].
Ω	See equation (11.2.3).

ANSWERS TO EXAMPLES

1.1.1 3, 1, 3, 1, 7, 3, 1, 3, 6.

1.2.1
$$\omega_1^2 = \left(\frac{g}{l} + \frac{kh^2}{ml^2}\right).$$

1.2.3 607 lb.in.^{2}.

1.2.5
$$\frac{k}{M}\left[1 - \frac{\delta + \delta_{st}}{a}\right] < \omega^2 < \frac{k}{M}\left[1 + \frac{\delta + \delta_{st}}{a}\right],$$

where
$$\delta_{st} = Mg/k.$$

1.2.6 $\sqrt{(2E/M)}$; $\sqrt{(2E/k)}$. Ellipses become circles.

1.2.7
$$\omega_1^2 = \frac{3[8l\lambda \sin^3\theta - Mg\cos\theta(1 + 2\sin^2\theta)]}{8Ml\sin\theta}.$$

1.3.1
$$x = \sqrt{\left[x_0^2 + \left(\frac{\dot{x}_0}{\omega}\right)^2\right]} \exp i\left[\omega t - \tan^{-1}\left(\frac{\dot{x}_0}{\omega x_0}\right)\right],$$

$$x = \sqrt{\left[x_0^2 + \left(\frac{\dot{x}_0}{\omega}\right)^2\right]} \exp i\left[\omega t + \tan^{-1}\left(\frac{\omega x_0}{\dot{x}_0}\right)\right].$$

1.4.1 $(k_1 + k_3 - M_1\omega^2)(k_2 + k_3 - M_2\omega^2) - k_3^2 = 0.$

1.4.2 $\omega_1^2 = k/M$ when $x_1/x_2 = \chi_1 = 1,$

$$\omega_2^2 = \frac{k + 2k_3}{M} \quad \text{when} \quad x_1/x_2 = \chi_2 = -1.$$

1.4.3 $\omega^2 = k_2/I_2.$

1.4.4 $[(k_1 + k_2) - I_1\omega^2](k_2 - I_2\omega^2) = k_2^2.$ In the following diagrams Θ_1 and Θ_2 are the amplitudes of the motions of the disks I_1 and I_2 respectively.

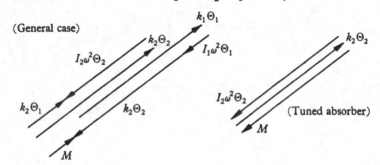

(General case)

(Tuned absorber)

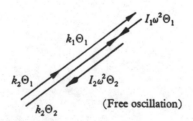

(Free oscillation)

Ans. to Ex. 1.4.4

1.4.5 $\omega_1^2 = 27\cdot6$ rad.^{2}/sec.^{2};

$\omega_2^2 = 148\cdot4$ rad.^{2}/sec.^{2};

$\chi_1 = 0\cdot16$; $\chi_2 = -0\cdot79$.

1.5.1 $\dfrac{1}{mgl - ml^2\omega^2}$, $\omega_1 = \surd(g/l)$.

1.6.1 $-M_1\omega^2 - \dfrac{kM_2\omega^2}{k - M_2\omega^2} = 0$.

1.6.6 $\dfrac{1}{k - M\omega^2 - \dfrac{mg\omega^2}{g - l\omega^2}}$.

1.6.8 $\omega_1^2 = 0$, $\omega_2^2 = 3k/2M$, $\omega_3^2 = 5k/2M$.

1.6.9 $35\cdot8$,   62 rad./sec.

1.7.1 1660 rad./sec.

1.7.2 0, $0\cdot02$, $-0\cdot01$.

1.7.3 $0\cdot20$, $-3\cdot0$.

1.7.4 $\dfrac{\pi k}{105}$, $\tfrac{2}{7}$ deg., $\tfrac{1}{7}$ deg.

1.8.3 $\omega^2 = \dfrac{k(m \pm n)^2}{m^2 I_n + n^2 I_m}$.

1.8.5 $22\cdot4$ r.p.s.

1.9.1 $\omega_1 = 71$ rad./sec., $\omega_2 = 100$ rad./sec., $\omega_3 = 129$ rad./sec.

1.9.5 $\omega_1 = 233$ rad./sec.

1.9.6 Approx. 3720 rad./sec.

1.9.7 $48700 < k < 139000$ lb.in./rad.

1.10.2 $x_3/x_5 = -0\cdot779$.

2.1.1 $z_1 = z_2 = z_3 = 1$,   6; $(x_2 - x_1)^2 + (y_2 - y_1)^2 + (z_2 - z_1)^2 = l^2$,   8.

2.1.2 $4M\ddot{q}_1 + 7M\ddot{q}_2 + 2(2k - mg)\, q_1 + (7k + 6mg)\, q_2 = 38F e^{i\omega t}$,

$2(2 + l)\, \ddot{q}_1 + (7 - 6l)\, \ddot{q}_2 + 2gq_1 - 6gq_2 = 0$.

2.2.2 $\tfrac{1}{2}Mga$.

2.3.1 850 lb.,   750 lb.,   1300 lb.;   1 in.

2.3.2 $9\cdot52$ lb.

2.3.4 Rotations of B, C, D are

$3mgr/2k$, $2mgr/k$, $3mgr/2k$.

2.4.1 (a) See diagram.

(b) $Mga \sin\theta + M\alpha a \cos\theta - \dfrac{4Ma^2}{3}\, \ddot{\theta}$.

(c) $\ddot{\theta} = \dfrac{3(\alpha \cos\theta + g \sin\theta)}{4a}$.

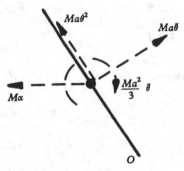

Ans. to Ex. 2.4.1

2.4.3 P; $\frac{4}{27}Pl$.

2.4.4 $[M(R^2+K^2)+m(R^2+r^2-2Rr\cos\theta)]\ddot{\theta}+mrR\dot{\theta}^2\sin\theta+mgr\sin\theta=0$ (assuming that contact with the surface is never lost).

2.5.1 $(m_1l^2+m_2h^2\sec^4\theta)\ddot{\theta}+2m_2h^2\sec^5\theta.\sin\theta.\dot{\theta}^2+m_1gl\sin\theta=0$.

2.5.3 $\omega^2=g/a$, $\omega^4-g\left(\dfrac{1}{l}+\dfrac{1}{a}\right)\left(1+\dfrac{2m}{M}\right)\omega^2+\dfrac{g^2}{la}\left(1+\dfrac{2m}{M}\right)=0$.

2.6.3 (a) $M_2bc[\ddot{\phi}\cos(\phi-\psi)-\dot{\phi}^2\sin(\phi-\psi)]+M_2(c^2+K_2^2)\ddot{\psi}$.

(b) $\dfrac{\partial T}{\partial\psi}=M_2bc\dot{\phi}\dot{\psi}\sin(\phi-\psi)$.

3.1.1 $20r\ddot{\theta}+3\cdot2g\theta=0$.

3.1.2
$$\frac{M}{3}\ddot{q}_1+\frac{M}{6}\ddot{q}_2+\frac{SE}{a}\left(\frac{1+2\sqrt{2}}{2\sqrt{2}}\right)q_1-\frac{SE}{a\sqrt{8}}q_3=0,$$
$$\frac{M}{6}\ddot{q}_1+\frac{M}{3}\ddot{q}_2+\frac{SE}{a}q_2=0,$$
$$M\ddot{q}_3-\frac{SE}{a\sqrt{8}}q_1+\frac{SE}{a\sqrt{8}}q_3=0.$$

3.2.1 $\alpha_{12}=k_1(k_2-I_3\omega^2)/\Delta$, where $\Delta=k_1I_2I_3\omega^4-(I_2+I_3)k_1k_2\omega^2$.

3.2.2 $\alpha_{11}=\alpha_{22}=\dfrac{mgl+kh^2-ml^2\omega^2}{\Delta}$, $\alpha_{12}=\dfrac{kh^2}{\Delta}$, $\alpha_{vv}=l^2\alpha_{11}$, $\alpha_{vz}=l^2\alpha_{12}$,

where
$$\Delta=\begin{vmatrix} mgl+kh^2-ml^2\omega^2 & -kh^2 \\ -kh^2 & mgl+kh^2-ml^2\omega^2 \end{vmatrix}.$$

3.2.3 $\dfrac{1}{M\omega^2}=\dfrac{l^2}{2\Delta}[mgl+2kh^2-ml\omega^2]$, where Δ is that of Ex. 3.2.2.

3.2.4
$$\alpha_{11}=\left[\frac{M^2\omega^4}{3}-\frac{M\omega^2SE}{a}\left(\frac{1+6\sqrt{2}}{6\sqrt{2}}\right)+\frac{S^2E^2}{a^2\sqrt{8}}\right]\Big/\Delta,\quad \alpha_{12}=\left[\frac{M\omega^2}{6}\left(\frac{SE}{a\sqrt{8}}-M\omega^2\right)\right]\Big/\Delta,$$

where
$$\Delta=\begin{vmatrix} \dfrac{SE(1+2\sqrt{2})}{a\sqrt{8}}-\dfrac{M\omega^2}{3} & -\dfrac{M\omega^2}{6} & -\dfrac{SE}{a\sqrt{8}} \\[2ex] -\dfrac{M\omega^2}{6} & \dfrac{SE}{a}-\dfrac{M\omega^2}{3} & 0 \\[2ex] -\dfrac{SE}{a\sqrt{8}} & 0 & \dfrac{SE}{a\sqrt{8}}-M\omega^2 \end{vmatrix}.$$

3.2.5 $F(\alpha_{11}+2\alpha_{12}+\alpha_{22}-4\alpha_{33})e^{i\omega t}/8$.

3.2.7
$$\begin{vmatrix} \beta_{11}+\gamma_{11} & \beta_{12}+\gamma_{12} \\ \beta_{12}+\gamma_{12} & \beta_{22}+\gamma_{22} \end{vmatrix}=0,$$

where
$$\beta_{11}=\frac{\cos^2\theta}{k_1-M\omega^2}+\frac{\sin^2\theta}{k_2-M\omega^2},$$
$$\beta_{22}=\frac{\sin^2\theta}{k_1-M\omega^2}+\frac{\cos^2\theta}{k_2-M\omega^2},$$
$$\beta_{12}=\frac{-\cos\theta\sin\theta}{k_1-M\omega^2}+\frac{\cos\theta\sin\theta}{k_2-M\omega^2},$$
$$\gamma_{11}=\frac{k_3-m\omega^2}{-k_3m\omega^2},\quad \gamma_{22}=\frac{-1}{m\omega^2},\quad \gamma_{12}=0.$$

3.4.1 $\omega_1 = \sqrt{(k_1/M_1)}$ giving $q_1:q_2::b_2+b_3:b_3,$

$\omega_2 = \sqrt{(k_2/M_2)}$ giving $q_1:q_2::b_1:b_1+b_2,$

$$\alpha_{11} = \frac{(b_2+b_3)^2}{b^2(k_1-M_1\omega^2)} + \frac{b_1^2}{b^2(k_2-M_2\omega^2)},$$

$$\alpha_{22} = \frac{b_3^2}{b^2(k_1-M_1\omega^2)} + \frac{(b_1+b_2)^2}{b^2(k_2-M_2\omega^2)},$$

$$\alpha_{12} = \frac{b_3(b_2+b_3)}{b^2(k_1-M_1\omega^2)} + \frac{b_1(b_1+b_2)}{b^2(k_2-M_2\omega^2)},$$

where $$b = b_1+b_2+b_3.$$

3.4.2 (a) $\omega_1^2 = 0.219k/M$ giving $q_1/q_2 = 0.562,$

$\omega_2^2 = 2.28k/M$ giving $q_1/q_2 = -3.56.$

(b) $\Psi_1 = 0.522, \quad \Psi_2 = 0.929.$

(c) $Q_1/Q_2 = 0.28, \quad Q_1/Q_2 = -1.78.$

3.4.3 $$\omega_1^2 = \frac{2k}{M} \quad \text{giving} \quad q_1:q_2:q_3::1:1:1,$$

$$\omega_2^2 = \frac{3k}{M} \quad \text{giving} \quad q_1:q_2:q_3::1:0:-1,$$

$$\omega_3^2 = \frac{6k}{M} \quad \text{giving} \quad q_1:q_2:q_3::1:-1:1.$$

3.4.7 (c), (a), (h).

3.5.1 $$q_1 = 4e^{i\omega_1 t}+2e^{i\omega_2 t}, \quad q_2 = 8e^{i\omega_1 t}-2e^{i\omega_2 t},$$

where $$\omega_1^2 = k/2M \quad \text{and} \quad \omega_2^2 = 2k/M.$$

3.5.2 $$p_1 = \left(\frac{\tfrac{2}{3}}{k-2M\omega^2}\right)\Phi_2\,e^{i\omega t}, \quad p_2 = \left(\frac{-\tfrac{1}{3}}{2k-M\omega^2}\right)\Phi_2\,e^{i\omega t}.$$

3.5.4 $$\Omega_1^2 = \frac{k}{(2+\sqrt{2})I}, \quad \Omega_2^2 = \frac{k}{(2-\sqrt{2})I}.$$

3.5.6 3.66 rad./sec.

3.7.1 $$\frac{6}{3k-2M\omega^2}+\frac{1}{4k-M\omega^2}.$$

3.7.2 $$\frac{1}{2k}+\frac{6}{3k-2M\omega^2}+\frac{1}{4k-M\omega^2} = 0,$$

$$\omega_1^2 = 3.21k/M, \quad \omega_2^2 = 10.29k/M,$$

$$\frac{p_1}{p_2} = \frac{1.39}{l}, \quad \frac{p_1}{p_2} = -\frac{2.15}{l},$$

where p_1 is a clockwise rotation about the mid-point of the bar and p_2 is the downward displacement at that point.

3.7.4 $$\frac{3.84}{3840-\omega^2}\pm\frac{2.88}{6480-\omega^2} \text{ in./lb.wt.}, \quad \omega = 17.2, 60, 81 \text{ rad./sec.}$$

3.7.6 $\dfrac{4}{45(200^2-\omega^2)}+\dfrac{4}{27(300^2-\omega^2)}+\dfrac{1}{63(150^2-\omega^2)}+\dfrac{1}{36(250^2-\omega^2)}-\dfrac{1}{6\omega^2}=0.$

3.7.7 Natural frequencies are given by the equation

$$\frac{1}{384}-\frac{1}{\omega^2}+\frac{1}{6150-\omega^2}+\frac{5\cdot07}{15\,900-\omega^2}=0.$$

3.8.1 $\quad _1Q_1=\dfrac{Q_1}{3}+\dfrac{2Q_2}{3},\quad _1Q_2=\dfrac{Q_1}{3}+\dfrac{2Q_2}{3},\quad _2Q_1=\dfrac{2Q_1}{3}-\dfrac{2Q_2}{3},\quad _2Q_2=-\dfrac{Q_1}{3}+\dfrac{Q_2}{3}.$

3.8.3 $\quad \alpha_{11}=\dfrac{0\cdot53}{2\cdot5\times10^4-\omega^2}+\dfrac{1\cdot33}{12\times10^4-\omega^2}+\dfrac{1\cdot04}{70\times10^4-\omega^2}$ in./lb.wt.,

$$\alpha_{12}=\frac{1\cdot05}{2\cdot5\times10^4-\omega^2}+\frac{0\cdot67}{12\times10^4-\omega^2}-\frac{0\cdot83}{70\times10^4-\omega^2}\ \text{in./lb.wt.,}$$

$$\alpha_{22}=\frac{2\cdot11}{2\cdot5\times10^4-\omega^2}+\frac{0\cdot33}{12\times10^4-\omega^2}+\frac{0\cdot66}{70\times10^4-\omega^2}\ \text{in./lb.wt.;}$$

$$\omega_1=96\ \text{rad./sec.}$$

3.9.1 $$\omega_c^2=\frac{k(1-2x+2x^2)}{I(1+x^2)}.$$

3.9.2 $$\omega_c^2=\frac{0\cdot0947k}{I},\quad \omega_1^2=\frac{0\cdot0945k}{I}.$$

3.10.1 $\quad \alpha_{12}=\alpha_{23}=k(k+2\lambda-2M\omega^2)/\Delta,$

$$\alpha_{22}=(k-2M\omega^2)\,(k+2\lambda-2M\omega^2)/\Delta,$$

where $\qquad \Delta=-M\omega^2(k+2\lambda-2M\omega^2)\,(5k-2M\omega^2).$

3.10.2 No.

$$\alpha_{11}=[(k^2+2k\lambda)-(5k+\lambda)M\omega^2+2M^2\omega^4]/\Delta,$$

$$\alpha_{13}=(k^2+2k\lambda-\lambda M\omega^2)/\Delta,$$

where Δ is that of Ex. 3.10.1.

No.

3.11.2 $1/\sqrt{3}$. Rotation about a point distant $\tfrac13 l$ from, and on the opposite side of, the mid-point.

4.1.2 $$\frac{Sv_1^2l}{2h(l-h)},\quad \frac{Sv_1l}{h(l-h)}.$$

4.2.4 $\quad \dfrac{1}{\lambda^2}\tan^2\left[\dfrac{\lambda l(1-r)}{2}\right]-\dfrac{2}{\lambda\lambda'}\tan\left[\dfrac{\lambda l(1-r)}{2}\right]\cot\lambda'lr-\dfrac{1}{\lambda'^2}=0,$

where $$\lambda'=\omega\sqrt{\frac{\mu'}{S}}.$$

4.2.5 $\qquad (a)\ \tan\lambda l=\dfrac{4S\lambda}{M\omega^2},\quad (b)\ \tan\lambda l=\dfrac{2S\lambda}{M\omega^2}.$

4.4.1 $$v_1=\sqrt{2}.p_1\sin\frac{\pi x}{2l}.$$

4.4.2 $$v_5=-2p_5\sin\frac{5\pi x}{2l}.$$

4.6.1
$$v_r(x) = \frac{2Fl}{Sr^2\pi^2} \sin \frac{r\pi h}{l} \sin \frac{r\pi x}{l}.$$

4.6.2
$$v(x, t) = \frac{2w\, e^{i\omega t}}{\mu\pi} \sum_{r=1}^{\infty} \left\{ \frac{\left(\cos \dfrac{r\pi}{4} - \cos \dfrac{3r\pi}{4}\right)}{r(\omega_r^2 - \omega^2)} \sin \frac{r\pi x}{l} \right\},$$

where ω_r is the rth natural frequency of a taut fixed-fixed string.

4.6.3
$$v(x, t) = \frac{2wa\, e^{i\omega t}}{\mu l} \sum_{r=1}^{\infty} \frac{\phi_r(x)}{\omega_r(\omega_r^2 - \omega^2)},$$

where ω_r is the rth natural frequency of a taut fixed-free string and $a = \sqrt{(S/\mu)}$.

4.7.1
$$v_3(x, t) = \frac{8Y}{25\pi^2} \cos \omega_3 t . \sin \frac{5\pi x}{2l},$$

where
$$\omega_3 = \frac{5\pi}{2l} \sqrt{\left(\frac{S}{\mu}\right)}.$$

4.8.1
$$(a)\quad f_r(x) = \frac{2w}{r\pi} \left(\cos \frac{r\pi}{4} - \cos \frac{3r\pi}{4}\right) \sin \frac{r\pi x}{l}.$$

$$(b)\quad v_r(x) = \frac{2wl^2}{Sr^3\pi^3} \left(\cos \frac{r\pi}{4} - \cos \frac{3r\pi}{4}\right) \sin \frac{r\pi x}{l}.$$

4.8.2
$$(a)\quad v = -\frac{wx^2}{2S} \quad (0 \leqslant x \leqslant \tfrac{1}{2}l).$$

$$(b)\quad v = \frac{2wl^2}{S\pi^2} \sum_{r=1}^{\infty} \left\{ \frac{1}{r^2} \cos \frac{r\pi}{2} . \left[1 - \cos \frac{r\pi x}{l}\right] \right\},$$

4.8.5 $2/l.$

5.2.1 (a) 0·233 in., 0·214 in., 0·221 in.; (b) 0·218 in.

5.2.2 $(a)\quad -\dfrac{wl^2}{9S},\ -\dfrac{5wl^2}{18S},\ -\dfrac{wl^2}{3S};\quad (b)\ v = -K\dfrac{wl^2}{S},$

where K is given in the following table.

Deflexion	Values of the constant K					
	Exact	Number of terms in series approximation				
		1	2	3	4	5
$v(\tfrac{1}{3}l) - v(o)$	0·1111	0·0838	0·1152	0·1113	0·1106	0·1109
$v(\tfrac{2}{3}l) - v(o)$	0·2778	0·2513	0·2828	0·2788	0·2768	0·2776
$v(l) - v(o)$	0·3333	0·3351	0·3351	0·3351	0·3325	0·3334

5.3.1 (a) 248 rad./sec. (b) 237 rad./sec., 3410 rad./sec.
  (c) 224 rad./sec. (d) 224 rad./sec., 3150 rad./sec.

6.2.1 153·3 cyc./sec., 2686 cyc./sec.

6.2.2 155 cyc./sec.

6.2.6
$$J_L \tan \frac{\omega l_L}{a} + r^2 J_N \tan \frac{\omega l_N}{a} = 0,$$

where $a^2 = G/\rho$.

6.2.7 $GJ\lambda \cot \lambda l + k = 0$ in the notation of this section.

6.2.8 (a) $-\dfrac{1}{\sin \lambda l [I\omega^2 \cot \lambda l + GJ\lambda]} F e^{i\omega t}$, where $\lambda = \omega/a$, $J = I'/\rho l$.

(b) $-\dfrac{2l^2}{I'\pi a^2} F e^{ia\pi t/2l}$.

6.3.2 $\theta_r = A_r \sin \lambda_r x$, where A_r is a constant and λ_r is the rth root of the equation
$$GJ\lambda \cot \lambda l + k = 0.$$

6.3.3
$$a_r = \frac{J\rho}{2\lambda_r} (\lambda_r l - \sin \lambda_r l \cos \lambda_r l), \quad c_r = \frac{GJ\lambda_r^2 l + k \sin^2 \lambda_r l}{2}.$$

6.4.1 170 cyc./sec.

6.5.1
$$\omega_1 = \frac{1 \cdot 84}{l} \sqrt{\left(\frac{G}{\rho}\right)}, \quad -0 \cdot 27.$$

6.5.2 $\omega_1 = 16180$ rad./sec., $\omega_2 = 36560$ rad./sec.

6.5.3 $\omega_1 = 5360$ rad./sec., $0 \cdot 00115$ rad.

6.6.1 See the curve given.

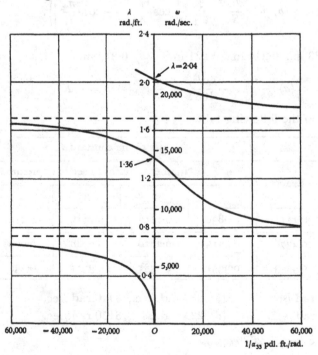

Ans. Ex. 6.6.1

6.6.2 See the curves.

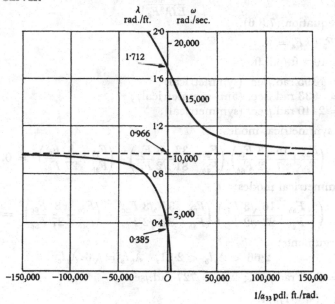

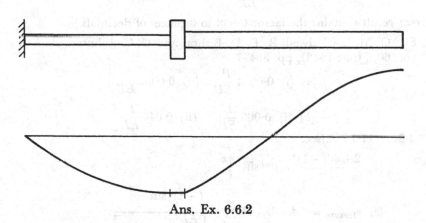

Ans. Ex. 6.6.2

6.7.1 (a) $\omega_1 = 0{\cdot}785 \sqrt{(k/I)}$, $\omega_2 = 1{\cdot}571 \sqrt{(k/I)}$.

(b) $\omega_1 = 0{\cdot}766 \sqrt{(k/I)}$, $\omega_2 = 1{\cdot}414 \sqrt{(k/I)}$.

6.7.2 See the reference given in the text.

6.8.1
$$\frac{-1}{A\rho l \omega^2} + \frac{2}{A\rho l} \sum_{r=1}^{\infty} \frac{1}{\omega_r^2 - \omega^2},$$

where $\omega_r^2 = r^2 \pi^2 a^2 / l^2$ in the notation of this section.

6.9.1 17 000 ft./sec.

7.1.1
$$S = -EI \frac{\partial^3 v}{\partial x^3} + I\rho \frac{\partial^3 v}{\partial x \partial t^2}.$$

7.2.1
$$\frac{F_8}{EI\lambda^2 F_6}.$$

7.3.1 See equation (7.3.6).

7.3.2 $EI\lambda F_5 + kF_4 = 0.$

7.4.1 $2 \cdot 4 \times 10^{-6}$ ft./pdl.ft.

7.4.2 $\omega_1 = 195$ rad./sec. (symmetrical);
$\omega_2 = 833$ rad./sec. (anti-symmetrical);
$\omega_3 = 2440$ rad./sec. (symmetrical).

7.4.3 For symmetrical modes:

$$\left(\frac{F_{5b}}{F_{4b}} + \frac{16\sqrt{3}}{27\sqrt{2}}\frac{F_{2c}}{F_{6c}}\right)\left(\frac{F_{6b}}{F_{4b}} + \frac{32\sqrt{3}}{81\sqrt{2}}\frac{F_{1c}}{F_{6c}}\right) + \left(\frac{F_{1b}}{F_{4b}} + \frac{8}{27}\frac{F_{5c}}{F_{6c}}\right)^2 = 0.$$

For anti-symmetrical modes:

$$\left(-\frac{F_{5b}}{F_{4b}} + \frac{16\sqrt{3}}{27\sqrt{2}}\frac{F_{1c}}{F_{5c}}\right)\left(\frac{F_{6b}}{F_{4b}} + \frac{32\sqrt{3}}{81\sqrt{2}}\frac{F_{2c}}{F_{5c}}\right) - \left(\frac{F_{1b}}{F_{4b}} - \frac{8}{27}\frac{F_{6c}}{F_{5c}}\right)^2 = 0.$$

Range of arguments:

$$2 \cdot 66 < \lambda_b l_b < 2 \cdot 91, \quad \lambda_b l_b = \sqrt{6} . \lambda_c l_c.$$

7.5.1 $\omega_1 = 68 \cdot 0$ rad./sec.; $\omega_2 = 727$ rad./sec.

7.6.1 $1 \cdot 0$; $(4r-1)\pi/4$.

7.7.1 $0 \cdot 125 \dfrac{wl^4}{EI}$, $0 \cdot 165 \dfrac{wl^3}{EI}$ (the exact numerical factors are $0 \cdot 1250$ and $0 \cdot 1667$).

7.7.2 $\quad -0 \cdot 009 \dfrac{Wl^3}{EI}$ (from 2 terms).

(The correct result contains the factor $0 \cdot 008$ to 3 places of decimals.)

7.7.3. See G. M. L. Gladwell, R. E. D. Bishop and D. C. Johnson, *J. Roy. Aero. Soc.*, vol. 66, (June 1962), pp. 394–7.

7.8.1 (a) (i) $0 \cdot 0054 \dfrac{l^3}{EI}$; (ii) $0 \cdot 0053 \dfrac{l^3}{EI}$.

(b) (i) $0 \cdot 063 \dfrac{l}{EI}$; (ii) $0 \cdot 044 \dfrac{l}{EI}$.

7.8.2 $4 \cdot 24$, $17 \cdot 7$ cyc./sec.

7.8.3 (b) $\dfrac{2A\rho\omega^2(-1)^r}{r\pi} v_l e^{i\omega t} \sin\dfrac{r\pi x}{l}.$

(c) $v_{\text{flexure}} = 2A\rho\omega^2 v_l e^{i\omega t} \displaystyle\sum_{r=1}^{\infty} \dfrac{(-1)^r \sin\dfrac{r\pi x}{l}}{r\pi\left[\dfrac{EIr^4\pi^4}{l^4} - A\rho\omega^2\right]}.$

(d) $\ddot{v}_{\text{total}} = -\omega^2 v_l e^{i\omega t}\left\{\dfrac{x}{l} + 2A\rho\omega^2 \displaystyle\sum_{r=1}^{\infty} \dfrac{(-1)^r \sin\dfrac{r\pi x}{l}}{r\pi\left[\dfrac{EIr^4\pi^4}{l^4} - A\rho\omega^2\right]}\right\}.$

(e) If the terminal force is $Fe^{i\omega t}$,

$$\alpha_{ll} = \frac{F}{v_l} = -A\rho l\omega^2\left\{\frac{1}{3} + \frac{2A\rho\omega^2}{\pi^2}\sum_{r=1}^{\infty}\frac{1}{r^2\left[\dfrac{EIr^4\pi^4}{l^4} - A\rho\omega^2\right]}\right\}.$$

Note that this series converges very rapidly.

7.8.4 $\omega_1 \doteqdot 54\cdot8$ cyc./sec., $\omega_2 \doteqdot 181$ cyc./sec.

7.10.2 The exact result is 195 rad./sec. (see Ex. 7.4.2); if the central portion is assumed to remain undistorted and the remainder to form a sine curve, the resulting estimate is 202 rad./sec.

7.10.3 The distortion shapes shown in the figure are composed of quarter cosine curves, in (a) the amplitudes are equal, in (b) the outer portion has half the amplitude of the clamped portion and in (c) the outer portion is straight; the corresponding Rayleigh estimates are (a) 99·5 rad./sec., (b) 87·5 rad./sec., (c) 87·0 rad./sec. The true frequency is 83·2 rad./sec. (see §7.5).

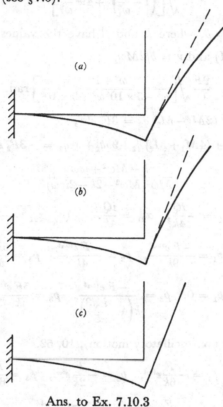

(a)

(b)

(c)

Ans. to Ex. 7.10.3

7.10.5 10·8 cyc./sec.

7.11.1
$$M = -\frac{H}{2}\left[e^{i(\omega t - \lambda l)} + e^{-\lambda l}\,e^{i\omega t}\right].$$

$$S = -\frac{H\lambda}{2}\left[e^{i(\omega t - \lambda l + \pi/2)} + e^{-\lambda l}\,e^{i\omega t}\right].$$

8.1.1 $(255 - 66i) \times 10^{-6}$ ft./pdl. $263 \times 10^{-6}e^{-0\cdot25i}$ ft./pdl.

8.1.2 (a) $1/k \to 1/ib\omega$.
(b) $0 \to \frac{1}{2}\pi$.
(c) $(128 - 16i) \times 10^{-6}$ ft./pdl., $129 \times 10^{-6}e^{-0\cdot124i}$ ft./pdl.

8.1.3 (a) $1/ib\omega \to -1/M\omega^2$.
(b) $\frac{1}{2}\pi \to \pi$.
(c) $(-236 - 57i) \times 10^{-6}$ ft./pdl., $244 \times 10^{-6}e^{-2\cdot91i}$ ft./pdl.

8.2.1 $\ddot{x}+2\nu\omega_1\dot{x}+\omega_1^2 x = \omega_1^2\phi+2\nu\omega_1\dot{\phi}, \quad \ddot{y}+2\nu\omega_1\dot{y}+\omega_1^2 y = -\ddot{\phi}.$

8.2.2 $\qquad R = \dfrac{\Phi\sqrt{\left(1+4\nu^2\dfrac{\omega^2}{\omega_1^2}\right)}}{\sqrt{\left[\left(1-\dfrac{\omega^2}{\omega_1^2}\right)^2+\dfrac{4\nu^2\omega^2}{\omega_1^2}\right]}}, \quad \alpha = \tan^{-1}\left[\dfrac{2\nu\dfrac{\omega}{\omega_1}}{4\nu^2-1+\dfrac{\omega_1^2}{\omega^2}}\right].$

8.2.4 $S = L\Phi,$ where $L = \dfrac{\dfrac{\omega^2}{\omega_1^2}}{\sqrt{\left[\left(1-\dfrac{\omega^2}{\omega_1^2}\right)^2+4\nu^2\dfrac{\omega^2}{\omega_1^2}\right]}}; \quad \beta = \tan^{-1}\left[\dfrac{2\nu\dfrac{\omega}{\omega_1}}{1-\dfrac{\omega^2}{\omega_1^2}}\right].$

8.2.6 $R = \dfrac{me}{M}L,$ $\gamma = \beta,$ where L and β have the values given in the answer of Ex. 8.2.4 if $\omega_1 = \sqrt{(k/M)}$ and $\nu = b/2M\omega_1.$

8.3.1 $\qquad \dfrac{9F}{\omega}\sqrt{\left[\dfrac{\omega^2+10^4}{\omega^4-5\times10^4\omega^2+9\times10^8}\right]}$ rad.

8.4.1 $(M_1+M_2)\ddot{q}_1+(2M_1-M_2)\ddot{q}_2 = 3F_2\,e^{i\omega t},$

$\qquad (2M_1-M_2)\ddot{q}_1+(4M_1+M_2)\ddot{q}_2+9b\dot{q}_2+9kq_2 = -3F_2\,e^{i\omega t}.$

8.5.1 $\qquad \dfrac{k-M\omega^2+ib\omega}{M\omega^2(M\omega^2-2k-2ib\omega)}.$

8.5.2 $\qquad$ (i) $\alpha_{11} = -\dfrac{iQ}{4k},$ $\alpha_{21} = \dfrac{iQ}{4k}.$ $\qquad$ (ii) $\alpha_{11} = -\dfrac{i}{kQ\sqrt{2}}.$

8.7.1 $\qquad$ (a) $p_1 = \dfrac{-F\,e^{i\omega_2 t}}{9k},$ $\quad p_2 = \dfrac{-F\,e^{i\omega_2 t}}{4k},$ $\quad p_3 = \dfrac{5F\,e^{i\omega_2 t}}{ik\sqrt{3}}.$

$\qquad$ (b) $p_1 = 0,$ $\quad p_2 = \dfrac{-F\,e^{i\omega_2 t}}{k\left(4-\dfrac{i\sqrt{3}}{5}\right)},$ $\quad p_3 = \dfrac{5F\,e^{i\omega_2 t}}{ik\sqrt{3}}.$

$Q = $ (not defined for non-oscillatory motion), 10, 52.

8.7.2 $\qquad$ (a) $p_1 = -\dfrac{F}{6k}e^{i\omega t},$ $\quad p_2 = -\dfrac{F}{2k}e^{i\omega t},$ $\quad p_3 = \dfrac{F}{6k}e^{i\omega t}.$

$\qquad$ (b) $p_1 = -\dfrac{F}{6k}e^{i\omega t},$ $\quad p_2 = \dfrac{-F\,e^{i\omega t}}{k\left(2-\dfrac{i\sqrt{2}}{5}\right)},$ $\quad p_3 = \dfrac{F}{6k}e^{i\omega t}.$

There is no error in p_1 or p_3; the error in the amplitude of p_2 is 1 % and the error in the phase is approximately 8°.

9.1.2 In the notation of this section

$$W_{12} = \frac{kR^2}{2}+\frac{\pi hR^2}{4},$$

$$W_{23} = -\frac{kR^2}{2(1+\mu^2)}+\frac{\pi hR^2}{4}-\frac{hR^2}{2}\left(\tan^{-1}\mu+\frac{\mu}{1+\mu^2}\right),$$

$$W_{34} = -\frac{kR^2\mu^2}{2(1+\mu^2)}+\frac{hR^2}{2}\left(\tan^{-1}\mu+\frac{\mu}{1+\mu^2}\right).$$

9.2.2 (a) $\alpha = \dfrac{1}{k - M\omega^2 + i\omega^2 f}$.

(b) If $\sigma = f/M$ and $\omega_1^2 = k/M$,

$$N_a = \frac{1}{\sqrt{\left[\left(1 - \dfrac{\omega^2}{\omega_1^2}\right)^2 + \sigma^2 \dfrac{\omega^4}{\omega^4}\right]}}.$$

$$\xi = \tan^{-1} \left\{ \frac{\dfrac{\omega^2}{\omega_1^2}\sigma}{1 - \dfrac{\omega^2}{\omega_1^2}} \right\}.$$

(c)

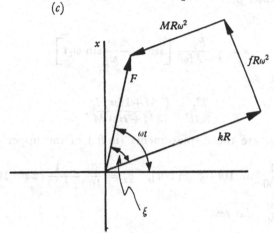

Ans. to Ex. 9.2.2 (c)

(d) $Q = \dfrac{M}{f} = \dfrac{1}{\sigma}$.

9.2.3 3·2 ft.pdl./rad.

9.4.1 (a) $S = \dfrac{i}{2}\left[hL^2 q_1^2 + hL^2 q_2^2 - 2hL^2 q_1 q_2\right]$.

(b) $U = \frac{1}{2}\left[(mgl + kL^2 + ihL^2)\, q_1^2 + (mgl + kL^2 + ihL^2)\, q_2^2 - 2L^2(k + ih)\, q_1 q_2\right]$.

(c) $ml^2 \ddot{q}_1 + (mgl + kL^2 + ihL^2)\, q_1 - L^2(k + ih)\, q_2 = Fe^{i\omega t},$
  $ml^2 \ddot{q}_2 - L^2(k + ih)\, q_1 + (mgl + kL^2 + ihL^2)\, q_2 = 0.$ $\Big\}$

9.5.1 $\alpha_{xx} = \dfrac{1}{2(k + ih) - M\omega^2} + \dfrac{b^2}{2b^2(k + ih) + \lambda - M(R^2 + a^2)\, \omega^2}$.

9.5.2 $\dfrac{1}{\alpha_{xx}} = ih + \dfrac{(2k - M\omega^2)\,[2b^2 k + \lambda - M\omega^2(R^2 + a^2)]}{4b^2 k + \lambda - M\omega^2(R^2 + a^2 + b^2)}$.

9.6.1 $Q = \dfrac{k}{h}$ in both modes.

9.6.3 $51 + 10i, \quad 199 + 20i$.

9.6.4 $\dfrac{q_1}{q_2} = -15i, \quad \dfrac{q_1}{q_2} = -0\cdot13i$.

9.8.1 About 12 ft.

10.3.1
$$q_j = \mathcal{R}\left\{\sum_{m=1}^{n} \left[{}_mC_j \, e^{i \, m\theta_j} u_m\right]\right\},$$

where the co-ordinate u_m has the form

$$u_m = U_m \, e^{(\rho m + i\sigma m) \, t},$$

U_m being complex.

10.6.1
$$x = \frac{F}{2\omega_1^2 M} \left[(1 + \omega_1 t) \, e^{-\omega_1 t} - \cos \omega_1 t\right].$$

10.7.3
$$u = \frac{8Fl}{\pi^2 AE} \sum_{r=1}^{\infty} \frac{1}{r^2} \sin \frac{r\pi}{2} \cos \frac{r\pi at}{2l} \sin \frac{r\pi x}{2l}$$

in the notation of Chapter 6.

11.2.1 180 ft.

11.2.3
$$x = \frac{F_0}{k - M\omega^2} \left[\sin \omega t - \frac{\omega}{\omega_1} \sin \omega_1 t\right].$$

11.2.4 5·8 ft./sec.².

11.3.2
$$\frac{7I_1}{3M}, \quad \frac{(7M + 16m) I_1}{(3M + 7m) M}.$$

11.5.2 If q_1 and q_2 are the displacements (in ft.) of the upper and lower masses respectively,

(a) $q_1 = \dfrac{3t}{100} - \dfrac{1}{500} \left[\sin 10t + \tfrac{1}{4} \sin 20t\right], \quad q_2 = \dfrac{3t}{100} - \dfrac{1}{500} \left[2 \sin 10t - \tfrac{1}{4} \sin 20t\right].$

(b) $q_1 = q_2 = \dfrac{3\pi}{100}$ (at rest).

INDEX

Printed in the United States
By Bookmasters